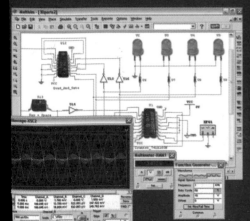

PRINCIPLES OF ELECTRIC CIRCUITS

Conventional Current Version

Eighth Edition

Thomas L. Floyd

PEARSON
Prentice
Hall

Upper Saddle River, New Jersey
Columbus, Ohio

Library of Congress Cataloging-in-Publication Data

Floyd, Thomas L.
  Principles of electric circuits: conventional current version / Thomas L. Floyd.— 8th ed.
    p. cm.
  Includes index.
  ISBN 0-13-170179-7 (alk. paper)
  I. Electric circuits.   I. Title.
TK454.F56   2007b
621.319′2—dc22   2005057487

Acquisitions Editor: Kate Linsner
Production Editor: Rex Davidson
Design Coordinator: Diane Ernsberger
Editorial Assistant: Lara Dimmick
Cover Designer: Candace Rowley
Cover art: Getty
Production Manager: Matt Ottenweller
Senior Marketing Manager: Ben Leonard
Marketing Assistant: Les Roberts
Senior Marketing Coordinator: Liz Farrell

This book was set in Times Roman by TechBooks/GTS and was printed and bound by Courier Kendallville, Inc. The cover was printed by Coral Graphic Services, Inc.

Pearson Education Ltd.
Pearson Education Singapore Pte. Ltd.
Pearson Education Canada, Ltd.
Pearson Education—Japan

Pearson Education Australia Pty. Limited
Pearson Education North Asia Ltd.
Pearson Educación de Mexico, S. A. de C. V.
Pearson Education Malaysia Pte. Ltd.

10 9 8 7 6 5 4 3 2 1
ISBN: 0-13-170179-7

DEDICATION

Once again, to Sheila
With love

PREFACE

The eighth edition of *Principles of Electric Circuits: Conventional Current Version* provides a complete and straightforward coverage of the basics of electrical components and circuits. Fundamental circuit laws and analysis methods are explained and applied in a variety of basic circuits. Applications, many of which are new to this edition, are emphasized, and most chapters have a special feature called *A Circuit Application*. Troubleshooting continues to be an important part of this edition, and many chapters have a separate section devoted to the topic.

New in This Edition

- Text layout and design
- Multisim files for selected examples
- Multisim 8 files, in addition to Multisim 2001 and Multisim 7, for Troubleshooting and Analysis problems
- Phasor coverage has been moved to Chapter 11, Introduction to Alternating Current and Voltage
- Complex number coverage has been moved to Chapter 15, *RC* Circuits
- New problems in most chapters
- Numerous miscellaneous improvements throughout
- Innovative PowerPoint® slides for each chapter available on CD-ROM

Features

- Full-color format
- Chapter openers with a chapter outline, introduction, chapter objectives, key terms list, and website reference
- An introduction and objectives at the beginning of each section within a chapter
- *A Circuit Application* feature at the end of most chapters
- Abundance of high-quality illustrations
- Short biographies of key figures in the history of electricity in several chapters
- *Safety Notes* located at appropriate points throughout the text and identified by a special logo
- Many worked examples
- A Related Problem in each worked example with answers at the end of the chapter
- Section Reviews with answers at the end of the chapter

- Troubleshooting section in many chapters

- Summary at the end of each chapter

- Key terms defined at the end of the chapter and in the comprehensive glossary at the end of the book

- Formula list at the end of each chapter

- Self-test at the end of each chapter with answers at the end of the chapter

- A Circuit Dynamics Quiz that tests the student's grasp of what happens in a circuit as a result of certain changes or faults. Answers are at the end of the chapter.

- Sectionalized problem set for each chapter, with the more difficult problems indicated by an asterisk. Answers to odd-numbered problems are at the end of the book.

- A comprehensive glossary at the end of the book that defines all boldface and key terms in the textbook

- The conventional direction of current is used. (An alternate version of this text uses electron-flow direction.)

Accompanying Student Resources

Experiments in Basic Circuits, Eighth Edition: lab manual by David Buchla (ISBN: 0-13-170181-9). Solutions are provided in the Instructor's Resource Manual.

Experiments in Electric Circuits, Eighth Edition: lab manual by Brian Stanley (ISBN: 0-13-170180-0). Solutions are provided in the Instructor's Resource Manual.

Multisim® CD-ROM: Packaged with each text, this CD contains a set of Multisim circuit files referenced in the text. Many of these circuits have hidden faults. All circuit files are provided on the CD-ROM in Multisim 2001®, Multisim 7®, and Multisim 8®. Circuit files in later versions of Multisim will be posted to the Companion Website at *www.prenhall.com/floyd* as subsequent versions of the software are developed by the manufacturer, Electronics Workbench.

These Multisim circuit files are provided for use by anyone who has Multisim software. Anyone who does not have Multisim software and wishes to purchase it in order to use the circuit files may do so by ordering it from *www.prenhall.com/ewb*. However, although the circuit files are intended to complement classroom, textbook, and laboratory study, it is not necessary to use these files in order to successfully study dc/ac circuits or use Floyd's *Principles of Electric Circuits, Eighth Edition*.

Companion Website (www.prenhall.com/floyd):   For the student, this website offers the opportunity to test his or her own progress and practice answering sample test questions.

Instructor Resources

To access supplementary materials online, instructors need to request an instructor access code. Go to **www.prenhall.com,** click the **Instructor Resource Center** link, and then click **Register Today** for an instructor access code. Within 48 hours after registering you will receive a confirming e-mail including an instructor access code. Once you have received your code, go to the site and log on for full instructions on downloading the materials you wish to use.

PowerPoint® Slides   A completely new set of innovative PowerPoint® slides, created by David M. Buchla, dynamically illustrates key concepts in the text. Each slide contains a summary with examples, key term definitions, and a quiz for each chapter. This is an excellent tool for classroom presentation to supplement the textbook. Another folder of PowerPoint® slides contains all figures from the text. The PowerPoints® are available on both CD and the Internet.

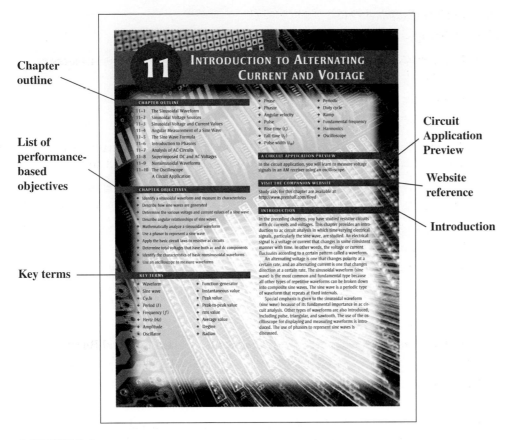

▲ FIGURE P–1

A typical chapter opener.

Instructor's Resource Manual   Includes solutions to chapter problems, solutions to *A Circuit Application* features, a test item file, Multisim circuit file summary, and solutions to both lab manuals. Available in print and online.

Prentice Hall TestGen   This is a computerized test bank. Available on CD-ROM and online.

Illustration of Chapter Features

Chapter Opener   Each chapter begins as shown in Figure P–1. Each chapter opener includes the chapter number and title, a brief introduction, lists of text sections and chapter objectives, a key terms list, *A Circuit Application* preview, and a website reference for study aids and supplementary materials.

Section Opener   Each section in a chapter begins with a brief introduction that includes a general overview and section objectives. An illustration is given in Figure P–2.

Section Review   Each section in a chapter ends with a review consisting of questions or exercises that emphasize the main concepts covered in the section. An example is shown in Figure P-2. Answers to the Section Reviews are at the end of the chapter.

Worked Examples and Related Problems   Numerous worked examples throughout each chapter help to illustrate and clarify basic concepts or specific procedures. Each example ends with a Related Problem that reinforces or expands on the example by requiring the student to work through a problem similar to the example. Selected examples have a Multisim circuit exercise. A typical worked example with a Related Problem is shown in Figure P–3.

▶ FIGURE P–2

A typical section opener and section review.

Section review questions end each section.

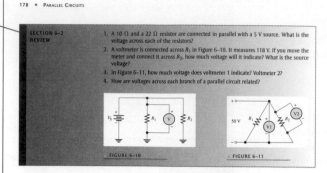

Introductory paragraph begins each section.

Performance-based section objectives.

▶ FIGURE P–3

A typical worked example and related problem.

Examples are set off from text.

Each example contains a related problem relevant to the example.

Selected examples provide a related Multisim circuit file reference.

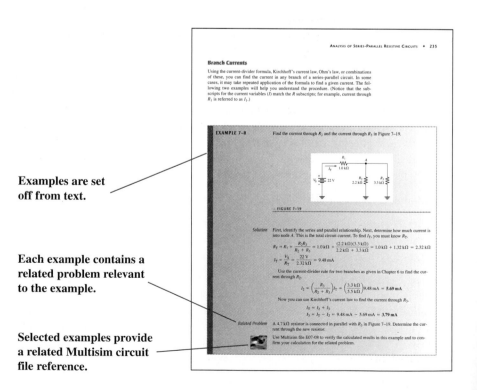

A *Circuit Application* is set off from text.   **A series of activities relates theory to practice.**

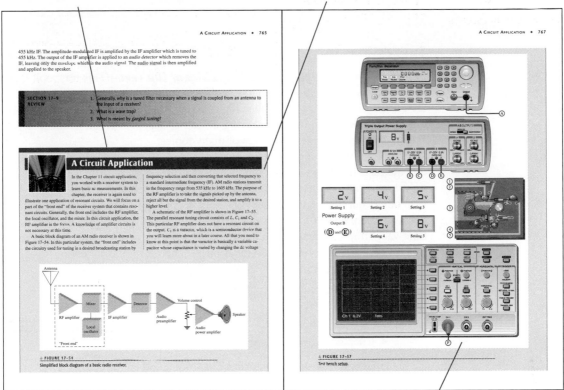

Realistic instrument and circuit board graphics

▲ **FIGURE P–4**

A portion of a typical *A Circuit Application* feature.

Troubleshooting Sections   Many chapters include a troubleshooting section that relates to the topics covered in the chapter and emphasizes logical thinking as well as a structured approach called APM (analysis, planning, and measurement) where applicable. Particular troubleshooting methods, such as half-splitting, are applied when appropriate.

A Circuit Application   This special feature at the end of each chapter (except Chapters 1 and 21) presents a practical application of certain topics covered in the chapter. Each of these features includes a series of activities, many of which involve comparing circuit board layouts with schematics, analyzing circuits, using measurements to determine circuit operation, and in some cases, developing simple test procedures. Results and answers are found in the Instructor's Resource Manual (IRM). A portion of a representative *A Circuit Application* feature is illustrated in Figure P–4.

Chapter End Matter   The following pedagogical features are found at the end of each chapter:

- ◆ Summary
- ◆ Key terms glossary
- ◆ Formula list
- ◆ Self-Test
- ◆ Circuit Dynamics Quiz
- ◆ Problems

◆ Answers to section reviews, related problems for examples, self-test, and the circuit dynamics quiz

Suggestions for Teaching with *Principles of Electric Circuits*

Selected Course Emphasis and Flexibility of the Text   This textbook is designed primarily for use in a two-term course sequence in which dc topics (Chapters 1 through 10) are covered in the first term and ac topics (Chapters 11 through 21) are covered in the second term. A one-term course covering dc and ac topics is possible but would require very selective and abbreviated coverage of many topics.

If time limitations or course emphasis restrict the topics that can be covered, as is usually the case, there are several options for selective coverage. The following suggestions for light treatment or omission do not necessarily imply that a certain topic is less important than others but that, in the context of a specific program, the topic may not require the emphasis that the more fundamental topics do. Because course emphasis, level, and available time vary from one program to another, the omission or abbreviated treatment of selected topics must be made on an individual basis. Therefore, the following suggestions are intended only as a general guide.

1. Chapters that may be considered for omission or selective coverage:

   ◆ Chapter 8, Circuit Theorems and Conversions

   ◆ Chapter 9, Branch, Loop, and Node Analyses

   ◆ Chapter 10, Magnetism and Electromagnetism

   ◆ Chapter 18, Passive Filters

   ◆ Chapter 19, Circuit Theorems in AC Analysis

   ◆ Chapter 20, Time Response of Reactive Circuits

   ◆ Chapter 21, Three-Phase Systems in Power Applications

2. *A Circuit Application* features and troubleshooting sections can be omitted without affecting other material.

3. Other specific topics may be omitted or covered lightly on a section-by-section basis at the discretion of the instructor.

The order in which certain topics appear in the text can be altered at the instructor's discretion. For example, the topics of capacitors and inductors (Chapters 12 and 13) can be covered at the end of the dc course in the first term by delaying coverage of the ac topics in Sections 12–6, 12–7, 13–5, and 13–6 until the ac course in the second term. Another possibility is to cover Chapters 12 and 13 in the second term but cover Chapter 15 (*RC* Circuits) immediately after Chapter 12 (Capacitors) and cover Chapter 16 (*RL* Circuits) immediately after Chapter 13 (Inductors).

A Circuit Application   These features are useful for motivation and for introducing applications of basic concepts and components. Suggestions for using these sections are:

   ◆ As an integral part of the chapter to illustrate how the concepts and components can be applied in a practical situation. The activities can be assigned for homework.

   ◆ As extra credit assignments.

   ◆ As in-class activities to promote discussion and interaction and to help students understand why they need to know the material.

Coverage of Reactive Circuits   Chapters 15, 16, and 17 have been designed to provide two approaches to teaching these topics on reactive circuits.

The first option is to cover the topics on the basis of components. That is, first cover all of Chapter 15 (*RC* Circuits), then all of Chapter 16 (*RL* Circuits), and, finally, all of Chapter 17 (*RLC* Circuits and Resonance).

The second option is to cover the topics on the basis of circuit type. That is, first cover all topics related to series reactive circuits, then all topics related to parallel reactive circuits, and finally, all topics related to series-parallel reactive circuits. To facilitate this second approach, each of the chapters has been divided into the following parts: *Part 1: Series Circuits, Part 2: Parallel Circuits, Part 3: Series-Parallel Circuits,* and *Part 4: Special Topics.* So, for series reactive circuits, cover Part 1 of all three chapters in sequence. For parallel reactive circuits, cover Part 2 of all three chapters in sequence. For series-parallel reactive circuits, cover Part 3 of all three chapters in sequence. Finally, cover Part 4 of all three chapters.

To the Student

Any career training requires hard work, and electronics is no exception. The best way to learn new material is by reading, thinking, and doing. This text is designed to help you along the way by providing an overview and objectives for each section, numerous worked-out examples, exercises, and review questions.

Read each section of the text carefully and think about what you have read. Sometimes you may need to read the section more than once. Work through each example problem step by step before you try the related problem that goes with the example. After each section, answer the review questions. Answers to the related problems and the section review questions are at the end of the chapter.

Review the chapter summary, the key term definitions, and the formula list. Take the multiple choice self test and the Circuit Dynamics Quiz. Check your answers against those at the end of the chapter. Finally, work the problems. Working problems is the most important way to check your comprehension and solidify concepts. Verify your answers to the odd-numbered problems with those provided at the end of the book.

Careers in Electronics

The field of electronics is very diverse, and career opportunities are available in many areas. Because electronics is currently found in so many different applications and new technology is being developed at a fast rate, its future appears limitless. There is hardly an area of our lives that is not enhanced to some degree by electronics technology. Those who acquire a sound, basic knowledge of electrical and electronic principles and are willing to continue learning will always be in demand.

The importance of obtaining a thorough understanding of the basic principles contained in this text cannot be overemphasized. Most employers prefer to hire people who have both a thorough grounding in the basics and the ability and eagerness to grasp new concepts and techniques. If you have a good training in the basics, an employer will train you in the specifics of the job to which you are assigned.

There are many types of job classifications for which a person with training in electronics technology may qualify. A few of the most common job functions are discussed briefly in the following paragraphs.

Service Shop Technician   Technical personnel in this category are involved in the repair or adjustment of both commercial and consumer electronic equipment that is returned to the dealer or manufacturer for service. Specific areas include TVs, VCRs, CD and DVD players, stereo equipment, CB radios, and computer hardware. This area also offers opportunities for self-employment.

Industrial Manufacturing Technician   Manufacturing personnel are involved in the testing of electronic products at the assembly-line level or in the maintenance and troubleshooting of electronic and electromechanical systems used in the testing and manufacturing of products. Virtually every type of manufacturing plant, regardless of its product, uses automated equipment that is electronically controlled.

Laboratory Technician   These technicians are involved in breadboarding, prototyping, and testing new or modified electronic systems in research and development laboratories. They generally work closely with engineers during the development phase of a product.

Field Service Technician   Field service personnel service and repair electronic equipment—for example, computer systems, radar installations, automatic banking equipment, and security systems—at the user's location.

Engineering Assistant/Associate Engineer   Personnel in this category work closely with engineers in the implementation of a concept and in the basic design and development of electronic systems. Engineering assistants are frequently involved in a project from its initial design through the early manufacturing stages.

Technical Writer   Technical writers compile technical information and then use the information to write and produce manuals and audiovisual materials. A broad knowledge of a particular system and the ability to clearly explain its principles and operation are essential.

Technical Sales   Technically trained people are in demand as sales representatives for high-technology products. The ability both to understand technical concepts and to communicate the technical aspects of a product to a potential customer is very valuable. In this area, as in technical writing, competency in expressing yourself orally and in writing is essential. Actually, being able to communicate well is very important in any technical job category because you must be able to record data clearly and explain procedures, conclusions, and actions taken so that others can readily understand what you are doing.

Milestones in Electronics

Before you begin your study of electric circuits, let's briefly look at some of the important developments that led to the electronics technology we have today. The names of many of the early pioneers in electricity and electromagnetics still live on in terms of familiar units and quantities. Names such as Ohm, Ampere, Volta, Farad, Henry, Coulomb, Oersted, and Hertz are some of the better known examples. More widely known names such as Franklin and Edison are also significant in the history of electricity and electronics because of their tremendous contributions. Short biographies of some of these pioneers, like shown here, are located throughout the text.

The Beginning of Electronics   Early experiments with electronics involved electric currents in vacuum tubes. Heinrich Geissler (1814–1879) removed most of the air from a glass tube and found that the tube glowed when there was current through it. Later, Sir William Crookes (1832–1919) found the current in vacuum tubes seemed to consist of particles. Thomas Edison (1847–1931) experimented with carbon filament bulbs with plates and discovered that there was a current from the hot filament to a positively charged plate. He patented the idea but never used it.

Other early experimenters measured the properties of the particles that flowed in vacuum tubes. Sir Joseph Thompson (1856–1940) measured properties of these particles, later called *electrons*.

Although wireless telegraphic communication dates back to 1844, electronics is basically a 20th century concept that began with the invention of the vacuum tube amplifier. An early vacuum tube that allowed current in only one direction was constructed by John A. Fleming in 1904. Called the Fleming valve, it was the forerunner of vacuum tube diodes. In 1907, Lee deForest added a grid to the vacuum tube. The new device, called the audiotron, could amplify a weak signal. By adding the control element, deForest ushered in the electronics revolution. It was with an improved version of his device that made transcontinental telephone service and radios possible. In 1912, a radio amateur in San Jose, California, was regularly broadcasting music!

In 1921, the secretary of commerce, Herbert Hoover, issued the first license to a broadcast radio station; within two years over 600 licenses were issued. By the end of the 1920s radios were in many homes. A new type of radio, the superheterodyne radio, invented by

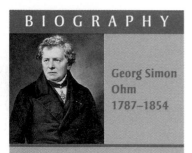

Edwin Armstrong, solved problems with high-frequency communication. In 1923, Vladimir Zworykin, an American researcher, invented the first television picture tube, and in 1927 Philo T. Farnsworth applied for a patent for a complete television system.

The 1930s saw many developments in radio, including metal tubes, automatic gain control, "midget sets," directional antennas, and more. Also started in this decade was the development of the first electronic computers. Modern computers trace their origins to the work of John Atanasoff at Iowa State University. Beginning in 1937, he envisioned a binary machine that could do complex mathematical work. By 1939, he and graduate student Clifford Berry had constructed a binary machine called ABC, (for Atanasoff-Berry Computer) that used vacuum tubes for logic and condensers (capacitors) for memory. In 1939, the magnetron, a microwave oscillator, was invented in Britain by Henry Boot and John Randall. In the same year, the klystron microwave tube was invented in America by Russell and Sigurd Varian.

During World War II, electronics developed rapidly. Radar and very high-frequency communication were made possible by the magnetron and klystron. Cathode ray tubes were improved for use in radar. Computer work continued during the war. By 1946, John von Neumann had developed the first stored program computer, the Eniac, at the University of Pennsylvania. The decade ended with one of the most important inventions ever, the transistor.

Solid-State Electronics   The crystal detectors used in early radios were the forerunners of modern solid-state devices. However, the era of solid-state electronics began with the invention of the transistor in 1947 at Bell Labs. The inventors were Walter Brattain, John Bardeen, and William Shockley. PC (printed circuit) boards were introduced in 1947, the year the transistor was invented. Commercial manufacturing of transistors began in Allentown, Pennsylvania, in 1951.

The most important invention of the 1950s was the integrated circuit. On September 12, 1958, Jack Kilby, at Texas Instruments, made the first integrated circuit. This invention literally created the modern computer age and brought about sweeping changes in medicine, communication, manufacturing, and the entertainment industry. Many billions of "chips"—as integrated circuits came to be called—have since been manufactured.

The 1960s saw the space race begin and spurred work on miniaturization and computers. The space race was the driving force behind the rapid changes in electronics that followed. The first successful "op-amp" was designed by Bob Widlar at Fairchild Semiconductor in 1965. Called the μA709, it was very successful but suffered from "latch-up" and other problems. Later, the most popular op-amp ever, the 741, was taking shape at Fairchild. This op-amp became the industry standard and influenced design of op-amps for years to come.

By 1971, a new company that had been formed by a group from Fairchild introduced the first microprocessor. The company was Intel and the product was the 4004 chip, which had the same processing power as the Eniac computer. Later in the same year, Intel announced the first 8-bit processor, the 8008. In 1975, the first personal computer was introduced by Altair, and Popular Science magazine featured it on the cover of the January, 1975, issue. The 1970s also saw the introduction of the pocket calculator and new developments in optical integrated circuits.

By the 1980s, half of all U.S. homes were using cable hookups instead of television antennas. The reliability, speed, and miniaturization of electronics continued throughout the 1980s, including automated testing and calibrating of PC boards. The computer became a part of instrumentation and the virtual instrument was created. Computers became a standard tool on the workbench.

The 1990s saw a widespread application of the Internet. In 1993, there were 130 websites, and now there are millions. Companies scrambled to establish a home page and many of the early developments of radio broadcasting had parallels with the Internet. In 1995, the FCC allocated spectrum space for a new service called Digital Audio Radio Service. Digital television standards were adopted in 1996 by the FCC for the nation's next generation of broadcast television.

The 21st century dawned in January 2001. One of the major technology stories has been the continuous and explosive growth of the Internet. Internet usage in North America has increased by over 100% from 2000 to 2005. The rest of the world experienced almost 200%

growth during the same period. The processing speed of computers is increasing at a steady rate and data storage media capacity is increasing at an amazing pace. Carbon nanotubes are seen to be the next step forward for computer chips, eventually replacing transistor technology.

Acknowledgments

Many capable people have been part of this revision of *Principles of Electric Circuits*. It has been thoroughly reviewed and checked for both content and accuracy. Those at Prentice Hall who have contributed greatly to this project throughout the many phases of development and production include Rex Davidson and Kate Linsner. Lois Porter, whose attention to details is unbelievable, has once more done an outstanding job of editing the manuscript. Jane Lopez has again provided the excellent illustrations and beautiful graphics work used in the text. David Buchla contributed a significant amount of material for this revision and provided many recommendations. As with the previous edition, Gary Snyder created the circuit files for the Multisim features in this edition.

I wish to express my appreciation to those already mentioned as well as those who provided many valuable suggestions and constructive criticism that have greatly influenced this textbook. The following individuals served as reviewers and provided insightful feedback for this edition: Eldon E. Brown, Jr., Cape Fear Community College; Montie Fleshman, New River Community College; James Jennings, Community College of Southern Nevada; Ronald J. LaSpisa, University of Oklahoma; E. Ed Margaff, Marion Technical College; David Misner, Hutchinson Community College; and Gerald Schickman, Miami Dade Community College. Special thanks are owed to David Heiserman for his extensive evaluation of the text.

Other instructors using the prior edition contributed significantly by participating in an online survey:

Hamid Allamehzadeh ENMU
Tim Baker John A. Logan College
Walter Banzhaf University of Hartford
Joseph Baumert NICC
Kenneth D. Belk Marion Technical College
Seddik Benhamida DeVry University
Rick Buffaloe Idaho State University
Robert Cannella Jr. College of DuPage
Ken Carpenter University of New Mexico
Dan Celenti NYCCT
James Diehl Del Mar College
James Dunn Boise State University
Anthony Edwards Rockingham Community College
Tom Eppes University of Hartford
Larry Gazaway Spokane Community College
David Grant Renton Technical College
Mark Gray Cape Fear Community College
Victor Greenwood Northwest Technical Institute
Osman Gurdal JCSU
Joshua Guttman Bergen Community College
Robert Heffner North Harris College
Christopher Henggeler Westwood College
H. Randolph Holt Northern Kentucky University
Andy Huertaz Albuquerque TVI
Mark Hughes Cleveland Community College
Osama Hussein New York City College of Technology
James Jazdzewski Gateway Technical College
David Jones Lenoir Community College
Benjamin Jun Ivy Tech State College
Lynn Kelly New Mexico State University
Ron LaSpisa U of Oklahoma
George Lee Massasoit Community College
Erik Mayer Bowling Green State University
Paul Mayer Eastern Maine Community College
Stan Middlebrooks Herzing College

Dave Misner Hutchinson Community College
Jim Nutt NMSU-A
Larry Patterson Belmont Technical College
David Phillips Linn State Technical College
James Reardon New England Institute of Technology
Rick Reardon Eastern Maine Community College
Steven Rice C.O.T. University of Montana
Bob Romano Cincinnati Technical and Community College
Jimmie Russell DeVry University
Manavi Sallick Miami Dade College
Joseph Santaniello Spartanburg Technical College
Robert Scoff The University of Memphis
S.T. (Tom) Sharar Asheville-Buncombe Technical College
James Smith Central Washington University
James Stack Boise State University
Richard Sturtevant STCC
Tony Suranno Hagerstown Community College
Greg Szepanski Holyoke Community College
Pratap Reddy Talusani Houston Community College
Ralph Tanner Western Michigan University
Calvin Taylor Centralia College
Ron Tinckham Santa Fe Community College
Don Tosh Evangel University
David Tyree Vincennes University
Ramon Vigil TVI
Paul Vonderwell Vincennes University
Harold Wiebe Northern Kentucky University
Steven Wilson Spokane Community College
Venancio Ybarra, Jr. NHMCCD Cy-Fair College
Steve Yelton Cincinnati State Technical and Community College
Tim Yoxtheimer Central Washington University

Tom Floyd

CONTENTS

QUANTITIES AND UNITS

1

CHAPTER OBJECTIVES

◆ Discuss the SI standard
◆ Use scientific notation (powers of ten) to represent quantities
◆ Use engineering notation and metric prefixes to represent large and small quantities
◆ Convert from one unit with a metric prefix to another

KEY TERMS

◆ SI
◆ Scientific notation
◆ Power of ten
◆ Exponent
◆ Engineering notation
◆ Metric prefix

VISIT THE COMPANION WEBSITE

Study aids for this chapter are available at
http://www.prenhall.com/floyd

INTRODUCTION

You must be familiar with the units used in electronics and know how to express electrical quantities in various ways using metric prefixes. Scientific notation and engineering notation are indispensable tools whether you use a computer, a calculator, or do computations the old-fashioned way.

SAFETY NOTE

When you work with electricity, you must always consider safety first. Safety notes throughout the book remind you of the importance of safety and provide tips for a safe workplace. Basic safety precautions are introduced in Chapter 2.

1–1 UNITS OF MEASUREMENT

In the 19th century, the principal weight and measurement units dealt with commerce. As technology advanced, scientists and engineers saw the need for international standard measurement units. In 1875, at a conference called by the French, representatives from eighteen nations signed a treaty that established international standards. Today, all engineering and scientific work use an improved international system of units, Le Système International d'Unités, abbreviated **SI***.

After completing this section, you should be able to

 ♦ **Discuss the SI standard**
   ♦ Specify the fundamental SI units
   ♦ Specify the supplementary units
   ♦ Explain what derived units are

Fundamental and Derived Units

The SI system is based on seven fundamental units (sometimes called *base units*) and two supplementary units. All measurements can be expressed as some combination of fundamental and supplementary units. Table 1–1 lists the fundamental units, and Table 1–2 lists the supplementary units.

The fundamental electrical unit, the ampere, is the unit for electrical current. Current is abbreviated with the letter I (for intensity) and uses the symbol A (for ampere). The ampere is unique in that it uses the fundamental unit of time (t) in its definition (second). All other electrical and magnetic units (such as voltage, power, and magnetic flux) use various combinations of fundamental units in their definitions and are called *derived units*.

For example, the derived unit of voltage, which is the volt (V), is defined in terms of fundamental units as $m^2 \cdot kg \cdot s^{-3} \cdot A^{-1}$. As you can see, this combination of fundamental units is very cumbersome and impractical. Therefore, volt is used as the derived unit.

▶ TABLE 1–1

SI fundamental units.

QUANTITY	UNIT	SYMBOL
Length	Meter	m
Mass	Kilogram	kg
Time	Second	s
Electric current	Ampere	A
Temperature	Kelvin	K
Luminous intensity	Candela	cd
Amount of substance	Mole	mol

▶ TABLE 1–2

SI supplementary units.

QUANTITY	UNIT	SYMBOL
Plane angle	Radian	r
Solid angle	Steradian	sr

*All bold terms are in the end-of-book glossary. The bold terms in color are key terms and are also defined at the end of the chapter.

Letter symbols are used to represent both quantities and their units. One symbol is used to represent the name of the quantity, and another symbol is used to represent the unit of measurement of that quantity. For example, *P* stands for *power,* and W stands for *watt,* which is the unit of power. Another example is voltage. In this case, the same letter stands for both the quantity and its unit. Italic *V* represents voltage and nonitalic V represents the volt, which is the unit of voltage. As a rule, italic letters stand for the quantity and nonitalic letters represent the unit of that quantity.

Table 1–3 lists the most important electrical quantities, along with their derived SI units and symbols. Table 1–4 lists magnetic quantities, along with their derived SI units and symbols.

◀ TABLE 1–3

Electrical quantities and derived units with SI symbols.

QUANTITY	SYMBOL	SI UNIT	SYMBOL
Capacitance	*C*	Farad	F
Charge	*Q*	Coulomb	C
Conductance	*G*	Siemens	S
Energy	*W*	Joule	J
Frequency	*f*	Hertz	Hz
Impedance	*Z*	Ohm	Ω
Inductance	*L*	Henry	H
Power	*P*	Watt	W
Reactance	*X*	Ohm	Ω
Resistance	*R*	Ohm	Ω
Voltage	*V*	Volt	V

◀ TABLE 1–4

Magnetic quantities and derived units with SI symbols.

QUANTITY	SYMBOL	SI UNIT	SYMBOL
Magnetic field intensity	*H*	Ampere-turns/meter	At/m
Magnetic flux	ϕ	Weber	Wb
Magnetic flux density	*B*	Tesla	T
Magnetomotive force	F_m	Ampere-turn	At
Permeability	μ	Webers/ampere-turn · meter	Wb/At · m
Reluctance	$\mathcal{R}$	Ampere-turns/weber	At/Wb

SECTION 1–1 REVIEW
Answers are at the end of the chapter.

1. How does a fundamental unit differ from a derived unit?
2. What is the fundamental electrical unit?
3. What does *SI* stand for?
4. Without referring to Table 1–3, list as many electrical quantities as possible, including their symbols, units, and unit symbols.
5. Without referring to Table 1–4, list as many magnetic quantities as possible, including their symbols, units, and unit symbols.

1–2 SCIENTIFIC NOTATION

In electrical and electronics fields, you will encounter both very small and very large quantities. For example, it is common to have electrical current values of only a few thousandths or even a few millionths of an ampere and to have resistance values ranging up to several thousand or several million ohms.

After completing this section, you should be able to

◆ **Use scientific notation (powers of ten) to represent quantities**

    ◆ Express any number using a power of ten

    ◆ Perform calculations with powers of ten

Scientific notation provides a convenient method to represent large and small numbers and to perform calculations involving such numbers. In scientific notation, a quantity is expressed as a product of a number between 1 and 10 and a power of ten. For example, the quantity 150,000 is expressed in scientific notation as 1.5×10^5, and the quantity 0.00022 is expressed as 2.2×10^{-4}.

Powers of Ten

Table 1–5 lists some powers of ten, both positive and negative, and the corresponding decimal numbers. The **power of ten** is expressed as an exponent of the base 10 in each case (10^x). An **exponent** is a number to which a base number is raised. It indicates the number of places that the decimal point is moved to the right or left to produce the decimal number. For a positive power of ten, move the decimal point to the right to get the equivalent decimal number. For example, for an exponent of 4,

$$10^4 = 1 \times 10^4 = 1.0000. = 10,000$$

For a negative the power of ten, move the decimal point to the left to get the equivalent decimal number. For example, for an exponent of −4,

$$10^{-4} = 1 \times 10^{-4} = .0001. = 0.0001$$

▼ TABLE 1–5

Some positive and negative powers of ten.

$10^6 = 1,000,000$	$10^{-6} = 0.000001$
$10^5 = 100,000$	$10^{-5} = 0.00001$
$10^4 = 10,000$	$10^{-4} = 0.0001$
$10^3 = 1,000$	$10^{-3} = 0.001$
$10^2 = 100$	$10^{-2} = 0.01$
$10^1 = 10$	$10^{-1} = 0.1$
$10^0 = 1$	

EXAMPLE 1–1

Express each number in scientific notation.

    (a) 200   **(b)** 5000   **(c)** 85,000   **(d)** 3,000,000

Solution   In each case, move the decimal point an appropriate number of places to the left to determine the positive power of ten.

    (a) $200 = \mathbf{2 \times 10^2}$ **(b)** $5000 = \mathbf{5 \times 10^3}$

    (c) $85{,}000 = \mathbf{8.5 \times 10^4}$ **(d)** $3{,}000{,}000 = \mathbf{3 \times 10^6}$

Related Problem*   Express 4750 in scientific notation.

*Answers are at the end of the chapter.

EXAMPLE 1–2

Express each number in scientific notation.

    (a) 0.2   **(b)** 0.005   **(c)** 0.00063   **(d)** 0.000015

Solution   In each case, move the decimal point an appropriate number of places to the right to determine the negative power of ten.

    (a) $0.2 = \mathbf{2 \times 10^{-1}}$ **(b)** $0.005 = \mathbf{5 \times 10^{-3}}$

    (c) $0.00063 = \mathbf{6.3 \times 10^{-4}}$ **(d)** $0.000015 = \mathbf{1.5 \times 10^{-5}}$

Related Problem   Express 0.00738 in scientific notation.

EXAMPLE 1–3

Express each of the following as a regular decimal number:

    (a) 1×10^5 **(b)** 2×10^3 **(c)** 3.2×10^{-2} **(d)** 2.50×10^{-6}

Solution   Move the decimal point to the right or left a number of places indicated by the positive or the negative power of ten respectively.

    (a) $1 \times 10^5 = \mathbf{100{,}000}$ **(b)** $2 \times 10^3 - \mathbf{2000}$

    (c) $3.2 \times 10^{-2} = \mathbf{0.032}$ **(d)** $2.5 \times 10^{-6} = \mathbf{0.0000025}$

Related Problem   Express 9.12×10^3 as a regular decimal number.

Calculations Using Powers of Ten

The advantage of scientific notation is in addition, subtraction, multiplication, and division of very small or very large numbers.

Addition   The steps for adding numbers in powers of ten are as follows:

1. Express the numbers to be added in the same power of ten.

2. Add the numbers without their powers of ten to get the sum.

3. Bring down the common power of ten, which is the power of ten of the sum.

EXAMPLE 1–4   Add 2×10^6 and 5×10^7 and express the result in scientific notation.

Solution   1. Express both numbers in the same power of ten: $(2 \times 10^6) + (50 \times 10^6)$.

2. Add $2 + 50 = 52$.

3. Bring down the common power of ten (10^6); the sum is $52 \times 10^6 = \mathbf{5.2 \times 10^7}$.

Related Problem   Add 3.1×10^3 and 5.5×10^4.

Subtraction   The steps for subtracting numbers in powers of ten are as follows:

1. Express the numbers to be subtracted in the same power of ten.

2. Subtract the numbers without their powers of ten to get the difference.

3. Bring down the common power of ten, which is the power of ten of the difference.

EXAMPLE 1–5   Subtract 2.5×10^{-12} from 7.5×10^{-11} and express the result in scientific notation.

Solution   1. Express each number in the same power of ten: $(7.5 \times 10^{-11}) - (0.25 \times 10^{-11})$.

2. Subtract $7.5 - 0.25 = 7.25$.

3. Bring down the common power of ten (10^{-11}); the difference is $\mathbf{7.25 \times 10^{-11}}$.

Related Problem   Subtract 3.5×10^{-6} from 2.2×10^{-5}.

Multiplication   The steps for multiplying numbers in powers of ten are as follows:

1. Multiply the numbers directly without their powers of ten.

2. Add the powers of ten algebraically (the exponents do not have to be the same).

EXAMPLE 1–6   Multiply 5×10^{12} and 3×10^{-6} and express the result in scientific notation.

Solution   Multiply the numbers, and algebraically add the powers.

$$(5 \times 10^{12})(3 \times 10^{-6}) = 15 \times 10^{12+(-6)} = 15 \times 10^6 = \mathbf{1.5 \times 10^7}$$

Related Problem   Multiply 3.2×10^6 and 1.5×10^{-3}.

Division   The steps for dividing numbers in powers of ten are as follows:

1. Divide the numbers directly without their powers of ten.

2. Subtract the power of ten in the denominator from the power of ten in the numerator (the powers do not have to be the same).

EXAMPLE 1–7

Divide 5.0×10^8 by 2.5×10^3 and express the result in scientific notation.

Solution   Write the division problem with a numerator and denominator as

$$\frac{5.0 \times 10^8}{2.5 \times 10^3}$$

Divide the numbers and subtract the powers of ten (3 from 8).

$$\frac{5.0 \times 10^8}{2.5 \times 10^3} = 2 \times 10^{8-3} = \mathbf{2 \times 10^5}$$

Related Problem   Divide 8×10^{-6} by 2×10^{-10}.

**SECTION 1–2
REVIEW**

1. Scientific notation uses powers of ten.   (True or False)
2. Express 100 as a power of ten.
3. Express the following numbers in scientific notation:
   (a) 4350   **(b)** 12,010   **(c)** 29,000,000
4. Express the following numbers in scientific notation:
   (a) 0.760   **(b)** 0.00025   **(c)** 0.000000597
5. Do the following operations:
   (a) $(1 \times 10^5) + (2 \times 10^5)$ **(b)** $(3 \times 10^6)(2 \times 10^4)$
   (c) $(8 \times 10^3) \div (4 \times 10^2)$ **(d)** $(2.5 \times 10^{-6}) - (1.3 \times 10^{-7})$

1–3   ENGINEERING NOTATION AND METRIC PREFIXES

Engineering notation, a specialized form of scientific notation, is used widely in technical fields to represent large and small quantities. In electronics, engineering notation is used to represent values of voltage, current, power, resistance, capacitance, inductance, and time, to name a few. Metric prefixes are used in conjunction with engineering notation as a "short hand" for the certain powers of ten that are multiples of three.

After completing this section, you should be able to

♦ **Use engineering notation and metric prefixes to represent large and small quantities**

   ♦ List the metric prefixes

   ♦ Change a power of ten in engineering notation to a metric prefix

   ♦ Use metric prefixes to express electrical quantities

   ♦ Convert one metric prefix to another

Engineering Notation

Engineering notation is similar to scientific notation. However, in **engineering notation** a number can have from one to three digits to the left of the decimal point and the power-of-ten exponent must be a multiple of three. For example, the number 33,000 expressed in engineering notation is 33×10^3. In scientific notation, it is expressed as 3.3×10^4. As another example, the number 0.045 expressed in engineering notation is 45×10^{-3}. In scientific notation, it is expressed as 4.5×10^{-2}.

EXAMPLE 1–8

Express the following numbers in engineering notation:

(a) 82,000   **(b)** 243,000   **(c)** 1,956,000

Solution

In engineering notation,

(a) 82,000 is expressed as **82×10^3**.

(b) 243,000 is expressed as **243×10^3**.

(c) 1,956,000 is expressed as **1.956×10^6**.

Related Problem   Express 36,000,000,000 in engineering notation.

EXAMPLE 1–9

Convert each of the following numbers to engineering notation:

(a) 0.0022   **(b)** 0.000000047   **(c)** 0.00033

Solution

In engineering notation,

(a) 0.0022 is expressed as **2.2×10^{-3}**.

(b) 0.000000047 is expressed as **47×10^{-9}**.

(c) 0.00033 is expressed as **330×10^{-6}**.

Related Problem   Express 0.0000000000056 in engineering notation.

Metric Prefixes

In engineering notation **metric prefixes** represent each of the most commonly used powers of ten. These metric prefixes are listed in Table 1–6 with their symbols and corresponding powers of ten.

Metric prefixes are used only with numbers that have a unit of measure, such as volts, amperes, and ohms, and precede the unit symbol. For example, 0.025 amperes can be expressed in engineering notation as 25×10^{-3} A. This quantity expressed using a metric prefix is 25 mA, which is read 25 milliamps. Note that the metric prefix *milli* has replaced 10^{-3}.

▶ TABLE 1–6

Metric prefixes with their symbols and corresponding powers of ten and values.

METRIC PREFIX	SYMBOL	POWER OF TEN	VALUE
femto	f	10^{-15}	one-quadrillionth
pico	p	10^{-12}	one-trillionth
nano	n	10^{-9}	one-billionth
micro	μ	10^{-6}	one-millionth
milli	m	10^{-3}	one-thousandth
kilo	k	10^3	one thousand
mega	M	10^6	one million
giga	G	10^9	one billion
tera	T	10^{12}	one trillion

As another example, 100,000,000 ohms can be expressed as $100 \times 10^6 \, \Omega$. This quantity expressed using a metric prefix is $100 \, M\Omega$, which is read 100 megohms. The metric prefix *mega* has replaced 10^6.

EXAMPLE 1–10

Express each quantity using a metric prefix:

(a) 50,000 V   (b) 25,000,000 Ω (c) 0.000036 A

Solution

(a) $50,000 \, V = 50 \times 10^3 \, V = \mathbf{50 \, kV}$

(b) $25,000,000 \, \Omega = 25 \times 10^6 \, \Omega = \mathbf{25 \, M\Omega}$

(c) $0.000036 \, A = 36 \times 10^{-6} \, A = \mathbf{36 \, \mu A}$

Related Problem

Express using metric prefixes:

(a) 56,000,000 Ω (b) 0.000470 A

Calculator Tip

All scientific and graphing calculators provide features for entering and displaying numbers in various formats. Scientific and engineering notation are special cases of exponential (power of ten) notation. Most calculators have a key labeled EE (or EXP) that is used to enter the exponent of numbers. To enter a number in exponential notation, enter the base number first, including the sign, and then press the EE key, followed by the exponent, including the sign.

Scientific and graphing calculators have displays for showing the power of ten. Some calculators display the exponent as a small raised number on the right side of the display.

47.0 ⁰³

Other calculators display the number with a small E followed by the exponent.

47.0E03

Notice that the base 10 is not generally shown, but it is implied or represented by the E. When you write the number out, you need to include the base 10. The displayed number shown above is written out as 47.0×10^3.

Some calculators are placed in the scientific or engineering notation mode using a secondary or tertiary function, such as SCI or ENG. Then numbers are entered in regular decimal form. The calculator automatically converts them to the proper format. Other calculators provide for mode selection using a menu.

Always check the owner's manual for your particular calculator to determine how to use the exponential notation features.

SECTION 1–3 REVIEW

1. Express the following numbers in engineering notation:
   (a) 0.0056   (b) 0.0000000283   (c) 950,000   (d) 375,000,000,000

2. List the metric prefix for each of the following powers of ten:
   10^6, 10^3, 10^{-3}, 10^{-6}, 10^{-9}, and 10^{-12}

3. Use an appropriate metric prefix to express 0.000001 A.

4. Use an appropriate metric prefix to express 250,000 W.

1–4 METRIC UNIT CONVERSIONS

It is sometimes necessary or convenient to convert a quantity from one unit with a metric prefix to another, such as from milliamperes (mA) to microamperes (μA). Moving the decimal point in the number an appropriate number of places to the left or to the right, depending on the particular conversion, results in a metric unit conversion.

After completing this section, you should be able to

♦ **Convert from one unit with a metric prefix to another**

   ♦ Convert between milli, micro, nano, and pico

   ♦ Convert between kilo and mega

The following basic rules apply to metric unit conversions:

1. When converting from a larger unit to a smaller unit, move the decimal point to the right.

2. When converting from a smaller unit to a larger unit, move the decimal point to the left.

3. Determine the number of places to move the decimal point by finding the difference in the powers of ten of the units being converted.

For example, when converting from milliamperes (mA) to microamperes (μA), move the decimal point three places to the right because there is a three-place difference between the two units (mA is 10^{-3} A and μA is 10^{-6} A). The following examples illustrate a few conversions.

EXAMPLE 1–11

Convert 0.15 milliampere (0.15 mA) to microamperes (μA).

Solution   Move the decimal point three places to the right.

$$0.15\,\text{mA} = 0.15 \times 10^{-3}\,\text{A} = 150 \times 10^{-6}\,\text{A} = \mathbf{150\,\mu A}$$

Related Problem   Convert 1 mA to microamperes.

EXAMPLE 1–12

Convert 4500 microvolts (4500 μV) to millivolts (mV).

Solution   Move the decimal point three places to the left.

$$4500\,\mu V = 4500 \times 10^{-6}\,\text{V} = 4.5 \times 10^{-3}\,\text{V} = \mathbf{4.5\,mV}$$

Related Problem   Convert 1000 μV to millivolts.

EXAMPLE 1–13

Convert 5000 nanoamperes (5000 nA) to microamperes (μA).

Solution   Move the decimal point three places to the left.

$$5000\,\text{nA} = 5000 \times 10^{-9}\,\text{A} = 5 \times 10^{-6}\,\text{A} = \mathbf{5\,\mu A}$$

Related Problem   Convert 893 nA to microamperes.

EXAMPLE 1–14

Convert 47,000 picofarads (47,000 pF) to microfarads (μF).

Solution Move the decimal point six places to the left.

$$47{,}000 \text{ pF} = 47{,}000 \times 10^{-12}\,\text{F} = 0.047 \times 10^{-6}\,\text{F} = \mathbf{0.047\ \mu F}$$

Related Problem Convert 10,000 pF to microfarads.

EXAMPLE 1–15

Convert 0.00022 microfarad (0.00022 μF) to picofarads (pF).

Solution Move the decimal point six places to the right.

$$0.00022\ \mu\text{F} = 0.00022 \times 10^{-6}\,\text{F} = 220 \times 10^{-12}\,\text{F} = \mathbf{220\ pF}$$

Related Problem Convert 0.0022 μF to picofarads.

EXAMPLE 1–16

Convert 1800 kilohms (1800 kΩ) to megohms (MΩ).

Solution Move the decimal point three places to the left.

$$1800\ \text{k}\Omega = 1800 \times 10^{3}\,\Omega = 1.8 \times 10^{6}\,\Omega = \mathbf{1.8\ M\Omega}$$

Related Problem Convert 2.2 kΩ to megohms.

When adding (or subtracting) quantities with different metric prefixes, first convert one of the quantities to the same prefix as the other quantity.

EXAMPLE 1–17

Add 15 mA and 8000 μA and express the sum in milliamperes.

Solution Convert 8000 μA to 8 mA and add.

$$15\ \text{mA} + 8000\ \mu\text{A} = 15 \times 10^{-3}\,\text{A} + 8000 \times 10^{-6}\,\text{A}$$
$$= 15 \times 10^{-3}\,\text{A} + 8 \times 10^{-3}\,\text{A} = 15\ \text{mA} + 8\ \text{mA} = \mathbf{23\ mA}$$

Related Problem Add 2873 mA to 10,000 μA; express the sum in milliamperes.

SECTION 1–4 REVIEW

1. Convert 0.01 MV to kilovolts (kV).
2. Convert 250,000 pA to milliamperes (mA).
3. Add 0.05 MW and 75 kW and express the result in kW.
4. Add 50 mV and 25,000 μV and express the result in mV.

SUMMARY

- ◆ SI is an abbreviation for Le Système International d'Unités and is a standardized system of units.
- ◆ A fundamental unit is an SI unit from which other SI units are derived. There are seven fundamental units.
- ◆ Scientific notation is a method for representing very large and very small numbers as a number between one and ten (one digit to left of decimal point) times a power of ten.
- ◆ Engineering notation is a form of scientific notation in which quantities are represented with one, two, or three digits to the left of the decimal point times a power of ten that is a multiple of three.
- ◆ Metric prefixes represent powers of ten in numbers expressed in engineering notation.

KEY TERMS

These key terms are also defined in the end-of-book glossary.

Engineering notation   A system for representing any number as a one-, two-, or three-digit number times a power of ten with an exponent that is a multiple of 3.

Exponent   The number to which a base number is raised.

Metric prefix   An affix that represents a power-of-ten number expressed in engineering notation.

Power of ten   A numerical representation consisting of a base of 10 and an exponent; the number 10 raised to a power.

Scientific notation   A system for representing any number as a number between 1 and 10 times an appropriate power of ten.

SI   Standardized internationalized system of units used for all engineering and scientific work; abbreviation for French *Le Système International d'Unités*.

SELF-TEST

Answers are at the end of the chapter.

1. Which of the following is not an electrical quantity?
   (a) current   (b) voltage   (c) time   (d) power
2. The unit of current is
   (a) volt   (b) watt   (c) ampere   (d) joule
3. The unit of voltage is
   (a) ohm   (b) watt   (c) volt   (d) farad
4. The unit of resistance is
   (a) ampere   (b) henry   (c) hertz   (d) ohm
5. Hertz is the unit of
   (a) power   (b) inductance   (c) frequency   (d) time
6. 15,000 W is the same as
   (a) 15 mW   (b) 15 kW   (c) 15 MW   (d) 15 μW
7. The quantity 4.7×10^3 is the same as
   (a) 470   (b) 4700   (c) 47,000   (d) 0.0047
8. The quantity 56×10^{-3} is the same as
   (a) 0.056   (b) 0.560   (c) 560   (d) 56,000
9. The number 3,300,000 can be expressed in engineering notation as
   (a) 3300×10^3 (b) 3.3×10^{-6} (c) 3.3×10^6 (d) either answer (a) or (c)
10. Ten milliamperes can be expressed as
    (a) 10 MA   (b) 10 μA   (c) 10 kA   (d) 10 mA
11. Five thousand volts can be expressed as
    (a) 5000 V   (b) 5 MV   (c) 5 kV   (d) either answer (a) or (c)
12. Twenty million ohms can be expressed as
    (a) 20 mΩ   (b) 20 MW   (c) 20 MΩ   (d) 20 μΩ

PROBLEMS   Answers to odd-numbered problems are at the end of the book.

SECTION 1–2   Scientific Notation

1. Express each of the following numbers in scientific notation:
   (a) 3000   (b) 75,000   (c) 2,000,000

2. Express each fractional number in scientific notation:
   (a) 1/500   (b) 1/2000   (c) 1/5,000,000

3. Express each of the following numbers in scientific notation:
   (a) 8400   (b) 99,000   (c) 0.2×10^6

4. Express each of the following numbers in scientific notation:
   (a) 0.0002   (b) 0.6   (c) 7.8×10^{-2}

5. Express each of the following numbers in scientific notation:
   (a) 32×10^3 (b) 6800×10^{-6} (c) 870×10^8

6. Express each of the following as a regular decimal number:
   (a) 2×10^5 (b) 5.4×10^{-9} (c) 1.0×10^1

7. Express each of the following as a regular decimal number:
   (a) 2.5×10^{-6} (b) 5.0×10^2 (c) 3.9×10^{-1}

8. Express each number in regular decimal form:
   (a) 4.5×10^{-6} (b) 8×10^{-9} (c) 4.0×10^{-12}

9. Add the following numbers:
   (a) $(9.2 \times 10^6) + (3.4 \times 10^7)$ (b) $(5 \times 10^3) + (8.5 \times 10^{-1})$
   (c) $(5.6 \times 10^{-8}) + (4.6 \times 10^{-9})$

10. Perform the following subtractions:
    (a) $(3.2 \times 10^{12}) - (1.1 \times 10^{12})$ (b) $(2.6 \times 10^8) - (1.3 \times 10^7)$
    (c) $(1.5 \times 10^{-12}) - (8 \times 10^{-13})$

11. Perform the following multiplications:
    (a) $(5 \times 10^3)(4 \times 10^5)$ (b) $(1.2 \times 10^{12})(3 \times 10^2)$
    (c) $(2.2 \times 10^{-9})(7 \times 10^{-6})$

12. Divide the following:
    (a) $(1.0 \times 10^3) \div (2.5 \times 10^2)$ (b) $(2.5 \times 10^{-6}) \div (5.0 \times 10^{-8})$
    (c) $(4.2 \times 10^8) \div (2 \times 10^{-5})$

SECTION 1–3   Engineering Notation and Metric Prefixes

13. Express each of the following numbers in engineering notation:
    (a) 89,000   (b) 450,000   (c) 12,040,000,000,000

14. Express each number in engineering notation:
    (a) 2.35×10^5 (b) 7.32×10^7 (c) 1.333×10^9

15. Express each number in engineering notation:
    (a) 0.000345   (b) 0.025   (c) 0.00000000129

16. Express each number in engineering notation:
    (a) 9.81×10^{-3} (b) 4.82×10^{-4} (c) 4.38×10^{-7}

17. Add the following numbers and express each result in engineering notation:
    (a) $(2.5 \times 10^{-3}) + (4.6 \times 10^{-3})$ (b) $(68 \times 10^6) + (33 \times 10^6)$
    (c) $(1.25 \times 10^6) + (250 \times 10^3)$

18. Multiply the following numbers and express each result in engineering notation:
    (a) $(32 \times 10^{-3})(56 \times 10^3)$ (b) $(1.2 \times 10^{-6})(1.2 \times 10^{-6})$
    (c) $100(55 \times 10^{-3})$

19. Divide the following numbers and express each result in engineering notation:
    (a) $50 \div (2.2 \times 10^3)$ (b) $(5 \times 10^3) \div (25 \times 10^{-6})$
    (c) $560 \times 10^3 \div (660 \times 10^3)$
20. Express each number in Problem 13 in ohms using a metric prefix.
21. Express each number in Problem 15 in amperes using a metric prefix.
22. Express each of the following as a quantity having a metric prefix:
    (a) 31×10^{-3} A   (b) 5.5×10^3 V   (c) 20×10^{-12} F
23. Express the following using metric prefixes:
    (a) 3×10^{-6} F   (b) 3.3×10^6 Ω   (c) 350×10^{-9} A
24. Express the following using metric prefixes:
    (a) 2.5×10^{-12} A   (b) 8×10^9 Hz   (c) 4.7×10^3 Ω
25. Express each quantity by converting the metric prefix to a power-of-10:
    (a) 7.5 pA   (b) 3.3 GHz   (c) 280 nW
26. Express each quantity in engineering notation:
    (a) 5 μA   (b) 43 mV   (c) 275 kΩ   (d) 10 MW

SECTION 1–4   Metric Unit Conversions

27. Perform the indicated conversions:
    (a) 5 mA to microamperes   (b) 3200 μW to milliwatts
    (c) 5000 kV to megavolts   (d) 10 MW to kilowatts
28. Determine the following:
    (a) The number of microamperes in 1 milliampere
    (b) The number of millivolts in 0.05 kilovolt
    (c) The number of megohms in 0.02 kilohm
    (d) The number of kilowatts in 155 milliwatts
29. Add the following quantities:
    (a) 50 mA + 680 μA   (b) 120 kΩ + 2.2 MΩ   (c) 0.02 μF + 3300 pF
30. Do the following operations:
    (a) $10\,kΩ \div (2.2\,kΩ + 10\,kΩ)$ (b) $250\,mV \div 50\,μV$ (c) $1\,MW \div 2\,kW$

ANSWERS

SECTION REVIEWS

SECTION 1–1   Units of Measurement

1. Fundamental units define derived units.
2. Ampere
3. SI is the abbreviation for Système International.
4. Refer to Table 1–3 after you have compiled your list of electrical quantities.
5. Refer to Table 1–4 after you have compiled your list of magnetic quantities.

SECTION 1–2   Scientific Notation

1. True
2. 10^2
3. (a) 4.35×10^3 (b) 1.201×10^4 (c) 2.9×10^7
4. (a) 7.6×10^{-1} (b) 2.5×10^{-4} (c) 5.97×10^{-7}
5. (a) 3×10^5 (b) 6×10^{10} (c) 2×10^1 (d) 2.37×10^{-6}

SECTION 1–3 **Engineering Notation and Metric Prefixes**

 1. (a) 5.6×10^{-3} (b) 28.3×10^{-9} (c) 950×10^{3} (d) 375×10^{9}
 2. Mega (M), kilo (k), milli (m), micro (μ), nano (n), and pico (p)
 3. $1\ \mu$A (one microampere)
 4. 250 kW (250 kilowatts)

SECTION 1–4 **Metric Unit Conversions**

 1. $0.01\ \text{MV} = 10\ \text{kV}$
 2. $250{,}000\ \text{pA} = 0.00025\ \text{mA}$
 3. $0.05\ \text{MW} + 75\ \text{kW} = 50\ \text{kW} + 75\ \text{kW} = 125\ \text{kW}$
 4. $50\ \text{mV} + 25{,}000\ \mu\text{V} = 50\ \text{mV} + 25\ \text{mV} = 75\ \text{mV}$

RELATED PROBLEMS FOR EXAMPLES

1–1 4.75×10^{3}
1–2 7.38×10^{-3}
1–3 9120
1–4 5.81×10^{4}
1–5 1.85×10^{-5}
1–6 4.8×10^{3}
1–7 4×10^{4}
1–8 36×10^{9}
1–9 5.6×10^{-12}
1–10 (a) $56\ \text{M}\Omega$ (b) $470\ \mu$A
1–11 $1000\ \mu$A
1–12 1 mV
1–13 $0.893\ \mu$A
1–14 $0.01\ \mu$F
1–15 2200 pF
1–16 $0.0022\ \text{M}\Omega$
1–17 2883 mA

SELF-TEST

 1. (c)   2. (c)   3. (c)   4. (d)   5. (c)   6. (b)
 7. (b)   8. (a)   9. (d)   10. (d)   11. (d)   12. (c)

2

VOLTAGE, CURRENT, AND RESISTANCE

CHAPTER OBJECTIVES

◆ Describe the basic structure of atoms

◆ Explain the concept of electrical charge

◆ Define *voltage*, *current*, and *resistance* and discuss the characteristics of each

◆ Discuss a voltage source and a current source

◆ Recognize and discuss various types and values of resistors

◆ Describe a basic electric circuit

◆ Make basic circuit measurements

◆ Recognize electrical hazards and practice proper safety procedures

KEY TERMS

◆ Atom
◆ Electron
◆ Free electron
◆ Conductor
◆ Semiconductor
◆ Insulator
◆ Charge
◆ Coulomb
◆ Voltage
◆ Volt
◆ Current
◆ Ampere
◆ Resistance
◆ Ohm
◆ Conductance
◆ Siemens

◆ Voltage source
◆ Current source
◆ Resistor
◆ Potentiometer
◆ Rheostat
◆ Circuit
◆ Load
◆ Closed circuit
◆ Open circuit
◆ AWG
◆ Ground
◆ Voltmeter
◆ Ammeter
◆ Ohmmeter
◆ DMM
◆ Electrical shock

A CIRCUIT APPLICATION PREVIEW

In the circuit application, you will see how the theory presented in this chapter is applied to a practical circuit that simulates part of a car's lighting system. An automobile's lights are examples of simple types of electric circuits. When you turn on the headlights and taillights, you are connecting the light bulbs to the battery, which provides the voltage and produces current through each bulb. The current causes the bulbs to emit light. The light bulbs themselves have resistance that limits the amount of current. The instrument panel light in most cars can be adjusted for brightness. By turning a knob, you actually change the resistance in the circuit, thereby causing the current to change. The amount of current through the light bulb determines its brightness.

VISIT THE COMPANION WEBSITE

Study aids for this chapter are available at
http://www.prenhall.com/floyd

INTRODUCTION

The useful application of electronics technology to practical situations requires that you first understand the theory that is the basis of a given application. Once you have mastered the theory, you can learn to apply it in practice. In this chapter and throughout the rest of the book, you will learn to put technology theory into practice in circuit applications.

The theoretical concepts of electrical current, voltage, and resistance are introduced in this chapter. You will learn how to express each of these quantities in the proper units and how each quantity is measured. The essential elements that form a basic electric circuit and how they are put together are covered.

You will be introduced to the types of devices that generate voltage and current. In addition, you will see a variety of components that are used to introduce resistance into electric circuits. The operation of protective devices such as fuses and circuit breakers are discussed, and mechanical switches that are commonly used in electric circuits are introduced. Also, you will learn how to control and measure voltage, current, and resistance using measuring instruments.

Voltage is essential in any kind of electric circuit. Voltage is the potential energy of electrical charge required to make the circuit work. Current is also necessary for electric circuits to operate, but it takes voltage to produce the current. Current is the movement of electrons through a circuit. Resistance in a circuit limits the amount of current. A water system can be used as an analogy for a simple circuit. Voltage can be considered analogous to the pressure required to force water through the pipes. Current through wires can be thought of as analogous to the water moving through the pipes. Resistance can be thought of as analogous to the restriction on the water flow produced by adjusting a valve.

2–1   ATOMIC STRUCTURE

All matter is made of atoms; and all atoms consist of electrons, protons, and neutrons. In this section, you will learn about the structure of an atom, including electron shells and orbits, valence electrons, ions, and energy levels. The configuration of certain electrons in an atom is the key factor in determining how well a given conductive or semiconductive material conducts electric current.

After completing this section, you should be able to

◆ **Describe the basic structure of atoms**

    ◆ Define *nucleus, proton, neutron,* and *electron*

    ◆ Define *atomic number*

    ◆ Define *shell*

    ◆ Explain what a valence electron is

    ◆ Describe ionization

    ◆ Explain what a free electron is

    ◆ Define *conductor, semiconductor,* and *insulator*

An **atom** is the smallest particle of an **element** that retains the characteristics of that element. Each of the known 109 elements has atoms that are different from the atoms of all other elements. This gives each element a unique atomic structure. According to the classic Bohr model, an atom is visualized as having a planetary type of structure that consists of a central nucleus surrounded by orbiting electrons, as illustrated in Figure 2–1. The **nucleus** consists of positively charged particles called **protons** and uncharged particles called **neutrons**. The basic particles of negative charge are called **electrons**.

◀ **FIGURE 2–1**

The Bohr model of an atom showing electrons in circular orbits around the nucleus. The "tails" on the electrons indicate they are moving.

⊖ Electron   ⊕ Proton   ◯ Neutron

Each type of atom has a certain number of electrons and protons that distinguishes it from the atoms of all other elements. For example, the simplest atom is that of hydrogen, which has one proton and one electron, as pictured in Figure 2–2(a). As another example, the helium atom, shown in Figure 2–2(b), has two protons and two neutrons in the nucleus and two electrons orbiting the nucleus.

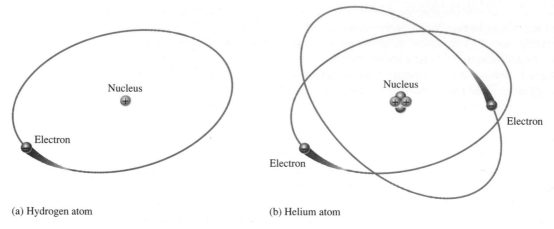

(a) Hydrogen atom   (b) Helium atom

▲ FIGURE 2–2

The two simplest atoms, hydrogen and helium.

Atomic Number

All elements are arranged in the periodic table of the elements in order according to their **atomic number**. The atomic number equals the number of protons in the nucleus. For example, hydrogen has an atomic number of 1 and helium has an atomic number of 2. In their normal (or neutral) state, all atoms of a given element have the same number of electrons as protons; the positive charges cancel the negative charges, and the atom has a net charge of zero, making it electrically balanced.

Shells, Orbits, and Energy Levels

As you have seen in the Bohr model, electrons orbit the nucleus of an atom at certain distances from the nucleus and are restricted to these specific orbits. Each orbit corresponds to a different energy level within the atom known as a **shell**. The shells are designated 1, 2, 3, and so on, with 1 being closest to the nucleus. Electrons further from the nucleus are at higher energy levels.

The line spectrums of hydrogen from the Bohr model of the atom shows that the electrons can only absorb or emit a specific amount of energy that represents the exact difference between the levels. Figure 2–3 shows the energy levels within the hydrogen

▶ FIGURE 2–3

Energy levels in hydrogen.

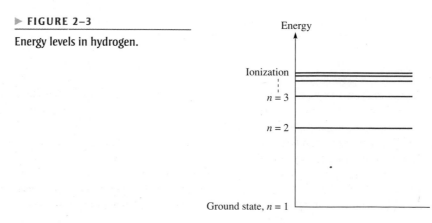

atom. The lowest level ($n = 1$) is called the *ground state* and represents the most stable atom with a single electron in the first shell. If this electron acquires a specific amount of energy by absorbing a photon, it can be raised to one of the higher energy levels. In this higher state, it can emit a photon with exactly the same energy and return to the ground state. Transitions between the levels account for various phenomena we see in electronics, such as the color of light from a light-emitting diode.

After Bohr's work, Erin Schroedinger (1887–1961) proposed a mathematical theory for the atom that explained more complicated atoms. He suggested that the electron has a wavelike property, and he considered the simplest case as having a three-dimensional standing wave pattern due to vibrations. Schroedinger theorized the standing wave of an electron with a spherical shape could have only certain wavelengths. This wave-mechanics model of the atom gave the same equation for the electron energy in hydrogen as Bohr's model, but in the wave-mechanics model, more complicated atoms could be explained by involving shapes other than spheres and adding a designation for the orientation of a given shape within the atom. In both models, electrons near the nucleus have less energy than those further out, which was the basic concept of the energy levels.

The idea of discrete energy levels within the atom is still a foundation for understanding the atom, and the wave-mechanics model has been very successful at predicting the energy levels for various atoms. The wave-mechanics model of the atom used the shell number, called the *principal quantum number,* in the energy equation. Three other quantum numbers describe each electron within the atom. All electrons in an atom have a unique set of quantum numbers.

When an atom is part of a large group, as in a crystal, the discrete energy levels broaden into energy bands, which is an important idea in solid-state electronics. The bands also differentiate between conductors, semiconductors, and insulators.

Valence Electrons

Electrons that are in orbits farther from the nucleus have higher energy and are less tightly bound to the atom than those closer to the nucleus. This is because the force of attraction between the positively charged nucleus and the negatively charged electron decreases with increasing distance from the nucleus. Electrons with the highest energy levels exist in the outermost shell of an atom and are relatively loosely bound to the atom. This outermost shell is known as the **valence** shell, and electrons in this shell are called **valence electrons**. These valence electrons contribute to chemical reactions and bonding within the structure of a material, and they determine the material's electrical properties.

Energy Levels and Ionization Energy

If an electron absorbs a photon with sufficient energy, it escapes from the atom and becomes a **free electron**. This is indicated by the ionization energy level in Figure 2–3. Any time an atom or group of atoms is left with a net charge, it is called an **ion**. When an electron escapes from the neutral hydrogen atom (designated H), the atom is left with a net positive charge and becomes a *positive ion* (designated H^+). In some cases, an atom or group of atoms can acquire an electron, in which case it is called a *negative ion*.

The Copper Atom

Copper is the most commonly used metal in **electrical** applications. The copper atom has 29 electrons that orbit the nucleus in four shells. The number of electrons in each shell follows a predictable pattern according to the formula, $2N^2$, where N is the number of the shell. The first shell of any atom can have up to 2 electrons, the second shell up to 8 electrons, the third shell up to 18 electrons, and the fourth shell up to 32 electrons.

A copper atom is represented in Figure 2–4. Notice that the fourth or outermost shell, the valence shell, has only 1 valence electron. When the valence electron in the outer shell of the copper atom gains sufficient thermal energy, it can break away from the parent atom

▶ FIGURE 2–4

The copper atom.

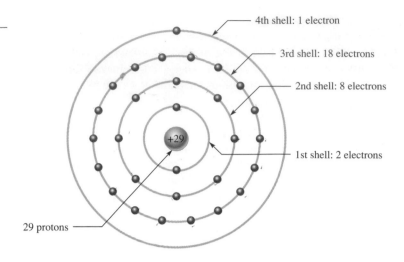

4th shell: 1 electron

3rd shell: 18 electrons

2nd shell: 8 electrons

1st shell: 2 electrons

29 protons

and become a free electron. In a piece of copper at room temperature, a "sea" of these free electrons is present. These electrons are not bound to a given atom but are free to move in the copper material. Free electrons make copper an excellent conductor and make electrical current possible.

Categories of Materials

Three categories of materials are used in electronics: conductors, semiconductors, and insulators.

Conductors   **Conductors** are materials that readily allow current. They have a large number of free electrons and are characterized by one to three valence electrons in their structure. Most metals are good conductors. Silver is the best conductor, and copper is next. Copper is the most widely used conductive material because it is less expensive than silver. Copper wire is commonly used as a conductor in electric circuits.

Semiconductors   **Semiconductors** are classed below the conductors in their ability to carry current because they have fewer free electrons than do conductors. Semiconductors have four valence electrons in their atomic structures. However, because of their unique characteristics, certain semiconductor materials are the basis for **electronic** devices such as the diode, transistor, and integrated circuit. Silicon and germanium are common semiconductive materials.

Insulators   **Insulators** are materials that are poor conductors of electric current. In fact, insulators are used to prevent current where it is not wanted. Compared to conductive materials, insulators have very few free electrons and are characterized by more than four valence electrons in their atomic structures.

**SECTION 2–1
REVIEW**
Answers are at the end of the chapter.

1. What is the basic particle of negative charge?
2. Define *atom*.
3. What does an atom consist of?
4. Define *atomic number*.
5. Do all elements have the same types of atoms?
6. What is a free electron?
7. What is a shell in the atomic structure?
8. Name two conductive materials.

2–2 ELECTRICAL CHARGE

As you know, an electron is the smallest particle that exhibits negative electrical charge. When an excess of electrons exists in a material, there is a net negative electrical charge. When a deficiency of electrons exists, there is a net positive electrical charge.

After completing this section, you should be able to

♦ **Explain the concept of electrical charge**

  ♦ Name the unit of charge

  ♦ Name the types of charge

  ♦ Discuss attractive and repulsive forces

  ♦ Determine the amount of charge on a given number of electrons

The charge of an electron and that of a proton are equal in magnitude. Electrical **charge**, an electrical property of matter that exists because of an excess or deficiency of electrons, is symbolized by Q. Static electricity is the presence of a net positive or negative charge in a material. Everyone has experienced the effects of static electricity from time to time, for example, when attempting to touch a metal surface or another person or when the clothes in a dryer cling together.

Materials with charges of opposite polarity are attracted to each other, and materials with charges of the same polarity are repelled, as indicated in Figure 2–5. A force acts between charges, as evidenced by the attraction or repulsion. This force, called an *electric field,* consists of invisible lines of force, as represented in Figure 2–6.

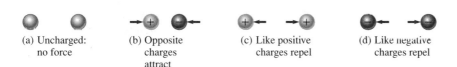

(a) Uncharged: no force (b) Opposite charges attract (c) Like positive charges repel (d) Like negative charges repel

▲ **FIGURE 2–5**

Attraction and repulsion of electrical charges.

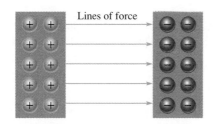

Lines of force

◀ **FIGURE 2–6**

Electric field between two oppositely charged surfaces.

BIOGRAPHY

Charles Augustin Coulomb, 1736–1806

Coulomb, a Frenchman, spent many years as a military engineer. When bad health forced him to retire, he devoted his time to scientific research. He is best known for his work on electricity and magnetism due to his development of the inverse square law for the force between two charges. The unit of electrical charge is named in his honor. (Photo credit: Courtesy of the Smithsonian Institution. Photo number 52,597.)

Coulomb: The Unit of Charge

Electrical charge (Q) is measured in coulombs, symbolized by C.

One coulomb is the total charge possessed by 6.25×10^{18} electrons.

A single electron has a charge of 1.6×10^{-19} C. The total charge Q, expressed in coulombs, for a given number of electrons is stated in the following formula:

$$Q = \frac{\text{number of electrons}}{6.25 \times 10^{18} \text{ electrons/C}}$$

Equation 2–1

Positive and Negative Charge

Consider a neutral atom—that is, one that has the same number of electrons and protons and thus has no net charge. As you know, when a valence electron is pulled away from the atom by the application of energy, the atom is left with a net positive charge (more protons than electrons) and becomes a positive ion. If an atom acquires an extra electron in its outer shell, it has a net negative charge and becomes a negative ion.

The amount of energy required to free a valence electron is related to the number of electrons in the outer shell. An atom can have up to eight valence electrons. The more complete the outer shell, the more stable the atom and thus the more energy is required to release an electron. Figure 2–7 illustrates the creation of a positive ion and a negative ion when a hydrogen atom gives up its single valence electron to a chlorine atom, forming gaseous hydrogen chloride (HCl). When the gaseous HCl is dissolved in water, hydrochloric acid is formed.

▶ FIGURE 2–7

Example of the formation of positive and negative ions.

 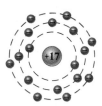

Hydrogen atom
(1 proton, 1 electron)

Chlorine atom
(17 protons, 17 electrons)

(a) The neutral hydrogen atom has a single valence electron.

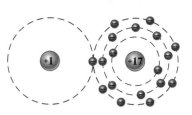

(b) The atoms combine by sharing the valence electron to form gaseous hydrogen chloride (HCl).

Positive hydrogen ion
(1 proton, no electrons)

Negative chloride ion
(17 protons, 18 electrons)

(c) When dissolved in water, hydrogen chloride gas separates into positive hydrogen ions and negative chloride ions. The chlorine atom retains the electron given up by the hydrogen atom forming both positive and negative ions in the same solution.

EXAMPLE 2–1

How many coulombs do 93.8×10^{16} electrons represent?

Solution $Q = \dfrac{\text{number of electrons}}{6.25 \times 10^{18} \text{ electrons/C}} = \dfrac{93.8 \times 10^{16} \text{ electrons}}{6.25 \times 10^{18} \text{ electrons/C}} = 15 \times 10^{-2} \text{ C} = \mathbf{0.15\ C}$

Related Problem* How many electrons does it take to have 3 C of charge?

---

*Answers are at the end of the chapter.

SECTION 2–2 REVIEW

1. What is the symbol for charge?
2. What is the unit of charge, and what is the unit symbol?
3. What causes positive and negative charge?
4. How much charge, in coulombs, is there in 10×10^{12} electrons?

2–3 VOLTAGE, CURRENT, AND RESISTANCE

Voltage, current, and resistance are the basic quantities present in all electrical circuits. Voltage is necessary to produce current, and resistance limits the amount of current in a circuit. The relationship of these three quantities is described by Ohm's law in Chapter 3.

After completing this section, you should be able to

- ◆ **Define *voltage, current*, and *resistance* and discuss the characteristics of each**
  - ◆ State the formula for voltage and name its unit
  - ◆ State the formula for current and name its unit
  - ◆ Explain the movement of electrons
  - ◆ Name the unit of resistance

Voltage

As you have seen, a force of attraction exists between a positive and a negative charge. A certain amount of energy must be exerted, in the form of work, to overcome the force and move the charges a given distance apart. All opposite charges possess a certain potential energy because of the separation between them. The difference in potential energy per charge is the potential difference or **voltage**. Voltage is the driving force in electric circuits and is what establishes current.

As an analogy, consider a water tank that is supported several feet above the ground. A given amount of energy must be exerted in the form of work to pump water up to fill the tank. Once the water is stored in the tank, it has a certain potential energy which, if released, can be used to perform work.

Voltage, symbolized by V, is defined as energy or work per unit charge.

$$V = \frac{W}{Q} \qquad \textbf{Equation 2–2}$$

where: V = voltage in volts (V)
W = energy in joules (J)
Q = charge in coulombs (C)

The unit of voltage is the volt, symbolized by V.

One volt is the potential difference (voltage) between two points when one joule of energy is used to move one coulomb of charge from one point to the other.

BIOGRAPHY

Alessandro Volta
1745–1827

Volta, an Italian, invented a device to generate static electricity and he also discovered methane gas. Volta investigated reactions between dissimilar metals and developed the first battery in 1800. Electrical potential, more commonly known as voltage, and the unit of voltage, the volt, are named in his honor. (Photo credit: AIP Emilio Segrè Visual Archives, Lande Collection.)

EXAMPLE 2–2

If 50 J of energy are available for every 10 C of charge, what is the voltage?

Solution

$$V = \frac{W}{Q} = \frac{50\,\text{J}}{10\,\text{C}} = \textbf{5 V}$$

Related Problem

How much energy is used to move 50 C from one point to another when the voltage between the two points is 12 V?

Current

Voltage provides energy to electrons, allowing them to move through a circuit. This movement of electrons is the current, which results in work being done in an electrical circuit.

As you have learned, free electrons are available in all conductive and semiconductive materials. These electrons drift randomly in all directions, from atom to atom, within the structure of the material, as indicated in Figure 2–8.

▶ **FIGURE 2–8**

Random motion of free electrons in a material.

If a voltage is placed across a conductive or semiconductive material, one end becomes positive and the other negative, as indicated in Figure 2–9. The repulsive force produced by the negative voltage at the left end causes the free electrons (negative charges) to move toward the right. The attractive force produced by the positive voltage at the right end pulls the free electrons to the right. The result is a net movement of the free electrons from the negative end of the material to the positive end, as shown in Figure 2–9.

▶ **FIGURE 2–9**

Electrons flow from negative to positive when a voltage is applied across a conductive or semiconductive material.

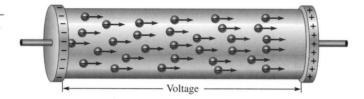

Voltage

The movement of these free electrons from the negative end of the material to the positive end is the electrical current, symbolized by *I*.

Electrical current is the rate of flow of charge.

Current in a conductive material is determined by the number of electrons (amount of charge) that flow past a point in a unit of time.

Equation 2–3

$$I = \frac{Q}{t}$$

where: I = current in amperes (A)

Q = charge in coulombs (C)

t = time in seconds (s)

One ampere (1 A) is the amount of current that exists when a number of electrons having a total charge of one coulomb (1 C) move through a given cross-sectional area in one second (1 s).

See Figure 2–10. Remember, one coulomb is the charge carried by 6.25×10^{18} electrons.

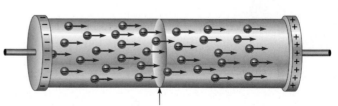

When a number of electrons having a total charge of 1 C pass through a cross-sectional area in 1 s, there is 1 A of current.

▲ **FIGURE 2–10**

Illustration of 1 A of current (1 C/s) in a material.

EXAMPLE 2–3	Ten coulombs of charge flow past a given point in a wire in 2 s. What is the current in amperes?
Solution	$$I = \frac{Q}{t} = \frac{10\,C}{2\,s} = 5\,A$$
Related Problem	If there are 8 A of current through the filament of a lamp, how many coulombs of charge move through the filament in 1.5 s?

Resistance

When there is current through a material, the free electrons move through the material and occasionally collide with atoms. These collisions cause the electrons to lose some of their energy, thus restricting their movement. The more collisions, the more the flow of electrons is restricted. This restriction varies and is determined by the type of material. The property of a material to restrict or oppose the flow of electrons is called resistance, R.

Resistance is the opposition to current.

Resistance is expressed in ohms, symbolized by the Greek letter omega (Ω).

One ohm (1 Ω) of resistance exists if there is one ampere (1 A) of current in a material when one volt (1 V) is applied across the material.

The schematic symbol for resistance is shown in Figure 2–11.

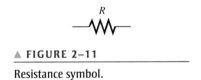

▲ FIGURE 2–11

Resistance symbol.

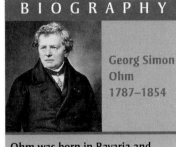

BIOGRAPHY

Georg Simon Ohm
1787–1854

Ohm was born in Bavaria and struggled for years to gain recognition for his work in formulating the relationship of current, voltage, and resistance. This mathematical relationship is known today as Ohm's law and the unit of resistance is named in his honor. (Photo credit: Library of Congress, LC-USZ62-40943.)

Conductance The reciprocal of resistance is **conductance**, symbolized by G. It is a measure of the ease with which current is established. The formula is

$$G = \frac{1}{R}$$

Equation 2–4

The unit of conductance is the **siemens**, abbreviated S. For example, the conductance of a 22 kΩ resistor is

$$G = \frac{1}{22\ k\Omega} = 45.5\ \mu S$$

The obsolete unit of *mho* (ohm spelled backwards) was previously used for conductance.

SECTION 2–3 REVIEW	1. Define *voltage*.
	2. What is the unit of voltage?
	3. How much is the voltage when there are 24 joules of energy for 10 coulombs of charge?
	4. Define *current* and state its unit.
	5. How many electrons make up one coulomb of charge?
	6. What is the current in amperes when 20 C flow past a point in a wire in 4 s?
	7. Define *resistance*.
	8. Name the unit of resistance.
	9. Define one ohm.

2–4 VOLTAGE AND CURRENT SOURCES

A **voltage source** provides electrical energy or electromotive force (emf), more commonly known as voltage. Voltage is produced by means of chemical energy, light energy, and magnetic energy combined with mechanical motion. A current source provides a constant current to a load.

After completing this section, you should be able to

◆ **Discuss a voltage source and a current source**

   ◆ List six categories of voltage sources

   ◆ Describe the basic operation of a battery

   ◆ Explain how a solar cell creates voltage

   ◆ Discuss the principle of a generator

   ◆ Describe what an electronic power supply does

BIOGRAPHY

Ernst Werner
von Siemens
1816–1872

Siemens was born in Prussia. While in prison for acting as a second in a duel, he began to experiment with chemistry, which led to his invention of the first electroplating system. In 1837, Siemens began making improvements in the early telegraph and contributed greatly to the development of telegraphic systems. The unit of conductance is named in his honor. (Photo credit: AIP Emilio Segrè Visual Archives, E. Scott Barr Collection.)

The Voltage Source

The Ideal Voltage Source   An ideal voltage source can provide a constant voltage for any current required by a circuit. The ideal voltage source does not exist but can be closely approximated in practice. We will assume ideal unless otherwise specified.

Voltage sources can be either dc or ac. A common symbol for a dc voltage source is shown in Figure 2–12(a) and one for an ac voltage source is shown in part (b). AC voltage sources will be used later in the book.

▶ **FIGURE 2–12**

Symbols for voltage sources.

(a) DC voltage source   (b) AC voltage source

A graph showing voltage versus current for an ideal dc voltage source is called the *VI* characteristic and is illustrated in Figure 2–13. As you can see, the voltage is constant for any current (within limits) from the source. For a practical voltage source connected in a circuit, the voltage decreases slightly as the current increases. Current is always drawn from a voltage source when a load such as a resistance is connected to it.

▶ **FIGURE 2–13**

VI characteristic of an ideal voltage source.

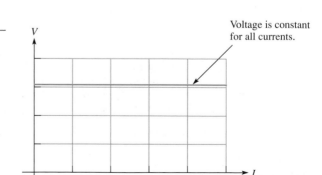

Voltage is constant for all currents.

Types of DC Voltage Sources

Batteries A **battery** is a type of voltage source that converts chemical energy into electrical energy. A battery consists of one or more electro-chemical cells that are electrically connected. A cell consists of four basic components: a positive electrode, a negative electrode, an electrolyte, and a porous separator. The *positive electrode* has a deficiency of electrons due to chemical reaction, the *negative electrode* has a surplus of electrons due to chemical reaction, the *electrolyte* provides a mechanism for charge flow between positive and negative electrodes, and the *separator* electrically isolates the positive and negative electrodes. A basic diagram of a battery cell is shown in Figure 2–14.

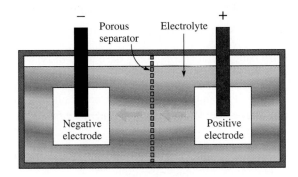

◀ **FIGURE 2–14**

Diagram of a battery cell.

The materials used in a battery cell determine the voltage that it produces. The chemical reaction at each of the electrodes produces a fixed potential at each electrode. For example, in a lead-acid cell, a potential of −1.685 V is produced at the positive electrode and a potential of +0.365 V is produced at the negative electrode. This means that the voltage between the two electrodes of a cell is 2.05 V, which is the standard lead-acid electrode potential. Factors such as acid concentration will affect this value to some degree so that the typical voltage of a commercial lead-acid cell is 2.15 V. The voltage of any battery cell depends on the cell chemistry. Nickel-cadmium cells are about 1.2 V and lithium cells can be as high as almost 4 V.

Although the voltage of a battery cell is fixed by its chemistry, the capacity is variable and depends on the quantity of materials in the cell. Essentially, the *capacity* of a cell is the number of electrons that can be obtained from it and is measured by the amount of current that can be supplied over time.

Batteries normally consist of multiple cells that are electrically connected together internally. The way that the cells are connected and the type of cells determine the voltage and capacity of the battery. If the positive electrode of one cell is connected to the negative electrode of the next and so on, as illustrated in Figure 2–15(a), the battery voltage is the sum of the individual cell voltages. This is called a series connection. To increase battery capacity, the positive electrodes of several cells are connected together and all the negative electrodes are connected together, as illustrated in Figure 2–15(b). This is called a parallel connection. Also, by using larger cells, which have a greater quantity of material, the ability to supply current can be increased but the voltage is not affected.

(a) Series-connected battery

(b) Parallel-connected battery

▲ **FIGURE 2–15**

Cells connected to form batteries.

Batteries are divided into two major classes, primary and secondary. Primary batteries are used once and discarded because their chemical reactions are irreversible. Secondary batteries can be recharged and reused many times because they are characterized by reversible chemical reactions.

There are many types, shapes, and sizes of batteries. Some of the sizes that you are most familiar with are AAA, AA, C, D, and 9 V. There is also a less common size called AAAA, which is smaller than the AAA. Batteries for hearing aids, watches, and other miniature applications are usually in a flat round configuration and are often called button batteries or coin batteries. Large multicell batteries are used in lanterns and industrial applications and, of course, there is the familiar automotive battery.

In addition to the many sizes and shapes, batteries are usually classified according to their chemical makeup as follows. Each of these classifications are typically available in several physical configurations.

- *Alkaline-MnO$_2$* This is a primary battery that is commonly used in palm-type computers, photographic equipment, toys, radios, and recorders.

- *Lithium-MnO$_2$* This is a primary battery that is commonly used in photographic and electronic equipment, smoke alarms, personal organizers, memory backup, and communications equipment.

- *Zinc air* This is a primary battery that is commonly used in hearing aids, medical monitoring instruments, pagers, and other frequency-use applications.

- *Silver oxide* This is a primary battery that is commonly used in watches, photographic equipment, hearing aids, and electronics requiring high-capacity batteries.

- *Nickel-metal hydride* This is a secondary (rechargable) battery that is commonly used in portable computers, cell phones, camcorders, and other portable consumer electronics.

- *Lead-acid* This is a secondary (rechargable) battery that is commonly used in automotive, marine, and other similar applications.

Solar Cells The operation of solar cells is based on the **photovoltaic effect**, which is the process whereby light energy is converted directly into electrical energy. A basic solar cell consists of two layers of different types of semiconductive materials joined together to form a junction. When one layer is exposed to light, many electrons acquire enough energy to break away from their parent atoms and cross the junction. This process forms negative ions on one side of the junction and positive ions on the other, and thus a potential difference (voltage) is developed. Figure 2–16 shows the construction of a basic solar cell.

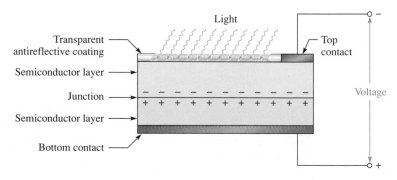

▲ **FIGURE 2–16**

Construction of a basic solar cell.

Generator Electrical **generators** convert mechanical energy into electrical energy using a principle called *electromagnetic induction* (see Chapter 10). A conductor is rotated through a magnetic field, and a voltage is produced across the conductor. A typical generator is pictured in Figure 2–17.

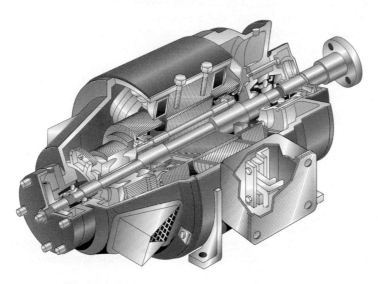

▲ FIGURE 2–17

Cutaway view of a dc voltage generator.

The Electronic Power Supply **Electronic power supplies** convert the ac voltage from a wall outlet to a constant (dc) voltage that is available across two terminals, as indicated in Figure 2–18(a). Typical commercial power supplies are shown in Figure 2–18(b).

Thermocouples The **thermocouple** is a thermoelectric type of voltage source that is commonly used to sense temperature. A thermocouple is formed by the junction of two dissimilar metals, and its operation is based on the **Seebeck effect** that describes the voltage generated at the junction of the metals as a function of temperature.

Standard types of thermocouple are characterized by the specific metals used. These standard thermocouples produce predictable output voltages for a range of temperatures. The most common is type K, made of chromel and alumel. Other types are also designated by letters as E, J, N, B, R, and S. Most thermocouples are available in wire or probe form.

Piezoelectric Sensors These sensors act as voltage sources and are based on the **piezoelectric effect** where a voltage is generated when a piezoelectric material is mechanically deformed by an external force. Quartz and ceramic are two types of piezoelectric material. Piezoelectric sensors are used in applications such as pressure sensors, force sensors, accelerometers, microphones, ultrasonic devices, and many others.

The Current Source

The Ideal Current Source As you know, an ideal voltage source can provide a constant voltage for any load. An ideal **current source** can provide a constant current in any load. Just as in the case of a voltage source, the ideal current source does not exist but can be approximated in practice. We will assume ideal unless otherwise specified.

The symbol for a current source is shown in Figure 2–19(a). The *IV* characteristic for an ideal current source is a horizontal line as illustrated in Figure 2–19(b). Notice that the current is constant for any voltage across the current source.

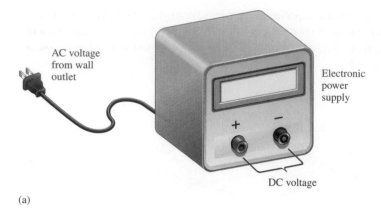

(a)

(b)

▲ FIGURE 2–18

Electronic power supplies. (Courtesy of B+K Precision)

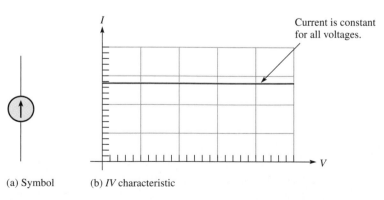

(a) Symbol   (b) *IV* characteristic

▲ FIGURE 2–19

The current source.

Actual Current Sources Power supplies are normally thought of as voltage sources because they are the most common source in the laboratory. However, current sources can also be considered a type of power supply. Typical commercial constant-current sources are illustrated in Figure 2–20.

(a)

(b)

◄ **FIGURE 2–20**

Typical commercial current sources. (Courtesy of Lake Shore Cryotronics)

In most transistor circuits, the transistor acts as a current source because part of the *IV* characteristic curve is a horizontal line as shown by the transistor characteristic in Figure 2–21. The flat part of the graph indicates where the transistor current is constant over a range of voltages. The constant-current region is used to form a constant-current source.

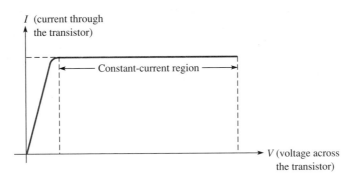

◄ **FIGURE 2–21**

Characteristic curve of a transistor showing the constant-current region.

One common application of a constant-current source is in constant-current battery chargers, as illustrated in a simplified way in Figure 2–22. The rectifier is a circuit that acts as a dc voltage source by converting the ac voltage from a standard wall outlet to a constant dc voltage. This voltage is effectively applied in parallel with a battery that is to be charged and in series with a constant-current source. The battery voltage is initially low but increases over time due to the constant charging current. The total voltage across the current source is the voltage from the rectifier minus the voltage of the battery, which increases as the battery charges.

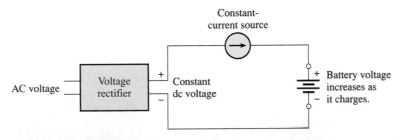

◄ **FIGURE 2–22**

Battery charger as an example of a current source application.

SECTION 2–4 REVIEW	1. Define a voltage source.
	2. Explain how a battery produces voltage.
	3. Describe how a solar cell produces voltage.
	4. Discuss how an electrical generator creates voltage.

5. Explain what an electronic power supply does.

6. Define a current source.

7. Name an electronic component that is used as a current source.

2–5 RESISTORS

A component that is specifically designed to have a certain amount of resistance is called a **resistor**. The principal applications of resistors are to limit current in a circuit, to divide voltage, and, in certain cases, to generate heat. Although resistors come in many shapes and sizes, they can all be placed in one of two main categories: fixed and variable.

After completing this section, you should be able to

◆ **Recognize and discuss various types and values of resistors**

   ◆ Distinguish between fixed resistors and variable resistors

   ◆ Know how the physical size of a resistor determines its ability to dissipate power

   ◆ Read a color code or other designation to determine the resistance value

   ◆ Describe how certain resistors are constructed

Fixed Resistors

Fixed resistors are available with a large selection of resistance values that are set during manufacturing and cannot be changed easily. They are constructed using various methods and materials. Figure 2–23 shows several common types.

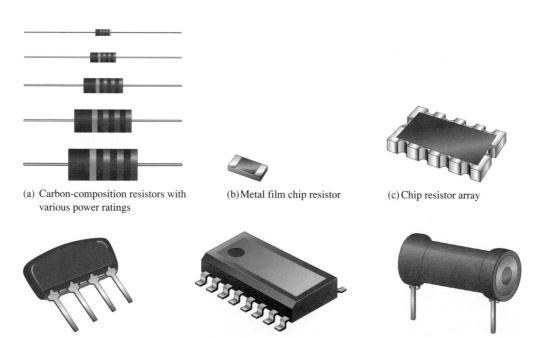

(a) Carbon-composition resistors with various power ratings

(b) Metal film chip resistor

(c) Chip resistor array

(d) Resistor network (simm)

(e) Resistor network (surface mount)

(f) Radial-lead for PC board insertion

▲ FIGURE 2–23

Typical fixed resistors.

One common fixed resistor is the carbon-composition type, which is made with a mixture of finely ground carbon, insulating filler, and a resin binder. The ratio of carbon to insulating filler sets the resistance value. The mixture is formed into rods, and conductive lead connections are made. The entire resistor is then encapsulated in an insulated coating for protection. Figure 2–24(a) shows the construction of a typical carbon-composition resistor.

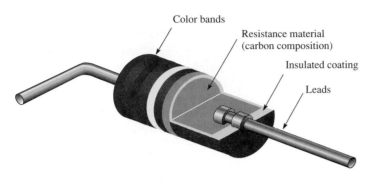
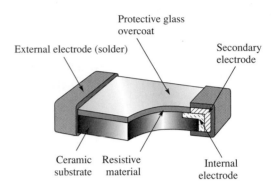

(a) Cutaway view of a carbon-composition resistor   (b) Cutaway view of a tiny chip resistor

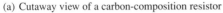

▲ FIGURE 2–24

Two types of fixed resistors (not to scale).

The chip resistor is another type of fixed resistor and is in the category of SMT (surface mount technology) components. It has the advantage of a very small size for compact assemblies. Figure 2–24(b) shows the construction of a chip resistor.

Other types of fixed resistors include carbon film, metal film, and wirewound. In film resistors, a resistive material is deposited evenly onto a high-grade ceramic rod. The resistive film may be carbon (carbon film) or nickel chromium (metal film). In these types of resistors, the desired resistance value is obtained by removing part of the resistive material in a helical pattern along the rod using a spiraling technique, as shown in Figure 2–25(a). Very close **tolerance** can be achieved with this method. Film resistors are also available in the form of resistor networks, as shown in Figure 2–25(b).

◀ FIGURE 2–25

Construction views of typical film resistors.

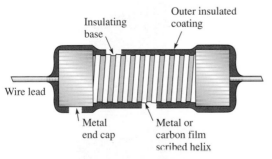
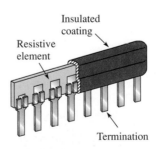

(a) Film resistor showing spiraling technique   (b) Resistor network

Wirewound resistors are constructed with resistive wire wound around an insulating rod and then sealed. Normally, wirewound resistors are used in applications that require higher power ratings. Since they are constructed with a coil of wire, wirewound resistors have significant inductance and are not used at higher frequencies. Some typical wirewound resistors are shown in Figure 2–26.

▲ FIGURE 2–26

Typical wirewound power resistors.

Resistor Color Codes

Fixed resistors with value tolerances of 5% or 10% are color coded with four bands to indicate the resistance value and the tolerance. This color-code band system is shown in Figure 2–27, and the color code is listed in Table 2–1. The bands are always closer to one end.

▶ FIGURE 2–27

Color-code bands on a 4-band resistor.

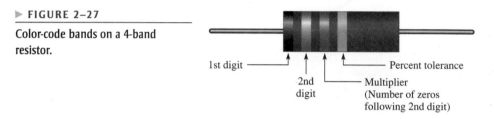

1st digit — Percent tolerance
2nd digit — Multiplier (Number of zeros following 2nd digit)

The color code is read as follows:

1. Start with the band closest to one end of the resistor. The first band is the first digit of the resistance value. If it is not clear which is the banded end, start from the end that does not begin with a gold or silver band.

▶ TABLE 2–1

Resistor 4-band color code.

	Digit	Color
	0	Black
	1	Brown
	2	Red
	3	Orange
Resistance value, first three bands:	4	Yellow
First band—1st digit		
Second band—2nd digit	5	Green
Third band—multiplier (number of zeros following the 2nd digit)	6	Blue
	7	Violet
	8	Gray
	9	White
Fourth band—tolerance	±5%	Gold
	±10%	Silver

2. The second band is the second digit of the resistance value.

3. The third band is the number of zeros following the second digit, or the multiplier.

4. The fourth band indicates the percent tolerance and is usually gold or silver.

For example, a 5% tolerance means that the *actual* resistance value is within ±5% of the color-coded value. Thus, a 100 Ω resistor with a tolerance of ±5% can have an acceptable range of values from a minimum of 95 Ω to a maximum of 105 Ω.

For resistance values less than 10 Ω, the third band is either gold or silver. Gold represents a multiplier of 0.1, and silver represents 0.01. For example, a color code of red, violet, gold, and silver represents 2.7 Ω with a tolerance of ±10%. A table of standard resistance values is in Appendix A.

EXAMPLE 2–4

Find the resistance value in ohms and the percent tolerance for each of the color-coded resistors shown in Figure 2–28.

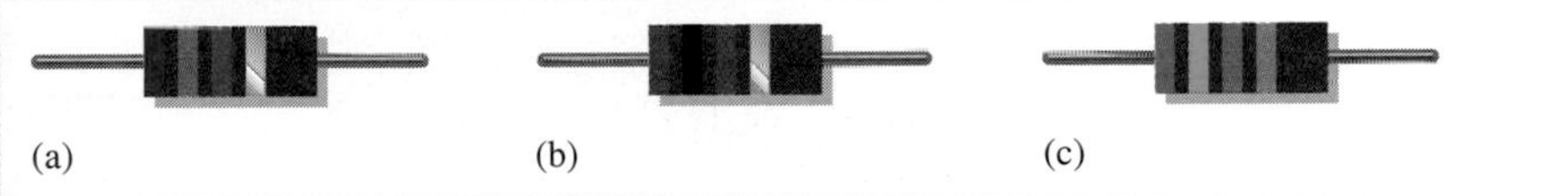

(a)   (b)   (c)

▲ **FIGURE 2–28**

Solution   **(a)** First band is red = 2, second band is violet = 7, third band is orange = 3 zeros, fourth band is silver = 10% tolerance.

$$R = 27,000\ \Omega \pm 10\%$$

(b) First band is brown = 1, second band is black = 0, third band is brown = 1 zero, fourth band is silver = 10% tolerance.

$$R = 100\ \Omega \pm 10\%$$

(c) First band is green = 5, second band is blue = 6, third band is green = 5 zeros, fourth band is gold = 5% tolerance.

$$R = 5,600,000\ \Omega \pm 5\%$$

Related Problem   A certain resistor has a yellow first band, a violet second band, a red third band, and a gold fourth band. Determine its value in ohms and its percent tolerance.

Five-Band Color Code   Certain precision resistors with tolerances of 2%, 1%, or less are generally color coded with five bands, as shown in Figure 2–29. Begin at the band closest to one end. The first band is the first digit of the resistance value, the second band is the second digit, the third band is the third digit, the fourth band is the multiplier (number of zeros after the third digit), and the fifth band indicates the percent tolerance. Table 2–2 shows the 5-band color code.

◀ **FIGURE 2–29**

Color-code bands on a 5-band resistor.

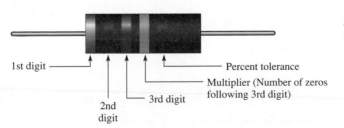

1st digit

2nd digit

3rd digit

Multiplier (Number of zeros following 3rd digit)

Percent tolerance

▶ TABLE 2–2

Resistor 5-band color code.

	DIGIT	COLOR
Resistance value, first three bands: First band—1st digit Second band—2nd digit Third band—3rd digit Fourth band—multiplier (number of zeros following 3rd digit)	0	Black
	1	Brown
	2	Red
	3	Orange
	4	Yellow
	5	Green
	6	Blue
	7	Violet
	8	Gray
	9	White
Fourth band—multiplier	0.1	Gold
	0.01	Silver
Fifth band—tolerance	±2%	Red
	±1%	Brown
	±0.5%	Green
	±0.25%	Blue
	±0.1%	Violet

Resistor Reliability Band   An extra band on some color-coded resistors indicates the resistor's reliability in percent of failures per 1000 hours (1000 h) of use. The reliability color code is listed in Table 2–3. For example, a brown fifth band on a 4-band color-coded resistor means that if a group of like resistors is operated under standard conditions for 1000 h, 1% of the resistors in that group will fail.

Resistors, as well as other components, should be operated substantially below their rated values to enhance their reliability.

▶ TABLE 2–3

Reliability color code.

COLOR	FAILURES DURING 1000 h OF OPERATION
Brown	1.0%
Red	0.1%
Orange	0.01%
Yellow	0.001%

EXAMPLE 2–5

Find the resistance value in ohms and the percent tolerance for each of the color-coded resistors shown in Figure 2–30.

(a)   (b)   (c)

▲ FIGURE 2–30

Solution   **(a)** First band is red = 2, second band is violet = 7, third band is black = 0, fourth band is gold = ×0.1, fifth band is red = ±2% tolerance.

$$R = 270 \times 0.1 = \mathbf{27\ \Omega \pm 2\%}$$

(b) First band is yellow = 4, second band is black = 0, third band is red = 2, fourth band is black = 0, fifth band is brown = ±1% tolerance.

$$R = 402\ \Omega \pm 1\%$$

(c) First band is orange = 3, second band is orange = 3, third band is red = 2, fourth band is orange = 3, fifth band is green = ±0.5% tolerance.

$$R = 332{,}000\ \Omega \pm 0.5\%$$

Related Problem A certain resistor has a yellow first band, a violet second band, a green third band, a gold fourth band, and a red fifth band. Determine its value in ohms and its percent tolerance.

Resistor Label Codes

Not all types of resistors are color coded. Many, including surface-mount resistors, use typographical marking to indicate the resistance value and tolerance. These label codes consist of either all numbers (numeric) or a combination of numbers and letters (alphanumeric). In some cases when the body of the resistor is large enough, the entire resistance value and tolerance are stamped on it in standard form.

Numeric Labeling This type of marking uses three digits to indicate the resistance value, as shown in Figure 2–31 using a specific example. The first two digits give the first two digits of the resistance value, and the third digit gives the multiplier or number of zeros that follow the first two digits. This code is limited to values of 10 Ω or greater.

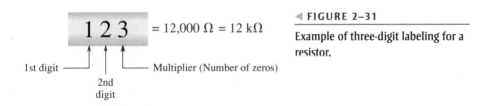

◀ FIGURE 2–31

Example of three-digit labeling for a resistor.

Alphanumeric Labeling Another common type of marking is a three- or four-character label that uses both digits and letters. This type of label typically consists of only three digits or two or three digits and one of the letters R, K, or M. The letter is used to indicate the multiplier, and the position of the letter indicates the decimal point placement. The letter R indicates a multiplier of 1 (no zeros after the digits), the K indicates a multiplier of 1000 (three zeros after the digits), and the M indicates a multiplier of 1,000,000 (six zeros after the digits). In this format, values from 100 to 999 consist of three digits and no letter to represent the three digits in the resistance value. Figure 2–32 shows three examples of this type of resistor label.

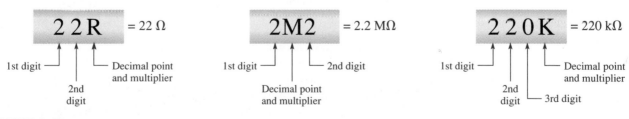

▲ FIGURE 2–32

Examples of the alphanumeric resistor label.

EXAMPLE 2–6

Interpret the following alphanumeric resistor labels:

(a) 470   (b) 5R6   (c) 68K   (d) 10M   (e) 3M3

Solution   (a) 470 = **470 Ω**   (b) 5R6 = **5.6 Ω**   (c) 68K = **68 kΩ**

(d) 10M = **10 MΩ**   (e) 3M3 = **3.3 MΩ**

Related Problem   What is the resistance indicated by 1K25?

One system of labels for resistance tolerance values uses the letters F, G, and J:

$$F = \pm 1\% \qquad G = \pm 2\% \qquad J = \pm 5\%$$

For example, 620F indicates a 620 Ω resistor with a tolerance of ±1%, 4R6G is a 4.6 Ω ±2% resistor, and 56KJ is a 56 kΩ ±5% resistor.

Variable Resistors

Variable resistors are designed so that their resistance values can be changed easily with a manual or an automatic adjustment.

Two basic uses for variable resistors are to divide voltage and to control current. The variable resistor used to divide voltage is called a **potentiometer**. The variable resistor used to control current is called a **rheostat**. Schematic symbols for these types are shown in Figure 2–33. The potentiometer is a three-terminal device, as indicated in part (a). Terminals 1 and 2 have a fixed resistance between them, which is the total resistance. Terminal 3 is connected to a moving contact (**wiper**). You can vary the resistance between 3 and 1 or between 3 and 2 by moving the contact up or down.

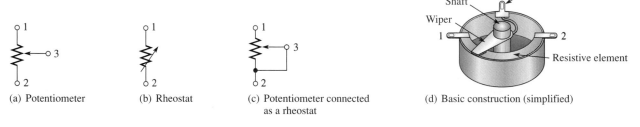

(a) Potentiometer   (b) Rheostat   (c) Potentiometer connected as a rheostat   (d) Basic construction (simplified)

▲ **FIGURE 2–33**

Potentiometer and rheostat symbols and basic construction of one type of potentiometer.

Figure 2–33(b) shows the rheostat as a two-terminal variable resistor. Part (c) shows how you can use a potentiometer as a rheostat by connecting terminal 3 to either terminal 1 or terminal 2. Parts (b) and (c) are equivalent symbols. Part (d) shows a simplified construction diagram of a potentiometer (which can also be configured as a rheostat). Some typical potentiometers are pictured in Figure 2–34.

Potentiometers and rheostats can be classified as linear or tapered, as shown in Figure 2–35, where a potentiometer with a total resistance of 100 Ω is used as an example. As shown in part (a), in a linear potentiometer, the resistance between either terminal and the moving contact varies linearly with the position of the moving contact. For example, one-half of the total contact movement results in one-half the total resistance. Three-quarters of the total movement results in three-quarters of the total resistance between the moving contact and one terminal, or one-quarter of the total resistance between the other terminal and the moving contact.

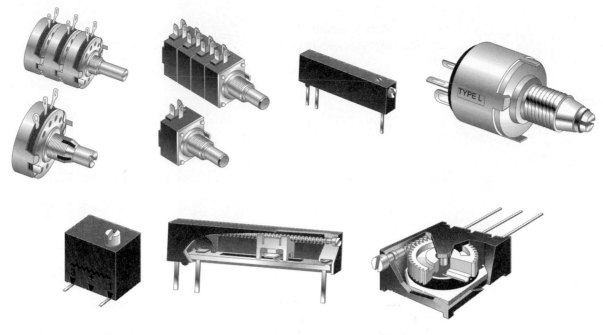

▲ FIGURE 2–34

Typical potentiometers and two construction views.

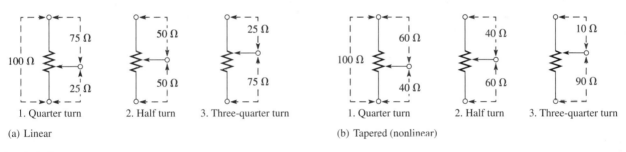

(a) Linear   (b) Tapered (nonlinear)

▲ FIGURE 2–35

Examples of linear and tapered potentiometers.

In the **tapered** potentiometer, the resistance varies nonlinearly with the position of the moving contact, so that one-half of a turn does not necessarily result in one-half the total resistance. This concept is illustrated in Figure 2–35(b), where the nonlinear values are arbitrary.

The potentiometer is used as a voltage-control device because when a fixed voltage is applied across the end terminals, a variable voltage is obtained at the wiper contact with respect to either end terminal. The rheostat is used as a current-control device because the current can be changed by changing the wiper position.

Two Types of Automatically Variable Resistors   A **thermistor** is a type of variable resistor that is temperature sensitive. When its temperature coefficient is negative, the resistance changes inversely with temperature. When its temperature coefficient is positive, the resistance changes directly with temperature.

The resistance of a **photoconductive cell** changes with a change in light intensity. This cell also has a negative temperature coefficient. Symbols for both of these devices are shown in Figure 2–36. Sometimes the Greek letter lambda (λ) is used in conjunction with the photoconductive cell symbol.

▶ FIGURE 2–36

Symbols for resistive devices with sensitivities to temperature and light.

(a) Thermistor

(b) Photoconductive cell

SECTION 2–5 REVIEW

1. What are the two main categories of resistors? Briefly explain the difference between them.
2. In the 4-band resistor color code, what does each band represent?
3. Determine the resistance and percent tolerance for each of the resistors in Figure 2–37.

▶ FIGURE 2–37

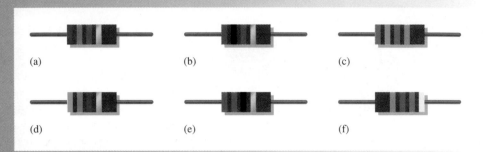

(a) (b) (c)

(d) (e) (f)

4. From the selection of resistors in Figure 2–38, select the following values: 330 Ω, 2.2 kΩ, 56 kΩ, 100 kΩ, and 39 kΩ.

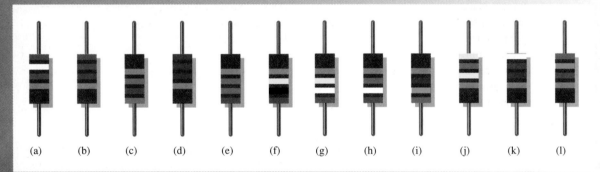

(a) (b) (c) (d) (e) (f) (g) (h) (i) (j) (k) (l)

△ FIGURE 2–38

5. What resistance value is indicated by each alphanumeric label:
   (a) 33R (b) 5K6 (c) 900 (d) 6M8
6. What is the basic difference between a rheostat and a potentiometer?
7. What is a thermistor?

2–6 THE ELECTRIC CIRCUIT

A basic electric circuit is an arrangement of physical components that use voltage, current, and resistance to perform some useful function.

After completing this section, you should be able to

◆ **Describe a basic electric circuit**

   ◆ Relate a schematic to a physical circuit

   ◆ Define *open circuit* and *closed circuit*

- ◆ Describe various types of protective devices
- ◆ Describe various types of switches
- ◆ Explain how wire sizes are related to gauge numbers
- ◆ Define *ground* or *common*

Direction of Current

For a few years after the discovery of electricity, people assumed all current consisted of moving positive charges. However, in the 1890s, the electron was identified as the charge carrier in solid conductors.

Today, there are two accepted conventions for the direction of electrical current. *Electron flow direction,* preferred by many in the fields of electrical and electronics technology, assumes for analysis purposes that current is out of the negative terminal of a voltage source, through the circuit, and into the positive terminal of the source. *Conventional current direction* assumes for analysis purposes that current is out of the positive terminal of a voltage source, through the circuit, and into the negative terminal of the source. By following the direction of conventional current, there is a rise in voltage across a source (negative to positive) and a drop in voltage across a resistor (positive to negative).

Since you cannot actually see current, only its effects, it actually makes no difference which direction of current is assumed as long as it is used *consistently.* The results of electric circuit analysis are not affected by the direction of current that is assumed for analytical purposes. The direction used for analysis is largely a matter of preference, and there are many proponents for each approach.

Conventional current direction is also used in electronics technology and is used almost exclusively at the engineering level. Conventional current direction is used throughout this text. An alternate version of this text that uses electron flow direction is also available.

SAFETY NOTE

To avoid electrical shock, never touch a circuit while it is connected to a voltage source. If you need to handle a circuit, remove a component, or change a component, first make sure the voltage source is disconnected.

The Basic Circuit

Basically, an electric **circuit** consists of a voltage source, a load, and a path for current between the source and the load. Figure 2–39 shows in pictorial form an example of a simple electric circuit: a battery connected to a lamp with two conductors (wires). The battery is the voltage source, the lamp is the **load** on the battery because it draws current from the battery, and the two wires provide the current path from the positive terminal of the battery to the lamp and back to the negative terminal of the battery. Current goes through the filament of the lamp (which has a resistance), causing it to emit visible light. Current through the battery occurs by chemical action.

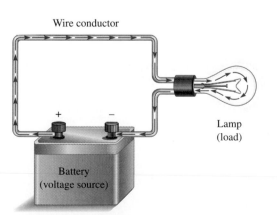

Wire conductor

Lamp (load)

Battery (voltage source)

◄ FIGURE 2–39

A simple electric circuit.

In many practical cases, one terminal of the battery is connected to a common or ground point. For example, in most automobiles, the negative battery terminal is connected to the metal chassis of the car. The chassis is the ground for the automobile electrical system and acts as a conductor that completes the circuit.

The Electric Circuit Schematic   An electric circuit can be represented by a **schematic** using standard symbols for each element, as shown in Figure 2–40 for the simple circuit in Figure 2–39. A schematic, in an organized manner, shows how the various components in a given circuit are interconnected so that the operation of the circuit can be determined.

▲ **FIGURE 2–40**

Schematic for the circuit in Figure 2–39.

Circuit Current Control and Protection

The example circuit in Figure 2–39 illustrated a **closed circuit**—that is, a circuit in which the current has a complete path. When the current path is broken, the circuit is called an **open circuit**.

Mechanical Switches   **Switches** are commonly used for controlling the opening or closing of circuits. For example, a switch is used to turn a lamp on or off, as illustrated in Figure 2–41. Each circuit pictorial is shown with its associated schematic. The type of switch indicated is a single-pole–single-throw (SPST) toggle switch. The term *pole* refers to the movable arm in a switch, and the term *throw* indicates the number of contacts that are affected (either opened or closed) by a single switch action (a single movement of a pole).

Figure 2–42 shows a somewhat more complicated circuit using a single-pole–double-throw (SPDT) type of switch to control the current to two different lamps. When one lamp

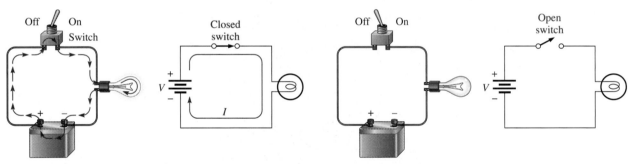

(a) There is current in a *closed* circuit because there is a complete current path (switch is ON or in the *closed* position). Current is almost always indicated by a red arrow in this text.

(b) There is no current in an *open* circuit because the path is broken (switch is OFF or in the *open* position).

▲ **FIGURE 2–41**

Illustration of closed and open circuits using an SPST switch for control.

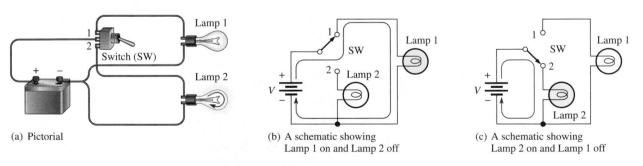

(a) Pictorial

(b) A schematic showing Lamp 1 on and Lamp 2 off

(c) A schematic showing Lamp 2 on and Lamp 1 off

▲ **FIGURE 2–42**

An example of an SPDT switch controlling two lamps.

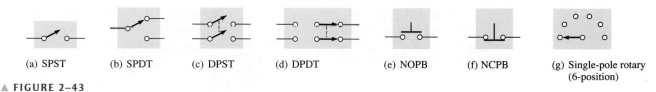

▲ FIGURE 2–43

Switch symbols.

is on, the other is off, and vice versa, as illustrated by the two schematics in parts (b) and (c), which represent each of the switch positions.

In addition to the SPST and the SPDT switches (symbols are shown in Figure 2–43(a) and (b)), the following other types are important:

- *Double-pole–single-throw (DPST)*   The DPST switch permits simultaneous opening or closing of two sets of contacts. The symbol is shown in Figure 2–43(c). The dashed line indicates that the contact arms are mechanically linked so that both move with a single switch action.

- *Double-pole–double-throw (DPDT)*   The DPDT switch provides connection from one set of contacts to either of two other sets. The schematic symbol is shown in Figure 2–43(d).

- *Push-button (PB)*   In the normally open push-button switch (NOPB), shown in Figure 2–43(e), connection is made between two contacts when the button is depressed, and connection is broken when the button is released. In the normally closed push-button switch (NCPB), shown in Figure 2–43(f), connection between the two contacts is broken when the button is depressed.

- *Rotary*   In a rotary switch, connection between one contact and any one of several others is made by turning a knob. A symbol for a simple six-position rotary switch is shown in Figure 2–43(g).

Figure 2–44 shows several varieties of mechanical switches, and Figure 2–45 shows the construction view of a typical toggle switch.

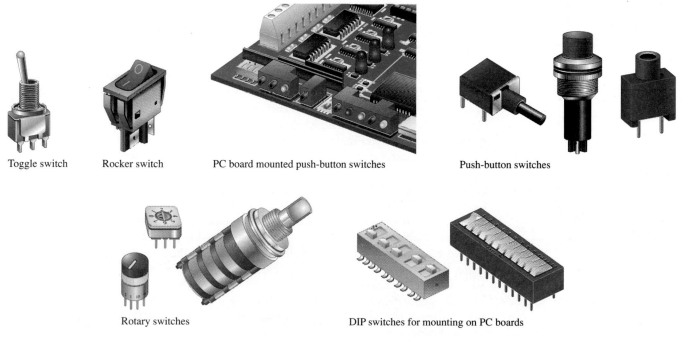

Toggle switch   Rocker switch   PC board mounted push-button switches   Push-button switches

Rotary switches   DIP switches for mounting on PC boards

▲ FIGURE 2–44

Typical mechanical switches.

▶ FIGURE 2–45

▶ FIGURE 2–45

Construction view of a typical toggle switch.

Semiconductor Switches   Transistors are widely used as switches in many applications. The transistor can be used as the equivalent of a single-pole–single-throw switch. You can open and close a circuit path by controlling the state of the transistor. Two types of transistor symbols are shown in Figure 2–46 with their mechanical switch equivalents.

▶ FIGURE 2–46

Transistor switches.

Current creates a closed switch.

No current creates an open switch.

Voltage creates a closed switch.

No voltage creates an open switch.

(a) Bipolar transistor

(b) Field-effect transistor

Here is a greatly simplified description of operation. One type, called the *bipolar transistor,* is controlled by current. When there is current at a specific terminal, the transistor acts as a closed switch; when there is no current at that terminal, the transistor acts as an open switch, as illustrated in Figure 2–46(a). Another type, called the *field-effect transistor,* is controlled by voltage. When there is voltage at a specific terminal, the transistor acts as a closed switch; when there is no voltage at that terminal, the transistor acts as an open switch, as illustrated in part (b).

Protective Devices   **Fuses** and **circuit breakers** are used to deliberately create an open circuit when the current exceeds a specified number of amperes due to a malfunction or other abnormal condition in a circuit. For example, a 20 A fuse or circuit breaker will open a circuit when the current exceeds 20 A.

The basic difference between a fuse and a circuit breaker is that when a fuse is "blown," it must be replaced; but when a circuit breaker opens, it can be reset and reused repeatedly. Both of these devices protect against damage to a circuit due to excess current or prevent a hazardous condition created by the overheating of wires and other components when the current is too great. Several typical fuses and circuit breakers, along with their schematic symbols, are shown in Figure 2–47.

Two basic categories of fuses in terms of their physical configuration are cartridge type and plug type (screw in). Cartridge-type fuses have various-shaped housings with leads or other types of contacts, as shown in Figure 2–47(a). A typical plug-type fuse is shown in part (b). Fuse operation is based on the melting temperature of a wire or other metal element. As current increases, the fuse element heats up and when the rated current is exceeded, the element reaches its melting temperature and opens, thus removing power from the circuit.

Two common types of fuses are the fast-acting and the time-delay (slow-blow). Fast-acting fuses are type F and time-delay fuses are type T. In normal operation, most fuses are subjected to intermittent current surges the exceed the rated current, such as when power to a circuit is turned on. Over time, this reduces the fuse's ability to withstand short surges or even current at the rated value. A slow-blow fuse can tolerate greater and longer duration surges of current than the typical fast-acting fuse. A fuse symbol is shown in Figure 2–47(c).

Typical circuit breakers are shown in Figure 2–47(d) and the symbol is shown in part (e). Generally, a circuit breaker detects excess current either by the heating effect of the current

(a) Cartridge fuses   (b) Plug fuse   (c) Fuse symbol

(d) Circuit breakers   (e) Circuit breaker symbol

▲ **FIGURE 2–47**

Typical fuses and circuit breakers and their symbols.

or by the magnetic field it creates. In a circuit breaker based on the heating effect, a bimetallic spring opens the contacts when the rated current is exceeded. Once opened, the contact is held open by mechanical means until manually reset. In a circuit breaker based on a magnetic field, the contacts are opened by a sufficient magnetic force created by excess current and must be mechanically reset.

Wires

Wires are the most common form of conductive material used in electrical applications. They vary in diameter and are arranged according to standard gauge numbers, called **AWG** (American Wire Gauge) sizes. As the gauge number increases, the wire diameter decreases. The size of a wire is also specified in terms of its cross-sectional area, as illustrated in Figure 2–48. A unit of cross-sectional area used for wires is the **circular mil**, abbreviated CM. One circular mil is the area of a wire with a diameter of 0.001 inch (1 mil). You can

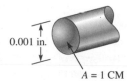

◀ **FIGURE 2–48**

Cross-sectional area of a wire.

d

Cross-sectional area, A

0.001 in.

$A = 1$ CM

find the cross-sectional area by expressing the diameter in thousandths of an inch (mils) and squaring it, as follows:

Equation 2–5

$$A = d^2$$

where A is the cross-sectional area in circular mils and d is the diameter in mils. Table 2–4 lists the AWG sizes with their corresponding cross-sectional area and resistance in ohms per 1000 ft at 20°C.

▼ TABLE 2–4

American Wire Gauge (AWG) sizes and resistances for solid round copper.

AWG #	AREA (CM)	RESISTANCE (Ω/1000 FT AT 20°C)	AWG #	AREA (CM)	RESISTANCE (Ω/1000 FT AT 20°C)
0000	211,600	0.0490	19	1,288.1	8.051
000	167,810	0.0618	20	1,021.5	10.15
00	133,080	0.0780	21	810.10	12.80
0	105,530	0.0983	22	642.40	16.14
1	83,694	0.1240	23	509.45	20.36
2	66,373	0.1563	24	404.01	25.67
3	52,634	0.1970	25	320.40	32.37
4	41,742	0.2485	26	254.10	40.81
5	33,102	0.3133	27	201.50	51.47
6	26,250	0.3951	28	159.79	64.90
7	20,816	0.4982	29	126.72	81.83
8	16,509	0.6282	30	100.50	103.2
9	13,094	0.7921	31	79.70	130.1
10	10,381	0.9989	32	63.21	164.1
11	8,234.0	1.260	33	50.13	206.9
12	6,529.0	1.588	34	39.75	260.9
13	5,178.4	2.003	35	31.52	329.0
14	4,106.8	2.525	36	25.00	414.8
15	3,256.7	3.184	37	19.83	523.1
16	2,582.9	4.016	38	15.72	659.6
17	2,048.2	5.064	39	12.47	831.8
18	1,624.3	6.385	40	9.89	1049.0

EXAMPLE 2–7

What is the cross-sectional area of a wire with a diameter of 0.005 inch?

Solution

$$d = 0.005 \text{ in.} = 5 \text{ mils}$$
$$A = d^2 = 5^2 = \textbf{25 CM}$$

Related Problem What is the cross-sectional area of a 0.0015 in. diameter wire?

Wire Resistance Although copper wire conducts electricity extremely well, it still has some resistance, as do all conductors. The resistance of a wire depends on three physical characteristics: (a) type of material, (b) length of wire, and (c) cross-sectional area. In addition, temperature can also affect the resistance.

Each type of conductive material has a characteristic called its *resistivity, ρ*. For each material, ρ is a constant value at a given temperature. The formula for the resistance of a wire of length *l* and cross-sectional area *A* is

$$R = \frac{\rho l}{A}$$

Equation 2–6

This formula shows that resistance increases with an increase in resistivity and length and decreases with an increase in cross-sectional area. For resistance to be calculated in ohms, the length must be in feet, the cross-sectional area in circular mils, and the resistivity in CM-Ω/ft.

EXAMPLE 2–8

Find the resistance of a 100 ft length of copper wire with a cross-sectional area of 810.1 CM. The resistivity of copper is 10.37 CM-Ω/ft.

Solution

$$R = \frac{\rho l}{A} = \frac{(10.37 \text{ CM-}\Omega\text{/ft})(100 \text{ ft})}{810.1 \text{ CM}} = \textbf{1.280 } \boldsymbol{\Omega}$$

Related Problem

Use Table 2–4 to determine the resistance of 100 ft of copper wire with a cross-sectional area of 810.1 CM. Compare with the calculated result.

As mentioned, Table 2–4 lists the resistance of the various standard wire sizes in ohms per 1000 feet at 20°C. For example, a 1000 ft length of 14 gauge copper wire has a resistance of 2.525 Ω. A 1000 ft length of 22 gauge wire has a resistance of 16.14 Ω. For a given length, the smaller wire has more resistance. Thus, for a given voltage, larger wires can carry more current than smaller ones.

Ground

Ground is the reference point in electric circuits. The term *ground* originated from the fact that one conductor of a circuit was typically connected with an 8-foot long metal rod driven into the earth itself. Today, this type of connection is referred to as an *earth ground*. In household wiring, earth ground is indicated with a green or bare copper wire. Earth ground is normally connected to the metal chassis of an appliance or a metal electrical box for safety. Unfortunately, there have been exceptions to this rule, which can present a safety hazard if a metal chassis is not at earth ground. It is a good idea to confirm that a metal chassis is actually at earth ground potential before doing any work on an instrument or appliance.

Another type of ground is called a *reference ground*. Voltages are always specified with respect to another point. If that point is not stated explicitly, the reference ground is understood. Reference ground defines 0 V for the circuit. The reference ground can be at a completely different potential than the earth ground. Reference ground is also called **common** and labeled COM or COMM because it represents a common conductor. When you are wiring a protoboard in the laboratory, you will normally reserve one of the bus strips (a long line along the length of the board) for this common conductor.

Three ground symbols are shown in Figure 2–49. Unfortunately, there is not a separate symbol to distinguish between earth ground and reference ground. The symbol in (a) represents either an earth ground or a reference ground, (b) shows a chassis ground, and (c) is an alternate reference symbol typically used when there is more than one common connection (such as analog and digital ground in the same circuit). In this book, the symbol in part (a) will be used throughout.

An instrument such as a laboratory power supply may have a green terminal that is labeled as earth ground. Figure 2–50 shows a triple output power supply. Each of the three

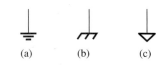

(a) (b) (c)

▲ **FIGURE 2–49**

Commonly used ground symbols.

▶ **FIGURE 2–50**

A triple output power supply.
(Courtesy of B+K Precision)

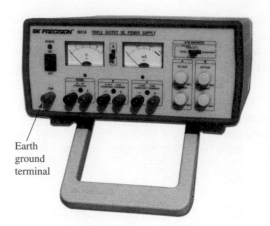

Earth ground terminal

power supplies in the same chassis is isolated from the earth ground. An earth ground connection is brought out to the front panel on the separate green connector on the left side, for cases when earth ground is required. Internally, this is tied to the center (round) pin on the ac plug.

If you want to connect a positive supply to a circuit, the earth ground is not used and the ground reference (common) for the circuit is the (−) terminal of the supply. If you require a negative voltage, the (+) terminal is the ground reference. Many circuits require both positive and negative supplies, so in this case the (+) terminal of one supply can be connected to the negative terminal of the other supply which becomes the reference. Figure 2–51 illustrates this type of connection. The earth ground is not used for this application. Notice that for the example circuit board, the amplifier circuit requires a common connection between two points which is connected to the power supply reference ground or common. As illustrated in the figure, a schematic drawing can show the common reference with a ground symbol or symbols. When a ground symbol is not shown on a circuit schematic, the common or reference point is generally assumed to be either the negative side or the positive of the voltage source, depending on the circuit configuration.

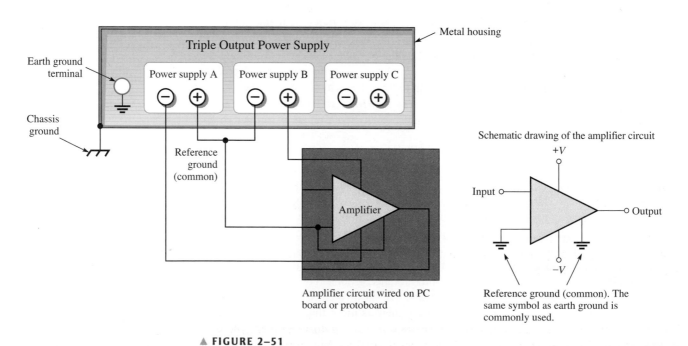

Amplifier circuit wired on PC board or protoboard

▲ **FIGURE 2–51**

Example of using grounds in a circuit application.

Figure 2–52 illustrates a simple circuit with ground connections. The current is from the positive terminal of the 12 V source, through the lamp, and back to the negative terminal of the source through the ground connection. Ground provides a return path for the current back to the source because all of the ground points are electrically the same point. The voltage at the top of the circuit is +12 V with respect to ground.

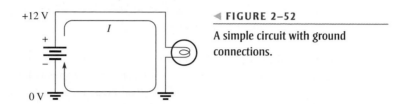

◀ FIGURE 2–52

A simple circuit with ground connections.

SECTION 2–6 REVIEW	1. What are the basic elements of an electric circuit?
	2. What is an open circuit?
	3. What is a closed circuit?
	4. What is the difference between a fuse and a circuit breaker?
	5. Which wire is larger in diameter, AWG 3 or AWG 22?
	6. What is ground (common) in an electric circuit?

2–7 BASIC CIRCUIT MEASUREMENTS

An electronics technician cannot function without knowing how to measure voltage, current, and resistance.

After completing this section, you should be able to

◆ **Make basic circuit measurements**

   ◆ Properly measure voltage in a circuit

   ◆ Properly measure current in a circuit

   ◆ Properly measure resistance

   ◆ Set up and read basic meters

Voltage, current, and resistance measurements are commonly required in electronics work. The instrument used to measure voltage is a **voltmeter**, the instrument used to measure current is an **ammeter**, and the instrument used to measure resistance is an **ohmmeter**. Commonly, all three instruments are combined into a single instrument called a **multimeter**, in which you can choose what specific quantity to measure by selecting the appropriate function with a switch.

Meter Symbols

Throughout this book, certain symbols will be used in circuits to represent meters, as shown in Figure 2–53. You may see any of four types of symbols for voltmeters, ammeters, or ohmmeters, depending on which symbol most effectively conveys the information

| (a) Digital | (b) Bar graph | (c) Analog | (d) Generic |

▲ FIGURE 2–53

Examples of meter symbols used in this book. Each of the symbols can be used to represent either an ammeter (A), a voltmeter (V), or an ohmmeter (Ω).

required. The digital meter symbol is used when specific values are to be indicated in a circuit. The bar graph meter symbol and sometimes the analog meter symbol are used to illustrate the operation of a circuit when *relative* measurements or changes in quantities, rather than specific values, need to be depicted. A changing quantity may be indicated by an arrow in the display showing an increase or decrease. The generic symbol is used to indicate placement of meters in a circuit when no values or value changes need to be shown.

Measuring Current

Figure 2–54 illustrates how to measure current with an ammeter. Part (a) shows a simple circuit in which the current through the resistor is to be measured. First make sure the range setting of the ammeter is greater than the expected current and then connect the ammeter in the current path by first opening the circuit, as shown in part (b). Then insert the meter as shown in part (c). Such a connection is a series connection. The polarity of the meter must be such that the current is in at the positive terminal and out at the negative terminal.

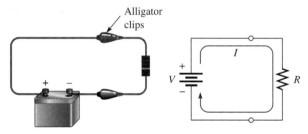

(a) Circuit in which the current is to be measured

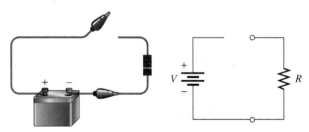

(b) Open the circuit either between the resistor and the positive terminal or between the resistor and the negative terminal of source.

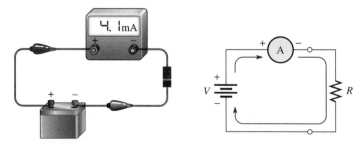

(c) Install the ammeter in the current path with polarity as shown (negative to negative, positive to positive).

▲ FIGURE 2–54

Example of an ammeter connection in a simple circuit to measure current.

Measuring Voltage

To measure voltage, connect the voltmeter across the component for which the voltage is to be found. Such a connection is a parallel connection. The negative terminal of the meter must

be connected to the negative side of the circuit, and the positive terminal of the meter must be connected to the positive side of the circuit. Figure 2–55 shows a voltmeter connected to measure the voltage across the resistor.

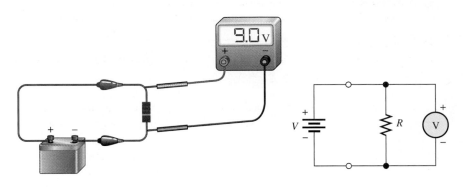

▲ FIGURE 2–55

Example of a voltmeter connection in a simple circuit to measure voltage.

Measuring Resistance

To measure resistance, first turn off the power and disconnect one end or both ends of the resistor from the circuit; then connect the ohmmeter across the resistor. This procedure is shown in Figure 2–56.

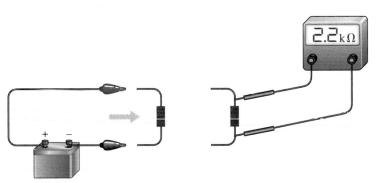

◀ FIGURE 2–56

Example of using an ohmmeter to measure resistance.

(a) Disconnect the resistor from the circuit to avoid damage to the meter and/or incorrect measurement.

(b) Measure the resistance. (Polarity is not important.)

Digital Multimeters

A **DMM** (digital multimeter) is an electronic instrument that combines meters for the measurement of voltage, current, and resistance. DMMs are the most widely used type of electronic measuring instrument. Generally, DMMs provide more functions, better accuracy, greater ease of reading, and greater reliability than do many analog meters. Analog meters have at least one advantage over DMMs, however. They can track short-term variations and trends in a measured quantity that many DMMs are too slow to respond to. Figure 2–57 shows typical DMMs.

DMM Functions The basic functions found on most DMMs include the following:

- ◆ Ohms
- ◆ DC voltage and current
- ◆ AC voltage and current

Some DMMs provide special functions such as transistor or diode tests, power measurement, and decibel measurement for audio amplifier tests. Some meters require manual

▶ FIGURE 2–57

Typical digital multimeters (DMMs).
(Courtesy of B+K Precision)

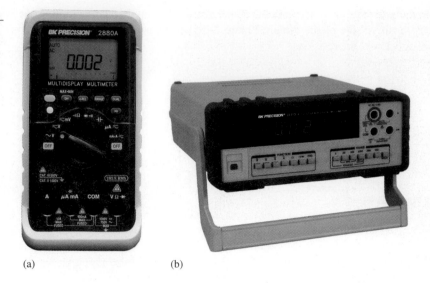

(a)   (b)

selection of the ranges for the functions. Many meters provide automatic range selection and are called *autoranging*.

DMM Displays   DMMs are available with either LCD (liquid-crystal display) or LED (light-emitting diode) readouts. The LCD is the most commonly used readout in battery-powered instruments because it requires only very small amounts of current. A typical battery-powered DMM with an LCD readout operates on a 9 V battery that will last from a few hundred hours to 2000 hours and more. The disadvantages of LCD readouts are that (a) they are difficult or impossible to see in low-light conditions and (b) they are relatively slow to respond to measurement changes. LEDs, on the other hand, can be seen in the dark and respond quickly to changes in measured values. LED displays require much more current than LCDs, and, therefore, battery life is shortened when LEDs are used in portable equipment.

Both LCD and LED DMM displays are in a 7-segment format. Each digit in a display consists of seven separate segments as shown in Figure 2–58(a). Each of the ten decimal digits is formed by the activation of appropriate segments, as illustrated in Figure 2–58(b). In addition to the seven segments, there is also a decimal point.

▶ FIGURE 2–58

Seven-segment display.

(a)   (b)

Resolution   The **resolution** of a DMM is the smallest increment of a quantity that the meter can measure. The smaller the increment, the better the resolution. One factor that determines the resolution of a meter is the number of digits in the display.

Because many DMMs have 3½ digits in their display, we will use this case for illustration. A 3½-digit multimeter has three digit positions that can indicate from 0 through 9, and one digit position that can indicate only a value of 0 or 1. This latter digit, called the *half-digit,* is always the most significant digit in the display. For example, suppose that a DMM is reading 0.999 V, as shown in Figure 2–59(a). If the voltage increases by 0.001 V to 1 V, the display correctly shows 1.000 V, as shown in part (b). The "1" is the half-digit. Thus, with 3½ digits, a variation of 0.001 V, which is the resolution, can be observed.

Now, suppose that the voltage increases to 1.999 V. This value is indicated on the meter as shown in Figure 2–59(c). If the voltage increases by 0.001 V to 2 V, the half-digit cannot display the "2," so the display shows 2.00. The half-digit is blanked and only three digits are active, as indicated in part (d). With only three digits active, the resolution is 0.01 V rather than 0.001 V

(a) Resolution: 0.001 V   (b) Resolution: 0.001 V   (c) Resolution: 0.001 V   (d) Resolution: 0.01 V

▲ **FIGURE 2–59**

A 3½-digit DMM illustrates how the resolution changes with the number of digits in use.

as it is with 3½ active digits. The resolution remains 0.01 V up to 19.99 V. The resolution goes to 0.1 V for readings of 20.0 V to 199.9 V. At 200 V, the resolution goes to 1 V, and so on.

The resolution capability of a DMM is also determined by the internal circuitry and the rate at which the measured quantity is sampled. DMMs with displays of 4½ through 8½ digits are also available.

Accuracy   The **accuracy** is the degree to which a measured value represents the true or accepted value of a quantity. The accuracy of a DMM is established strictly by its internal circuitry and calibration. For typical DMMs, accuracies range from 0.01% to 0.5%, with some precision laboratory-grade DMMs going to 0.002%.

Reading Analog Multimeters

Although the DMM is the predominate type of multimeter, you may occasionally have to use an analog meter. A representation of a typical analog multimeter is shown in Figure 2–60. This particular instrument can be used to measure both direct current (dc) and alternating current (ac) quantities as well as resistance values. It has four selectable

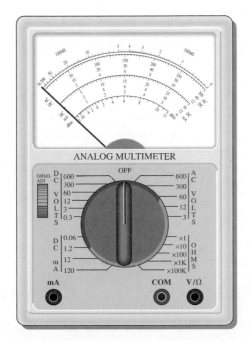

◄ **FIGURE 2–60**

A typical analog multimeter.

functions: dc volts (DC VOLTS), dc milliamperes (DC mA), ac volts (AC VOLTS), and OHMS. Most analog multimeters are similar to this one.

Within each function there are several ranges, as indicated by the brackets around the selector switch. For example, the DC VOLTS function has 0.3 V, 3 V, 12 V, 60 V, 300 V, and 600 V ranges. Thus, dc voltages from 0.3 V full-scale to 600 V full-scale can be measured. On the DC mA function, direct currents from 0.06 mA full-scale to 120 mA full-scale can be measured. On the ohm scale, the settings are ×1, ×10, ×100, ×1000, and ×100,000.

The Ohm Scale   Ohms are read on the top scale of the meter. This scale is nonlinear; that is, the values represented by each division (large or small) vary as you go across the scale. In Figure 2–60, notice how the scale becomes more compressed as you go from right to left.

To read the actual value in ohms, multiply the number on the scale as indicated by the pointer by the factor selected by the switch. For example, when the switch is set at ×100 and the pointer is at 20, the reading is 20 × 100 = 2000 Ω.

As another example, assume that the switch is at ×10 and the pointer is at the seventh small division between the 1 and 2 marks, indicating 17 Ω (1.7 × 10). Now, if the meter remains connected to the same resistance and the switch setting is changed to ×1, the pointer will move to the second small division between the 15 and 20 marks. This, of course, is also a 17 Ω reading, illustrating that a given resistance value can often be read at more than one range switch setting. However, the meter should be *zeroed* each time the range is changed by touching the leads together and adjusting the needle.

The AC-DC and DC mA Scales   The second, third, and fourth scales from the top, labeled "AC" and "DC," are used in conjunction with the DC VOLTS and AC VOLTS functions. The upper ac-dc scale ends at the 300 mark and is used with range settings, such as 0.3, 3, and 300. For example, when the switch is at 3 on the DC VOLTS function, the 300 scale has a full-scale value of 3 V; at the range setting of 300, the full-scale value is 300 V. The middle ac-dc scale ends at 60. This scale is used in conjunction with range settings, such as 0.06, 60, and 600. For example, when the switch is at 60 on the DC VOLTS function, the full-scale value is 60 V. The lower ac-dc scale ends at 12 and is used in conjunction with switch settings, such as 1.2, 12, and 120. The three DC mA scales are used in a similar way to measure current.

EXAMPLE 2–9

Determine the quantity (voltage, current, or resistance) that is being measured and its value for each specified switch setting on the meter in Figure 2–61.

(a) The switch is set on the DC VOLTS function and the 60 V range.

(b) The switch is set on the DC mA function and the 12 mA range.

(c) The switch is set on the OHMS function and the ×1K range.

FIGURE 2–61

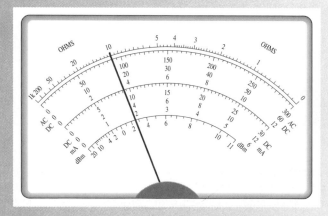

Solution   (a) The reading taken from the middle AC-DC scale is **18 V**.

(b) The reading taken from the lower AC-DC scale is **3.8 mA**.

(c) The reading taken from the ohms scale (top) is **10 kΩ**.

Related Problem   In Figure 2–61 the switch is moved to the ×100 ohms setting. Assuming that the same resistance is being measured as in part (c), what will the needle do?

SECTION 2–7 REVIEW

1. Name the meters for measurement of (a) current, (b) voltage, and (c) resistance.
2. Place two ammeters in the circuit of Figure 2–42 to measure the current through either lamp (be sure to observe the polarities). How can the same measurements be accomplished with only one ammeter?
3. Show how to place a voltmeter to measure the voltage across Lamp 2 in Figure 2–42.
4. List two common types of DMM displays, and discuss the advantages and disadvantages of each.
5. Define *resolution* in a DMM.
6. The multimeter in Figure 2–60 is set on the 3 V range to measure dc voltage. The pointer is at 150 on the upper ac-dc scale. What voltage is being measured?
7. How do you set up the meter in Figure 2–60 to measure 275 V dc, and on what scale do you read the voltage?
8. If you expect to measure a resistance in excess of 20 kΩ, where do you set the switch?

2–8   ELECTRICAL SAFETY

Safety is a major concern when working with electricity. The possibility of an electric shock or a burn is always present, so caution should always be used. You provide a current path when voltage is applied across two points on your body, and current produces electrical shock. Electrical components often operate at high temperatures, so you can sustain skin burns when you come in contact with them. Also, the presence of electricity creates a potential fire hazard.

After completing this section, you should be able to

♦ **Recognize electrical hazards and practice proper safety procedures**

  ♦ Describe the cause of electrical shock

  ♦ List the groups of current paths through the body

  ♦ Discuss the effects of current on the human body

  ♦ List the safety precautions that you should observe when you work with electricity

Electric Shock

Current through your body, not the voltage, is the cause of **electrical shock**. Of course, it takes voltage across a resistance to produce current. When a point on your body comes in contact with a voltage and another point comes in contact with a different voltage or with ground, such as a metal chassis, there will be current through your body from one point to the other. The path of the current depends on the points across which the voltage occurs. The severity of the resulting electrical shock depends on the amount of voltage and the path that the current takes through your body. The current path through the body determines which tissues and organs will be affected.

Effects of Current on the Human Body   The amount of current is dependent on voltage and resistance. The human body has resistance that depends on many factors, which include body mass, skin moisture, and points of contact of the body with a voltage potential. Table 2–5 shows the effects for various values of current in milliamperes.

▶ **TABLE 2–5**

Physical effects of electrical current. Values vary depending on body mass.

CURRENT (mA)	PHYSICAL EFFECT
0.4	Slight sensation
1.1	Perception threshold
1.8	Shock, no pain, no loss of muscular control
9	Painful shock, no loss of muscular control
16	Painful shock, let-go threshold
23	Severe painful shock, muscular contractions, breathing difficulty
75	Ventricular fibrillation, threshold
235	Ventricular fibrillation, usually fatal for duration of 5 seconds or more
4,000	Heart paralysis (no ventricular fibrillation)
5,000	Tissue burn

Body Resistance   Resistance of the human body is typically between $10\,\text{k}\Omega$ and $50\,\text{k}\Omega$ and depends on the two points between which it is measured. The moisture of the skin also affects the resistance between two points. The resistance determines the amount of voltage required to produce each of the effects listed in Table 2–5. For example, if you have a resistance of $10\,\text{k}\Omega$ between two given points on your body, 90 V across those two points will produce enough current (9 mA) to cause painful shock.

Safety Precautions

There are many practical things that you should do when you work with electrical and electronic equipment. Some important precautions are listed here.

- ◆ Avoid contact with any voltage source. Turn power off before you work on circuits when touching circuit parts is required.
- ◆ Do not work alone. A telephone should be available for emergencies.
- ◆ Do not work when tired or taking medications that make you drowsy.
- ◆ Remove rings, watches, and other metallic jewelry when you work on circuits.
- ◆ Do not work on equipment until you know proper procedures and are aware of potential hazards.
- ◆ Use equipment with three-wire power cords (three-prong plug).
- ◆ Make sure power cords are in good condition and grounding pins are not missing or bent.
- ◆ Keep your tools properly maintained. Make sure the insulation on metal tool handles is in good condition.
- ◆ Handle tools properly and maintain a neat work area.
- ◆ Wear safety glasses when appropriate, particularly when soldering and clipping wires.
- ◆ Always shut off power and discharge capacitors before you touch any part of a circuit with your hands.

◆ Know the location of the emergency power-off switch and emergency exits.

◆ Never try to override or tamper with safety devices such as an interlock switch.

◆ Always wear shoes and keep them dry. Do not stand on metal or wet floors.

◆ Never handle instruments when your hands are wet.

◆ Never assume that a circuit is off. Double-check it with a reliable meter before handling.

◆ Set the limiter on electronic power supplies to prevent currents larger than necessary to supply the circuit under test.

◆ Some devices such as capacitors can store a lethal charge for long periods after power is removed. They must be properly discharged before you work with them.

◆ When making circuit connections, always make the connection to the point with the highest voltage as your last step.

◆ Avoid contact with the terminals of power supplies.

◆ Always use wires with insulation and connectors or clips with insulating shrouds.

◆ Keep cables and wires as short as possible. Connect polarized components properly.

◆ Report any unsafe condition.

◆ Be aware of and follow all workplace and laboratory rules. Do not have drinks or food near equipment.

◆ If another person cannot let go of an energized conductor, switch the power off immediately. If that is not possible, use any available nonconductive material to try to separate the body from the contact.

SECTION 2 8 REVIEW

1. What causes physical pain and/or damage to the body when electrical contact is made?
2. It's OK to wear a ring when working on an electrical circuit.   (T or F)
3. Standing on a wet floor presents no safety hazard when working with electricity. (T or F)
4. A circuit can be rewired without removing the power if you are careful.   (T or F)
5. Electrical shock can be extremely painful or even fatal.   (T or F)

A Circuit Application

In this application, a dc voltage is applied to a circuit in order to produce current through a lamp and produce light. You will see how the current is controlled by resistance. The circuit that you will be working with simulates the instrument illumination circuit in a car, which allows you to increase or decrease the amount of light on the instruments.

The instrument panel illumination circuit in an automobile operates from the 12 V battery that is the voltage source for the circuit. The circuit uses a potentiometer connected as a rheostat, controlled by a knob on the instrument panel, which is used to

set the amount of current through the lamp to back-light the instruments. The brightness of the lamp is proportional to the amount of current through the lamp. The switch used to turn the lamp on and off is the same one used for the headlights. There is a fuse for circuit protection in case of a short circuit.

Figure 2–62 shows the schematic for the illumination circuit. Figure 2–63 shows a breadboarded circuit which simulates the illumination circuit by using components that are functionally equivalent but not physically the same as those in a car. A laboratory dc power supply is used in the place of an actual automobile battery. The protoboard in Figure 2–63 is a type that is commonly used for constructing circuits on the test bench.

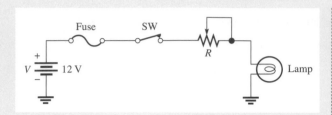

▲ FIGURE 2–62

Basic panel illumination circuit schematic.

The Test Bench

Figure 2–63 shows the breadboarded circuit, a dc power supply, and a digital multimeter. The power supply is connected to provide 12 V to the circuit. The multimeter is used to measure current, voltage, and resistance in the circuit.

◆ Identify each component in the circuit and check the breadboarded circuit in Figure 2–63 to make sure it is connected as the schematic in Figure 2–62 indicates.

◆ Explain the purpose of each component in the circuit.

As shown in Figure 2–64, the typical protoboard consists of rows of small sockets into which component leads and wires are inserted. In this particular configuration, all five sockets in each row are connected together and are effectively one electrical point as shown in the bottom view. All sockets arranged on the outer edges of the board are typically connected together as shown.

Measuring Current with the Multimeter

Set the DMM to the ammeter function to measure current. You must break the circuit in order to connect the ammeter in series to measure current. Refer to Figure 2–65.

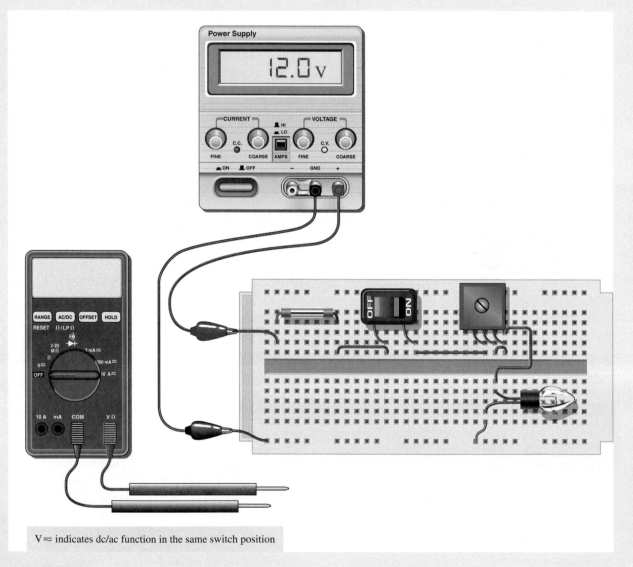

V≈ indicates dc/ac function in the same switch position

▲ FIGURE 2–63

Test bench setup for simulating the panel illumination circuit.

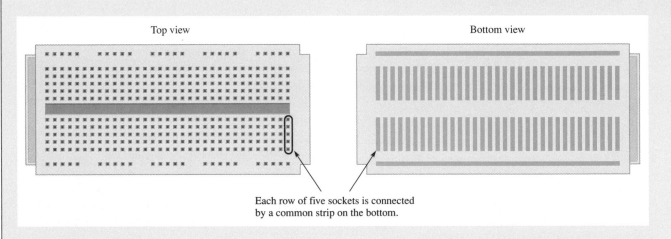

Top view

Bottom view

Each row of five sockets is connected
by a common strip on the bottom.

▲ **FIGURE 2–64**

A typical protoboard used for breadboarding.

◀ **FIGURE 2–65**

Current measurements. The
circled numbers indicate the
meter-to-circuit connections.

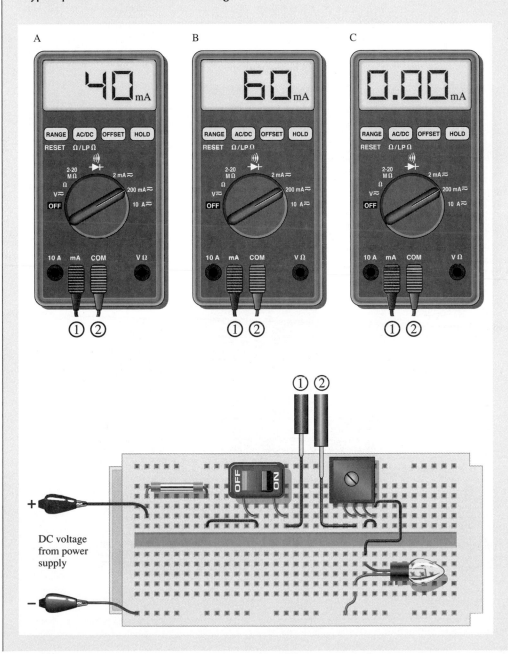

59

◆ Redraw the schematic in Figure 2–62 to include the ammeter.

◆ For which measurement (A, B, or C) is the lamp brightest? Explain.

◆ List the change(s) in the circuit that can cause the ammeter reading to go from A to B.

◆ List the circuit condition(s) that will produce the ammeter reading in C.

Measuring Voltage with the Multimeter

Set the DMM to the voltmeter function to measure voltage. You must connect the voltmeter to the two points across which you are measuring the voltage. Refer to Figure 2–66.

◆ Across which component is the voltage measured?

◆ Redraw the schematic in Figure 2–62 to include the voltmeter.

◆ For which measurement (A or B) is the lamp brighter? Explain.

◆ List the change(s) in the circuit that can cause the voltmeter reading to go from A to B.

Measuring Resistance with the Multimeter

Set the DMM to the ohmmeter function to measure resistance. Before you connect the ohmmeter, you must disconnect the resistance to be measured from the circuit. Before you disconnect any component, first turn the power supply off. Refer to Figure 2–67.

◆ For which component is the resistance measured?

◆ For which measurement (A or B) will the lamp be brighter when the circuit is reconnected and the power turned on? Explain.

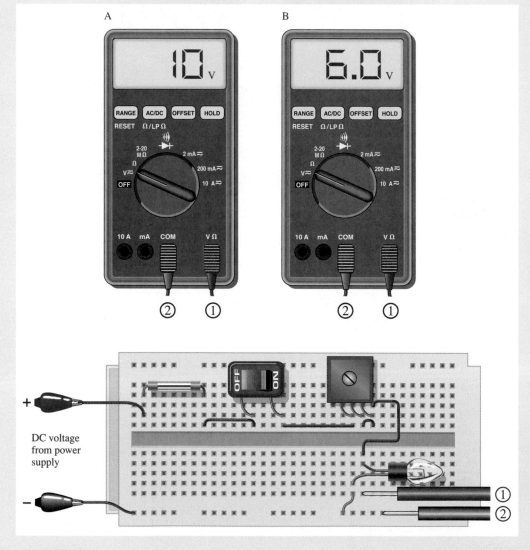

▲ FIGURE 2–66

Voltage measurements.

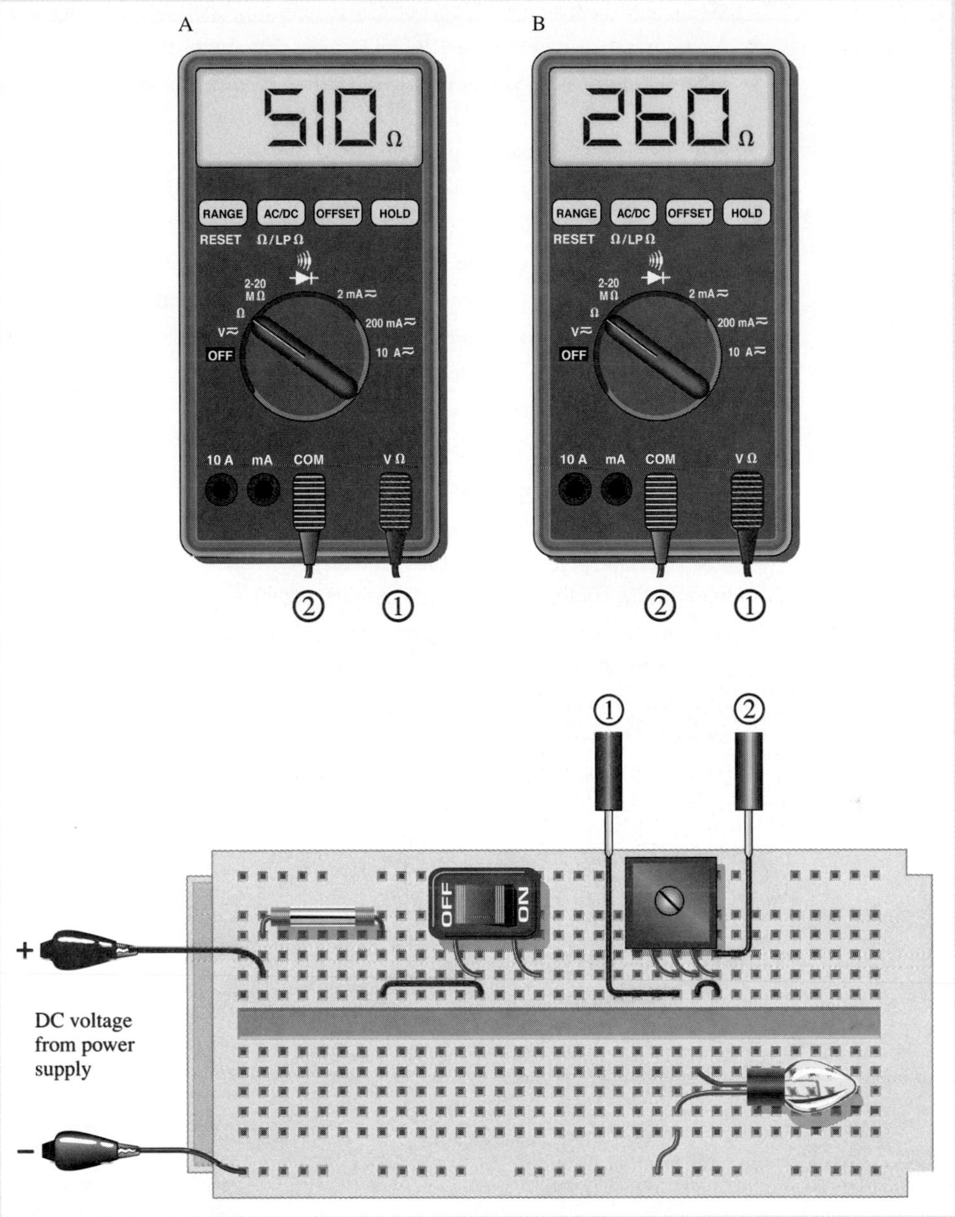

▲ FIGURE 2–67

Resistance measurements.

Review

1. If the dc supply voltage in the panel illumination circuit is reduced, how is the amount of light produced by the lamp affected? Explain.

2. Should the potentiometer be adjusted to a higher or lower resistance for the circuit to produce more light?

SUMMARY

◆ An atom is the smallest particle of an element that retains the characteristics of that element.

◆ When electrons in the outer orbit of an atom (valence electrons) break away, they become free electrons.

◆ Free electrons make current possible.

◆ Like charges repel each other, and opposite charges attract each other.

◆ Voltage must be applied to a circuit to produce current.

◆ Resistance limits the current.

◆ Basically, an electric circuit consists of a source, a load, and a current path.

◆ An open circuit is one in which the current path is broken.

◆ A closed circuit is one which has a complete current path.

◆ An ammeter is connected in line with the current path.

◆ A voltmeter is connected across the current path.

◆ An ohmmeter is connected across a resistor (resistor must be disconnected from circuit).

◆ One coulomb is the charge of 6.25×10^{18} electrons.

◆ One volt is the potential difference (voltage) between two points when one joule of energy is used to move one coulomb from one point to the other.

◆ One ampere is the amount of current that exists when one coulomb of charge moves through a given cross-sectional area of a material in one second.

◆ One ohm is the resistance when there is one ampere of current in a material with one volt applied across the material.

◆ Figure 2–68 shows the electrical symbols introduced in this chapter.

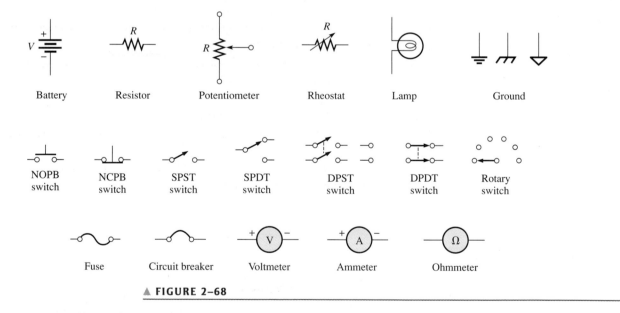

▲ FIGURE 2–68

KEY TERMS

Key terms and other bold terms in the chapter are defined in the end-of-book glossary.

Ammeter An electrical instrument used to measure current.

Ampere (A) The unit of electrical current.

Atom The smallest particle of an element possessing the unique characteristics of that element.

AWG American wire gauge; a standardization based on wire diameter.

Charge   An electrical property of matter that exists because of an excess or a deficiency of electrons. Charge can be either positive or negative.

Circuit   An interconnection of electrical components designed to produce a desired result. A basic circuit consists of a source, a load, and an interconnecting current path.

Closed circuit   A circuit with a complete current path.

Conductance   The ability of a circuit to allow current. The unit is the siemens (S).

Conductor   A material in which electric current is easily established. An example is copper.

Coulomb (C)   The unit of electrical charge; the total charge possessed by 6.25×10^{18} electrons.

Current   The rate of flow of charge (electrons).

Current source   A device that produces a constant current for a varying load.

DMM   Digital multimeter; an electronic instrument that combines meters for measurement of voltage, current, and resistance.

Electrical shock   The physical sensation resulting from electrical current through the body.

Electron   A basic particle of electrical charge in matter. The electron possesses negative charge.

Free electron   A valence electron that has broken away from its parent atom and is free to move from atom to atom within the atomic structure of a material.

Ground   The common or reference point in a circuit.

Insulator   A material that does not allow current under normal conditions.

Load   An element connected across the output terminals of a circuit that draws current from the source and upon which work is done.

Ohm (Ω)   The unit of resistance.

Ohmmeter   An instrument for measuring resistance.

Open circuit   A circuit in which there is not a complete current path.

Potentiometer   A three-terminal variable resistor.

Resistance   Opposition to current. The unit is the ohm (Ω).

Resistor   An electrical component specifically designed to have a certain amount of resistance.

Rheostat   A two-terminal variable resistor.

Semiconductor   A material that has a conductance value between that of a conductor and an insulator. Silicon and germanium are examples.

Siemens (S)   The unit of conductance.

Volt   The unit of voltage or electromotive force.

Voltage   The amount of energy per charge available to move electrons from one point to another in an electric circuit.

Voltage source   A device that produces a constant voltage for a varying load.

Voltmeter   An instrument used to measure voltage.

FORMULAS

2–1	$Q = \dfrac{\text{number of electrons}}{6.25 \times 10^{18} \text{ electrons/C}}$	Charge
2–2	$V = \dfrac{W}{Q}$	Voltage equals energy divided by charge.
2–3	$I = \dfrac{Q}{t}$	Current equals charge divided by time.
2–4	$G = \dfrac{1}{R}$	Conductance is the reciprocal of resistance.
2–5	$A = d^2$	Cross-sectional area equals the diameter squared.
2–6	$R = \dfrac{\rho l}{A}$	Resistance is resistivity times length divided by cross-sectional area.

Answers are at the end of the chapter.

1. A neutral atom with an atomic number of three has how many electrons?
   (a) 1   (b) 3   (c) none   (d) depends on the type of atom
2. Electron orbits are called
   (a) shells   (b) nuclei   (c) waves   (d) valences
3. Materials in which there is no current when voltage is applied are called
   (a) filters   (b) conductors   (c) insulators   (d) semiconductors
4. When placed close together, a positively charged material and a negatively charged material will
   (a) repel   (b) become neutral   (c) attract   (d) exchange charges
5. The charge on a single electron is
   (a) 6.25×10^{-18} C   (b) 1.6×10^{-19} C   (c) 1.6×10^{-19} J   (d) 3.14×10^{-6} C
6. *Potential difference* is another term for
   (a) energy   (b) voltage   (c) distance of an electron from the nucleus   (d) charge
7. The unit of energy is the
   (a) watt   (b) coulomb   (c) joule   (d) volt
8. Which one of the following is not a type of energy source?
   (a) battery   (b) solar cell   (c) generator   (d) potentiometer
9. Which one of the following is not a possible condition in an electric circuit?
   (a) voltage and no current   (b) current and no voltage
   (c) voltage and current   (d) no voltage and no current
10. Electrical current is defined as
    (a) free electrons
    (b) the rate of flow of free electrons
    (c) the energy required to move electrons
    (d) the charge on free electrons
11. There is no current in a circuit when
    (a) a switch is closed   (b) a switch is open   (c) there is no voltage
    (d) answers (a) and (c)   (e) answers (b) and (c)
12. The primary purpose of a resistor is to
    (a) increase current   (b) limit current
    (c) produce heat   (d) resist current change
13. Potentiometers and rheostats are types of
    (a) voltage sources   (b) variable resistors
    (c) fixed resistors   (d) circuit breakers
14. The current in a given circuit is not to exceed 22 A. Which value of fuse is best?
    (a) 10 A   (b) 25 A   (c) 20 A   (d) a fuse is not necessary

More difficult problems are indicated by an asterisk (*).
Answers to odd-numbered problems are at the end of the book.

SECTION 2–2   Electrical Charge

1. What is the charge in coulombs of the nucleus of a copper atom?
2. What is the charge in coulombs of the nucleus of a chlorine atom?
3. How many coulombs of charge do 50×10^{31} electrons possess?
4. How many electrons does it take to make 80 μC (microcoulombs) of charge?

SECTION 2–3   Voltage, Current, and Resistance

5. Determine the voltage in each of the following cases:

   (a) 10 J/C   (b) 5 J/2 C   (c) 100 J/25 C

6. Five hundred joules of energy are used to move 100 C of charge through a resistor. What is the voltage across the resistor?

7. What is the voltage of a battery that uses 800 J of energy to move 40 C of charge through a resistor?

8. How much energy does a 12 V battery use to move 2.5 C through a circuit?

9. If a resistor with a current of 2 A through it converts 1000 J of electrical energy into heat energy in 15 s, what is the voltage across the resistor?

10. Determine the current in each of the following cases:

    (a) 75 C in 1 s   (b) 10 C in 0.5 s   (c) 5 C in 2 s

11. Six-tenths coulomb passes a point in 3 s. What is the current in amperes?

12. How long does it take 10 C to flow past a point if the current is 5 A?

13. How many coulombs pass a point in 0.1 s when the current is 1.5 A?

14. 5.74×10^{17} electrons flow through a wire in 250 ms. What is the current in amperes?

15. Find the conductance for each of the following resistance values:

    (a) 5 Ω   (b) 25 Ω   (c) 100 Ω

16. Find the resistance corresponding to the following conductances:

    (a) 0.1 S   (b) 0.5 S   (c) 0.02 S

SECTION 2–4   Voltage and Current Sources

17. List four common sources of voltage.

18. Upon what principle is electrical generators based?

19. How does an electronic power supply differ from the other sources of voltage?

20. A certain current source provides 100 mA to a 1 kΩ load. If the resistance is decreased to 500 Ω, what the current in the load?

SECTION 2–5   Resistors

21. Determine the resistance values and tolerance for the following 4-band resistors:

    (a) red, violet, orange, gold   (b) brown, gray, red, silver

22. Find the minimum and the maximum resistance within the tolerance limits for each resistor in Problem 21.

23. Determine the color bands for each of the following 4-band, 5% values: 330 Ω, 2.2 kΩ, 56 kΩ, 100 kΩ, and 39 kΩ.

24. Determine the resistance and tolerance of each of the following 4-band resistors:

    (a) brown, black, black, gold

    (b) green, brown, green, silver

    (c) blue, gray, black, gold

25. Determine the color bands for each of the following 4-band resistors. Assume each has a 5% tolerance.

    (a) 0.47 Ω   (b) 270 kΩ   (c) 5.1 MΩ

26. Determine the resistance and tolerance of each of the following 5-band resistors:

    (a) red, gray, violet, red, brown

    (b) blue, black, yellow, gold, brown

    (c) white, orange, brown, brown, brown

27. Determine the color bands for each of the following 5-band resistors. Assume each has a 1% tolerance.

    (a) 14.7 kΩ   (b) 39.2 Ω   (c) 9.76 kΩ

28. The adjustable contact of a linear potentiometer is set at the mechanical center of its adjustment. If the total resistance is 1000 Ω, what is the resistance between each end terminal and the adjustable contact?

29. What resistance is indicated by 4K7?

30. Determine the resistance and tolerance of each resistor labeled as follows:

   (a) 4R7J   (b) 5602M   (c) 1501F

SECTION 2–6   The Electric Circuit

31. Trace the current path in Figure 2–69(a) with the switch in position 2.

32. With the switch in either position, redraw the circuit in Figure 2–69(d) with a fuse connected to protect the circuit against excessive current.

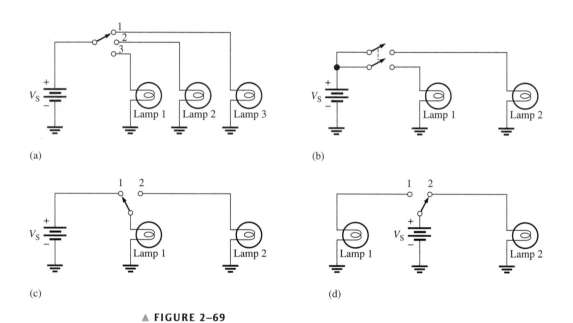

▲ FIGURE 2–69

33. There is only one circuit in Figure 2–69 in which it is possible to have all lamps on at the same time. Determine which circuit it is.

34. Through which resistor in Figure 2–70 is there always current, regardless of the position of the switches?

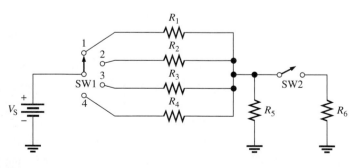

▲ FIGURE 2–70

*35. Devise a switch arrangement whereby two voltage sources (V_{S1} and V_{S2}) can be connected simultaneously to either of two resistors (R_1 and R_2) as follows:

$$V_{S1} \text{ connected to } R_1 \text{ and } V_{S2} \text{ connected to } R_2$$

or $\qquad\qquad V_{S1} \text{ connected to } R_2 \text{ and } V_{S2} \text{ connected to } R_1$

36. The different sections of a stereo system are represented by the blocks in Figure 2–71. Show how a single switch can be used to connect the phonograph, the CD (compact disk) player, the tape deck, the AM tuner, or the FM tuner to the amplifier by a single knob control. Only one section can be connected to the amplifier at any time.

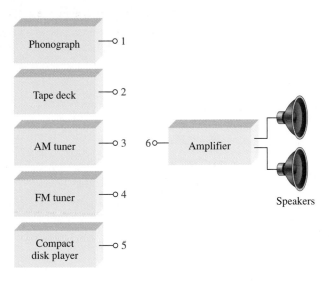

▲ FIGURE 2–71

SECTION 2–7   Basic Circuit Measurements

37. Show the placement of an ammeter and a voltmeter to measure the current and the source voltage in Figure 2–72.

▶ FIGURE 2–72

38. Explain how you would measure the resistance of R_2 in Figure 2–72.

39. In Figure 2–73, how much voltage does each meter indicate when the switch is in position 1? In position 2?

▶ FIGURE 2–73

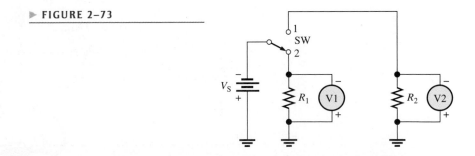

40. In Figure 2–73, indicate how to connect an ammeter to measure the current from the voltage source regardless of the switch position.

41. In Figure 2–70, show the proper placement of ammeters to measure the current through each resistor and the current out of the battery.

42. Show the proper placement of voltmeters to measure the voltage across each resistor in Figure 2–70.

43. What is the voltage reading of the meter in Figure 2–74?

▶ **FIGURE 2–74**

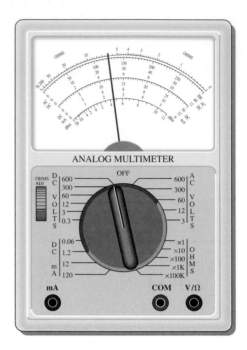

44. How much resistance is the ohmmeter in Figure 2–75 measuring?

▶ **FIGURE 2–75**

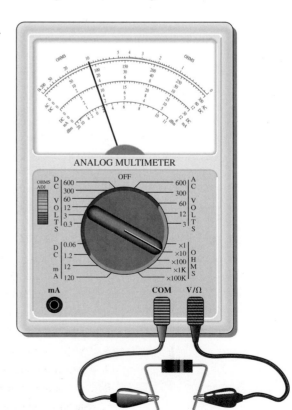

45. Determine the resistance indicated by each of the following ohmmeter readings and range settings:

   (a) pointer at 2, range setting at ×10

   (b) pointer at 15, range setting at ×100,000

   (c) pointer at 45, range setting at ×100

46. What is the maximum resolution of a 4½-digit DMM?

47. Indicate how you would connect the multimeter in Figure 2–75 to the circuit in Figure 2–76 to measure each of the following quantities. In each case indicate the appropriate function and range.

   (a) I_1 **(b)** V_1 **(c)** R_1

▶ **FIGURE 2–76**

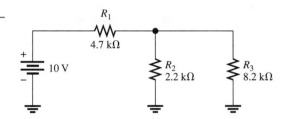

ANSWERS

SECTION REVIEWS

SECTION 2–1   Atomic Structure

   1. The electron is the basic particle of negative charge.

   2. An atom is the smallest particle of an element that retains the unique characteristics of the element.

   3. An atom is a positively charged nucleus surrounded by orbiting electrons.

   4. Atomic number is the number of protons in a nucleus.

   5. No, each element has a different type of atom.

   6. A free electron is an outer-shell electron that has drifted away from the parent atom.

   7. Shells are energy bands in which electrons orbit the nucleus of an atom.

   8. Copper and silver

SECTION 2–2   Electrical Charge

   1. Q = charge

   2. Unit of charge is the coulomb; C

   3. Positive or negative charge is caused by the loss or acquisition respectively of an outer-shell (valence) electron.

   4. $Q = \dfrac{10 \times 10^{12} \text{ electrons}}{6.25 \times 10^{18} \text{ electrons/C}} = 1.6 \times 10^{-6}\,\text{C} = 1.6\,\mu\text{C}$

SECTION 2–3   Voltage, Current, and Resistance

   1. Voltage is energy per unit charge.

   2. Volt is the unit of voltage.

   3. $V = W/Q = 24\,\text{J}/10\,\text{C} = 2.4\,\text{V}$

   4. Current is the rate of flow of electrons; its unit is the ampere (A).

   5. electrons/coulomb = 6.25×10^{18}

   6. $I = Q/t = 20\,\text{C}/4\,\text{s} = 5\,\text{A}$

   7. Resistance is opposition to current.

   8. The unit of resistance is the ohm (Ω).

   9. One ohm exists when 1 V produces 1 A.

SECTION 2–4 Voltage and Current Sources

1. A voltage source is a device that produces a constant voltage for a varying load.

2. A battery converts chemical energy to electrical energy.

3. A solar cell uses the photovoltaic effect to convert light energy to electrical energy.

4. A generator produces voltage by rotating a conductor through a magnetic field based on the principle of electromagnetic induction.

5. An electronic power supply converts the commercially available ac voltage to dc voltage.

6. A current source is a device that produces a constant current for a varying load.

7. Transistor

SECTION 2–5 Resistors

1. Two resistor categories are fixed and variable. The value of a fixed resistor cannot be changed, but that of a variable resistor can.

2. *First band:* first digit of resistance value. *Second band:* second digit of resistance value. *Third band:* multiplier (number of zeros following the second digit). *Fourth band:* % tolerance.

3. **(a)** $27\,k\Omega \pm 10\%$ **(b)** $100\,\Omega \pm 10\%$
   (c) $5.6\,M\Omega \pm 5\%$ **(d)** $6.8\,k\Omega \pm 10\%$
   (e) $33\,\Omega \pm 10\%$ **(f)** $47\,k\Omega \pm 5\%$

4. $330\,\Omega$: (b); $2.2\,k\Omega$: (d); $56\,k\Omega$: (e); $100\,k\Omega$: (f); $39\,k\Omega$: (a)

5. **(a)** 33R $= 33\,\Omega$ **(b)** 5K6 $= 5.6\,k\Omega$
   (c) 900 $= 900\,\Omega$ **(d)** 6M8 $= 6.8\,M\Omega$

6. A rheostat has two terminals; a potentiometer has three terminals.

7. A thermistor is a temperature-sensitive resistor.

SECTION 2–6 The Electric Circuit

1. An electric circuit consists of a source, load, and current path between source and load.

2. An open circuit is one that has no path for current.

3. A closed circuit is one that has a complete path for current.

4. A fuse is not resettable, a circuit breaker is.

5. AWG 3 is larger.

6. Ground is the common or reference point.

SECTION 2–7 Basic Circuit Measurements

1. **(a)** An ammeter measures current.
   (b) A voltmeter measures voltage.
   (c) An ohmmeter measures resistance.

2. See Figure 2–77.

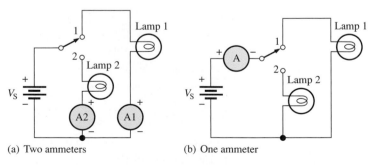

(a) Two ammeters   (b) One ammeter

▲ **FIGURE 2–77**

3. See Figure 2–78.

▶ **FIGURE 2–78**

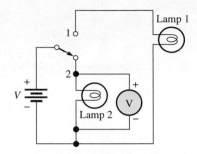

4. Two types of DMM displays are liquid-crystal display (LCD) and light-emitting display (LED). The LCD requires little current, but it is difficult to see in low light and is slow to respond. The LED can be seen in the dark, and it responds quickly. However, it requires much more current than does the LCD.

5. Resolution is the smallest increment of a quantity that the meter can measure.

6. 1.5 V

7. Set the range switch to 300 and read on the upper ac-dc scale.

8. ×1000 range

SECTION 2–8 Electrical Safety

1. Current

2. F

3. F

4. F

5. T

A Circuit Application

1. Less voltage causes less light because the current is reduced.

2. A lower resistance will result in more light.

RELATED PROBLEMS FOR EXAMPLES

2–1 1.88×10^{19} electrons

2–2 600 J

2–3 12 C

2–4 4700 Ω ± 5%

2–5 47.5 Ω ± 2%

2–6 1.25 kΩ

2–7 2.25 CM

2–8 1.280 Ω; same as calculated result

2–9 The needle will move left to the "100" mark.

SELF-TEST

1. (b)	**2.** (a)	**3.** (c)	**4.** (c)	**5.** (b)	**6.** (b)	**7.** (c)	**8.** (d)
9. (b)	**10.** (b)	**11.** (e)	**12.** (b)	**13.** (b)	**14.** (c)		

3

OHM'S LAW

CHAPTER OBJECTIVES

◆ Explain Ohm's law

◆ Calculate current in a circuit

◆ Calculate voltage in a circuit

◆ Calculate resistance in a circuit

◆ Describe a basic approach to troubleshooting

KEY TERMS

◆ Ohm's law

◆ Linear

◆ Troubleshooting

A CIRCUIT APPLICATION PREVIEW

In this application, you will see how Ohm's law is used in a practical circuit. You are assigned the task of modifying an existing test fixture for use in a new application. The test fixture is a resistance box with an array of switch-selectable resistors of various values. You will determine and specify the changes necessary in the existing circuit for the new application. Also, you will develop a test procedure once the modifications have been made.

VISIT THE COMPANION WEBSITE

Study aids for this chapter are available at
http://www.prenhall.com/floyd

INTRODUCTION

In Chapter 2, you studied the concepts of voltage, current, and resistance. You also were introduced to a basic electric circuit. In this chapter, you will learn how voltage, current, and resistance are interrelated. You will also learn how to analyze a simple electric circuit.

Ohm's law is perhaps the single most important tool for the analysis of electric circuits, and you *must* know how to apply it.

In 1826 Georg Simon Ohm found that current, voltage, and resistance are related in a specific and predictable way. Ohm expressed this relationship with a formula that is known today as Ohm's law. In this chapter, you will learn Ohm's law and how to use it in solving circuit problems. A general approach to troubleshooting using an analysis, planning, and measurement (APM) approach is also introduced.

3–1 THE RELATIONSHIP OF CURRENT, VOLTAGE, AND RESISTANCE

Ohm's law describes mathematically how voltage, current, and resistance in a circuit are related. Ohm's law is expressed in three equivalent forms depending on which quantity you need to determine. As you will learn, current and voltage are linearly proportional. However, current and resistance are inversely proportional.

After completing this section, you should be able to

◆ **Explain Ohm's law**

    ◆ Describe how V, I, and R are related

    ◆ Express I as a function of V and R

    ◆ Express V as a function of I and R

    ◆ Express R as a function of V and I

    ◆ Show graphically that I and V are directly proportional

    ◆ Show graphically that I and R are inversely proportional

    ◆ Explain why I and V are linearly proportional

Ohm determined experimentally that if the voltage across a resistor is increased, the current through the resistor will also increase; and, likewise, if the voltage is decreased, the current will decrease. For example, if the voltage is doubled, the current will double. If the voltage is halved, the current will also be halved. This relationship is illustrated in Figure 3–1, with relative meter indications of voltage and current.

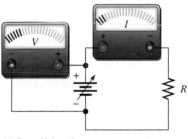
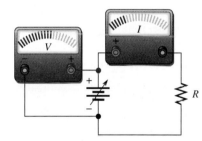

(a) Less V, less I (b) More V, more I

◀ **FIGURE 3–1**

Effect on the current of changing the voltage with the resistance at a constant value.

Ohm also determined that if the voltage is kept constant, less resistance results in more current, and, also, more resistance results in less current. For example, if the resistance is halved, the current doubles. If the resistance is doubled, the current is halved. This concept is illustrated by the meter indications in Figure 3–2, where the resistance is increased and the voltage is held constant.

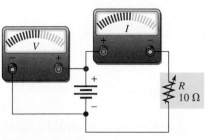
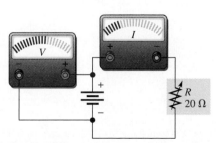

(a) Less R, more I (b) More R, less I

◀ **FIGURE 3–2**

Effect on the current of changing the resistance with the voltage at a constant value.

Ohm's law states that current is directly proportional to voltage and inversely proportional to resistance. The circuits in Figures 3–1 and 3–2 illustrate Ohm's law, which is given in the following formula:

Equation 3–1

$$I = \frac{V}{R}$$

where: I = current in amperes (A)

V = voltage in volts (V)

R = resistance in ohms (Ω)

For a constant value of R, if the value of V is increased, the value of I increases; if V is decreased, I decreases. If V is constant and R is increased, I decreases. Similarly, if V is constant and R is decreased, I increases.

Using Equation 3–1, you can calculate the current if you know the values of voltage and resistance. By manipulating Equation 3–1, you can obtain expressions for voltage and resistance.

Equation 3–2

$$V = IR$$

Equation 3–3

$$R = \frac{V}{I}$$

With Equation 3–2, you can calculate voltage if you know the values of current and resistance. With Equation 3–3, you can calculate resistance if you know the values of voltage and current.

The three expressions—Equations 3–1, 3–2 and 3–3—are all equivalent. They are simply three ways of using Ohm's law.

The Linear Relationship of Current and Voltage

In resistive circuits, current and voltage are linearly proportional. **Linear** means that if one of the quantities is increased or decreased by a certain percentage, the other will increase or decrease by the same percentage, assuming that the resistance is constant in value. For example, if the voltage across a resistor is tripled, the current will triple.

EXAMPLE 3–1

Show that if the voltage in the circuit of Figure 3–3 is increased to three times its present value, the current will triple in value.

▶ **FIGURE 3–3**

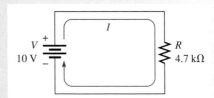

Solution   With 10 V, the current is

$$I = \frac{V}{R} = \frac{10\,\text{V}}{4.7\,\text{k}\Omega} = \mathbf{2.13\,mA}$$

If the voltage is increased to 30 V, the current will be

$$I = \frac{V}{R} = \frac{30\,\text{V}}{4.7\,\text{k}\Omega} = \mathbf{6.38\,mA}$$

The current went from 2.13 mA to 6.38 mA when the voltage was tripled to 30 V.

*Related Problem**   If the voltage in Figure 3–3 is quadrupled, will the current also quadruple?

*Answers are at the end of the chapter.

Let's take a constant value of resistance, for example, 10 Ω, and calculate the current for several values of voltage ranging from 10 V to 100 V in the circuit in Figure 3–4(a). The current values obtained are shown in Figure 3–4(b). The graph of the *I* values versus the *V* values is shown in Figure 3–4(c). Note that it is a straight line graph. This graph tells us that a change in voltage results in a linearly proportional change in current. No matter what value *R* is, assuming that *R* is constant, the graph of *I* versus *V* will always be a straight line.

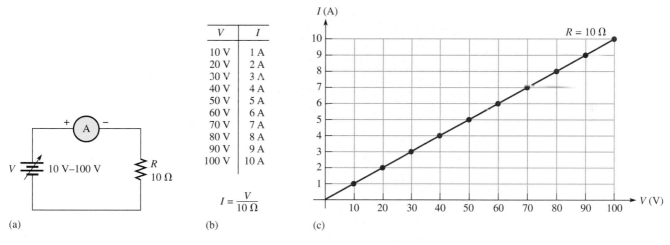

(a)   (b)   (c)

▲ **FIGURE 3–4**

Graph of current versus voltage for the circuit in part (a).

EXAMPLE 3–2

Assume that you are measuring the current in a circuit that is operating with 25 V. The ammeter reads 50 mA. Later, you notice that the current has dropped to 40 mA. Assuming that the resistance did not change, you must conclude that the voltage source has changed. How much has the voltage changed, and what is its new value?

Solution   The current has dropped from 50 mA to 40 mA, which is a decrease of 20%. Since the voltage is linearly proportional to the current, the voltage has decreased by the same percentage that the current did. Taking 20% of 25 V, you get

$$\text{Change in voltage} = (0.2)(25\ \text{V}) = \mathbf{5\ V}$$

Subtract this change from the original voltage to get the new voltage.

$$\text{New voltage} = 25\ \text{V} - 5\ \text{V} = \mathbf{20\ V}$$

Notice that you did not need the resistance value in order to find the new voltage.

Related Problem   If the current drops to 0 A under the same conditions stated in the example, what is the voltage?

The Inverse Relationship of Current and Resistance

As you have seen, current varies inversely with resistance as expressed by Ohm's law, $I = V/R$. When the resistance is reduced, the current goes up; when the resistance is increased, the current goes down. For example, if the source voltage is held constant and the resistance is halved, the current doubles in value; when the resistance is doubled, the current is reduced by half.

Let's take a constant value of voltage, for example, 10 V, and calculate the current for several values of resistance ranging from 10 Ω to 100 Ω in the circuit in Figure 3–5(a). The values obtained are shown in Figure 3–5(b). The graph of the I values versus the R values is shown in Figure 3–5 (c).

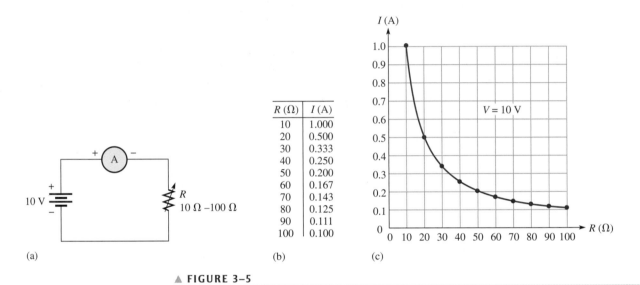

R (Ω)	I (A)
10	1.000
20	0.500
30	0.333
40	0.250
50	0.200
60	0.167
70	0.143
80	0.125
90	0.111
100	0.100

(a) (b) (c)

▲ FIGURE 3–5

Graph of current versus resistance for the circuit in part (a).

1. Ohm's law defines how three basic quantities are related. What are these quantities?
2. Write the Ohm's law formula for current.
3. Write the Ohm's law formula for voltage.
4. Write the Ohm's law formula for resistance.
5. If the voltage across a fixed-value resistor is tripled, does the current increase or decrease, and by how much?
6. If the voltage across a fixed resistor is cut in half, how much will the current change?
7. There is a fixed voltage across a resistor, and you measure a current of 1 A. If you replace the resistor with one that has twice the resistance value, how much current will you measure?
8. In a circuit the voltage is doubled and the resistance is cut in half. Would the current increase or decrease and if so, by how much?
9. In a circuit, $V = 2$ V and $I = 10$ mA. If V is changed to 1 V, what will I equal?
10. If $I = 3$ A at a certain voltage, what will it be if the voltage is doubled?

3–2 CALCULATING CURRENT

Examples in this section illustrate the Ohm's law formula $I = V/R$.

After completing this section, you should be able to

◆ **Calculate current in a circuit**

   ◆ Use Ohm's law to find current when you know voltage and resistance values

   ◆ Use voltage and resistance values expressed with metric prefixes

The following examples use the formula $I = V/R$. In order to get current in amperes, you must express the value of voltage in volts and the value of resistance in ohms.

EXAMPLE 3–3

How many amperes of current are in the circuit of Figure 3–6?

▶ **FIGURE 3–6**

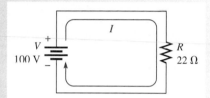

Solution   Use the formula $I = V/R$, and substitute 100 V for V and 22 Ω for R.

$$I = \frac{V}{R} = \frac{100\,V}{22\,\Omega} = \textbf{4.55 A}$$

Related Problem   If R is changed to 33 Ω in Figure 3–6, what is the current?

 Use Multisim file E03-03 to verify the calculated results in this example and your calculation for the related problem.

EXAMPLE 3–4

If the resistance in Figure 3–6 is changed to 47 Ω and the voltage to 50 V, what is the new value of current?

Solution   Substitute $V = 50$ V and $R = 47$ Ω into the formula $I = V/R$.

$$I = \frac{V}{R} = \frac{50\,V}{47\,\Omega} = \textbf{1.06 A}$$

Related Problem   If $V = 5$ V and $R = 1000$ Ω, what is the current?

Units with Metric Prefixes

In electronics, resistance values of thousands of ohms or even millions of ohms are common. The metric prefixes *kilo* (k) and *mega* (M) are used to indicate large values. Thus, thousands of ohms are expressed in kilohms (kΩ), and millions of ohms are expressed in megohms (MΩ). The following four examples illustrate how to use kilohms and megohms to calculate current. Volts (V) divided by kilohms (kΩ) results in milliamperes (mA). Volts (V) divided by megohms (MΩ) results in microamperes (μA).

EXAMPLE 3–5

Calculate the current in Figure 3–7.

▶ **FIGURE 3–7**

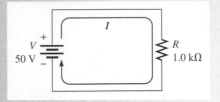

Solution Remember that 1.0 kΩ is the same as $1 \times 10^3 \, \Omega$. Use the formula $I = V/R$ and substitute 50 V for V and $1 \times 10^3 \, \Omega$ for R.

$$I = \frac{V}{R} = \frac{50 \, \text{V}}{1.0 \, \text{k}\Omega} = \frac{50 \, \text{V}}{1 \times 10^3 \, \Omega} = 50 \times 10^{-3} \, \text{A} = \textbf{50 mA}$$

Related Problem Calculate the current in Figure 3–7 if R is changed to 10 kΩ.

In Example 3–5, 50×10^{-3} A is expressed as 50 milliamperes (50 mA). This can be used to advantage when you divide volts by kilohms. The current will be in milliamperes, as Example 3–6 illustrates.

EXAMPLE 3–6

How many milliamperes are in the circuit of Figure 3–8?

▶ **FIGURE 3–8**

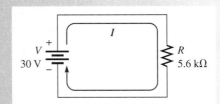

Solution When you divide volts by kilohms, you get current in milliamperes.

$$I = \frac{V}{R} = \frac{30 \, \text{V}}{5.6 \, \text{k}\Omega} = \textbf{5.36 mA}$$

Related Problem What is the current in milliamperes if R is changed to 2.2 kΩ?

 Use Multisim file E03-06 to verify the calculated results in this example and to confirm your calculation for the related problem.

If you apply volts when resistance values are in megohms, the current is in microamperes (μA), as Examples 3–5 and 3–6 show.

EXAMPLE 3–7

Determine the amount of current in the circuit of Figure 3–9.

▶ **FIGURE 3–9**

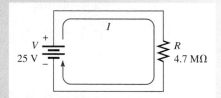

Solution   Recall that 4.7 MΩ equals 4.7×10^6 Ω. Substitute 25 V for V and 4.7×10^6 Ω for R.

$$I = \frac{V}{R} = \frac{25\text{ V}}{4.7\text{ M}\Omega} = \frac{25\text{ V}}{4.7 \times 10^6\,\Omega} = 5.32 \times 10^{-6}\text{ A} = \mathbf{5.32\,\mu A}$$

Related Problem   What is the current if V is increased to 100 V in Figure 3–6?

EXAMPLE 3–8

Change the value of R in Figure 3–9 to 1.8 MΩ. What is the new value of current?

Solution   When you divide volts by megohms, you get current in microamperes.

$$I = \frac{V}{R} = \frac{25\text{ V}}{1.8\text{ M}\Omega} = \mathbf{13.9\,\mu A}$$

Related Problem   If R is doubled in the circuit of Figure 3–6, what is the new value of current?

Small voltages, usually less than 50 V are common in semiconductor circuits. Occasionally, however, large voltages are encountered. For example, the high-voltage supply in some television receivers is around 20,000 V (20 kilovolts, or 20 kV), and transmission voltages generated by the power companies may be as high as 345,000 V (345 kV). The following two examples illustrate how to use voltage values in the kilovolt range to calculate current.

EXAMPLE 3–9

How much current is produced by a voltage of 24 kV across a 12 kΩ resistor?

Solution   Since kilovolts are divided by kilohms, the prefixes cancel; therefore, the current is in amperes.

$$I = \frac{V}{R} = \frac{24\text{ kV}}{12\text{ k}\Omega} = \frac{24 \times 10^3\text{ V}}{12 \times 10^3\,\Omega} = \mathbf{2\text{ A}}$$

Related Problem   What is the current in mA produced by 1 kV across a 27 kΩ resistor?

EXAMPLE 3–10

How much current is there through a 100 MΩ resistor when 50 kV are applied?

Solution   In this case, divide 50 kV by 100 MΩ to get the current. Substitute 50×10^3 V for 50 kV and 100×10^6 Ω for 100 MΩ.

$$I = \frac{V}{R} = \frac{50\text{ kV}}{100\text{ M}\Omega} = \frac{50 \times 10^3\text{ V}}{100 \times 10^6\,\Omega} = 0.5 \times 10^{-3}\text{ A} = \mathbf{0.5\text{ mA}}$$

Remember that the power of ten in the denominator is subtracted from the power of ten in the numerator. So 50 was divided by 100, giving 0.5, and 6 was subtracted from 3, giving 10^{-3}.

Related Problem   How much current is there through a 6.8 MΩ resistor when 10 kV are applied?

SECTION 3–2
REVIEW

In Problems 1–4, calculate *I*.
1. *V* = 10 V and *R* = 5.6 Ω.
2. *V* = 100 V and *R* = 560 Ω.
3. *V* = 5 V and *R* = 2.2 kΩ.
4. *V* = 15 V and *R* = 4.7 MΩ.
5. If a 4.7 MΩ resistor has 20 kV across it, how much current is there?
6. How much current will 10 kV across a 2.2 MΩ resistor produce?

3–3   CALCULATING VOLTAGE

Examples in this section illustrate the Ohm's law expression *V* = *IR*.

After completing this section, you should be able to

◆ **Calculate voltage in a circuit**

   ◆ Use Ohm's law to find voltage when you know current and resistance values

   ◆ Use current and resistance values expressed with metric prefixes

The following examples use the formula *V* = *IR*. To obtain voltage in volts, you must express the value of *I* in amperes and the value of *R* in ohms.

EXAMPLE 3–11

In the circuit of Figure 3–10, how much voltage is needed to produce 5 A of current?

▶ **FIGURE 3–10**

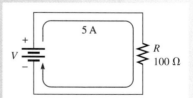

Solution   Substitute 5 A for *I* and 100 Ω for *R* into the formula *V* = *IR*.

$$V = IR = (5\text{ A})(100\ \Omega) = \textbf{500 V}$$

Thus, 500 V are required to produce 5 A of current through a 100 Ω resistor.

Related Problem   In Figure 3–10, how much voltage is required to produce 12 A of current?

Units with Metric Prefixes

The following two examples illustrate how to use current values in the milliampere (mA) and microampere (μA) ranges to calculate voltage.

EXAMPLE 3–12

How much voltage will be measured across the resistor in Figure 3–11?

▶ **FIGURE 3–11**

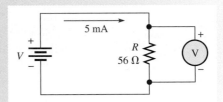

Solution   Five milliamperes equals 5×10^{-3} A. Substitute the values for I and R into the formula $V = IR$.

$$V = IR = (5\,\text{mA})(56\,\Omega) = (5 \times 10^{-3}\,\text{A})(56\,\Omega) = 280 \times 10^{-3}\,\text{V} = \mathbf{280\,mV}$$

When you multiply milliamperes by ohms, you get millivolts.

Related Problem   How much voltage is measured across R if $R = 33\,\Omega$ and $I = 1.5\,\text{mA}$ in Figure 3–11?

 Use Multisim file E03-12 to verify the calculated results in this example and to confirm your calculation for the related problem.

EXAMPLE 3–13

Suppose that there is a current of 8 μA through a 10 Ω resistor. How much voltage is across the resistor?

Solution   Eight microamperes equals 8×10^{-6} A. Substitute the values for I and R into the formula $V = IR$.

$$V = IR = (8\,\mu\text{A})(10\,\Omega) = (8 \times 10^{-6}\,\text{A})(10\,\Omega) = 80 \times 10^{-6}\,\text{V} = \mathbf{80\,\mu V}$$

When you multiply microamperes by ohms, you get microvolts.

Related Problem   If there are 3.2 μA through a 47 Ω resistor, what is the voltage across the resistor?

The following two examples illustrate how to use resistance values in the kilohm (kΩ) and megohm (MΩ) ranges to calculate voltage.

EXAMPLE 3–14

The circuit in Figure 3–12 has a current of 10 mA. What is the voltage?

▶ **FIGURE 3–12**

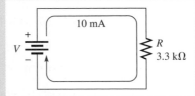

Solution Ten milliamperes equals 10×10^{-3} A and 3.3 kΩ equals 3.3×10^3 Ω. Substitute these values into the formula $V = IR$.

$$V = IR = (10\,\text{mA})(3.3\,\text{k}\Omega) = (10 \times 10^{-3}\,\text{A})(3.3 \times 10^3\,\Omega) = \mathbf{33\ V}$$

Notice that 10^{-3} and 10^3 cancel. Therefore, milliamperes cancel kilohms when multiplied, and the result is volts.

Related Problem If the current in Figure 3–12 is 25 mA, what is the voltage?

 Use Multisim file E03-14 to verify the calculated results in this example and to confirm your calculation for the related problem.

EXAMPLE 3–15 If there is a current of 50 μA through a 4.7 MΩ resistor, what is the voltage?

Solution Fifty microamperes equals 50×10^{-6} A and 4.7 MΩ is 4.7×10^6 Ω. Substitute these values into the formula $V = IR$.

$$V = IR = (50\,\mu\text{A})(4.7\,\text{M}\Omega) = (50 \times 10^{-6}\,\text{A})(4.7 \times 10^6\,\Omega) = \mathbf{235\ V}$$

Notice that 10^{-6} and 10^6 cancel. Therefore, microamperes cancel megohms when multiplied, and the result is volts.

Related Problem If there are 450 μA through a 3.9 MΩ resistor, what is the voltage?

SECTION 3–3 REVIEW

In Problems 1–7, calculate V.
1. $I = 1$ A and $R = 10$ Ω.
2. $I = 8$ A and $R = 470$ Ω.
3. $I = 3$ mA and $R = 100$ Ω.
4. $I = 25$ μA and $R = 56$ Ω.
5. $I = 2$ mA and $R = 1.8$ kΩ.
6. $I = 5$ mA and $R = 100$ MΩ.
7. $I = 10$ μA and $R = 2.2$ MΩ.
8. How much voltage is required to produce 100 mA through a 4.7 kΩ resistor?
9. What voltage do you need to cause 3 mA of current in a 3.3 kΩ resistor?
10. A battery produces 2 A through a 6.8 Ω resistive load. What is the battery voltage?

3–4 Cᴀʟᴄᴜʟᴀᴛɪɴɢ Rᴇꜱɪꜱᴛᴀɴᴄᴇ

Examples in this section illustrate the Ohm's law expression $R = V/I$.

After completing this section, you should be able to

◆ **Calculate resistance in a circuit**
   ◆ Use Ohm's law to find resistance when you know voltage and current values
   ◆ Use current and voltage values expressed with metric prefixes

The following examples use the formula $R = V/I$. To get resistance in ohms, you must express the value of I in amperes and the value of V in volts.

EXAMPLE 3–16

In the circuit of Figure 3–13, how much resistance is needed to draw 3.08 A of current from the battery?

▶ **FIGURE 3–13**

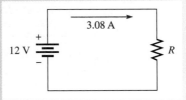

Solution   Substitute 12 V for V and 3.08 A for I into the formula $R = V/I$.

$$R = \frac{V}{I} = \frac{12\,\text{V}}{3.08\,\text{A}} = \textbf{3.90}\ \boldsymbol{\Omega}$$

Related Problem   In Figure 3–13, to what value must R be changed for a current of 5.45 A?

Units with Metric Prefixes

The following two examples illustrate how to use current values in the milliampere (mA) and microampere (μA) ranges to calculate resistance.

EXAMPLE 3–17

Suppose that the ammeter in Figure 3–14 indicates 4.55 mA of current and the voltmeter reads 150 V. What is the value of R?

▶ **FIGURE 3–14**

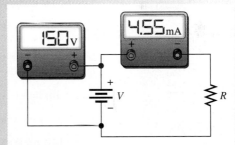

Solution   4.55 mA equals 4.55×10^{-3} A. Substitute the voltage and current values into the formula $R = V/I$.

$$R = \frac{V}{I} = \frac{150\,\text{V}}{4.55\,\text{mA}} = \frac{150\,\text{V}}{4.55 \times 10^{-3}\,\text{A}} = 33 \times 10^{3}\,\Omega = \textbf{33 k}\boldsymbol{\Omega}$$

When volts are divided by milliamperes, the resistance is in kilohms.

Related Problem   If the ammeter indicates 1.10 mA and the voltmeter reads 75 V, what is the value of R?

   Use Multisim file E03-17 to verify the calculated results in this example and to confirm your calculation for the related problem.

EXAMPLE 3–18

Suppose that the value of the resistor in Figure 3–14 is changed. If the battery voltage is still 150 V and the ammeter reads 68.2 μA, what is the new resistor value?

Solution 68.2 μA equals 68.2×10^{-6} A. Substitute V and I values into the equation for R.

$$R = \frac{V}{I} = \frac{150 \text{ V}}{68.2 \, \mu\text{A}} = \frac{150 \text{ V}}{68.2 \times 10^{-6} \text{ A}} = 2.2 \times 10^6 \, \Omega = \mathbf{2.2 \, M\Omega}$$

When volts are divided by microamperes, the resistance has units of megohms.

Related Problem If the resistor is changed in Figure 3–14 so that the ammeter reads 48.5 μA, what is the new resistor value? Assume $V = 150$ V.

SECTION 3–4 REVIEW

In Problems, 1–5, calculate R.

1. $V = 10$ V and $I = 2.13$ A.
2. $V = 270$ V and $I = 10$ A.
3. $V = 20$ kV and $I = 5.13$ A.
4. $V = 15$ V and $I = 2.68$ mA.
5. $V = 5$ V and $I = 2.27$ μA.
6. You have a resistor across which you measure 25 V, and your ammeter indicates 53.2 mA of current. What is the resistor's value in kilohms? In ohms?

3–5   INTRODUCTION TO TROUBLESHOOTING

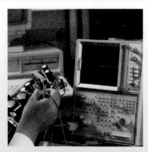

Technicians must be able to diagnose and repair malfunctioning circuits and systems. In this section, you learn a general approach to troubleshooting using a simple example. Troubleshooting coverage is an important part of this textbook, so you will find a troubleshooting section in many of the chapters and troubleshooting problems, including Multisim circuits, for skill building.

After completing this section, you should be able to

◆ **Describe a basic approach to troubleshooting**

  ◆ List three steps in troubleshooting

  ◆ Explain what is meant by half-splitting

  ◆ Discuss and compare the three basic measurements of voltage, current, and resistance

Troubleshooting is the application of logical thinking combined with a thorough knowledge of circuit or system operation to correct a malfunction. The basic approach to troubleshooting consists of three steps: *analysis, planning,* and *measuring.* We will refer to this 3-step approach as APM.

Analysis

The first step in troubleshooting a circuit is to analyze clues or symptoms of the failure. The analysis can begin by determining the answer to certain questions:

1. Has the circuit ever worked?

2. If the circuit once worked, under what conditions did it fail?

3. What are the symptoms of the failure?

4. What are the possible causes of failure?

Planning

The second step in the troubleshooting process, after analyzing the clues, is formulating a logical plan of attack. Much time can be saved by proper planning. A working knowledge of the circuit is a prerequisite to a plan for troubleshooting. If you are not certain how the circuit is supposed to operate, take time to review circuit diagrams (schematics), operating instructions, and other pertinent information. A schematic with proper voltages marked at various test points is particularly useful. Although logical thinking is perhaps the most important tool in troubleshooting, it rarely can solve the problem by itself.

Measuring

The third step is to narrow the possible failures by making carefully thought out measurements. These measurements usually confirm the direction you are taking in solving the problem, or they may point to a new direction that you should take. Occasionally, you may find a totally unexpected result.

An APM Example

The thought process that is part of the APM approach can be illustrated with a simple example. Suppose you have a string of 8 decorative 12 V bulbs connected in series to a 120 V source V_S, as shown in Figure 3–15. Assume that this circuit worked properly at one time but stopped working after it was moved to a new location. When plugged in at the new location, the lamps fail to turn on. How do you go about finding the trouble?

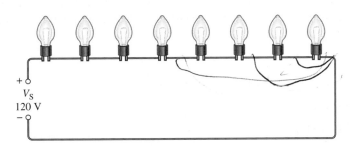

▲ **FIGURE 3–15**

A string of bulbs connected to a voltage source.

The Analysis Thought Process   You may think like this as you proceed to analyze the situation:

◆ Since the circuit worked before it was moved, the problem could be that there is no voltage at the new location.

◆ Perhaps the wiring was loose and pulled apart when moved.

◆ It is possible that a bulb is burned out or loose in its socket.

With this reasoning, you have considered possible causes and failures that may have occurred. The thought process continues:

♦ The fact that the circuit once worked eliminates the possibility that the original circuit was improperly wired.

♦ If the fault is due to an open path, it is unlikely that there is more than one break which could be either a bad connection or a burned out bulb.

You have now analyzed the problem and are ready to plan the process of finding the fault in the circuit.

The Planning Thought Process   The first part of your plan is to measure for voltage at the new location. If the voltage is present, then the problem is in the string of lights. If voltage is not present, check the circuit breaker at the distribution box in the house. Before resetting breakers, you should think about why a breaker may be tripped. Let's assume that you find the voltage is present. This means that the problem is in the string of lights.

The second part of your plan is to measure either the resistance in the string of lights or to measure voltages across the bulbs. The decision whether to measure resistance or voltage is a toss-up and can be made based on the ease of making the test. Seldom is a troubleshooting plan developed so completely that all possible contingencies are included. You will frequently need to modify the plan as you go along.

The Measurement Process   You proceed with the first part of your plan by using a multimeter to check the voltage at the new location. Assume the measurement shows a voltage of 120 V. Now you have eliminated the possibility of no voltage. You know that, since you have voltage across the string and there is no current because no bulb is on, there must be an open in the current path. Either a bulb is burned out, a connection at the lamp socket is broken, or the wire is broken.

Next, you decide to locate the break by measuring resistance with your multimeter. Applying logical thinking, you decide to measure the resistance of each half of the string instead of measuring the resistance of each bulb. By measuring the resistance of half the bulbs at once, you can usually reduce the effort required to find the open. This technique is a type of troubleshooting procedure called **half-splitting**.

Once you have identified the half in which the open occurs, as indicated by an infinite resistance, you use half-splitting again on the faulty half and continue until you narrow the fault down to a faulty bulb or connection. This process is shown in Figure 3–16, assuming for purposes of illustration that the seventh bulb is burned out.

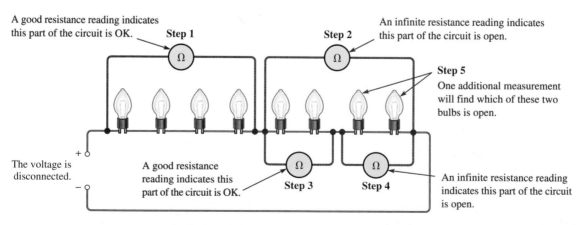

▲ FIGURE 3–16

Illustration of the half-splitting method of troubleshooting. The numbered steps indicate the sequence in which the multimeter is moved from one position to another.

As you can see in the figure, the half-splitting approach in this particular case takes a maximum of five measurements to identify the open bulb. If you had decided to measure each bulb individually and had started at the left, you would have needed seven measurements. Sometimes half-splitting saves steps; sometimes it doesn't. The number of steps required depends on where you make your measurements and in what sequence.

Unfortunately, most troubleshooting is more difficult than this example. However, analysis and planning are essential for effective troubleshooting in any situation. As measurements are made, the plan is often modified; the experienced troubleshooter narrows the search by fitting the symptoms and measurements into a probable cause. In some cases, low-cost equipment is simply discarded when troubleshooting and repair costs are comparable to replacement costs.

Comparison of *V*, *R*, and *I* Measurements

As you know from Section 2–7, you can measure voltage, current, or resistance in a circuit. To measure voltage, place the voltmeter in parallel across the component; that is, place one lead on each side of the component. This makes voltage measurements the easiest of the three types of measurements.

To measure resistance, connect the ohmmeter across a component; however, the voltage must be first disconnected, and sometimes the component itself must be removed from the circuit. Therefore, resistance measurements are generally more difficult than voltage measurements.

To measure current, place the ammeter in series with the component; that is, the ammeter must be in line with the current path. To do this you must disconnect a component lead or a wire before you connect the ammeter. This usually makes a current measurement the most difficult to perform.

SECTION 3–5 REVIEW

1. Name the three steps in the APM approach to troubleshooting.
2. Explain the basic idea of the half-splitting technique.
3. Why are voltages easier to measure than currents in a circuit?

A Circuit Application

In this application, an existing resistance box that is to be used as part of a test setup in the lab is to be checked out and modified. Your task is to modify the circuit so that it will meet the requirements of the new application. You will have to apply your knowledge of Ohm's law in order to complete this assignment.

The specifications are as follows:

1. Each resistor is switch selectable and only one resistor is selected at a time.

2. The lowest resistor value is to be 10 Ω.

3. Each successively higher resistance in the switch sequence must be a decade (10 times) increase over the previous one.

4. The maximum resistor value must be 1.0 MΩ.

5. The maximum voltage across any resistor in the box will be 4 V.

6. Two additional resistors are required, one to limit the current to 10 mA ± 10% and the other to limit the current to 5 mA ± 10% with a 4 V drop.

The Existing Resistor Circuit

The existing resistance box is shown in both top and bottom views in Figure 3–17. The switch is a rotary type.

The Schematic

◆ From Figure 3–17, determine the resistor values and draw the schematic for the existing circuit. Determine the resistor numbering from the *R* labels on the top view.

▶ FIGURE 3–17

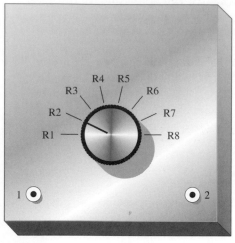

(a) Top view

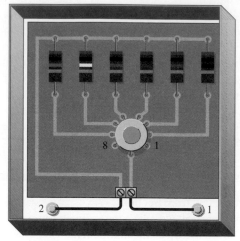

(b) Bottom view

The Schematic for the New Requirements

◆ Draw the schematic for a circuit that will accomplish the following:

1. One resistor at a time is to be connected by the switch between terminals 1 and 2 of the box.

2. Provide switch-selectable resistor values beginning with $10 \, \Omega$ and increasing in decade increments to $1.0 \, M\Omega$.

3. Each of the resistors must be selectable by a sequence of adjacent switch positions in ascending order.

4. There must be two switch-selectable resistors, one is in switch position 1 (shown in Figure 3–17, bottom view) and must limit the current to $10 \, mA \pm 10\%$ with a 4 V drop and the other is in switch position 8 and must limit the current to $5 \, mA \pm 10\%$ with a 4 V drop.

5. All the resistors must be standard values with 10% tolerance. See Appendix A for standard resistor values.

◆ Determine the modifications that must be made to the existing circuit board to meet the specifications and develop a detailed list of the changes including resistance values, wiring, and new components. You should number each point in the schematic for easy reference.

A Test Procedure

◆ After the resistance box has been modified to meet the new specifications, it must be tested to see if it is working properly. Determine how you would test the resistance box and what instruments you would use. Then detail your test procedure in a step-by-step format.

Troubleshooting the Circuit

◆ An ohmmeter is connected across terminals 1 and 2 of the resistance box. Determine the most likely fault in each of the following cases:

1. The ohmmeter shows an infinitely high resistance when the switch is in position 3.

2. The ohmmeter shows an infinitely high resistance in all switch positions.

3. The ohmmeter shows an incorrect value of resistance when the switch is in position 6.

Review

1. Explain how you applied Ohm's law to this application.

2. Determine the current through each resistor when 4 V is applied across it.

SUMMARY

◆ Voltage and current are linearly proportional.

◆ Ohm's law gives the relationship of voltage, current, and resistance.

◆ Current is inversely proportional to resistance.

◆ A kilohm ($k\Omega$) is one thousand ohms.

◆ A megohm ($M\Omega$) is one million ohms.

◆ A microampere (μA) is one-millionth of an ampere.

◆ A milliampere (mA) is one-thousandth of an ampere.

◆ Use $I = V/R$ to calculate current.

◆ Use $V = IR$ to calculate voltage.

◆ Use $R = V/I$ to calculate resistance.

◆ APM is a 3-step troubleshooting approach, consisting of analysis, planning, and measuring.

◆ Half-splitting is a troubleshooting technique that can be used to reduce the number of measurements required to find a problem.

KEY TERMS

Key terms and other bold terms in the chapter are defined in the end-of-book glossary.

Linear   Characterized by a straight-line relationship.

Ohm's law   A law stating that current is directly proportional to voltage and inversely proportional to resistance.

Troubleshooting   A systematic process of isolating, identifying, and correcting a fault in a circuit or system.

FORMULAS

3–1	$I = \dfrac{V}{R}$	Form of Ohm's law for calculating current
3–2	$V = IR$	Form of Ohm's law for calculating voltage
3–3	$R = \dfrac{V}{I}$	Form of Ohm's law for calculating resistance

SELF-TEST

Answers are at the end of the chapter.

1. Ohm's law states that
   (a) current equals voltage times resistance
   (b) voltage equals current times resistance
   (c) resistance equals current divided by voltage
   (d) voltage equals current squared times resistance

2. When the voltage across a resistor is doubled, the current will
   (a) triple (b) halve (c) double (d) not change

3. When 10 V are applied across a 20 Ω resistor, the current is
   (a) 10 A (b) 0.5 A (c) 200 A (d) 2 A

4. When there are 10 mA of current through 1.0 kΩ resistor, the voltage across the resistor is
   (a) 100 V (b) 0.1 V (c) 10 kV (d) 10 V

5. If 20 V are applied across a resistor and there are 6.06 mA of current, the resistance is
   (a) 3.3 kΩ (b) 33 kΩ (c) 330 kΩ (d) 3.03 kΩ

6. A current of 250 μA through a 4.7 kΩ resistor produces a voltage drop of
   (a) 53.2 V (b) 1.18 mV (c) 18.8 V (d) 1.18 V

7. A resistance of 2.2 MΩ is connected across a 1 kV source. The resulting current is approximately
   (a) 2.2 mA (b) 0.455 mA (c) 45.5 μA (d) 0.455 A

8. How much resistance is required to limit the current from a 10 V battery to 1 mA?
   (a) 100 Ω (b) 1.0 kΩ (c) 10 Ω (d) 10 kΩ

9. An electric heater draws 2.5 A from a 110 V source. The resistance of the heating element is
   (a) 275 Ω (b) 22.7 mΩ (c) 44 Ω (d) 440 Ω

10. The current through a flashlight bulb is 20 mA and the total battery voltage is 4.5 V. The resistance of the bulb is
    (a) 90 Ω (b) 225 Ω (c) 4.44 Ω (d) 45 Ω

CIRCUIT DYNAMICS QUIZ

Answers are at the end of the chapter.

1. If the current through a fixed resistor goes from 10 mA to 12 mA, the voltage across the resistor
   (a) increases   (b) decreases   (c) stays the same

2. If the voltage across a fixed resistor goes from 10 V to 7 V, the current through the resistor
   (a) increases   (b) decreases   (c) stays the same

3. A variable resistor has 5 V across it. If you reduce the resistance, the current through it
   (a) increases   (b) decreases   (c) stays the same

4. If the voltage across a resistor increases from 5 V to 10 V and the current increases from 1 mA to 2 mA, the resistance
   (a) increases   (b) decreases   (c) stays the same

Refer to Figure 3–14.

5. If the voltmeter reading changes to 175 V, the ammeter reading
   (a) increases   (b) decreases   (c) stays the same

6. If R is changed to a larger value and the voltmeter reading stays at 150 V, the current
   (a) increases   (b) decreases   (c) stays the same

7. If the resistor is removed from the circuit leaving an open, the ammeter reading
   (a) increases   (b) decreases   (c) stays the same

8. If the resistor is removed from the circuit leaving an open, the voltmeter reading
   (a) increases   (b) decreases   (c) stays the same

Refer to Figure 3–21.

9. If the rheostat is adjusted to increase the resistance, the current through the heating element
   (a) increases   (b) decreases   (c) stays the same

10. If the rheostat is adjusted to increase the resistance, the source voltage
    (a) increases   (b) decreases   (c) stays the same

11. If the fuse opens, the voltage across the heating element
    (a) increases   (b) decreases   (c) stays the same

12. If the source voltage increases, the voltage across the heating element
    (a) increases   (b) decreases   (c) stays the same

13. If the fuse is changed to one with a higher rating, the current through the rheostat
    (a) increases   (b) decreases   (c) stays the same

Refer to Figure 3–23.

14. If the lamp burns out (opens), the current
    (a) increases   (b) decreases   (c) stays the same

15. If the lamp burns out, the voltage across it
    (a) increases   (b) decreases   (c) stays the same

PROBLEMS

More difficult problems are indicated by an astrisk (*).
Answers to odd-numbered problems are at the end of the book.

SECTION 3–1   The Relationship of Current, Voltage, and Resistance

1. In a circuit consisting of a voltage source and a resistor, describe what happens to the current when
   (a) the voltage is tripled
   (b) the voltage is reduced by 75%
   (c) the resistance is doubled

    (d) the resistance is reduced by 35%

    (e) the voltage is doubled and the resistance is cut in half

    (f) the voltage is doubled and the resistance is doubled

2. State the formula used to find I when the values of V and R are known.

3. State the formula used to find V when the values of I and R are known.

4. State the formula used to find R when the values of V and I are known.

5. A variable voltage source is connected to the circuit of Figure 3–18. Start at 0 V and increase the voltage in 10 V steps up to 100 V. Determine the current at each voltage point, and plot a graph of V versus I. Is the graph a straight line? What does the graph indicate?

▶ FIGURE 3–18

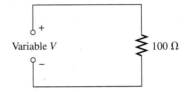

6. In a certain circuit, $I = 5$ mA when $V = 1$ V. Determine the current for each of the following voltages in the same circuit:

    (a) $V = 1.5$ V   (b) $V = 2$ V   (c) $V = 3$ V

    (d) $V = 4$ V   (e) $V = 10$ V

7. Figure 3–19 is a graph of current versus voltage for three resistance values. Determine R_1, R_2, and R_3.

▶ FIGURE 3–19

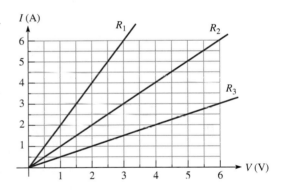

8. Plot the current-voltage relationship for a four-band resistor with the color code gray, red, red, gold.

9. Plot the current-voltage relationship for a five-band resistor with the color code brown, green, gray, brown, red.

10. Which circuit in Figure 3–20 has the most current? The least current?

▶ FIGURE 3–20

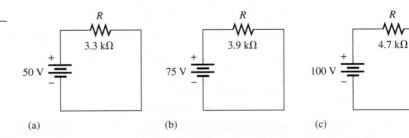

(a)   (b)   (c)

*11. You are measuring the current in a circuit that is operated on a 10 V battery. The ammeter reads 50 mA. Later, you notice that the current has dropped to 30 mA. Eliminating the possibility of a resistance change, you must conclude that the voltage has changed. How much has the voltage of the battery changed, and what is its new value?

*12. If you wish to increase the amount of current in a resistor from 100 mA to 150 mA by changing the 20 V source, by how many volts should you change the source? To what new value should you set it?

13. Plot a graph of current versus voltage for voltage values ranging from 10 V to 100 V in 10 V steps for each of the following resistance values:

   (a) $1.0 \, \Omega$ (b) $5.0 \, \Omega$ (c) $20 \, \Omega$ (d) $100 \, \Omega$

14. Does the graph in Problem 13 indicate a linear relationship between voltage and current? Explain.

SECTION 3–2   Calculating Current

15. Determine the current in each case:

   (a) $V = 5 \, V, R = 1.0 \, \Omega$ (b) $V = 15 \, V, R = 10 \, \Omega$
   (c) $V = 50 \, V, R = 100 \, \Omega$ (d) $V = 30 \, V, R = 15 \, k\Omega$
   (e) $V = 250 \, V, R = 5.6 \, M\Omega$

16. Determine the current in each case:

   (a) $V = 9 \, V, R = 2.7 \, k\Omega$ (b) $V = 5.5 \, V, R = 10 \, k\Omega$
   (c) $V = 40 \, V, R = 68 \, k\Omega$ (d) $V = 1 \, kV, R = 2.2 \, k\Omega$
   (e) $V = 66 \, kV, R = 10 \, M\Omega$

17. A $10 \, \Omega$ resistor is connected across a 12 V battery. What is the current through the resistor?

18. A certain resistor has the following color code: orange, orange, red, gold. Determine the maximum and minimum currents you should expect to measure when a 12 V source is connected across the resistor.

19. A 4-band resistor is connected across the terminals of a 25 V source. Determine the current in the resistor if the color code is yellow, violet, orange, silver.

20. A 5-band resistor is connected across a 12 V source. Determine the current if the color code is orange, violet, yellow, gold, brown.

21. If the voltage in Problem 20 is doubled, will a 0.5 A fuse blow? Explain your answer.

*22. The potentiometer connected as a rheostat in Figure 3–21 is used to control the current to a heating element. When the rheostat is adjusted to a value of $8 \, \Omega$ or less, the heating element can burn out. What is the rated value of the fuse needed to protect the circuit if the voltage across the heating element at the point of maximum current is 100 V and the voltage across the rheostat is the difference between the heating element voltage and the source voltage?

▶ FIGURE 3–21

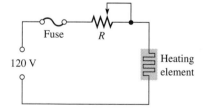

SECTION 3–3   Calculating Voltage

23. Calculate the voltage for each value of I and R:

   (a) $I = 2 \, A, R = 18 \, \Omega$ (b) $I = 5 \, A, R = 56 \, \Omega$
   (c) $I = 2.5 \, A, R = 680 \, \Omega$ (d) $I = 0.6 \, A, R = 47 \, \Omega$
   (e) $I = 0.1 \, A, R = 560 \, \Omega$

24. Calculate the voltage for each value of I and R:

   (a) $I = 1 \, mA, R = 10 \, \Omega$ (b) $I = 50 \, mA, R = 33 \, \Omega$
   (c) $I = 3 \, A, R = 5.6 \, k\Omega$ (d) $I = 1.6 \, mA, R = 2.2 \, k\Omega$

(e) $I = 250\,\mu A, R = 1.0\,k\Omega$ **(f)** $I = 500\,mA, R = 1.5\,M\Omega$

(g) $I = 850\,\mu A, R = 10\,M\Omega$ **(h)** $I = 75\,\mu A, R = 47\,\Omega$

25. Three amperes of current are measured through a 27 Ω resistor connected across a voltage source. How much voltage does the source produce?

26. Assign a voltage value to each source in the circuits of Figure 3–22 to obtain the indicated amounts of current.

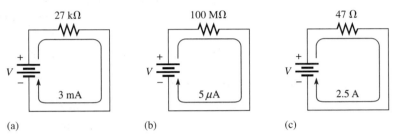

(a)   (b)   (c)

▲ **FIGURE 3–22**

***27.** A 6 V source is connected to a 100 Ω resistor by two 12 ft lengths of 18 gauge copper wire. The total resistance is the resistance of both wires added to the 100 Ω resistor. Determine the following:

    (a) Current

    (b) Resistor voltage drop

    (c) Voltage drop across each length of wire

SECTION 3–4   Calculating Resistance

28. Calculate the resistance of a rheostat for each value of V and I:

    (a) $V = 10\,V, I = 2\,A$ **(b)** $V = 90\,V, I = 45\,A$

    (c) $V = 50\,V, I = 5\,A$ **(d)** $V = 5.5\,V, I = 10\,A$

    (e) $V = 150\,V, I = 0.5\,A$

29. Calculate the resistance of a rheostat for each set of V and I values:

    (a) $V = 10\,kV, I = 5\,A$ **(b)** $V = 7\,V, I = 2\,mA$

    (c) $V = 500\,V, I = 250\,mA$ **(d)** $V = 50\,V, I = 500\,\mu A$

    (e) $V = 1\,kV, I = 1\,mA$

30. Six volts are applied across a resistor. A current of 2 mA is measured. What is the value of the resistor?

31. The filament of a lamp in the circuit of Figure 3–23(a) has a certain amount of resistance, represented by an equivalent resistance in Figure 3–23(b). If the lamp operates with 120 V and 0.8 A of current, what is the resistance of its filament when it is on?

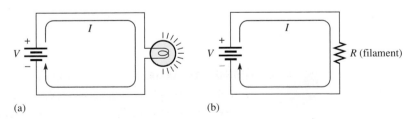

(a)   (b)

▲ **FIGURE 3–23**

32. A certain electrical device has an unknown resistance. You have available a 12 V battery and an ammeter. How would you determine the value of the unknown resistance? Draw the necessary circuit connections.

33. By varying the rheostat (variable resistor) in the circuit of Figure 3–24, you can change the amount of current. The setting of the rheostat is such that the current is 750 mA. What is the resistance value of this setting? To adjust the current to 1 A, to what resistance value must you set the rheostat? What is the problem with this circuit?

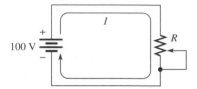

▶ **FIGURE 3–24**

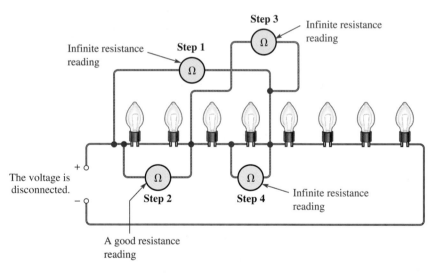

*34.** A 120 V lamp-dimming circuit is controlled by a rheostat and protected from excessive current by a 2 A fuse. To what minimum resistance value can the rheostat be set without blowing the fuse? Assume a lamp resistance of 15 Ω.

35. Repeat Problem 34 for a 110 V circuit and a 1 A fuse.

SECTION 3–5   Introduction to Troubleshooting

36. In the light circuit of Figure 3–25, identify the faulty bulb based on the series of ohmmeter readings shown.

▲ **FIGURE 3–25**

37. Assume you have a 32-light string and one of the bulbs is burned out. Using the half-splitting approach and starting in the left half of the circuit, how many resistance measurements will it take to find the faulty bulb if it is seventeenth from the left?

Multisim Troubleshooting and Analysis

These problems require your Multisim CD-ROM.

38. Open file P03-38 on your CD-ROM and determine which one of the three circuits is not working properly.

39. Open file P03-39 and measure the resistance values of the resistors.

40. Open file P03-40 and determine the values of the current and voltage.

41. Open file P03-41 and determine the value of the source voltage and the resistance.

42. Open file P03-42 and find the problem with the circuit.

ANSWERS

SECTION REVIEWS

SECTION 3–1   The Relationship of Current, Voltage, and Resistance

1. Current, voltage, and resistance
2. $I = V/R$
3. $V = IR$
4. $R = V/I$
5. When voltage is tripled, current increases by three times.
6. When voltage is halved, current reduces to one-half of original value.
7. 0.5 A
8. The current would increase by four times if the voltage doubles and the resistance is halved.
9. $I = 5\,\text{mA}$
10. $I = 6\,\text{A}$

SECTION 3–2   Calculating Current

1. $I = 10\,\text{V}/5.6\,\Omega = 1.79\,\text{A}$
2. $I = 100\,\text{V}/560\,\Omega = 179\,\text{mA}$
3. $I = 5\,\text{V}/2.2\,\text{k}\Omega = 2.27\,\text{mA}$
4. $I = 15\,\text{V}/4.7\,\text{M}\Omega = 3.19\,\mu\text{A}$
5. $I = 20\,\text{kV}/4.7\,\text{M}\Omega = 4.26\,\text{mA}$
6. $I = 10\,\text{kV}/2.2\,\text{k}\Omega = 4.55\,\text{A}$

SECTION 3–3   Calculating Voltage

1. $V = (1\,\text{A})(10\,\Omega) = 10\,\text{V}$
2. $V = (8\,\text{A})(470\,\Omega) = 3.76\,\text{kV}$
3. $V = (3\,\text{mA})(100\,\Omega) = 300\,\text{mV}$
4. $V = (25\,\mu\text{A})(56\,\Omega) = 1.4\,\text{mV}$
5. $V = (2\,\text{mA})(1.8\,\text{k}\Omega) = 3.6\,\text{V}$
6. $V = (5\,\text{mA})(100\,\text{M}\Omega) = 500\,\text{kV}$
7. $V = (10\,\mu\text{A})(2.2\,\text{M}\Omega) = 22\,\text{V}$
8. $V = (100\,\text{mA})(4.7\,\text{k}\Omega) = 470\,\text{V}$
9. $V = (3\,\text{mA})(3.3\,\text{k}\Omega) = 9.9\,\text{V}$
10. $V = (2\,\text{A})(6.8\,\Omega) = 13.6\,\text{V}$

SECTION 3–4   Calculating Resistance

1. $R = 10\,\text{V}/2.13\,\text{A} = 4.7\,\Omega$
2. $R = 270\,\text{V}/10\,\text{A} = 27\,\Omega$
3. $R = 20\,\text{kV}/5.13\,\text{A} = 3.9\,\text{k}\Omega$
4. $R = 15\,\text{V}/2.68\,\text{mA} = 5.6\,\text{k}\Omega$
5. $R = 5\,\text{V}/2.27\,\mu\text{A} = 2.2\,\text{M}\Omega$
6. $R = 25\,\text{V}/53.2\,\text{mA} = 0.47\,\text{k}\Omega = 470\,\Omega$

SECTION 3–5   Introduction to Troubleshooting

1. Analysis, planning, and measurement
2. Half-splitting identifies the fault by successively isolating half of the remaining circuit.
3. Voltage is measured across a component; current is measured in series with the component.

A Circuit Application

1. For the new resistors, $R = V/I$.
2. $I = 4\,\text{V}/10\,\Omega = 400\,\text{mA}; I = 4\,\text{V}/100\,\Omega = 40\,\text{mA}; I = 4\,\text{V}/1.0\,\text{k}\Omega = 4\,\text{mA};$
   $I = 4\,\text{V}/10\,\text{k}\Omega = 400\,\mu\text{A}; I = 4\,\text{V}/100\,\text{k}\Omega = 40\,\mu\text{A}; I = 4\,\text{V}/1.0\,\text{M}\Omega = 4\,\mu\text{A}.$

RELATED PROBLEMS FOR EXAMPLES

3–1   Yes
3–2   0 V
3–3   3.03 A
3–4   0.005 A
3–5   0.005 A
3–6   13.6 mA
3–7   21.3 μA
3–8   2.66 μA
3–9   37.0 mA
3–10 1.47 mA
3–11 1200 V
3–12 49.5 mV
3–13 0.150 mV
3–14 82.5 V
3–15 1755 V
3–16 2.20 Ω
3–17 68.2 kΩ
3–18 3.30 MΩ

SELF-TEST

1. (b)   2. (c)   3. (b)   4. (d)   5. (a)   6. (d)   7. (b)   8. (d)
9. (c)   10. (b)

CIRCUIT DYNAMICS QUIZ

1. (a)   2. (b)   3. (a)   4. (c)   5. (a)   6. (b)   7. (b)   8. (c)
9. (b)   10. (c)   11. (b)   12. (a)   13. (c)   14. (b)   15. (c)

ENERGY AND POWER

4

CHAPTER OBJECTIVES

◆ Define *energy* and *power*

◆ Calculate power in a circuit

◆ Properly select resistors based on power consideration

◆ Explain energy conversion and voltage drop

◆ Discuss power supplies and their characteristics

KEY TERMS

◆ Energy

◆ Power

◆ Joule (J)

◆ Watt (W)

◆ Kilowatt-hour

◆ Voltage drop

◆ Power supply

◆ Ampere-hour rating

◆ Efficiency

A CIRCUIT APPLICATION PREVIEW

In the application you will see how the theory learned in this chapter is applicable to the resistance box introduced in the last chapter. Suppose that the resistance box is to be used in testing a circuit in which there will be a maximum of 4 V across all the resistors. You will evaluate the power rating of each resistor and, if it is not sufficient, to replace the resistor with one that has an adequate power rating.

VISIT THE COMPANION WEBSITE

Study aids for this chapter are available at
http://www.prenhall.com/floyd

INTRODUCTION

From Chapter 3, you know the relationship of current, voltage, and resistance as stated by Ohm's law. The existence of these three quantities in an electric circuit results in the fourth basic quantity known as power. A specific relationship exists between power and I, V, and R.

Energy is the ability to do work, and power is the rate at which energy is used. Current carries electrical energy through a circuit. As the free electrons pass through the resistance of the circuit, they give up their energy when they collide with atoms in the resistive material. The electrical energy given up by the electrons is converted into heat energy. The rate at which the electrical energy is used is the power in the circuit.

4–1   ENERGY AND POWER

When there is current through a resistance, electrical energy is converted to heat or other form of energy, such as light. A common example of this is a light bulb that becomes too hot to touch. The current through the filament that produces light also produces unwanted heat because the filament has resistance. Electrical components must be able to dissipate a certain amount of energy in a given period of time.

After completing this section, you should be able to

- ◆ **Define *energy* and *power***
  - ◆ Express power in terms of energy
  - ◆ State the unit of power
  - ◆ State the common units of energy
  - ◆ Perform energy and power calculations

Energy is the ability to do work, and power is the rate at which energy is used.

Power (*P*) is a certain amount of energy (*W*) used in a certain length of time (*t*), expressed as follows:

Equation 4–1

$$P = \frac{W}{t}$$

where: P = power in watts (W)
W = energy in joules (J)
t = time in seconds (s)

Note that an italic *W* is used to represent energy in the form of work and a nonitalic W is used for watts, the unit of power. The **joule (J)** is the SI unit of energy.

Energy in joules divided by time in seconds gives power in watts. For example, if 50 J of energy are used in 2 s, the power is 50 J/2 s = 25 W. By definition,

One watt (W) is the amount of power when one joule of energy is used in one second.

Thus, the number of joules used in one second is always equal to the number of watts. For example, if 75 J are used in 1 s, the power is $P = W/t = 75$ J/1 s = 75 W.

Amounts of power much less than one watt are common in certain areas of electronics. As with small current and voltage values, metric prefixes are used to designate small amounts of power. Thus, milliwatts (mW), microwatts (μW), and even picowatts (pW) are commonly found in some applications.

In the electrical utilities field, kilowatts (kW) and megawatts (MW) are common units. Radio and television stations also use large amounts of power to transmit signals. Electric motors are commonly rated in horsepower (hp) where 1 hp = 746 W.

Since power is the rate at which energy is used, as expressed in Equation 4–1, power utilized over a period of time represents energy consumption. If you multiply power in watts and time in seconds, you have energy in joules, symbolized by *W*.

$$W = Pt$$

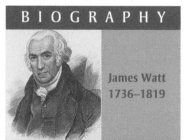

EXAMPLE 4–1

An amount of energy equal to 100 J is used in 5 s. What is the power in watts?

Solution

$$P = \frac{\text{energy}}{\text{time}} = \frac{W}{t} = \frac{100\,\text{J}}{5\,\text{s}} = \textbf{20 W}$$

Related Problem* If 100 W of power occurs for 30 s, how much energy, in joules, is used?

*Answers are at the end of the chapter.

EXAMPLE 4–2

Express the following values of electrical power using appropriate metric prefixes:

(a) 0.045 W (b) 0.000012 W (c) 3500 W (d) 10,000,000 W

Solution (a) 0.045 W = **45 mW** (b) 0.000012 W = **12 μW**

(c) 3500 W = **3.5 kW** (d) 10,000,000 W = **10 MW**

Related Problem Express the following amounts of power in watts without metric prefixes:

(a) 1 mW (b) 1800 μW (c) 1000 mW (d) 1 μW

The Kilowatt-hour (kWh) Unit of Energy

The joule has been defined as a unit of energy. However, there is another way to express energy. Since power is expressed in watts and time in seconds, units of energy called the watt-second (Ws), watt-hour (Wh), and kilowatt-hour (kWh) can be used.

When you pay your electric bill, you are charged on the basis of the amount of energy you use, not the power. Because power companies deal in huge amounts of energy, the most practical unit is the kilowatt-hour. You use a **kilowatt-hour** of energy when you use one thousand watts of power for one hour. For example, a 100 W light bulb burning for 10 h uses 1 kWh of energy.

$$W = Pt = (100\,\text{W})(10\,\text{h}) = 1000\,\text{Wh} = 1\,\text{kWh}$$

BIOGRAPHY

James Prescott Joule 1818–1889

Joule, a British physicist, is known for his research in electricity and thermodynamics. He formulated the relationship that states that the amount of heat energy produced by an electrical current in a conductor is proportional to the conductor's resistance and the time. The unit of energy is named in his honor. (Photo credit: Library of Congress.)

EXAMPLE 4–3

Determine the number of kilowatt-hours (kWh) for each of the following energy consumptions:

(a) 1400 W for 1 h (b) 2500 W for 2 h (c) 100,000 W for 5 h

Solution (a) 1400 W = 1.4 kW (b) 2500 W = 2.5 kW

$W = Pt = (1.4\,\text{kW})(1\,\text{h}) = \textbf{1.4 kWh}$ $W = (2.5\,\text{kW})(2\,\text{h}) = \textbf{5 kWh}$

(c) 100,000 W = 100 kW

$W = (100\,\text{kW})(5\,\text{h}) = \textbf{500 kWh}$

Related Problem How many kilowatt-hours are used by a 250 W bulb burning for 8 h?

SECTION 4–1
REVIEW
Answers are at the end of the chapter.

1. Define *power*.
2. Write the formula for power in terms of energy and time.
3. Define *watt*.
4. Express each of the following values of power in the most appropriate units:
   (a) 68,000 W   **(b)** 0.005 W   **(c)** 0.000025 W
5. If you use 100 W of power for 10 h, how much energy (in kWh) have you used?
6. Convert 2000 Wh to kilowatt-hours.
7. Convert 360,000 Ws to kilowatt-hours.

4–2   POWER IN AN ELECTRIC CIRCUIT

The generation of heat, which occurs when electrical energy is converted to heat energy, in an electric circuit is often an unwanted by-product of current through the resistance in the circuit. In some cases, however, the generation of heat is the primary purpose of a circuit as, for example, in an electric resistive heater. In any case, you must frequently deal with power in electrical and electronic circuits.

After completing this section, you should be able to

◆ **Calculate power in a circuit**

   ◆ Determine power when you know *I* and *R* values

   ◆ Determine power when you know *V* and *I* values

   ◆ Determine power when you know *V* and *R* values

When there is current through resistance, the collisions of the electrons produce heat as a result of the conversion of electrical energy, as indicated in Figure 4–1. The amount of power dissipated in an electric circuit is dependent on the amount of resistance and on the amount of current, expressed as follows:

Equation 4–2
$$P = I^2R$$

where: P = power in watts (W)
I = current in amperes (A)
R = resistance in ohms (Ω)

You can get an equivalent expression for power in terms of voltage and current by substituting *V* for *IR* (I^2 is $I \times I$).

$$P = I^2R = (I \times I)R = I(IR) = (IR)I$$

Equation 4–3
$$P = VI$$

▶ **FIGURE 4–1**

Power dissipation in an electric circuit results in heat energy given off by the resistance.

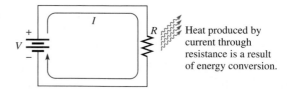

Heat produced by current through resistance is a result of energy conversion.

where P is in watts when V is in volts and I is in amperes. You can obtain another equivalent expression by substituting V/R for I (Ohm's law).

$$P = VI = V\left(\frac{V}{R}\right)$$

$$P = \frac{V^2}{R}$$

Equation 4–4

The relationships between power and current, voltage, and resistance expressed in the preceding formulas are known as **Watt's law**. In each case, I must be in amps, V in volts, and R in ohms. To calculate the power in a resistance, you can use any one of the three power formulas, depending on what information you have. For example, assume that you know the values of current and voltage. In this case calculate the power with the formula $P = VI$. If you know I and R, use the formula $P = I^2R$. If you know V and R, use the formula $P = V^2/R$.

EXAMPLE 4–4

Calculate the power in each of the three circuits of Figure 4–2.

FIGURE 4–2

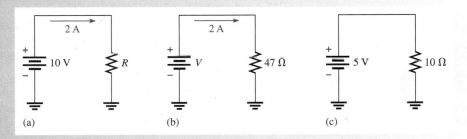

(a) (b) (c)

Solution In circuit (a), you know V and I. Therefore, use Equation 4–3.

$$P = VI = (10 \text{ V})(2 \text{ A}) = \mathbf{20\ W}$$

In circuit (b), you know I and R. Therefore, use Equation 4–2.

$$P = I^2R = (2 \text{ A})^2(47\ \Omega) = \mathbf{188\ W}$$

In circuit (c), you know V and R. Therefore, use Equation 4–4.

$$P = \frac{V^2}{R} = \frac{(5 \text{ V})^2}{10\ \Omega} = \mathbf{2.5\ W}$$

Related Problem Determine P in each circuit of Figure 4–2 for the following changes:

Circuit (a): I doubled and V remains the same

Circuit (b): R doubled and I remains the same

Circuit (c): V halved and R remains the same

EXAMPLE 4–5

A 100 W light bulb operates on 120 V. How much current does it require?

Solution Use the formula $P = VI$ and solve for I by first transposing the terms to get I on the left side in the equation.

$$VI = P$$

Rearranging,

$$I = \frac{P}{V}$$

Substituting 100 W for P and 120 V for V yields

$$I = \frac{P}{V} = \frac{100\ W}{120\ V} = 0.833\ A = \mathbf{833\ mA}$$

Related Problem   A light bulb draws 545 mA from a 110 V source. What is the power dissipated?

SECTION 4–2 REVIEW

1. If there are 10 V across a resistor and a current of 3 A through it, what is the power dissipated?
2. How much power does the source in Figure 4–3 generate? What is the power in the resistor? Are the two values the same? Why?

▶ **FIGURE 4–3**

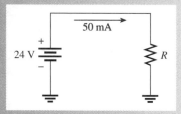

3. If there is a current of 5 A through a 56 Ω resistor, what is the power dissipated?
4. How much power is dissipated by 20 mA through a 4.7 kΩ resistor?
5. Five volts are applied to a 10 Ω resistor. What is the power dissipated?
6. How much power does a 2.2 kΩ resistor with 8 V across it dissipate?
7. What is the resistance of a 75 W bulb that takes 0.5 A?

4–3   RESISTOR POWER RATINGS

As you know, a resistor gives off heat when there is current through it. The limit to the amount of heat that a resistor can give off is specified by its power rating.

After completing this section, you should be able to

◆ **Properly select resistors based on power consideration**

   ◆ Define *power rating*

   ◆ Explain how physical characteristics of resistors determine their power rating

   ◆ Check for resistor failure with an ohmmeter

The **power rating** is the maximum amount of power that a resistor can dissipate without being damaged by excessive heat buildup. The power rating is not related to the ohmic value (resistance) but rather is determined mainly by the physical composition, size, and shape of the resistor. All else being equal, the larger the surface area of a resistor, the more power it can dissipate. *The surface area of a cylindrically shaped resistor is equal to the length (l) times the circumference (c),* as indicated in Figure 4–4. The area of the ends is not included.

The power rating of a resistor is directly related to its surface area.

Metal-film resistors are available in standard power ratings from ⅛ W to 1 W, as shown in Figure 4–5. Available power ratings for other types of resistors vary. For example, wirewound resistors have ratings up to 225 W or greater. Figure 4–6 shows some of these resistors.

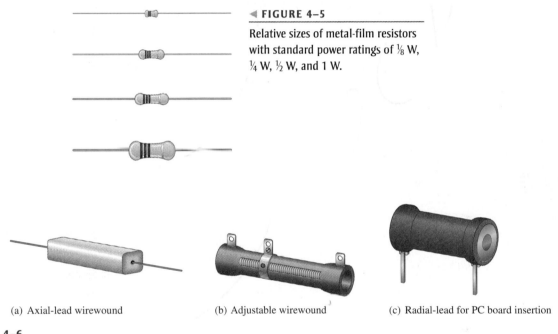

◀ FIGURE 4–5

Relative sizes of metal-film resistors with standard power ratings of ⅛ W, ¼ W, ½ W, and 1 W.

(a) Axial-lead wirewound (b) Adjustable wirewound (c) Radial-lead for PC board insertion

▲ FIGURE 4–6

Typical resistors with high power ratings.

When a resistor is used in a circuit, its power rating must be greater than the maximum power that it will have to handle. For example, if a resistor is to dissipate 0.75 W in a circuit application, its rating should be at least the next higher standard value which is 1 W. *A rating larger than the actual power should be used when possible as a safety margin.*

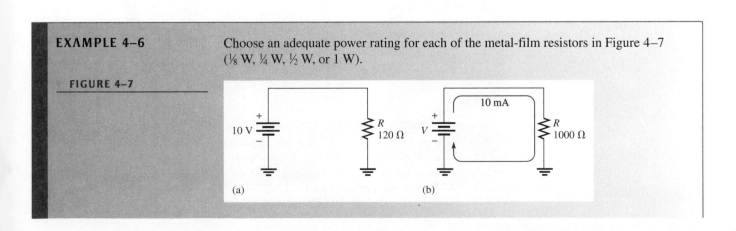

EXAMPLE 4–6

Choose an adequate power rating for each of the metal-film resistors in Figure 4–7 (⅛ W, ¼ W, ½ W, or 1 W).

FIGURE 4–7

Solution   In Figure 4–7(a), the actual power is

$$P = \frac{V^2}{R} = \frac{(10\ V)^2}{120\ \Omega} = \frac{100\ V^2}{120\ \Omega} = 0.833\ W$$

Select a resistor with a power rating higher than the actual power. In this case, a **1 W resistor** should be used.
In Figure 4–7(b), the actual power is

$$P = I^2R = (10\ mA)^2(1000\ \Omega) = (10 \times 10^{-3}\ A)^2(1000\ \Omega) = 0.1\ W$$

At least a **⅛ W (0.125 W) resistor** should be used in this case.

Related Problem   A certain resistor is required to dissipate 0.25 W. What standard rating should be used?

When the power in a resistor is greater than its rating, the resistor will become excessively hot. As a result, either the resistor will burn open or its resistance value will be greatly altered.

A resistor that has been damaged because of overheating can often be detected by the charred or altered appearance of its surface. If there is no visual evidence, a resistor that is suspected of being damaged can be checked with an ohmmeter for an open or incorrect resistance value. Recall that one or both leads of a resistor should be removed from a circuit to measure resistance.

Checking a Resistor with an Ohmmeter

A typical digital multimeter and an analog multimeter are shown in Figures 4–8(a) and 4–8(b), respectively. For the digital meter in Figure 4–8(a), you use the round function switch to select ohms (Ω). You do not have to manually select a range because this particular meter is autoranging and you have a direct digital readout of the resistance value. The large round switch on the analog meter is called a *range switch*. Notice the resistance (OHMS) settings on both meters.

For the analog meter in part (b), each setting indicates the amount by which the ohms scale (top scale) on the meter is to be multiplied. For example, if the pointer is at 50 on the

SAFETY NOTE

Resistors can become very hot in normal operation. To avoid a burn, do not touch a circuit component while the power is connected to the circuit. After power has been turned off, allow time for the components to cool down.

▶ **FIGURE 4–8**

Typical portable multimeters.
(a) Courtesy of Fluke Corporation. Reproduced with permission.
(b) Courtesy of B+K Precision.

(a) Digital multimeter   (b) Analog multimeter

ohms scale and the range switch is set at ×10, the resistance being measured is 50 × 10 Ω = 500 Ω. *If the resistor is open, the pointer will stay at full left scale (∞ means infinite) regardless of the range switch setting.*

EXAMPLE 4–7

Determine whether the resistor in each circuit of Figure 4–9 has possibly been damaged by overheating.

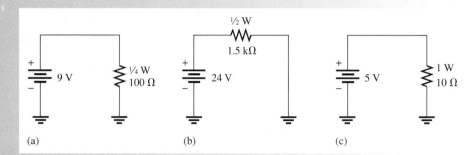

▲ **FIGURE 4–9**

Solution   In the circuit in Figure 4–9(a),

$$P = \frac{V^2}{R} = \frac{(9\text{ V})^2}{100\text{ }\Omega} = 0.810\text{ W} = 810\text{ mW}$$

The rating of the resistor is ¼ W (0.25 W), which is insufficient to handle the power. The resistor has been overheated and may be burned out, making it an open.
    In the circuit of Figure 4–9(b),

$$P = \frac{V^2}{R} = \frac{(24\text{ V})^2}{1.5\text{ k}\Omega} = 0.384\text{ W} = 384\text{ mW}$$

The rating of the resistor is ½ W (0.5 W), which is sufficient to handle the power.
    In the circuit of Figure 4–9(c),

$$P = \frac{V^2}{R} = \frac{(5\text{ V})^2}{10\text{ }\Omega} = 2.5\text{ W}$$

The rating of the resistor is 1 W, which is insufficient to handle the power. The resistor has been overheated and may be burned out, making it an open.

Related Problem   A 0.25 W, 1.0 kΩ resistor is connected across a 12 V battery. Is the power rating adequate?

**SECTION 4–3
REVIEW**

1. Name two important values associated with a resistor.
2. How does the physical size of a resistor determine the amount of power that it can handle?
3. List the standard power ratings of metal-film resistors.
4. A resistor must handle 0.3 W. What minimum power rating of a metal-film resistor should be used to dissipate the energy properly?

4–4 ENERGY CONVERSION AND VOLTAGE DROP IN RESISTANCE

As you have learned, when there is current through a resistance, electrical energy is converted to heat energy. This heat is caused by collisions of the free electrons within the atomic structure of the resistive material. When a collision occurs, heat is given off; and the electron gives up some of its acquired energy as it moves through the material.

After completing this section, you should be able to

- ◆ **Explain energy conversion and voltage drop**
  - ◆ Discuss the cause of energy conversion in a circuit
  - ◆ Define *voltage drop*
  - ◆ Explain the relationship between energy conversion and voltage drop

Figure 4–10 illustrates charge in the form of electrons flowing from the negative terminal of a battery, through a circuit, and back to the positive terminal. As they emerge from the negative terminal, the electrons are at their highest energy level. The electrons flow through each of the resistors that are connected together to form a current path (this type of connection is called series, as you will learn in Chapter 5). As the electrons flow through each resistor, some of their energy is given up in the form of heat. Therefore, the electrons have more energy when they enter a resistor than when they exit the resistor, as illustrated in the figure by the decrease in the intensity of the red color. When they have traveled through the circuit back to the positive terminal of the battery, the electrons are at their lowest energy level.

▶ **FIGURE 4–10**

A loss of energy by electrons (charge) as they flow through a resistance creates a voltage drop because voltage equals energy divided by charge.

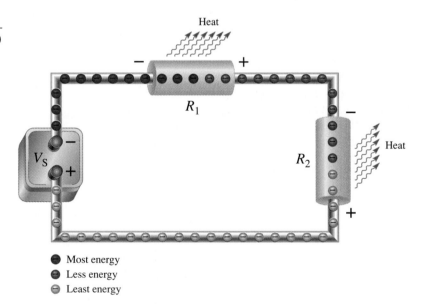

Recall that voltage equals energy per charge ($V = W/Q$) and charge is a property of electrons. Based on the voltage of the battery, a certain amount of energy is imparted to all of the electrons that flow out of the negative terminal. The same number of electrons flow at each point throughout the circuit, but their energy decreases as they move through the resistance of the circuit.

In Figure 4–10, the voltage at the left end of R_1 is equal to W_{enter}/Q, and the voltage at the right end of R_1 is equal to W_{exit}/Q. The same number of electrons that enter R_1 also exit R_1, so Q is constant. However, the energy W_{exit} is less than W_{enter}, so the voltage at the right end of R_1 is less than the voltage at the left end. This decrease in voltage across the resistor due to a loss of energy is called a **voltage drop**. The voltage at the right end of R_1 is less negative (or more positive) than the voltage at the left end. The voltage drop is indicated by − and + signs (the + implies a less negative or more positive voltage).

The electrons have lost some energy in R_1 and now they enter R_2 with a reduced energy level. As they flow through R_2, they lose more energy, resulting in another voltage drop across R_2.

SECTION 4–4 REVIEW	1. What is the basic reason for energy conversion in a resistor? 2. What is a voltage drop? 3. What is the polarity of a voltage drop in relation to conventional current direction?

4–5   POWER SUPPLIES

In general, a **power supply** is a device that provides power to a load. Recall that a load is any electrical device or circuit that is connected to the output of the power supply and draws current from the supply.

After completing this section, you should be able to

◆ **Discuss power supplies and their characteristics**

  ◆ Define *ampere-hour rating* of batteries

  ◆ Discuss electronic power supply efficiency

Figure 4–11 shows a representation of a power supply with a loading device connected to it. The load can be anything from a light bulb to a computer. The power supply produces a voltage across its two output terminals and provides current through the load, as indicated in the figure. The product IV_{OUT} is the amount of power produced by the supply and consumed by the load. For a given output voltage (V_{OUT}), more current drawn by the load means more power from the supply.

◀ **FIGURE 4–11**

Power supply and load.

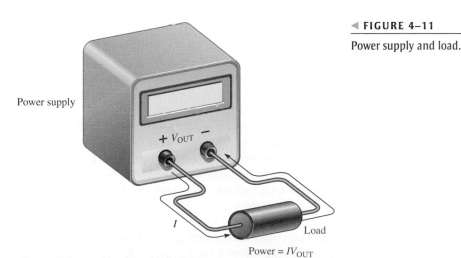

Power = IV_{OUT}

Power supplies range from simple batteries to regulated electronic circuits where an accurate output voltage is automatically maintained. A battery is a dc power supply that converts chemical energy into electrical energy. Electronic power supplies normally convert 110 V ac (alternating current) from a wall outlet into a regulated dc (direct current) voltage at a level suitable for electronic components.

Ampere-hour Ratings of Batteries

Batteries convert chemical energy into electrical energy. Because of their limited source of chemical energy, batteries have a certain capacity that limits the amount of time over which they can produce a given power level. This capacity is measured in ampere-hours. The **ampere-hour (Ah) rating** determines the length of time that a battery can deliver a certain amount of average current to a load at the rated voltage.

A rating of one ampere-hour means that a battery can deliver an average of one ampere of current to a load for one hour at the rated voltage output. This same battery can deliver an average of two amperes for one-half hour. The more current the battery is required to deliver, the shorter the life of the battery. In practice, a battery usually is rated for a specified current level and output voltage. For example, a 12 V automobile battery may be rated for 70 Ah at 3.5 A. This means that it can produce an average of 3.5 A for 20 h at the rated voltage.

EXAMPLE 4–8

For how many hours can a battery deliver 2 A if it is rated at 70 Ah?

Solution   The ampere-hour rating is the current times the number of hours (x).

$$70 \text{ Ah} = (2 \text{ A})(x \text{ h})$$

Solving for the number of hours, x, yields

$$x = \frac{70 \text{ Ah}}{2 \text{ A}} = \textbf{35 h}$$

Related Problem   A certain battery delivers 10 A for 6 h. What is its Ah rating?

Power Supply Efficiency

An important characteristic of electronic power supplies is efficiency. **Efficiency** is the ratio of the output power delivered to a load to the input power to a circuit.

Equation 4–5

$$\text{Efficiency} = \frac{P_{\text{OUT}}}{P_{\text{IN}}}$$

Efficiency is often expressed as a percentage. For example, if the input power is 100 W and the output power is 50 W, the efficiency is (50 W/100 W) $\times$ 100% = 50%.

All electronic power supplies require that power be put into them. For example, an electronic power supply generally uses the ac power from a wall outlet as its input. Its output is usually a regulated dc voltage. The output power is *always* less than the input power because some of the total power must be used internally to operate the power supply circuitry. This internal power dissipation is normally called the *power loss*. The output power is the input power minus the power loss.

Equation 4–6

$$P_{\text{OUT}} = P_{\text{IN}} - P_{\text{LOSS}}$$

High efficiency means that very little power is dissipated in the power supply and there is a higher proportion of output power for a given input power.

EXAMPLE 4–9

A certain electronic power supply requires 25 W of input power. It can produce an output power of 20 W. What is its efficiency, and what is the power loss?

Solution

$$\text{Efficiency} = \frac{P_{OUT}}{P_{IN}} = \frac{20 \text{ W}}{25 \text{ W}} = \mathbf{0.8}$$

Expressed as a percentage,

$$\text{Efficiency} = \left(\frac{20 \text{ W}}{25 \text{ W}}\right)100\% = 80\%$$

The power loss is

$$P_{LOSS} = P_{IN} - P_{OUT} = 25 \text{ W} - 20 \text{ W} = \mathbf{5 \text{ W}}$$

Related Problem A power supply has an efficiency of 92%. If P_{IN} is 50 W, what is P_{OUT}?

SECTION 4–5 REVIEW

1. When a loading device draws an increased amount of current from a power supply, does this change represent a greater or a smaller load on the supply?

2. A power supply produces an output voltage of 10 V. If the supply provides 0.5 A to a load, what is the power to a load?

3. If a battery has an ampere-hour rating of 100 Ah, how long can it provide 5 A to a load?

4. If the battery in Question 3 is a 12 V device, what is its power to a load for the specified value of current?

5. An electronic power supply used in the lab operates with an input power of 1 W. It can provide an output power of 750 mW. What is its efficiency? Determine the power loss.

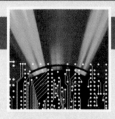

A Circuit Application

In this application, the resistance box that you modified in Chapter 3 is back. The last time, you verified that all the resistor values were correct. This time you must make sure each resistor has a sufficient power rating; and if the power rating is insufficient, replace the resistor with one that is adequate.

Power Ratings

Assume the power rating of each resistor in the resistance box as modified in Chapter 3 is ⅛ W. The box is shown in Figure 4–12.

◆ Determine if the power rating of each resistor is adequate for a maximum of 4 V.

◆ If a rating is not adequate, determine the lowest rating required to handle the maximum power. Choose from standard ratings of ⅛ W, ¼ W, ½ W, 1 W, 2 W, and 5 W.

◆ Add the power rating of each resistor to the schematic developed in Chapter 3.

Review

1. How many resistors were replaced because of inadequate power ratings?

2. If the resistance must operate with 10 V maximum, which resistors must be changed and to what minimum power ratings?

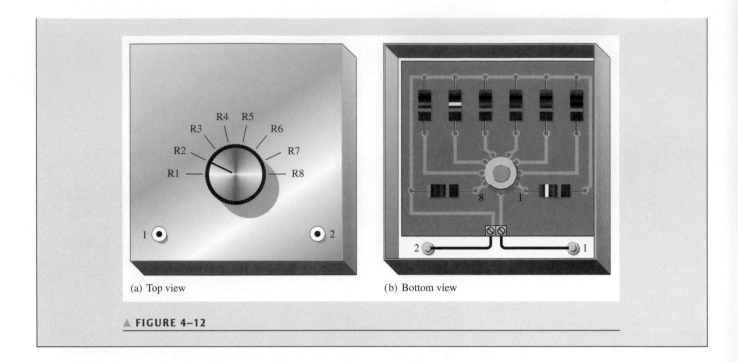

(a) Top view   (b) Bottom view

▲ FIGURE 4–12

SUMMARY

◆ The power rating in watts of a resistor determines the maximum power that it can handle safely.

◆ Resistors with a larger physical size can dissipate more power in the form of heat than smaller ones.

◆ A resistor should have a power rating higher than the maximum power that it is expected to handle in the circuit.

◆ Power rating is not related to resistance value.

◆ A resistor normally opens when it overheats and fails.

◆ Energy is the ability to do work and is equal to power multiplied by time.

◆ The kilowatt-hour is a unit of energy.

◆ One kilowatt-hour equals one thousand watts used for one hour or any other combination of watts and hours that has a product of one.

◆ A power supply is an energy source used to operate electrical and electronic devices.

◆ A battery is one type of power supply that converts chemical energy into electrical energy.

◆ An electronic power supply converts commercial energy (ac from the power company) to regulated dc at various voltage levels.

◆ The output power of a supply is the output voltage times the load current.

◆ A load is a device that draws current from the power supply.

◆ The capacity of a battery is measured in ampere-hours (Ah).

◆ One ampere-hour equals one ampere used for one hour, or any other combination of amperes and hours that has a product of one.

◆ A circuit with a high efficiency has a smaller percentage power loss than one with a lower efficiency.

KEY TERMS

Key terms and other bold terms in the chapter are defined in the end-of-book glossary.

Ampere-hour (Ah) rating   A number given in ampere-hours determined by multiplying the current (A) times the length of time (h) a battery can deliver that current to a load.

Efficiency   The ratio of the output power delivered to a load to the input power to a circuit, usually expressed as a percentage.

Energy The ability to do work.

Joule (J) The SI unit of energy.

Kilowatt-hour (kWh) A large unit of energy used mainly by utility companies.

Power The rate of energy usage.

Power supply A device that provides power to a load.

Voltage drop The decrease in voltage across a resistor due to a loss of energy.

Watt (W) The unit of power. One watt is the power when 1 J of energy is used in 1 s.

FORMULAS

4–1	$P = \dfrac{W}{t}$	Power equals energy divided by time.
4–2	$P = I^2 R$	Power equals current squared times resistance.
4–3	$P = VI$	Power equals voltage times current.
4–4	$P = \dfrac{V^2}{R}$	Power equals voltage squared divided by resistance.
4–5	$\text{Efficiency} = \dfrac{P_{OUT}}{P_{IN}}$	Power supply efficiency
4–6	$P_{OUT} = P_{IN} - P_{LOSS}$	Output power is input power less power loss.

SELF-TEST

Answers are at the end of the chapter.

1. Power can be defined as

    (a) energy   (b) heat

    (c) the rate at which energy is used   (d) the time required to use energy

2. Two hundred joules of energy are consumed in 10 s. The power is

    (a) 2000 W   (b) 10 W   (c) 20 W   (d) 2 W

3. If it takes 300 ms to use 10,000 J of energy, the power is

    (a) 33.3 kW   (b) 33.3 W   (c) 33.3 mW

4. In 50 kW, there are

    (a) 500 W   (b) 5000 W   (c) 0.5 MW   (d) 50,000 W

5. In 0.045 W, there are

    (a) 45 kW   (b) 45 mW   (c) 4,500 μW   (d) 0.00045 MW

6. For 10 V and 50 mA, the power is

    (a) 500 mW   (b) 0.5 W   (c) 500,000 μW   (d) answers (a), (b), and (c)

7. When the current through a 10 kΩ resistor is 10 mA, the power is

    (a) 1 W   (b) 10 W   (c) 100 mW   (d) 1000 μW

8. A 2.2 kΩ resistor dissipates 0.5 W. The current is

    (a) 15.1 mA   (b) 0.227 mA   (c) 1.1 mA   (d) 4.4 mA

9. A 330 Ω resistor dissipates 2 W. The voltage is

    (a) 2.57 V   (b) 660 V   (c) 6.6 V   (d) 25.7 V

10. If you used 500 W of power for 24 h, you have used

    (a) 0.5 kWh   (b) 2400 kWh   (c) 12,000 kWh   (d) 12 kWh

11. How many watt-hours represent 75 W used for 10 h?

    (a) 75 Wh   (b) 750 Wh   (c) 0.75 Wh   (d) 7500 Wh

12. A 100 Ω resistor must carry a maximum current of 35 mA. Its rating should be at least

    (a) 35 W   (b) 35 mW   (c) 123 mW   (d) 3500 mW

13. The power rating of a resistor that is to handle up to 1.1 W should be

   (a) 0.25 W   (b) 1 W   (c) 2 W   (d) 5 W

14. A 22 Ω half-watt resistor and a 220 Ω half-watt resistor are connected across a 10 V source. Which one(s) will overheat?

   (a) 22 Ω   (b) 220 Ω   (c) both   (d) neither

15. When the needle of an analog ohmmeter indicates infinity, the resistor being measured is

   (a) overheated   (b) shorted   (c) open   (d) reversed

16. A 12 V battery is connected to a 600 Ω load. Under these conditions, it is rated at 50 Ah. How long can it supply current to the load?

   (a) 2500 h   (b) 50 h   (c) 25 h   (d) 4.16 h

17. A given power supply is capable of providing 8 A for 2.5 h. Its ampere-hour rating is

   (a) 2.5 Ah   (b) 20 Ah   (c) 8 Ah

18. A power supply produces a 0.5 W output with an input of 0.6 W. Its percentage of efficiency is

   (a) 50%   (b) 60%   (c) 83.3%   (d) 45%

CIRCUIT DYNAMICS QUIZ

Answers are at the end of the chapter.

1. If the current through a fixed resistor goes from 10 mA to 12 mA, the power in the resistor

   (a) increases   (b) decreases   (c) stays the same

2. If the voltage across a fixed resistor goes from 10 V to 7 V, the power in the resistor

   (a) increases   (b) decreases   (c) stays the same

3. A variable resistor has 5 V across it. If you reduce the resistance, the power in the resistor

   (a) increases   (b) decreases   (c) stays the same

4. If the voltage across a resistor increases from 5 V to 10 V and the current increases from 1 mA to 2 mA, the power

   (a) increases   (b) decreases   (c) stays the same

5. If the resistance of a load connected to a battery is increased, the amount of time the battery can supply current

   (a) increases   (b) decreases   (c) stays the same

6. If the amount of time that a battery supplies current to a load is decreased, its ampere-hour rating

   (a) increases   (b) decreases   (c) stays the same

7. If the current that a battery supplies to a load is increased, the battery life

   (a) increases   (b) decreases   (c) stays the same

8. If there is no load connected to a battery, its ampere-hour rating

   (a) increases   (b) decreases   (c) stays the same

Refer to Figure 4–11.

9. If the output voltage of the power supply increases, the power to the constant load

   (a) increases   (b) decreases   (c) stays the same

10. For a constant output voltage, if the current to the load decreases, the load power

   (a) increases   (b) decreases   (c) stays the same

11. For a constant output voltage, if the resistance of the load increases, the power in the load

   (a) increases   (b) decreases   (c) stays the same

12. If the load is removed from the circuit leaving an open, ideally the power supply output voltage

   (a) increases   (b) decreases   (c) stays the same

PROBLEMS

More difficult problems are indicated by an asterisk (*).
Answers to odd-numbered problems are at the end of the book.

SECTION 4–1 Energy and Power

1. Prove that the unit for power (the watt) is equivalent to one volt × one amp.

2. Show that there are 3.6×10^6 joules in a kilowatt-hour.

3. What is the power when energy is consumed at the rate of 350 J/s?

4. How many watts are used when 7500 J of energy are consumed in 5 h?

5. How many watts does 1000 J in 50 ms equal?

6. Convert the following to kilowatts:
   (a) 1000 W (b) 3750 W (c) 160 W (d) 50,000 W

7. Convert the following to megawatts:
   (a) 1,000,000 W (b) 3×10^6 W (c) 15×10^7 W (d) 8700 kW

8. Convert the following to milliwatts:
   (a) 1 W (b) 0.4 W (c) 0.002 W (d) 0.0125 W

9. Convert the following to microwatts:
   (a) 2 W (b) 0.0005 W (c) 0.25 mW (d) 0.00667 mW

10. Convert the following to watts:
    (a) 1.5 kW (b) 0.5 MW (c) 350 mW (d) 9000 μW

11. A particular electronic device uses 100 mW of power. If it runs for 24 h, how many joules of energy does it consume?

*12. If a 300 W bulb is allowed to burn continuously for 30 days, how many kilowatt-hours of energy does it consume?

*13. At the end of a 31 day period, your utility bill shows that you have used 1500 kWh. What is your average daily power usage?

14. Convert 5×10^6 watt-minutes to kWh.

15. Convert 6700 watt-seconds to kWh.

16. For how many seconds must there be 5 A of current through a 47 Ω resistor in order to consume 25 J?

SECTION 4–2 Power in an Electric Circuit

17. If a 75 V source is supplying 2 A to a load, what is the resistance value of the load?

18. If a resistor has 5.5 V across it and 3 mA through it, what is the power?

19. An electric heater works on 120 V and draws 3 A of current. How much power does it use?

20. What is the power when there are 500 mA of current through a 4.7 kΩ resistor?

21. Calculate the power dissipated by a 10 kΩ resistor carrying 100 μA.

22. If there are 60 V across a 680 Ω resistor, what is the power?

23. A 56 Ω resistor is connected across the terminals of a 1.5 V battery. What is the power dissipation in the resistor?

24. If a resistor is to carry 2 A of current and handle 100 W of power, how many ohms must it be? Assume that the voltage can be adjusted to any required value.

25. A 12 V source is connected across a 10 Ω resistor.
    (a) How much energy is used in two minutes?
    (b) If the resistor is disconnected after one minute, is the power during the first minute greater than, less than, or equal to the power during a two minute interval?

SECTION 4–3 Resistor Power Ratings

26. A 6.8 kΩ resistor has burned out in a circuit. You must replace it with another resistor with the same resistance value. If the resistor carries 10 mA, what should its power rating be? Assume that you have available resistors in all the standard power ratings.

27. A certain type of power resistor comes in the following ratings: 3 W, 5 W, 8 W, 12 W, 20 W. Your particular application requires a resistor that can handle approximately 8 W. Which rating would you use for a minimum safety margin of 20% above the rated value? Why?

SECTION 4–4 Energy Conversion and Voltage Drop in Resistance

28. For each circuit in Figure 4–13, assign the proper polarity for the voltage drop across the resistor.

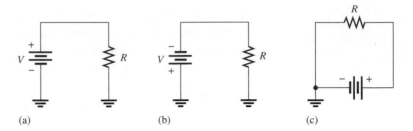

(a)   (b)   (c)

▲ FIGURE 4–13

SECTION 4–5 Power Supplies

29. A 50 Ω load uses 1 W of power. What is the output voltage of the power supply?

30. Assume that an alkaline D-cell battery can maintain an average voltage of 1.25 V for 90 hours in a 10 Ω load before becoming unusable. What average power is delivered to the load during the life of the battery?

31. What is the total energy in joules that is delivered during the 90 hours for the battery in Problem 30?

32. A battery can provide an average of 1.5 A of current for 24 h. What is its ampere-hour rating?

33. How much average current can be drawn from an 80 Ah battery for 10 h?

34. If a battery is rated at 650 mAh, how much average current will it provide for 48 h?

35. If the input power is 500 mW and the output power is 400 mW, how much power is lost? What is the efficiency of this power supply?

36. To operate at 85% efficiency, how much output power must a source produce if the input power is 5 W?

*37. A certain power supply provides a continuous 2 W to a load. It is operating at 60% efficiency. In a 24 h period, how many kilowatt-hours does the power supply use?

Multisim Troubleshooting and Analysis

These problems require your Multisim CD-ROM.

38. Open file P04-38 and determine the current, voltage, and resistance. Using the measured values, calculate the power.

39. Open file P04-39 and determine the current, voltage, and resistance. Calculate the power from these values.

40. Open file P04-40. Measure the current in the lamp and determine if the value agrees with that determined using the power and voltage rating of the lamp.

ANSWERS

SECTION REVIEWS

SECTION 4–1 Energy and Power

1. Power is the rate at which energy is used.
2. $P = W/t$
3. Watt is the unit of power. One watt is the power when 1 J of energy is used in 1 s.

4. **(a)** 68,000 W = 68 kW   **(b)** 0.005 W = 5 mW   **(c)** 0.000025 W = 25 μW

5. $W = (0.1 \text{ kW})(10 \text{ h}) = 1 \text{ kWh}$

6. 2000 Wh = 2 kWh

7. 360,000 Ws = 0.1 kWh

SECTION 4–2   Power in an Electric Circuit

1. $P = (10 \text{ V})(3 \text{ A}) = 30 \text{ W}$

2. $P = (24 \text{ V})(50 \text{ mA}) = 1.2 \text{ W}$; 1.2 W; the values are the same because all energy generated by the source is dissipated by the resistance.

3. $P = (5 \text{ A})^2(56 \text{ }\Omega) = 1400 \text{ W}$

4. $P = (20 \text{ mA})^2(4.7 \text{ k}\Omega) = 1.88 \text{ W}$

5. $P = (5 \text{ V})^2/10 \text{ }\Omega = 2.5 \text{ W}$

6. $P = (8 \text{ V})^2/2.2 \text{ k}\Omega = 29.1 \text{ mW}$

7. $R = 75 \text{ W}/(0.5 \text{ A})^2 = 300 \text{ }\Omega$

SECTION 4–3   Resistor Power Ratings

1. Resistors have resistance and a power rating.

2. A larger surface area of a resistor dissipates more power.

3. 0.125 W, 0.25 W, 0.5 W, 1 W

4. A 0.5 W rating should be used for 0.3 W.

SECTION 4–4   Energy Conversion and Voltage Drop in Resistance

1. Energy conversion in a resistor is caused by collisions of free electrons with the atoms in the material.

2. Voltage drop is a decrease in voltage across a resistor due to a loss of energy.

3. Voltage drop is positive to negative in the direction of conventional current.

SECTION 4–5   Power Supplies

1. More current means a greater load.

2. $P = (10 \text{ V})(0.5 \text{ A}) = 5 \text{ W}$

3. $t = 100 \text{ Ah}/5 \text{ A} = 20 \text{ h}$

4. $P = (12 \text{ V})(5 \text{ A}) = 60 \text{ W}$

5. Eff = $(0.75 \text{ W}/1 \text{ W})100\% = 75\%$; $P_{\text{LOSS}} = 1000 \text{ mW} - 750 \text{ mW} = 250 \text{ mW}$

A Circuit Application

1. Two

2. 10 Ω, 10 W; 100 Ω, 1 W; 400 Ω, ¼ W

RELATED PROBLEMS FOR EXAMPLES

4–1 3000 J

4–2 **(a)** 0.001 W   **(b)** 0.0018 W   **(c)** 1 W   **(d)** 0.000001 W

4–3 2 kWh

4–4 **(a)** 40 W   **(b)** 376 W   **(c)** 625 mW

4–5 60 W

4–6 0.5 W

4–7 Yes

4–8 60 Ah

4–9 46 W

SELF-TEST

1. (c)   2. (c)   3. (a)   4. (d)   5. (b)   6. (d)   7. (a)   8. (a)
9. (d)   10. (d)   11. (b)   12. (c)   13. (c)   14. (a)   15. (c)   16. (a)
17. (b)   18. (c)

CIRCUIT DYNAMICS QUIZ

1. (a)   2. (b)   3. (a)   4. (a)   5. (a)   6. (c)
7. (b)   8. (c)   9. (a)   10. (b)   11. (b)   12. (c)

SERIES CIRCUITS

5

CHAPTER OUTLINE

CHAPTER OBJECTIVES

◆ Identify a series resistive circuit

◆ Determine the current throughout a series circuit

◆ Determine total series resistance

◆ Apply Ohm's law in series circuits

◆ Determine the total effect of voltage sources connected in series

◆ Apply Kirchhoff's voltage law

◆ Use a series circuit as a voltage divider

◆ Determine power in a series circuit

◆ Measure voltage with respect to ground

◆ Troubleshoot series circuits

KEY TERMS

◆ Series

◆ Kirchhoff's voltage law

◆ Voltage divider

◆ Reference ground

◆ Open

◆ Short

A CIRCUIT APPLICATION PREVIEW

In the application, you will evaluate a voltage-divider circuit board connected to a 12 V battery to provide a selection of fixed reference voltages for use with an electronic instrument.

VISIT THE COMPANION WEBSITE

Study aids for this chapter are available at
http://www.prenhall.com/floyd

INTRODUCTION

In Chapter 3 you learned about Ohm's law, and in Chapter 4 you learned about power in resistors. In this chapter, those concepts are applied to circuits in which resistors are connected in a series arrangement.

Resistive circuits can be of two basic forms: series and parallel. In this chapter, series circuits are studied. Parallel circuits are covered in Chapter 6, and combinations of series and parallel resistors are examined in Chapter 7. In this chapter, you will see how Ohm's law is used in series circuits; and you will learn another important circuit law, Kirchhoff's voltage law. Also, several applications of series circuits, including voltage dividers, are presented.

When resistors are connected in series and a voltage is applied across the series connection, there is only one path for current; and, therefore, each resistor in series has the same amount of current through it. All of the resistances in series add together to produce a total resistance. The voltage drops across each of the resistors add up to the voltage applied across the entire series connection.

5–1 RESISTORS IN SERIES

When connected in series, resistors form a "string" in which there is only one path for current.

After completing this section, you should be able to

- ◆ **Identify a series resistive circuit**
  - ◆ Translate a physical arrangement of resistors into a schematic

The schematic in Figure 5–1(a) shows two resistors connected in series between point *A* and point *B*. Part (b) shows three resistors in series, and part (c) shows four in series. Of course, there can be any number of resistors in a series circuit.

(a)

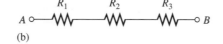

(b)

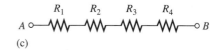

(c)

▲ **FIGURE 5–1**

Resistors in series.

When a voltage source is connected between point *A* and *B*, the only way for current to get from one point to the other in any of the connections of Figure 5–1 is to go through each of the resistors. The following statement describes a series circuit:

A series circuit provides only one path for current between two points so that the current is the same through each series resistor.

In an actual circuit diagram, a series circuit may not always be as easy to visually identify as those in Figure 5–1. For example, Figure 5–2 shows series resistors drawn in other ways with voltage applied. Remember, if there is only one current path between two points, the resistors between those two points are in series, no matter how they appear in a diagram.

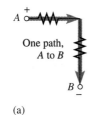

(a)

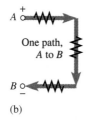

(b)

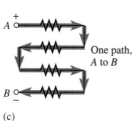

(c)

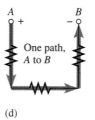

(d)

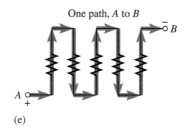

(e)

▲ **FIGURE 5–2**

Some examples of series circuits. Notice that the current is the same at all points because the current has only one path.

EXAMPLE 5–1

Suppose that there are five resistors positioned on a protoboard as shown in Figure 5–3. Wire them together in series so that, starting from the positive (+) terminal, R_1 is first, R_2 is second, R_3 is third, and so on. Draw a schematic showing this connection.

▶ FIGURE 5–3

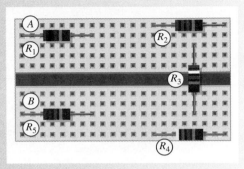

Solution   The wires are connected as shown in Figure 5–4(a), which is the assembly diagram. The schematic is shown in Figure 5–4(b). Note that the schematic does not necessarily show the actual physical arrangement of the resistors as does the assembly diagram. The schematic shows how components are connected electrically; the assembly diagram shows how components are arranged and interconnected physically.

FIGURE 5–4

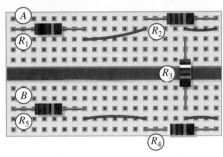

(a) Assembly diagram

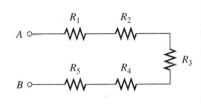

(b) Schematic

Related Problem*   **(a)** Show how you would rewire the protoboard in Figure 5–4(a) so that all the odd-numbered resistors come first followed by the even-numbered ones. **(b)** Determine the resistance value of each resistor.

*Answers are at the end of the chapter.

EXAMPLE 5–2   Describe how the resistors on the printed circuit (PC) board in Figure 5–5 are related electrically. Determine the resistance value of each resistor.

FIGURE 5–5

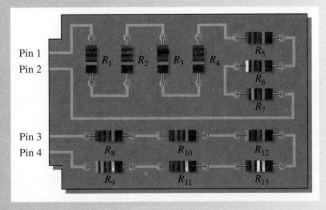

Solution   Resistors R_1 through R_7 are in series with each other. This series combination is connected between pins 1 and 2 on the PC board.

Resistors R_8 through R_{13} are in series with each other. This series combination is connected between pins 3 and 4 on the PC board.

The values of the resistors are $R_1 = 2.2\,\text{k}\Omega$, $R_2 = 3.3\,\text{k}\Omega$, $R_3 = 1.0\,\text{k}\Omega$, $R_4 = 1.2\,\text{k}\Omega$, $R_5 = 3.3\,\text{k}\Omega$, $R_6 = 4.7\,\text{k}\Omega$, $R_7 = 5.6\,\text{k}\Omega$, $R_8 = 12\,\text{k}\Omega$, $R_9 = 68\,\text{k}\Omega$, $R_{10} = 27\,\text{k}\Omega$, $R_{11} = 12\,\text{k}\Omega$, $R_{12} = 82\,\text{k}\Omega$, and $R_{13} = 270\,\text{k}\Omega$.

Related Problem   How is the circuit changed when pin 2 and pin 3 in Figure 5–5 are connected?

SECTION 5–1
REVIEW
Answers are at the end of the chapter.

1. How are the resistors connected in a series circuit?
2. How can you identify a series circuit?
3. Complete the schematics for the circuits in each part of Figure 5–6 by connecting each group of resistors in series in numerical order from terminal *A* to terminal *B*.
4. Connect each group of series resistors in Figure 5–6 in series with each other.

▶ FIGURE 5–6

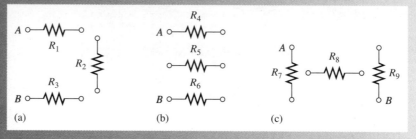

(a)　　　　　　(b)　　　　　　(c)

5–2   CURRENT IN A SERIES CIRCUIT

The current is the same through all points in a series circuit. The current through each resistor in a series circuit is the same as the current through all the other resistors that are in series with it.

After completing this section, you should be able to

◆ **Determine the current throughout a series circuit**

　◆ Show that the current is the same at all points in a series circuit

Figure 5–7 shows three resistors connected in series to a dc voltage source. *At any point in this circuit, the current into that point must equal the current out of that point,* as illustrated by the current directional arrows. Notice also that the current out of each resistor must equal the current into each resistor because there is no place where part of the current can branch off and go somewhere else. Therefore, the current in each section of the circuit is the same as the current in all other sections. It has only one path going from the positive (+) side of the source to the negative (−) side.

Let's assume that the battery in Figure 5–7 supplies one ampere of current to the series resistance. There is one ampere of current out of the battery's positive terminal. When ammeters are connected at several points in the circuit, as shown in Figure 5–8, each meter reads one ampere.

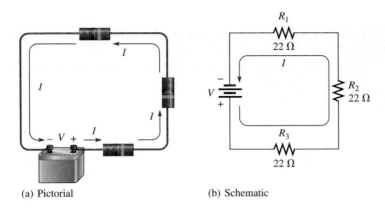

(a) Pictorial (b) Schematic

▲ FIGURE 5–7

Current into any point in a series circuit is the same as the current out of that point.

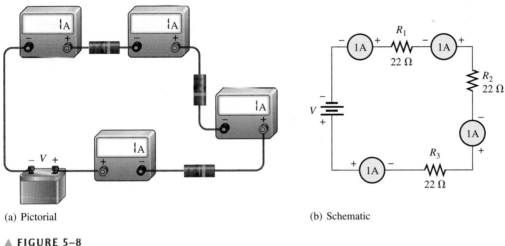

(a) Pictorial (b) Schematic

▲ FIGURE 5–8

Current is the same at all points in a series circuit.

SECTION 5–2 REVIEW

1. In a circuit with a 10 Ω and a 4.7 Ω resistor in series, there is 1 A of current through the 10 Ω resistor. How much current is through the 4.7 Ω resistor?

2. A milliammeter is connected between points A and B in Figure 5–9. It measures 50 mA. If you move the meter and connect it between points C and D, how much current will it indicate? Between E and F?

▶ FIGURE 5–9

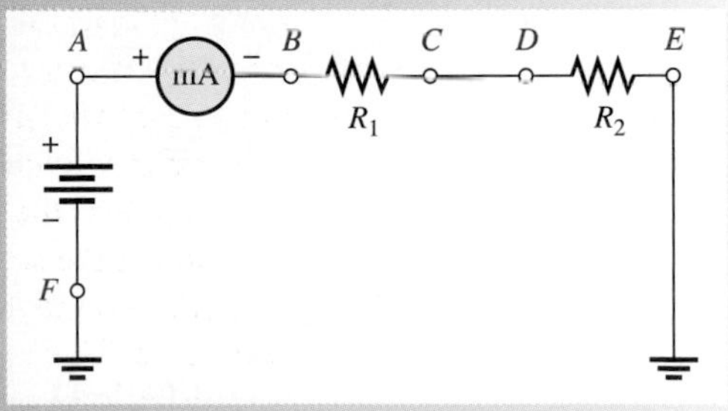

3. In Figure 5–10, how much current does ammeter 1 indicate? How much current does ammeter 2 indicate?

4. Describe current in a series circuit?

▶ FIGURE 5–10

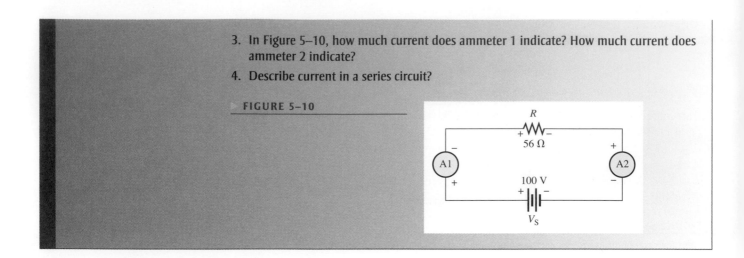

5–3 TOTAL SERIES RESISTANCE

The total resistance of a series circuit is equal to the sum of the resistances of each individual series resistor.

After completing this section, you should be able to

◆ **Determine total series resistance**

    ◆ Explain why resistance values add when resistors are connected in series

    ◆ Apply the series resistance formula

Series Resistor Values Add

When resistors are connected in series, the resistor values add because each resistor offers opposition to the current in direct proportion to its resistance. A greater number of resistors connected in series creates more opposition to current. More opposition to current implies a higher value of resistance. Thus, every time a resistor is added in series, the total resistance increases.

Figure 5–11 illustrates how series resistances add to increase the total resistance. Figure 5–11(a) has a single 10 Ω resistor. Figure 5–11(b) shows another 10 Ω resistor connected

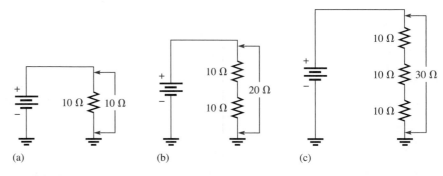

▲ FIGURE 5–11

Total resistance increases with each additional series resistor.

in series with the first one, making a total resistance of 20 Ω. If a third 10 Ω resistor is connected in series with the first two, as shown in Figure 5–11(c), the total resistance becomes 30 Ω.

Series Resistance Formula

For any number of individual resistors connected in series, the total resistance is the sum of each of the individual values.

$$R_T = R_1 + R_2 + R_3 + \cdots + R_n$$

Equation 5–1

where R_T is the total resistance and R_n is the last resistor in the series string (n can be any positive integer equal to the number of resistors in series). For example, if there are four resistors in series ($n = 4$), the total resistance formula is

$$R_T = R_1 + R_2 + R_3 + R_4$$

If there are six resistors in series ($n = 6$), the total resistance formula is

$$R_T = R_1 + R_2 + R_3 + R_4 + R_5 + R_6$$

To illustrate the calculation of total series resistance, let's determine R_T in the circuit of Figure 5–12, where V_S is the source voltage. The circuit has five resistors in series. To get the total resistance, simply add the values.

$$R_T = 56\ \Omega + 100\ \Omega + 27\ \Omega + 10\ \Omega + 47\ \Omega = 240\ \Omega$$

Note in Figure 5–12 that the order in which the resistances are added does not matter. You can physically change the positions of the resistors in the circuit without affecting the total resistance or the current.

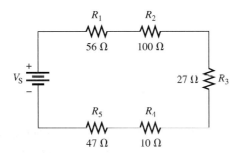

▲ **FIGURE 5–12**

Example of five resistors in series.

EXAMPLE 5–3

Connect the resistors in Figure 5–13 in series, and determine the total resistance, R_T.

▶ **FIGURE 5–13**

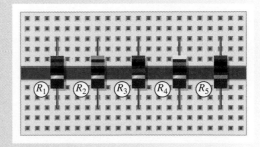

Solution   The resistors are connected as shown in Figure 5–14. Find the total resistance by adding all the values.

$$R_T = R_1 + R_2 + R_3 + R_4 + R_5 = 33\,\Omega + 68\,\Omega + 100\,\Omega + 47\,\Omega + 10\,\Omega = \mathbf{258\,\Omega}$$

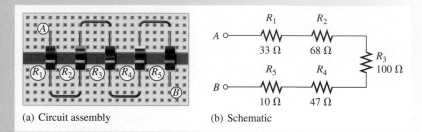

(a) Circuit assembly   (b) Schematic

▲ **FIGURE 5–14**

Related Problem   Determine the total resistance in Figure 5–14(a) if the positions of R_2 and R_4 are interchanged.

EXAMPLE 5–4   What is the total resistance (R_T) in the circuit of Figure 5–15?

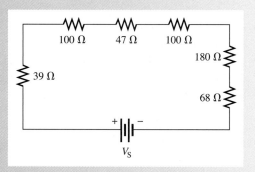

▲ **FIGURE 5–15**

Solution   Sum all the values.

$$R_T = 39\,\Omega + 100\,\Omega + 47\,\Omega + 100\,\Omega + 180\,\Omega + 68\,\Omega = \mathbf{534\,\Omega}$$

Related Problem   What is the total resistance for the following series resistors: 1.0 kΩ, 2.2 kΩ, 3.3 kΩ, and 5.6 kΩ?

EXAMPLE 5–5

Determine the value of R_4 in the circuit of Figure 5–16.

▶ **FIGURE 5–16**

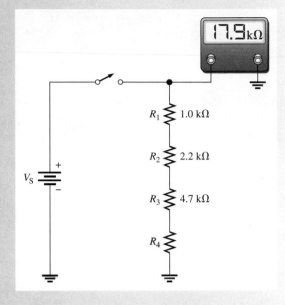

Solution   From the ohmmeter reading, $R_T = 17.9\,\text{k}\Omega$.

$$R_T = R_1 + R_2 + R_3 + R_4$$

Solving for R_4 yields

$$R_4 = R_T - (R_1 + R_2 + R_3) = 17.9\,\text{k}\Omega - (1.0\,\text{k}\Omega + 2.2\,\text{k}\Omega + 4.7\,\text{k}\Omega) = \mathbf{10\,k\Omega}$$

Related Problem   Determine the value of R_4 in Figure 5–16 if the ohmmeter reading is 14.7 kΩ.

Equal-Value Series Resistors

When a circuit has more than one resistor of the same value in series, there is a shortcut method to obtain the total resistance: Simply multiply the resistance value by the number of equal-value resistors that are in series. This method is essentially the same as adding the values. For example, five 100 Ω resistors in series have an R_T of 5(100 Ω) = 500 Ω. In general, the formula is expressed as

$$R_T = nR$$ **Equation 5–2**

where n is the number of equal-value resistors and R is the resistance value.

EXAMPLE 5–6

Find the R_T of eight 22 Ω resistors in series.

Solution   Find R_T by adding the values.

$$R_T = 22\,\Omega + 22\,\Omega + 22\,\Omega + 22\,\Omega + 22\,\Omega + 22\,\Omega + 22\,\Omega + 22\,\Omega = \mathbf{176\,\Omega}$$

However, it is much easier to multiply to get the same result.

$$R_T = 8(22\,\Omega) = \mathbf{176\,\Omega}$$

Related Problem   Find R_T for three 1.0 kΩ resistors and two 720 Ω resistors in series.

1. The following resistors (one each) are in series: 1.0 Ω, 2.2 Ω, 3.3 Ω, and 4.7 Ω. What is the total resistance?
2. The following resistors are in series: one 100 Ω, two 56 Ω, four 12 Ω, and one 330 Ω. What is the total resistance?
3. Suppose that you have one resistor each of the following values: 1.0 kΩ, 2.7 kΩ, 5.6 kΩ, and 560 kΩ. To get a total resistance of approximately 13.8 kΩ, you need one more resistor. What should its value be?
4. What is the R_T for twelve 56 Ω resistors in series?
5. What is the R_T for twenty 5.6 kΩ resistors and thirty 8.2 kΩ resistors in series?

5–4 APPLICATION OF OHM'S LAW

The basic concepts of series circuits and Ohm's law can be applied to series circuit analysis.

After completing this section, you should be able to

◆ **Apply Ohm's law in series circuits**

  ◆ Find the current in a series circuit

  ◆ Find the voltage across each resistor in series

The following are key points to remember when you analyze series circuits:

1. Current through any of the series resistors is the same as the total current.

2. If you know the total applied voltage and the total resistance, you can determine the total current by Ohm's law.

$$I_T = \frac{V_T}{R_T}$$

3. If you know the voltage drop across one of the series resistors (R_x), you can determine the total current by Ohm's law.

$$I_T = \frac{V_x}{R_x}$$

4. If you know the total current, you can find the voltage drop across any of the series resistors by Ohm's law.

$$V_R = I_T R_x$$

5. The polarity of a voltage drop across a resistor is positive at the end of the resistor that is closest to the positive terminal of the voltage source.

6. The current through a resistor is defined to be in a direction from the positive end of the resistor to the negative end.

7. An open in a series circuit prevents current; and, therefore, there is zero voltage drop across each series resistor. The total voltage appears across the points between which there is an open.

Now let's look at several examples that use Ohm's law for series circuit analysis.

EXAMPLE 5–7

Find the current in the circuit of Figure 5–17.

▶ **FIGURE 5–17**

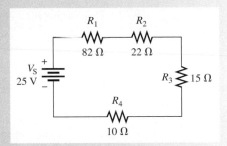

Solution   The current is determined by the source voltage V_S and the total resistance R_T. First, calculate the total resistance.

$$R_T = R_1 + R_2 + R_3 + R_4 = 82 \ \Omega + 22 \ \Omega + 15 \ \Omega + 10 \ \Omega = 129 \ \Omega$$

Next, use Ohm's law to calculate the current.

$$I = \frac{V_S}{R_T} = \frac{25 \text{ V}}{129 \ \Omega} = 0.194 \text{ A} = \mathbf{194 \text{ mA}}$$

where V_S is the total voltage and I is the total current. Remember, the same current exists at all points in the circuit. Thus, each resistor has 194 mA through it.

Related Problem   What is the current in the circuit of Figure 5–17 if R_4 is changed to 100 Ω?

Use Multisim file E05-07 to verify the calculated results in this example and to confirm your calculation for the related problem.

EXAMPLE 5–8

The current in the circuit of Figure 5–18 is 1 mA. For this amount of current, what must the source voltage V_S be?

▶ **FIGURE 5–18**

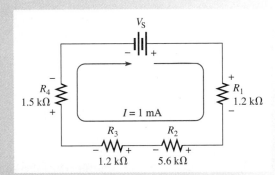

Solution   In order to calculate V_S, first determine R_T.

$$R_T = R_1 + R_2 + R_3 + R_4 = 1.2 \text{ k}\Omega + 5.6 \text{ k}\Omega + 1.2 \text{ k}\Omega + 1.5 \text{ k}\Omega = 9.5 \text{ k}\Omega$$

Next, use Ohm's law to determine V_S.

$$V_S = IR_T = (1 \text{ mA})(9.5 \text{ k}\Omega) = \mathbf{9.5 \text{ V}}$$

Related Problem Calculate V_S if the 5.6 kΩ resistor is changed to 3.9 kΩ with the current the same.

 Use Multisim file E05-08 to verify the calculated results in this example and to confirm your calculation for the related problem.

EXAMPLE 5–9 Calculate the voltage across each resistor in Figure 5–19, and find the value of V_S. To what maximum value can V_S be raised if the current is to be limited to 5 mA?

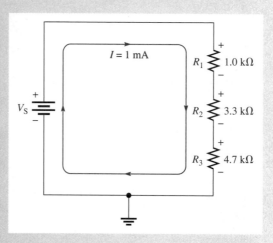

▲ **FIGURE 5–19**

Solution By Ohm's law, the voltage across each resistor is equal to its resistance multiplied by the current through it. Use the Ohm's law formula $V = IR$ to determine the voltage across each of the resistors. Keep in mind that there is the same current through each series resistor. The voltage across R_1 (designated V_1) is

$$V_1 = IR_1 = (1 \text{ mA})(1.0 \text{ k}\Omega) = \textbf{1 V}$$

The voltage across R_2 is

$$V_2 = IR_2 = (1 \text{ mA})(3.3 \text{ k}\Omega) = \textbf{3.3 V}$$

The voltage across R_3 is

$$V_3 = IR_3 = (1 \text{ mA})(4.7 \text{ k}\Omega) = \textbf{4.7 V}$$

To find the value of V_S, first determine R_T.

$$R_T = 1.0 \text{ k}\Omega + 3.3 \text{ k}\Omega + 4.7 \text{ k}\Omega = 9 \text{ k}\Omega$$

The source voltage V_S is equal to the current times the total resistance.

$$V_S = IR_T = (1 \text{ mA})(9 \text{ k}\Omega) = \textbf{9 V}$$

Notice that if you add the voltage drops of the resistors, they total 9 V, which is the same as the source voltage.

V_S can be increased to a value where I = 5 mA. Calculate the maximum value of V_S as follows:

$$V_{S(max)} = IR_T = (5\text{ mA})(9\text{ k}\Omega) = \mathbf{45\ V}$$

Related Problem Repeat the calculations for V_1, V_2, V_3, V_S, and $V_{S(max)}$ if $R_3 = 2.2\text{ k}\Omega$ and I is maintained at 1 mA.

Use Multisim file E05-09 to verify the calculated results in this example and to confirm your calculations for the related problem.

EXAMPLE 5–10

Some resistors are not color-coded with bands but have the values stamped on the resistor body. When the circuit board shown in Figure 5–20 was assembled, the resistors were erroneously mounted with the labels turned down, and there is no documentation showing the resistor values. Without removing the resistors from the board, use Ohm's law to determine the resistance of each one. Assume that a multimeter and a power supply are available but the ohmmeter function of the multimeter does not work.

▶ **FIGURE 5–20**

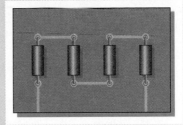

Solution The resistors are all in series, so the current is the same through each one. Measure the current by connecting a 12 V source (arbitrary value) and an ammeter as shown in Figure 5–21. Measure the voltage across each resistor. Start with the voltmeter across the first resistor, and then repeat this measurement for the other three resistors. For illustration, the voltage values indicated on the board are assumed to be the measured values.

▶ **FIGURE 5–21**

The voltmeter readings across each resistor are indicated.

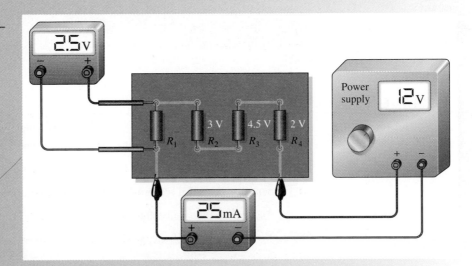

Determine the resistance of each resistor by substituting the measured values of current and voltage into the Ohm's law formula.

$$R_1 = \frac{V_1}{I} = \frac{2.5 \text{ V}}{25 \text{ mA}} = 100 \text{ } \Omega$$

$$R_2 = \frac{V_2}{I} = \frac{3 \text{ V}}{25 \text{ mA}} = 120 \text{ } \Omega$$

$$R_3 = \frac{V_3}{I} = \frac{4.5 \text{ V}}{25 \text{ mA}} = 180 \text{ } \Omega$$

$$R_4 = \frac{V_4}{I} = \frac{2 \text{ V}}{25 \text{ mA}} = 80 \text{ } \Omega$$

Notice that the largest-value resistor has the largest voltage drop across it.

Related Problem   What is an easier way to determine the resistance values?

**SECTION 5–4
REVIEW**

1. A 10 V battery is connected across three 100 Ω resistors in series. What is the current through each resistor?

2. How much voltage is required to produce 50 mA through the circuit of Figure 5–22?

   FIGURE 5–22

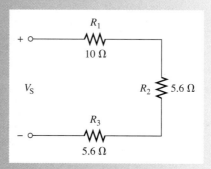

3. How much voltage is dropped across each resistor in Figure 5–22 when the current is 50 mA?

4. There are four equal-value resistors connected in series with a 5 V source. A current of 4.63 mA is measured. What is the value of each resistor?

5–5   VOLTAGE SOURCES IN SERIES

Recall that a voltage source is an energy source that provides a constant voltage to a load. Batteries and electronic power supplies are practical examples of dc voltage sources.

After completing this section, you should be able to

◆ **Determine the total effect of voltage sources connected in series**

   ◆ Determine the total voltage of series sources with the same polarities

   ◆ Determine the total voltage of series sources with opposite polarities

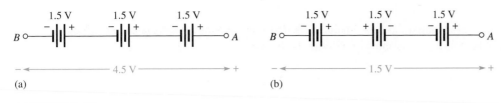

(a)   (b)

▲ FIGURE 5–23

Voltage sources in series add algebraically.

When two or more voltage sources are in series, the total voltage is equal to the algebraic sum of the individual source voltages. The algebraic sum means that the polarities of the sources must be included when the sources are combined in series. Sources with opposite polarities have voltages with opposite signs.

$$V_{S(tot)} = V_{S1} + V_{S2} + \cdots + V_{Sn}$$

When the voltage sources are all in the same direction in terms of their polarities, as in Figure 5–23(a), all of the voltages have the same sign when added; there is a total of 4.5 V from terminal A to terminal B with A more positive than B.

$$V_{AB} = 1.5\text{ V} + 1.5\text{ V} + 1.5\text{ V} = +4.5\text{ V}$$

The voltage has a double subscript, AB, to indicate that it is the voltage at point A with respect to point B.

In Figure 5–23(b), the middle voltage source is opposite to the other two; so its voltage has an opposite sign when added to the others. For this case the total voltage from A to B is

$$V_{AB} = +1.5\text{ V} - 1.5\text{ V} + 1.5\text{ V} = +1.5\text{ V}$$

Terminal A is 1.5 V more positive than terminal B.

A familiar example of voltage sources in series is the flashlight. When you put two 1.5 V batteries in your flashlight, they are connected in series, giving a total of 3 V. When connecting batteries or other voltage sources in series to increase the total voltage, always connect from the positive (+) terminal of one to the negative (−) terminal of another. Such a connection is illustrated in Figure 5–24.

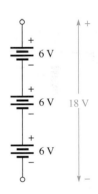

▲ FIGURE 5–24

Connection of three 6 V batteries to obtain 18 V.

EXAMPLE 5–11

What is the total source voltage ($V_{S(tot)}$) in Figure 5–25?

▶ FIGURE 5–25

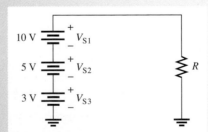

Solution The polarity of each source is the same (the sources are connected in the same direction in the circuit). Add the three voltages to get the total.

$$V_{S(tot)} = V_{S1} + V_{S2} + V_{S3} = 10\text{ V} + 5\text{ V} + 3\text{ V} = \textbf{18 V}$$

The three individual sources can be replaced by a single equivalent source of 18 V with its polarity as shown in Figure 5–26.

▶ FIGURE 5–26

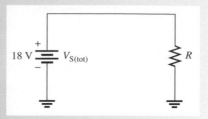

Related Problem   If V_{S3} in Figure 5–25 is reversed, what is the total source voltage?

    Use Multisim file E05-11 to verify the calculated results in this example and to confirm your calculation for the related problem.

EXAMPLE 5–12   Determine $V_{S(tot)}$ in Figure 5–27.

▶ FIGURE 5–27

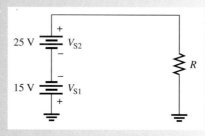

Solution   These sources are connected in opposing directions. If you go clockwise around the circuit, you go from plus to minus through V_{S1}, and minus to plus through V_{S2}. The total voltage is the difference of the two source voltages (algebraic sum of oppositely signed values). The total voltage has the same polarity as the larger-value source. Here we will choose V_{S2} to be positive.

$$V_{S(tot)} = V_{S2} - V_{S1} = 25\ V - 15\ V = \mathbf{10\ V}$$

The two sources in Figure 5–27 can be replaced by a single 10 V equivalent source with polarity as shown in Figure 5–28.

▶ FIGURE 5–28

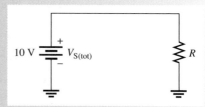

Related Problem   If an 8 V source in the direction of V_{S1} is added in series in Figure 5–27, what is $V_{S(tot)}$?

    Use Multisim file E05-12 to verify the calculated results in this example and to confirm your calculation for the related problem.

SECTION 5–5
REVIEW

1. Four 1.5 V flashlight batteries are connected in series plus to minus. What is the total voltage of all four cells?
2. How many 12 V batteries must be connected in series to produce 60 V? Draw a schematic that shows the battery connections.
3. The resistive circuit in Figure 5–29 is used to bias a transistor amplifier. Show how to connect two 15 V power supplies in order to get 30 V across the two resistors.
4. Determine the total source voltage in each circuit of Figure 5–30.
5. Draw the equivalent single-source circuit for each circuit of Figure 5–30.

▶ FIGURE 5–29

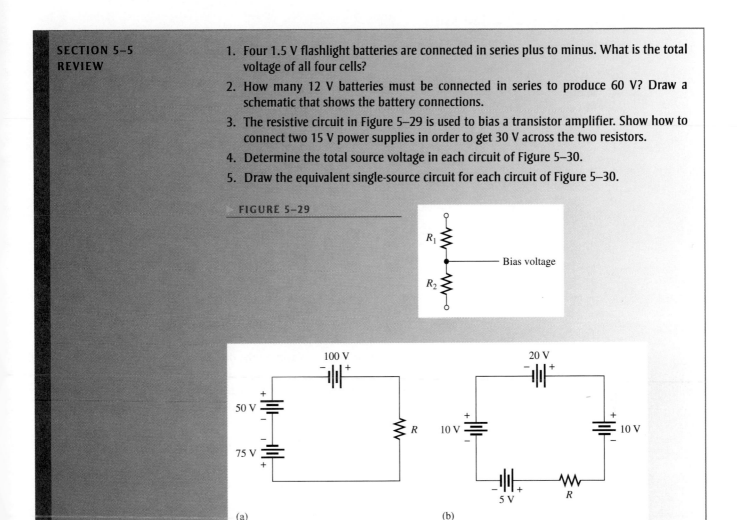

(a)   (b)

FIGURE 5–30

5–6   KIRCHHOFF'S VOLTAGE LAW

Kirchhoff's voltage law is a fundamental circuit law that states that the algebraic sum of all the voltages around a single closed path is zero or, in other words, the sum of the voltage drops equals the total source voltage.

After completing this section, you should be able to

◆ **Apply Kirchhoff's voltage law**

  ◆ State Kirchhoff's voltage law

  ◆ Determine the source voltage by adding the voltage drops

  ◆ Determine an unknown voltage drop

In an electric circuit, the voltages across the resistors (voltage drops) *always* have polarities opposite to the source voltage polarity. For example, in Figure 5–31, follow a clockwise loop around the circuit. Note that the source polarity is minus-to-plus and each

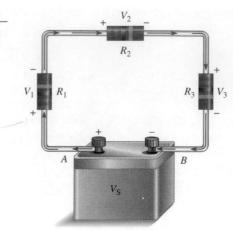

voltage drop is plus-to-minus. The voltage drops across resistors are designated as V_1, V_2, and so on.

In Figure 5–31, the current is out of the positive side of the source and through the resistors as the arrows indicate. The current is into the positive side of each resistor and out the negative side. The drop in energy level across a resistor creates a potential difference, or voltage drop, with a plus-to-minus polarity in the direction of the current.

The voltage from point A to point B in the circuit of Figure 5–31 is the source voltage, V_S. Also, the voltage from A to B is the sum of the series resistor voltage drops. Therefore, the source voltage is equal to the sum of the three voltage drops, as stated by **Kirchhoff's voltage law**.

> **The sum of all the voltage drops around a single closed path in a circuit is equal to the total source voltage in that loop.**

The general concept of Kirchhoff's voltage law is illustrated in Figure 5–32 and expressed by Equation 5–3.

Equation 5–3

$$V_S = V_1 + V_2 + V_3 + \cdots + V_n$$

where the subscript n represents the number of voltage drops.

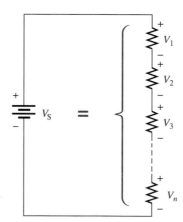

If all the voltage drops around a closed path are added and then this total is subtracted from the source voltage, the result is zero. This result occurs because the sum of the voltage drops always equals the source voltage.

> **The algebraic sum of all the voltages (both source and drops) around a single closed path is zero.**

Therefore, another way of expressing Kirchhoff's voltage law in equation form is

$$V_S - V_1 - V_2 - V_3 - \cdots - V_n = 0$$

Equation 5–4

You can verify Kirchhoff's voltage law by connecting a circuit and measuring each resistor voltage and the source voltage as illustrated in Figure 5–33. When the resistor voltages are added together, their sum will equal the source voltage. Any number of resistors can be added.

The three examples that follow use Kirchhoff's voltage law to solve circuit problems.

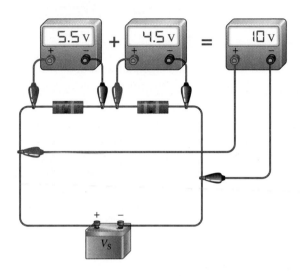

▲ FIGURE 5–33

Illustration of an experimental verification of Kirchhoff's voltage law.

EXAMPLE 5–13

Determine the source voltage V_S in Figure 5–34 where the two voltage drops are given. There is no voltage drop across the fuse.

▶ FIGURE 5–34

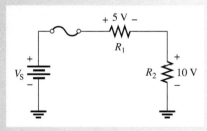

Solution   By Kirchhoff's voltage law, (Eq. 5–3), the source voltage (applied voltage) must equal the sum of the voltage drops. Adding the voltage drops gives the value of the source voltage.

$$V_S = 5\,\text{V} + 10\,\text{V} = \mathbf{15\,V}$$

Related Problem   If V_S is increased to 30 V, determine the two voltage drops. What is the voltage across each component (including the fuse) if the fuse is blown?

EXAMPLE 5–14

Determine the unknown voltage drop, V_3, in Figure 5–35.

▶ **FIGURE 5–35**

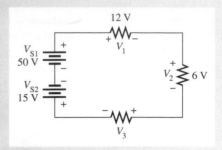

Solution By Kirchhoff's voltage law (Eq. 5–4), the algebraic sum of all the voltages around the circuit is zero. The value of each voltage drop except V_3 is known. Substitute these values into the equation.

$$-V_{S2} + V_{S1} - V_1 - V_2 - V_3 = 0$$
$$-15\,\text{V} + 50\,\text{V} - 12\,\text{V} - 6\,\text{V} - V_3 = 0\,\text{V}$$

Next, combine the known values, transpose 17 V to the right side of the equation, and cancel the minus signs.

$$17\,\text{V} - V_3 = 0\,\text{V}$$
$$-V_3 = -17\,\text{V}$$
$$V_3 = \mathbf{17\,V}$$

The voltage drop across R_3 is 17 V, and its polarity is as shown in Figure 5–35.

Related Problem Determine V_3 if the polarity of V_{S2} is reversed in Figure 5–35.

EXAMPLE 5–15

Find the value of R_4 in Figure 5–36.

▶ **FIGURE 5–36**

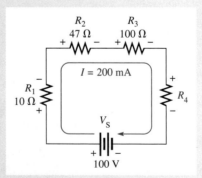

Solution In this problem you will use both Ohm's law and Kirchhoff's voltage law.

First, use Ohm's law to find the voltage drop across each of the known resistors.

$$V_1 = IR_1 = (200\,\text{mA})(10\,\Omega) = 2.0\,\text{V}$$
$$V_2 = IR_2 = (200\,\text{mA})(47\,\Omega) = 9.4\,\text{V}$$
$$V_3 = IR_3 = (200\,\text{mA})(100\,\Omega) = 20\,\text{V}$$

Next, use Kirchhoff's voltage law to find V_4, the voltage drop across the unknown resistor.

$$V_S - V_1 - V_2 - V_3 - V_4 = 0\text{ V}$$
$$100\text{ V} - 2.0\text{ V} - 9.4\text{ V} - 20\text{ V} - V_4 = 0\text{ V}$$
$$68.6\text{ V} - V_4 = 0\text{ V}$$
$$V_4 = 68.6\text{ V}$$

Now that you know V_4, use Ohm's law to calculate R_4.

$$R_4 = \frac{V_4}{I} = \frac{68.6\text{ V}}{200\text{ mA}} = \mathbf{343\ \Omega}$$

This is most likely a 330 Ω resistor because 343 Ω is within a standard tolerance range ($+5\%$) of 330 Ω.

Related Problem   Determine R_4 in Figure 5–36 for $V_S = 150$ V and $I = 200$ mA.

Use Multisim file E05-15 to verify the calculated results in the example and to confirm your calculation for the related problem.

**SECTION 5–6
REVIEW**

1. State Kirchhoff's voltage law in two ways.
2. A 50 V source is connected to a series resistive circuit. What is the sum of the voltage drops in this circuit?
3. Two equal-value resistors are connected in series across a 10 V battery. What is the voltage drop across each resistor?
4. In a series circuit with a 25 V source, there are three resistors. One voltage drop is 5 V, and the other is 10 V. What is the value of the third voltage drop?
5. The individual voltage drops in a series string are as follows: 1 V, 3 V, 5 V, 8 V, and 7 V. What is the total voltage applied across the series string?

5–7   VOLTAGE DIVIDERS

A series circuit acts as a voltage divider. The voltage divider is an important application of series circuits.

After completing this section, you should be able to

- **Use a series circuit as a voltage divider**
  - Apply the voltage-divider formula
  - Use a potentiometer as an adjustable voltage divider
  - Describe some voltage-divider applications

A circuit consisting of a series string of resistors connected to a voltage source acts as a **voltage divider**. Figure 5–37 shows a circuit with two resistors in series, although there can be any number. There are two voltage drops across the resistors: one across R_1 and one

▶ FIGURE 5–37

Two-resistor voltage divider.

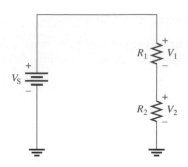

across R_2. These voltage drops are V_1 and V_2, respectively, as indicated in the schematic. Since each resistor has the same current, the voltage drops are proportional to the resistance values. For example, if the value of R_2 is twice that of R_1, then the value of V_2 is twice that of V_1.

The total voltage drop around a single closed path divides among the series resistors in amounts directly proportional to the resistance values. For example, in Figure 5–37, if V_S is 10 V, R_1 is 50 Ω, and R_2 is 100 Ω, then V_1 is one-third the total voltage, or 3.33 V, because R_1 is one-third the total resistance of 150 Ω. Likewise, V_2 is two-thirds V_S, or 6.67 V.

Voltage-Divider Formula

With a few calculations, you can develop a formula for determining how the voltages divide among series resistors. Assume a circuit with n resistors in series as shown in Figure 5–38, where n can be any number.

▶ FIGURE 5–38

Generalized voltage divider with n resistors.

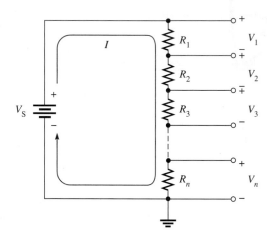

Let V_x represent the voltage drop across any one of the resistors and R_x represent the number of a particular resistor or combination of resistors. By Ohm's law, you can express the voltage drop across R_x as follows:

$$V_x = IR_x$$

The current through the circuit is equal to the source voltage divided by the total resistance $(I = V_S/R_T)$. In the circuit of Figure 5–38, the total resistance is $R_1 + R_2 + R_3 + \cdots + R_n$. By substitution of V_S/R_T for I in the expression for V_x,

$$V_x = \left(\frac{V_S}{R_T}\right)R_x$$

Rearranging the terms you get

Equation 5–5

$$V_x = \left(\frac{R_x}{R_T}\right)V_S$$

Equation 5–5 is the general voltage-divider formula, which can be stated as follows:

The voltage drop across any resistor or combination of resistors in a series circuit is equal to the ratio of that resistance value to the total resistance, multiplied by the source voltage.

EXAMPLE 5–16

Determine V_1 (the voltage across R_1) and V_2 (the voltage across R_2) in the voltage divider in Figure 5–39.

▶ **FIGURE 5–39**

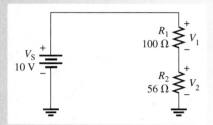

Solution To determine V_1, use the voltage-divider formula, $V_x = (R_x/R_T)V_S$, where $x = 1$. The total resistance is

$$R_T = R_1 + R_2 = 100 \ \Omega + 56 \ \Omega = 156 \ \Omega$$

R_1 is 100 Ω and V_S is 10 V. Substitute these values into the voltage-divider formula.

$$V_1 = \left(\frac{R_1}{R_T}\right)V_S = \left(\frac{100 \ \Omega}{156 \ \Omega}\right)10 \ V = \textbf{6.41 V}$$

There are two ways to find the value of V_2: Kirchhoff's voltage law or the voltage-divider formula. If you use Kirchhoff's voltage law ($V_S = V_1 + V_2$), substitute the values for V_S and V_1 as follows:

$$V_2 = V_S - V_1 = 10 \ V - 6.41 \ V = \textbf{3.59 V}$$

To determine V_2, use the voltage-divider formula where $x = 2$.

$$V_2 = \left(\frac{R_2}{R_T}\right)V_S = \left(\frac{56 \ \Omega}{156 \ \Omega}\right)10 \ V = \textbf{3.59 V}$$

Related Problem Find the voltages across R_1 and R_2 in Figure 5–39 if R_2 is changed to 180 Ω.

Use Multisim file E05-16 to verify the calculated results in this example and to confirm your calculations for the related problem.

EXAMPLE 5–17

Calculate the voltage drop across each resistor in the voltage divider of Figure 5–40.

▶ **FIGURE 5–40**

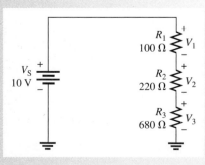

Solution Look at the circuit for a moment and consider the following: The total resistance is 1000 Ω. Ten percent of the total voltage is across R_1 because it is 10% of the total resistance (100 Ω is 10% of 1000 Ω). Likewise, 22% of the total voltage is dropped across R_2 because it is 22% of the total resistance (220 Ω is 22% of 1000 Ω). Finally, R_3 drops 68% of the total voltage because 680 Ω is 68% of 1000 Ω.

Because of the convenient values in this problem, it is easy to figure the voltages mentally. ($V_1 = 0.10 \times 10$ V = 1 V, $V_2 = 0.22 \times 10$ V = 2.2 V, and $V_3 = 0.68 \times 10$ V = 6.8 V). Such is not always the case, but sometimes a little thinking will produce a result more efficiently and eliminate some calculating. This is also a good way to estimate what your results should be so that you will recognize an unreasonable answer as a result of a calculation error.

Although you have already reasoned through this problem, the calculations will verify your results.

$$V_1 = \left(\frac{R_1}{R_T}\right)V_S = \left(\frac{100\ \Omega}{1000\ \Omega}\right)10\ \text{V} = \mathbf{1\ V}$$

$$V_2 = \left(\frac{R_2}{R_T}\right)V_S = \left(\frac{220\ \Omega}{1000\ \Omega}\right)10\ \text{V} = \mathbf{2.2\ V}$$

$$V_3 = \left(\frac{R_3}{R_T}\right)V_S = \left(\frac{680\ \Omega}{1000\ \Omega}\right)10\ \text{V} = \mathbf{6.8\ V}$$

Notice that the sum of the voltage drops is equal to the source voltage, in accordance with Kirchhoff's voltage law. This check is a good way to verify your results.

Related Problem If R_1 and R_2 in Figure 5–40 are changed to 680 Ω, what are the voltage drops?

 Use Multisim file E05-17 to verify the calculated results in this example and to confirm your calculations for the related problem.

EXAMPLE 5–18

Determine the voltages between the following points in the voltage divider of Figure 5–41:

(a) *A* to *B*   **(b)** *A* to *C*   **(c)** *B* to *C*   **(d)** *B* to *D*   **(e)** *C* to *D*

▶ **FIGURE 5–41**

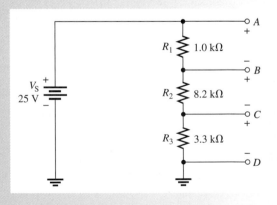

Solution First, determine R_T.

$$R_T = R_1 + R_2 + R_3 = 1.0\ \text{k}\Omega + 8.2\ \text{k}\Omega + 3.3\ \text{k}\Omega = 12.5\ \text{k}\Omega$$

Next, apply the voltage-divider formula to obtain each required voltage.

(a) The voltage A to B is the voltage drop across R_1.

$$V_{AB} = \left(\frac{R_1}{R_T}\right)V_S = \left(\frac{1.0\,k\Omega}{12.5\,k\Omega}\right)25\,V = \mathbf{2\,V}$$

(b) The voltage from A to C is the combined voltage drop across both R_1 and R_2. In this case, R_x in the general formula given in Equation 5–5 is $R_1 + R_2$.

$$V_{AC} = \left(\frac{R_1 + R_2}{R_T}\right)V_S = \left(\frac{9.2\,k\Omega}{12.5\,k\Omega}\right)25\,V = \mathbf{18.4\,V}$$

(c) The voltage from B to C is the voltage drop across R_2.

$$V_{BC} = \left(\frac{R_2}{R_T}\right)V_S = \left(\frac{8.2\,k\Omega}{12.5\,k\Omega}\right)25\,V = \mathbf{16.4\,V}$$

(d) The voltage from B to D is the combined voltage drop across both R_2 and R_3. In this case, R_x in the general formula is $R_2 + R_3$.

$$V_{BD} = \left(\frac{R_2 + R_3}{R_T}\right)V_S = \left(\frac{11.5\,k\Omega}{12.5\,k\Omega}\right)25\,V = \mathbf{23\,V}$$

(e) Finally, the voltage from C to D is the voltage drop across R_3.

$$V_{CD} = \left(\frac{R_3}{R_T}\right)V_S = \left(\frac{3.3\,k\Omega}{12.5\,k\Omega}\right)25\,V = \mathbf{6.6\,V}$$

If you connect this voltage divider, you can verify each of the calculated voltages by connecting a voltmeter between the appropriate points in each case.

Related Problem Determine each of the previously calculated voltages if V_S is doubled.

 Use Multisim file E05-18 to verify the calculated results in this example and to confirm your calculations for the related problem.

A Potentiometer as an Adjustable Voltage Divider

Recall from Chapter 2 that a potentiometer is a variable resistor with three terminals. A potentiometer connected to a voltage source is shown in Figure 5–42(a) with the schematic shown in part (b). Notice that the two end terminals are labeled 1 and 2. The adjustable terminal or wiper is labeled 3. The potentiometer functions as a voltage divider, which can be

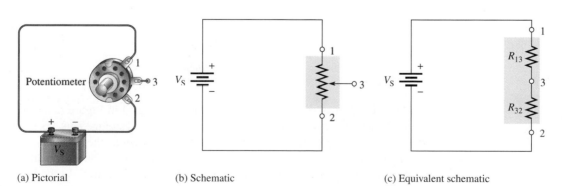

(a) Pictorial   (b) Schematic   (c) Equivalent schematic

▲ **FIGURE 5–42**

The potentiometer used as a voltage divider.

illustrated by separating the total resistance into two parts, as shown in Figure 5–42(c). The resistance between terminal 1 and terminal 3 (R_{13}) is one part, and the resistance between terminal 3 and terminal 2 (R_{32}) is the other part. So this potentiometer is equivalent to a two-resistor voltage divider that can be manually adjusted.

Figure 5–43 shows what happens when the wiper contact (3) is moved. In part (a), the wiper is exactly centered, making the two resistances equal. If you measure the voltage across terminals 3 to 2 as indicated by the voltmeter symbol, you have one-half of the total source voltage. When the wiper is moved up, as in part (b), the resistance between terminals 3 and 2 increases, and the voltage across it increases proportionally. When the wiper is moved down, as in part (c), the resistance between terminals 3 and 2 decreases, and the voltage decreases proportionally.

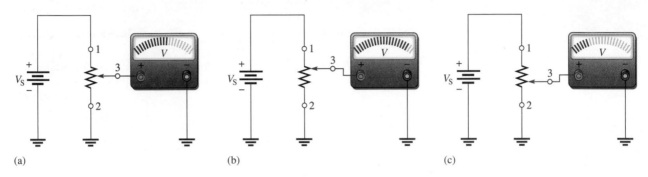

(a)   (b)   (c)

▲ **FIGURE 5–43**

Adjusting the voltage divider.

Applications

The volume control of radio or TV receivers is a common application of a potentiometer used as a voltage divider. Since the loudness of the sound is dependent on the amount of voltage associated with the audio signal, you can increase or decrease the volume by adjusting the potentiometer, that is, by turning the knob of the volume control on the set. The block diagram in Figure 5–44 shows how a potentiometer can be used for volume control in a typical receiver.

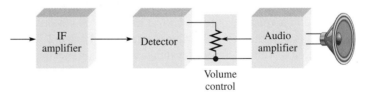

▲ **FIGURE 5–44**

A variable voltage divider used for volume control in a radio receiver.

Another application of a voltage divider is illustrated in Figure 5–45, which depicts a potentiometer voltage divider as a level sensor in a liquid storage tank. As shown in part (a), the float moves up as the tank is filled and moves down as the tank empties. The float is mechanically linked to the wiper arm of a potentiometer, as shown in part (b). The output voltage varies proportionally with the position of the wiper arm. As the liquid in the tank decreases, the sensor output voltage also decreases. The output voltage goes to the indicator circuitry, which controls a digital readout to show the amount of liquid in the tank. The schematic of this system is shown in part (c).

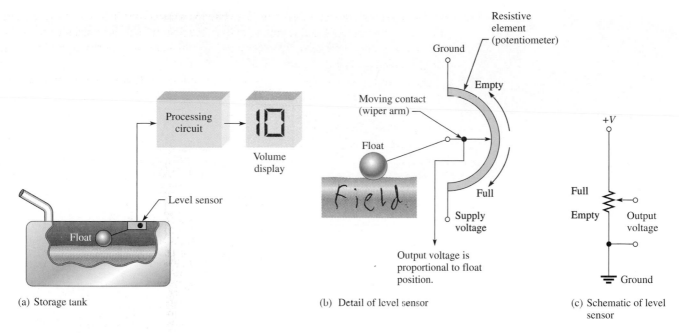

Ground

Resistive
element
(potentiometer)

Empty

Moving contact
(wiper arm)

Float

Field

Full

Supply
voltage

Output voltage is
proportional to float
position.

+V

Full

Empty   Output
voltage

Ground

Processing
circuit

Volume
display

Level sensor

Float

(a) Storage tank

(b) Detail of level sensor

(c) Schematic of level
sensor

▲ **FIGURE 5–45**

A potentiometer voltage divider used in a level sensor.

Still another application for voltage dividers is in setting the dc operating voltage (bias) in transistor amplifiers. Figure 5–46 shows a voltage divider used for this purpose. You will study transistor amplifiers and biasing in a later course, so it is important that you understand the basics of voltage dividers at this point.

These examples are only three out of many possible applications of voltage dividers.

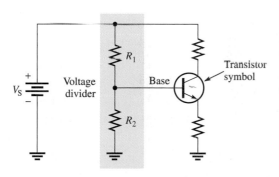

V_S Voltage
divider

R_1

Base

R_2

Transistor
symbol

◀ **FIGURE 5–46**

The voltage divider used as a bias circuit for a transistor amplifier, where the voltage at the base of the transistor is determined by the voltage divider as $V_{base} = (R_2/(R_1 + R_2))V_S$.

SECTION 5–7
REVIEW

1. What is a voltage divider?
2. How many resistors can there be in a series voltage-divider circuit?
3. Write the general formula for voltage dividers.
4. If two series resistors of equal value are connected across a 10 V source, how much voltage is there across each resistor?
5. A 47 kΩ resistor and an 82 kΩ resistor are connected as a voltage divider. The source voltage is 100 V. Draw the circuit, and determine the voltage across each of the resistors.

6. The circuit of Figure 5–47 is an adjustable voltage divider. If the potentiometer is linear, where would you set the wiper in order to get 5 V from A to B and 5 V from B to C?

▶ FIGURE 5–47

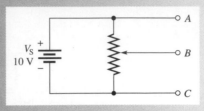

5–8 POWER IN SERIES CIRCUITS

The power dissipated by each individual resistor in a series circuit contributes to the total power in the circuit. The individual powers are additive.

After completing this section, you should be able to

- ◆ **Determine power in a series circuit**
  - ◆ Apply one of the power formulas

The total amount of power in a series resistive circuit is equal to the sum of the powers in each resistor in series.

Equation 5–6

$$P_T = P_1 + P_2 + P_3 + \cdots + P_n$$

where P_T is the total power and P_n is the power in the last resistor in series.

The power formulas that you learned in Chapter 4 are applicable to series circuits. Since there is the same current through each resistor in series, the following formulas are used to calculate the total power:

$$P_T = V_S I$$
$$P_T = I^2 R_T$$
$$P_T = \frac{V_S^2}{R_T}$$

where I is the current through the circuit, V_S is the total source voltage across the series connection, and R_T is the total resistance.

EXAMPLE 5–19

Determine the total amount of power in the series circuit in Figure 5–48.

▶ FIGURE 5–48

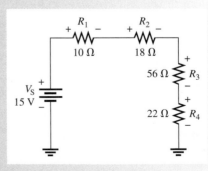

Solution The source voltage is 15 V. The total resistance is

$$R_T = R_1 + R_2 + R_3 + R_4 = 10\ \Omega + 18\ \Omega + 56\ \Omega + 22\ \Omega = 106\ \Omega$$

The easiest formula to use is $P_T = V_S^2/R_T$ since you know both V_S and R_T.

$$P_T = \frac{V_S^2}{R_T} = \frac{(15\ \text{V})^2}{106\ \Omega} = \frac{225\ \text{V}^2}{106\ \Omega} = 2.12\ \text{W}$$

If you determine the power in each resistor separately and all of these powers are added, you obtain the same result. First, find the current.

$$I = \frac{V_S}{R_T} = \frac{15\ \text{V}}{106\ \Omega} = 142\ \text{mA}$$

Next, calculate the power for each resistor using $P = I^2R$.

$$P_1 = I^2R_1 = (142\ \text{mA})^2(10\ \Omega) = 200\ \text{mW}$$
$$P_2 = I^2R_2 = (142\ \text{mA})^2(18\ \Omega) = 360\ \text{mW}$$
$$P_3 = I^2R_3 = (142\ \text{mA})^2(56\ \Omega) = 1.12\ \text{W}$$
$$P_4 = I^2R_4 = (142\ \text{mA})^2(22\ \Omega) = 441\ \text{mW}$$

Then, add these powers to get the total power.

$$P_T = P_1 + P_2 + P_3 + P_4 = 200\ \text{mW} + 360\ \text{mW} + 1.12\ \text{W} + 441\ \text{mW} = \mathbf{2.12\ W}$$

This result compares to the total power as determined previously by the formula $P_T = V_S^2/R_T$.

Related Problem What is the power in the circuit of Figure 5–48 if V_S is increased to 30 V?

The amount of power in a resistor is important because the power rating of the resistors must be high enough to handle the expected power in the circuit. The following example illustrates practical considerations relating to power in a series circuit.

EXAMPLE 5–20

Determine if the indicated power rating (½ W) of each resistor in Figure 5–49 is sufficient to handle the actual power. If a rating is not adequate, specify the required minimum rating.

▶ **FIGURE 5–49**

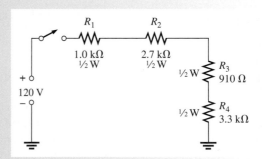

Solution First, determine the total resistance.

$$R_T = R_1 + R_2 + R_3 + R_4 = 1.0\ \text{k}\Omega + 2.7\ \text{k}\Omega + 910\ \Omega + 3.3\ \text{k}\Omega = 7.91\ \text{k}\Omega$$

Next, calculate the current.

$$I = \frac{V_S}{R_T} = \frac{120\,V}{7.91\,k\Omega} = 15\,mA$$

Then calculate the power in each resistor.

$$P_1 = I^2 R_1 = (15\,mA)^2 (1.0\,k\Omega) = \textbf{225 mW}$$
$$P_2 = I^2 R_2 = (15\,mA)^2 (2.7\,k\Omega) = \textbf{608 mW}$$
$$P_3 = I^2 R_3 = (15\,mA)^2 (910\,\Omega) = \textbf{205 mW}$$
$$P_4 = I^2 R_4 = (15\,mA)^2 (3.3\,k\Omega) = \textbf{743 mW}$$

R_2 and R_4 do not have a rating sufficient to handle the actual power, which exceeds ½ W in each of these two resistors, and they may burn out if the switch is closed. These resistors should be replaced by 1 W resistors.

Related Problem   Determine the minimum power rating required for each resistor in Figure 5–49 if the source voltage is increased to 240 V.

**SECTION 5–8
REVIEW**

1. If you know the power in each resistor in a series circuit, how can you find the total power?
2. The resistors in a series circuit dissipate the following powers: 2 W, 5 W, 1 W, and 8 W. What is the total power in the circuit?
3. A circuit has a 100 Ω, a 330 Ω, and a 680 Ω resistor in series. There is a current of 1 A through the circuit. What is the total power?

5–9   VOLTAGE MEASUREMENTS

Voltage is relative. That is, the voltage at one point in a circuit is always measured relative to another point. For example, if we say that there are +100 V at a certain point in a circuit, we mean that the point is 100 V more positive than some designated reference point in the circuit. This reference point is called the ground or common point.

After completing this section, you should be able to

♦ **Measure voltage with respect to ground**

  ♦ Determine and identify ground in a circuit

  ♦ Define the term *reference ground*

The concept of *ground* was introduced in Chapter 2. In most electronic equipment, a large conductive area on a printed circuit board or the metal housing is used as the **reference ground** or **common**, as illustrated in Figure 5–50.

Reference ground has a potential of zero volts (0 V) with respect to all other points in the circuit that are referenced to it, as illustrated in Figure 5–51. In part (a), the negative side of the source is grounded, and all voltages indicated are positive with respect to ground. In part (b), the positive side of the source is ground. The voltages at all other points are therefore negative with respect to ground. Recall that all points shown grounded in a circuit are connected together through ground and are effectively the same point electrically.

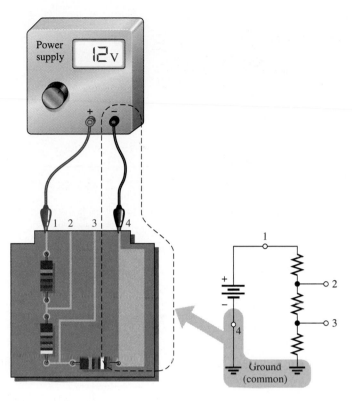

▲ FIGURE 5–50

Simple illustration of ground in a circuit.

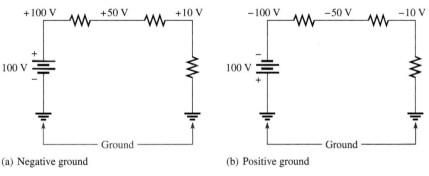

(a) Negative ground   (b) Positive ground

▲ FIGURE 5–51

Example of negative and positive grounds.

Measuring Voltages with Respect to Ground

When you measure voltages with respect to the reference ground in a circuit, connect one meter lead to the reference ground, and the other to the point at which the voltage is to be measured. In a negative ground circuit, as illustrated in Figure 5–52, the negative meter terminal is connected to the reference ground. The positive terminal of the voltmeter is then connected to the positive voltage point. The meter reads the positive voltage at point *A* with respect to ground.

For a circuit with a positive ground, the positive voltmeter lead is connected to reference ground, and the negative lead is connected to the negative voltage point, as indicated in Figure 5–53. Here the meter reads the negative voltage at point *A* with respect to ground.

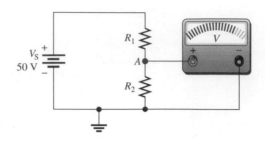

▲ FIGURE 5–52

Measuring a voltage with respect to negative ground.

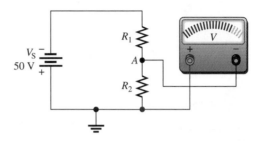

▲ FIGURE 5–53

Measuring a voltage with respect to positive ground.

When you must measure voltages at several points in a circuit, you can clip the ground lead to ground at one point in the circuit and leave it there. Then move the other lead from point to point as you measure the voltages. This method is illustrated pictorially in Figure 5–54 and in equivalent schematic form in Figure 5–55.

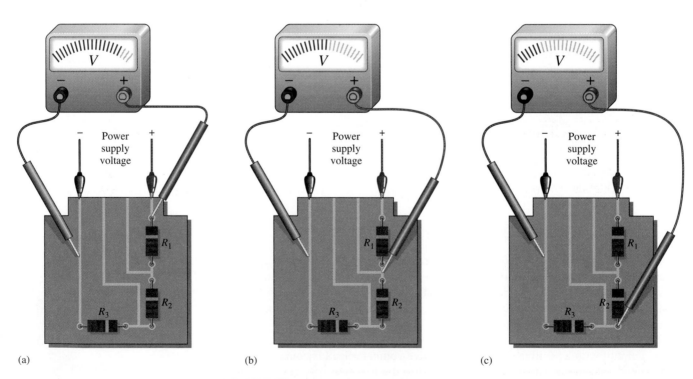

(a)   (b)   (c)

▲ FIGURE 5–54

Measuring voltages at several points in a circuit with respect to ground.

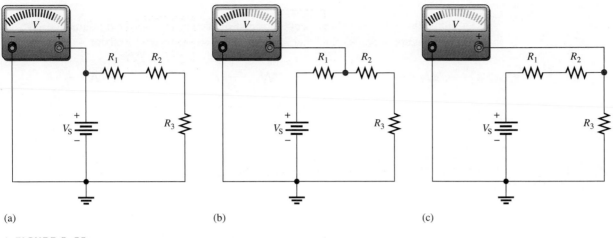

(a)　　　　　(b)　　　　　(c)

▲ FIGURE 5–55

Equivalent schematics for Figure 5–54.

Measuring Voltage Across an Ungrounded Resistor

Voltage can normally be measured across a resistor, as shown in Figure 5 56, even though neither side of the resistor is connected to ground. If the measuring instrument is not isolated from power line ground, the negative lead of the meter will ground one side of the resistor and alter the operation of the circuit. In this situation, another method must be used, as illustrated in Figure 5–57. The voltages on each side of the resistor are measured with respect to ground. The difference of these two measurements is the voltage drop across the resistor.

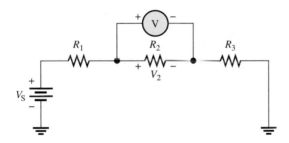

▲ FIGURE 5–56

Measuring voltage across a resistor.

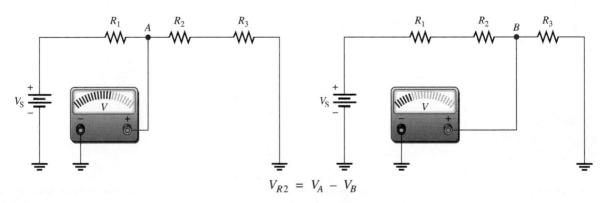

$$V_{R2} = V_A - V_B$$

▲ FIGURE 5–57

Measuring voltage across R_2 with two separate measurements to ground. Note $V_A - V_B$ can be expressed as V_{AB}, where the second subscript letter, B, is the reference.

EXAMPLE 5–21

Determine the voltages with respect to ground of each of the indicated points in each circuit of Figure 5–58. Assume that 25 V are dropped across each resistor.

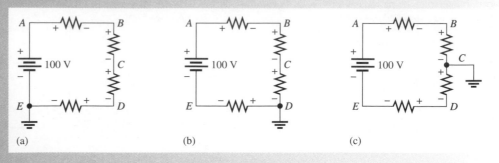

▲ FIGURE 5–58

Solution

In circuit (a), the voltage polarities are as shown. Point E is ground. Single-letter subscripts denote voltage at a point with respect to ground. The voltages with respect to ground are as follows:

$$V_E = \mathbf{0\,V}, \quad V_D = \mathbf{+25\,V}, \quad V_C = \mathbf{+50\,V}, \quad V_B = \mathbf{+75\,V}, \quad V_A = \mathbf{+100\,V}$$

In circuit (b), the voltage polarities are as shown. Point D is ground. The voltages with respect to ground are as follows:

$$V_E = \mathbf{-25\,V}, \quad V_D = \mathbf{0\,V}, \quad V_C = \mathbf{+25\,V}, \quad V_B = \mathbf{+50\,V}, \quad V_A = \mathbf{+75\,V}$$

In circuit (c), the voltage polarities are as shown. Point C is ground. The voltages with respect to ground are as follows:

$$V_E = \mathbf{-50\,V}, \quad V_D = \mathbf{-25\,V}, \quad V_C = \mathbf{0\,V}, \quad V_B = \mathbf{+25\,V}, \quad V_A = \mathbf{+50\,V}$$

Related Problem

If the ground is at point A in the circuit in Figure 5–58, what are the voltages at each of the points with respect to ground?

Use Multisim file E05-21 to verify the calculated results in this example and to confirm your calculation for the related problem.

SECTION 5–9 REVIEW

1. What is the reference point in a circuit called?
2. Voltages in a circuit are generally referenced with respect to ground. (T or F)
3. The housing or chassis is often used as reference ground. (T or F)

5–10 TROUBLESHOOTING

Open resistors or contacts and one point shorted to another are common problems in all circuits including series circuits.

After completing this section, you should be able to

◆ **Troubleshoot series circuits**

   ◆ Check for an open circuit

   ◆ Check for a short circuit

   ◆ Identify primary causes of opens and shorts

Open Circuit

The most common failure in a series circuit is an **open**. For example, when a resistor or a lamp burns out, it causes a break in the current path and creates an open circuit, as illustrated in Figure 5–59.

An open in a series circuit prevents current.

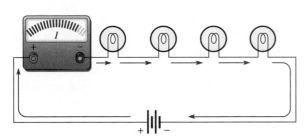

(a) A complete series circuit has current.

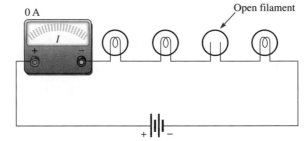

(b) An open series circuit has no current.

▲ **FIGURE 5–59**

Current ceases when an open occurs.

Troubleshooting an Open In Chapter 3, you were introduced to the analysis, planning, and measurement (APM) approach to troubleshooting. You also learned about the half-splitting method and saw an example using an ohmmeter. Now, the same principles will be applied using voltage measurements instead of resistance measurements. As you know, voltage measurements are generally the easiest to make because you do not have to disconnect anything.

As a beginning step, prior to analysis, it is a good idea to make a visual check of the faulty circuit. Occasionally, you can find a charred resistor, a broken lamp filament, a loose wire, or a loose connection this way. However, it is possible (and probably more common) for a resistor or other component to open without showing visible signs of damage. When a visual check reveals nothing, then proceed with the APM approach.

When an open occurs in a series circuit, all of the source voltage appears across the open. The reason for this is that the open condition prevents current through the series circuit. With no current, there can be no voltage drop across any of the other resistors (or other component). Since $IR = (0\,A)R = 0\,V$, the voltage on each end of a good resistor is the same. Therefore, the voltage applied across a series string also appears across the open component because there are no other voltage drops in the circuit, as illustrated in Figure 5–60. The source voltage will appear across the open resistor in accordance with Kirchhoff's voltage law as follows:

$$V_S = V_1 + V_2 + V_3 + V_4 + V_5 + V_6$$
$$V_4 = V_S - V_1 - V_2 - V_3 - V_5 - V_6$$
$$= 10\,V - 0\,V - 0\,V - 0\,V - 0\,V - 0\,V$$
$$V_4 = V_S = 10\,V$$

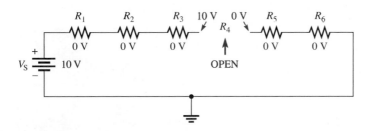

▲ **FIGURE 5–60**

The source voltage appears across the open series resistor.

Example of Half-Splitting Using Voltage Measurements   Suppose a circuit has four resistors in series. You have determined by *analyzing* the symptoms (there is voltage but no current) that one of the resistors is open, and you are *planning* to find the open resistor using a voltmeter for *measuring* by the half-splitting method. A sequence of measurements for this particular case is illustrated in Figure 5–61.

Step 1.   Measure across R_1 and R_2 (the left half of the circuit). A 0 V reading indicates that neither of these resistors is open.

Step 2.   Move meter to measure across R_3 and R_4; the reading is 10 V. This indicates there is an open in the right half of the circuit, so either R_3 or R_4 is the faulty resistor (assume no bad connections).

Step 3.   Move the meter to measure across R_3. A measurement of 10 V across R_3 identifies it as the open resistor. If you had measured across R_4, it would have indicated 0 V. This would have also identified R_3 as the faulty component because it would have been the only one left that could have 10 V across it.

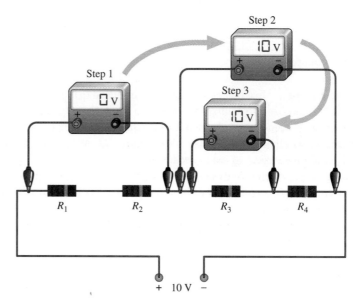

▲ **FIGURE 5–61**

Troubleshooting a series circuit for an open using half-splitting.

Short Circuit

Sometimes an unwanted short circuit occurs when two conductors touch or a foreign object such as solder or a wire clipping accidentally connects two sections of a circuit together. This situation is particularly common in circuits with a high component density. Several potential causes of short circuits are illustrated on the PC board in Figure 5–62.

▶ **FIGURE 5–62**

Examples of shorts on a PC board.

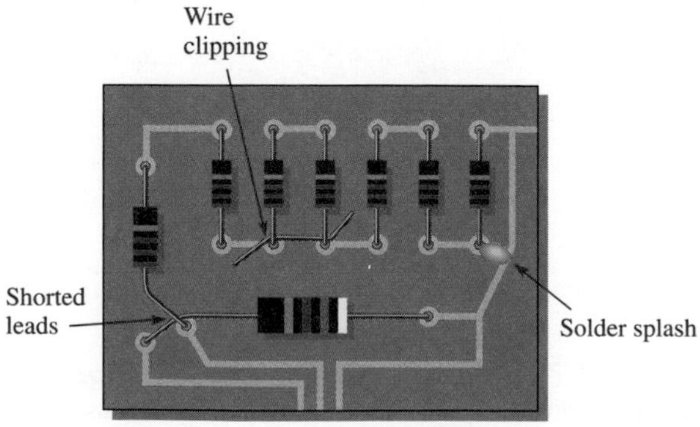

When there is a **short**, a portion of the series resistance is bypassed (all of the current goes through the short), thus reducing the total resistance as illustrated in Figure 5–63. Notice that the current increases as a result of the short.

A short in a series circuit causes more current than normal.

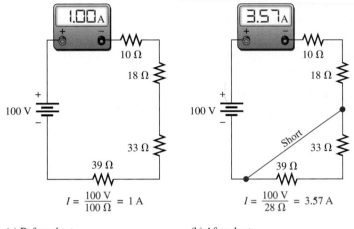

(a) Before short

$$I = \frac{100 \text{ V}}{100 \text{ }\Omega} = 1 \text{ A}$$

(b) After short

$$I = \frac{100 \text{ V}}{28 \text{ }\Omega} = 3.57 \text{ A}$$

◀ **FIGURE 5–63**

Example of the effect of a short in a series circuit.

Troubleshooting a Short A short is very difficult to troubleshoot. As in any troubleshooting situation, it is a good idea to make a visual check of the faulty circuit. In the case of a short in the circuit, a wire clipping, solder splash, or touching leads is often found to be the culprit. In terms of component failure, shorts are less common than opens in many types of components. Furthermore, a short in one part of a circuit can cause overheating in another part due to the higher current caused by the short. As a result two failures, an open and a short, may occur together.

When a short occurs in a series circuit, there is essentially no voltage across the shorted *part.* A short has zero or near zero resistance, although shorts with significant resistance values can occur from time to time. These are called *resistive shorts.* For purposes of illustration, zero resistance is assumed for all shorts.

In order to troubleshoot a short, measure the voltage across each resistor until you get a reading of 0 V. This is the straightforward approach and does not use half-splitting. In order to apply the half-splitting method, you must know the correct voltage at each point in the circuit and compare it to measured voltages. Example 5–22 illustrates using half-splitting to find a short.

EXAMPLE 5–22

Assume you have determined that there is a short in a circuit with four series resistors because the current is higher than it should be. You know that the voltage at each point in the circuit should be as shown in Figure 5–64 if the circuit is working properly. The voltages are shown relative to the negative terminal of the source. Find the location of the short.

FIGURE 5–64

Series circuit (without a short) with correct voltages indicated.

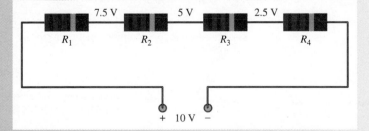

Solution   Use the half-splitting method to troubleshoot the short.

Step 1: Measure across R_1 and R_2. The meter shows a reading of 6.67 V, which is higher than the normal voltage (it should be 5 V). Look for a voltage that is lower than normal because a short will make the voltage less across that part of the circuit.

Step 2: Move the meter and measure across R_3 and R_4; the reading of 3.33 V is incorrect and lower than normal (it should be 5 V). This shows that the short is in the right half of the circuit and that either R_3 or R_4 is shorted.

Step 3: Again move the meter and measure across R_3. A reading of 3.3 V across R_3 tells you that R_4 is shorted because it must have 0 V across it. Figure 5–65 illustrates this troubleshooting technique.

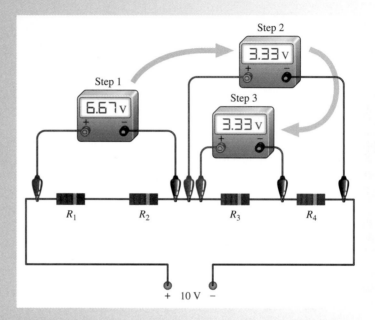

▲ **FIGURE 5–65**

Troubleshooting a series circuit for a short using half-splitting.

Related Problem   Assume that R_1 is shorted in Figure 5–65. What would the Step 1 measurement be?

SECTION 5–10 REVIEW

1. Define *short*.

2. Define *open*.

3. What happens when a series circuit opens?

4. Name two general ways in which an open circuit can occur in practice. What may cause a short circuit to occur?

5. When a resistor fails, it will normally fail open.   (T or F)

6. The total voltage across a string of series resistors is 24 V. If one of the resistors is open, how much voltage is there across it? How much is there across each of the good resistors?

A Circuit Application

For this application, you have a voltage-divider circuit board to evaluate and modify if necessary. You will use it to obtain five different voltage levels from a 12 V battery that has a 6.5 Ah rating. The voltage divider provides positive reference voltages to an electronic circuit in an analog-to-digital converter. Your job is to check the circuit to see if it provides the following voltages within a tolerance of ±5% with respect to the negative side of the battery: 10.4 V, 8.0 V, 7.3 V, 6.0 V, and 2.7 V. If the existing circuit does not provide the specified voltages, you will modify it so that it does. Also, you must make sure that the power ratings of the resistors are adequate for the application and determine how long the battery will last with the voltage divider connected to it.

The Schematic of the Circuit

◆ Use Figure 5–66 to determine the resistor values and draw the schematic of the voltage-divider circuit. All the resistors on the board are ¼ W.

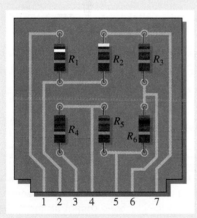

▲ **FIGURE 5–66**

The Voltages

◆ Determine the voltage at each pin on the circuit board with respect to the negative side of the battery when the positive side of the 12 V battery is connected to pin 3 and the negative side is connected to pin 1. Compare the existing voltages to the following specifications:

Pin 1: negative terminal of 12 V battery

Pin 2: 2.7 V ± 5%

Pin 3: positive terminal of 12 V battery

Pin 4: 10.4 V ± 5%

Pin 5: 8.0 V ± 5%

Pin 6: 7.3 V ± 5%

Pin 7: 6.0 V ± 5%

◆ If the output voltages of the existing circuit are not the same as those stated in the specifications, make the necessary changes in the circuit to meet the specifications. Draw a schematic of the modified circuit showing resistor values and adequate power ratings.

The Battery

◆ Find the total current drawn from the 12 V battery when the voltage-divider circuit is connected and determine how many days the 6.5 Ah battery will last.

A Test Procedure

◆ Determine how you would test the voltage-divider circuit board and what instruments you would use. Then detail your test procedure in a step-by-step format.

Troubleshooting

◆ Determine the most likely fault for each of the following cases. Voltages are with respect to the negative battery terminal (pin 1 on the circuit board).

1. No voltage at any of the pins on the circuit board

2. 12 V at pins 3 and 4. All other pins have 0 V.

3. 12 V at all pins except 0 V at pin 1

4. 12 V at pin 6 and 0 V at pin 7

5. 3.3 V at pin 2

Review

1. What is the total power dissipated by the voltage-divider circuit in Figure 5–66 with a 12 V battery?

2. What are the output voltages from the voltage divider if the positive terminal of a 6 V battery is connected to pin 3 and the negative terminal to pin 1?

3. When the voltage-divider board is connected to the electronic circuit to which it is providing positive reference voltages, which pin on the board should be connected to the ground of the electronic circuit?

SUMMARY

- The current is the same at all points in a series circuit.
- The total series resistance is the sum of all resistors in the series circuit.
- The total resistance between any two points in a series circuit is equal to the sum of all resistors connected in series between those two points.
- If all of the resistors in a series circuit are of equal value, the total resistance is the number of resistors multiplied by the resistance value of one resistor.
- Voltage sources in series add algebraically.
- Kirchhoff's voltage law: The sum of all the voltage drops around a single closed path in a circuit is equal to the total source voltage in that loop.
- Kirchhoff's voltage law: The algebraic sum of all the voltages (both source and drops) around a single closed path is zero.
- The voltage drops in a circuit are always opposite in polarity to the total source voltage.
- Conventional current is defined to be out of the positive side of a source and into the negative side.
- Conventional current is defined to be into the positive side of each resistor and out of the more negative (less positive) side.
- A voltage drop results from a decrease in energy level across a resistor.
- A voltage divider is a series arrangement of resistors connected to a voltage source.
- A voltage divider is so named because the voltage drop across any resistor in the series circuit is divided down from the total voltage by an amount proportional to that resistance value in relation to the total resistance.
- A potentiometer can be used as an adjustable voltage divider.
- The total power in a resistive circuit is the sum of all the individual powers of the resistors making up the series circuit.
- Ground (common) is zero volts with respect to all points referenced to it in the circuit.
- *Negative ground* is the term used when the negative side of the source is grounded.
- *Positive ground* is the term used when the positive side of the source is grounded.
- The voltage across an open component always equals the source voltage.
- The voltage across a shorted component is always 0 V.

KEY TERMS

These key terms are also in the end-of-book glossary.

Kirchhoff's voltage law   A law stating that (1) the sum of the voltage drops around a single closed path equals the source voltage in that loop or (2) the algebraic sum of all the voltages (drops and source) around a single closed path is zero.

Open   A circuit condition in which the current path is broken.

Reference ground   A method of grounding whereby the metal chassis that houses the assembly or a large conductive area on a printed circuit board is used as the common or reference point.

Series   In an electric circuit, a relationship of components in which the components are connected such that they provide a single current path between two points.

Short   A circuit condition in which there is a zero or abnormally low resistance path between two points; usually an inadvertent condition.

Voltage divider   A circuit consisting of series resistors across which one or more output voltages are taken.

FORMULAS

5–1	$R_T = R_1 + R_2 + R_3 + \cdots + R_n$	Total resistance of n resistors in series
5–2	$R_T = nR$	Total resistance of n equal-value resistors in series
5–3	$V_S = V_1 + V_2 + V_3 + \cdots + V_n$	Kirchhoff's voltage law

5–4 $V_S - V_1 - V_2 - V_3 - \cdots - V_n = 0$ Kirchhoff's voltage law stated another way

5–5 $V_x = \left(\dfrac{R_x}{R_T}\right)V_S$ Voltage-divider formula

5–6 $P_T = P_1 + P_2 + P_3 + \cdots + P_n$ Total power

SELF-TEST

Answers are at the end of the chapter.

1. Five equal-value resistors are connected in series and there is a current of 2 mA into the first resistor. The amount of current out of the second resistor is

   (a) equal to 2 mA   (b) less than 2 mA   (c) greater than 2 mA

2. To measure the current out of the third resistor in a circuit consisting of four series resistors, an ammeter can be placed

   (a) between the third and fourth resistors   (b) between the second and third resistors

   (c) at the positive terminal of the source   (d) at any point in the circuit

3. When a third resistor is connected in series with two series resistors, the total resistance

   (a) remains the same   (b) increases

   (c) decreases   (d) increases by one-third

4. When one of four series resistors is removed from a circuit and the circuit reconnected, the current

   (a) decreases by the amount of current through the removed resistor

   (b) decreases by one-fourth

   (c) quadruples

   (d) increases

5. A series circuit consists of three resistors with values of 100 Ω, 220 Ω, and 330 Ω. The total resistance is

   (a) less than 100 Ω   (b) the average of the values   (c) 550 Ω   (d) 650 Ω

6. A 9 V battery is connected across a series combination of 68 Ω, 33 Ω, 100 Ω, and 47 Ω resistors. The amount of current is

   (a) 36.3 mA   (b) 27.6 A   (c) 22.3 mA   (d) 363 mA

7. While putting four 1.5 V batteries in a flashlight, you accidentally put one of them in backward. The voltage across the bulb will be

   (a) 6 V   (b) 3 V   (c) 4.5 V   (d) 0 V

8. If you measure all the voltage drops and the source voltage in a series circuit and add them together, taking into consideration the polarities, you will get a result equal to

   (a) the source voltage   (b) the total of the voltage drops

   (c) zero   (d) the total of the source voltage and the voltage drops

9. There are six resistors in a given series circuit and each resistor has 5 V dropped across it. The source voltage is

   (a) 5 V   (b) 30 V

   (c) dependent on the resistor values   (d) dependent on the current

10. A series circuit consists of a 4.7 kΩ, a 5.6 kΩ, and a 10 kΩ resistor. The resistor that has the most voltage across it is

    (a) the 4.7 kΩ   (b) the 5.6 kΩ

    (c) the 10 kΩ   (d) impossible to determine from the given information

11. Which of the following series combinations dissipates the most power when connected across a 100 V source?

    (a) One 100 Ω resistor   (b) Two 100 Ω resistors

    (c) Three 100 Ω resistors   (d) Four 100 Ω resistors

12. The total power in a certain circuit is 1 W. Each of the five equal-value series resistors making up the circuit dissipates

    (a) 1 W   (b) 5 W   (c) 0.5 W   (d) 0.2 W

13. When you connect an ammeter in a series-resistive circuit and turn on the source voltage, the meter reads zero. You should check for

(a) a broken wire   (b) a shorted resistor

(c) an open resistor   (d) answers (a) and (c)

14. While checking out a series-resistive circuit, you find that the current is higher than it should be. You should look for

(a) an open circuit   (b) a short   (c) a low resistor value   (d) answers (b) and (c)

CIRCUIT DYNAMICS QUIZ

Answers are at the end of the chapter.

Refer to Figure 5–70.

1. If the current shown by one of the milliammeters increases, the current shown by the other two

(a) increases   (b) decreases   (c) stays the same

2. If the source voltage decreases, the current indicated by each milliammeter

(a) increases   (b) decreases   (c) stays the same

3. If the current through R_1 increases as a result of R_1 being replaced by a different resistor, the current indicated by each milliammeter

(a) increases   (b) decreases   (c) stays the same

Refer to Figure 5–73.

4. With a 10 V voltage source connected between points A and B, when the switches are thrown from position 1 to position 2, the total current from the source

(a) increases   (b) decreases   (c) stays the same

5. For the conditions described in Question 4, the current through R_3

(a) increases   (b) decreases   (c) stays the same

6. When the switches are in position 1 and a short develops across R_3, the current through R_2

(a) increases   (b) decreases   (c) stays the same

7. When the switches are in position 2 and a short develops across R_3, the current through R_5

(a) increases   (b) decreases   (c) stays the same

Refer to Figure 5–77.

8. If the switch is thrown from position A to position B, the ammeter reading

(a) increases   (b) decreases   (c) stays the same

9. If the switch is thrown from position B to position C, the voltage across R_4

(a) increases   (b) decreases   (c) stays the same

10. If the switch is thrown from position C to position D, the current through R_3

(a) increases   (b) decreases   (c) stays the same

Refer to Figure 5–84(b).

11. If R_1 is changed to 1.2 kΩ, the voltage from A to B

(a) increases   (b) decreases   (c) stays the same

12. If R_2 and R_3 are interchanged, the voltage from A to B

(a) increases   (b) decreases   (c) stays the same

13. If the source voltage increases from 8 V to 10 V, the voltage from A to B

(a) increases   (b) decreases   (c) stays the same

Refer to Figure 5–91.

14. If the 9 V source is reduced to 5 V, the current in the circuit

(a) increases   (b) decreases   (c) stays the same

15. If the 9 V source is reversed, the voltage at point B with respect to ground

(a) increases   (b) decreases   (c) stays the same

PROBLEMS

More difficult problems are indicated by an asterisk (*).
Answers to odd-numbered problems are at the end of the book.

SECTION 5–1 Resistors in Series

1. Connect each set of resistors in Figure 5–67 in series between points A and B.

◀ FIGURE 5–67

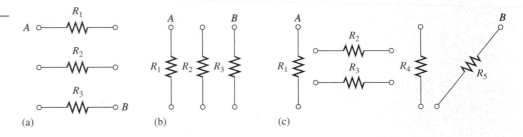

(a)   (b)   (c)

2. Determine which resistors in Figure 5–68 are in series. Show how to interconnect the pins to put all the resistors in series.

◀ FIGURE 5–68

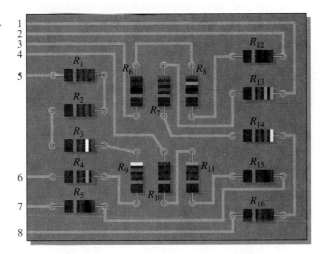

3. Determine the nominal resistance between pins 1 and 8 in the circuit board in Figure 5–68.

4. Determine the nominal resistance between pins 2 and 3 in the circuit board in Figure 5–68.

5. On the double-sided PC board in Figure 5–69, identify each group of series resistors. Note that many of the interconnections feed through the board from the top side to the bottom side.

◀ FIGURE 5–69

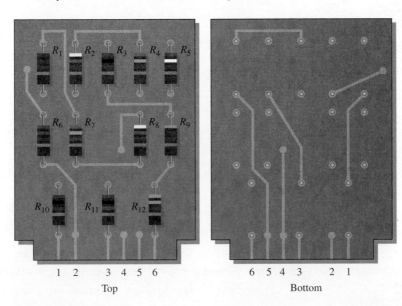

1 2   3 4 5 6   6 5 4 3   2 1
       Top   Bottom

SECTION 5–2 Current in a Series Circuit

6. What is the current through each resistor in a series circuit if the total voltage is 12 V and the total resistance is 120 Ω?

7. The current from the source in Figure 5–70 is 5 mA. How much current does each milliammeter in the circuit indicate?

▶ **FIGURE 5–70**

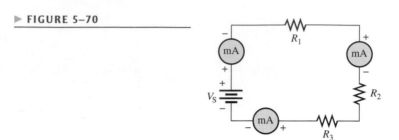

8. Show how to connect a voltage source and an ammeter to the PC board in Figure 5–68 to measure the current in R_1. Which other resistor currents are measured by this setup?

*9. Using 1.5 V batteries, a switch, and three lamps, devise a circuit to apply 4.5 V across either one lamp, two lamps in series, or three lamps in series with a single-control switch. Draw the schematic.

SECTION 5–3 Total Series Resistance

10. The following resistors (one each) are connected in a series circuit: 1.0 Ω, 2.2 Ω, 5.6 Ω, 12 Ω, and 22 Ω. Determine the total resistance.

11. Find the total resistance of each of the following groups of series resistors:

    (a) 560 Ω and 1000 Ω   **(b)** 47 Ω and 56 Ω

    (c) 1.5 kΩ, 2.2 kΩ, and 10 kΩ   **(d)** 1.0 MΩ, 470 kΩ, 1.0 kΩ, 2.2 MΩ

12. Calculate R_T for each circuit of Figure 5–71.

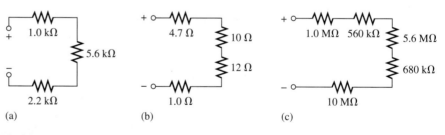

▲ **FIGURE 5–71**

13. What is the total resistance of twelve 5.6 kΩ resistors in series?

14. Six 56 Ω resistors, eight 100 Ω resistors, and two 22 Ω resistors are all connected in series. What is the total resistance?

15. If the total resistance in Figure 5–72 is 17.4 kΩ, what is the value of R_5?

▶ **FIGURE 5–72**

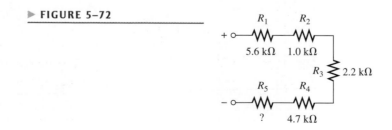

*16. You have the following resistor values available to you in the lab in unlimited quantities: $10\,\Omega$, $100\,\Omega$, $470\,\Omega$, $560\,\Omega$, $680\,\Omega$, $1.0\,k\Omega$, $2.2\,k\Omega$, and $5.6\,k\Omega$. All of the other standard values are out of stock. A project that you are working on requires an $18\,k\Omega$ resistance. What combinations of the available values would you use in series to achieve this total resistance?

17. Find the total resistance in Figure 5–71 if all three circuits are connected in series.

18. What is the total resistance from A to B for each switch position in Figure 5–73?

▶ FIGURE 5–73

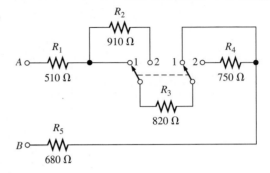

SECTION 5–4   Application of Ohm's Law

19. What is the current in each circuit of Figure 5–74?

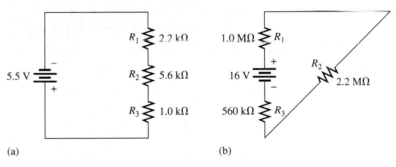

(a)   (b)

▲ FIGURE 5–74

20. Determine the voltage drop across each resistor in Figure 5–74.

21. Three $470\,\Omega$ resistors are connected in series with a 48 V source.

   (a) What is the current is in the circuit?

   (b) What is the voltage across each resistor?

   (c) What is the minimum power rating of the resistors?

22. Four equal-value resistors are in series with a 5 V battery, and 2.23 mA are measured. What is the value of each resistor?

23. What is the value of each resistor in Figure 5–75?

24. Determine V_{R1}, R_2, and R_3 in Figure 5–76.

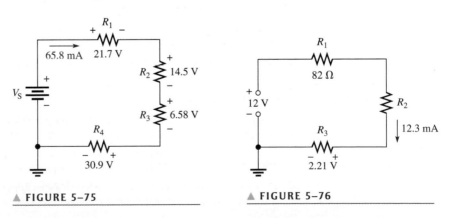

▲ FIGURE 5–75   ▲ FIGURE 5–76

25. For the circuit in Figure 5–77 the meter reads 7.84 mA when the switch is in position *A*.
    (a) What is the resistance of R_4?
    (b) What should be the meter reading for switch positions *B*, *C*, and *D*?
    (c) Will a ¼ A fuse blow in any position of the switch?

26. Determine the current measured by the meter in Figure 5–78 for each position of the ganged switch.

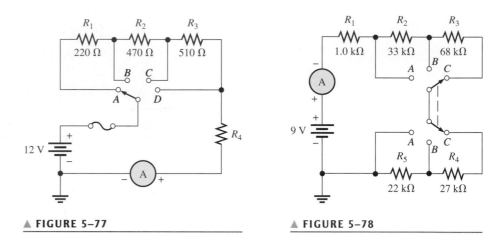

▲ FIGURE 5–77   ▲ FIGURE 5–78

SECTION 5–5 Voltage Sources in Series

27. *Series aiding* is a term sometimes used to describe voltage sources of the same polarity in series. If a 5 V and a 9 V source are connected in this manner, what is the total voltage?

28. The term *series opposing* means that sources are in series with opposite polarities. If a 12 V and a 3 V battery are series opposing, what is the total voltage?

29. Determine the total source voltage in each circuit of Figure 5–79.

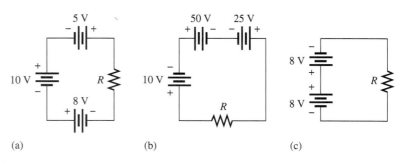

(a)   (b)   (c)

▲ FIGURE 5–79

SECTION 5–6 Kirchhoff's Voltage Law

30. The following voltage drops are measured across three resistors in series: 5.5 V, 8.2 V, and 12.3 V. What is the value of the source voltage to which these resistors are connected?

31. Five resistors are in series with a 20 V source. The voltage drops across four of the resistors are 1.5 V, 5.5 V, 3 V, and 6 V. How much voltage is dropped across the fifth resistor?

32. Determine the unspecified voltage drop(s) in each circuit of Figure 5–80. Show how to connect a voltmeter to measure each unknown voltage drop.

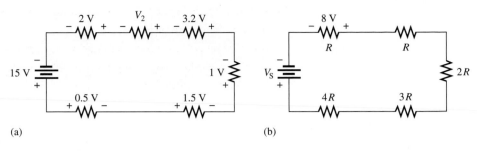

▲ FIGURE 5–80

33. In the circuit of Figure 5–81, determine the resistance of R_4.

34. Find R_1, R_2, and R_3 in Figure 5–82.

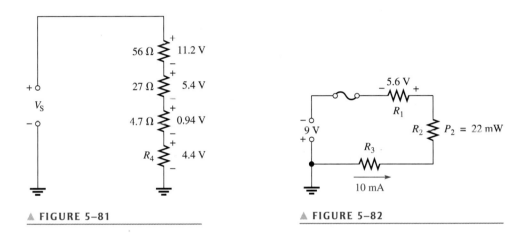

▲ FIGURE 5–81

▲ FIGURE 5–82

35. Determine the voltage across R_5 for each position of the switch in Figure 5–83. The current in each position is as follows: A, 3.35 mA; B, 3.73 mA; C, 4.50 mA; D, 6.00 mA.

36. Using the result of Problem 35, determine the voltage across each resistor in Figure 5–83 for each switch position.

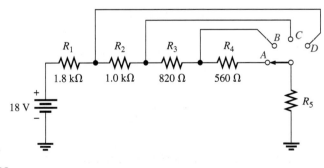

▲ FIGURE 5–83

SECTION 5–7   Voltage Dividers

*37. The total resistance of a circuit is 560 Ω. What percentage of the total voltage appears across a 27 Ω resistor that makes up part of the total series resistance?

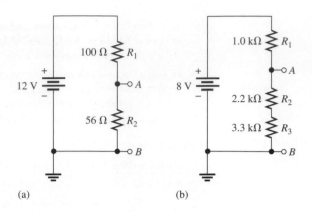

▲ FIGURE 5–84

38. Determine the voltage between points *A* and *B* in each voltage divider of Figure 5–84.

39. Determine the voltage with respect to ground for output *A*, *B*, and *C* in Figure 5–85(a).

40. Determine the minimum and maximum voltage from the voltage divider in Figure 5–85(b).

*41.** What is the voltage across each resistor in Figure 5–86? *R* is the lowest-value resistor, and all others are multiples of that value as indicated.

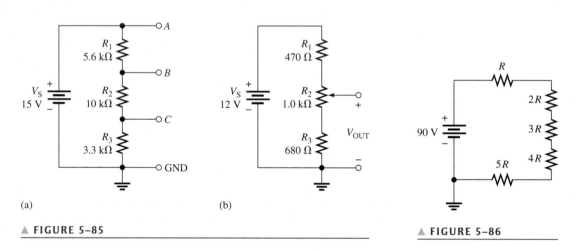

▲ FIGURE 5–85   ▲ FIGURE 5–86

42. Determine the voltage at each point in Figure 5–87 with respect to the negative side of the battery.

43. If there are 10 V across R_1 in Figure 5–88, what is the voltage across each of the other resistors?

*44.** With the table of standard resistor values given in Appendix A, design a voltage divider to provide the following approximate voltages with respect to ground using a 30 V source: 8.18 V,

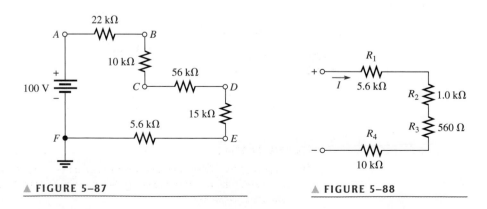

▲ FIGURE 5–87   ▲ FIGURE 5–88

14.7 V, and 24.6 V. The current drain on the source must be limited to no more than 1 mA. The number of resistors, their values, and their wattage ratings must be specified. A schematic showing the circuit arrangement and resistor placement must be provided.

*45. Design a variable voltage divider to provide an output voltage adjustable from a minimum of 10 V to a maximum of 100 V within ±1% using a 1 to 120 V source. The maximum voltage must occur at the maximum resistance setting of the potentiometer, and the minimum voltage must occur at the minimum resistance (zero) setting. The current is to be 10 mA.

SECTION 5–8   Power in Series Circuits

46. Five series resistors each handle 50 mW. What is the total power?

47. What is the total power in the circuit in Figure 5–88? Use the results of Problem 43.

48. The following ¼ W resistors are in series: 1.2 kΩ, 2.2 kΩ, 3.9 kΩ, and 5.6 kΩ. What is the maximum voltage that can be applied across the series resistors without exceeding a power rating? Which resistor will burn out first if excessive voltage is applied?

49. Find R_T in Figure 5–89.

50. A certain series circuit consists of a ⅛ W resistor, a ¼ W resistor and a ½ W resistor. The total resistance is 2400 Ω. If each of the resistors is operating in the circuit at its maximum power dissipation, determine the following:

    (a) I (b) V_T (c) The value of each resistor

▶ **FIGURE 5–89**

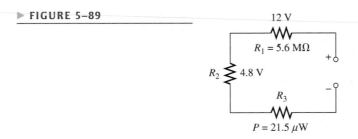

SECTION 5–9   Voltage Measurements

51. Determine the voltage at each point with respect to ground in Figure 5–90.

52. In Figure 5–91, how would you determine the voltage across R_2 by measuring, without connecting a meter directly across the resistor?

53. Determine the voltage at each point with respect to ground in Figure 5–91.

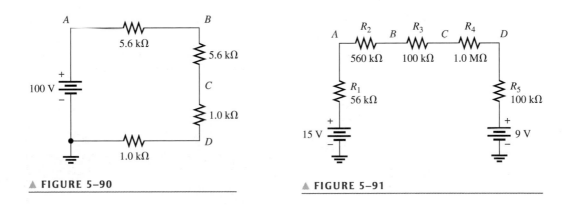

▲ FIGURE 5–90   ▲ FIGURE 5–91

SECTION 5–10   Troubleshooting

54. A string of five series resistors is connected across a 12 V battery. Zero volts is measured across all of the resistors except R_2. What is wrong with the circuit? What voltage will be measured across R_2?

▶ FIGURE 5–92

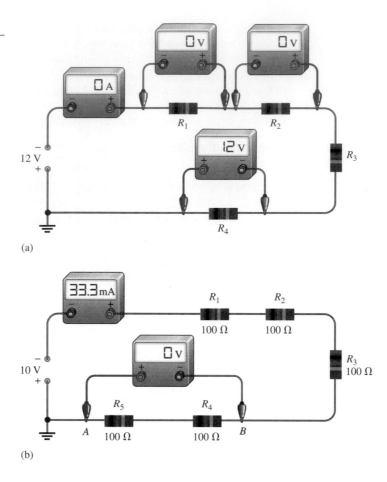

(a)

(b)

55. By observing the meters in Figure 5–92, determine the types of failures in the circuits and which components have failed.

56. What current would you measure in Figure 5–91(b) if only R_2 were shorted?

*57. Table 5–1 shows the results of resistance measurements on the PC board in Figure 5–93. Are these results correct? If not, identify the possible problems.

▶ TABLE 5–1

BETWEEN PINS	RESISTANCE
1 and 2	∞
1 and 3	∞
1 and 4	4.23 kΩ
1 and 5	∞
1 and 6	∞
2 and 3	23.6 kΩ
2 and 4	∞
2 and 5	∞
2 and 6	∞
3 and 4	∞
3 and 5	∞
3 and 6	∞
4 and 5	∞
4 and 6	∞
5 and 6	19.9 kΩ

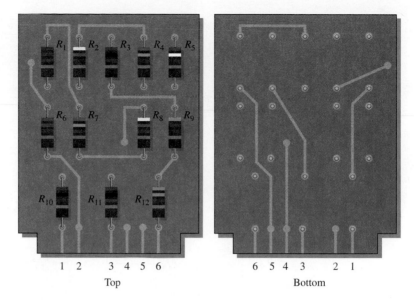

▲ FIGURE 5–93

*58. You measure 15 kΩ between pins 5 and 6 on the PC board in Figure 5–93. Does this indicate a problem? If so, identify it.

*59. In checking out the PC board in Figure 5–93, you measure 17.83 kΩ between pins 1 and 2. Also, you measure 13.6 kΩ between pins 2 and 4. Does this indicate a problem on the PC board? If so, identify the fault.

*60. The three groups of series resistors on the PC board in Figure 5–93 are connected in series with each other to form a single series circuit by connecting pin 2 to pin 4 and pin 3 to pin 5. A voltage source is connected across pins 1 and 6 and an ammeter is placed in series. As you increase the source voltage, you observe the corresponding increase in current. Suddenly, the current drops to zero and you smell smoke. All resistors are ½ W.

    (a) What has happened?

    (b) Specifically, what must you do to fix the problem?

    (c) At what voltage did the failure occur?

Multisim Troubleshooting and Analysis

These problems require your Multisim CD-ROM.

61. Open file P05-61 and measure the total series resistance.

62. Open file P05-62 and determine by measurement if there is an open resistor and, if so, which one.

63. Open file P05-63 and determine the unspecified resistance value.

64. Open file P05-64 and determine the unspecified source voltage.

65. Open file P05-65 and find the shorted resistor if there is one.

ANSWERS

SECTION REVIEWS

SECTION 5–1   Resistors in Series

1. Series resistors are connected end-to-end in a "string" with each lead of a given resistor connected to a different resistor.

2. There is a single current path in a series circuit.

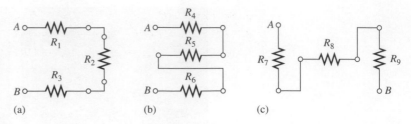

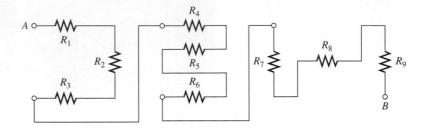

▲ FIGURE 5–94

▲ FIGURE 5–95

3. See Figure 5–94.

4. See Figure 5–95.

SECTION 5–2 Current in a Series Circuit

1. $I = 1$ A

2. The milliammeter measures 50 mA between C and D and 50 mA between E and F.

3. $I = 100$ V/56 Ω = 1.79 A; 1.79 A

4. In a series circuit, current is the same at all points.

SECTION 5–3 Total Series Resistance

1. $R_T = 1.0\ \Omega + 2.2\ \Omega + 3.3\ \Omega + 4.7\ \Omega = 11.2\ \Omega$

2. $R_T = 100\ \Omega + 2(56\ \Omega) + 4(12\ \Omega) + 330\ \Omega = 590\ \Omega$

3. $R_5 = 13.8\ k\Omega - (1.0\ k\Omega + 2.7\ k\Omega + 5.6\ k\Omega + 560\ \Omega) = 3.94\ k\Omega$

4. $R_T = 12(56\ \Omega) = 672\ \Omega$

5. $R_T = 20(5.6\ k\Omega) + 30(8.2\ k\Omega) = 358\ k\Omega$

SECTION 5–4 Application of Ohm's Law

1. $I = 10$ V/300 Ω = 33.3 mA

2. $V_S = (50\ mA)(21.2\ \Omega) = 1.06$ V

3. $V_1 = (50\ mA)(10\ \Omega) = 0.5$ V; $V_2 = (50\ mA)(5.6\ \Omega) = 0.28$ V; $V_3 = (50\ mA)(5.6\ \Omega) = 0.28$ V

4. $R = \frac{1}{4}(5\ V/4.63\ mA) = 270\ \Omega$

SECTION 5–5 Voltage Sources in Series

1. $V_T = 4(1.5\ V) = 6.0$ V

2. 60 V/12 V = 5; see Figure 5–96.

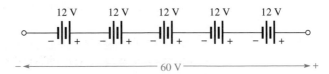

▲ FIGURE 5–96

3. See Figure 5–97.
4. (a) $V_{S(tot)} = 100\,V + 50\,V - 75\,V = 75\,V$
   (b) $V_{S(tot)} = 20\,V + 10\,V - 10\,V - 5\,V = 15\,V$
5. See Figure 5–98.

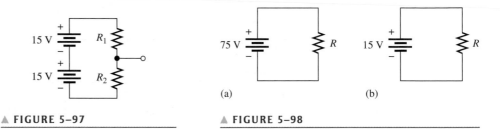

▲ FIGURE 5–97   ▲ FIGURE 5–98

SECTION 5–6 Kirchhoff's Voltage Law

1. (a) Kirchhoff's law states the algebraic sum of the voltages around a closed path is zero;
   (b) Kirchhoff's law states the sum of the voltage drops equals the total source voltage.
2. $V_T = V_S = 50\,V$
3. $V_1 - V_2 = 5\,V$
4. $V_3 = 25\,V - 10\,V - 5\,V = 10\,V$
5. $V_S = 1\,V + 3\,V + 5\,V + 8\,V + 7\,V = 24\,V$

SECTION 5–7 Voltage Dividers

1. A voltage divider is a circuit with two or more series resistors in which the voltage taken across any resistor or combination of resistors is proportional to the value of that resistance.
2. Two or more resistors form a voltage divider.
3. $V_x = (R_x/R_1)V_S$
4. $V_R = 10\,V/2 = 5\,V$
5. $V_{47} = (47\,k\Omega/129\,k\Omega)100\,V = 36.4\,V$; $V_{82} = (82\,k\Omega/129\,k\Omega)100\,V = 63.6\,V$; see Figure 5–99.
6. Set the wiper at the midpoint.

▶ FIGURE 5–99

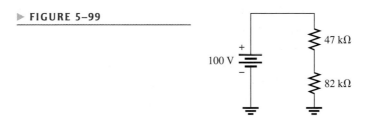

SECTION 5–8 Power in Series Circuits

1. Add the power in each resistor to get total power.
2. $P_T = 2\,W + 5\,W + 1\,W + 8\,W = 16\,W$
3. $P_T = (1\,A)^2(1110\,\Omega) = 1110\,W$

SECTION 5–9 Voltage Measurements

1. The reference point in a circuit is called ground or common.
2. True
3. True

SECTION 5–10 **Troubleshooting**

1. A short is a zero resistance path that bypasses a portion of a circuit.

2. An open is a break in the current path.

3. When a circuit opens, current ceases.

4. An open can be created by a switch or by a component failure. A short can be created by a switch or, unintentionally, by a wire clipping or solder splash.

5. True, a resistor normally fails open.

6. 24 V across the open R; 0 V across the other Rs

A Circuit Application

1. $P_T = (12\,V)^2/16.6\,k\Omega = 8.67\,mW$

2. Pin 2: 1.41 V; Pin 6: 3.65 V; Pin 5: 4.01 V; Pin 4: 5.20 V; Pin 7: 3.11 V

3. Pin 3 connects to ground.

RELATED PROBLEMS FOR EXAMPLES

5–1 (a) See Figure 5–100.

　　　(b) $R_1 = 1.0\,k\Omega$, $R_2 = 33\,k\Omega$, $R_3 = 39\,k\Omega$, $R_4 = 470\,\Omega$, $R_5 = 22\,k\Omega$

▶ FIGURE 5–100

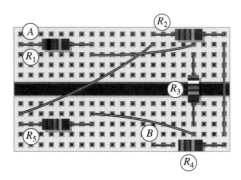

5–2 All resistors on the board are in series.

5–3 258 Ω

5–4 12.1 kΩ

5–5 6.8 kΩ

5–6 4440 Ω

5–7 114 mA

5–8 7.8 V

5–9 $V_1 = 1\,V$, $V_2 = 3.3\,V$, $V_3 = 2.2\,V$; $V_S = 6.5\,V$; $V_{S(max)} = 32.5\,V$

5–10 Use an ohmmeter.

5–11 12 V

5–12 2 V

5–13 10 V and 20 V; $V_{fuse} = V_S = 30\,V$; $V_{R1} = V_{R2} = 0\,V$

5–14 47 V

5–15 593 Ω. This is most likely a 560 Ω resistor because 593 Ω is within a standard tolerance (+10%) of 560 Ω.

5–16 $V_1 = 3.57\,V$; $V_2 = 6.43\,V$

5–17 $V_1 = V_2 = V_3 = 3.33\,V$

5–18 $V_{AB} = 4\,V$; $V_{AC} = 36.8\,V$; $V_{BC} = 32.8\,V$; $V_{BD} = 46\,V$; $V_{CD} = 13.2\,V$

5–19 8.49 W

5–20 $P_1 = 0.92 \text{ W (1 W)}; P_2 = 2.49 \text{ W (5 W)}; P_3 = 0.838 \text{ W (1 W)}; P_4 = 3.04 \text{ W (5 W)}$

5–21 $V_A = 0 \text{ V}; V_B = -25 \text{ V}; V_C = -50 \text{ V}; V_D = -75 \text{ V}; V_E = -100 \text{ V}$

5–22 3.33 V

SELF-TEST

1. (a) **2.** (d) **3.** (b) **4.** (d) **5.** (d) **6.** (a) **7.** b) **8.** (c)

9. (b) **10.** (c) **11.** (a) **12.** (d) **13.** (d) **14.** (d)

CIRCUIT DYNAMICS QUIZ

1. (a) **2.** (b) **3.** (a) **4.** (b) **5.** (b) **6.** (c)

7. (a) **8.** (a) **9.** (a) **10.** (b) **11.** (b) **12.** (c)

13. (a) **14.** (a) **15.** (b)

6

PARALLEL CIRCUITS

CHAPTER OBJECTIVES

◆ Identify a parallel resistive circuit
◆ Determine the voltage across each parallel branch
◆ Apply Kirchhoff's current law
◆ Determine total parallel resistance
◆ Apply Ohm's law in a parallel circuit
◆ Determine the total effect of current sources in parallel
◆ Use a parallel circuit as a current divider
◆ Determine power in a parallel circuit
◆ Describe some basic applications of parallel circuits
◆ Troubleshoot parallel circuits

KEY TERMS

◆ Branch
◆ Parallel
◆ Kirchhoff's current law
◆ Node
◆ Current divider

A CIRCUIT APPLICATION PREVIEW

In this application, a panel-mounted power supply will be modified by adding a milliammeter to indicate current to a load. Expansion of the meter for multiple current ranges using parallel (shunt) resistors will be demonstrated. The problem with very low-value resistors when a switch is used to select the current ranges will be introduced and the effect of switch contact resistance will be demonstrated. A way of eliminating the contact resistance problem will be presented. Finally, the ammeter circuit will be installed in the power supply. The knowledge of parallel circuits and of basic ammeters that you will acquire in this chapter plus your understanding of Ohm's law, current dividers, and the resistor color code will be put to good use.

VISIT THE COMPANION WEBSITE

Study aids for this chapter are available at
http://www.prenhall.com/floyd

INTRODUCTION

In Chapter 5, you learned about series circuits and how to apply Ohm's law and Kirchhoff's voltage law. You also saw how a series circuit can be used as a voltage divider to obtain several specified voltages from a single source voltage. The effects of opens and shorts in series circuits were also examined.

In this chapter, you will see how Ohm's law is used in parallel circuits; and you will learn Kirchhoff's current law. Also, several applications of parallel circuits, including automotive lighting, residential wiring, and the internal wiring of analog ammeters are presented. You will learn how to determine total parallel resistance and how to troubleshoot for open resistors.

When resistors are connected in parallel and a voltage is applied across the parallel circuit, each resistor provides a separate path for current. The total resistance of a parallel circuit is reduced as more resistors are connected in parallel. The voltage across each of the parallel resistors is equal to the voltage applied across the entire parallel circuit.

6–1 RESISTORS IN PARALLEL

When two or more resistors are individually connected between two separate points, they are in parallel with each other. A parallel circuit provides more than one path for current.

After completing this section, you should be able to

- **Identify a parallel resistive circuit**
  - Translate a physical arrangement of parallel resistors into a schematic

Each current path is called a **branch**, and a **parallel** circuit is one that has more than one branch. Two resistors connected in parallel are shown in Figure 6–1(a). As shown in part (b), the current out of the source (I_T) divides when it gets to point A. I_1 goes through R_1 and I_2 goes through R_2. If additional resistors are connected in parallel with the first two, more current paths are provided between point A and point B, as shown in Figure 6–1(c). All points along the top shown in blue are electrically the same as point A, and all points along the bottom shown in green are electrically the same as point B.

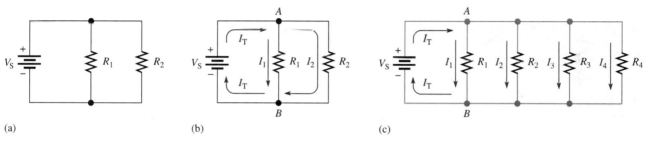

(a)　　　　　(b)　　　　　(c)

▲ FIGURE 6–1

Resistors in parallel.

In Figure 6–1, it is obvious that the resistors are connected in parallel. Often, in actual circuit diagrams, the parallel relationship is not as clear. It is important that you learn to recognize parallel circuits regardless of how they may be drawn.

A rule for identifying parallel circuits is as follows:

If there is more than one current path (branch) between two separate points and if the voltage between those two points also appears across each of the branches, then there is a parallel circuit between those two points.

Figure 6–2 shows parallel resistors drawn in different ways between two separate points labeled A and B. Notice that in each case, the current has two paths going from A to B, and

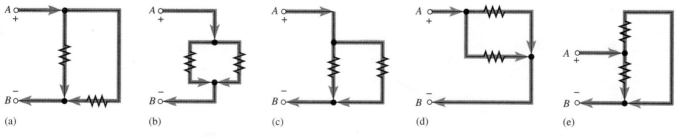

(a)　　　　　(b)　　　　　(c)　　　　　(d)　　　　　(e)

▲ FIGURE 6–2

Examples of circuits with two parallel paths.

the voltage across each branch is the same. Although these examples show only two parallel paths, there can be any number of resistors in parallel.

EXAMPLE 6–1

Five resistors are positioned on a protoboard as shown in Figure 6–3. Show the wiring required to connect all of the resistors in parallel between *A* and *B*. Draw a schematic and label each of the resistors with its value.

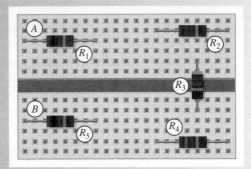

▲ **FIGURE 6–3**

Solution   Wires are connected as shown in the assembly diagram of Figure 6–4(a). The schematic is shown in Figure 6–4(b). Again, note that the schematic does not necessarily have to show the actual physical arrangement of the resistors. The schematic shows how components are connected electrically.

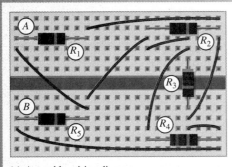

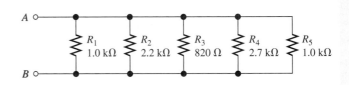

(a) Assembly wiring diagram

(b) Schematic

▲ **FIGURE 6–4**

Related Problem*   How would the circuit have to be rewired if R_2 is removed?

*Answers are at the end of the chapter.

EXAMPLE 6–2   Determine the parallel groupings in Figure 6–5 and the value of each resistor.

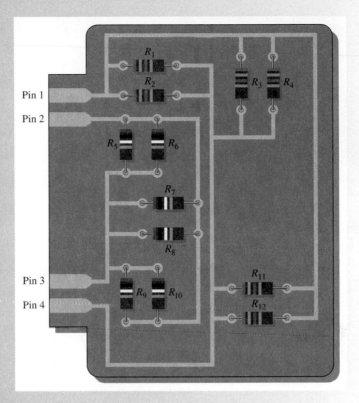

▲ FIGURE 6–5

Solution   Resistors R_1 through R_4 and R_{11} and R_{12} are all in parallel. This parallel combination is connected to pins 1 and 4. Each resistor in this group is 56 kΩ.

Resistors R_5 through R_{10} are all in parallel. This combination is connected to pins 2 and 3. Each resistor in this group is 100 kΩ.

Related Problem   How would you connect all of the resistors in Figure 6–5 in parallel?

**SECTION 6–1
REVIEW**
Answers are at the end of the chapter.

1. How are the resistors connected in a parallel circuit?
2. How do you identify a parallel circuit?
3. Complete the schematics for the circuits in each part of Figure 6–6 by connecting the resistors in parallel between points *A* and *B*.

4. Connect each group of parallel resistors in Figure 6–6 in parallel with each other.

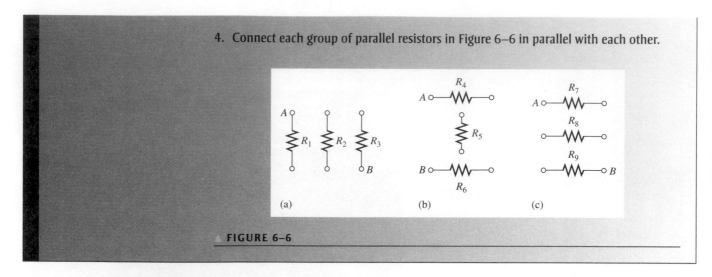

▲ FIGURE 6–6

6–2 VOLTAGE IN A PARALLEL CIRCUIT

The voltage across any given branch of a parallel circuit is equal to the voltage across each of the other branches in parallel. As you know, each current path in a parallel circuit is called a branch.

After completing this section, you should be able to

◆ **Determine the voltage across each parallel branch**

◆ Explain why the voltage is the same across all parallel resistors

To illustrate voltage in a parallel circuit, let's examine Figure 6–7(a). Points *A*, *B*, *C*, and *D* along the left side of the parallel circuit are electrically the same point because the voltage is the same along this line. You can think of all of these points as being connected by a single wire to the negative terminal of the battery. The points *E*, *F*, *G*, and *H* along the right side of the circuit are all at a voltage equal to that of the positive terminal of the source. Thus, voltage across each parallel resistor is the same, and each is equal to the source voltage. Note that the parallel circuit in Figure 6–7 resembles a ladder.

Figure 6–7(b) is the same circuit as in part (a), drawn in a slightly different way. Here the left side of each resistor is connected to a single point, which is the negative battery terminal. The right side of each resistor is connected to a single point, which is the positive battery terminal. The resistors are still all in parallel across the source.

▶ FIGURE 6–7

Voltage across parallel branches is the same.

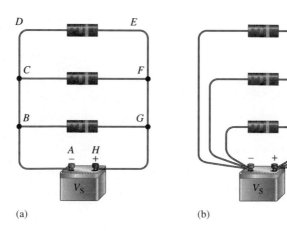

(a)   (b)

In Figure 6–8, a 12 V battery is connected across three parallel resistors. When the voltage is measured across the battery and then across each of the resistors, the readings are the same. As you can see, the same voltage appears across each branch in a parallel circuit.

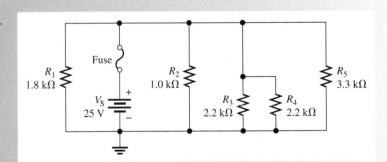

◀ FIGURE 6–8

The same voltage appears across each resistor in parallel.

(a) Pictorial

(b) Schematic

EXAMPLE 6–3

Determine the voltage across each resistor in Figure 6–9.

FIGURE 6–9

Solution   The five resistors are in parallel, so the voltage across each one is equal to the applied source voltage. There is no voltage across the fuse. The voltage across the resistors is

$$V_1 = V_2 = V_3 = V_4 = V_5 = V_S = \textbf{25 V}$$

Related Problem   If R_4 is removed from the circuit, what is the voltage across R_3?

Use Multisim file E06-03 to verify the calculated results in this example and to confirm your calculation for the related problem.

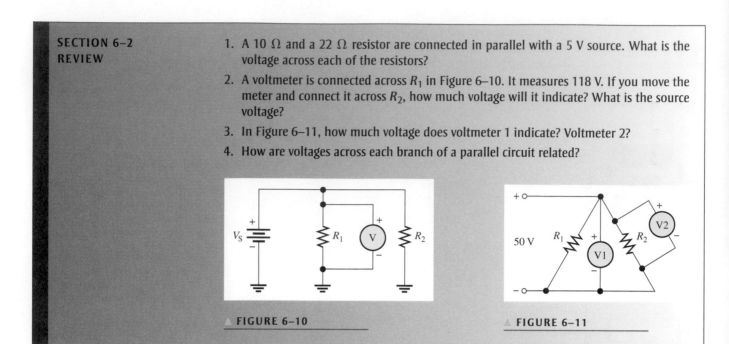

1. A 10 Ω and a 22 Ω resistor are connected in parallel with a 5 V source. What is the voltage across each of the resistors?
2. A voltmeter is connected across R_1 in Figure 6–10. It measures 118 V. If you move the meter and connect it across R_2, how much voltage will it indicate? What is the source voltage?
3. In Figure 6–11, how much voltage does voltmeter 1 indicate? Voltmeter 2?
4. How are voltages across each branch of a parallel circuit related?

▲ FIGURE 6–10   ▲ FIGURE 6–11

6–3   KIRCHHOFF'S CURRENT LAW

Kirchhoff's voltage law deals with voltages in a single closed path. Kirchhoff's current law applies to currents in multiple paths.

After completing this section, you should be able to

◆ **Apply Kirchhoff's current law**

   ◆ State Kirchhoff's current law

   ◆ Define *node*

   ◆ Determine the total current by adding the branch currents

   ◆ Determine an unknown branch current

Kirchhoff's current law, often abbreviated KCL, can be stated as follows:

The sum of the currents into a node (total current in) is equal to the sum of the currents out of that node (total current out).

A **node** is any point or junction in a circuit where two or more components are connected. In a parallel circuit, a node or junction is a point where the parallel branches come together. For example, in the circuit of Figure 6–12, point A is one node and point B is another. Let's start at the positive terminal of the source and follow the current. The total current I_T from the source is *into* node A. At this point, the current splits up among the three branches as indicated. Each of the three branch currents (I_1, I_2, and I_3) is *out of* node A. Kirchhoff's current law says that the total current into node A is equal to the total current out of node A; that is,

$$I_T = I_1 + I_2 + I_3$$

Now, following the currents in Figure 6–12 through the three branches, you see that they come back together at node B. Currents I_1, I_2, and I_3 are into node B, and

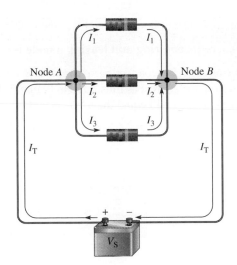

◀ **FIGURE 6–12**

Kirchhoff's current law: The current into a node equals the current out of that node.

I_T is out of node B. Kirchhoff's current law formula at node B is therefore the same as at node A.

$$I_T = I_1 + I_2 + I_3$$

Figure 6–13 shows a generalized circuit node where a number of branches are connected at a point in a circuit. Currents $I_{IN(1)}$ through $I_{IN(n)}$ are into the node (n can be any number). Currents $I_{OUT(1)}$ through $I_{OUT(m)}$ are out of the node (m can be any number, but not necessarily equal to n). By Kirchhoff's current law, the sum of the currents into a node must equal the sum of the currents out of the node. With reference to Figure 6–13, a general formula for Kirchhoff's current law is

$$I_{IN(1)} + I_{IN(2)} + \cdots + I_{IN(n)} = I_{OUT(1)} + I_{OUT(2)} + \cdots + I_{OUT(m)}$$

Equation 6–1

If all the terms on the right side of Equation 6–1 are brought over to the left side, their signs change to negative, and a zero is left on the right side as follows:

$$I_{IN(1)} + I_{IN(2)} + \cdots + I_{IN(n)} - I_{OUT(1)} - I_{OUT(2)} - \cdots - I_{OUT(m)} = 0$$

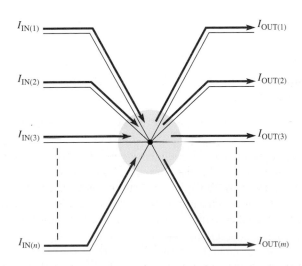

◀ **FIGURE 6–13**

Generalized circuit node illustrating Kirchhoff's current law.

Based on this last equation, Kirchhoff's current law can also be stated in this way:

The algebraic sum of all the currents entering and leaving a node is equal to zero.

You can verify Kirchhoff's current law by connecting a circuit and measuring each branch current and the total current from the source, as illustrated in Figure 6–14. When the branch currents are added together, their sum will equal the total current. This rule applies for any number of branches.

The following three examples illustrate use of Kirchhoff's current law.

▶ **FIGURE 6–14**

An illustration of Kirchhoff's current law.

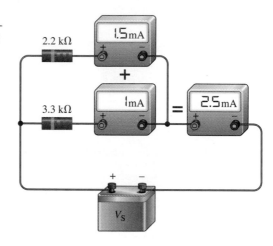

EXAMPLE 6–4

The branch currents are shown in the circuit of Figure 6–15. Determine the total current entering node A and the total current leaving node B.

▶ **FIGURE 6–15**

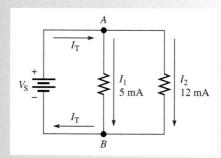

Solution   The total current out of node A is the sum of the two branch currents. So the total current into node A is

$$I_T = I_1 + I_2 = 5\,\text{mA} + 12\,\text{mA} = \mathbf{17\,mA}$$

The total current entering node B is the sum of the two branch currents. So the total current out of node B is

$$I_T = I_1 + I_2 = 5\,\text{mA} + 12\,\text{mA} = \mathbf{17\,mA}$$

Note that this equation can be equivalently expressed as $I_T - I_1 - I_2 = 0$.

Related Problem   If a third branch is added to the circuit in Figure 6–15 and its current is 3 mA, what is the total current into node A and out of node B?

EXAMPLE 6–5   Determine the current I_2 through R_2 in Figure 6–16.

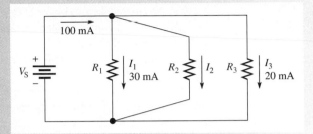

◢ **FIGURE 6–16**

Solution   The total current into the junction of the three branches is $I_T = I_1 + I_2 + I_3$. From Figure 6–16, you know the total current and the branch currents through R_1 and R_3. Solve for I_2 as follows:

$$I_2 = I_T - I_1 - I_3 = 100\,\text{mA} - 30\,\text{mA} - 20\,\text{mA} = \mathbf{50\,mA}$$

Related Problem   Determine I_T and I_2 if a fourth branch is added to the circuit in Figure 6–16 and it has 12 mA through it.

EXAMPLE 6–6   Use Kirchhoff's current law to find the current measured by ammeters A3 and A5 in Figure 6–17.

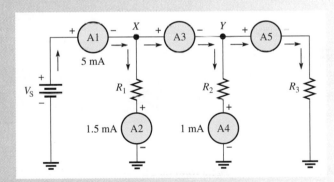

◢ **FIGURE 6–17**

Solution   The total current into node X is 5 mA. Two currents are out of node X: 1.5 mA through resistor R_1 and the current through A3. Kirchhoff's current law applied at node X gives

$$5\,\text{mA} = 1.5\,\text{mA} + I_{A3}$$

Solving for I_{A3} yields

$$I_{A3} = 5\,\text{mA} - 1.5\,\text{mA} = \mathbf{3.5\,mA}$$

The total current into node Y is $I_{A3} = 3.5\,\text{mA}$. Two currents are out of node Y: 1 mA through resistor R_2 and the current through A5 and R_3. Kirchhoff's current law applied at node Y gives

$$3.5\,\text{mA} = 1\,\text{mA} + I_{A5}$$

Solving for I_{A5} yields

$$I_{A5} = 3.5\,\text{mA} - 1\,\text{mA} = \mathbf{2.5\,mA}$$

Related Problem How much current will an ammeter measure when it is placed in the circuit right below R_3 in Figure 6–17? Below the negative battery terminal?

SECTION 6–3 REVIEW

1. State Kirchhoff's current law in two ways.
2. There is a total current of 2.5 mA into a node and the current out of the node divides into three parallel branches. What is the sum of all three branch currents?
3. In Figure 6–18, 100 mA and 300 mA are into the node. What is the amount of current out of the node?
4. Determine I_1 in the circuit of Figure 6–19.
5. Two branch currents enter a node, and two branch currents leave the same node. One of the currents entering the node is 1 mA, and one of the currents leaving the node is 3 mA. The total current entering and leaving the node is 8 mA. Determine the value of the unknown current entering the node and the value of the unknown current leaving the node.

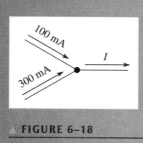

▲ FIGURE 6–18

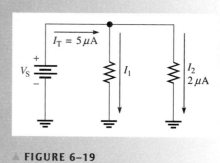

▲ FIGURE 6–19

6–4 TOTAL PARALLEL RESISTANCE

When resistors are connected in parallel, the total resistance of the circuit decreases. The total resistance of a parallel circuit is always less than the value of the smallest resistor. For example, if a 10 Ω resistor and a 100 Ω resistor are connected in parallel, the total resistance is less than 10 Ω.

After completing this section, you should be able to

◆ **Determine total parallel resistance**

  ◆ Explain why resistance decreases as resistors are connected in parallel

  ◆ Apply the parallel-resistance formula

As you know, when resistors are connected in parallel, the current has more than one path. The number of current paths is equal to the number of parallel branches.

In Figure 6–20(a), there is only one current path because it is a series circuit. There is a certain amount of current, I_1, through R_1. If resistor R_2 is connected in parallel with R_1, as shown in Figure 6–20(b), there is an additional amount of current, I_2, through R_2. The total current from the source has increased with the addition of the parallel resistor. Assuming that the source voltage is constant, an increase in the total current from the source means that the total resistance has decreased, in accordance with Ohm's law. Additional resistors connected in parallel will further reduce the resistance and increase the total current.

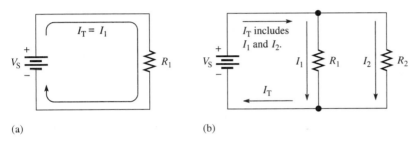

(a)   (b)

▲ FIGURE 6–20

Addition of resistors in parallel reduces total resistance and increases total current.

Formula for Total Parallel Resistance

The circuit in Figure 6–21 shows a general case of n resistors in parallel (n can be any number). From Kirchhoff's current law, the equation for current is

$$I_T = I_1 + I_2 + I_3 + \cdots + I_n$$

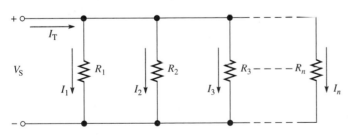

▲ FIGURE 6–21

Circuit with n resistors in parallel.

Since V_S is the voltage across each of the parallel resistors, by Ohm's law, $I_1 = V_S/R_1$, $I_2 = V_S/R_2$, and so on. By substitution into the equation for current,

$$\frac{V_S}{R_T} = \frac{V_S}{R_1} + \frac{V_S}{R_2} + \frac{V_S}{R_3} + \cdots + \frac{V_S}{R_n}$$

The term V_S can be factored out of the right side of the equation and canceled with V_S on the left side, leaving only the resistance terms.

$$\frac{1}{R_T} = \frac{1}{R_1} + \frac{1}{R_2} + \frac{1}{R_3} + \cdots + \frac{1}{R_n}$$

Recall that the reciprocal of resistance ($1/R$) is called *conductance,* which is symbolized by G. The unit of conductance is the siemens (S). The equation for $1/R_T$ can be expressed in terms of conductance as

$$G_T = G_1 + G_2 + G_3 + \cdots + G_n$$

Solve for R_T by taking the reciprocal of (that is, by inverting) both sides of the equation for $1/R_T$.

Equation 6–2

$$R_T = \frac{1}{\left(\dfrac{1}{R_1}\right) + \left(\dfrac{1}{R_2}\right) + \left(\dfrac{1}{R_3}\right) + \cdots + \left(\dfrac{1}{R_n}\right)}$$

Equation 6–2 shows that to find the total parallel resistance, add all the $1/R$ (or conductance, G) terms and then take the reciprocal of the sum.

$$R_T = \frac{1}{G_T}$$

EXAMPLE 6–7

Calculate the total parallel resistance between points A and B of the circuit in Figure 6–22.

▶ **FIGURE 6–22**

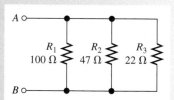

Solution   Use Equation 6–2 to calculate the total parallel resistance when you know the individual resistances. First, find the conductance, which is the reciprocal of the resistance, of each of the three resistors.

$$G_1 = \frac{1}{R_1} = \frac{1}{100\ \Omega} = 10\ \text{mS}$$

$$G_2 = \frac{1}{R_2} = \frac{1}{47\ \Omega} = 21.3\ \text{mS}$$

$$G_3 = \frac{1}{R_3} = \frac{1}{22\ \Omega} = 45.5\ \text{mS}$$

Next, calculate R_T by adding G_1, G_2, and G_3 and taking the reciprocal of the sum.

$$R_T = \frac{1}{G_T} = \frac{1}{10\ \text{mS} + 21.3\ \text{mS} + 45.5\ \text{mS}} = \frac{1}{76.8\ \text{mS}} = \textbf{13.0}\ \boldsymbol{\Omega}$$

For a quick accuracy check, notice that the value of R_T (13.0 Ω) is smaller than the smallest value in parallel, which is R_3 (22 Ω), as it should be.

Related Problem   If a 33 Ω resistor is connected in parallel in Figure 6–22, what is the new value of R_T?

Calculator Tip

The parallel-resistance formula is easily solved on a calculator using Equation 6–2. The general procedure is to enter the value of R_1 and then take its reciprocal by pressing the x^{-1} key. (The reciprocal is a secondary function on some calculators.) Next press the $+$ key; then enter the value of R_2 and take its reciprocal using the x^{-1} key and press the $+$ key. Repeat this procedure until all of the resistor values have been entered; then press ENTER. The final step is to press the x^{-1} key and the ENTER key to get R_T. The total parallel resistance is now on the display. The display format may vary, depending on the particular calculator. For example, the steps required for a typical calculator solution of Example 6–7 are as follows:

1. Enter 100. Display shows 100.

2. Press x^{-1} (or 2nd then x^{-1}). Display shows 100^{-1}.

3. Press $+$. Display shows $100^{-1} +$.

4. Enter 47. Display shows $100^{-1} + 47$.

5. Press x^{-1} (or 2nd then x^{-1}). Display shows $100^{-1} + 47^{-1}$.

6. Press $+$. Display shows $100^{-1} + 47^{-1} +$.

7. Enter 22. Display shows $100^{-1} + 47^{-1} + 22$.

8. Press x^{-1} (or 2nd then x^{-1}). Display shows $100^{-1} + 47^{-1} + 22^{-1}$.

9. Press ENTER. Display shows a result of $76.7311411992\text{E}{-3}$.

10. Press x^{-1} (or 2nd then x^{-1}) and then ENTER. Display shows a result of $13.0325182758\text{E}0$.

The number displayed in Step 10 is the total resistance in ohms. Round it to 13.0 Ω.

The Case of Two Resistors in Parallel

Equation 6–2 is a general formula for finding the total resistance for any number of resistors in parallel. The combination of two resistors in parallel occurs commonly in practice. Also, any number of resistors in parallel can be broken down into pairs as an alternate way to find the R_T. Based on Equation 6–2, the formula for the total resistance of two resistors in parallel is

$$R_T = \frac{1}{\left(\dfrac{1}{R_1}\right) + \left(\dfrac{1}{R_2}\right)}$$

Combining the terms in the denominator yields

$$R_T = \frac{1}{\left(\dfrac{R_1 + R_2}{R_1 R_2}\right)}$$

which can be rewritten as follows:

$$R_T = \frac{R_1 R_2}{R_1 + R_2}$$

Equation 6–3

Equation 6–3 states

The total resistance for two resistors in parallel is equal to the product of the two resistors divided by the sum of the two resistors.

This equation is sometimes referred to as the "product over the sum" formula.

EXAMPLE 6-8

Calculate the total resistance connected to the voltage source of the circuit in Figure 6–23.

▶ **FIGURE 6–23**

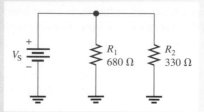

Solution Use Equation 6–3.

$$R_T = \frac{R_1R_2}{R_1 + R_2} = \frac{(680\ \Omega)(330\ \Omega)}{680\ \Omega + 330\ \Omega} = \frac{224{,}400\ \Omega^2}{1010\ \Omega} = \mathbf{222\ \Omega}$$

Related Problem Determine R_T if a 220 Ω replaces R_1 in Figure 6–23.

The Case of Equal-Value Resistors in Parallel

Another special case of parallel circuits is the parallel connection of several resistors each having the same resistance value. There is a shortcut method of calculating R_T when this case occurs.

If several resistors in parallel have the same resistance, they can be assigned the same symbol R. For example, $R_1 = R_2 = R_3 = \cdots = R_n = R$. Starting with Equation 6–2, you can develop a special formula for finding R_T.

$$R_T = \frac{1}{\left(\dfrac{1}{R}\right) + \left(\dfrac{1}{R}\right) + \left(\dfrac{1}{R}\right) + \cdots + \left(\dfrac{1}{R}\right)}$$

Notice that in the denominator, the same term, $1/R$, is added n times (n is the number of equal-value resistors in parallel). Therefore, the formula can be written as

$$R_T = \frac{1}{n/R}$$

or

Equation 6–4

$$R_T = \frac{R}{n}$$

Equation 6–4 says that when any number of resistors (n), all having the same resistance (R), are connected in parallel, R_T is equal to the resistance divided by the number of resistors in parallel.

EXAMPLE 6-9

Four 8 Ω speakers are connected in parallel to the output of an amplifier. What is the total resistance across the output of the amplifier?

Solution There are four 8 Ω resistors in parallel. Use Equation 6–4 as follows:

$$R_T = \frac{R}{n} = \frac{8\ \Omega}{4} = \mathbf{2\ \Omega}$$

Related Problem If two of the speakers are removed, what is the resistance across the output?

Determining an Unknown Parallel Resistor

Sometimes you need to determine the values of resistors that are to be combined to produce a desired total resistance. For example, you use two parallel resistors to obtain a known total resistance. If you know or arbitrarily choose one resistor value, then you can calculate the second resistor value using Equation 6–3 for two parallel resistors. The formula for determining the value of an unknown resistor R_x is developed as follows:

$$\frac{1}{R_T} = \frac{1}{R_A} + \frac{1}{R_x}$$

$$\frac{1}{R_x} = \frac{1}{R_T} - \frac{1}{R_A}$$

$$\frac{1}{R_x} = \frac{R_A - R_T}{R_A R_T}$$

$$R_x = \frac{R_A R_T}{R_A - R_T}$$

Equation 6–5

where R_x is the unknown resistor and R_A is the known or selected value.

EXAMPLE 6–10

Suppose that you wish to obtain a resistance as close to 150 Ω as possible by combining two resistors in parallel. There is a 330 Ω resistor available. What other value do you need?

Solution $R_T = 150\ \Omega$ and $R_A = 330\ \Omega$. Therefore,

$$R_x = \frac{R_A R_T}{R_A - R_T} = \frac{(330\ \Omega)(150\ \Omega)}{330\ \Omega - 150\ \Omega} = 275\ \Omega$$

The closest standard value is **270 Ω**.

Related Problem   If you need to obtain a total resistance of 130 Ω, what value can you add in parallel to the parallel combination of 330 Ω and 270 Ω? First find the value of 330 Ω and 270 Ω in parallel and treat that value as a single resistor.

Notation for Parallel Resistors

Sometimes, for convenience, parallel resistors are designated by two parallel vertical marks. For example, R_1 in parallel with R_2 can be written as $R_1 \| R_2$. Also, when several resistors are in parallel with each other, this notation can be used. For example,

$$R_1 \| R_2 \| R_3 \| R_4 \| R_5$$

indicates that R_1 through R_5 are all in parallel.

This notation is also used with resistance values. For example,

$$10\,\text{k}\Omega \| 5\,\text{k}\Omega$$

means that a 10 kΩ resistor is in parallel with a 5 kΩ resistor.

SECTION 6–4 REVIEW

1. Does the total resistance increase or decrease as more resistors are connected in parallel?
2. The total parallel resistance is always less than what value?
3. Write the general formula for R_T with any number of resistors in parallel.

4. Write the special formula for two resistors in parallel.
5. Write the special formula for any number of equal-value resistors in parallel.
6. Calculate R_T for Figure 6–24.
7. Determine R_T for Figure 6–25.
8. Find R_T for Figure 6–26.

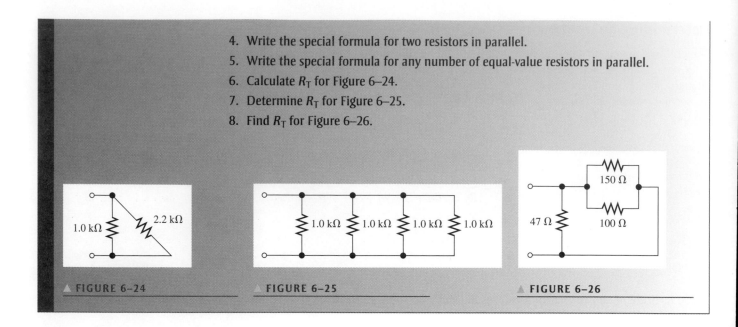

▲ FIGURE 6–24   ▲ FIGURE 6–25   ▲ FIGURE 6–26

6–5 APPLICATION OF OHM'S LAW

Ohm's law can be applied to parallel circuit analysis.

After completing this section, you should be able to

◆ **Apply Ohm's law in a parallel circuit**

   ◆ Find the total current in a parallel circuit

   ◆ Find each branch current in a parallel circuit

   ◆ Find the voltage across a parallel circuit

   ◆ Find the resistance of a parallel circuit

The following examples illustrate how to apply Ohm's law to determine the total current, branch currents, voltage, and resistance in parallel circuits.

EXAMPLE 6–11   Find the total current produced by the battery in Figure 6–27.

▶ FIGURE 6–27

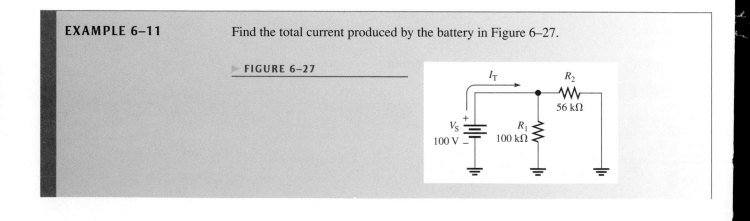

Solution The battery "sees" a total parallel resistance that determines the amount of current that it generates. First, calculate R_T.

$$R_T = \frac{R_1 R_2}{R_1 + R_2} = \frac{(100\,k\Omega)(56\,k\Omega)}{100\,k\Omega + 56\,k\Omega} = \frac{5600\,k\Omega^2}{156\,k\Omega} = 35.9\,k\Omega$$

The battery voltage is 100 V. Use Ohm's law to find I_T.

$$I_T = \frac{V_S}{R_T} = \frac{100\,V}{35.9\,k\Omega} = \textbf{2.79 mA}$$

Related Problem What is I_T in Figure 6–27 if R_2 is changed to 120 kΩ? What is the current through R_1?

Use Multisim file E06-11 to verify the calculated results in this example and to confirm your calculation for the related problem.

EXAMPLE 6–12

Determine the current through each resistor in the parallel circuit of Figure 6–28.

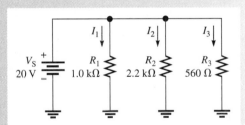

▲ **FIGURE 6–28**

Solution The voltage across each resistor (branch) is equal to the source voltage. That is, the voltage across R_1 is 20 V, the voltage across R_2 is 20 V, and the voltage across R_3 is 20 V. The current through each resistor is determined as follows:

$$I_1 = \frac{V_S}{R_1} = \frac{20\,V}{1.0\,k\Omega} = \textbf{20 mA}$$

$$I_2 = \frac{V_S}{R_2} = \frac{20\,V}{2.2\,k\Omega} = \textbf{9.09 mA}$$

$$I_3 = \frac{V_S}{R_3} = \frac{20\,V}{560\,\Omega} - \textbf{35.7 mA}$$

Related Problem If an additional resistor of 910 Ω is connected in parallel to the circuit in Figure 6–28, determine all of the branch currents.

Use Multisim file E06-12 to verify the calculated results in this example and to confirm your calculations for the related problem.

EXAMPLE 6–13

Find the voltage V_S across the parallel circuit in Figure 6–29.

▶ **FIGURE 6–29**

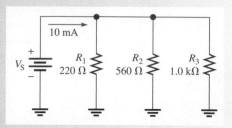

Solution   The total current into the parallel circuit is 10 mA. If you know the total resistance, then you can apply Ohm's law to get the voltage. The total resistance is

$$R_T = \frac{1}{G_1 + G_2 + G_3}$$

$$= \frac{1}{\left(\dfrac{1}{R_1}\right) + \left(\dfrac{1}{R_2}\right) + \left(\dfrac{1}{R_3}\right)}$$

$$= \frac{1}{\left(\dfrac{1}{220\ \Omega}\right) + \left(\dfrac{1}{560\ \Omega}\right) + \left(\dfrac{1}{1.0\ k\Omega}\right)}$$

$$= \frac{1}{4.55\ mS + 1.79\ mS + 1\ mS} = \frac{1}{7.34\ mS} = 136\ \Omega$$

Therefore, the source voltage is

$$V_S = I_T R_T = (10\ mA)(136\ \Omega) = \mathbf{1.36\ V}$$

Related Problem   Find the voltage if R_3 is decreased to 680 Ω in Figure 6–29 and I_T is 10 mA.

 Use Multisim file E06-13 to verify the calculated results in this example and to confirm your calculation for the related problem.

EXAMPLE 6–14

The circuit board in Figure 6–30 has three resistors in parallel. The values of two of the resistors are known from the color bands, but the top resistor is not clearly marked (maybe the bands are worn off from handling). Determine the value of the unknown resistor R_1 using only an ammeter and a dc power supply.

▶ **FIGURE 6–30**

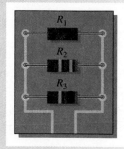

Solution   If you can determine the total resistance of the three resistors in parallel, then you can use the parallel-resistance formula to calculate the unknown resistance. You can use Ohm's law to find the total resistance if voltage and total current are known.

In Figure 6–31, a 12 V source (arbitrary value) is connected across the resistors, and the total current is measured. Using these measured values, find the total resistance.

$$R_T = \frac{V}{I_T} = \frac{12\ V}{24.1\ mA} = 498\ \Omega$$

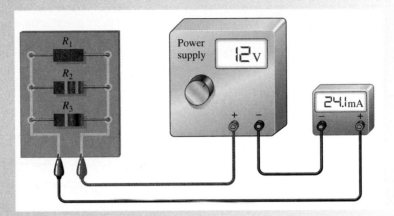

▲ FIGURE 6–31

Find the unknown resistance as follows:

$$\frac{1}{R_T} = \frac{1}{R_1} + \frac{1}{R_2} + \frac{1}{R_3}$$

$$\frac{1}{R_1} = \frac{1}{R_T} - \frac{1}{R_2} - \frac{1}{R_3} = \frac{1}{498\ \Omega} - \frac{1}{1.8\ k\Omega} - \frac{1}{1.0\ k\Omega} = 453\ \mu S$$

$$R_1 = \frac{1}{453\ \mu S} = \mathbf{2.21\ k\Omega}$$

Related Problem   Explain how to determine the value of R_1 using an ohmmeter and without removing R_1 from the circuit.

**SECTION 6–5
REVIEW**

1. A 10 V battery is connected across three 680 Ω resistors that are in parallel. What is the total current from the battery?

2. How much voltage is required to produce 20 mA of current through the circuit of Figure 6–32?

▶ FIGURE 6–32

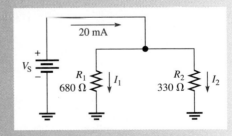

3. How much current is there through each resistor of Figure 6–32?
4. There are four equal-value resistors in parallel with a 12 V source, and 5.85 mA of current from the source. What is the value of each resistor?
5. A 1.0 kΩ and a 2.2 kΩ resistor are connected in parallel. There is a total of 100 mA through the parallel combination. How much voltage is dropped across the resistors?

6–6 CURRENT SOURCES IN PARALLEL

As you learned in Chapter 2, a current source is a type of energy source that provides a constant current to a load even if the resistance of that load changes. A transistor can be used as a current source; therefore, current sources are important in electronic circuits. Although the study of transistors is beyond the scope of this book, you should understand how current sources act in parallel.

After completing this section, you should be able to

◆ **Determine the total effect of current sources in parallel**

  ◆ Determine the total current from parallel sources having the same direction

  ◆ Determine the total current from parallel sources having opposite directions

In general, the total current produced by current sources in parallel is equal to the algebraic sum of the individual current sources. The algebraic sum means that you must consider the direction of current when you combine the sources in parallel. For example, in Figure 6–33(a), the three current sources in parallel provide current in the same direction (into node A). So the total current into node A is

$$I_T = 1\,A + 2\,A + 2\,A = 5\,A$$

In Figure 6–33(b), the 1 A source provides current in a direction opposite to the other two. The total current into node A in this case is

$$I_T = 2\,A + 2\,A - 1\,A = 3\,A$$

▶ FIGURE 6–33

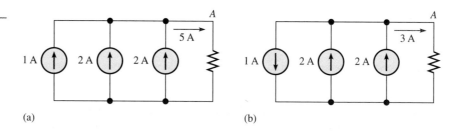

(a)   (b)

EXAMPLE 6–15   Determine the current through R_L in Figure 6–34.

▶ FIGURE 6–34

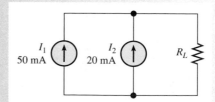

Solution The two current sources are in the same direction; so the current through R_L is

$$I_{R_L} = I_1 + I_2 = 50\,\text{mA} + 20\,\text{mA} = \textbf{70\,mA}$$

Related Problem Determine the current through R_L if the direction of I_2 is reversed.

**SECTION 6–6
REVIEW**

1. Four 0.5 A current sources are connected in parallel in the same direction. What current will be produced through a load resistor?
2. How many 100 mA current sources must be connected in parallel to produce a total current output of 300 mA? Draw a schematic showing the sources connected.
3. In a certain transistor amplifier circuit, the transistor can be represented by a 10 mA current source, as shown in Figure 6–35. In a certain transistor amplifier, two transistors act in parallel. How much current is there through the resistor R_E?

► **FIGURE 6–35**

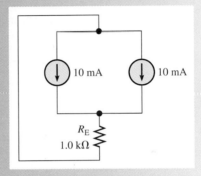

6–7 CURRENT DIVIDERS

A parallel circuit acts as a current divider because the current entering the junction of parallel branches "divides" up into several individual branch currents.

After completing this section, you should be able to

◆ **Use a parallel circuit as a current divider**

  ◆ Apply the current-divider formula

  ◆ Determine an unknown branch current

In a parallel circuit, the total current into the junction of the parallel branches divides among the branches. Thus, a parallel circuit acts as a **current divider**. This current-divider principle is illustrated in Figure 6–36 for a two-branch parallel circuit in which part of the total current I_T goes through R_1 and part through R_2.

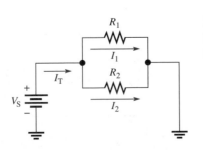

◄ **FIGURE 6–36**

Total current divides between the two branches.

Since the same voltage is across each of the resistors in parallel, the branch currents are inversely proportional to the values of the resistors. For example, if the value of R_2 is twice that of R_1, then the value of I_2 is one-half that of I_1. In other words,

The total current divides among parallel resistors into currents with values inversely proportional to the resistance values.

The branches with higher resistance have less current, and the branches with lower resistance have more current, in accordance with Ohm's law. If all the branches have the same resistance, the branch currents are all equal.

Figure 6–37 shows specific values to demonstrate how the currents divide according to the branch resistances. Notice that in this case the resistance of the upper branch is one-tenth the resistance of the lower branch, but the upper branch current is ten times the lower branch current.

▶ FIGURE 6–37

The branch with the lower resistance has more current, and the branch with the higher resistance has less current.

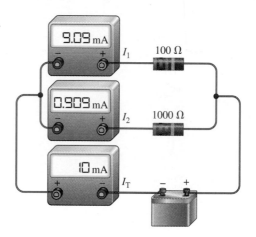

Current-Divider Formula

You can develop a formula for determining how currents divide among any number of parallel resistors as shown in Figure 6–38, where n is the total number of resistors.

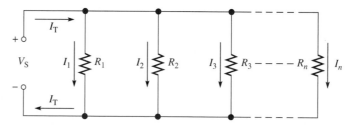

▲ FIGURE 6–38

A parallel circuit with n branches.

The current through any one of the parallel resistors is I_x, where x represents the number of a particular resistor (1, 2, 3, and so on). By Ohm's law, you can express the current through any one of the resistors in Figure 6–38 as follows:

$$I_x = \frac{V_S}{R_x}$$

The source voltage, V_S, appears across each of the parallel resistors, and R_x represents any one of the parallel resistors. The total source voltage, V_S, is equal to the total current times the total parallel resistance.

$$V_S = I_T R_T$$

Substituting $I_T R_T$ for V_S in the expression for I_x results in

$$I_x = \frac{I_T R_T}{R_x}$$

Rearranging terms yields

$$I_x = \left(\frac{R_T}{R_x}\right) I_T \qquad\qquad \text{Equation 6–6}$$

where $x = 1, 2, 3$, etc.

Equation 6–6 is the general current-divider formula and applies to a parallel circuit with any number of branches.

The current (I_x) through any branch equals the total parallel resistance (R_T) divided by the resistance (R_x) of that branch, and then multiplied by the total current (I_T) into the junction of parallel branches.

EXAMPLE 6–16   Determine the current through each resistor in the circuit of Figure 6–39.

▶ **FIGURE 6–39**

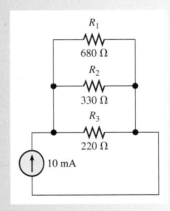

Solution   First calculate the total parallel resistance.

$$R_T = \cfrac{1}{\left(\cfrac{1}{R_1}\right) + \left(\cfrac{1}{R_2}\right) + \left(\cfrac{1}{R_3}\right)} = \cfrac{1}{\left(\cfrac{1}{680\ \Omega}\right) + \left(\cfrac{1}{330\ \Omega}\right) + \left(\cfrac{1}{220\ \Omega}\right)} = 111\ \Omega$$

The total current is 10 mA. Use Equation 6–6 to calculate each branch current.

$$I_1 = \left(\frac{R_T}{R_1}\right) I_T = \left(\frac{111\ \Omega}{680\ \Omega}\right) 10\ \text{mA} = \textbf{1.63 mA}$$

$$I_2 = \left(\frac{R_T}{R_2}\right) I_T = \left(\frac{111\ \Omega}{330\ \Omega}\right) 10\ \text{mA} = \textbf{3.36 mA}$$

$$I_3 = \left(\frac{R_T}{R_3}\right) I_T = \left(\frac{111\ \Omega}{220\ \Omega}\right) 10\ \text{mA} = \textbf{5.05 mA}$$

Related Problem   Determine the current through each resistor in Figure 6–39 if R_3 is removed.

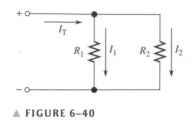

▲ FIGURE 6–40

Current-Divider Formulas for Two Branches   Two parallel resistors are common in practical circuits, as shown in Figure 6–40. As you know from Equation 6–3,

$$R_T = \frac{R_1 R_2}{R_1 + R_2}$$

Using the general current-divider formula in Equation 6–6, the formulas for I_1 and I_2 can be written as follows:

$$I_1 = \left(\frac{R_T}{R_1}\right) I_T \qquad \text{and} \qquad I_2 = \left(\frac{R_T}{R_2}\right) I_T$$

Substituting $R_1 R_2/(R_1 + R_2)$ for R_T and canceling terms result in

$$I_1 = \frac{\left(\dfrac{\cancel{R}_1 R_2}{R_1 + R_2}\right)}{\cancel{R}_1} I_T \qquad \text{and} \qquad I_2 = \frac{\left(\dfrac{R_1 \cancel{R}_2}{R_1 + R_2}\right)}{\cancel{R}_2} I_T$$

Therefore, the current-divider formulas for the special case of two branches are

Equation 6–7
$$I_1 = \left(\frac{R_2}{R_1 + R_2}\right) I_T$$

Equation 6–8
$$I_2 = \left(\frac{R_1}{R_1 + R_2}\right) I_T$$

Note that in Equations 6–7 and 6–8, the current in one of the branches is equal to the opposite branch resistance divided by the sum of the two resistors, all multiplied by the total current. In all applications of the current-divider equations, you must know the total current into the parallel branches.

EXAMPLE 6–17

Find I_1 and I_2 in Figure 6–41.

▶ FIGURE 6–41

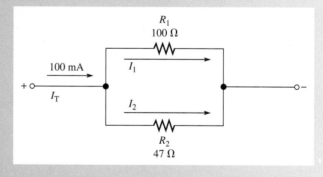

Solution   Use Equation 6–7 to determine I_1.

$$I_1 = \left(\frac{R_2}{R_1 + R_2}\right) I_T = \left(\frac{47\ \Omega}{147\ \Omega}\right) 100\ \text{mA} = \textbf{32.0 mA}$$

Use Equation 6–8 to determine I_2.

$$I_2 = \left(\frac{R_1}{R_1 + R_2}\right) I_T = \left(\frac{100\ \Omega}{147\ \Omega}\right) 100\ \text{mA} = \textbf{68.0 mA}$$

Related Problem   If $R_1 = 56\ \Omega$, and $R_2 = 82\ \Omega$ in Figure 6–41 and I_T stays the same, what will each branch current be?

SECTION 6–7
REVIEW

1. Write the general current-divider formula.
2. Write the two special formulas for calculating each branch current for a two-branch circuit.
3. A circuit has the following resistors in parallel with a voltage source: 220 kΩ, 100 kΩ, 82 kΩ, 47 kΩ, and 22 kΩ. Which resistor has the most current through it? The least current?
4. Find I_1 and I_2 in the circuit of Figure 6–42.
5. Determine the current through R_3 in Figure 6–43.

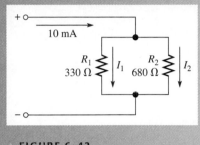

▲ FIGURE 6–42

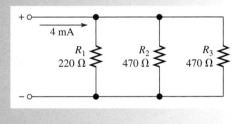

▲ FIGURE 6–43

6–8 POWER IN PARALLEL CIRCUITS

Total power in a parallel circuit is found by adding up the powers of all the individual resistors, the same as for series circuits.

After completing this section, you should be able to

- ◆ **Determine power in a parallel circuit**

Equation 6–9 states the formula for finding total power in a concise way for any number of resistors in parallel.

$$P_T = P_1 + P_2 + P_3 + \cdots + P_n$$

Equation 6–9

where P_T is the total power and P_n is the power in the last resistor in parallel. As you can see, the powers are additive, just as in a series circuit.

The power formulas in Chapter 4 are directly applicable to parallel circuits. The following formulas are used to calculate the total power P_T:

$$P_T = VI_T$$
$$P_T = I_T^2 R_T$$
$$P_T = \frac{V^2}{R_T}$$

where V is the voltage across the parallel circuit, I_T is the total current into the parallel circuit, and R_T is the total resistance of the parallel circuit. Examples 6–18 and 6–19 show how total power can be calculated in a parallel circuit.

EXAMPLE 6–18

Determine the total amount of power in the parallel circuit in Figure 6–44.

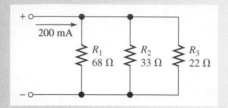

▲ **FIGURE 6–44**

Solution The total current is 200 mA. The total resistance is

$$R_T = \frac{1}{\left(\dfrac{1}{68\,\Omega}\right) + \left(\dfrac{1}{33\,\Omega}\right) + \left(\dfrac{1}{22\,\Omega}\right)} = 11.1\,\Omega$$

The easiest power formula to use is $P_T = I_T^2 R_T$ because you know both I_T and R_T.

$$P_T = I_T^2 R_T = (200\,\text{mA})^2(11.1\,\Omega) = \mathbf{444\,mW}$$

Let's demonstrate that if you determine the power in each resistor and if you add all of these values together, you will get the same result. First, find the voltage across each branch of the circuit.

$$V = I_T R_T = (200\,\text{mA})(11.1\,\Omega) = 2.22\,\text{V}$$

Remember that the voltage across all branches is the same.

Next, use $P = V^2/R$ to calculate the power for each resistor.

$$P_1 = \frac{(2.22\,\text{V})^2}{68\,\Omega} = 72.5\,\text{mW}$$

$$P_2 = \frac{(2.22\,\text{V})^2}{33\,\Omega} = 149\,\text{mW}$$

$$P_3 = \frac{(2.22\,\text{V})^2}{22\,\Omega} = 224\,\text{mW}$$

Add these powers to get the total power.

$$P_T = 72.5\,\text{mW} + 149\,\text{mW} + 224\,\text{mW} = 446\,\text{mW}$$

This calculation shows that the sum of the individual powers is equal (approximately) to the total power as determined by one of the power formulas. Rounding to three significant figures accounts for the difference.

Related Problem Find the total power in Figure 6–44 if the total current is doubled.

EXAMPLE 6–19

The amplifier in one channel of a stereo system as shown in Figure 6–45 drives two speakers. If the maximum voltage* to the speakers is 15 V, how much power must the amplifier be able to deliver to the speakers?

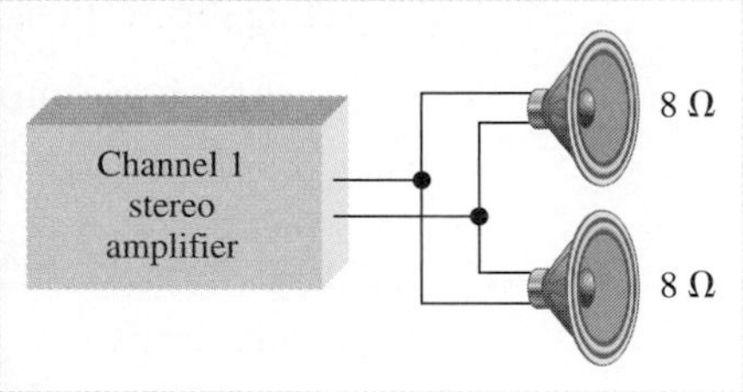

▲ FIGURE 6–45

Solution　The speakers are connected in parallel to the amplifier output, so the voltage across each is the same. The maximum power to each speaker is

$$P_{max} = \frac{V_{max}^2}{R} = \frac{(15 \text{ V})^2}{8 \ \Omega} = 28.1 \text{ W}$$

The total power that the amplifier must be capable of delivering to the speaker system is twice the power in an individual speaker because the total power is the sum of the individual powers.

$$P_{T(max)} = P_{(max)} + P_{(max)} = 2P_{(max)} = 2(28.1 \text{ W}) = \textbf{56.2 W}$$

Related Problem　If the amplifier can produce a maximum of 18 V, what is the maximum total power to the speakers?

*Voltage is ac in this case; but power is determined the same for ac voltage as for dc voltage, as you will see later.

SECTION 6–8 REVIEW

1. If you know the power in each resistor in a parallel circuit, how can you find the total power?
2. The resistors in a parallel circuit dissipate the following powers: 238 mW, 512 mW, 109 mW, and 876 mW. What is the total power in the circuit?
3. A circuit has a 1.0 kΩ, a 2.7 kΩ, and a 3.9 kΩ resistor in parallel. There is a total current of 1 A into the parallel circuit. What is the total power?

6–9 PARALLEL CIRCUIT APPLICATIONS

Parallel circuits are found in some form in virtually every electronic system. In many of these applications, the parallel relationship of components may not be obvious until you have covered some advanced topics that you will study later. For now, let's look at some examples of common and familiar applications of parallel circuits.

After completing this section, you should be able to

◆ **Describe some basic applications of parallel circuits**

   ◆ Discuss the lighting system in automobiles

   ◆ Discuss residential wiring

   ◆ Explain basically how a multiple-range ammeter works

Automotive

One advantage of a parallel circuit over a series circuit is that when one branch opens, the other branches are not affected. For example, Figure 6–46 shows a simplified diagram of an automobile lighting system. When one headlight on a car goes out, it does not cause the other lights to go out because they are all in parallel.

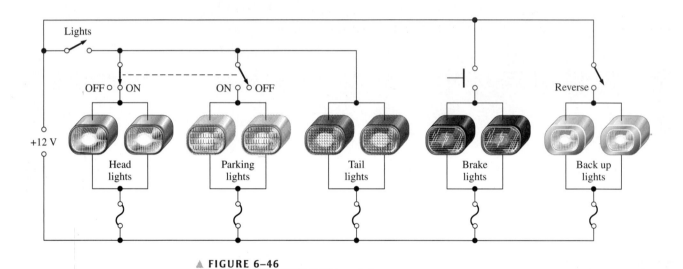

▲ **FIGURE 6–46**

Simplified diagram of the exterior light system of an automobile.

    Notice that the brake lights are switched on independently of the headlights and taillights. They come on only when the driver closes the brake light switch by depressing the brake pedal. When the lights switch is closed, both headlights and both taillights are on. When the headlights are on, the parking lights are off and vice versa. If any one of the lights burns out (opens), there is still current in each of the other lights. The back-up lights are switched on when the reverse gear is engaged.

Residential

Another common use of parallel circuits is in residential electrical systems. All the lights and appliances in a home are wired in parallel. Figure 6–47 shows a typical room wiring arrangement with two switch-controlled lights and three wall outlets in parallel.

Analog Ammeters

Parallel circuits are used in the analog (needle-type) ammeter or milliammeter. Although analog meters are not as common as they once were, they are still used as panel meters in certain applications and analog multimeters are still available. Parallel circuits are an

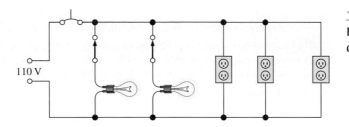

Example of parallel circuits in residential wiring.

important part of analog ammeter operation because they allow the selection of various ranges in order to measure many different current values.

The mechanism in an ammeter that causes the pointer to move in proportion to the current is called the *meter movement,* which is based on a magnetic principle that you will learn later. Right now, it is sufficient to know that a given meter movement has a certain resistance and a maximum current. This maximum current, called the *full-scale deflection current,* causes the pointer to go all the way to the end of the scale. For example, a certain meter movement has a 50 Ω resistance and a full-scale deflection current of 1 mA. A meter with this particular movement can measure currents of 1 mA or less as indicated in Figure 6–48(a) and (b). Currents greater than 1 mA will cause the pointer to "peg" (or stop) slightly past the full scale mark as indicated in part (c), which can damage the meter.

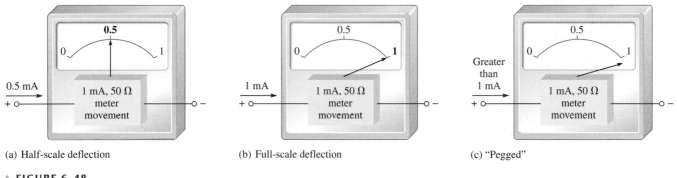

(a) Half-scale deflection   (b) Full-scale deflection   (c) "Pegged"

▲ FIGURE 6–48

A 1 mA analog ammeter.

Figure 6–49 shows a simple ammeter with a resistor in parallel with the 1 mA meter movement; this resistor is called a *shunt resistor.* Its purpose is to bypass a portion of current around the meter movement to extend the range of current that can be measured. The

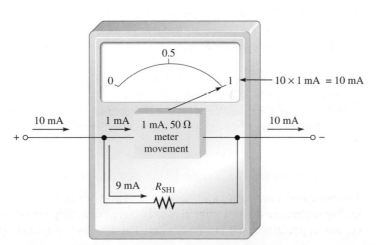

A 10 mA analog ammeter.

figure specifically shows 9 mA through the shunt resistor and 1 mA through the meter movement. Thus, up to 10 mA can be measured. To find the actual current value, simply multiply the reading on the scale by 10.

A multiple-range ammeter has a range switch that permits the selection of several full-scale current settings. In each switch position, a certain amount of current is bypassed through a parallel resistor as determined by the resistance value. In our example, the current though the movement is never greater than 1 mA.

Figure 6–50 illustrates a meter with three ranges: 1 mA, 10 mA, and 100 mA. When the range switch is in the 1 mA position, all of the current into the meter goes through the meter movement. In the 10 mA setting, up to 9 mA goes through R_{SH1} and up to 1 mA through the movement. In the 100 mA setting, up to 99 mA goes through R_{SH2}, and the movement can still have only 1 mA for full-scale.

The scale reading is interpreted based on the range setting. For example, in Figure 6–50, if 50 mA of current are being measured, the needle points at the 0.5 mark on the scale; you must multiply 0.5 by 100 to find the current value. In this situation, 0.5 mA is through the movement (half-scale deflection) and 49.5 mA are through R_{SH2}.

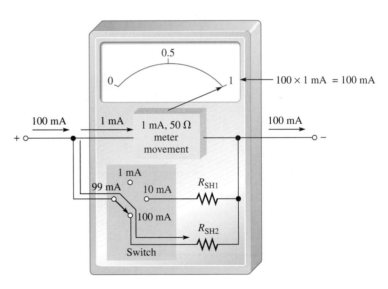

▲ FIGURE 6–50

An analog ammeter with three ranges.

Effect of the Ammeter on a Circuit   As you know, an ammeter is connected in series to measure the current in a circuit. Ideally, the meter should not alter the current that it is intended to measure. In practice, however, the meter unavoidably has some effect on the circuit because its internal resistance is connected in series with the circuit resistance. However, in most cases, the meter's internal resistance is so small compared to the circuit resistance that it can be neglected.

For example, if a meter has a 50 Ω movement (R_M) and a 0.1 mA full-scale current (I_M), the maximum voltage dropped across the movement is

$$V_M = I_M R_M = (0.1 \text{ mA})(50 \text{ Ω}) = 5 \text{ mV}$$

The shunt resistance (R_{SH}) for the 10 mA range, for example, is

$$R_{SH} = \frac{V_M}{I_{SH}} = \frac{5 \text{ mV}}{9.9 \text{ mA}} = 0.505 \text{ Ω}$$

As you can see, the total resistance of the ammeter on the 10 mA range is the resistance of the meter movement in parallel with the shunt resistance.

$$R_{M(tot)} = R_M \| R_{SH} = 50\ \Omega \| 0.505\ \Omega = 0.5\ \Omega$$

EXAMPLE 6–20

How much does a 10 mA ammeter with a 0.1 mA, 50 Ω movement affect the current in the circuit of Figure 6–51?

FIGURE 6–51

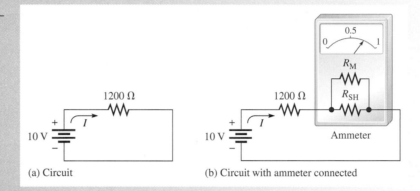

(a) Circuit

(b) Circuit with ammeter connected

Solution   The original current in the circuit (with no meter) is

$$I_{orig} = \frac{10\ V}{1200\ \Omega} = 8.3333\ mA$$

The meter is set on the 10 mA range in order to measure this particular amount of current. The meter's resistance on the 10 mA range is 0.5 Ω. When the meter is connected in the circuit, its resistance is in series with the 1200 Ω resistor. Thus, there is a total of 1200.5 Ω.

The current in the circuit is reduced slightly by inserting the meter.

$$I_{meas} = \frac{10\ V}{1200.5\ \Omega} = 8.3299\ mA$$

The current with the presence of the meter differs from the original circuit current by only **3.4 μA** or **0.04%**.

Therefore, the meter does not significantly alter the current value, a situation which, of course, is necessary because the measuring instrument should not change the quantity that is to be measured accurately.

Related Problem   How much will the measured current differ from the original current if the circuit resistance in Figure 6–51 is 12 kΩ rather than 1200 Ω?

SECTION 6–9
REVIEW

1. For the ammeter in Figure 6–51, what is the maximum resistance that the meter will have when connected in a circuit? What is the maximum current that can be measured at the setting?
2. Do the shunt resistors have resistance values considerably less than or more than that of the meter movement? Why?

6–10 TROUBLESHOOTING

Recall that an open circuit is one in which the current path is interrupted and there is no current. In this section we examine what happens when a branch of a parallel circuit opens.

After completing this section, you should be able to

◆ **Troubleshoot parallel circuits**

    ◆ Check for an open in a circuit

Open Branches

If a switch is connected in a branch of a parallel circuit, as shown in Figure 6–52, an open or a closed path can be made by the switch. When the switch is closed, as in Figure 6–52(a), R_1 and R_2 are in parallel. The total resistance is 50 Ω (two 100 Ω resistors in parallel). Current is through both resistors. If the switch is opened, as in Figure 6–52(b), R_1 is effectively removed from the circuit, and the total resistance is 100 Ω. Current is now only through R_2. In general,

> **When an open occurs in a parallel branch, the total resistance increases, the total current decreases, and the same current continues through each of the remaining parallel paths.**

The decrease in total current equals the amount of current that was previously in the open branch. The other branch currents remain the same.

▶ **FIGURE 6–52**

When switch opens, total current decreases and current through R_2 remains unchanged.

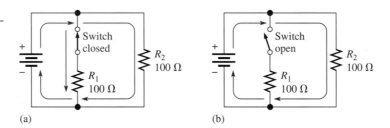

(a)   (b)

Consider the lamp circuit in Figure 6–53. There are four bulbs in parallel with a 12 V source. In part (a), there is current through each bulb. Now suppose that one of the bulbs burns out, creating an open path as shown in Figure 6–53(b). This light will go out because there is no current through the open path. Notice, however, that current continues through all the other parallel bulbs, and they continue to glow. The open branch does not change the

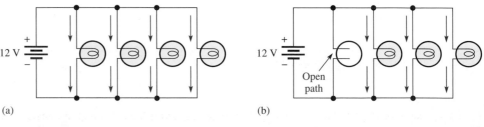

(a)   (b)

▲ **FIGURE 6–53**

When one lamp opens, total current decreases and other branch currents remain unchanged.

voltage across the parallel branches; it remains at 12 V, and the current through each branch remains the same.

You can see that a parallel circuit has an advantage over a series circuit in lighting systems because if one or more of the parallel bulbs burn out, the others will stay on. In a series circuit, when one bulb goes out, all of the others go out also because the current path is completely interrupted.

When a resistor in a parallel circuit opens, the open resistor cannot be located by measurement of the voltage across the branches because the same voltage exists across all the branches. Thus, there is no way to tell which resistor is open by simply measuring voltage. The good resistors will always have the same voltage as the open one, as illustrated in Figure 6–54 (note that the middle resistor is open).

If a visual inspection does not reveal the open resistor, it must be located by current measurements. In practice, measuring current is more difficult than measuring voltage because you must insert the ammeter in series to measure the current. Thus, a wire or a PC board connection must be cut or disconnected, or one end of a component must be lifted off the circuit board, in order to connect the ammeter in series. This procedure, of course, is not required when voltage measurements are made because the meter leads are simply connected across a component.

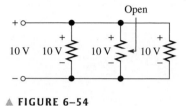

▲ **FIGURE 6–54**

Parallel branches (open or not) have the same voltage.

Finding an Open Branch by Current Measurement

In a parallel circuit with a suspected open branch, the total current can be measured to find the open. *When a parallel resistor opens, the total current, I_T, is always less than its normal value.* Once you know I_T and the voltage across the branches, a few calculations will determine the open resistor when all the resistors are of different resistance values.

Consider the two-branch circuit in Figure 6–55(a). If one of the resistors opens, the total current will equal the current in the good resistor. Ohm's law quickly tells you what the current in each resistor should be.

$$I_1 = \frac{50 \text{ V}}{560 \text{ }\Omega} = 89.3 \text{ mA}$$

$$I_2 = \frac{50 \text{ V}}{100 \text{ }\Omega} = 500 \text{ mA}$$

$$I_T = I_1 + I_2 = 589.3 \text{ mA}$$

If R_2 is open, the total current is 89.3 mA, as indicated in Figure 6–55(b). If R_1 is open, the total current is 500 mA, as indicated in Figure 6–55(c).

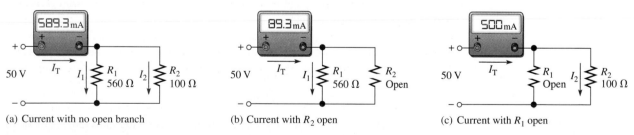

(a) Current with no open branch (b) Current with R_2 open (c) Current with R_1 open

▲ **FIGURE 6–55**

Finding an open path by current measurement.

This procedure can be extended to any number of branches having unequal resistances. If the parallel resistances are all equal, the current in each branch must be checked until a branch is found with no current. This is the open resistor.

EXAMPLE 6–21

In Figure 6–56, there is a total current of 31.09 mA, and the voltage across the parallel branches is 20 V. Is there an open resistor, and, if so, which one is it?

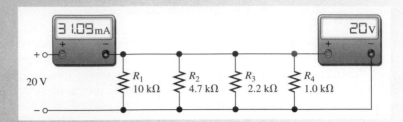

▲ FIGURE 6–56

Solution Calculate the current in each branch.

$$I_1 = \frac{V}{R_1} = \frac{20\,V}{10\,k\Omega} = 2\,mA$$

$$I_2 = \frac{V}{R_2} = \frac{20\,V}{4.7\,k\Omega} = 4.26\,mA$$

$$I_3 = \frac{V}{R_3} = \frac{20\,V}{2.2\,k\Omega} = 9.09\,mA$$

$$I_4 = \frac{V}{R_4} = \frac{20\,V}{1.0\,k\Omega} = 20\,mA$$

The total current should be

$$I_T = I_1 + I_2 + I_3 + I_4 = 2\,mA + 4.26\,mA + 9.09\,mA + 20\,mA = 35.35\,mA$$

The actual measured current is 31.09 mA, as stated, which is 4.26 mA less than normal, indicating that the branch carrying 4.26 mA is open. Thus, **R_2 must be open.**

Related Problem What is the total current measured in Figure 6–56 if R_4 and not R_2 is open?

Finding an Open Branch by Resistance Measurement

If the parallel circuit to be checked can be disconnected from its voltage source and from any other circuit to which it may be connected, a measurement of the total resistance can be used to locate an open branch.

Recall that conductance, G, is the reciprocal of resistance ($1/R$) and its unit is the siemens (S). The total conductance of a parallel circuit is the sum of the conductances of all the resistors.

$$G_T = G_1 + G_2 + G_3 + \cdots + G_n$$

To locate an open branch, do the following steps:

1. Calculate what the total conductance should be using the individual resistor values.

$$G_{T(calc)} = \frac{1}{R_1} + \frac{1}{R_2} + \frac{1}{R_3} + \cdots + \frac{1}{R_n}$$

2. Measure the total resistance with an ohmmeter and calculate the total measured conductance.

$$G_{T(meas)} = \frac{1}{R_{T(meas)}}$$

3. Subtract the measured total conductance (Step 2) from the calculated total conductance (Step 1). The result is the conductance of the open branch and the resistance is obtained by taking its reciprocal ($R = 1/G$).

$$R_{\text{open}} = \frac{1}{G_{\text{T(calc)}} - G_{\text{T(meas)}}}$$

Equation 6–10

EXAMPLE 6–22

Check the PC board in Figure 6–57 for open branches.

FIGURE 6–57

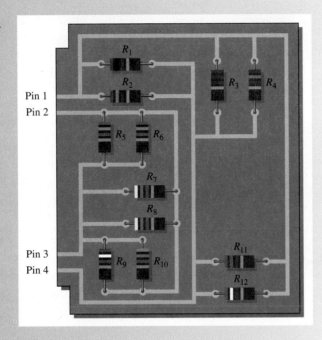

Solution There are two separate parallel circuits on the board. The circuit between pin 1 and pin 4 is checked as follows (we will assume one of the resistors is open):

1. Calculate what the total conductance should be using the individual resistor values.

$$G_{\text{T(calc)}} = \frac{1}{R_1} + \frac{1}{R_2} + \frac{1}{R_3} + \frac{1}{R_4} + \frac{1}{R_{11}} + \frac{1}{R_{12}}$$

$$= \frac{1}{1.0\,\text{k}\Omega} + \frac{1}{1.8\,\text{k}\Omega} + \frac{1}{2.2\,\text{k}\Omega} + \frac{1}{2.7\,\text{k}\Omega} + \frac{1}{3.3\,\text{k}\Omega} + \frac{1}{3.9\,\text{k}\Omega} = 2.94\,\text{mS}$$

2. Measure the total resistance with an ohmmeter and calculate the total measured conductance. Assume that your ohmmeter measures 402 Ω.

$$G_{\text{T(meas)}} = \frac{1}{402\,\Omega} = 2.49\,\text{mS}$$

3. Subtract the measured total conductance (Step 2) from the calculated total conductance (Step 1). The result is the conductance of the open branch and the resistance is obtained by taking its reciprocal.

$$G_{\text{open}} = G_{\text{T(calc)}} - G_{\text{T(meas)}} = 2.94\,\text{mS} - 2.49\,\text{mS} = 0.45\,\text{mS}$$

$$R_{\text{open}} = \frac{1}{G_{\text{open}}} = \frac{1}{0.45\,\text{mS}} = 2.2\,\text{k}\Omega$$

Resistor R_3 is open and must be replaced.

Related Problem   Your ohmmeter indicates 9.6 kΩ between pin 2 and pin 3 on the PC board in Figure 6–57. Determine if this is correct and, if not, which resistor is open.

Shorted Branches

When a branch in a parallel circuit shorts, the current increases to an excessive value, causing a fuse or circuit breaker to blow. This results in a difficult troubleshooting problem because it is hard to isolate the shorted branch.

A pulser and a current tracer are tools often used to find shorts in a circuit. They are not restricted to use in digital circuits but can be effective in any type of circuit. The pulser is a pen-shaped tool that applies pulses to a selected point in a circuit, causing pulses of current to flow through the shorted path. The current tracer is also a pen-shaped tool that senses pulses of current. By following the current with the tracer, the current path can be identified.

SECTION 6–10 REVIEW

1. If a parallel branch opens, what changes can be detected in the circuit's voltage and the currents, assuming that the parallel circuit is across a constant-voltage source?
2. What happens to the total resistance if one branch opens?
3. If several light bulbs are connected in parallel and one of the bulbs opens (burns out), will the others continue to glow?
4. There is 100 mA of current in each branch of a parallel circuit. If one branch opens, what is the current in each of the remaining branches?
5. A three-branch circuit normally has the following branch currents: 100 mA, 250 mA, and 120 mA. If the total current measures 350 mA, which branch is open?

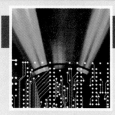

A Circuit Application

In this application, a dc power supply is modified by adding a 3-range ammeter to indicate current to the load. As you have learned, parallel resistances can be used to extend the range of an ammeter. These parallel resistors, called *shunts,* bypass current around the meter movement, allowing for the meter to effectively measure higher currents than the maximum current for which the meter movement is designed.

The Power Supply

A rack-mounted power supply is shown in Figure 6–58. The voltmeter indicates the output voltage, which can be adjusted from 0 V

▶ **FIGURE 6–58**

Front panel view of a rack-mounted power supply.

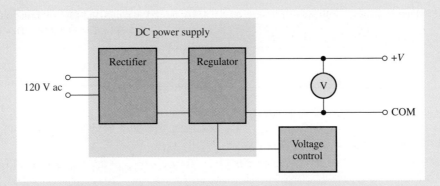

▲ FIGURE 6–59

Basic block diagram of the dc power supply.

to 10 V using the voltage control. The power supply is capable of providing up to 2 A to a load. A basic block diagram of the power supply is shown in Figure 6–59. It consists of a rectifier circuit that converts ac voltage from the wall outlet to dc voltage and a regulator circuit that keeps the output voltage at a constant value.

It is required that the power supply be modified by adding an ammeter with three switch-selected current ranges of 25 mA, 250 mA, and 2.5 A. To accomplish this, two shunt resistances are used that can each be switched into a parallel connection with the meter movement. This approach works fine as long as the required values of the shunt resistors are not too small. However, there are problems at very low values of shunt resistance and you will see why next.

The Shunt Circuit

An ammeter is selected that has a full-scale deflection of 25 mA and a resistance of 6 Ω*. Two shunt resistors must be added—

one for 250 mA and one for 2.5 A full-scale deflections. The internal meter movement provides the 25 mA range. This is shown in Figure 6–60. The range selection is provided by a 1-pole, 3-position rotary switch with a contact resistance of 50 mΩ. Contact resistance of switches can be from less than 20 mΩ to about 100 mΩ. The contact resistance of a given switch can vary with temperature, current, and usage and, therefore, cannot be relied upon to remain within a reasonable tolerance of the specified value. Also, the switch is a make-before-break type, which means that contact with the previous position is not broken until contact with the new position is made.

The shunt resistance value for the 2.5 A range is determined as follows where the voltage across the meter movement is

$$V_M = I_M R_M = (25 \text{ mA})(6 \text{ Ω}) = 150 \text{ mV}$$

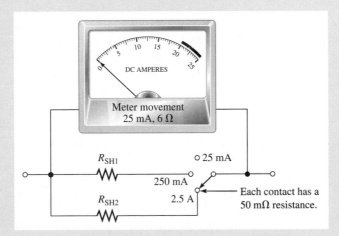

▲ FIGURE 6–60

Ammeter modified to provide three current ranges.

*See the Simpson model 1227 milliammeter at www.simpsonelectric.com.

The current through the shunt resistor for full-scale deflection is

$$I_{SH2} = I_{FULL\ SCALE} - I_M = 2.5\ \text{A} - 25\ \text{mA} = 2.475\ \text{A}$$

The total shunt resistance is

$$R_{SH2(tot)} = \frac{V_M}{I_{SH2}} = \frac{150\ \text{mV}}{2.475\ \text{A}} = 60.6\ \text{m}\Omega$$

Low-ohm precision resistors are generally available in values from 1 mΩ to 10 Ω or greater from various manufacturers.

Notice in Figure 6–60 that the contact resistance, R_{CONT}, of the switch appears in series with R_{SH2}. Since the total shunt resistance must be 60.6 mΩ, the value of the shunt resistor R_{SH2} is

$$R_{SH2} = R_{SH2(tot)} - R_{CONT} = 60.6\ \text{m}\Omega - 50\ \text{m}\Omega = 10.6\ \text{m}\Omega$$

Although this value, or one close to it, may be available, the problem in this case is that the switch contact resistance is almost twice that of R_{SH2} and any variation in it would create a significant inaccuracy in the meter. As you can see, this approach is not acceptable for these particular requirements.

Another Approach

A variation of the standard shunt resistance circuit is shown in Figure 6–61. The shunt resistor, R_{SH}, is connected in parallel for the two higher current range settings and disconnected for the 25 mA setting using a 2-pole, 3-position switch. This circuit avoids dependency on the switch contact resistance by using resistor values that are large enough to make it insignificant. The disadvantages of this meter circuit are that it requires a more complex switch and the voltage drops from input to output are greater than in the previous shunt circuit.

For the 250 mA range, the current through the meter movement for full-scale deflection is 25 mA. The voltage across the meter movement is 150 mV.

$$I_{SH} = 250\ \text{mA} - 25\ \text{mA} = 225\ \text{mA}$$

$$R_{SH} = \frac{150\ \text{mV}}{225\ \text{mA}} = 0.67\ \Omega = 670\ \text{m}\Omega$$

This value of R_{SH} is more than thirty times the expected switch contact resistance of 20 mΩ, thus minimizing the effect of the contact resistance.

For the 2.5 A range, the current through the meter movement for full-scale deflection is still 25 mA. This is also the current through R_1.

$$I_{SH} = 2.5\ \text{A} - 25\ \text{mA} = 2.475\ \text{A}$$

The voltage across the meter circuit from A to B is

$$V_{AB} = I_{SH}R_{SH} = (2.475\ \text{A})(670\ \text{m}\Omega) = 1.66\ \text{V}$$

Applying Kirchhoff's voltage law and Ohm's law to find R_1,

$$V_{R1} + V_M = V_{AB}$$

$$V_{R1} = V_{AB} - V_M = 1.66\ \text{V} - 150\ \text{mV} = 1.51\ \text{V}$$

$$R_1 = \frac{V_{R1}}{I_M} = \frac{1.51\ \text{V}}{25\ \text{mA}} = 60.4\ \Omega$$

This value is much greater than the contact resistance of the switch.

◆ Determine the maximum power dissipated by R_{SH} in Figure 6–61 for each range setting.

◆ How much voltage is there from A to B in Figure 6–61 when the switch is set to the 2.5 A range and the current is 1 A?

◆ The meter indicates 250 mA. How much does the voltage across the meter circuit from A to B change when the switch is moved from the 250 mA position to the 2.5 A position?

◆ Assume the meter movement has a resistance of 4 Ω instead of 6 Ω. Specify any changes necessary in the circuit of Figure 6–61.

Implementing the Power Supply Modification

Once the proper values are obtained, the resistors are placed on a board which is then mounted in the power supply. The resistors and the range switch are connected to the power supply as shown in Figure 6–62. The ammeter circuit is connected between the rectifier circuit in the power supply and the regulator circuit in order to reduce the impact of the voltage drop across the meter circuit on the output voltage. The regulator maintains, within certain limits, a constant dc output voltage even though its input voltage coming through the meter circuit may change.

▶ **FIGURE 6–61**

Meter circuit redesigned to eliminate or minimize the effect of switch contact resistance. The switch is a 2-pole, 3-position make-before-break rotary type.

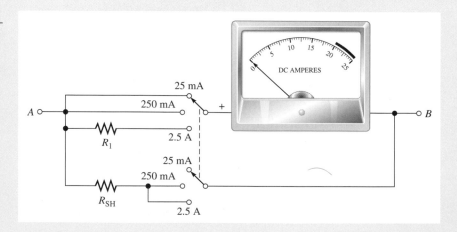

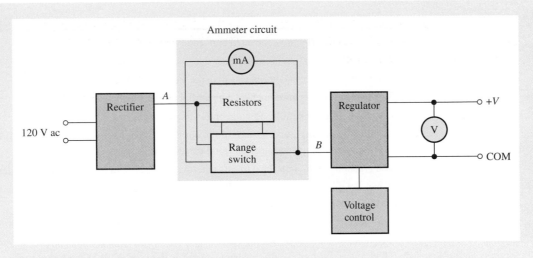

Ammeter circuit

▲ FIGURE 6–62

Block diagram of dc power supply with 3-range milliammeter.

◀ FIGURE 6–63

The power supply with the addition of the milliammeter and the current range selection switch.

Figure 6–63 shows the modified power supply front panel with the rotary range switch and milliammeter installed. The red portion of the scale indicates excess current for the 2.5 A range since the power supply has a maximum current of 2 A for safe operation.

Review

1. When the meter is set to the 250 mA range, which resistance has the most current through it?

2. Determine the total resistance from *A* to *B* of the meter circuit in Figure 6–61 for each of the three current ranges.

3. Explain why the circuit in Figure 6–61 was used instead of the one in Figure 6–60.

4. If the pointer is at the 15 and the range switch is set to 250 mA, what is the current?

5. How much current is indicated by the ammeter in Figure 6–64 for each of the three range switch settings in Figure 6–61?

◀ FIGURE 6–64

SUMMARY

♦ Resistors in parallel are connected between two points (nodes).

♦ A parallel combination has more than one path for current.

♦ The total parallel resistance is less than the lowest-value resistor.

♦ The voltages across all branches of a parallel circuit are the same.

♦ Current sources in parallel add algebraically.

♦ Kirchhoff's current law: The sum of the currents into a junction (total current in) equals the sum of the currents out of the junction (total current out).

♦ The algebraic sum of all the currents entering and leaving a junction is equal to zero.

♦ A parallel circuit is a current divider, so called because the total current entering the junction of parallel branches divides up into each of the branches.

♦ If all of the branches of a parallel circuit have equal resistance, the currents through all of the branches are equal.

♦ The total power in a parallel-resistive circuit is the sum of all of the individual powers of the resistors making up the parallel circuit.

♦ The total power for a parallel circuit can be calculated with the power formulas using values of total current, total resistance, or total voltage.

♦ If one of the branches of a parallel circuit opens, the total resistance increases, and therefore the total current decreases.

♦ If a branch of a parallel circuit opens, there is no change in current through the remaining branches.

KEY TERMS

These key terms are also in the end-of-book glossary.

Branch One current path in a parallel circuit.

Current divider A parallel circuit in which the currents divide inversely proportional to the parallel branch resistances.

Kirchhoff's current law A circuit law stating that the total current into a node equals the total current out of the node. Equivalently, the algebraic sum of all the currents entering and leaving a node is zero.

Node A point in a circuit at which two or more components are connected; also known as a *junction*.

Parallel The relationship in electric circuits in which two or more current paths are connected between two separate nodes.

FORMULAS

6–1 $I_{IN(1)} + I_{IN(2)} + \cdots + I_{IN(n)} = I_{OUT(1)} + I_{OUT(2)} + \cdots + I_{OUT(m)}$ Kirchhoff's current law

6–2 $R_T = \dfrac{1}{\left(\dfrac{1}{R_1}\right) + \left(\dfrac{1}{R_2}\right) + \left(\dfrac{1}{R_3}\right) + \cdots + \left(\dfrac{1}{R_n}\right)}$ Total parallel resistance

6–3 $R_T = \dfrac{R_1 R_2}{R_1 + R_2}$ Special case for two resistors in parallel

6–4 $R_T = \dfrac{R}{n}$ Special case for n equal-value resistors in parallel

6–5 $R_x = \dfrac{R_A R_T}{R_A - R_T}$ Unknown parallel resistor

6–6 $I_x = \left(\dfrac{R_T}{R_x}\right) I_T$ General current-divider formula

6–7 $I_1 = \left(\dfrac{R_2}{R_1 + R_2}\right) I_T$ Two-branch current-divider formula

6–8 $I_2 = \left(\dfrac{R_1}{R_1 + R_2}\right) I_T$ Two-branch current-divider formula

6–9 $P_T = P_1 + P_2 + P_3 + \cdots + P_n$ Total power

6–10 $R_{open} = \dfrac{1}{G_{T(calc)} - G_{T(meas)}}$ Open branch resistance

SELF-TEST

Answers are at the end of the chapter.

1. In a parallel circuit, each resistor has
   (a) the same current   (b) the same voltage
   (c) the same power   (d) all of the above

2. When a 1.2 kΩ resistor and a 100 Ω resistor are connected in parallel, the total resistance is
   (a) greater than 1.2 kΩ
   (b) greater than 100 Ω but less than 1.2 kΩ
   (c) less than 100 Ω but greater than 90 Ω
   (d) less than 90 Ω

3. A 330 Ω resistor, a 270 Ω resistor, and a 68 Ω resistor are all in parallel. The total resistance is approximately
   (a) 668 Ω   (b) 47 Ω   (c) 68 Ω   (d) 22 Ω

4. Eight resistors are in parallel. The two lowest-value resistors are both 1.0 kΩ. The total resistance
   (a) is less than 8 kΩ   (b) is greater than 1.0 kΩ
   (c) is less than 1.0 kΩ   (d) is less than 500 Ω

5. When an additional resistor is connected across an existing parallel circuit, the total resistance
   (a) decreases   (b) increases
   (c) remains the same   (d) increases by the value of the added resistor

6. If one of the resistors in a parallel circuit is removed, the total resistance
   (a) decreases by the value of the removed resistor   (b) remains the same
   (c) increases   (d) doubles

7. One current into a junction is 500 mA and the other current into the same junction is 300 mA. The total current out of the junction is
   (a) 200 mA   (b) unknown   (c) 800 mA   (d) the larger of the two

8. The following resistors are in parallel across a voltage source: 390 Ω, 560 Ω, and 820 Ω. The resistor with the least current is
   (a) 390 Ω   (b) 560 Ω
   (c) 820 Ω   (d) impossible to determine without knowing the voltage

9. A sudden decrease in the total current into a parallel circuit may indicate
   (a) a short   (b) an open resistor
   (c) a drop in source voltage (d) either (b) or (c)

10. In a four-branch parallel circuit, there are 10 mA of current in each branch. If one of the branches opens, the current in each of the other three branches is
    (a) 13.3 mA   (b) 10 mA   (c) 0 A   (d) 30 mA

11. In a certain three-branch parallel circuit, R_1 has 10 mA through it, R_2 has 15 mA through it, and R_3 has 20 mA through it. After measuring a total current of 35 mA, you can say that
    (a) R_1 is open   (b) R_2 is open
    (c) R_3 is open   (d) the circuit is operating properly

12. If there are a total of 100 mA into a parallel circuit consisting of three branches and two of the branch currents are 40 mA and 20 mA, the third branch current is

(a) 60 mA   (b) 20 mA   (c) 160 mA   (d) 40 mA

13. A complete short develops across one of five parallel resistors on a PC board. The most likely result is

(a) the shorted resistor will burn out

(b) one or more of the other resistors will burn out

(c) the fuse in the power supply will blow

(d) the resistance values will be altered

14. The power dissipation in each of four parallel branches is 1 W. The total power dissipation is

(a) 1 W   (b) 4 W   (c) 0.25 W   (d) 16 W

CIRCUIT DYNAMICS QUIZ

Answers are at the end of the chapter.

Refer to Figure 6–68.

1. If R_1 opens with the switch in the position shown, the voltage at terminal A with respect to ground

(a) increases   (b) decreases   (c) stays the same

2. If the switch is thrown from position A to position B, the total current

(a) increases   (b) decreases   (c) stays the same

3. If R_4 opens with the switch in position C, the total current

(a) increases   (b) decreases   (c) stays the same

4. If a short develops between B and C while the switch is in position B, the total current

(a) increases   (b) decreases   (c) stays the same

Refer to Figure 6–74(b).

5. If R_2 opens, the current through R_1

(a) increases   (b) decreases   (c) stays the same

6. If R_3 opens, the voltage across it

(a) increases   (b) decreases   (c) stays the same

7. If R_1 opens, the voltage across it

(a) increases   (b) decreases   (c) stays the same

Refer to Figure 6–75.

8. If the resistance of the rheostat R_2 is increased, the current through R_1

(a) increases   (b) decreases   (c) stays the same

9. If the fuse opens, the voltage across the rheostat R_2

(a) increases   (b) decreases   (c) stays the same

10. If the rheostat R_2 develops a short between the wiper and ground, the current through it

(a) increases   (b) decreases   (c) stays the same

Refer to Figure 6–79.

11. If the 2.25 mA source opens while the switch is in position C, the current through R

(a) increases   (b) decreases   (c) stays the same

12. If the 2.25 mA source opens while the switch is in position B, the current through R

(a) increases   (b) decreases   (c) stays the same

Refer to Figure 6–87.

13. If pins 4 and 5 are shorted together, the resistance between pins 3 and 6

(a) increases   (b) decreases   (c) stays the same

14. If the bottom connection of R_1 is shorted to the top connection of R_5, the resistance between pins 1 and 2

    (a) increases   **(b)** decreases   **(c)** stays the same

15. If R_7 opens, the resistance between pins 5 and 6

    (a) increases   **(b)** decreases   **(c)** stays the same

PROBLEMS

More difficult problems are indicated by an asterisk (*).
Answers to odd-numbered problems are at the end of the book.

SECTION 6–1   Resistors in Parallel

1. Show how to connect the resistors in Figure 6–65(a) in parallel across the battery.

2. Determine whether or not all the resistors in Figure 6–65(b) are connected in parallel on the printed circuit (PC) board.

*3. Identify which groups of resistors are in parallel on the double-sided PC board in Figure 6–66.

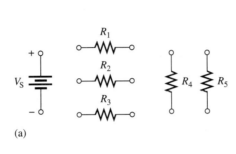

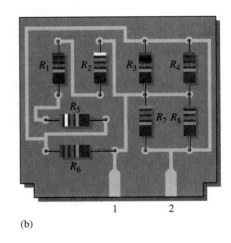

(a)

(b)

▲ FIGURE 6–65

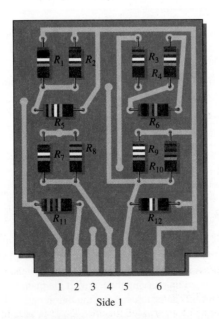

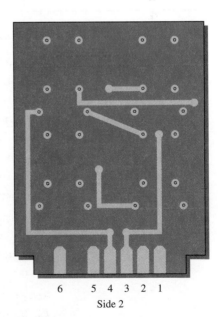

Side 1

Side 2

▲ FIGURE 6–66

SECTION 6–2 Voltage in a Parallel Circuit

4. What is the voltage across and the current through each parallel resistor if the total voltage is 12 V and the total resistance is 550 Ω? There are four resistors, all of equal value.

5. The source voltage in Figure 6–67 is 100 V. How much voltage does each of the meters read?

▶ FIGURE 6–67

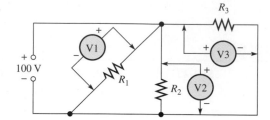

6. What is the total resistance of the circuit as seen from the voltage source for each position of the switch in Figure 6–68?

7. What is the voltage across each resistor in Figure 6–68 for each switch position?

8. What is the total current from the voltage source in Figure 6–68 for each switch position?

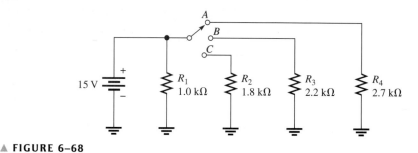

▲ FIGURE 6–68

SECTION 6–3 Kirchhoff's Current Law

9. The following currents are measured in the same direction in a three-branch parallel circuit: 250 mA, 300 mA, and 800 mA. What is the value of the current into the junction of these three branches?

10. There is a total of 500 mA of current into five parallel resistors. The currents through four of the resistors are 50 mA, 150 mA, 25 mA, and 100 mA. What is the current through the fifth resistor?

11. In the circuit of Figure 6–69, determine the resistance R_2, R_3, and R_4.

▶ FIGURE 6–69

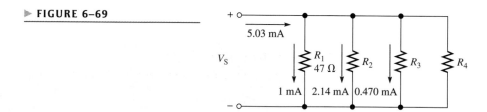

*12. The electrical circuit in a room has a ceiling lamp that draws 1.25 A and four wall outlets. Two table lamps that each draw 0.833 A are plugged into two outlets, and an electric heater that draws 10 A is connected to the third outlet. When all of these items are in use, how much current is in the main line serving the room? If the main line is protected by a 15 A circuit breaker, how much current can be drawn from the fourth outlet? Draw a schematic of this wiring.

*13. The total resistance of a parallel circuit is 25 Ω. What is the current through a 220 Ω resistor that makes up part of the parallel circuit if the total current is 100 mA?

SECTION 6–4 Total Parallel Resistance

14. The following resistors are connected in parallel: 1.0 MΩ, 2.2 MΩ, 5.6 MΩ, 12 MΩ, and 22 MΩ. Determine the total resistance.

15. Find the total resistance for each of the following groups of parallel resistors:
    (a) 560 Ω and 1000 Ω
    (b) 47 Ω and 56 Ω
    (c) 1.5 kΩ, 2.2 kΩ, 10 kΩ
    (d) 1.0 MΩ, 470 kΩ, 1.0 kΩ, 2.7 MΩ

16. Calculate R_T for each circuit in Figure 6–70.

▶ FIGURE 6–70

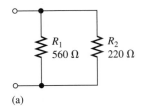
(a)

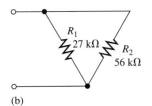
(b)

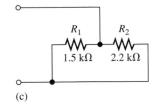
(c)

17. What is the total resistance of twelve 6.8 kΩ resistors in parallel?

18. Five 470 Ω, ten 1000 Ω, and two 100 Ω resistors are all connected in parallel. What is the total resistance for each of the three groupings?

19. Find the total resistance for the entire parallel circuit in Problem 18.

20. If the total resistance in Figure 6–71 is 389.2 Ω, what is the value of R_2?

▶ FIGURE 6–71

R_1 680 Ω R_2

21. What is the total resistance between point A and ground in Figure 6–72 for the following conditions?
    (a) SW1 and SW2 open
    (b) SW1 closed, SW2 open
    (c) SW1 open, SW2 closed
    (d) SW1 and SW2 closed

▶ FIGURE 6–72

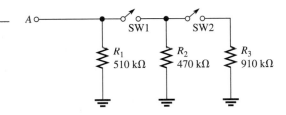

SECTION 6–5 Application of Ohm's Law

22. What is the total current in each circuit of Figure 6–73?

▶ FIGURE 6–73

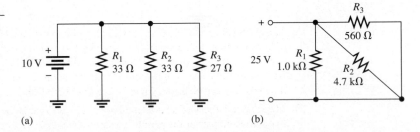
(a)   (b)

23. Three 33 Ω resistors are connected in parallel with a 110 V source. What is the current from the source?

24. Four equal-value resistors are connected in parallel. Five volts are applied across the parallel circuit, and 1.11 mA are measured from the source. What is the value of each resistor?

25. Many types of decorative lights are connected in parallel. If a set of lights is connected to a 110 V source and the filament of each bulb has a hot resistance of 2.2 kΩ, what is the current through each bulb? Why is it better to have these bulbs in parallel rather than in series?

26. Find the values of the unspecified labeled quantities in each circuit of Figure 6–74.

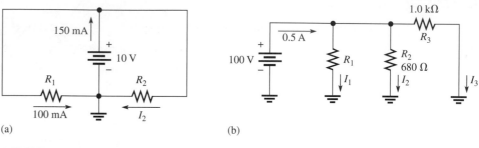

(a) (b)

▲ **FIGURE 6–74**

27. To what minimum value can the 100 Ω rheostat in Figure 6–75 be adjusted before the 0.5 A fuse blows?

▶ **FIGURE 6–75**

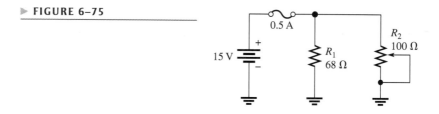

28. Determine the total current from the source and the current through each resistor for each switch position in Figure 6–76.

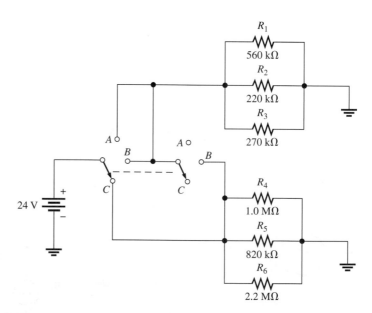

▲ **FIGURE 6–76**

29. Find the values of the unspecified quantities in Figure 6–77.

▶ **FIGURE 6–77**

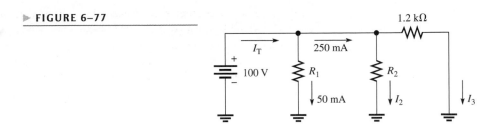

SECTION 6–6 Current Sources in Parallel

30. Determine the current through R_L in each circuit in Figure 6–78.

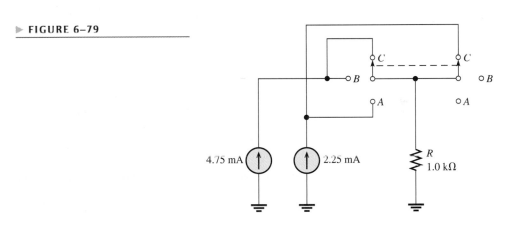

▲ **FIGURE 6–78**

31. Find the current through the resistor for each position of the ganged switch in Figure 6–79.

▶ **FIGURE 6–79**

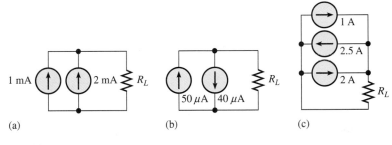

SECTION 6–7 Current Dividers

32. How much branch current should each meter in Figure 6–80 indicate?

▶ **FIGURE 6–80**

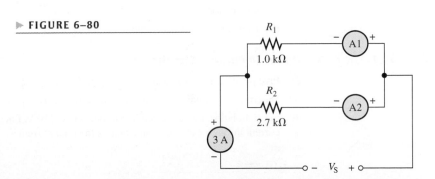

33. Determine the current in each branch of the current dividers of Figure 6–81.

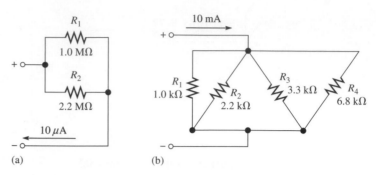

(a)   (b)

▲ FIGURE 6–81

34. What is the current through each resistor in Figure 6–82? R is the lowest-value resistor, and all others are multiples of that value as indicated.

▶ FIGURE 6–82

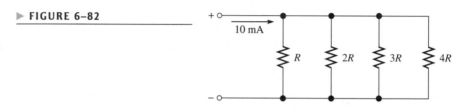

35. Determine all of the resistor values in Figure 6–83. $R_T = 773 \ \Omega$.

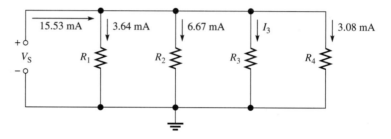

▲ FIGURE 6–83

*36. (a) Determine the required value of the shunt resistor R_{SH1} in the ammeter of Figure 6–49 if the resistance of the meter movement is 50 Ω.

(b) Find the required value for R_{SH2} in the meter circuit of Figure 6–50 ($R_M = 50 \ \Omega$).

*37. Special shunt resistors designed to drop 50 mV in high current-measuring applications are available from manufacturers. A 50 mV, 10 kΩ full-scale voltmeter is connected across the shunt to make the measurement.

(a) What value of shunt resistance is required to use a 50 mV meter in a 50 A measurement application?

(b) How much current is through the meter?

SECTION 6–8   Power in Parallel Circuits

38. Five parallel resistors each handle 250 mW. What is the total power?

39. Determine the total power in each circuit of Figure 6–81.

40. Six light bulbs are connected in parallel across 110 V. Each bulb is rated at 75 W. What is the current through each bulb, and what is the total current?

*41. Find the values of the unspecified quantities in Figure 6–84.

▶ FIGURE 6–84

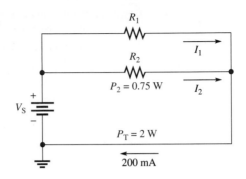

*42. A certain parallel circuit consists of only ½ W resistors. The total resistance is 1.0 kΩ, and the total current is 50 mA. If each resistor is operating at one-half its maximum power level, determine the following:

(a) The number of resistors   (b) The value of each resistor

(c) The current in each branch   (d) The applied voltage

SECTION 6–10 **Troubleshooting**

43. If one of the bulbs burns out in Problem 40, how much current will be through each of the remaining bulbs? What will the total current be?

44. In Figure 6–85, the current and voltage measurements are indicated. Has a resistor opened, and, if so, which one?

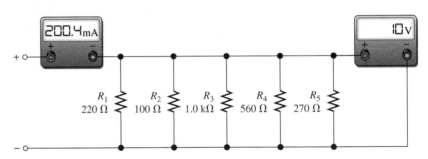

▲ FIGURE 6–85

45. What is wrong with the circuit in Figure 6–86?

46. What is wrong with the circuit in Figure 6–86 if the meter reads 5.55 mA?

▶ FIGURE 6–86

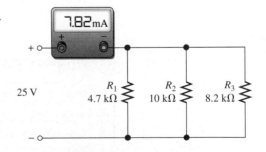

*47. Develop a test procedure to check the circuit board in Figure 6–87 to make sure that there are no open components. You must do this test without removing a component from the board. List the procedure in a detailed step-by-step format.

*48. For the circuit board shown in Figure 6–88, determine the resistance between the following pins if there is a short between pins 2 and 4:

(a) 1 and 2   (b) 2 and 3   (c) 3 and 4   (d) 1 and 4

*49. For the circuit board shown in Figure 6–88, determine the resistance between the following pins if there is a short between pins 3 and 4:

(a) 1 and 2   (b) 2 and 3   (c) 2 and 4   (d) 1 and 4

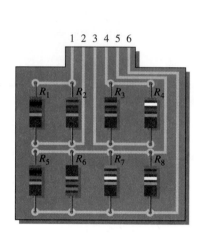

▲ FIGURE 6–87

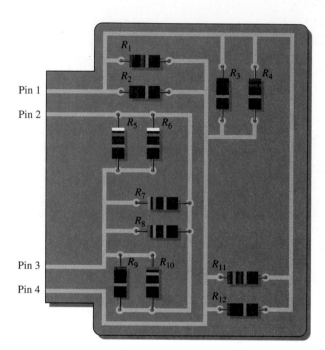

▲ FIGURE 6–88

Multisim Troubleshooting and Analysis

These problems require your Multisim CD-ROM.

50. Open file P06-50 and measure the total parallel resistance.

51. Open file P06-51. Determine by measurement if there is an open resistor and, if so, which one.

52. Open file P06-52 and determine the unspecified resistance value.

53. Open file P06-53 and determine the unspecified source voltage.

54. Open file P06-54 and find the fault if there is one.

ANSWERS

SECTION REVIEWS

SECTION 6–1   Resistors in Parallel

1. Parallel resistors are connected between the same two separate points.

2. A parallel circuit has more than one current path between two given points.

3. See Figure 6–89.

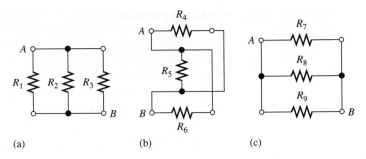

▲ FIGURE 6–89

4. See Figure 6–90.

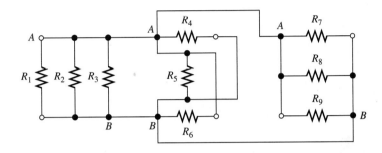

▲ FIGURE 6–90

SECTION 6–2   Voltage in a Parallel Circuit

1. $V_{10\Omega} = V_{22\Omega} = 5\,\text{V}$
2. $V_{R2} = 118\,\text{V}; V_S = 118\,\text{V}$
3. $V_{R1} = 50\,\text{V}$ and $V_{R2} = 50\,\text{V}$
4. Voltage is the same across all parallel branches.

SECTION 6–3   Kirchhoff's Current Law

1. Kirchhoff's law: The algebraic sum of all the currents at a junction is zero; The sum of the currents entering a junction equals the sum of the currents leaving that junction.
2. $I_1 = I_2 = I_3 = I_T = 2.5\,\text{mA}$
3. $I_{OUT} = 100\,\text{mA} + 300\,\text{mA} = 400\,\text{mA}$
4. $I_1 = I_T - I_2 = 3\,\mu\text{A}$
5. $I_{IN} = 8\,\text{mA} - 1\,\text{mA} = 7\,\text{mA}; I_{OUT} = 8\,\text{mA} - 3\,\text{mA} = 5\,\text{mA}$

SECTION 6–4   Total Parallel Resistance

1. R_T decreases with more resistors in parallel.
2. The total parallel resistance is less than the smallest branch resistance.
3. $R_T = \dfrac{1}{(1/R_1) + (1/R_2) + \cdots + (1/R_n)}$
4. $R_T = R_1 R_2/(R_1 + R_2)$
5. $R_T = R/n$
6. $R_T = (1.0\,\text{k}\Omega)(2.2\,\text{k}\Omega)/3.2\,\text{k}\Omega = 688\,\Omega$
7. $R_T = 1.0\,\text{k}\Omega/4 = 250\,\Omega$
8. $R_T = \dfrac{1}{1/47\,\Omega + 1/150\,\Omega + 1/100\,\Omega} = 26.4\,\Omega$

SECTION 6–5 Application of Ohm's Law

1. $I_T = 10\,\text{V}/22.7\,\Omega = 44.1\,\text{mA}$
2. $V_S = (20\,\text{mA})(222\,\Omega) = 4.44\,\text{V}$
3. $I_1 = 4.44\,\text{V}/680\,\Omega = 6.53\,\text{mA}; I_2 = 4.44\,\text{V}/330\,\Omega = 13.5\,\text{mA}$
4. $R_T = 12\,\text{V}/5.85\,\text{mA} = 2.05\,\text{k}\Omega; R = (2.05\,\text{k}\Omega)(4) = 8.2\,\text{k}\Omega$
5. $V = (100\,\text{mA})(688\,\Omega) = 68.8\,\text{V}$

SECTION 6–6 Current Sources in Parallel

1. $I_T = 4(0.5\,\text{A}) = 2\,\text{A}$
2. Three sources; See Figure 6–91.
3. $I_{R_E} = 10\,\text{mA} + 10\,\text{mA} = 20\,\text{mA}$

▶ **FIGURE 6–91**

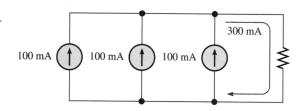

100 mA   100 mA   100 mA   300 mA

SECTION 6–7 Current Dividers

1. $I_x = (R_T/R_x)I_T$
2. $I_1 = \left(\dfrac{R_2}{R_1 + R_2}\right)I_T \qquad I_2 = \left(\dfrac{R_1}{R_1 + R_2}\right)I_T$
3. The 22 kΩ has the most current; the 220 kΩ has the least current.
4. $I_1 = (680\,\Omega/1010\,\Omega)10\,\text{mA} = 6.73\,\text{mA}; I_2 = (330\,\Omega/1010\,\Omega)10\,\text{mA} = 3.27\,\text{mA}$
5. $I_3 = (114\,\Omega/470\,\Omega)4\,\text{mA} = 970\,\mu\text{A}$

SECTION 6–8 Power in Parallel Circuits

1. Add the power of each resistor to get total power.
2. $P_T = 238\,\text{mW} + 512\,\text{mW} + 109\,\text{mW} + 876\,\text{mW} = 1.74\,\text{W}$
3. $P_T = (1\,\text{A})^2(615\,\Omega) = 615\,\text{W}$

SECTION 6–9 Parallel Circuit Applications

1. $R_{max} = 50\,\Omega; I_{max} = 1\,\text{mA}$
2. R_{SH} is less than R_M because the shunt resistors must allow currents much greater than the current through the meter movement.

SECTION 6–10 Troubleshooting

1. When a branch opens, there is no change in voltage; the total current decreases.
2. If a branch opens, total parallel resistance increases.
3. The remaining bulbs continue to glow.
4. All remaining branch currents are 100 mA.
5. The branch with 120 mA is open.

A Circuit Application

1. R_{SH} has the most current.
2. 25 mA range: $R_{AB} = R_M = 6\,\Omega$
   250 mA range: $R_{AB} = R_M \| R_{SH} = 6\,\Omega \| 670\,\text{m}\Omega = 603\,\text{m}\Omega$
   2.5 A range: $R_{AB} = (R_1 + R_M) \| R_{SH} = (60.4\,\Omega + 6\,\Omega) \| 670\,\text{m}\Omega =$
         $66.4\,\Omega \| 670\,\text{m}\Omega = 663\,\text{m}\Omega$

3. The meter circuit in Figure 6–61 negates the effect of the switch contact resistance.

4. 150 mA

5. 25 mA range: 7.5 mA

    250 mA range: 75 mA

    2.5 A range: 750 mA

RELATED PROBLEMS FOR EXAMPLES

6–1 See Figure 6–92.

▶ **FIGURE 6–92**

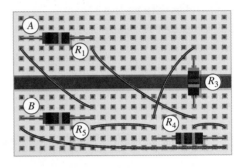

6–2 Connect pin 1 to pin 2 and pin 3 to pin 4.

6–3 25 V

6–4 20 mA into node A and out of node B

6–5 $I_T = 112\,\text{mA}, I_2 = 50\,\text{mA}$

6–6 2.5 mA; 5 mA

6–7 9.33 Ω

6–8 132 Ω

6–9 4 Ω

6–10 1044 Ω

6–11 1.83 mA; 1 mA

6–12 $I_1 = 20\,\text{mA}; I_2 = 9.09\,\text{mA}; I_3 = 35.7\,\text{mA}, I_4 = 22.0\,\text{mA}$

6–13 1.28 V

6–14 Measure R_T with an ohmmeter and calculate R_1 using $R_1 = 1/[(1/R_T) - (1/R_2) - (1/R_3)]$

6–15 30 mA

6–16 $I_1 = 3.27\,\text{mA}; I_2 = 6.73\,\text{mA}$

6–17 $I_1 = 59.4\,\text{mA}; I_2 = 40.6\,\text{mA}$

6–18 1.78 W

6–19 81 W

6–20 0.0347 μA

6–21 15.4 mA

6–22 Not correct, R_{10} (68 kΩ) must be open.

SELF-TEST

  1. (b)   **2.** (c)   **3.** (b)   **4.** (d)   **5.** (a)   **6.** (c)   **7.** (c)   **8.** (c)

  9. (d)   **10.** (b)   **11.** (a)   **12.** (d)   **13.** (c)   **14.** (b)

CIRCUIT DYNAMICS QUIZ

  1. (c)   **2.** (a)   **3.** (c)   **4.** (a)   **5.** (c)   **6.** (c)   **7.** (c)   **8.** (c)

  9. (b)   **10.** (c)   **11.** (b)   **12.** (c)   **13.** (a)   **14.** (c)   **15.** (a)

7

SERIES-PARALLEL CIRCUITS

CHAPTER OBJECTIVES

◆ Identify series-parallel relationships

◆ Analyze series-parallel circuits

◆ Analyze loaded voltage dividers

◆ Determine the loading effect of a voltmeter on a circuit

◆ Analyze ladder networks

◆ Analyze and apply a Wheatstone bridge

◆ Troubleshoot series-parallel circuits

KEY TERMS

◆ Bleeder current

◆ Wheatstone bridge

◆ Balanced bridge

◆ Unbalanced bridge

A CIRCUIT APPLICATION PREVIEW

In the circuit application, you will learn how a Wheatstone bridge in conjunction with a thermistor can be used in a temperature-control application. The circuit in this application is designed to turn a heating element on and off in order to keep the temperature of a liquid in a tank at a desired level.

VISIT THE COMPANION WEBSITE

Study aids and supplementary materials for this chapter are available at http://www.prenhall.com/floyd

INTRODUCTION

In Chapters 5 and 6, series circuits and parallel circuits were studied individually. In this chapter, both series and parallel resistors are combined into series-parallel circuits. In many practical situations, you will have both series and parallel combinations within the same circuit, and the analysis methods you learned for series circuits and for parallel circuits will apply.

Important types of series-parallel circuits are introduced in this chapter. These circuits include the voltage divider with a resistive load, the ladder network, and the Wheatstone bridge.

The analysis of series-parallel circuits requires the use of Ohm's law, Kirchhoff's voltage and current laws, and the methods for finding total resistance and power that you learned in the last two chapters. The topic of loaded voltage dividers is important because this type of circuit is found in many practical situations. One example is the voltage-divider bias circuit for a transistor amplifier, which you will study in a later course. Ladder networks are important in several areas, including a major type of digital-to-analog conversion, which you will study in a digital fundamentals course. The Wheatstone bridge is used in many types of systems for the measurement of unknown parameters, including most electronic scales.

7–1   IDENTIFYING SERIES-PARALLEL RELATIONSHIPS

A series-parallel circuit consists of combinations of both series and parallel current paths. It is important to be able to identify how the components in a circuit are arranged in terms of their series and parallel relationships.

After completing this section, you should be able to

◆ **Identify series-parallel relationships**

   ◆ Recognize how each resistor in a given circuit is related to the other resistors

   ◆ Determine series and parallel relationships on a PC board

Figure 7–1(a) shows an example of a simple series-parallel combination of resistors. Notice that the resistance from point A to point B is R_1. The resistance from point B to point C is R_2 and R_3 in parallel ($R_2 \parallel R_3$). The total resistance from point A to point C is R_1 in series with the parallel combination of R_2 and R_3, as indicated in Figure 7–1(b).

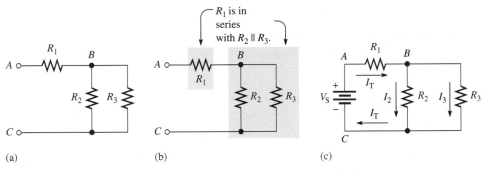

(a)   (b)   (c)

▲ **FIGURE 7–1**

A simple series-parallel resistive circuit.

When the circuit of Figure 7–1(a) is connected to a voltage source as shown in Figure 7–1(c), the total current is through R_1 and divides at point B into the two parallel paths. These two branch currents then recombine, and the total current is into the negative source terminal as shown.

Now, to illustrate series-parallel relationships, let's increase the complexity of the circuit in Figure 7–1(a) step-by-step. In Figure 7–2(a), another resistor (R_4) is connected in series with R_1. The resistance between points A and B is now $R_1 + R_4$, and this combination is in series with the parallel combination of R_2 and R_3, as illustrated in Figure 7–2(b).

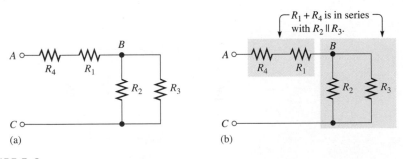

(a)   (b)

▲ **FIGURE 7–2**

R_4 is added to the circuit in series with R_1.

In Figure 7–3(a), R_5 is connected in series with R_2. The series combination of R_2 and R_5 is in parallel with R_3. This entire series-parallel combination is in series with the series combination of R_1 and R_4, as illustrated in Figure 7–3(b).

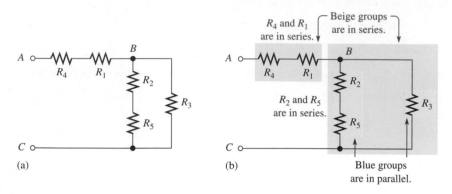

▲ FIGURE 7–3

R_5 is added to the circuit in series with R_2.

In Figure 7–4(a), R_6 is connected in parallel with the series combination of R_1 and R_4. The series-parallel combination of R_1, R_4, and R_6 is in series with the series-parallel combination of R_2, R_3, and R_5, as indicated in Figure 7–4(b).

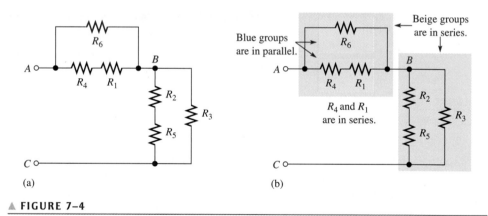

▲ FIGURE 7–4

R_6 is added to the circuit in parallel with the series combination of R_1 and R_4.

EXAMPLE 7–1

Identify the series-parallel relationships in Figure 7–5.

▶ FIGURE 7–5

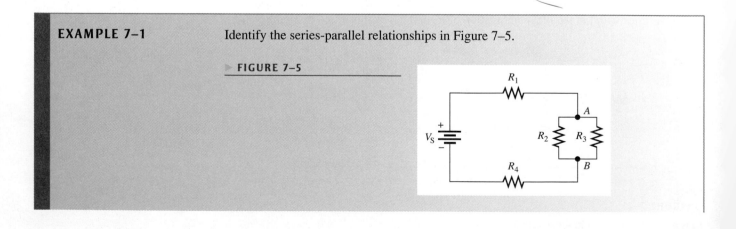

Solution Starting at the positive terminal of the source, follow the current paths. All of the current produced by the source must go through R_1, which is in series with the rest of the circuit.

The total current takes two paths when it gets to node A. Part of it is through R_2, and part of it is through R_3. Resistors R_2 and R_3 are in parallel with each other, and this parallel combination is in series with R_1.

At node B, the currents through R_2 and R_3 come together again. Thus, the total current is through R_4. Resistor R_4 is in series with R_1 and the parallel combination of R_2 and R_3. The currents are shown in Figure 7–6, where I_T is the total current.

▶ **FIGURE 7–6**

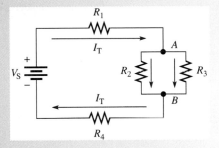

In summary, R_1 and R_4 are in series with the parallel combination of R_2 and R_3 as stated by the following expression:

$$R_1 + R_2 \parallel R_3 + R_4$$

Related Problem* If another resistor, R_5, is connected from node A to the negative side of the source in Figure 7–6, what is its relationship to the other resistors?

*Answers are at the end of the chapter.

EXAMPLE 7–2 Identify the series-parallel relationships in Figure 7–7.

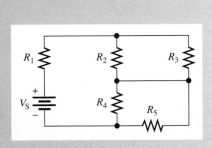

▲ **FIGURE 7–7**

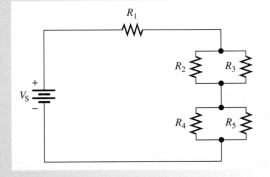

▲ **FIGURE 7–8**

Solution Sometimes it is easier to see a particular circuit arrangement if it is drawn in a different way. In this case, the circuit schematic is redrawn in Figure 7–8, which better illustrates the series-parallel relationships. Now you can see that R_2 and R_3 are in parallel with each other and also that R_4 and R_5 are in parallel with each other. Both parallel combinations are in series with each other and with R_1 as stated by the following expression:

$$R_1 + R_2 \| R_3 + R_4 \| R_5$$

Related Problem If a resistor is connected from the bottom end of R_3 to the top end of R_5 in Figure 7–8, what effect does it have on the circuit?

EXAMPLE 7–3

Describe the series-parallel combination between terminals A and D in Figure 7–9.

▶ **FIGURE 7–9**

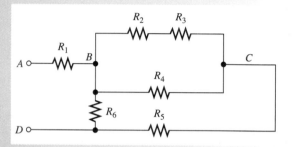

Solution Between nodes B and C, there are two parallel paths. The lower path consists of R_4, and the upper path consists of a series combination of R_2 and R_3. This parallel combination is in series with R_5. The R_2, R_3, R_4, R_5 combination is in parallel with R_6. Resistor R_1 is in series with this entire combination as stated by the following expression:

$$R_1 + R_6 \| (R_5 + R_4 \| (R_2 + R_3))$$

Related Problem If a resistor is connected from C to D in Figure 7–9, describe its parallel relationship.

EXAMPLE 7–4

Describe the total resistance between each pair of terminals in Figure 7–10.

▶ **FIGURE 7–10**

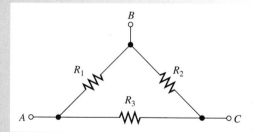

Solution 1. From A to B: R_1 is in parallel with the series combination of R_2 and R_3.

$$R_1 \| (R_2 + R_3)$$

2. From A to C: R_3 is in parallel with the series combination of R_1 and R_2.

$$R_3 \| (R_1 + R_2)$$

3. From B to C: R_2 is in parallel with the series combination of R_1 and R_3.

$$R_2 \| (R_1 + R_3)$$

> *Related Problem* In Figure 7–10, describe the total resistance between each terminal and an added ground if a new resistor, R_4, is connected from C to ground. None of the existing resistors connect directly to the ground.

Usually, the physical arrangement of components on a PC or protoboard bears no resemblance to the actual circuit relationships. By tracing out the circuit and rearranging the components on paper into a recognizable form, you can determine the series-parallel relationships.

EXAMPLE 7–5   Determine the relationships of the resistors on the PC board in Figure 7–11.

> ▶ **FIGURE 7–11**

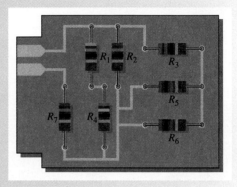

Solution   In Figure 7–12(a), the schematic is drawn in the same arrangement as that of the resistors on the board. In part (b), the resistors are rearranged so that the series-parallel relationships are more obvious. Resistors R_1 and R_4 are in series; $R_1 + R_4$ is in parallel with R_2; R_5 and R_6 are in parallel and this combination is in series with R_3. The R_3, R_5, and R_6 series-parallel combination is in parallel with both R_2 and the $R_1 + R_4$ combination. This entire series-parallel combination is in series with R_7. Figure 7–12(c) illustrates these relationships. Summarizing in equation form,

$$R_{AB} = (R_5 \parallel R_6 + R_3) \parallel R_2 \parallel (R_1 + R_4) + R_7$$

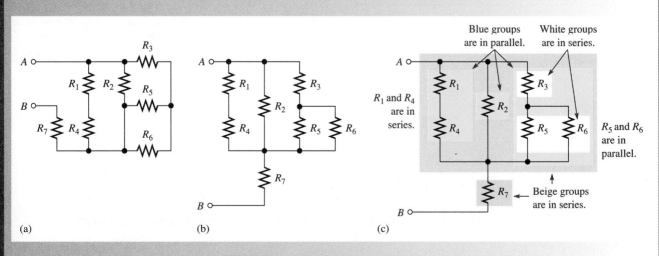

▲ **FIGURE 7–12**

> *Related Problem* If R_5 were removed from the circuit, what would be the relationship of R_3 and R_6?

1. Define *series-parallel resistive circuit*.
2. A certain series-parallel circuit is described as follows: R_1 and R_2 are in parallel. This parallel combination is in series with another parallel combination of R_3 and R_4. Draw the circuit.
3. In the circuit of Figure 7–13, describe the series-parallel relationships of the resistors.
4. Which resistors are in parallel in Figure 7–14?

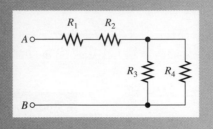

▲ FIGURE 7–13

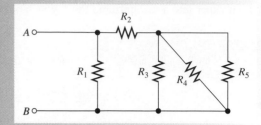

▲ FIGURE 7–14

5. Describe the parallel arrangements in Figure 7–15.
6. Are the parallel combinations in Figure 7–15 in series?

▶ FIGURE 7–15

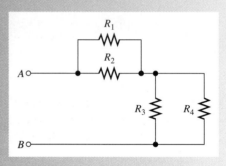

7–2 ANALYSIS OF SERIES-PARALLEL RESISTIVE CIRCUITS

The analysis of series-parallel circuits can be approached in many ways, depending on what information you need and what circuit values you know. The examples in this section do not represent an exhaustive coverage, but they give you an idea of how to approach series-parallel circuit analysis.

After completing this section, you should be able to

◆ **Analyze series-parallel circuits**

  ◆ Determine total resistance

  ◆ Determine all the currents

  ◆ Determine all the voltage drops

If you know Ohm's law, Kirchhoff's laws, the voltage-divider formula, and the current-divider formula, and if you know how to apply these laws, you can solve most resistive circuit analysis problems. The ability to recognize series and parallel combinations is, of course, essential. A few circuits, such as the unbalanced Wheatstone bridge, do not have basic series and parallel combinations. Other methods are needed for these cases, as we will discuss later.

Total Resistance

In Chapter 5, you learned how to determine total series resistance. In Chapter 6, you learned how to determine total parallel resistance. To find the total resistance (R_T) of a series-parallel combination, simply define the series and parallel relationships; then perform the calculations that you have previously learned. The following two examples illustrate this general approach.

EXAMPLE 7–6

Determine R_T of the circuit in Figure 7–16 between terminals A and B.

▶ **FIGURE 7–16**

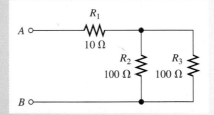

Solution   First, calculate the equivalent parallel resistance of R_2 and R_3. Since R_2 and R_3 are equal in value, you can use Equation 6–4.

$$R_{2\|3} = \frac{R}{n} = \frac{100 \ \Omega}{2} = 50 \ \Omega$$

Notice that the term $R_{2\|3}$ is used here to designate the total resistance of a portion of a circuit in order to distinguish it from the total resistance, R_T, of the complete circuit. Now, since R_1 is in series with $R_{2\|3}$, add their values as follows:

$$R_T = R_1 + R_{2\|3} = 10 \ \Omega + 50 \ \Omega = \mathbf{60 \ \Omega}$$

Related Problem   Determine R_T in Figure 7–16 if R_3 is changed to 82 Ω.

EXAMPLE 7–7

Find the total resistance between the positive and negative terminals of the battery in Figure 7–17.

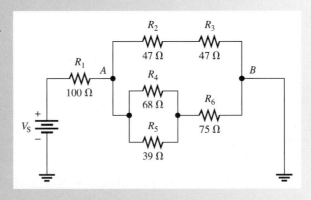

▲ **FIGURE 7–17**

Solution In the upper branch, R_2 is in series with R_3. This series combination is designated R_{2+3} and is equal to $R_2 + R_3$.

$$R_{2+3} = R_2 + R_3 = 47\ \Omega + 47\ \Omega = 94\ \Omega$$

In the lower branch, R_4 and R_5 are in parallel with each other. This parallel combination is designated $R_{4\|5}$.

$$R_{4\|5} = \frac{R_4 R_5}{R_4 + R_5} = \frac{(68\ \Omega)(39\ \Omega)}{68\ \Omega + 39\ \Omega} = 24.8\ \Omega$$

Also in the lower branch, the parallel combination of R_4 and R_5 is in series with R_6. This series-parallel combination is designated $R_{4\|5+6}$.

$$R_{4\|5+6} = R_6 + R_{4\|5} = 75\ \Omega + 24.8\ \Omega = 99.8\ \Omega$$

Figure 7–18 shows the original circuit in a simplified equivalent form.

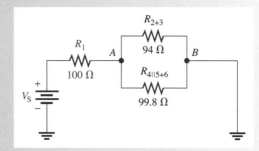

▲ **FIGURE 7–18**

Now you can find the equivalent resistance between *A* and *B*. It is R_{2+3} in parallel with $R_{4\|5+6}$. Calculate the equivalent resistance as follows:

$$R_{AB} = \frac{1}{\dfrac{1}{R_{2+3}} + \dfrac{1}{R_{4\|5+6}}} = \frac{1}{\dfrac{1}{94\ \Omega} + \dfrac{1}{99.8\ \Omega}} = 48.4\ \Omega$$

Finally, the total resistance is R_1 in series with R_{AB}.

$$R_T = R_1 + R_{AB} = 100\ \Omega + 48.4\ \Omega = \mathbf{148.4\ \Omega}$$

Related Problem Determine R_T if a 68 Ω resistor is added in parallel from *A* to *B* in Figure 7–17.

Total Current

Once you know the total resistance and the source voltage, you can apply Ohm's law to find the total current in a circuit. Total current is the source voltage divided by the total resistance.

$$I_T = \frac{V_S}{R_T}$$

For example, assuming that the source voltage is 30 V, the total current in the circuit of Example 7–7 (Figure 7–17) is

$$I_T = \frac{V_S}{R_T} = \frac{30\ \text{V}}{148.4\ \Omega} = 202\ \text{mA}$$

Branch Currents

Using the current-divider formula, Kirchhoff's current law, Ohm's law, or combinations of these, you can find the current in any branch of a series-parallel circuit. In some cases, it may take repeated application of the formula to find a given current. The following two examples will help you understand the procedure. (Notice that the subscripts for the current variables (I) match the R subscripts; for example, current through R_1 is referred to as I_1.)

EXAMPLE 7–8

Find the current through R_2 and the current through R_3 in Figure 7–19.

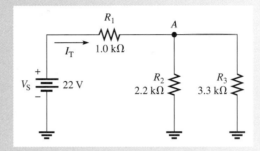

▲ **FIGURE 7–19**

Solution

First, identify the series and parallel relationship. Next, determine how much current is into node A. This is the total circuit current. To find I_T, you must know R_T.

$$R_T = R_1 + \frac{R_2 R_3}{R_2 + R_3} = 1.0\,\text{k}\Omega + \frac{(2.2\,\text{k}\Omega)(3.3\,\text{k}\Omega)}{2.2\,\text{k}\Omega + 3.3\,\text{k}\Omega} = 1.0\,\text{k}\Omega + 1.32\,\text{k}\Omega = 2.32\,\text{k}\Omega$$

$$I_T = \frac{V_S}{R_T} = \frac{22\,\text{V}}{2.32\,\text{k}\Omega} = 9.48\,\text{mA}$$

Use the current-divider rule for two branches as given in Chapter 6 to find the current through R_2.

$$I_2 = \left(\frac{R_3}{R_2 + R_3}\right) I_T = \left(\frac{3.3\,\text{k}\Omega}{5.5\,\text{k}\Omega}\right) 9.48\,\text{mA} = \mathbf{5.69\,mA}$$

Now you can use Kirchhoff's current law to find the current through R_3.

$$I_T = I_2 + I_3$$
$$I_3 = I_T - I_2 = 9.48\,\text{mA} - 5.69\,\text{mA} = \mathbf{3.79\,mA}$$

Related Problem

A 4.7 kΩ resistor is connected in parallel with R_3 in Figure 7–19. Determine the current through the new resistor.

Use Multisim file E07-08 to verify the calculated results in this example and to confirm your calculation for the related problem.

EXAMPLE 7–9

Determine the current through R_4 in Figure 7–20 if $V_S = 50$ V.

▶ **FIGURE 7–20**

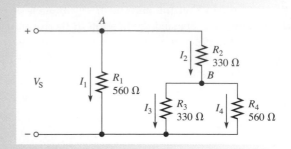

Solution First, find the current (I_2) into node B. Once you know this current, use the current-divider formula to find I_4, the current through R_4.

Notice that there are two main branches in the circuit. The left-most branch consists of only R_1. The right-most branch has R_2 in series with the parallel combination of R_3 and R_4. The voltage across both of these main branches is the same and equal to 50 V. Calculate the equivalent resistance ($R_{2+3\|4}$) of the right-most main branch and then apply Ohm's law; I_2 is the total current through this main branch. Thus,

$$R_{2+3\|4} = R_2 + \frac{R_3 R_4}{R_3 + R_4} = 330\ \Omega + \frac{(330\ \Omega)(560\ \Omega)}{890\ \Omega} = 538\ \Omega$$

$$I_2 = \frac{V_S}{R_{2+3\|4}} = \frac{50\ \text{V}}{538\ \Omega} = 93\ \text{mA}$$

Use the two-resistor current-divider formula to calculate I_4.

$$I_4 = \left(\frac{R_3}{R_3 + R_4}\right) I_2 = \left(\frac{330\ \Omega}{890\ \Omega}\right) 93\ \text{mA} = \textbf{34.5 mA}$$

Related Problem Determine the current through R_1 and R_3 in Figure 7–20 if $V_S = 20$ V.

Voltage Drops

To find the voltages across certain parts of a series-parallel circuit, you can use the voltage-divider formula given in Chapter 5, Kirchhoff's voltage law, Ohm's law, or combinations of each. The following three examples illustrate use of the formulas. (The subscripts for V match the subscripts for the corresponding R: V_1 is the voltage across R_1; V_2 is the voltage across R_2, etc.)

EXAMPLE 7–10

Determine the voltage drop from node A to ground in Figure 7–21. Then find the voltage (V_1) across R_1.

▶ **FIGURE 7–21**

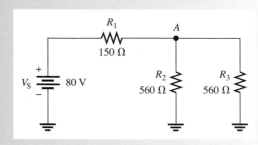

Solution Note that R_2 and R_3 are in parallel in this circuit. Since they are equal in value, their equivalent resistance from node A to ground is

$$R_A = \frac{560 \ \Omega}{2} = 280 \ \Omega$$

In the equivalent circuit shown in Figure 7–22, R_1 is in series with R_A. The total circuit resistance as seen from the source is

$$R_T = R_1 + R_A = 150 \ \Omega + 280 \ \Omega = 430 \ \Omega$$

► **FIGURE 7–22**

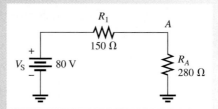

Use the voltage-divider formula to find the voltage across the parallel combination of Figure 7–21 (between node A and ground).

$$V_A = \left(\frac{R_A}{R_T}\right)V_S = \left(\frac{280 \ \Omega}{430 \ \Omega}\right)80 \ V = \mathbf{52.1 \ V}$$

Now use Kirchhoff's voltage law to find V_1.

$$V_S = V_1 + V_A$$
$$V_1 = V_S - V_A = 80 \ V - 52.1 \ V = \mathbf{27.9 \ V}$$

Related Problem Determine V_A and V_1 if R_1 is changed to $220 \ \Omega$ in Figure 7–21.

Use Multisim file E07-10 to verify the calculated results in this example and to confirm your calculation for the related problem.

EXAMPLE 7–11 Determine the voltage drop across each resistor in the circuit of Figure 7–23.

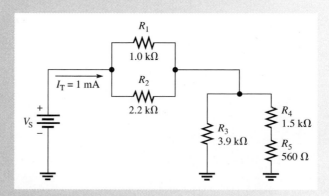

▲ **FIGURE 7–23**

Solution The source voltage is not given, but you know the total current from the figure. Since R_1 and R_2 are in parallel, they each have the same voltage. The current through R_1 is

$$I_1 = \left(\frac{R_2}{R_1 + R_2}\right)I_T = \left(\frac{2.2\,k\Omega}{3.2\,k\Omega}\right)1\,mA = 688\,\mu A$$

The voltages across R_1 and R_2 are

$$V_1 = I_1 R_1 = (688\,\mu A)(1.0\,k\Omega) = \mathbf{688\,mV}$$
$$V_2 = V_1 = \mathbf{688\,mV}$$

The series combination of R_4 and R_5 form the branch resistance, R_{4+5}. Apply the current-divider formula to determine the current through R_3.

$$I_3 = \left(\frac{R_{4+5}}{R_3 + R_{4+5}}\right)I_T = \left(\frac{2.06\,k\Omega}{5.96\,k\Omega}\right)1\,mA = 346\,\mu A$$

The voltage across R_3 is

$$V_3 = I_3 R_3 = (346\,\mu A)(3.9\,k\Omega) = \mathbf{1.35\,V}$$

The currents through R_4 and R_5 are the same because these resistors are in series.

$$I_4 = I_5 = I_T - I_3 = 1\,mA - 346\,\mu A = 654\,\mu A$$

Calculate the voltages across R_4 and R_5 as follows:

$$V_4 = I_4 R_4 = (654\,\mu A)(1.5\,k\Omega) = \mathbf{981\,mV}$$
$$V_5 = I_5 R_5 = (654\,\mu A)(560\,\Omega) = \mathbf{366\,mV}$$

Related Problem What is the source voltage, V_S, in the circuit of Figure 7–23?

 Use Multisim file E07-11 to verify the calculated results in this example and to confirm your calculation for the related problem.

EXAMPLE 7–12 Determine the voltage drop across each resistor in Figure 7–24.

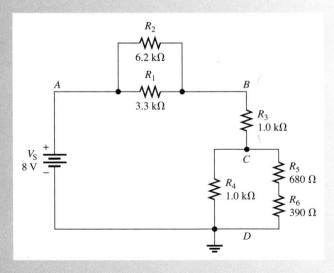

▲ FIGURE 7–24

Solution Because the total voltage is given in the figure, you can solve this problem using the voltage-divider formula. First, you need to reduce each parallel combination to an equivalent resistance. Since R_1 and R_2 are in parallel between A and B, combine their values.

$$R_{AB} = \frac{R_1 R_2}{R_1 + R_2} = \frac{(3.3\,k\Omega)(6.2\,k\Omega)}{9.5\,k\Omega} = 2.15\,k\Omega$$

Since R_4 is in parallel with the R_5 and R_6 series combination (R_{5+6}) between C and D, combine these values.

$$R_{CD} = \frac{R_4 R_{5+6}}{R_4 + R_{5+6}} = \frac{(1.0\,k\Omega)(1.07\,k\Omega)}{2.07\,k\Omega} = 517\,\Omega$$

The equivalent circuit is drawn in Figure 7–25. The total circuit resistance is

$$R_T = R_{AB} + R_3 + R_{CD} = 2.15\,k\Omega + 1.0\,k\Omega + 517\,\Omega = 3.67\,k\Omega$$

▶ **FIGURE 7–25**

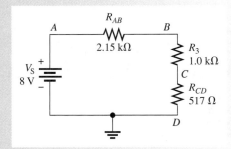

Next, use the voltage-divider formula to determine the voltages in the equivalent circuit.

$$V_{AB} = \left(\frac{R_{AB}}{R_T}\right)V_S = \left(\frac{2.15\,k\Omega}{3.67\,k\Omega}\right)8\,V = 4.69\,V$$

$$V_{CD} = \left(\frac{R_{CD}}{R_T}\right)V_S = \left(\frac{517\,\Omega}{3.67\,k\Omega}\right)8\,V = 1.13\,V$$

$$V_3 = \left(\frac{R_3}{R_T}\right)V_S = \left(\frac{1.0\,k\Omega}{3.67\,k\Omega}\right)8\,V = \mathbf{2.18\,V}$$

Refer to Figure 7–24. V_{AB} equals the voltage across both R_1 and R_2, so

$$V_1 = V_2 = V_{AB} = \mathbf{4.69\,V}$$

V_{CD} is the voltage across R_4 and across the series combination of R_5 and R_6. Therefore,

$$V_4 = V_{CD} = \mathbf{1.13\,V}$$

Now apply the voltage-divider formula to the series combination of R_5 and R_6 to get V_5 and V_6.

$$V_5 = \left(\frac{R_5}{R_5 + R_6}\right)V_{CD} = \left(\frac{680\,\Omega}{1070\,\Omega}\right)1.13\,V = \mathbf{718\,mV}$$

$$V_6 = \left(\frac{R_6}{R_5 + R_6}\right)V_{CD} = \left(\frac{390\,\Omega}{1070\,\Omega}\right)1.13\,V = \mathbf{412\,mV}$$

Related Problem R_2 is removed from the circuit in Figure 7–24. Calculate V_{AB}, V_{BC}, and V_{CD}.

Use Multisim file E07-12 to verify the calculated results in this example and to confirm your calculation for the related problem.

SECTION 7–2
REVIEW

1. List four circuit laws and formulas that may be necessary in the analysis of series-parallel circuits.
2. Find the total resistance between A and B in the circuit of Figure 7–26.
3. Find the current through R_3 in Figure 7–26.
4. Find the voltage drop across R_2 in Figure 7–26.
5. Determine R_T and I_T in Figure 7–27 as "seen" by the source.

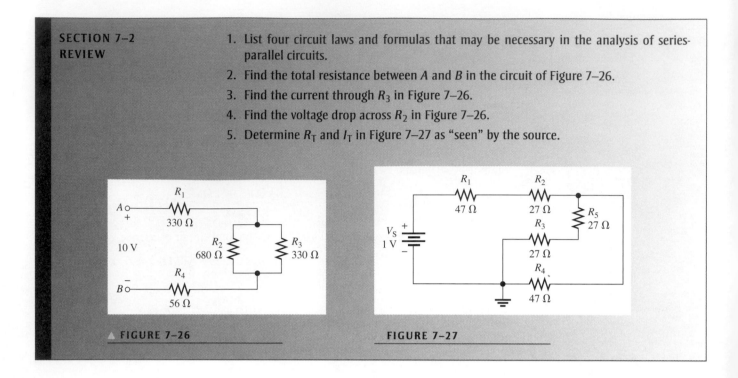

▲ FIGURE 7–26   FIGURE 7–27

7–3   VOLTAGE DIVIDERS WITH RESISTIVE LOADS

Voltage dividers were introduced in Chapter 5. In this section, you will learn how resistive loads affect the operation of voltage-divider circuits.

After completing this section, you should be able to

◆ **Analyze loaded voltage dividers**

  ◆ Determine the effect of a resistive load on a voltage-divider circuit

  ◆ Define *bleeder current*

The voltage divider in Figure 7–28(a) produces an output voltage (V_{OUT}) of 5 V because the two resistors are of equal value. This voltage is the *unloaded output voltage.* When a load resistor, R_L, is connected from the output to ground as shown in Figure 7–28(b), the output voltage is reduced by an amount that depends on the value of R_L. The load resistor is in parallel with R_2, reducing the resistance from node A to ground and, as a result, also

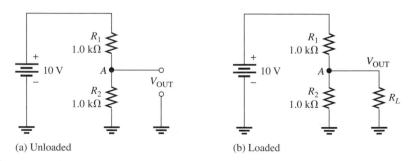

(a) Unloaded   (b) Loaded

▲ FIGURE 7–28

A voltage divider with both unloaded and loaded outputs.

reducing the voltage across the parallel combination. This is one effect of loading a voltage divider. Another effect of a load is that more current is drawn from the source because the total resistance of the circuit is reduced.

The larger R_L is, compared to R_2, the less the output voltage is reduced from its unloaded value, as illustrated in Figure 7–29. When two resistors are connected in parallel and one of the resistors is much greater than the other, the total resistance is close to the value of the smaller resistance.

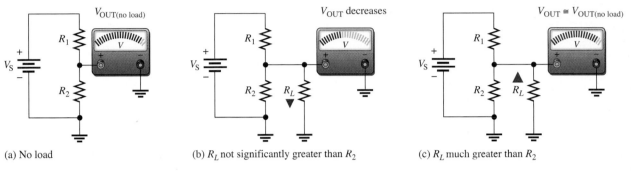

(a) No load   (b) R_L not significantly greater than R_2 (c) R_L much greater than R_2

▲ FIGURE 7–29

The effect of a load resistor.

EXAMPLE 7–13

(a) Determine the unloaded output voltage of the voltage divider in Figure 7–30.

(b) Find the loaded output voltages of the voltage divider in Figure 7–30 for the following two values of load resistance: $R_L = 10\,k\Omega$ and $R_L = 100\,k\Omega$.

▷ FIGURE 7–30

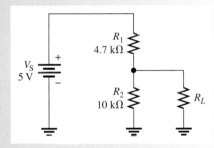

Solution   (a) The unloaded output voltage is

$$V_{OUT(unloaded)} = \left(\frac{R_2}{R_1 + R_2}\right)V_S = \left(\frac{10\,k\Omega}{14.7\,k\Omega}\right)5\,V = \textbf{3.40 V}$$

(b) With the $10\,k\Omega$ load resistor connected, R_L is in parallel with R_2, which gives

$$R_2 \parallel R_L = \frac{R_2 R_L}{R_2 + R_L} = \frac{100\,M\Omega}{20\,k\Omega} = 5\,k\Omega$$

The equivalent circuit is shown in Figure 7–31(a). The loaded output voltage is

$$V_{OUT(loaded)} = \left(\frac{R_2 \parallel R_L}{R_1 + R_2 \parallel R_L}\right)V_S = \left(\frac{5\,k\Omega}{9.7\,k\Omega}\right)5\,V = \textbf{2.58 V}$$

With the $100\,k\Omega$ load, the resistance from output to ground is

$$R_2 \parallel R_L = \frac{R_2 R_L}{R_2 + R_L} = \frac{(10\,k\Omega)(100\,k\Omega)}{110\,k\Omega} = 9.1\,k\Omega$$

The equivalent circuit is shown in Figure 7–31(b). The loaded output voltage is

$$V_{\text{OUT(loaded)}} = \left(\frac{R_2 \parallel R_L}{R_1 + R_2 \parallel R_L}\right)V_S = \left(\frac{9.1\text{ k}\Omega}{13.8\text{ k}\Omega}\right)5\text{ V} = \mathbf{3.30\text{ V}}$$

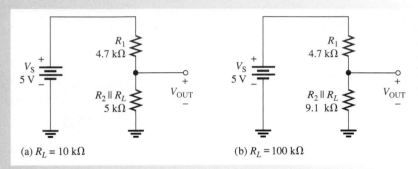

(a) $R_L = 10$ kΩ (b) $R_L = 100$ kΩ

FIGURE 7–31

For the smaller value of R_L, the reduction in V_{OUT} is

$$3.40\text{ V} - 2.58\text{ V} = 0.82\text{ V}$$

For the larger value of R_L, the reduction in V_{OUT} is

$$3.40\text{ V} - 3.30\text{ V} = 0.10\text{ V}$$

This illustrates the loading effect of R_L on the voltage divider.

Related Problem   Determine V_{OUT} in Figure 7–30 for a 1.0 MΩ load resistance.

Use Multisim file E07-13 to verify the calculated results in this example and to confirm your calculation for the related problem.

Load Current and Bleeder Current

In a multiple-tap loaded voltage-divider circuit, the total current drawn from the source consists of currents through the load resistors, called *load currents*, and the divider resistors. Figure 7–32 shows a voltage divider with two voltage outputs or two taps. Notice that the total current, I_T, through R_1 enters node A where the current divides into I_{RL1} through R_{L1} and into I_2 through R_2. At node B, the current I_2 divides into I_{RL2} through R_{RL2} and into

▶ **FIGURE 7–32**

Currents in a two-tap loaded voltage divider.

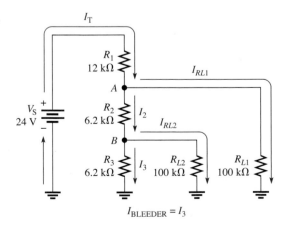

I_3 through R_3. Current I_3 is called the **bleeder current**, which is the current left after the total load current is subtracted from the total current in the circuit.

$$I_{\text{BLEEDER}} = I_{\text{T}} - I_{RL1} - I_{RL2}$$

Equation 7–1

EXAMPLE 7–14

Determine the load currents I_{RL1} and I_{RL2} and the bleeder current I_3 in the two-tap loaded voltage divider in Figure 7–32.

Solution The equivalent resistance from node A to ground is the $100\,\text{k}\Omega$ load resistor R_{L1} in parallel with the combination of R_2 in series with the parallel combination of R_3 and R_{L2}. Determine the resistance values first. R_3 in parallel with R_{L2} is designated R_B. The resulting equivalent circuit is shown in Figure 7–33(a).

$$R_B = \frac{R_3 R_{L2}}{R_3 + R_{L2}} = \frac{(6.2\,\text{k}\Omega)(100\,\text{k}\Omega)}{106.2\,\text{k}\Omega} = 5.84\,\text{k}\Omega$$

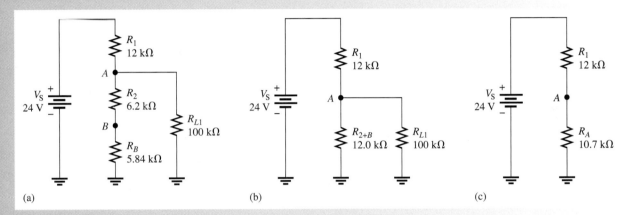

(a)   (b)   (c)

▲ **FIGURE 7–33**

R_2 in series with R_B is designated R_{2+B}. The resulting equivalent circuit is shown in Figure 7–33(b).

$$R_{2+B} = R_2 + R_B = 6.2\,\text{k}\Omega + 5.84\,\text{k}\Omega = 12.0\,\text{k}\Omega$$

R_{L1} in parallel with R_{2+B} is designated R_A. The resulting equivalent circuit is shown in Figure 7–33(c).

$$R_A = \frac{R_{L1} R_{2+B}}{R_{L1} + R_{2+B}} = \frac{(100\,\text{k}\Omega)(12.0\,\text{k}\Omega)}{112\,\text{k}\Omega} = 10.7\,\text{k}\Omega$$

R_A is the total resistance from node A to ground. The total resistance for the circuit is

$$R_{\text{T}} = R_A + R_1 = 10.7\,\text{k}\Omega + 12\,\text{k}\Omega = 22.7\,\text{k}\Omega$$

Determine the voltage across R_{L1} as follows, using the equivalent circuit in Figure 7–33(c):

$$V_{RL1} = V_A = \left(\frac{R_A}{R_{\text{T}}}\right) V_{\text{S}} = \left(\frac{10.7\,\text{k}\Omega}{22.7\,\text{k}\Omega}\right) 24\,\text{V} = 11.3\,\text{V}$$

The load current through R_{L1} is

$$I_{RL1} = \frac{V_{RL1}}{R_{L1}} = \left(\frac{11.3\,\text{V}}{100\,\text{k}\Omega}\right) = \mathbf{113\,\mu A}$$

Determine the voltage at node B by using the equivalent circuit in Figure 7–33(a) and the voltage at node A.

$$V_B = \left(\frac{R_B}{R_{2+B}}\right)V_A = \left(\frac{5.84\text{ k}\Omega}{12.0\text{ k}\Omega}\right)11.3\text{ V} = 5.50\text{ V}$$

The load current through R_{L2} is

$$I_{RL2} = \frac{V_{RL2}}{R_{L2}} = \frac{V_B}{R_{L2}} = \frac{5.50\text{ V}}{100\text{ k}\Omega} = \textbf{55 } \boldsymbol{\mu}\textbf{A}$$

The bleeder current is

$$I_3 = \frac{V_B}{R_3} = \frac{5.50\text{ V}}{6.2\text{ k}\Omega} = \textbf{887}\boldsymbol{\mu}\textbf{A}$$

Related Problem How can the bleeder current in Figure 7–32 be reduced without affecting the load currents?

Use Multisim file E07-14 to verify the calculated results in this example.

Bipolar Voltage Dividers

An example of a voltage divider that produces both positive and negative voltages from a single source is shown in Figure 7–34. Notice that neither the positive nor the negative terminal of the source is connected to reference ground or common. The voltages at nodes A and B are positive with respect to reference ground, and the voltages at nodes C and D are negative with respect to reference ground.

▶ **FIGURE 7–34**

A bipolar voltage divider. The positive and negative voltages are with respect to reference ground.

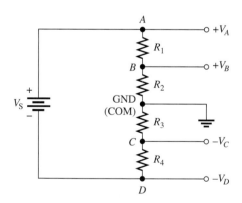

1. A load resistor is connected to an output tap on a voltage divider. What effect does the load resistor have on the output voltage at this tap?

2. A larger-value load resistor will cause the output voltage to change less than a smaller-value one will.   (T or F)

3. For the voltage divider in Figure 7–35, determine the unloaded output voltage with respect to ground. Also determine the output voltage with a 10 kΩ load resistor connected across the output.

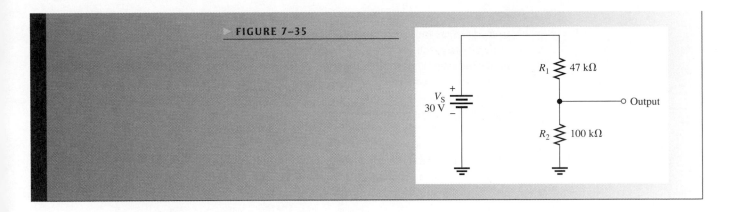

▶ FIGURE 7–35

R_1 47 kΩ

V_S 30 V

Output

R_2 100 kΩ

7–4 LOADING EFFECT OF A VOLTMETER

As you have learned, voltmeters must be connected in parallel with a resistor in order to measure the voltage across the resistor. Because of its internal resistance, a voltmeter puts a load on the circuit and will affect, to a certain extent, the voltage that is being measured. Until now, we have ignored the loading effect because the internal resistance of a voltmeter is very high, and normally it has negligible effect on the circuit that is being measured. However, if the internal resistance of the voltmeter is not sufficiently greater than the circuit resistance across which it is connected, the loading effect will cause the measured voltage to be less than its actual value. You should always be aware of this effect.

After completing this section, you should be able to

◆ **Determine the loading effect of a voltmeter on a circuit**

  ◆ Explain why a voltmeter can load a circuit

  ◆ Discuss the internal resistance of a voltmeter

When a voltmeter is connected to a circuit as shown, for example, in Figure 7–36(a), its internal resistance appears in parallel with R_3, as shown in part (b). The resistance from A

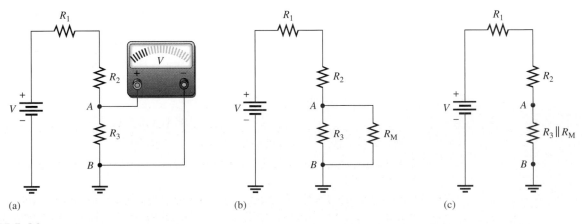

(a) (b) (c)

▲ FIGURE 7–36

The loading effect of a voltmeter.

to B is altered by the loading effect of the voltmeter's internal resistance, R_M, and is equal to $R_3 \parallel R_M$, as indicated in part (c).

If R_M is much greater than R_3, the resistance from A to B changes very little, and the meter indicates the actual voltage. If R_M is not sufficiently greater than R_3, the resistance from A to B is reduced significantly, and the voltage across R_3 is altered by the loading effect of the meter. A good rule of thumb is that *if the loading effect is less than 10%, it can usually be neglected, depending on the accuracy required.*

Two categories of voltmeters are the electromagnetic analog voltmeter (commonly called VOM), whose internal resistance is determined by its sensitivity factor, and the digital voltmeter (the most commonly used type and commonly called DMM), whose internal resistance is also typically at least 10 MΩ. The digital voltmeter presents fewer loading problems than the electromagnetic type because the internal resistances of DMMs are much higher.

EXAMPLE 7–15

How much does the digital voltmeter affect the voltage being measured for each circuit indicated in Figure 7–37? Assume the meter has an input resistance (R_M) of 10 MΩ.

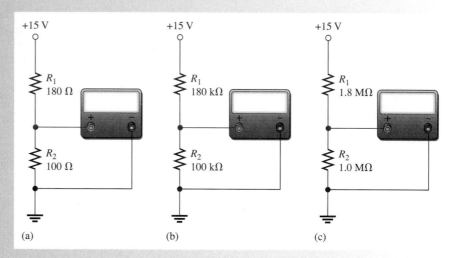

(a) (b) (c)

▲ **FIGURE 7–37**

Solution To show the small differences more clearly, the results are expressed in more than three significant figures in this example.

(a) Refer to Figure 7–37(a). The unloaded voltage across R_2 in the voltage-divider circuit is

$$V_2 = \left(\frac{R_2}{R_1 + R_2}\right)V_S = \left(\frac{100 \ \Omega}{280 \ \Omega}\right)15 \text{ V} = 5.357 \text{ V}$$

The meter's resistance in parallel with R_2 is

$$R_2 \parallel R_M = \left(\frac{R_2 R_M}{R_2 + R_M}\right) = \frac{(100 \ \Omega)(10 \text{ M}\Omega)}{10.0001 \text{ M}\Omega} = 99.999 \ \Omega$$

The voltage actually measured by the meter is

$$V_2 = \left(\frac{R_2 \parallel R_M}{R_1 + R_2 \parallel R_M}\right)V_S = \left(\frac{99.999 \ \Omega}{279.999 \ \Omega}\right)15 \text{ V} = 5.357 \text{ V}$$

The voltmeter has no measurable loading effect.

(b) Refer to Figure 7–37(b).

$$V_2 = \left(\frac{R_2}{R_1 + R_2}\right)V_S = \left(\frac{100\,k\Omega}{280\,k\Omega}\right)15\,V = 5.357\,V$$

$$R_2 \parallel R_M = \frac{R_2 R_M}{R_2 + R_M} = \frac{(100\,k\Omega)(10\,M\Omega)}{10.1\,M\Omega} = 99.01\,k\Omega$$

The voltage actually measured by the meter is

$$V_2 = \left(\frac{R_2 \parallel R_M}{R_1 + R_2 \parallel R_M}\right)V_S = \left(\frac{99.01\,k\Omega}{279.01\,k\Omega}\right)15\,V = 5.323\,V$$

The loading effect of the voltmeter reduces the voltage by a very small amount.

(c) Refer to Figure 7–37(c).

$$V_2 = \left(\frac{R_2}{R_1 + R_2}\right)V_S = \left(\frac{1.0\,M\Omega}{2.8\,M\Omega}\right)15\,V = 5.357\,V$$

$$R_2 \parallel R_M = \frac{R_2 R_M}{R_2 + R_M} = \frac{(1.0\,M\Omega)(10\,M\Omega)}{11\,M\Omega} = 909.09\,k\Omega$$

The voltage actually measured is

$$V_2 = \left(\frac{R_2 \parallel R_M}{R_1 + R_2 \parallel R_M}\right)V_S = \left(\frac{909.09\,k\Omega}{2.709\,M\Omega}\right)15\,V = 5.034\,V$$

The loading effect of the voltmeter reduces the voltage by a noticeable amount. As you can see, the higher the resistance across which a voltage is measured, the more the loading effect.

Related Problem   Calculate the voltage across R_2 in Figure 7–37(c) if the meter resistance is 20 MΩ.

SECTION 7–4 REVIEW

1. Explain why a voltmeter can potentially load a circuit.
2. If a voltmeter with a 10 MΩ internal resistance is measuring the voltage across a 1.0 kΩ resistor, should you normally be concerned about the loading effect?
3. If a voltmeter with a 10 MΩ resistance is measuring the voltage across a 3.3 MΩ resistor, should you be concerned about the loading effect?

7–5 LADDER NETWORKS

A resistive ladder network is a special type of series-parallel circuit. The R/2R ladder network is commonly used to scale down voltages to certain weighted values for digital-to-analog conversion, which is a process that you will study in another course.

After completing this section, you should be able to

♦ **Analyze ladder networks**

   ♦ Determine the voltages in a three-step ladder network

   ♦ Analyze an R/2R ladder

One approach to the analysis of a ladder network such as the one shown in Figure 7–38 is to simplify it one step at a time, starting at the side farthest from the source. In this way, you can determine the current in any branch or the voltage at any node, as illustrated in Example 7–16.

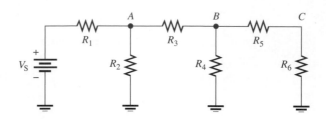

▲ FIGURE 7–38

Basic three-step ladder network.

EXAMPLE 7–16

Determine the current through each resistor and the voltage at each labeled node with respect to ground in the ladder network of Figure 7–39.

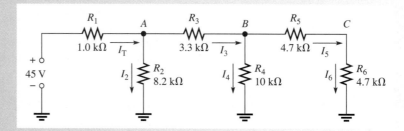

▲ FIGURE 7–39

Solution   To find the current through each resistor, you must know the total current from the source (I_T). To obtain I_T, you must find the total resistance "seen" by the source.

Determine R_T in a step-by-step process, starting at the right of the circuit diagram. First, notice that R_5 and R_6 are in series across R_4. Neglecting the circuit to the left of node B, the resistance from node B to ground is

$$R_B = \frac{R_4(R_5 + R_6)}{R_4 + (R_5 + R_6)} = \frac{(10\,\text{k}\Omega)(9.4\,\text{k}\Omega)}{19.4\,\text{k}\Omega} = 4.85\,\text{k}\Omega$$

Using R_B, you can draw the equivalent circuit as shown in Figure 7–40.

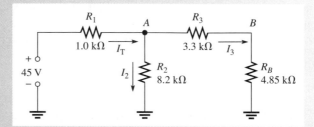

▲ FIGURE 7–40

Next, neglecting the circuit to the left of node A, the resistance from node A to ground (R_A) is R_2 in parallel with the series combination of R_3 and R_B. Calculate resistance R_A.

$$R_A = \frac{R_2(R_3 + R_B)}{R_2 + (R_3 + R_B)} = \frac{(8.2\,\text{k}\Omega)(8.15\,\text{k}\Omega)}{16.35\,\text{k}\Omega} = 4.09\,\text{k}\Omega$$

Using R_A, you can further simplify the equivalent circuit of Figure 7–40 as shown in Figure 7–41.

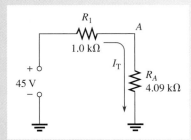

▲ FIGURE 7–41

Finally, the total resistance "seen" by the source is R_1 in series with R_A.

$$R_T = R_1 + R_A = 1.0\,\text{k}\Omega + 4.09\,\text{k}\Omega = 5.09\,\text{k}\Omega$$

The total circuit current is

$$I_T = \frac{V_S}{R_T} = \frac{45\,\text{V}}{5.09\,\text{k}\Omega} = \textbf{8.84 mA}$$

As indicated in Figure 7–40, I_T is into node A and divides between R_2 and the branch containing $R_3 + R_B$. Since the branch resistances are approximately equal in this particular example, half the total current is through R_2 and half into node B. So the currents through R_2 and R_3 are

$$I_2 = \textbf{4.42 mA}$$
$$I_3 = \textbf{4.42 mA}$$

If the branch resistances are not equal, use the current-divider formula. As indicated in Figure 7–39, I_3 is into node B and is divided between R_4 and the branch containing $R_5 + R_6$. Therefore, the currents through R_4, R_5, and R_6 can be calculated.

$$I_4 = \left(\frac{R_5 + R_6}{R_4 + (R_5 + R_6)}\right)I_3 = \left(\frac{9.4\,\text{k}\Omega}{19.4\,\text{k}\Omega}\right)4.42\,\text{mA} = \textbf{2.14 mA}$$

$$I_5 = I_6 = I_3 - I_4 = 4.42\,\text{mA} - 2.14\,\text{mA} = \textbf{2.28 mA}$$

To determine V_A, V_B, and V_C, apply Ohm's law.

$$V_A = I_2 R_2 = (4.42\,\text{mA})(8.2\,\text{k}\Omega) = \textbf{36.2 V}$$
$$V_B = I_4 R_4 = (2.14\,\text{mA})(10\,\text{k}\Omega) = \textbf{21.4 V}$$
$$V_C = I_6 R_6 = (2.28\,\text{mA})(4.7\,\text{k}\Omega) = \textbf{10.7 V}$$

Related Problem   Recalculate the currents through each resistor and the voltages at each node in Figure 7–39 if R_1 is increased to 2.2 kΩ.

Use Multisim file E07-16 to verify the calculated results in this example and to confirm your calculations for the related problem.

The *R/2R* Ladder Network

A basic *R/2R* ladder network is shown in Figure 7–42. As you can see, the name comes from the relationship of the resistor values. *R* represents a common value, and one set of resistors has twice the value of the others. This type of ladder network is used in applications where digital codes are converted to speech, music, or other types of analog signals as found, for example, in the area of digital recording and reproduction. This application is called *digital-to-analog (D/A) conversion.*

▶ **FIGURE 7–42**

A basic four-step *R/2R* ladder network.

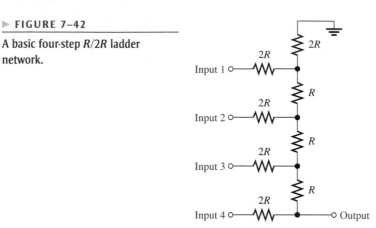

Let's examine the general operation of a basic *R/2R* ladder using the four-step circuit in Figure 7–43. In a later course in digital fundamentals, you will learn specifically how this type of circuit is used in D/A conversion.

▶ **FIGURE 7–43**

R/2R ladder with switch inputs to simulate a two-level (digital) code.

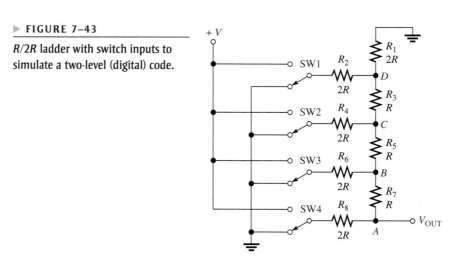

The switches used in this illustration simulate the digital (two-level) inputs. One switch position is connected to ground (0 V), and the other position is connected to a positive voltage (*V*). The analysis is as follows: Start by assuming that switch SW4 in Figure 7–43 is at the *V* position and the others are at ground so that the inputs are as shown in Figure 7–44(a).

The total resistance from node *A* to ground is found by first combining R_1 and R_2 in parallel from node *D* to ground. The simplified circuit is shown in Figure 7–44(b).

$$R_1 \parallel R_2 = \frac{2R}{2} = R$$

$R_1 \parallel R_2$ is in series with R_3 from node *C* to ground as illustrated in part (c).

$$R_1 \parallel R_2 + R_3 = R + R = 2R$$

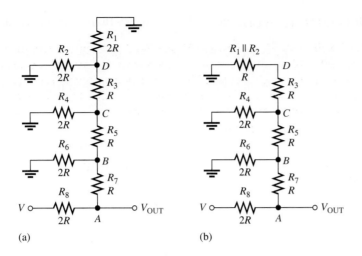

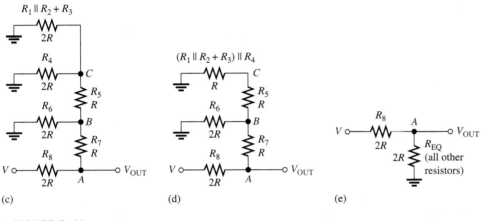

FIGURE 7–44

Simplification of R/2R ladder for analysis.

Next, this combination is in parallel with R_4 from node C to ground as shown in part (d).

$$(R_1 \parallel R_2 + R_3) \parallel R_4 = 2R \parallel 2R = \frac{2R}{2} = R$$

Continuing this simplification process results in the circuit in part (e) in which the output voltage can be expressed using the voltage-divider formula as

$$V_{OUT} = \left(\frac{2R}{4R}\right)V = \frac{V}{2}$$

A similar analysis, except with switch SW3 in Figure 7–43 connected to V and the other switches connected to ground, results in the simplified circuit shown in Figure 7–45.

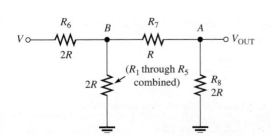

◀ **FIGURE 7–45**

Simplified ladder with only V input at SW3 in Figure 7–43.

The analysis for this case is as follows: The resistance from node B to ground is

$$R_B = (R_7 + R_8) \parallel 2R = 3R \parallel 2R = \frac{6R}{5}$$

Using the voltage-divider formula, we can express the voltage at node B with respect to ground as

$$V_B = \left(\frac{R_B}{R_6 + R_B}\right)V = \left(\frac{6R/5}{2R + 6R/5}\right)V = \left(\frac{6R/5}{10R/5 + 6R/5}\right)V = \left(\frac{6R/5}{16R/5}\right)V$$

$$= \left(\frac{6R}{16R}\right)V = \frac{3V}{8}$$

The output voltage is, therefore,

$$V_{OUT} = \left(\frac{R_8}{R_7 + R_8}\right)V_B = \left(\frac{2R}{3R}\right)\left(\frac{3V}{8}\right) = \frac{V}{4}$$

Notice that the output voltage in this case ($V/4$) is one half the output voltage ($V/2$) for the case where V is connected at switch SW4.

A similar analysis for each of the remaining switch inputs in Figure 7–43 results in output voltages as follows: For SW2 connected to V and the other switches connected to ground,

$$V_{OUT} = \frac{V}{8}$$

For SW1 connected to V and the other switches connected to ground,

$$V_{OUT} = \frac{V}{16}$$

When more than one input at a time are connected to V, the total output is the sum of the individual outputs, according to the superposition theorem that is covered in Section 8–4. These particular relationships among the output voltages for the various levels of inputs are important in the application of $R/2R$ ladder networks to digital-to-analog conversion.

**SECTION 7–5
REVIEW**

1. Draw a basic four-step ladder network.
2. Determine the total circuit resistance presented to the source by the ladder network of Figure 7–46.
3. What is the total current in Figure 7–46?
4. What is the current through R_2 in Figure 7–46?
5. What is the voltage at node A with respect to ground in Figure 7–46?

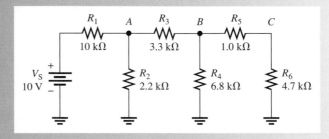

▲ FIGURE 7–46

7–6 THE WHEATSTONE BRIDGE

The Wheatstone bridge circuit can be used to precisely measure resistance. However, the bridge is most commonly used in conjunction with transducers to measure physical quantities such as strain, temperature, and pressure. Transducers are devices that sense a change in a physical parameter and convert that change into an electrical quantity such as a change in resistance. For example, a strain gauge exhibits a change in resistance when it is exposed to mechanical factors such as force, pressure, or displacement. A thermistor exhibits a change in its resistance when it is exposed to a change in temperature. The Wheatstone bridge can be operated in a balanced or an unbalanced condition. The condition of operation depends on the type of application.

After completing this section, you should be able to

♦ **Analyze and apply a Wheatstone bridge**

  ♦ Determine when a bridge is balanced

  ♦ Determine an unknown resistance with a balanced bridge

  ♦ Determine when a bridge is unbalanced

  ♦ Discuss measurements using an unbalanced bridge

A **Wheatstone bridge** circuit is shown in its most common "diamond" configuration in Figure 7–47(a). It consists of four resistors and a dc voltage source connected across the top and bottom points of the "diamond." The output voltage is taken across the left and right points of the "diamond" between A and B. In part (b), the circuit is drawn in a slightly different way to more clearly show its series-parallel configuration.

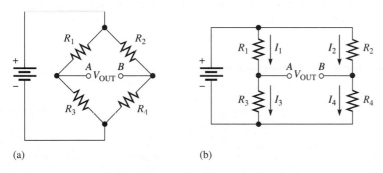

(a)   (b)

▲ **FIGURE 7–47**

Wheatstone bridge.

The Balanced Wheatstone Bridge

The Wheatstone bridge in Figure 7–47 is in the **balanced bridge** condition when the output voltage (V_{OUT}) between terminals A and B is equal to zero.

$$V_{OUT} = 0 \text{ V}$$

When the bridge is balanced, the voltages across R_1 and R_2 are equal ($V_1 = V_2$) and the voltages across R_3 and R_4 are equal ($V_3 = V_4$). Therefore, the voltage ratios can be written as

$$\frac{V_1}{V_3} = \frac{V_2}{V_4}$$

Substituting IR for V by Ohm's law gives

$$\frac{I_1 R_1}{I_3 R_3} = \frac{I_2 R_2}{I_4 R_4}$$

Since $I_1 = I_3$ and $I_2 = I_4$, all the current terms cancel, leaving the resistor ratios.

$$\frac{R_1}{R_3} = \frac{R_2}{R_4}$$

Solving for R_1 results in the following formula:

$$R_1 = R_3 \left(\frac{R_2}{R_4} \right)$$

This formula allows you to find the value of resistor R_1 in terms of the other resistor values when the bridge is balanced. You can also find the value of any other resistor in a similar way.

Using the Balanced Wheatstone Bridge to Find an Unknown Resistance
Assume that R_1 in Figure 7–47 has an unknown value, which we call R_X. Resistors R_2 and R_4 have fixed values so that their ratio, R_2/R_4, also has a fixed value. Since R_X can be any value, R_3 must be adjusted to make $R_1/R_3 = R_2/R_4$ in order to create a balanced condition. Therefore, R_3 is a variable resistor, which we will call R_V. When R_X is placed in the bridge, R_V is adjusted until the bridge is balanced as indicated by a zero output voltage. Then, the unknown resistance is found as

Equation 7–2

$$R_X = R_V \left(\frac{R_2}{R_4} \right)$$

The ratio R_2/R_4 is the scale factor.

An older type of measuring instrument called a galvanometer can be connected between the output terminals A and B to detect a balanced condition. The galvanometer is essentially a very sensitive ammeter that senses current in either direction. It differs from a regular ammeter in that the midscale point is zero. In modern instruments, an amplifier connected across the bridge output indicates a balanced condition when its output is 0 V.

From Equation 7–2, the value of R_V at balance multiplied by the scale factor R_2/R_4 is the actual resistance value of R_X. If $R_2/R_4 = 1$, then $R_X = R_V$, if $R_2/R_4 = 0.5$, then $R_X = 0.5R_V$, and so on. In a practical bridge circuit, the position of the R_V adjustment can be calibrated to indicate the actual value of R_X on a scale or with some other method of display.

EXAMPLE 7–17

Determine the value of R_X in the balanced bridge shown in Figure 7–48.

▶ **FIGURE 7–48**

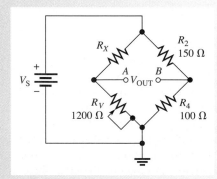

Solution   The scale factor is

$$\frac{R_2}{R_4} = \frac{150 \ \Omega}{100 \ \Omega} = 1.5$$

The bridge is balanced ($V_{OUT} = 0$ V) when R_V is set at 1200 Ω, so the unknown resistance is

$$R_X = R_V \left(\frac{R_2}{R_4} \right) = (1200 \ \Omega)(1.5) = \mathbf{1800 \ \Omega}$$

Related Problem   If R_V must be adjusted to 2.2 kΩ to balance the bridge in Figure 7–48, what is R_X?

Use Multisim file E07-17 to verify the calculated results in this example and to confirm your calculation for the related problem.

The Unbalanced Wheatstone Bridge

An **unbalanced bridge** condition occurs when V_{OUT} is not equal to zero. The unbalanced bridge is used to measure several types of physical quantities such as mechanical strain, temperature, or pressure. This can be done by connecting a transducer in one leg of the bridge, as shown in Figure 7–49. The resistance of the transducer changes proportionally to the changes in the parameter that it is measuring. If the bridge is balanced at a known point, then the amount of deviation from the balanced condition, as indicated by the output voltage, indicates the amount of change in the parameter being measured. Therefore, the value of the parameter being measured can be determined by the amount that the bridge is unbalanced.

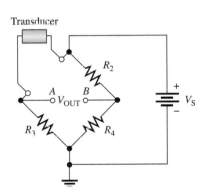

◀ **FIGURE 7–49**

A bridge circuit for measuring a physical parameter using a transducer.

A Bridge Circuit for Measuring Temperature   If temperature is to be measured, the transducer can be a thermistor, which is a temperature-sensitive resistor. The thermistor resistance changes in a predictable way as the temperature changes. A change in temperature causes a change in thermistor resistance, which causes a corresponding change in the output voltage of the bridge as it becomes unbalanced. The output voltage is proportional to the temperature; therefore, either a voltmeter connected across the output can be calibrated to show the temperature or the output voltage can be amplified and converted to digital form to drive a readout display of the temperature.

A bridge circuit used to measure temperature is designed so that it is balanced at a reference temperature and becomes unbalanced at a measured temperature. For example, let's say the bridge is to be balanced at 25°C. A thermistor will have a known value of resistance at 25°C. For simplicity, let's assume the other three bridge resistors are equal to the thermistor resistance at 25°C, so $R_{therm} = R_2 = R_3 = R_4$. For this particular case, the change in output voltage (ΔV_{OUT}) can be shown to be related to the change in R_{therm} by the following formula:

Equation 7–3

$$\Delta V_{\text{OUT}} \cong \Delta R_{\text{therm}}\left(\frac{V_S}{4R}\right)$$

The Δ (Greek letter delta) in front of a variable means a change in the variable. This formula applies only to the case where all resistances in the bridge are equal when the bridge is balanced. A derivation is provided in Appendix B. Keep in mind that the bridge can be initially balanced without having all the resistors equal as long as $R_1 = R_2$ and $R_3 = R_4$ (see Figure 7–47), but the formula for ΔV_{OUT} would be more complicated.

EXAMPLE 7–18

Determine the output voltage of the temperature-measuring bridge circuit in Figure 7–50 if the thermistor is exposed to a temperature of 50°C and its resistance at 25°C is 1.0 kΩ. Assume the resistance of the thermistor decreases to 900 Ω at 50°C.

▶ **FIGURE 7–50**

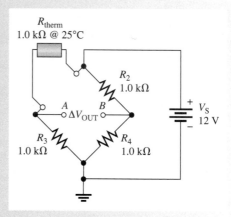

Solution

$$\Delta R_{\text{therm}} = 1.0\,\text{k}\Omega - 900\,\Omega = 100\,\Omega$$

$$\Delta V_{\text{OUT}} \cong \Delta R_{\text{therm}}\left(\frac{V_S}{4R}\right) = 100\,\Omega\left(\frac{12\,\text{V}}{4\,\text{k}\Omega}\right) = 0.3\,\text{V}$$

Since $V_{\text{OUT}} = 0\,\text{V}$ when the bridge is balanced at 25°C and it changes 0.3 V, then

$$V_{\text{OUT}} = \mathbf{0.3\,V}$$

when the temperature is 50°C.

Related Problem

If the temperature is increased to 60°C, causing the thermistor resistance in Figure 7–50 to decrease to 850 Ω, what is V_{OUT}?

Other Unbalanced Wheatstone Bridge Applications A Wheatstone bridge with a strain gauge can be used to measure certain forces. A strain gauge is a device that exhibits a change in resistance when it is compressed or stretched by the application of an external force. As the resistance of the strain gauge changes, the previously balanced bridge becomes unbalanced. This unbalance causes the output voltage to change from zero, and this change can be measured to determine the amount of strain. In strain gauges, the resistance change is extremely small. This tiny change unbalances a Wheatstone bridge because of its high sensitivity. For example, Wheatstone bridges with strain gauges are commonly used in weight scales.

Some resistive transducers have extremely small resistance changes, and these changes are difficult to measure accurately with a direct measurement. In particular, strain gauges are one of the most useful resistive transducers that convert the stretching or compression of a fine wire into a change in resistance. When strain causes the wire in the gauge to stretch, the resistance increases a small amount; and when it compresses, the resistance of the wire decreases.

Strain gauges are used in many types of scales, from those that are used for weighing small parts to those for weighing huge trucks. Typically, the gauges are mounted on a special block of aluminum that deforms when a weight is on the scale. The strain gauges are extremely delicate and must be mounted properly, so the entire assembly is generally prepared as a single unit called a load cell. A wide variety of load cells with different shapes and sizes are available from manufacturers depending on the application. A typical S-shaped load cell for a weighing application that has four strain gauges is illustrated in Figure 7–51(a). The gauges are mounted so that two of the gauges stretch (tension) when a load is placed on the scale and two of the gauges compress.

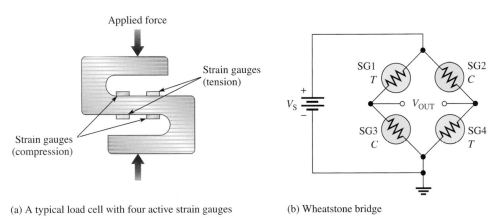

(a) A typical load cell with four active strain gauges   (b) Wheatstone bridge

▲ **FIGURE 7–51**

Load cells are usually connected to a Wheatstone bridge as shown in Figure 7–51(b) with strain gauges (SG) in tension (T) and compression (C) in opposite diagonal legs as shown. The output of the bridge is normally digitized and converted to a reading for a display or sent to a computer for processing. The major advantage of the Wheatstone bridge circuit is that it is capable of accurately measuring very small differences in resistance. The use of four active transducers increases the sensitivity of the measurement and makes the bridge the ideal circuit for instrumentation. The Wheatstone bridge circuit has the added benefit of compensating for temperature variations and wire resistance of connecting wires that would otherwise contribute to inaccuracies.

In addition to scales, strain gauges are used with Wheatstone bridges in other types of measurements including pressure measurements, displacement and acceleration measurements to name a few. In pressure measurements, the strain gauges are bonded to a flexible diaphragm that stretches when pressure is applied to the transducer. The amount of flexing is related to the pressure, which again converts to a very small resistance change.

SECTION 7–6 **REVIEW**	1. Draw a basic Wheatstone bridge circuit.
	2. Under what condition is a bridge balanced?
	3. What is the unknown resistance in Figure 7–48 when $R_V = 3.3\,\text{k}\Omega$, $R_2 = 10\,\text{k}\Omega$, and $R_4 = 2.2\,\text{k}\Omega$?
	4. How is a Wheatstone bridge used in the unbalanced condition?

7–7 TROUBLESHOOTING

As you know, troubleshooting is the process of identifying and locating a failure or problem in a circuit. Some troubleshooting techniques and the application of logical thought have already been discussed in relation to both series circuits and parallel circuits. A basic premise of troubleshooting is that you must know what to look for before you can successfully troubleshoot a circuit.

After completing this section, you should be able to

◆ **Troubleshoot series-parallel circuits**

   ◆ Determine the effects of an open in a circuit

   ◆ Determine the effects of a short in a circuit

   ◆ Locate opens and shorts

Opens and shorts are typical problems that occur in electric circuits. As mentioned in Chapter 5, if a resistor burns out, it will normally produce an open. Bad solder connections, broken wires, and poor contacts can also be causes of open paths. Pieces of foreign material, such as solder splashes, broken insulation on wires, and so on, can often lead to shorts in a circuit. A short is considered to be a zero resistance path between two points.

In addition to complete opens or shorts, partial opens or partial shorts can develop in a circuit. A partial open would be a much higher than normal resistance, but not infinitely large. A partial short would be a much lower than normal resistance, but not zero.

The following three examples illustrate troubleshooting series-parallel circuits.

EXAMPLE 7–19

From the indicated voltmeter reading in Figure 7–52, determine if there is a fault by applying the APM approach. If there is a fault, identify it as either a short or an open.

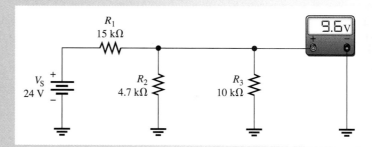

▲ **FIGURE 7–52**

Solution **Step 1:** Analysis

Determine what the voltmeter should be indicating as follows. Since R_2 and R_3 are in parallel, their combined resistance is

$$R_{2\|3} = \frac{R_2 R_3}{R_2 + R_3} = \frac{(4.7\,\text{k}\Omega)(10\,\text{k}\Omega)}{14.7\,\text{k}\Omega} = 3.20\,\text{k}\Omega$$

Determine the voltage across the parallel combination by the voltage-divider formula.

$$V_{2\|3} = \left(\frac{R_{2\|3}}{R_1 + R_{2\|3}}\right) V_S = \left(\frac{3.2\,\text{k}\Omega}{18.2\,\text{k}\Omega}\right) 24\,\text{V} = 4.22\,\text{V}$$

This calculation shows that 4.22 V is the voltage reading that you should get on the meter. However, the meter reads 9.6 V across $R_{2\|3}$. This value is incorrect, and, because it is higher than it should be, either R_2 or R_3 is probably open. Why? Because if either of these two resistors is open, the resistance across which the meter is connected is larger than expected. A higher resistance will drop a higher voltage in this circuit.

Step 2: Planning

Start trying to find the open resistor by assuming that R_2 is open. If it is, the voltage across R_3 is

$$V_3 = \left(\frac{R_3}{R_1 + R_3}\right)V_S = \left(\frac{10\,\text{k}\Omega}{25\,\text{k}\Omega}\right)24\,\text{V} = 9.6\,\text{V}$$

Since the measured voltage is also 9.6 V, this calculation shows that R_2 is probably open.

Step 3: Measurement

Disconnect power and remove R_2. Measure its resistance to verify it is open. If it is not, inspect the wiring, solder, or connections around R_2, looking for the open.

Related Problem   What would be the voltmeter reading if R_3 were open in Figure 7–52? If R_1 were open?

EXAMPLE 7–20

Suppose that you measure 24 V with the voltmeter in Figure 7–53. Determine if there is a fault, and, if there is, identify it.

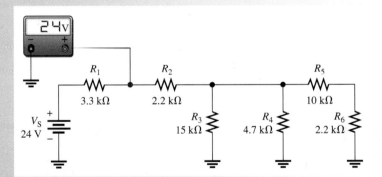

▲ **FIGURE 7–53**

Solution   **Step 1:** Analysis

There is no voltage drop across R_1 because both sides of the resistor are at +24 V. Either there is no current through R_1 from the source, which tells you that R_2 is open in the circuit, or R_1 is shorted.

Step 2: Planning

The most probable failure is an open R_2. If it is open, then there will be no current from the source. To verify this, measure across R_2 with the voltmeter. If R_2 is open, the meter will indicate 24 V. The right side of R_2 will be at zero volts because there is no current through any of the other resistors to cause a voltage drop across them.

Step 3: Measurement

The measurement to verify that R_2 is open is shown in Figure 7–54.

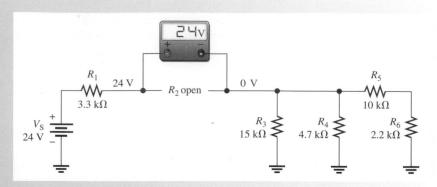

FIGURE 7–54

Related Problem   What would be the voltage across an open R_5 in Figure 7–53 assuming no other faults?

EXAMPLE 7–21   The two voltmeters in Figure 7–55 indicate the voltages shown. Apply logical thought and your knowledge of circuit operation to determine if there are any opens or shorts in the circuit and, if so, where they are located.

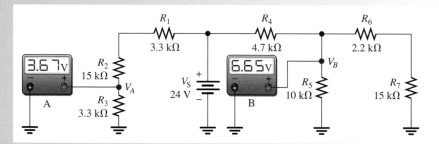

FIGURE 7–55

Solution   **Step 1:** Determine if the voltmeter readings are correct. R_1, R_2, and R_3 act as a voltage divider. Calculate the voltage (V_A) across R_3 as follows:

$$V_A = \left(\frac{R_3}{R_1 + R_2 + R_3} \right) V_S = \left(\frac{3.3\,\text{k}\Omega}{21.6\,\text{k}\Omega} \right) 24\,\text{V} = 3.67\,\text{V}$$

The voltmeter A reading is correct. This indicates that R_1, R_2, and R_3 are connected and are not faulty.

Step 2: See if the voltmeter B reading is correct. $R_6 + R_7$ is in parallel with R_5. The series-parallel combination of R_5, R_6, and R_7 is in series with R_4. Calculate the resistance of the R_5, R_6, and R_7 combination as follows:

$$R_{5\|(6+7)} = \frac{R_5(R_6 + R_7)}{R_5 + R_6 + R_7} = \frac{(10\,\text{k}\Omega)(17.2\,\text{k}\Omega)}{27.2\,\text{k}\Omega} = 6.32\,\text{k}\Omega$$

$R_{5\|(6+7)}$ and R_4 form a voltage divider, and voltmeter B measures the voltage across $R_{5\|(6+7)}$. Is it correct? Check as follows:

$$V_B = \left(\frac{R_{5\|(6+7)}}{R_4 + R_{5\|(6+7)}} \right) V_S = \left(\frac{6.32\,\text{k}\Omega}{11\,\text{k}\Omega} \right) 24\,\text{V} = 13.8\,\text{V}$$

Thus, the actual measured voltage (6.65 V) at this point is incorrect. Some logical thinking will help to isolate the problem.

Step 3: R_4 is not open, because if it were, the meter would read 0 V. If there were a short across it, the meter would read 24 V. Since the actual voltage is much less than it should be, $R_{5\|(6+7)}$ must be less than the calculated value of 6.32 kΩ. The most likely problem is a short across R_7. If there is a short from the top of R_7 to ground, R_6 is effectively in parallel with R_5. In this case,

$$R_5 \| R_6 = \frac{R_5 R_6}{R_5 + R_6} = \frac{(10\,\text{k}\Omega)(2.2\,\text{k}\Omega)}{12.2\,\text{k}\Omega} = 1.80\,\text{k}\Omega$$

Then V_B is

$$V_B = \left(\frac{1.80\,\text{k}\Omega}{6.5\,\text{k}\Omega}\right)24\,\text{V} = 6.65\,\text{V}$$

This value for V_B agrees with the voltmeter B reading. So there is a short across R_7. If this were an actual circuit, you would try to find the physical cause of the short.

Related Problem   If the only fault in Figure 7–55 is that R_2 is shorted, what will voltmeter A read? What will voltmeter B read?

**SECTION 7–7
REVIEW**

1. Name two types of common circuit faults.
2. In Figure 7–56, one of the resistors in the circuit is open. Based on the meter reading, determine which is the open resistor.

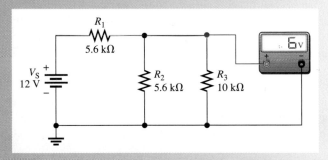

▲ **FIGURE 7–56**

3. For the following faults in Figure 7–57, what voltage would be measured at node *A* with respect to ground?
   (a) No faults   (b) R_1 open   (c) Short across R_5 (d) R_3 and R_4 open
   (e) R_2 open

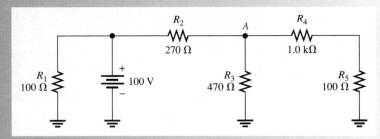

▲ **FIGURE 7–57**

A Circuit Application

The Wheatstone bridge is widely used in measurement applications that use sensors to convert a physical parameter to a change in resistance. Modern Wheatstone bridges are automated; with intelligent interface modules, the output can be conditioned and converted to any desired unit for display or processing (for example, the output might be displayed in pounds for a scale application).

The Wheatstone bridge provides a null measurement, which enables it to have great sensitivity. It can also be designed to compensate for changes in temperature, a great advantage for many resistive measurements, particularly when the resistance change of the sensor is very small. Usually, the output voltage of the bridge is increased by an amplifier that has a minimum loading effect on the bridge.

Temperature Controller

In this application, a Wheatstone bridge circuit is used in a temperature controller. The sensor is a thermistor ("thermal resistor"), which is a resistive sensor that changes resistance as temperature changes. Thermistors are available with positive or negative resistance characteristics as a function of temperature. The thermistor in this circuit is one of the resistors in a Wheatstone bridge but is located a short distance from the circuit board for sensing the temperature at a point off the board.

The threshold voltage for the output to change is controlled by the $10 \text{ k}\Omega$ potentiometer, R_3. The amplifier in this case is constructed using an operational amplifier ("op-amp"), which is an integrated circuit. The term *amplifier* is used in electronics to describe a device that produces a larger replica of the input voltage or current at its output. The term *gain* refers to the amount

of amplification. The op-amp in this circuit is configured as a *comparator,* which is be used to compare the voltage on one side of the bridge with the voltage on the other.

The advantage of a comparator is that it is extremely sensitive to an unbalanced bridge and produces a large output when the bridge is unbalanced. In fact, it is so sensitive, it is virtually impossible to adjust the bridge for perfect balance. Even the tiniest imbalance will cause the output to go to a voltage near either the maximum or minimum possible (the power supply voltages). This is handy for turning on a heater or other device based on the temperature.

The Control Circuit

This application has a tank containing a liquid that needs to be held at a warm temperature, as illustrated in Figure 7–58(a). The circuit board for the temperature controller is shown in Figure 7–58(b). The circuit board controls a heating unit (through an interface, which is not shown) when the temperature is too cold. The thermistor, which is located in the tank, is connected between one of the amplifier inputs and ground as illustrated.

The amplifier is a 741C op-amp, an inexpensive and popular device that will work fine for this application. The op-amp has two inputs, an output, and connections for positive and negative supply voltages. The schematic symbol for the op-amp with light-emitting diodes (LEDs) connected to the output is shown in Figure 7–59. The red LED is on when the op-amp output is a positive voltage, indicating the heater is on. The green LED is on when the output is a negative voltage, indicating the heater is off.

◆ Using the circuit board as a guide, complete the schematic in Figure 7–59. The op-amp inputs are connected to a Wheatstone bridge. Show the values for all resistors.

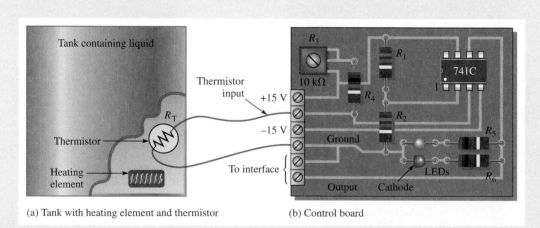

(a) Tank with heating element and thermistor

(b) Control board

▲ FIGURE 7–58

A CIRCUIT APPLICATION ◆ 263

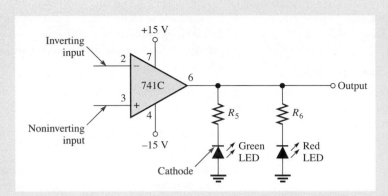

The op-amp and output LED indicators.

The Thermistor

The thermistor is a mixture of two metal-oxides, which exhibit a large resistance change as a function of temperature. The thermistor in the temperature controller circuit is located off the board near the point where temperature is to be sensed in the tank and is connected between the thermistor input and ground.

Thermistors have a nonlinear resistance-temperature characteristic described by the exponential equation:

$$R_T = R_0 e^{\beta\left(\frac{T_0 - T}{T_0 T}\right)}$$

where:

R_T = the resistance at a given temperature

R_0 = the resistance at a reference temperature

T_0 = the reference temperature in Kelvin (K), typically 298 K, which is 25°C

T = temperature in K

β = a constant (K) provided by the manufacturer

This exponential equation where e is the base of natural logarithms can be solved easily on a scientific calculator. Exponential equations are studied in later chapters.

The thermistor in this application is a Thermometrics RL2006-13.3K-140-D1 thermistor with a specified resistance of 25 kΩ at 25°C and a β of 4615 K. For convenience, the resistance of this thermistor is plotted as a function of temperature in Figure 7–60. Notice that the negative slope indicates this thermistor has a negative temperature coefficient (NTC); that is, its resistance decreases as the temperature increases.

As an example, the calculation for finding the resistance at $T = 50°C$ is shown. First, convert 50°C to K.

$$T = °C + 273 = 50°C + 273 = 323 \text{ K}$$

Also,

$$T_0 = °C + 273 = 25°C + 273 = 298 \text{ K}$$

$$R_0 = 25 \text{ k}\Omega$$

$$R_T = R_0 e^{\beta\left(\frac{T_0 - T}{T_0 T}\right)}$$

$$= (25 \text{ k}\Omega)e^{4615\left(\frac{298 - 323}{298 \times 323}\right)}$$

$$= (25 \text{ k}\Omega)e^{-1.198}$$

$$= (25 \text{ k}\Omega)(0.302)$$

$$= 7.54 \text{ k}\Omega$$

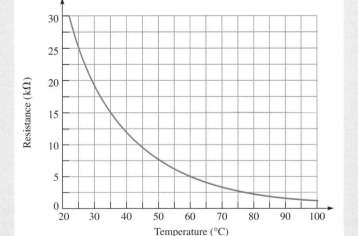

Using your calculator, first determine the value of the exponent $\beta(T_0 - T)/(T_0 T)$. Next determine the value of the term $e^{\beta\left(\frac{T_0 - T}{T_0 T}\right)}$. Finally, multiply by R_0. On many calculators, e^x is a secondary function.

◆ Calculate the resistance of the thermistor at a temperature of 40°C using the exponential equation and confirm that your calculation is correct by comparing your result with Figure 7–60. Remember that temperatures in the equation are in kelvin. (K = °C + 273).

◆ Calculate the resistance setting of R_3 to balance the bridge at 25°C.

◆ Calculate the output voltage of the bridge (input to the op-amp) when the temperature of the thermistor is 40°C. Assume the bridge was balanced at 25°C and that the only change is the resistance of the thermistor.

◆ If you needed to set the reference temperature to 0°C, what simple change would you make to the circuit? Show with a calculation that your change will work and draw the revised schematic.

Review

1. At 25°C, the thermistor will have about 7.5 V across it. Calculate the power it dissipates. Is there any loading effect on a temperature measurement due to this?

2. As the temperature increases, does the loading effect go up, down, or remain the same? Explain your answer.

3. Can ⅛ W resistors be used in this application? Explain your answer.

4. Why is only one LED on at a time at the output?

SUMMARY

◆ A series-parallel circuit is a combination of both series and parallel current paths.

◆ To determine total resistance in a series-parallel circuit, identify the series and parallel relationships, and then apply the formulas for series resistance and parallel resistance from Chapters 5 and 6.

◆ To find the total current, apply Ohm's law and divide the total voltage by the total resistance.

◆ To determine branch currents, apply the current-divider formula, Kirchhoff's current law, or Ohm's law. Consider each circuit problem individually to determine the most appropriate method.

◆ To determine voltage drops across any portion of a series-parallel circuit, use the voltage-divider formula, Kirchhoff's voltage law, or Ohm's law. Consider each circuit problem individually to determine the most appropriate method.

◆ When a load resistor is connected across a voltage-divider output, the output voltage decreases.

◆ The load resistor should be large compared to the resistance across which it is connected, in order that the loading effect may be minimized.

◆ To find total resistance of a ladder network, start at the point farthest from the source and reduce the resistance in steps.

◆ A balanced Wheatstone bridge can be used to measure an unknown resistance.

◆ A bridge is balanced when the output voltage is zero. The balanced condition produces zero current through a load connected across the output terminals of the bridge.

◆ An unbalanced Wheatstone bridge can be used to measure physical quantities using transducers.

◆ Opens and shorts are typical circuit faults.

◆ Resistors normally open when they burn out.

KEY TERMS

These key terms are also in the end-of-book glossary.

Balanced bridge A bridge circuit that is in the balanced state is indicated by 0 V across the output.

Bleeder current The current left after the total load current is subtracted from the total current into the circuit.

Unbalanced bridge   A bridge circuit that is in the unbalanced state as indicated by a voltage across the output that is proportional to the amount of deviation from the balanced state.

Wheatstone bridge   A 4-legged type of bridge circuit with which an unknown resistance can be accurately measured using the balanced state. Deviations in resistance can be measured using the unbalanced state.

FORMULAS

7–1 $I_{\text{BLEEDER}} = I_{\text{T}} - I_{RL1} - I_{RL2}$ Bleeder current

7–2 $R_X = R_V \left(\dfrac{R_2}{R_4} \right)$ Unknown resistance in a Wheatstone bridge

7–3 $\Delta V_{\text{OUT}} = \Delta R_{\text{therm}} \left(\dfrac{V_S}{4R} \right)$ Thermistor bridge output

SELF–TEST   Answers are at the end of the chapter.

1. Which of the following statements are true concerning Figure 7–61?

    (a) R_1 and R_2 are in series with R_3, R_4, and R_5.

    (b) R_1 and R_2 are in series.

    (c) R_3, R_4, and R_5 are in parallel.

    (d) The series combination of R_1 and R_2 is in parallel with the series combination of R_3, R_4, and R_5.

    (e) answers (b) and (d)

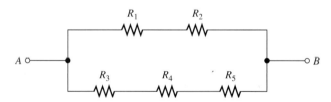

▲ FIGURE 7–61

2. The total resistance of Figure 7–61 can be found with which of the following formulas?

    (a) $R_1 + R_2 + R_3 \parallel R_4 \parallel R_5$ (b) $R_1 \parallel R_2 + R_3 \parallel R_4 \parallel R_5$

    (c) $(R_1 + R_2) \parallel (R_3 + R_4 + R_5)$ (d) none of these answers

3. If all of the resistors in Figure 7–61 have the same value, when voltage is applied across terminals A and B, the current is

    (a) greatest in R_5 (b) greatest in R_3, R_4, and R_5

    (c) greatest in R_1 and R_2 (d) the same in all the resistors

4. Two 1.0 kΩ resistors are in series and this series combination is in parallel with a 2.2 kΩ resistor. The voltage across one of the 1.0 kΩ resistors is 6 V. The voltage across the 2.2 kΩ resistor is

    (a) 6 V   (b) 3 V   (c) 12 V   (d) 13.2 V

5. The parallel combination of a 330 Ω resistor and a 470 Ω resistor is in series with the parallel combination of four 1.0 kΩ resistors. A 100 V source is connected across the circuit. The resistor with the most current has a value of

    (a) 1.0 kΩ   (b) 330 Ω   (c) 470 Ω

6. In the circuit described in Question 5, the resistor(s) with the most voltage has (have) a value of

    (a) 1.0 kΩ   (b) 470 Ω   (c) 330 Ω

7. In the circuit of Question 5, the percentage of the total current through any single $1.0\,k\Omega$ resistor is

   (a) 100%   (b) 25%   (c) 50%   (d) 31.3%

8. The output of a certain voltage divider is 9 V with no load. When a load is connected, the output voltage

   (a) increases   (b) decreases   (c) remains the same   (d) becomes zero

9. A certain voltage divider consists of two $10\,k\Omega$ resistors in series. Which of the following load resistors will have the most effect on the output voltage?

   (a) $1.0\,M\Omega$ (b) $20\,k\Omega$ (c) $100\,k\Omega$ (d) $10\,k\Omega$

10. When a load resistance is connected to the output of a voltage-divider circuit, the current drawn from the source

    (a) decreases   (b) increases   (c) remains the same   (d) is cut off

11. In a ladder network, simplification should begin at

    (a) the source   (b) the resistor farthest from the source

    (c) the center   (d) the resistor closest to the source

12. In a certain four-step $R/2R$ ladder network, the smallest resistor value is $10\,k\Omega$. The largest value is

    (a) indeterminable   (b) $20\,k\Omega$ (c) $50\,k\Omega$ (d) $100\,k\Omega$

13. The output voltage of a balanced Wheatstone bridge is

    (a) equal to the source voltage

    (b) equal to zero

    (c) dependent on all of the resistor values in the bridge

    (d) dependent on the value of the unknown resistor

14. A certain Wheatstone bridge has the following resistor values: $R_V = 8\,k\Omega$, $R_2 = 680\,\Omega$, and $R_4 = 2.2\,k\Omega$. The unknown resistance is

    (a) $2473\,\Omega$ (b) $25.9\,k\Omega$ (c) $187\,\Omega$ (d) $2890\,\Omega$

15. You are measuring the voltage at a given point in a circuit that has very high resistance values and the measured voltage is a little lower than it should be. This is possibly because of

    (a) one or more of the resistance values being off

    (b) the loading effect of the voltmeter

    (c) the source voltage is too low

    (d) all of these answers

CIRCUIT DYNAMICS
QUIZ

Answers are at the end of the chapter.

Refer to Figure 7–62(b).

1. If R_2 opens, the total current

   (a) increases   (b) decreases   (c) stays the same

2. If R_3 opens, the current in R_2

   (a) increases   (b) decreases   (c) stays the same

3. If R_4 opens, the voltage across it

   (a) increases   (b) decreases   (c) stays the same

4. If R_4 is shorted, the total current

   (a) increases   (b) decreases   (c) stays the same

Refer to Figure 7–64.

5. If R_{10} opens, with 10 V applied between terminals A and B, the total current

   (a) increases   (b) decreases   (c) stays the same

6. If R_1 opens with 10 V applied between terminals A and B, the voltage across R_1

   (a) increases (b) decreases (c) stays the same

7. If there is a short between the left contact of R_3 and the bottom contact of R_5, the total resistance between A and B

   (a) increases (b) decreases (c) stays the same

Refer to Figure 7–68.

8. If R_4 opens, the voltage at point C

   (a) increases (b) decreases (c) stays the same

9. If there is a short from point D to ground, the voltage from A to B

   (a) increases (b) decreases (c) stays the same

10. If R_5 opens, the current through R_1

    (a) increases (b) decreases (c) stays the same

Refer to Figure 7–74.

11. If a 10 kΩ load resistor is connected across the output terminals A and B, the output voltage

    (a) increases (b) decreases (c) stays the same

12. If the 10 kΩ load resistor mentioned in Question 11 is replaced by a 100 kΩ load resistor, V_{OUT}

    (a) increases (b) decreases (c) stays the same

Refer to Figure 7–75.

13. If there is a short between the V_2 and V_3 terminals of the switch, the voltage V_1 with respect to ground

    (a) increases (b) decreases (c) stays the same

14. If the switch is in the position shown and if the V_3 terminal of the switch is shorted to ground, the voltage across R_L

    (a) increases (b) decreases (c) stays the same

15. If R_4 opens with the switch in the position shown, the voltage across R_L

    (a) increases (b) decreases (c) stays the same

Refer to Figure 7–80.

16. If R_4 opens, V_{OUT}

    (a) increases (b) decreases (c) stays the same

17. If R_7 is shorted to ground, V_{OUT}

    (a) increases (b) decreases (c) stays the same

PROBLEMS

More difficult problems are indicated by an asterisk (*).
Answers to odd-numbered problems are at the end of the book.

SECTION 7–1 Identifying Series-Parallel Relationships

1. Visualize and draw the following series-parallel combinations:

   (a) R_1 in series with the parallel combination of R_2 and R_3

   (b) R_1 in parallel with the series combination of R_2 and R_3

   (c) R_1 in parallel with a branch containing R_2 in series with a parallel combination of four other resistors

2. Visualize and draw the following series-parallel circuits:

   (a) A parallel combination of three branches, each containing two series resistors

   (b) A series combination of three parallel circuits, each containing two resistors

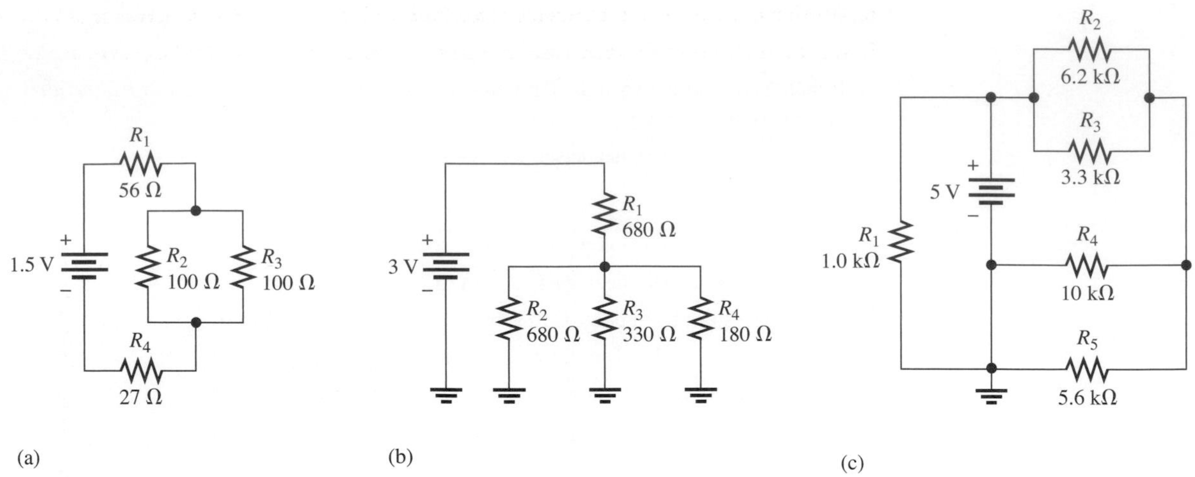

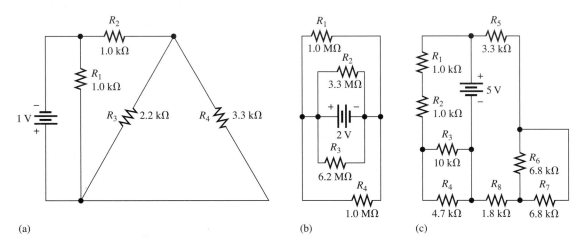

▲ FIGURE 7–62

3. In each circuit of Figure 7–62, identify the series and parallel relationships of the resistors viewed from the source.

4. For each circuit in Figure 7–63, identify the series and parallel relationships of the resistors viewed from the source.

▲ FIGURE 7–63

5. Draw the schematic of the PC board layout in Figure 7–64 showing resistor values and identify the series-parallel relationships.

▶ FIGURE 7–64

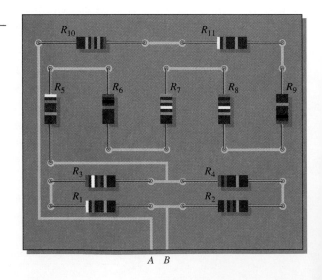

*6. Develop a schematic for the double-sided PC board in Figure 7–65 and label the resistor values.

*7. Lay out a PC board for the circuit in Figure 7–63(c). The battery is to be connected external to the board.

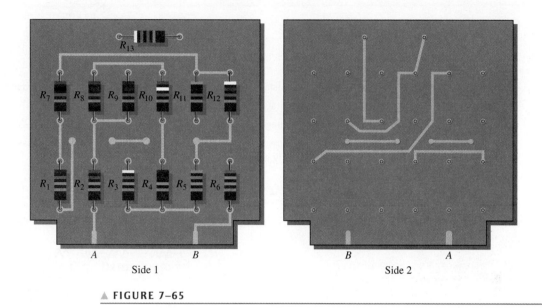

Side 1   Side 2

▲ FIGURE 7–65

SECTION 7–2   Analysis of Series-Parallel Resistive Circuits

8. A certain circuit is composed of two parallel resistors. The total resistance is 667 Ω. One of the resistors is 1.0 kΩ. What is the other resistor?

9. For each circuit in Figure 7–62, determine the total resistance presented to the source.

10. Repeat Problem 9 for each circuit in Figure 7–63.

11. Determine the current through each resistor in each circuit in Figure 7–62; then calculate each voltage drop.

12. Determine the current through each resistor in each circuit in Figure 7–63; then calculate each voltage drop.

13. Find R_T for all combinations of the switches in Figure 7–66.

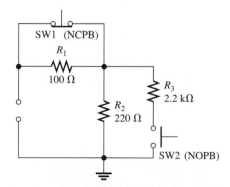

▲ FIGURE 7–66

▶ FIGURE 7–67

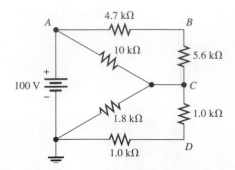

14. Determine the resistance between *A* and *B* in Figure 7–67 with the source removed.

15. Determine the voltage at each node with respect to ground in Figure 7–67.

16. Determine the voltage at each node with respect to ground in Figure 7–68.

17. In Figure 7–68, how would you determine the voltage across R_2 by measuring without connecting a meter directly across the resistor?

▶ FIGURE 7–68

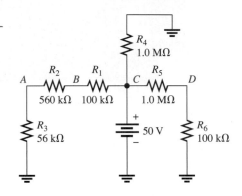

18. Determine the resistance of the circuit in Figure 7–67 as seen from the voltage source.

19. Determine the resistance of the circuit in Figure 7–68 as seen from the voltage source.

20. Determine the voltage, V_{AB}, in Figure 7–69.

▶ FIGURE 7–69

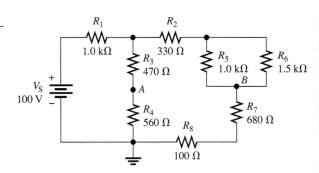

*21. (a) Find the value of R_2 in Figure 7–70. (b) Determine the power in R_2.

▶ FIGURE 7–70

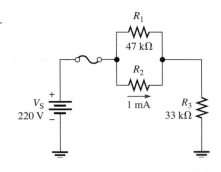

*22. Find the resistance between node A and each of the other nodes (R_{AB}, R_{AC}, R_{AD}, R_{AE}, R_{AF}, and R_{AG}) in Figure 7–71.

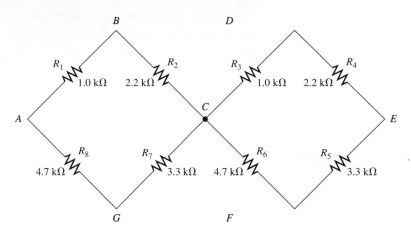

▲ FIGURE 7–71

*23. Find the resistance between each of the following sets of nodes in Figure 7–72: AB, BC, and CD.

▶ FIGURE 7–72

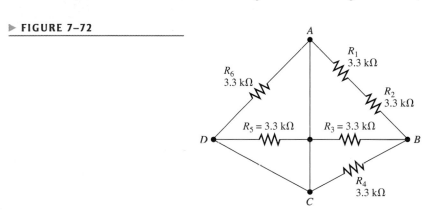

*24. Determine the value of each resistor in Figure 7–73.

▶ FIGURE 7–73

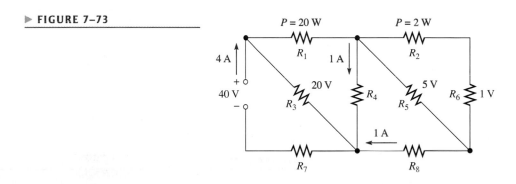

SECTION 7–3   Voltage Dividers with Resistive Loads

25. A voltage divider consists of two 56 kΩ resistors and a 15 V source. Calculate the unloaded output voltage. What will the output voltage be if a load resistor of 1.0 MΩ is connected to the output?

26. A 12 V battery output is divided down to obtain two output voltages. Three 3.3 kΩ resistors are used to provide the two taps. Determine the output voltages. If a 10 kΩ load is connected to the higher of the two outputs, what will its loaded value be?

27. Which will cause a smaller decrease in output voltage for a given voltage divider, a 10 kΩ load or a 47 kΩ load?

28. In Figure 7–74, determine the output voltage with no load across the output terminals. With a 100 kΩ load connected from A to B, what is the output voltage?

29. In Figure 7–74, determine the output voltage with a 33 kΩ load connected between A and B.

30. In Figure 7–74, determine the continuous current drawn from the source with no load across the output terminals. With a 33 kΩ load, what is the current drain?

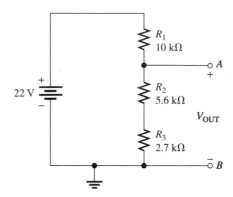

▲ FIGURE 7–74

*31. Determine the resistance values for a voltage divider that must meet the following specifications: The current drawn from the source under unloaded condition is not to exceed 5 mA. The source voltage is to be 10 V, and the required outputs are to be 5 V and 2.5 V. Sketch the circuit. Determine the effect on the output voltages if a 1.0 kΩ load is connected to each tap one at a time.

32. The voltage divider in Figure 7–75 has a switched load. Determine the voltage at each tap (V_1, V_2, and V_3) for each position of the switch.

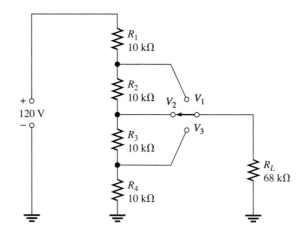

▲ FIGURE 7–75

*33. Figure 7–76 shows a dc biasing arrangement for a field-effect transistor amplifier. Biasing is a common method for setting up certain dc voltage levels required for proper amplifier operation. Although you are not expected to be familiar with transistor amplifiers at this point, the dc voltages and currents in the circuit can be determined using methods that you already know.

    (a) Find V_G and V_S **(b)** Determine I_1, I_2, I_D, and I_S **(c)** Find V_{DS} and V_{DG}

*34. Design a voltage divider to provide a 6 V output with no load and a minimum of 5.5 V across a 1.0 kΩ load. The source voltage is 24 V, and the unloaded current drain is not to exceed 100 mA.

▶ FIGURE 7–76

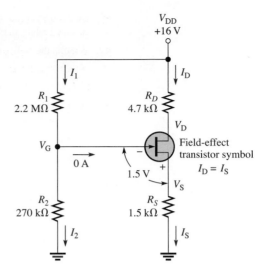

SECTION 7–4 Loading Effect of a Voltmeter

35. On which one of the following voltage range settings will a voltmeter present the minimum load on a circuit?

   (a) 1 V   (b) 10 V   (c) 100 V   (d) 1000 V

36. Determine the internal resistance of a 20,000 Ω/V voltmeter on each of the following range settings.

   (a) 0.5 V   (b) 1 V   (c) 5 V   (d) 50 V   (e) 100 V   (f) 1000 V

37. The voltmeter described in Problem 36 is used to measure the voltage across R_4 in Figure 7–62(a).

   (a) What range should be used?

   (b) How much less is the voltage measured by the meter than the actual voltage?

38. Repeat Problem 37 if the voltmeter is used to measure the voltage across R_1 in the circuit of Figure 7–62(b).

SECTION 7–5 Ladder Networks

39. For the circuit shown in Figure 7–77, calculate the following:

   (a) Total resistance across the source   (b) Total current from the source

   (c) Current through the 910 Ω resistor   (d) Voltage from A to point B

40. Determine the total resistance and the voltage at nodes A, B, and C in the ladder network of Figure 7–78.

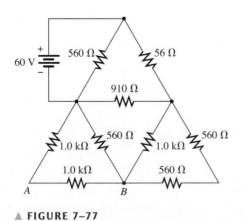

▲ FIGURE 7–77

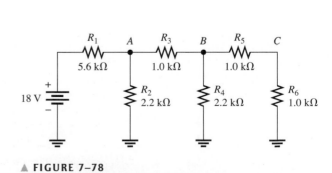

▲ FIGURE 7–78

*41. Determine the total resistance between terminals *A* and *B* of the ladder network in Figure 7–79. Also calculate the current in each branch with 10 V between *A* and *B*.

42. What is the voltage across each resistor in Figure 7–79 with 10 V between *A* and *B*?

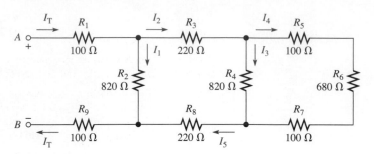

▲ FIGURE 7–79

*43. Find I_T and V_{OUT} in Figure 7–80.

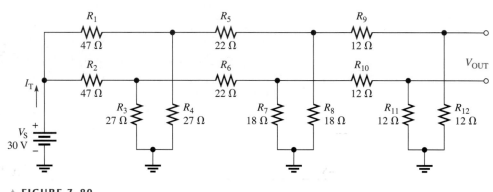

▲ FIGURE 7–80

44. Determine V_{OUT} for the *R/2R* ladder network in Figure 7–81 for the following conditions:
   (a) Switch SW2 connected to +12 V and the others connected to ground
   (b) Switch SW1 connected to +12 V and the others connected to ground

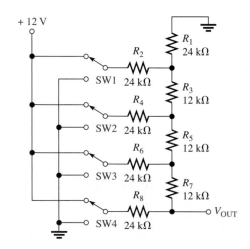

▲ FIGURE 7–81

45. Repeat Problem 44 for the following conditions:

    (a) SW3 and SW4 to +12 V, SW1 and SW2 to ground

    (b) SW3 and SW1 to +12 V, SW2 and SW4 to ground

    (c) All switches to +12 V

SECTION 7–6   The Wheatstone Bridge

46. A resistor of unknown value is connected to a Wheatstone bridge circuit. The bridge parameters for a balanced condition are set as follows: $R_V = 18\text{ k}\Omega$ and $R_2/R_4 = 0.02$. What is R_X?

47. A load cell has four identical strain gauges with an unstrained resistance of 120.000 Ω for each gauge (a standard value). When a load is added, the gauges in tension increase their resistance by 60 mΩ to 120.060 Ω and the gauges in compression decrease their resistance by 60 mΩ to 119.940 Ω as shown in Figure 7–82. What is the output voltage under load?

48. Determine the output voltage for the unbalanced bridge in Figure 7–83 for a temperature of 60°C. The temperature resistance characteristic for the thermistor is shown in Figure 7–60.

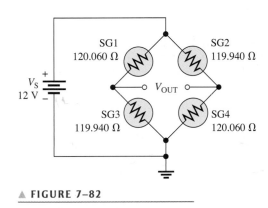

▲ FIGURE 7–82

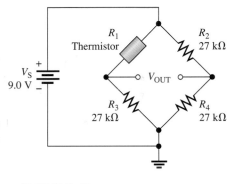

▲ FIGURE 7–83

SECTION 7–7   Troubleshooting

49. Is the voltmeter reading in Figure 7–84 correct?

50. Are the meter readings in Figure 7–85 correct?

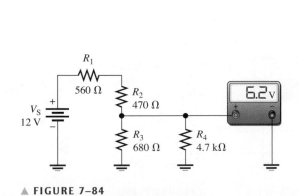

▲ FIGURE 7–84

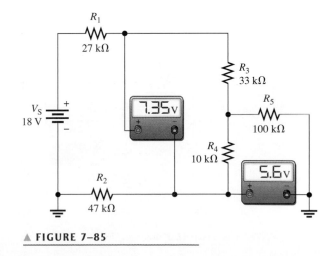

▲ FIGURE 7–85

51. There is one fault in Figure 7–86. Based on the meter indications, determine what the fault is.

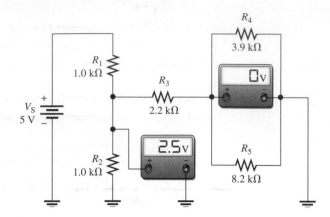

▲ FIGURE 7–86

52. Look at the meters in Figure 7–87 and determine if there is a fault in the circuit. If there is a fault, identify it.

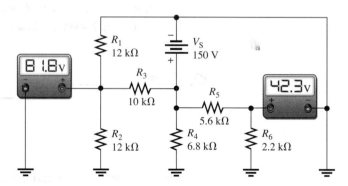

▲ FIGURE 7–87

53. Check the meter readings in Figure 7–88 and locate any fault that may exist.

54. If R_2 in Figure 7–89 opens, what voltages will be read at points A, B, and C?

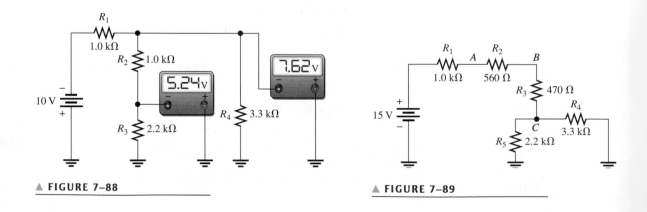

▲ FIGURE 7–88   ▲ FIGURE 7–89

Multisim Troubleshooting and Analysis

These problems require your Multisim CD-ROM.

55. Open file P07-55 and measure the total resistance.

56. Open file P07-56. Determine by measurement if there is an open resistor and, if so, which one.

57. Open file P07-57 and determine the unspecified resistance value.

58. Open file P07-58 and determine how much the load resistance affects each of the resistor voltages.

59. Open file P07-59 and find the shorted resistor, if there is one.

60. Open file P07-60 and adjust the value of R_X until the bridge is approximately balanced.

ANSWERS

SECTION REVIEWS

SECTION 7–1 Identifying Series-Parallel Relationships

1. A series-parallel resistive circuit is a circuit consisting of both series and parallel connections.
2. See Figure 7–90.
3. Resistors R_1 and R_2 are in series with the parallel combination of R_3 and R_4.
4. R_3, R_4, and R_5 are in parallel. Also the series-parallel combination $R_2 + (R_3 \parallel R_4 \parallel R_5)$ is in parallel with R_1.
5. Resistors R_1 and R_2 are in parallel; R_3 and R_4 are in parallel.
6. Yes, the parallel combinations are in series.

▶ **FIGURE 7–90**

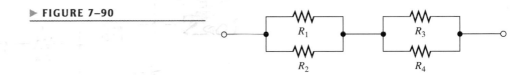

SECTION 7–2 Analysis of Series-Parallel Resistive Circuits

1. Voltage-divider and current-divider formulas, Kirchhoff's laws, and Ohm's law can be used in series-parallel analysis.
2. $R_T = R_1 + R_2 \parallel R_3 + R_4 = 608\ \Omega$
3. $I_3 = [R_2/(R_2 + R_3)]I_T = 11.1\ \text{mA}$
4. $V_2 = I_2R_2 = 3.65\ \text{V}$
5. $R_T = 47\ \Omega + 27\ \Omega + (27\ \Omega + 27\ \Omega) \parallel 47\ \Omega = 99.1\ \Omega;\ I_T = 1\ \text{V}/99.1\ \Omega = 10.1\ \text{mA}$

SECTION 7–3 Voltage Dividers with Resistive Loads

1. The load resistor decreases the output voltage.
2. True
3. $V_{\text{OUT(unloaded)}} = (100\ \text{k}\Omega/147\ \text{k}\Omega)30\ \text{V} = 20.4\ \text{V};\ V_{\text{OUT(loaded)}} = (9.1\ \text{k}\Omega/56.1\ \text{k}\Omega)30\ \text{V} = 4.87\ \text{V}$

SECTION 7–4 Loading Effect of a Voltmeter

1. A voltmeter loads a circuit because the internal resistance of the meter appears in parallel with the circuit resistance across which it is connected, reducing the resistance between those two points of the circuit and drawing current from the circuit.
2. No, because the meter resistance is much larger than $1.0\ \text{k}\Omega$.
3. Yes.

SECTION 7–5   Ladder Networks

1. See Figure 7–91.
2. $R_T = 11.6\,\text{k}\Omega$
3. $I_T = 10\,\text{V}/11.6\,\text{k}\Omega = 859\,\mu\text{A}$
4. $I_2 = 640\,\mu\text{A}$
5. $V_A = 1.41\,\text{V}$

▶ FIGURE 7–91

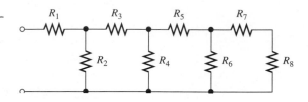

SECTION 7–6   The Wheatstone Bridge

1. See Figure 7–92.
2. The bridge is balanced when $V_A = V_B$; that is, when $V_{OUT} = 0$
3. $R_X = 15\,\text{k}\Omega$
4. An unbalanced bridge is used to measure transducer-sensed quantities.

▶ FIGURE 7–92

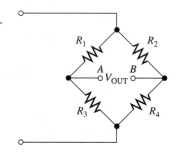

SECTION 7–7   Troubleshooting

1. Common circuit faults are opens and shorts.
2. The $10\,\text{k}\Omega$ resistor (R_3) is open.
3. (a) $V_A = 55\,\text{V}$ (b) $V_A = 55\,\text{V}$ (c) $V_A = 54.2\,\text{V}$ (d) $V_A = 100\,\text{V}$ (e) $V_A = 0\,\text{V}$

A Circuit Application

1. $P = 2.25\,\text{W}$; yes, a very tiny effect
2. The loading decreases.
3. Yes, worst case is $R_{THERM} = 0$, so 15 V is across $R_3 + R_4$, making power less than ⅛ W.
4. The output voltage of the op-amp can only be at a maximum or a minimum level, causing one LED or the other to be on at a time, but not both.

RELATED PROBLEMS FOR EXAMPLES

7–1   The new resistor is in parallel with $R_4 + R_2 \parallel R_3$.
7–2   The resistor has no effect because it is shorted.
7–3   The new resistor is in parallel with R_5.
7–4   A to gnd: $R_T = R_4 + R_3 \parallel (R_1 + R_2)$
       B to gnd: $R_T = R_4 + R_2 \parallel (R_1 + R_3)$
       C to gnd: $R_T = R_4$

7–5 R_3 and R_6 are in series.

7–6 55.1 Ω

7–7 128.3 Ω

7–8 2.38 mA

7–9 $I_1 = 35.7\,\text{mA}; I_3 = 23.4\,\text{mA}$

7–10 $V_A = 44.8\,\text{V}; V_1 = 35.2\,\text{V}$

7–11 2.04 V

7–12 $V_{AB} = 5.48\,\text{V}; V_{BC} = 1.66\,\text{V}; V_{CD} = 0.86\,\text{V}$

7–13 3.39 V

7–14 Increase R_1, R_2, and R_3 proportionally.

7–15 5.19 V

7–16 $I_1 = 7.16\,\text{mA}; I_2 = 3.57\,\text{mA}; I_3 = 3.57\,\text{mA}; I_4 = 1.74\,\text{mA}; I_5 = 1.85\,\text{mA};$
$I_6 = 1.85\,\text{mA}; V_A = 29.3\,\text{V}; V_B = 17.4\,\text{V}; V_C = 8.70\,\text{V}$

7–17 3.3 kΩ

7–18 0.45 V

7–19 5.73 V; 0 V

7–20 9.46 V

7–21 $V_A = 12\,\text{V}; V_B = 13.8\,\text{V}$

SELF-TEST

1. (e) **2.** (c) **3.** (c) **4.** (c) **5.** (b) **6.** (a) **7.** (b) **8.** (b)

9. (d) **10.** (b) **11.** (b) **12.** (b) **13.** (b) **14.** (a) **15.** (d)

CIRCUIT DYNAMICS QUIZ

1. (b) **2.** (a) **3.** (a) **4.** (a) **5.** (b) **6.** (a)

7. (b) **8.** (c) **9.** (c) **10.** (c) **11.** (b) **12.** (a)

13. (b) **14.** (b) **15.** (a) **16.** (a) **17.** (a)

8

CIRCUIT THEOREMS
AND CONVERSIONS

CHAPTER OBJECTIVES

◆ Describe the characteristics of a dc voltage source
◆ Describe the characteristics of a current source
◆ Perform source conversions
◆ Apply the superposition theorem to circuit analysis
◆ Apply Thevenin's theorem to simplify a circuit for analysis
◆ Apply Norton's theorem to simplify a circuit
◆ Apply the maximum power transfer theorem
◆ Perform Δ-to-Y and Y-to-Δ conversions

KEY TERMS

◆ Terminal equivalency
◆ Superposition theorem
◆ Thevenin's theorem
◆ Norton's theorem
◆ Maximum power transfer

A CIRCUIT APPLICATION PREVIEW

In the circuit application, a temperature-measurement and control circuit uses a Wheatstone bridge like the one you studied in Chapter 7. You will utilize Thevenin's theorem as well as other techniques in the evaluation of this circuit.

VISIT THE COMPANION WEBSITE

Study aids for this chapter are available at http://www. prenhall.com/floyd

INTRODUCTION

In previous chapters, you analyzed various types of circuits using Ohm's law and Kirchhoff's laws. Some types of circuits are difficult to analyze using only those basic laws and require additional methods in order to simplify the analysis.

The theorems and conversions in this chapter make analysis easier for certain types of circuits. These methods do not replace Ohm's law and Kirchhoff's laws, but they are normally used in conjunction with the laws in certain situations.

Because all electric circuits are driven by either voltage sources or current sources, it is important to understand how to work with these elements. The superposition theorem will help you to deal with circuits that have multiple sources. Thevenin's and Norton's theorems provide methods for reducing a circuit to a simple equivalent form for ease of analysis. The maximum power transfer theorem is used in applications where it is important for a given circuit to provide maximum power to a load. An example of this is an audio amplifier that provides maximum power to a speaker. Delta-to-wye and wye-to-delta conversions are sometimes useful when you analyze bridge circuits that are commonly found in systems that measure physical parameters such as temperature, pressure, and strain.

8–1 THE DC VOLTAGE SOURCE

As you learned in Chapter 2, the dc voltage source is one of the principal types of energy source in electronic applications, so it is important to understand its characteristics. The dc voltage source ideally provides constant voltage to a load even when the load resistance varies.

After completing this section, you should be able to

◆ **Describe the characteristics of a dc voltage source**

  ◆ Compare a practical voltage source to an ideal source

  ◆ Discuss the effect of loading on a practical voltage source

Figure 8–1(a) is the familiar symbol for an ideal dc voltage source. The voltage across its terminals, A and B, remains fixed regardless of the value of load resistance that may be connected across its output. Figure 8–1(b) shows a load resistor, R_L, connected. All of the source voltage, V_S, is dropped across R_L. Ideally, R_L can be changed to any value except zero, and the voltage will remain fixed. The ideal voltage source has an internal resistance of zero.

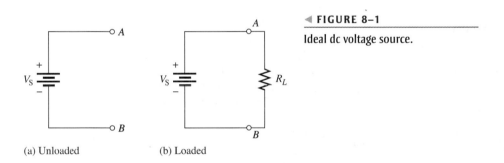

◀ **FIGURE 8–1**

Ideal dc voltage source.

(a) Unloaded   (b) Loaded

In reality, no voltage source is ideal; however, regulated power supplies can approach ideal when operated within the specified output current. All voltage sources have some inherent internal resistance as a result of their physical and/or chemical makeup, which can be represented by a resistor in series with an ideal source, as shown in Figure 8–2(a). R_S is the internal source resistance and V_S is the source voltage. With no load, the output voltage (voltage from A to B) is V_S. This voltage is sometimes called the *open circuit voltage.*

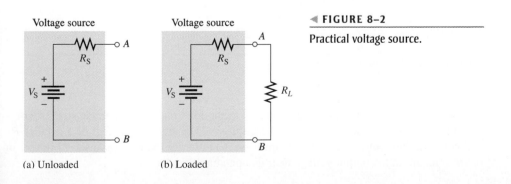

◀ **FIGURE 8–2**

Practical voltage source.

(a) Unloaded   (b) Loaded

Loading of the Voltage Source

When a load resistor is connected across the output terminals, as shown in Figure 8–2(b), all of the source voltage does not appear across R_L. Some of the voltage is dropped across R_S because R_S and R_L are in series.

If R_S is very small compared to R_L, the source approaches ideal because almost all of the source voltage, V_S, appears across the larger resistance, R_L. Very little voltage is dropped across the internal resistance, R_S. If R_L changes, most of the source voltage remains across the output as long as R_L is much larger than R_S. As a result, very little change occurs in the output voltage. The larger R_L is, compared to R_S, the less change there is in the output voltage. Example 8–1 illustrates the effect of changes in R_L on the output voltage when R_L is much greater than R_S.

EXAMPLE 8–1

Calculate the voltage output of the source in Figure 8–3 for the following values of R_L: 100 Ω, 560 Ω, and 1.0 kΩ.

▶ **FIGURE 8–3**

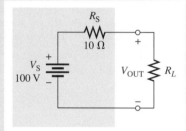

Solution For $R_L = 100$ Ω, the voltage output is

$$V_{OUT} = \left(\frac{R_L}{R_S + R_L}\right)V_S = \left(\frac{100\ \Omega}{110\ \Omega}\right)100\ V = \mathbf{90.9\ V}$$

For $R_L = 560$ Ω,

$$V_{OUT} = \left(\frac{560\ \Omega}{570\ \Omega}\right)100\ V = \mathbf{98.2\ V}$$

For $R_L = 1.0$ kΩ,

$$V_{OUT} = \left(\frac{1000\ \Omega}{1010\ \Omega}\right)100\ V = \mathbf{99.0\ V}$$

Notice that the output voltage is within 10% of the source voltage, V_S, for all three values of R_L because R_L is at least ten times R_S.

Related Problem* Determine V_{OUT} in Figure 8–3 if $R_S = 50$ Ω and $R_L = 10$ kΩ.

Use Multisim file E08-01 to verify the calculated results in this example and to confirm your calculation for the related problem.

---

*Answers are at the end of the chapter.

The output voltage decreases significantly as the load resistance is made smaller compared to the internal source resistance. Example 8–2 illustrates the effect of a smaller R_L and confirms the requirement that R_L must be much larger than R_S (at least 10 times) in order to maintain the output voltage near its open circuit value.

EXAMPLE 8–2

Determine V_{OUT} for $R_L = 10 \, \Omega$ and for $R_L = 1.0 \, \Omega$ in Figure 8–3.

Solution   For $R_L = 10 \, \Omega$, the voltage output is

$$V_{OUT} = \left(\frac{R_L}{R_S + R_L} \right) V_S = \left(\frac{10 \, \Omega}{20 \, \Omega} \right) 100 \text{ V} = \textbf{50 V}$$

For $R_L = 1.0 \, \Omega$,

$$V_{OUT} = \left(\frac{1.0 \, \Omega}{11 \, \Omega} \right) 100 \text{ V} = \textbf{9.09 V}$$

Related Problem   What is V_{OUT} with no load resistor in Figure 8–3?

SECTION 8–1 REVIEW
Answers are at the end of the chapter.

1. What is the symbol for the ideal voltage source?
2. Draw a practical voltage source.
3. What is the internal resistance of the ideal voltage source?
4. What effect does the load have on the output voltage of the practical voltage source?

8–2   THE CURRENT SOURCE

As you learned in Chapter 2, the current source is another type of energy source that ideally provides a constant current to a load even when the resistance of the load varies. The concept of the current source is important in certain types of transistor circuits.

After completing this section, you should be able to

◆ **Describe the characteristics of a current source**

  ◆ Compare a practical current source to an ideal source

  ◆ Discuss the effect of loading on a practical current source

Figure 8–4(a) shows a symbol for the ideal current source. The arrow indicates the direction of source current, I_S. An ideal current source produces a constant value of current through a load, regardless of the value of the load. This concept is illustrated in Figure 8–4(b), where a load resistor is connected to the current source between terminals A and B. The ideal current source has an infinitely large internal parallel resistance.

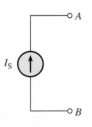

(a) Unloaded

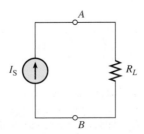

(b) Loaded

◀ **FIGURE 8–4**

Ideal current source.

Transistors act basically as current sources, and for this reason, knowledge of the current source concept is important. You will find that the equivalent model of a transistor does contain a current source.

Although the ideal current source can be used in most analysis work, no actual device is ideal. A practical current source representation is shown in Figure 8–5. Here the internal resistance appears in parallel with the ideal current source.

▶ **FIGURE 8–5**

Practical current source with load.

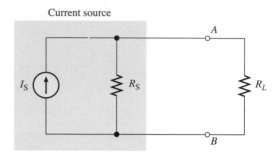

If the internal source resistance, R_S, is much larger than a load resistor, the practical source approaches ideal. The reason is illustrated in the practical current source shown in Figure 8–5. Part of the current, I_S, is through R_S, and part is through R_L. The internal source resistance, R_S, and the load resistor, R_L, act as a current divider. If R_S is much larger than R_L, most of the current is through R_L and very little through R_S. As long as R_L remains much smaller than R_S, the current through R_L will stay almost constant, no matter how much R_L changes.

If there is a constant-current source, you can normally assume that R_S is so much larger than the load resistance that R_S can be neglected. This simplifies the source to ideal, making the analysis easier.

Example 8–3 illustrates the effect of changes in R_L on the load current when R_L is much smaller than R_S. Generally, R_L should be at least ten times smaller than R_S ($10R_L \leq R_S$) for a source to act as a reasonable current source.

EXAMPLE 8–3

Calculate the load current (I_L) in Figure 8–6 for the following values of R_L: 100 Ω, 560 Ω, and 1.0 kΩ.

▶ **FIGURE 8–6**

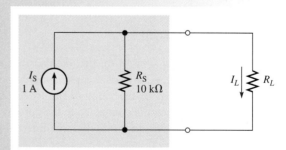

Solution For $R_L = 100$ Ω, the load current is

$$I_L = \left(\frac{R_S}{R_S + R_L}\right)I_S = \left(\frac{10\,\text{k}\Omega}{10.1\,\text{k}\Omega}\right)1\,\text{A} = \mathbf{990\,mA}$$

For $R_L = 560$ Ω,

$$I_L = \left(\frac{10\,\text{k}\Omega}{10.56\,\text{k}\Omega}\right)1\,\text{A} = \mathbf{947\,mA}$$

For $R_L = 1.0\,k\Omega$,

$$I_L = \left(\frac{10\,k\Omega}{11\,k\Omega}\right)1\,A = \textbf{909 mA}$$

Notice that the load current, I_L, is within 10% of the source current for each value of R_L because R_L is at least ten times smaller than R_S in each case.

Related Problem At what value of R_L in Figure 8–6 will the load current equal 750 mA?

SECTION 8–2 REVIEW

1. What is the symbol for an ideal current source?
2. Draw the practical current source.
3. What is the internal resistance of the ideal current source?
4. What effect does the load have on the load current of the practical current source?

8–3 SOURCE CONVERSIONS

In circuit analysis, it is sometimes useful to convert a voltage source to an equivalent current source, or vice versa.

After completing this section, you should be able to

◆ **Perform source conversions**

  ◆ Convert a voltage source to a current source

  ◆ Convert a current source to a voltage source

  ◆ Define *terminal equivalency*

Converting a Voltage Source to a Current Source

The source voltage, V_S, divided by the internal source resistance, R_S, gives the value of the equivalent source current.

$$I_S = \frac{V_S}{R_S}$$

The value of R_S is the same for both the voltage and current sources. As illustrated in Figure 8–7, the directional arrow for the current points from minus to plus. The equivalent current source is in parallel with R_S.

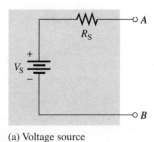

(a) Voltage source   (b) Current source

◀ **FIGURE 8–7**

Conversion of voltage source to equivalent current source.

Equivalency of two sources means that for any given load resistance connected to the two sources, the same load voltage and load current are produced by both sources. This concept is called **terminal equivalency**.

You can show that the voltage source and the current source in Figure 8–7 are equivalent by connecting a load resistor to each, as shown in Figure 8–8, and then calculating the load current. For the voltage source, the load current is

$$I_L = \frac{V_S}{R_S + R_L}$$

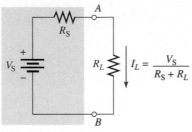

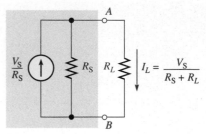

(a) Loaded voltage source   (b) Loaded current source

▲ FIGURE 8–8

Equivalent sources with loads.

For the current source,

$$I_L = \left(\frac{R_S}{R_S + R_L}\right)\frac{V_S}{R_S} = \frac{V_S}{R_S + R_L}$$

As you see, both expressions for I_L are the same. These equations show that the sources are equivalent as far as the load or terminals A and B are concerned.

EXAMPLE 8–4

Convert the voltage source in Figure 8–9 to an equivalent current source and show the equivalent circuit.

▶ FIGURE 8–9

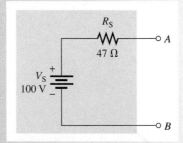

Solution   The value of the internal resistance, R_S, of the equivalent current is the same as the internal resistance of the voltage source. Therefore, the equivalent current source is

$$I_S = \frac{V_S}{R_S} = \frac{100\ \text{V}}{47\ \Omega} = 2.13\ \text{A}$$

Figure 8–10 shows the equivalent circuit.

▶ FIGURE 8–10

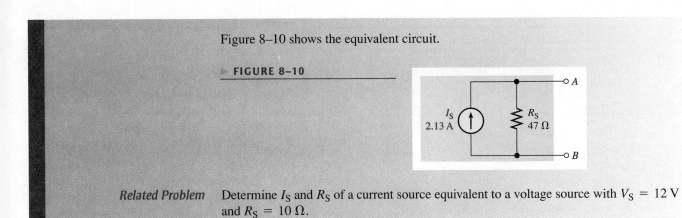

Related Problem   Determine I_S and R_S of a current source equivalent to a voltage source with $V_S = 12$ V and $R_S = 10\ \Omega$.

Converting a Current Source to a Voltage Source

The source current, I_S, multiplied by the internal source resistance, R_S, gives the value of the equivalent source voltage.

$$V_S = I_S R_S$$

Again, R_S remains the same. The polarity of the voltage source is minus to plus in the direction of the current. The equivalent voltage source is the voltage in series with R_S, as illustrated in Figure 8–11.

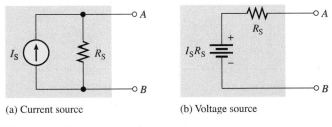

(a) Current source   (b) Voltage source

▲ **FIGURE 8–11**

Conversion of current source to equivalent voltage source.

EXAMPLE 8–5

Convert the current source in Figure 8–12 to an equivalent voltage source and show the equivalent circuit.

▶ FIGURE 8–12

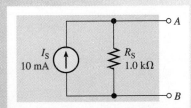

Solution   The value of R_S is the same as with a current source. Therefore, the equivalent voltage source is

$$V_S = I_S R_S = (10\ \text{mA})(1.0\ \text{k}\Omega) = 10\ \text{V}$$

Figure 8–13 shows the equivalent circuit.

▶ **FIGURE 8–13**

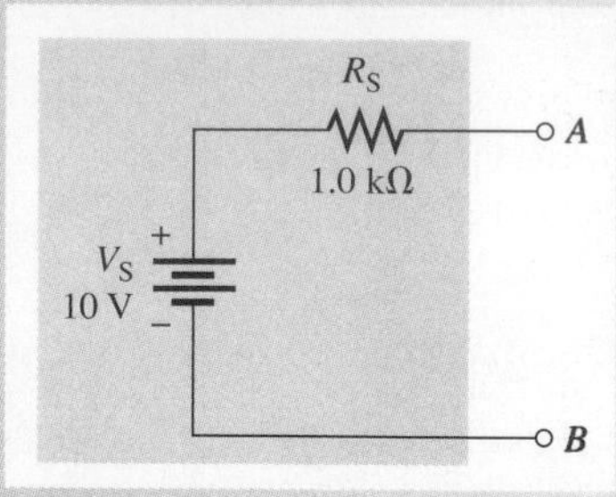

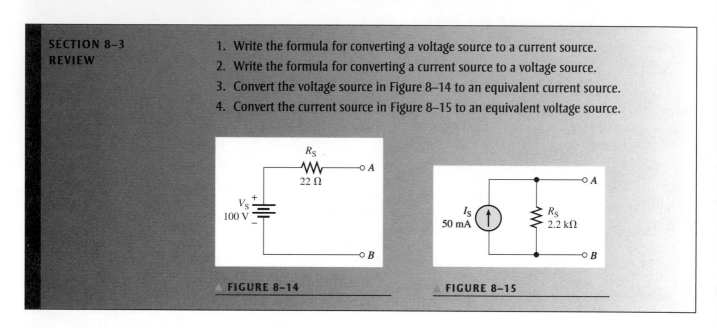

Related Problem   Determine V_S and R_S of a voltage source equivalent to a current source with $I_S = 500$ mA and $R_S = 600\ \Omega$.

SECTION 8–3 REVIEW

1. Write the formula for converting a voltage source to a current source.
2. Write the formula for converting a current source to a voltage source.
3. Convert the voltage source in Figure 8–14 to an equivalent current source.
4. Convert the current source in Figure 8–15 to an equivalent voltage source.

▲ **FIGURE 8–14**   ▲ **FIGURE 8–15**

8–4   THE SUPERPOSITION THEOREM

Some circuits require more than one voltage or current source. For example, most amplifiers operate with two voltage sources: an ac and a dc source. Additionally, some amplifiers require both a positive and a negative dc voltage source for proper operation. When multiple sources are used in a circuit, the superposition theorem provides a method for analysis.

After completing this section, you should be able to

◆ **Apply the superposition theorem to circuit analysis**

  ◆ State the superposition theorem

  ◆ List the steps in applying the theorem

The superposition method is a way to determine currents in a circuit with multiple sources by leaving one source at a time and replacing the other sources by their internal resistances.

Recall that an ideal voltage source has a zero internal resistance and an ideal current source has infinite internal resistance. All sources will be treated as ideal in order to simplify the coverage.

A general statement of the **superposition theorem** is as follows:

The current in any given branch of a multiple-source circuit can be found by determining the currents in that particular branch produced by each source acting alone, with all other sources replaced by their internal resistances. The total current in the branch is the algebraic sum of the individual currents in that branch.

The steps in applying the superposition method are as follows:

Step 1. Leave one voltage (or current) source at a time in the circuit and replace each of the other voltage (or current) sources with its internal resistance. For ideal sources a short represents zero internal resistance and an open represents infinite internal resistance.

Step 2. Determine the particular current (or voltage) that you want just as if there were only one source in the circuit.

Step 3. Take the next source in the circuit and repeat Steps 1 and 2. Do this for each source.

Step 4. To find the actual current in a given branch, algebraically sum the currents due to each individual source. (If the currents are in the same direction, they are added. If the currents are in opposite directions, they are subtracted with the direction of the resulting current the same as the larger of the original quantities.) Once you find the current, you can determine the voltage using Ohm's law.

The approach to superposition is demonstrated in Figure 8–16 for a series-parallel circuit with two ideal voltage sources. Study the steps in this figure.

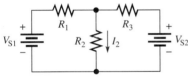

(a) Problem: Find I_2.

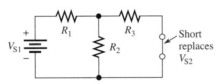

(b) Replace V_{S2} with zero resistance (short).

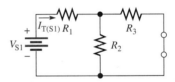

(c) Find R_T and I_T looking from V_{S1}:
$R_{T(S1)} = R_1 + R_2 \| R_3$
$I_{T(S1)} = V_{S1}/R_{T(S1)}$

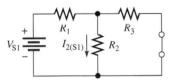

(d) Find I_2 due to V_{S1} (current divider):

$$I_{2(S1)} = \left(\frac{R_3}{R_2 + R_3} \right) I_{T(S1)}$$

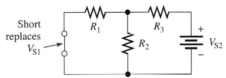

(e) Replace V_{S1} with zero resistance (short).

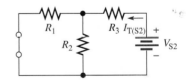

(f) Find R_T and I_T looking from V_{S2}:
$R_{T(S2)} = R_3 + R_1 \| R_2$
$I_{T(S2)} = V_{S2}/R_{T(S2)}$

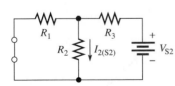

(g) Find I_2 due to V_{S2}:

$$I_{2(S2)} = \left(\frac{R_1}{R_1 + R_2} \right) I_{T(S2)}$$

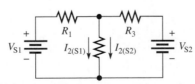

(h) Restore the original sources. Add $I_{2(S1)}$ and $I_{2(S2)}$ to get the actual I_2 (they are in same direction):
$I_2 = I_{2(S1)} + I_{2(S2)}$

▲ **FIGURE 8–16**

Demonstration of the superposition method.

EXAMPLE 8–6

Use the superposition theorem to find the current through R_2 of Figure 8–17.

▶ FIGURE 8–17

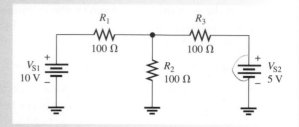

Solution **Step 1:** Replace V_{S2} with a short and find the current through R_2 due to voltage source V_{S1}, as shown in Figure 8–18. To find I_2, use the current-divider formula (Equation 6–6). Looking from V_{S1},

$$R_{T(S1)} = R_1 + \frac{R_3}{2} = 100 \ \Omega + 50 \ \Omega = 150 \ \Omega$$

$$I_{T(S1)} = \frac{V_{S1}}{R_{T(S1)}} = \frac{10 \ V}{150 \ \Omega} = 66.7 \ \text{mA}$$

The current through R_2 due to V_{S1} is

$$I_{2(S1)} = \left(\frac{R_3}{R_2 + R_3}\right) I_{T(S1)} = \left(\frac{100 \ \Omega}{200 \ \Omega}\right) 66.7 \ \text{mA} = 33.3 \ \text{mA}$$

Note that this current is downward through R_2.

▶ FIGURE 8–18

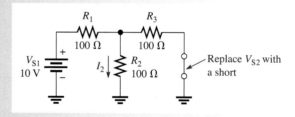

Step 2: Find the current through R_2 due to voltage source V_{S2} by replacing V_{S1} with a short, as shown in Figure 8–19. Looking from V_{S2},

$$R_{T(S2)} = R_3 + \frac{R_1}{2} = 100 \ \Omega + 50 \ \Omega = 150 \ \Omega$$

$$I_{T(S2)} = \frac{V_{S2}}{R_{T(S2)}} = \frac{5 \ V}{150 \ \Omega} = 33.3 \ \text{mA}$$

The current through R_2 due to V_{S2} is

$$I_{2(S2)} = \left(\frac{R_1}{R_1 + R_2}\right) I_{T(S2)} = \left(\frac{100 \ \Omega}{200 \ \Omega}\right) 33.3 \ \text{mA} = 16.7 \ \text{mA}$$

Note that this current is downward through R_2.

▶ FIGURE 8–19

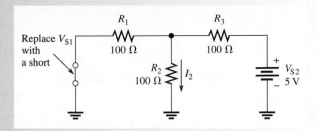

Step 3: Both component currents are downward through R_2, so they have the same algebraic sign. Therefore, add the values to get the total current through R_2.

$$I_{2(\text{tot})} = I_{2(S1)} + I_{2(S2)} = 33.3 \text{ mA} + 16.7 \text{mA} = \textbf{50 mA}$$

Related Problem   Determine the total current through R_2 if the polarity of V_{S2} in Figure 8–17 is reversed.

 Use Multisim file E08-06 to verify the calculated results in this example and to confirm your calculation for the related problem.

EXAMPLE 8–7   Find the current through R_2 in the circuit of Figure 8–20.

FIGURE 8–20

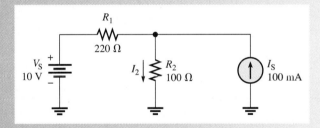

Solution   **Step 1:** Find the current through R_2 due to V_S by replacing I_S with an open, as shown in Figure 8–21.

FIGURE 8–21

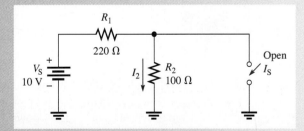

Notice that all of the current produced by V_S is through R_2. Looking from V_S,

$$R_T = R_1 + R_2 = 320 \text{ } \Omega$$

The current through R_2 due to V_S is

$$I_{2(V_S)} = \frac{V_S}{R_T} = \frac{10 \text{ V}}{320 \text{ } \Omega} = 31.2 \text{ mA}$$

Note that this current is downward through R_2.

Step 2: Find the current through R_2 due to I_S by replacing V_S with a short, as shown in Figure 8–22.

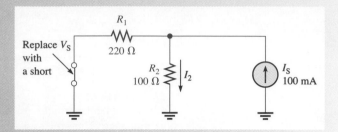

▲ **FIGURE 8–22**

Use the current-divider formula to determine the current through R_2 due to I_S.

$$I_{2(I_S)} = \left(\frac{R_1}{R_1 + R_2}\right)I_S = \left(\frac{220\ \Omega}{320\ \Omega}\right)100\ \text{mA} = 68.8\ \text{mA}$$

Note that this current also is downward through R_2.

Step 3: Both currents are in the same direction through R_2, so add them to get the total.

$$I_{2(\text{tot})} = I_{2(V_S)} + I_{2(I_S)} = 31.2\ \text{mA} + 68.8\ \text{mA} = \mathbf{100\ mA}$$

Related Problem   If the polarity of V_S in Figure 8–20 is reversed, how is the value of I_S affected?

EXAMPLE 8–8

Find the current through the 100 Ω resistor in Figure 8–23.

▶ **FIGURE 8–23**

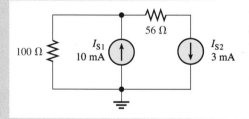

Solution   **Step 1:** Find the current through the 100 Ω resistor due to current source I_{S1} by replacing source I_{S2} with an open, as shown in Figure 8–24. As you can see, the entire 10 mA from the current source I_{S1} is downward through the 100 Ω resistor.

▶ **FIGURE 8–24**

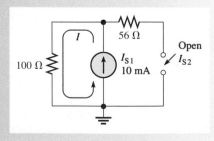

Step 2: Find the current through the 100 Ω resistor due to source I_{S2} by replacing source I_{S1} with an open, as indicated in Figure 8–25. Notice that all of the 3 mA from source I_{S2} is upward through the 100 Ω resistor.

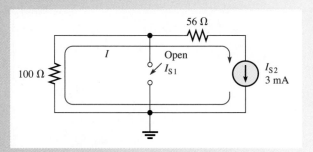

▲ FIGURE 8–25

Step 3: To get the total current through the 100 Ω resistor, subtract the smaller current from the larger because they are in opposite directions. The resulting total current is in the direction of the larger current from source I_{S1}.

$$I_{100\Omega(tot)} = I_{100\Omega(I_{S1})} - I_{100\Omega(I_{S2})}$$
$$= 10\,\text{mA} - 3\,\text{mA} = \mathbf{7\,mA}$$

The resulting current is downward through the resistor.

Related Problem If the 100 Ω resistor in Figure 8–23 is changed to 68 Ω, what will be the current through it?

EXAMPLE 8–9

Find the total current through R_3 in Figure 8–26.

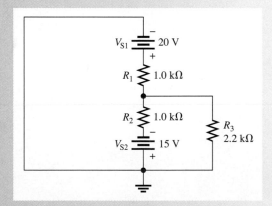

▲ FIGURE 8–26

Solution **Step 1:** Find the current through R_3 due to source V_{S1} by replacing source V_{S2} with a short, as shown in Figure 8–27.

▶ FIGURE 8–27

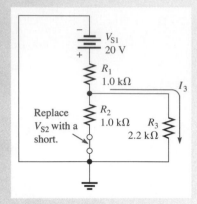

Looking from V_{S1},

$$R_{T(S1)} = R_1 + \frac{R_2 R_3}{R_2 + R_3} = 1.0\,k\Omega + \frac{(1.0\,k\Omega)(2.2\,k\Omega)}{3.2\,k\Omega} = 1.69\,k\Omega$$

$$I_{T(S1)} = \frac{V_{S1}}{R_{T(S1)}} = \frac{20\,V}{1.69\,k\Omega} = 11.8\,mA$$

Now apply the current-divider formula to get the current through R_3 due to source V_{S1}.

$$I_{3(S1)} = \left(\frac{R_2}{R_2 + R_3}\right) I_{T(S1)} = \left(\frac{1.0\,k\Omega}{3.2\,k\Omega}\right) 11.8\,mA = 3.69\,mA$$

Notice that this current is downward through R_3.

Step 2: Find I_3 due to source V_{S2} by replacing source V_{S1} with a short, as shown in Figure 8–28.

▶ FIGURE 8–28

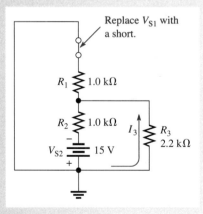

Looking from V_{S2},

$$R_{T(S2)} = R_2 + \frac{R_1 R_3}{R_1 + R_3} = 1.0\,k\Omega + \frac{(1.0\,k\Omega)(2.2\,k\Omega)}{3.2\,k\Omega} = 1.69\,k\Omega$$

$$I_{T(S2)} = \frac{V_{S2}}{R_{T(S2)}} = \frac{15\,V}{1.69\,k\Omega} = 8.88\,mA$$

Now apply the current-divider formula to find the current through R_3 due to source V_{S2}.

$$I_{3(S2)} = \left(\frac{R_1}{R_1 + R_3}\right) I_{T(S2)} = \left(\frac{1.0\,k\Omega}{3.2\,k\Omega}\right) 8.88\,mA = 2.78\,mA$$

Notice that this current is upward through R_3.

Step 3: Calculate the total current through R_3.

$$I_{3(\text{tot})} = I_{3(S1)} - I_{3(S2)} = 3.69 \text{ mA} - 2.78 \text{ mA} = 0.91 \text{ mA} = \mathbf{910 \ \mu A}$$

This current is downward through R_3.

Related Problem Find $I_{3(\text{tot})}$ in Figure 8–26 if V_{S1} is changed to 12 V and its polarity reversed.

Use Multisim file E08-09 to verify the calculated results in this example and to confirm your calculation for the related problem.

Although regulated dc power supplies are close to ideal voltage sources, many ac sources are not. For example, function generators generally have 50 Ω or 600 Ω of internal resistance, which appears as a resistance in series with an ideal source. Also, batteries can look ideal when they are fresh; but as they age, the internal resistance increases. When applying the superposition theorem, it is important to recognize when a source is not ideal and replace it with its equivalent internal resistance.

Current sources are not as common as voltage sources and are also not always ideal. If a current source is not ideal, as in the case of many transistors, it should be replaced by its equivalent internal resistance when the superposition theorem is applied.

**SECTION 8–4
REVIEW**

1. State the superposition theorem.
2. Why is the superposition theorem useful for analysis of multiple-source circuits?
3. Why is an ideal voltage source shorted and an ideal current source opened when the superposition theorem is applied?
4. Using the superposition theorem, find the current through R_1 in Figure 8–29.
5. If, as a result of applying the superposition theorem, two currents are in opposing directions through a branch of a circuit, in which direction is the net current?

▶ **FIGURE 8–29**

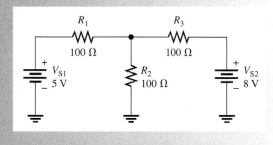

8–5 THEVENIN'S THEOREM

Thevenin's theorem provides a method for simplifying a circuit to a standard equivalent form. This theorem can be used to simplify the analysis of complex circuits.

After completing this section, you should be able to

◆ **Apply Thevenin's theorem to simplify a circuit for analysis**

   ◆ Describe the form of a Thevenin equivalent circuit

   ◆ Obtain the Thevenin equivalent voltage source

   ◆ Obtain the Thevenin equivalent resistance

◆ Explain terminal equivalency in the context of Thevenin's theorem

◆ Thevenize a portion of a circuit

◆ Thevenize a bridge circuit

The Thevenin equivalent form of any two-terminal resistive circuit consists of an equivalent voltage source (V_{TH}) and an equivalent resistance (R_{TH}), arranged as shown in Figure 8–30. The values of the equivalent voltage and resistance depend on the values in the original circuit. Any resistive circuit can be simplified regardless of its complexity with respect to two output terminals.

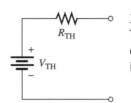

▶ FIGURE 8–30

The general form of a Thevenin equivalent circuit is a voltage source in series with a resistance.

The equivalent voltage, V_{TH}, is one part of the complete Thevenin equivalent circuit. The other part is R_{TH}.

The Thevenin equivalent voltage (V_{TH}) is the open circuit (no-load) voltage between two output terminals in a circuit.

Any component connected between these two terminals effectively "sees" V_{TH} in series with R_{TH}. As defined by **Thevenin's theorem**,

The Thevenin equivalent resistance (R_{TH}) is the total resistance appearing between two terminals in a given circuit with all sources replaced by their internal resistances.

Although a Thevenin equivalent circuit is not the same as its original circuit, it acts the same in terms of the output voltage and current. Try the following demonstration as illustrated in Figure 8–31. Place a resistive circuit of any complexity in a box with only the

▶ FIGURE 8–31

Which box contains the original circuit and which contains the Thevenin equivalent circuit? You cannot tell by observing the meters.

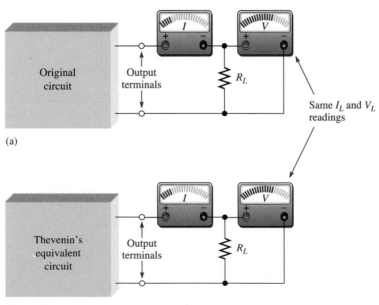

output terminals exposed. Then place the Thevenin equivalent of that circuit in an identical box with, again, only the output terminals exposed. Connect identical load resistors across the output terminals of each box. Next connect a voltmeter and an ammeter to measure the voltage and current for each load as shown in the figure. The measured values will be identical (neglecting tolerance variations), and you will not be able to determine which box contains the original circuit and which contains the Thevenin equivalent. That is, in terms of your observations based on any electrical measurements, both circuits appear to be the same. This condition is sometimes known as *terminal equivalency* because both circuits look the same from the "viewpoint" of the two output terminals.

To find the Thevenin equivalent of any circuit, determine the equivalent voltage, V_{TH}, and the equivalent resistance, R_{TH}, looking from the output terminals. As an example, the Thevenin equivalent for the circuit between terminals A and B is developed in Figure 8–32.

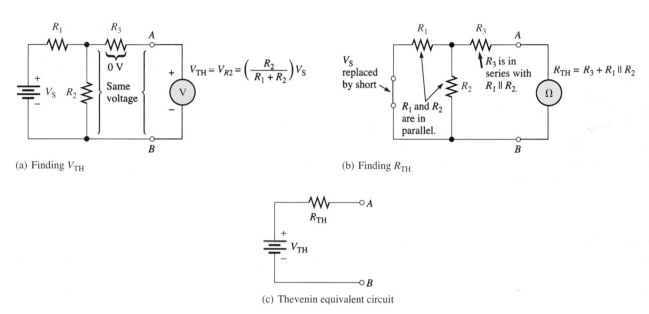

(a) Finding V_{TH}

(b) Finding R_{TH}

(c) Thevenin equivalent circuit

▲ **FIGURE 8–32**

Example of the simplification of a circuit by Thevenin's theorem.

In Figure 8–32(a), the voltage across the designated terminals A and B is the Thevenin equivalent voltage. In this particular circuit, the voltage from A to B is the same as the voltage across R_2 because there is no current through R_3 and, therefore, no voltage drop across it. The Thevenin voltage is expressed as follows for this particular example:

$$V_{TH} = \left(\frac{R_2}{R_1 + R_2}\right)V_S$$

In Figure 8–32(b), the resistance between terminals A and B with the source replaced by a short (zero internal resistance) is the Thevenin equivalent resistance. In this particular circuit, the resistance from A to B is R_3 in series with the parallel combination of R_1 and R_2. Therefore, R_{TH} is expressed as follows:

$$R_{TH} = R_3 + \frac{R_1 R_2}{R_1 + R_2}$$

The Thevenin equivalent circuit is shown in Figure 8–32(c).

EXAMPLE 8–10

Find the Thevenin equivalent circuit between *A* and *B* of the circuit in Figure 8–33.

▶ **FIGURE 8–33**

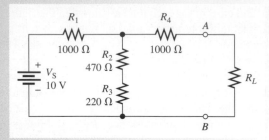

Solution First, remove R_L. Then V_{TH} equals the voltage across $R_2 + R_3$, as shown in Figure 8–34(a), because $V_4 = 0$ V since there is no current through it.

$$V_{TH} = \left(\frac{R_2 + R_3}{R_1 + R_2 + R_3}\right)V_S = \left(\frac{690\ \Omega}{1690\ \Omega}\right)10\ V = \textbf{4.08 V}$$

To find R_{TH}, first replace the source with a short to simulate a zero internal resistance. Then R_1 appears in parallel with $R_2 + R_3$, and R_4 is in series with the series-parallel combination of R_1, R_2, and R_3, as indicated in Figure 8–34(b).

$$R_{TH} = R_4 + \frac{R_1(R_2 + R_3)}{R_1 + R_2 + R_3} = 1000\ \Omega + \frac{(1000\ \Omega)(690\ \Omega)}{1690\ \Omega} = \textbf{1410 }\Omega$$

The resulting Thevenin equivalent circuit is shown in Figure 8–34(c).

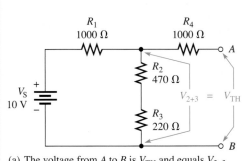

(a) The voltage from *A* to *B* is V_{TH} and equals V_{2+3}.

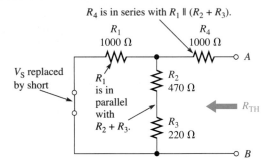

(b) Looking from terminals *A* and *B*, R_4 appears in series with the combination of R_1 in parallel with $(R_2 + R_3)$.

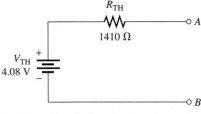

(c) Thevenin equivalent circuit

▲ **FIGURE 8–34**

Related Problem Determine V_{TH} and R_{TH} if a 560 Ω resistor is connected in parallel across R_2 and R_3.

Use Multisim file E08-10 to verify the calculated results in this example and to confirm your calculations for the related problem.

Thevenin Equivalency Depends on the Viewpoint

The Thevenin equivalent for any circuit depends on the location of the two output terminals from which the circuit is "viewed." In Figure 8–33, you viewed the circuit from between the two terminals labeled A and B. Any given circuit can have more than one Thevenin equivalent, depending on how the output terminals are designated. For example, if you view the circuit in Figure 8–35 from between terminals A and C, you obtain a completely different result than if you viewed it from between terminals A and B or from between terminals B and C.

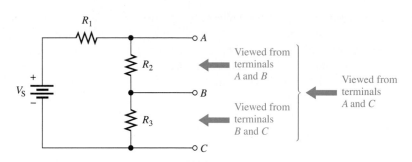

Thevenin's equivalent depends on the output terminals from which the circuit is viewed.

In Figure 8–36(a), when viewed from between terminals A and C, V_{TH} is the voltage across $R_2 + R_3$ and can be expressed using the voltage-divider formula as

$$V_{TH(AC)} = \left(\frac{R_2 + R_3}{R_1 + R_2 + R_3}\right)V_S$$

Also, as shown in Figure 8–36(b), the resistance between terminals A and C is $R_2 + R_3$ in parallel with R_1 (the source is replaced by a short) and can be expressed as

$$R_{TH(AC)} = \frac{R_1(R_2 + R_3)}{R_1 + R_2 + R_3}$$

The resulting Thevenin equivalent circuit is shown in Figure 8–36(c).

When viewed from between terminals B and C as indicated in Figure 8–36(d), V_{TH} is the voltage across R_3 and can be expressed as

$$V_{TH(BC)} = \left(\frac{R_3}{R_1 + R_2 + R_3}\right)V_S$$

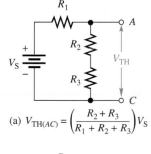

(a) $V_{TH(AC)} = \left(\dfrac{R_2 + R_3}{R_1 + R_2 + R_3}\right)V_S$

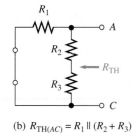

(b) $R_{TH(AC)} = R_1 \parallel (R_2 + R_3)$

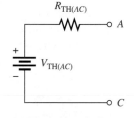

(c) Thevenin equivalent

(d) $V_{TH(BC)} = \left(\dfrac{R_3}{R_1 + R_2 + R_3}\right)V_S$

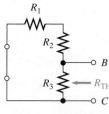

(e) $R_{TH(BC)} = R_3 \parallel (R_1 + R_2)$

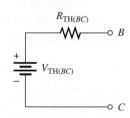

(f) Thevenin equivalent

Example of a circuit thevenized from two different sets of terminals. Parts (a), (b), and (c) illustrate one set of terminals and parts (d), (e), and (f) illustrate another set of terminals. (The V_{TH} and R_{TH} values are different for each case.)

As shown in Figure 8–36(e), the resistance between terminals B and C is R_3 in parallel with the series combination of R_1 and R_2.

$$R_{TH(BC)} = \frac{R_3(R_1 + R_2)}{R_1 + R_2 + R_3}$$

The resulting Thevenin equivalent is shown in Figure 8–36 (f).

Thevenizing a Portion of a Circuit

In many cases, it helps to thevenize only a portion of a circuit. For example, when you need to know the equivalent circuit as viewed by one particular resistor in the circuit, you can remove the resistor and apply Thevenin's theorem to the remaining part of the circuit as viewed from the points between which that resistor was connected. Figure 8–37 illustrates the thevenizing of part of a circuit.

▶ FIGURE 8–37

Example of thevenizing a portion of a circuit. In this case, the circuit is thevenized from the viewpoint of the load resistor, R_3.

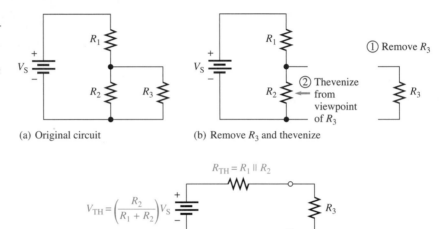

(a) Original circuit   (b) Remove R_3 and thevenize

$$V_{TH} = \left(\frac{R_2}{R_1 + R_2}\right)V_S$$

(c) Thevenin equivalent of original circuit with R_3 connected

Using this type of approach, you can easily find the voltage and current for a specified resistor for any number of resistor values using only Ohm's law. This method eliminates the necessity of reanalyzing the original circuit for each different resistance value.

Thevenizing a Bridge Circuit

The usefulness of Thevenin's theorem is perhaps best illustrated when it is applied to a Wheatstone bridge circuit. For example, when a load resistor is connected to the output terminals of a Wheatstone bridge, as shown in Figure 8–38, the circuit is difficult to analyze because it is not a straightforward series-parallel arrangement. There are no resistors that are in series or in parallel with another resistor.

▶ FIGURE 8–38

A Wheatstone bridge with a load resistor connected between the output terminals is not a straightforward series-parallel circuit.

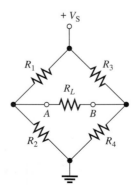

Using Thevenin's theorem, you can simplify the bridge circuit to an equivalent circuit viewed from the load resistor as shown step-by-step in Figure 8–39. Study carefully the steps in this figure. Once the equivalent circuit for the bridge is found, the voltage and current for any value of load resistor can easily be determined.

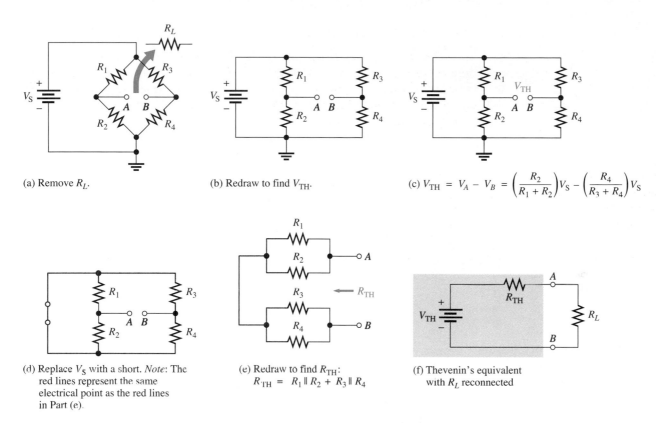

(a) Remove R_L.

(b) Redraw to find V_{TH}.

(c) $V_{TH} = V_A - V_B = \left(\dfrac{R_2}{R_1 + R_2} \right) V_S - \left(\dfrac{R_4}{R_3 + R_4} \right) V_S$

(d) Replace V_S with a short. *Note:* The red lines represent the same electrical point as the red lines in Part (e).

(e) Redraw to find R_{TH}:
$R_{TH} = R_1 \| R_2 + R_3 \| R_4$

(f) Thevenin's equivalent with R_L reconnected

▲ **FIGURE 8–39**

Simplifying a Wheatstone bridge with Thevenin's theorem.

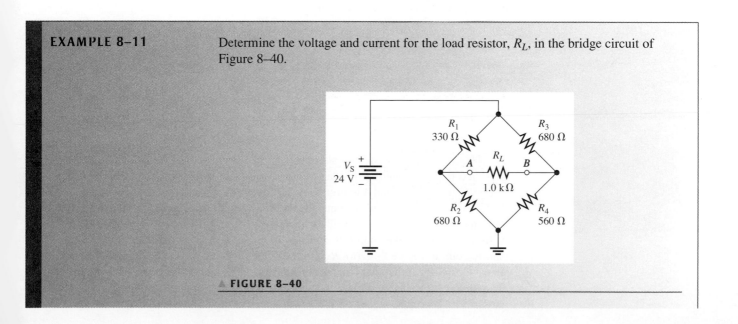

EXAMPLE 8–11

Determine the voltage and current for the load resistor, R_L, in the bridge circuit of Figure 8–40.

▲ **FIGURE 8–40**

Solution

Step 1: Remove R_L.

Step 2: To thevenize the bridge as viewed from between terminals A and B, as was shown in Figure 8–39, first determine V_{TH}.

$$V_{TH} = V_A - V_B = \left(\frac{R_2}{R_1 + R_2}\right)V_S - \left(\frac{R_4}{R_3 + R_4}\right)V_S$$

$$= \left(\frac{680\ \Omega}{1010\ \Omega}\right)24\ V - \left(\frac{560\ \Omega}{1240\ \Omega}\right)24\ V = 16.16\ V - 10.84\ V = 5.32\ V$$

Step 3: Determine R_{TH}.

$$R_{TH} = \frac{R_1 R_2}{R_1 + R_2} + \frac{R_3 R_4}{R_3 + R_4}$$

$$= \frac{(330\ \Omega)(680\ \Omega)}{1010\ \Omega} + \frac{(680\ \Omega)(560\ \Omega)}{1240\ \Omega} = 222\ \Omega + 307\ \Omega = 529\ \Omega$$

Step 4: Place V_{TH} and R_{TH} in series to form the Thevenin equivalent circuit.

Step 5: Connect the load resistor between terminals A and B of the equivalent circuit, and determine the load voltage and current as illustrated in Figure 8–41.

$$V_L = \left(\frac{R_L}{R_L + R_{TH}}\right)V_{TH} = \left(\frac{1.0\ k\Omega}{1.529\ k\Omega}\right)5.32\ V = \textbf{3.48 V}$$

$$I_L = \frac{V_L}{R_L} = \frac{3.48\ V}{1.0\ k\Omega} = \textbf{3.48 mA}$$

▶ **FIGURE 8–41**

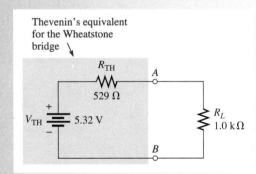

Thevenin's equivalent for the Wheatstone bridge

Related Problem Calculate I_L for $R_1 = 2.2\ k\Omega$, $R_2 = 3.3\ k\Omega$, $R_3 = 3.9\ k\Omega$, and $R_4 = 2.7\ k\Omega$.

Use Multisim file E08-11 to verify the calculated results in this example and to confirm your calculation for the related problem.

An Alternate Approach An alternate way to thevenize the Wheatstone bridge is to consider a different viewpoint. Instead of viewing it from between the A and B terminals, you can view it from terminal A to ground and from terminal B to ground, as illustrated in Figure 8–42(a) and (b). The resulting equivalent circuit is simplified into two facing Thevenin circuits that still include ground, as illustrated in Figure 8–42(c). When calculating the Thevenin resistance, the voltage source is replaced by a short; thus, two of the bridge resistors are shorted out. In Figure 8–42(a), R_3 and R_4 are shorted and in part (b) R_1 and R_2 are shorted. In each case, the remaining two resistors appear in parallel to form the Thevenin resistance. The load resistor can be replaced as in Figure 8–42(d), which is seen to be a simple series circuit with two opposing sources. The advantage of this method is that ground is still shown

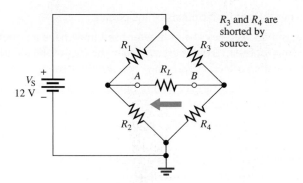
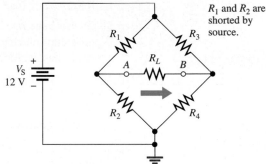

(a) Viewed between A and ground

(b) Viewed between B and ground

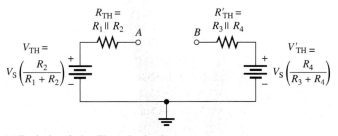
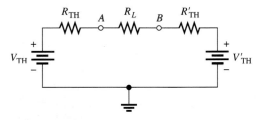

(c) Equivalent facing Thevenin circuits

(d) Load resistor connected

▲ **FIGURE 8–42**

in this equivalent circuit, so it is easy to find the voltage at terminal A or B with respect to ground by applying the superposition theorem to the equivalent circuit.

Summary of Thevenin's Theorem

Remember, the Thevenin equivalent circuit is *always* in the form of an equivalent voltage source in series with an equivalent resistance regardless of the original circuit that it replaces. The significance of Thevenin's theorem is that the equivalent circuit can replace the original circuit as far as any external load is concerned. Any load resistor connected between the terminals of a Thevenin equivalent circuit will have the same current through it and the same voltage across it as if it were connected to the terminals of the original circuit.

A summary of steps for applying Thevenin's theorem is as follows:

Step 1. Open the two terminals (remove any load) between which you want to find the Thevenin equivalent circuit.

Step 2. Determine the voltage (V_{TH}) across the two open terminals.

Step 3. Determine the resistance (R_{TH}) between the two open terminals with all sources replaced with their internal resistances (ideal voltage sources shorted and ideal current sources opened).

Step 4. Connect V_{TH} and R_{TH} in series to produce the complete Thevenin equivalent for the original circuit.

Step 5. Replace the load removed in Step 1 across the terminals of the Thevenin equivalent circuit. You can now calculate the load current and load voltage using only Ohm's law. They have the same value as the load current and load voltage in the original circuit.

Determining V_{TH} and R_{TH} by Measurement

Thevenin's theorem is largely an analytical tool that is applied theoretically in order to simplify circuit analysis. However, you can find Thevenin's equivalent for an actual circuit by the following general measurement methods. These steps are illustrated in Figure 8–43.

Step 1. Remove any load from the output terminals of the circuit.

Step 2. Measure the open terminal voltage. The voltmeter used must have an internal resistance much greater (at least 10 times greater) than the R_{TH} of the circuit so that it has negligible loading effect. (V_{TH} is the open terminal voltage.)

Step 3. Connect a variable resistor (rheostat) across the output terminals. Set it at its maximum value, which must be greater than R_{TH}.

Step 4. Adjust the rheostat until the terminal voltage equals $0.5V_{TH}$. At this point, the resistance of the rheostat is equal to R_{TH}.

Step 5. Disconnect the rheostat from the terminals and measure its resistance with an ohmmeter. This measured resistance is equal to R_{TH}.

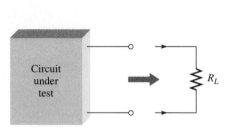

Step 1: Open the output terminals (remove load).

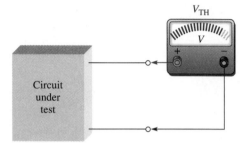

Step 2: Measure V_{TH}.

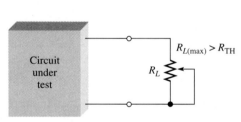

Step 3: Connect variable load resistance set to its maximum value across the terminals.

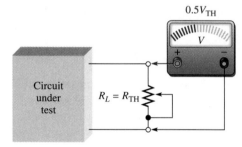

Step 4: Adjust R_L until $V_L = 0.5V_{TH}$. When $V_L = 0.5V_{TH}$, $R_L = R_{TH}$.

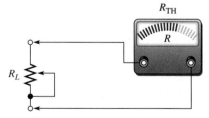

Step 5: Remove R_L from the circuit under test and measure its resistance to get R_{TH}.

▲ **FIGURE 8–43**

Determination of Thevenin's equivalent by measurement.

This procedure for determining R_{TH} differs from the theoretical procedure because it is impractical to short voltage sources or open current sources in an actual circuit. Also, when measuring R_{TH}, be certain that the circuit is capable of providing the required current to the variable resistor load and that the variable resistor can handle the required power. These considerations may make the procedure impractical in some cases.

An Example of a Practical Application

Although you have not studied transistor circuits, a basic amplifier can be used to illustrate the usefulness of Thevenin's equivalent circuit. A transistor circuit can be modeled with basic components including a dependent current source and a Thevenin equivalent circuit. Modeling is generally a mathematical simplification of a complex circuit, retaining only the most important parts of the circuit and eliminating those that have only a minimum effect.

A typical dc model of a transistor is shown in Figure 8–44. This type of transistor (bipolar junction transistor) has three terminals, labeled base (B), collector (C), and emitter (E). In this case, the emitter terminal is both an input and an output, so it is common. The dependent current source (diamond-shaped symbol) is controlled by the base current, I_B. For this example, the current from the dependent source is 200 times larger than the base current as expressed by the term βI_B, where β is a transistor gain parameter and, in this case, $\beta = 200$.

The transistor is part of a dc amplifier circuit, and you can use the basic model to predict the output current. The output current is larger than the input circuit can provide by itself. For example, the source can represent a small transducer, such as a solar cell with an internal resistance of 6.8 kΩ. It is shown as an equivalent Thevenin voltage and Thevenin resistance. The load could be any device that requires higher current than the source can provide directly.

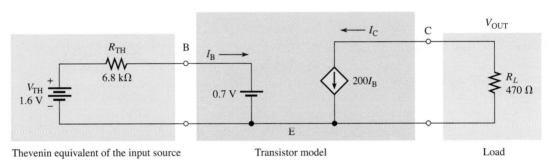

▲ **FIGURE 8–44**

DC transistor circuit. The diamond-shaped symbol indicates a dependent current source.

EXAMPLE 8–12

(a) Write KVL around the left part of the circuit in Figure 8–44. Solve for I_B.

(b) Determine the current from the dependent current source. This current is I_C.

(c) Calculate the output voltage and the power in the load resistor R_L.

(d) Compare the power determined in (c) with the power delivered to the load if the load resistor were connected directly to the Thevenin circuit.

Solution (a) $V_{TH} - R_{TH}I_B - 0.7\,\text{V} = 0$

$$I_B = \frac{V_{TH} - 0.7\,\text{V}}{R_{TH}} = \frac{1.6\,\text{V} - 0.7\,\text{V}}{6.8\,\text{k}\Omega} = \mathbf{132\,\mu A}$$

(b) $I_C = \beta I_B = 200(132\,\mu\text{A}) = \mathbf{26.5\,mA}$

(c) $V_{OUT} = I_C R_L = (26.5\,\text{mA})(470\,\Omega) = \mathbf{12.4\,V}$

$$P_L = \frac{V_{OUT}^2}{R_L} = \frac{(12.4\,\text{V})^2}{470\,\Omega} = \mathbf{327\,mW}$$

(d) $P_L = I_B^2 R_L = (132\,\mu\text{A})^2(470\,\Omega) = \mathbf{8.19\,\mu W}$

The power in the load resistor is 327 mW/8.19 μW = 39,927 times greater than the power that the Thevenin input circuit could deliver to the same load. This illustrates that the transistor can operate as a power amplifier.

Related Problem   Determine the input voltage at the base (B) of the transistor. Compare this value to V_{OUT}. By how much does the amplifier increase the input voltage?

SECTION 8–5
REVIEW

1. What are the two components of a Thevenin equivalent circuit?
2. Draw the general form of a Thevenin equivalent circuit.
3. How is V_{TH} defined?
4. How is R_{TH} defined?
5. For the original circuit in Figure 8–45, draw the Thevenin equivalent circuit as viewed from the output terminals A and B.

▶ FIGURE 8–45

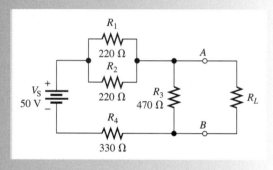

8–6 NORTON'S THEOREM

Like Thevenin's theorem, Norton's theorem provides a method of reducing a more complex circuit to a simpler equivalent form. The basic difference is that Norton's theorem results in an equivalent current source in parallel with an equivalent resistance.

After completing this section, you should be able to

◆ **Apply Norton's theorem to simplify a circuit**

   ◆ Describe the form of a Norton equivalent circuit

   ◆ Obtain the Norton equivalent current source

   ◆ Obtain the Norton equivalent resistance

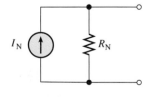

▲ FIGURE 8–46

Form of Norton's equivalent circuit.

Norton's theorem is a method for simplifying a two-terminal linear circuit to an equivalent circuit with only a current source in parallel with a resistor. The form of Norton's equivalent circuit is shown in Figure 8–46. Regardless of how complex the original two-terminal circuit is, it can always be reduced to this equivalent form. The equivalent current source is designated I_N, and the equivalent resistance is designated R_N. To apply Norton's theorem, you must know how to find the two quantities I_N and R_N. Once you know them for a given circuit, simply connect them in parallel to get the complete Norton circuit.

Norton's Equivalent Current (I_N)

Norton's equivalent current (I_N) is the short-circuit current between two output terminals in a circuit.

Any component connected between these two terminals effectively "sees" a current source I_N in parallel with R_N. To illustrate, suppose that a resistive circuit of some kind has a resistor (R_L) connected between two output terminals in the circuit, as shown in Figure 8–47(a). You want to find the Norton circuit that is equivalent to the one shown as "seen" by R_L. To find I_N, calculate the current between terminals A and B with these two terminals shorted, as shown in Figure 8–47(b). Example 8–13 demonstrates how to find I_N.

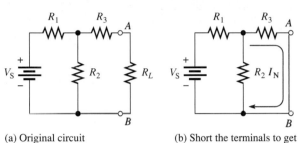

(a) Original circuit   (b) Short the terminals to get I_N.

◀ **FIGURE 8–47**

Determining the Norton equivalent current, I_N.

EXAMPLE 8–13

Determine I_N for the circuit within the beige area in Figure 8–48(a).

FIGURE 8–48

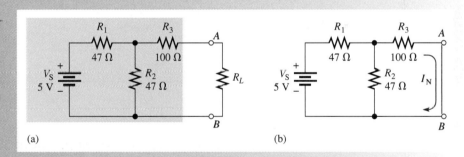

(a)   (b)

Solution   Short terminals A and B as shown in Figure 8–48(b). I_N is the current through the short. First, the total resistance seen by the voltage source is

$$R_T = R_1 + \frac{R_2 R_3}{R_2 + R_3} = 47\ \Omega + \frac{(47\ \Omega)(100\ \Omega)}{147\ \Omega} = 79\ \Omega$$

The total current from the source is

$$I_T = \frac{V_S}{R_T} = \frac{5\ V}{79\ \Omega} = 63.3\ mA$$

Now apply the current-divider formula to find I_N (the current through the short).

$$I_N = \left(\frac{R_2}{R_2 + R_3}\right)I_T = \left(\frac{47\ \Omega}{147\ \Omega}\right)63.3\ mA = \mathbf{20.2\ mA}$$

This is the value for the equivalent Norton current source.

Related Problem   Determine I_N in Figure 8–48(a) if the value of R_2 is doubled.

Use Multisim file E08-13 to verify the calculated results in this example and to confirm your calculation for the related problem.

Norton's Equivalent Resistance (R_N)

Norton's equivalent resistance (R_N) is defined in the same way as R_{TH}.

The Norton equivalent resistance, R_N, is the total resistance appearing between two output terminals in a given circuit with all sources replaced by their internal resistances.

Example 8–14 demonstrates how to find R_N.

EXAMPLE 8–14

Find R_N for the circuit within the beige area of Figure 8–48(a) (see Example 8–13).

Solution   First reduce V_S to zero by shorting it, as shown in Figure 8–49. Looking in at terminals A and B, you can see that the parallel combination of R_1 and R_2 is in series with R_3. Thus,

$$R_N = R_3 + \frac{R_1}{2} = 100 \ \Omega + \frac{47 \ \Omega}{2} = \mathbf{124 \ \Omega}$$

▶ **FIGURE 8–49**

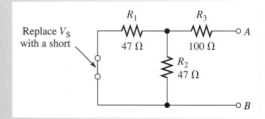

Related Problem   Determine R_N in Figure 8–48(a) if the value of R_2 is doubled.

Examples 8–13 and 8–14 have shown how to find the two equivalent components of a Norton equivalent circuit, I_N and R_N. Keep in mind that these values can be found for any linear circuit. Once these are known, they must be connected in parallel to form the Norton equivalent circuit, as illustrated in Example 8–15.

EXAMPLE 8–15

Draw the complete Norton equivalent circuit for the original circuit in Figure 8–48(a) (Example 8–13).

Solution   In Examples 8–13 and 8–14 you found that $I_N = 20.2 \ \text{mA}$ and $R_N = 124 \ \Omega$. The Norton equivalent circuit is shown in Figure 8–50.

▶ **FIGURE 8–50**

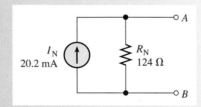

Related Problem   Find R_N for the circuit in Figure 8–48(a) if all the resistor values are doubled.

Summary of Norton's Theorem

Any load resistor connected between the output terminals of a Norton equivalent circuit will have the same current through it and the same voltage across it as if it were connected to the output terminals of the original circuit. A summary of steps for theoretically applying Norton's theorem is as follows:

Step 1. Short the two terminals between which you want to find the Norton equivalent circuit.

Step 2. Determine the current (I_N) through the shorted terminals.

Step 3. Determine the resistance (R_N) between the two open terminals with all sources replaced with their internal resistances (ideal voltage sources shorted and ideal current sources opened). $R_N = R_{TH}$.

Step 4. Connect I_N and R_N in parallel to produce the complete Norton equivalent for the original circuit.

Norton's equivalent circuit can also be derived from Thevenin's equivalent circuit by use of the source conversion method discussed in Section 8–3.

An Example of a Practical Application

A voltage amplifier in a digital light meter is modeled using a Norton equivalent circuit and a dependent voltage source. A block diagram of the light meter is shown in Figure 8–51. The light meter uses a photocell as a sensor. The photocell is a current source that produces a very small current proportional to the incident light. Because it is a current source, a Norton circuit can be used to model the photocell. The very small amount of current from the photocell is converted to a small input voltage across R_N. A dc amplifier is used to increase the voltage to a level sufficient for driving the analog-to-digital converter.

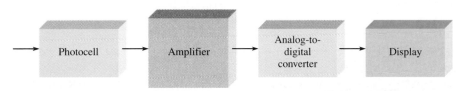

◀ **FIGURE 8–51**

Light meter block diagram.

In this application, only with the first two blocks in the light meter diagram are of interest. These have been modeled as shown in Figure 8–52. The photocell has been modeled as a Norton circuit on the input. The output of the Norton circuit is fed to the amplifier's input resistance, which converts the current I_N to a small voltage V_{IN}. The amplifier increases this voltage by 33 to drive the analog-to-digital converter which, for simplicity, is modeled simply as a load resistor, R_L. The value of 33 is the gain of this particular amplifier.

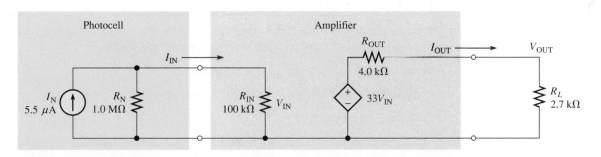

▲ **FIGURE 8–52**

Photocell and amplifier model. The diamond-shaped symbol indicates a dependent voltage source.

EXAMPLE 8–16   Refer to Figure 8–52.

(a) Apply the current-divider rule to the input Norton circuit to calculate I_{IN}.

(b) Use Ohm's law to calculate V_{IN}.

(c) Determine the voltage from the dependent voltage source. This gain is 33.

(d) Apply the voltage-divider rule to calculate V_{OUT}.

Solution   (a) $I_{IN} = I_N\left(\dfrac{R_N}{R_N + R_{IN}}\right) = (5.5\,\mu A)\left(\dfrac{1.0\,M\Omega}{1.1\,M\Omega}\right) = \mathbf{5\,\mu A}$

(b) $V_{IN} = I_{IN}R_{IN} = (5\,\mu A)(100\,k\Omega) = \mathbf{0.5\,V}$

(c) $33V_{IN} = (33)(0.5\,V) = \mathbf{16.5\,V}$

(d) $V_{OUT} = (33\,V_{IN})\left(\dfrac{R_L}{R_L + R_{OUT}}\right) = (16.5\,V)(0.403) = \mathbf{6.65\,V}$

Related Problem   If the photocell is replaced by one having the same current but a Norton equivalent resistance of 2.0 MΩ, what is the output voltage?

SECTION 8–6
REVIEW

1. What are the two components of a Norton equivalent circuit?
2. Draw the general form of a Norton equivalent circuit.
3. How is I_N defined?
4. How is R_N defined?
5. Find the Norton circuit as seen by R_L in Figure 8–53.

▶ **FIGURE 8–53**

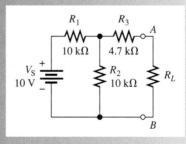

8–7   MAXIMUM POWER TRANSFER THEOREM

The maximum power transfer theorem is important when you need to know the value of the load at which the most power is delivered from the source.

After completing this section, you should be able to

◆ **Apply the maximum power transfer theorem**

  ◆ State the theorem

  ◆ Determine the value of load resistance for which maximum power is transferred from a given circuit

The **maximum power transfer** theorem is stated as follows:

For a given source voltage, maximum power is transferred from a source to a load when the load resistance is equal to the internal source resistance.

The source resistance, R_S, of a circuit is the equivalent resistance as viewed from the output terminals using Thevenin's theorem. A Thevenin equivalent circuit with its output resistance and load is shown in Figure 8–54. When $R_L = R_S$, the maximum power possible is transferred from the voltage source to R_L for a given value of V_S.

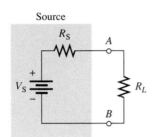

◀ **FIGURE 8–54**

Maximum power is transferred to the load when $R_L = R_S$.

Practical applications of the maximum power transfer theorem include audio systems such as stereo, radio, and public address. In these systems the resistance of the speaker is the load. The circuit that drives the speaker is a power amplifier. The systems are typically optimized for maximum power to the speakers. Thus, the resistance of the speaker must equal the internal source resistance of the amplifier.

Example 8–17 shows that maximum power occurs when $R_L = R_S$.

EXAMPLE 8–17

The source in Figure 8–55 has an internal source resistance of 75 Ω. Determine the load power for each of the following values of load resistance:

(a) 0 Ω   (b) 25 Ω   (c) 50 Ω   (d) 75 Ω   (e) 100 Ω   (f) 125 Ω

Draw a graph showing the load power versus the load resistance.

▶ **FIGURE 8–55**

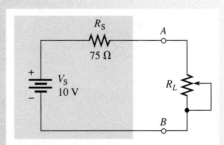

Solution   Use Ohm's law ($I = V/R$) and the power formula ($P = I^2R$) to find the load power, P_L, for each value of load resistance.

(a) For $R_L = 0\,\Omega$,

$$I = \frac{V_S}{R_S + R_L} = \frac{10\,\text{V}}{75\,\Omega + 0\,\Omega} = 133\,\text{mA}$$

$$P_L = I^2R_L = (133\,\text{mA})^2(0\,\Omega) = \mathbf{0\,mW}$$

(b) For $R_L = 25\,\Omega$,

$$I = \frac{V_S}{R_S + R_L} = \frac{10\,\text{V}}{75\,\Omega + 25\,\Omega} = 100\,\text{mA}$$
$$P_L = I^2 R_L = (100\,\text{mA})^2(25\,\Omega) = \textbf{250\,mW}$$

(c) For $R_L = 50\,\Omega$,

$$I = \frac{V_S}{R_S + R_L} = \frac{10\,\text{V}}{125\,\Omega} = 80\,\text{mA}$$
$$P_L = I^2 R_L = (80\,\text{mA})^2(50\,\Omega) = \textbf{320\,mW}$$

(d) For $R_L = 75\,\Omega$,

$$I = \frac{V_S}{R_S + R_L} = \frac{10\,\text{V}}{150\,\Omega} = 66.7\,\text{mA}$$
$$P_L = I^2 R_L = (66.7\,\text{mA})^2(75\,\Omega) = \textbf{334\,mW}$$

(e) For $R_L = 100\,\Omega$,

$$I = \frac{V_S}{R_S + R_L} = \frac{10\,\text{V}}{175\,\Omega} = 57.1\,\text{mA}$$
$$P_L = I^2 R_L = (57.1\,\text{mA})^2(100\,\Omega) = \textbf{326\,mW}$$

(f) For $R_L = 125\,\Omega$,

$$I = \frac{V_S}{R_S + R_L} = \frac{10\,\text{V}}{200\,\Omega} = 50\,\text{mA}$$
$$P_L = I^2 R_L = (50\,\text{mA})^2(125\,\Omega) = \textbf{313\,mW}$$

Notice that the load power is greatest when $R_L = 75\,\Omega$, which is the same as the internal source resistance. When the load resistance is less than or greater than this value, the power drops off, as the curve in Figure 8–56 graphically illustrates.

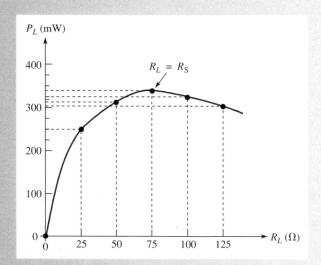

▲ **FIGURE 8–56**

Curve showing that the load power is maximum when $R_L = R_S$.

Related Problem   If the source resistance in Figure 8–55 is 600 Ω, what is the maximum power that can be delivered to a load?

SECTION 8–7 REVIEW

1. State the maximum power transfer theorem.

2. When is maximum power delivered from a source to a load?

3. A given circuit has an internal source resistance of 50 Ω. What will be the value of the load to which the maximum power is delivered?

8–8 DELTA-TO-WYE (Δ-TO-Y) AND WYE-TO-DELTA (Y-TO-Δ) CONVERSIONS

Conversions between delta-type and wye-type circuit arrangements are useful in certain specialized three-terminal applications. One example is in the analysis of a loaded Wheatstone bridge circuit.

After completing this section, you should be able to

◆ **Perform Δ-to-Y and Y-to-Δ conversions**

 ◆ Apply Δ-to-Y conversion to a bridge circuit

A resistive delta (Δ) circuit is a three-terminal arrangement as shown in Figure 8–57(a). A wye (Y) circuit is shown in Figure 8–57(b). Notice that letter subscripts are used to designate resistors in the delta circuit and that numerical subscripts are used to designate resistors in the wye circuit.

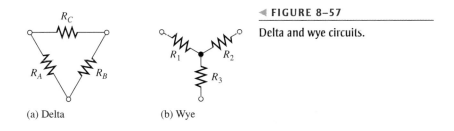

◀ FIGURE 8–57

Delta and wye circuits.

(a) Delta   (b) Wye

Δ-to-Y Conversion

It is convenient to think of the wye positioned within the delta, as shown in Figure 8–58. To convert from delta to wye, you need R_1, R_2, and R_3 in terms of R_A, R_B, and R_C. The conversion rule is as follows:

Each resistor in the wye is equal to the product of the resistors in two adjacent delta branches, divided by the sum of all three delta resistors.

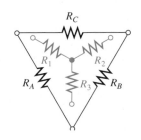

◀ FIGURE 8–58

"Y within Δ" aid for conversion formulas.

In Figure 8–58, R_A and R_C are adjacent to R_1; therefore,

Equation 8–1
$$R_1 = \frac{R_A R_C}{R_A + R_B + R_C}$$

Also, R_B and R_C are adjacent to R_2, so

Equation 8–2
$$R_2 = \frac{R_B R_C}{R_A + R_B + R_C}$$

and R_A and R_B are adjacent to R_3, so

Equation 8–3
$$R_3 = \frac{R_A R_B}{R_A + R_B + R_C}$$

Y-to-Δ Conversion

To convert from wye to delta, you need R_A, R_B, and R_C in terms of R_1, R_2, and R_3. The conversion rule is as follows:

Each resistor in the delta is equal to the sum of all possible products of wye resistors taken two at a time, divided by the opposite wye resistor.

In Figure 8–58, R_2 is opposite to R_A; therefore,

Equation 8–4
$$R_A = \frac{R_1 R_2 + R_1 R_3 + R_2 R_3}{R_2}$$

Also, R_1 is opposite to R_B, so

Equation 8–5
$$R_B = \frac{R_1 R_2 + R_1 R_3 + R_2 R_3}{R_1}$$

and R_3 is opposite to R_C, so

Equation 8–6
$$R_C = \frac{R_1 R_2 + R_1 R_3 + R_2 R_3}{R_3}$$

EXAMPLE 8–18

Convert the delta circuit in Figure 8–59 to a wye circuit.

▶ **FIGURE 8–59**

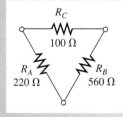

Solution Use Equations 8–1, 8–2, and 8–3.

$$R_1 = \frac{R_A R_C}{R_A + R_B + R_C} = \frac{(220\ \Omega)(100\ \Omega)}{220\ \Omega + 560\ \Omega + 100\ \Omega} = \mathbf{25\ \Omega}$$

$$R_2 = \frac{R_B R_C}{R_A + R_B + R_C} = \frac{(560\ \Omega)(100\ \Omega)}{880\ \Omega} = \mathbf{63.6\ \Omega}$$

$$R_3 = \frac{R_A R_B}{R_A + R_B + R_C} = \frac{(220\ \Omega)(560\ \Omega)}{880\ \Omega} = \mathbf{140\ \Omega}$$

The resulting wye circuit is shown in Figure 8–60.

▶ FIGURE 8–60

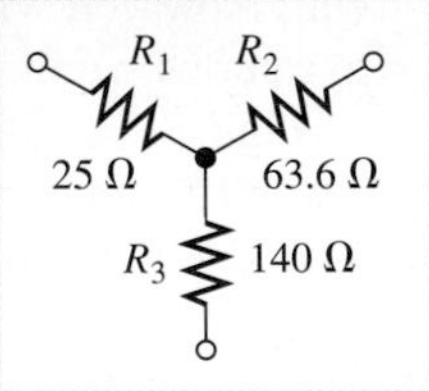

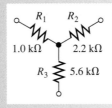

Related Problem Convert the delta circuit to a wye network for $R_A = 2.2\,\text{k}\Omega$, $R_B = 1.0\,\text{k}\Omega$, and $R_C = 1.8\,\text{k}\Omega$.

EXAMPLE 8–19

Convert the wye circuit in Figure 8–61 to a delta circuit.

▶ FIGURE 8–61

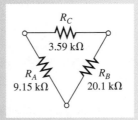

Solution Use Equations 8–4, 8–5, and 8–6.

$$R_A = \frac{R_1R_2 + R_1R_3 + R_2R_3}{R_2}$$

$$= \frac{(1.0\,\text{k}\Omega)(2.2\,\text{k}\Omega) + (1.0\,\text{k}\Omega)(5.6\,\text{k}\Omega) + (2.2\,\text{k}\Omega)(5.6\,\text{k}\Omega)}{2.2\,\text{k}\Omega} = \mathbf{9.15\,k\Omega}$$

$$R_B = \frac{R_1R_2 + R_1R_3 + R_2R_3}{R_1}$$

$$= \frac{(1.0\,\text{k}\Omega)(2.2\,\text{k}\Omega) + (1.0\,\text{k}\Omega)(5.6\,\text{k}\Omega) + (2.2\,\text{k}\Omega)(5.6\,\text{k}\Omega)}{1.0\,\text{k}\Omega} = \mathbf{20.1\,k\Omega}$$

$$R_C = \frac{R_1R_2 + R_1R_3 + R_2R_3}{R_3}$$

$$= \frac{(1.0\,\text{k}\Omega)(2.2\,\text{k}\Omega) + (1.0\,\text{k}\Omega)(5.6\,\text{k}\Omega) + (2.2\,\text{k}\Omega)(5.6\,\text{k}\Omega)}{5.6\,\text{k}\Omega} = \mathbf{3.59\,k\Omega}$$

The resulting delta circuit is shown in Figure 8–62.

▶ FIGURE 8–62

Related Problem Convert the wye circuit to a delta circuit for $R_1 = 100\,\Omega$, $R_2 = 330\,\Omega$, and $R_3 = 470\,\Omega$.

Application of Δ-to-Y Conversion to a Bridge Circuit

In Section 8–5 you learned how Thevenin's theorem can be used to simplify a bridge circuit. Now you will see how Δ-to-Y conversion can be used for converting a bridge circuit to a series-parallel form for easier analysis.

Figure 8–63 illustrates how the delta (Δ) formed by R_A, R_B, and R_C can be converted to a wye (Y), thus creating an equivalent series-parallel circuit. Equations 8–1, 8–2, and 8–3 are used in this conversion.

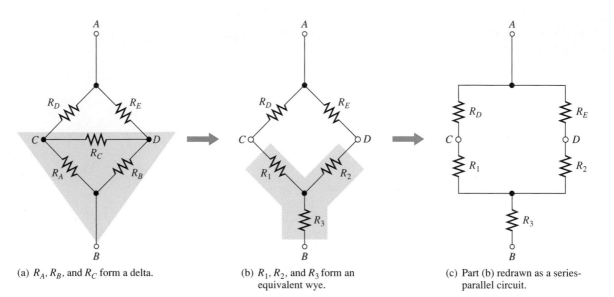

(a) R_A, R_B, and R_C form a delta.

(b) R_1, R_2, and R_3 form an equivalent wye.

(c) Part (b) redrawn as a series-parallel circuit.

▲ FIGURE 8–63

Conversion of a bridge circuit to a series-parallel configuration.

In a bridge circuit, the load is connected between terminals C and D. In Figure 8–63(a), R_C represents the load resistor. When voltage is applied across terminals A and B, the voltage from C to D (V_{CD}) can be determined using the equivalent series-parallel circuit in Figure 8–63(c) as follows. The total resistance from terminal A to terminal B is

$$R_T = \frac{(R_1 + R_D)(R_2 + R_E)}{(R_1 + R_D) + (R_2 + R_E)} + R_3$$

Then,

$$I_T = \frac{V_{AB}}{R_T}$$

The resistance of the parallel portion of the circuit in Figure 8–63(c) is

$$R_{T(p)} = \frac{(R_1 + R_D)(R_2 + R_E)}{(R_1 + R_D) + (R_2 + R_E)}$$

The current through the left branch is

$$I_{AC} = \left(\frac{R_{T(p)}}{R_1 + R_D}\right) I_T$$

The current through the right branch is

$$I_{AD} = \left(\frac{R_{T(p)}}{R_2 + R_E}\right) I_T$$

The voltage at terminal C with respect to terminal A is

$$V_{CA} = V_A - I_{AC}R_D$$

The voltage at terminal D with respect to terminal A is

$$V_{DA} = V_A - I_{AD}R_E$$

The voltage from terminal C to terminal D is

$$\begin{aligned} V_{CD} &= V_{CA} - V_{DA} \\ &= (V_A - I_{AC}R_D) - (V_A - I_{AD}R_E) = I_{AD}R_E - I_{AC}R_D \end{aligned}$$

V_{CD} is the voltage across the load (R_C) in the bridge circuit of Figure 8–63(a).
The load current through R_C can be found by Ohm's law.

$$I_{R_C} = \frac{V_{CD}}{R_C}$$

EXAMPLE 8–20

Determine the voltage across the load resistor and the current through the load resistor in the bridge circuit in Figure 8–64. Notice that the resistors are labeled for convenient conversion using Equations 8–1, 8–2, and 8–3. R_C is the load resistor.

▶ **FIGURE 8–64**

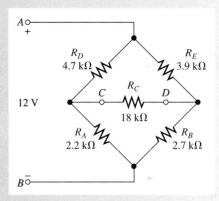

Solution   First, convert the delta formed by R_A, R_B, and R_C to a wye.

$$R_1 = \frac{R_A R_C}{R_A + R_B + R_C} = \frac{(2.2\,\text{k}\Omega)(18\,\text{k}\Omega)}{2.2\,\text{k}\Omega + 2.7\,\text{k}\Omega + 18\,\text{k}\Omega} = 1.73\,\text{k}\Omega$$

$$R_2 = \frac{R_B R_C}{R_A + R_B + R_C} = \frac{(2.7\,\text{k}\Omega)(18\,\text{k}\Omega)}{22.9\,\text{k}\Omega} = 2.12\,\text{k}\Omega$$

$$R_3 = \frac{R_A R_B}{R_A + R_B + R_C} = \frac{(2.2\,\text{k}\Omega)(2.7\,\text{k}\Omega)}{22.9\,\text{k}\Omega} = 259\,\Omega$$

The resulting equivalent series-parallel circuit is shown in Figure 8–65.
Next, determine R_T and the branch currents in Figure 8–65.

$$\begin{aligned} R_T &= \frac{(R_1 + R_D)(R_2 + R_E)}{(R_1 + R_D) + (R_2 + R_E)} + R_3 \\ &= \frac{(6.43\,\text{k}\Omega)(6.02\,\text{k}\Omega)}{6.43\,\text{k}\Omega + 6.02\,\text{k}\Omega} + 259\,\Omega = 3.11\,\text{k}\Omega + 259\,\Omega = 3.37\,\text{k}\Omega \end{aligned}$$

$$I_T = \frac{V_{AB}}{R_T} = \frac{12\,\text{V}}{3.37\,\text{k}\Omega} = 3.56\,\text{mA}$$

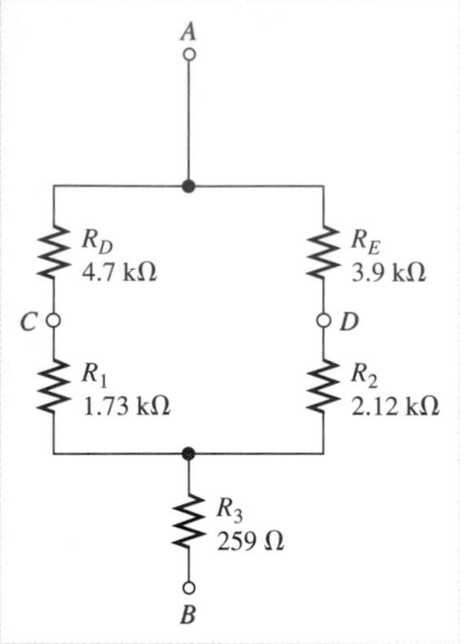

A

R_D
4.7 kΩ

R_E
3.9 kΩ

C

D

R_1
1.73 kΩ

R_2
2.12 kΩ

R_3
259 Ω

B

▲ **FIGURE 8–65**

The total resistance of the parallel part of the circuit, $R_{T(p)}$, is 3.11 kΩ.

$$I_{AC} = \left(\frac{R_{T(p)}}{R_1 + R_D}\right)I_T = \left(\frac{3.11\,\text{k}\Omega}{1.73\,\text{k}\Omega + 4.7\,\text{k}\Omega}\right)3.56\,\text{mA} = 1.72\,\text{mA}$$

$$I_{AD} = \left(\frac{R_{T(p)}}{R_2 + R_E}\right)I_T = \left(\frac{3.11\,\text{k}\Omega}{2.12\,\text{k}\Omega + 3.9\,\text{k}\Omega}\right)3.56\,\text{mA} = 1.84\,\text{mA}$$

The voltage from terminal C to terminal D is

$$V_{CD} = I_{AD}R_E - I_{AC}R_D = (1.84\,\text{mA})(3.9\,\text{k}\Omega) - (1.72\,\text{mA})(4.7\,\text{k}\Omega)$$
$$= 7.18\,\text{V} - 8.08\,\text{V} = -0.9\,\text{V}$$

V_{CD} is the voltage across the load (R_C) in the bridge circuit shown in Figure 8–64. The load current through R_C is

$$I_{R_C} = \frac{V_{CD}}{R_C} = \frac{-0.9\,\text{V}}{18\,\text{k}\Omega} = -50\,\mu\text{A}$$

Related Problem Determine the load current, I_{R_C}, in Figure 8–64 for the following resistor values: $R_A = 27\,\text{k}\Omega$, $R_B = 33\,\text{k}\Omega$, $R_D = 39\,\text{k}\Omega$, $R_E = 47\,\text{k}\Omega$, and $R_C = 100\,\text{k}\Omega$.

 Use Multisim file E08-20 to verify the calculated results in this example and to confirm your calculation for the related problem.

**SECTION 8–8
REVIEW**

1. Draw a delta circuit.
2. Draw a wye circuit.
3. Write the formulas for delta-to-wye conversion.
4. Write the formulas for wye-to-delta conversion.

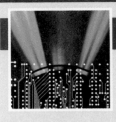

A Circuit Application

The Wheatstone bridge circuit was introduced in Chapter 7 and expanded in this chapter to include the use of Thevenin's theorem. In Chapter 7, the circuit application used a thermistor in one arm of the bridge to sense temperature. The bridge was used to compare the resistance of the thermistor with the resistance of a rheostat, which set the temperature at which the output would switch from one polarity to the opposite for turning on a heater in a tank filled with a liquid. In this circuit application, you will work with a similar circuit, but this time it will be used to monitor the temperature in the tank to provide a visual indication that the temperature is within a specified range.

The Temperature Monitor

The basic measuring circuit in the temperature monitor is a Wheatstone bridge with an ammeter and a series resistor acting as the load. The meter is an analog panel meter with a sensitivity of 50 μA full scale. The Wheatstone bridge temperature-measuring circuit is shown in Figure 8–66(a), and the meter panel is shown in Figure 8–66(b).

The Thermistor

The thermistor is the same one used in the circuit application of Chapter 7, specifically a Thermometrics RL2006-13.3K-140-D1 thermistor with a specified resistance of 25 kΩ at 25°C and a β of 4615K. Recall that β is a constant supplied by the manufacturer that indicates the shape of the temperature-resistance characteristic. As given earlier, the exponential equation for the resistance of a thermistor is approximated by

$$R_T = R_0 e^{\beta\left(\frac{T_0 - T}{T_0 T}\right)}$$

where:

R_T = the resistance at a given temperature

R_0 = the resistance at a reference temperature

T_0 = the reference temperature in K (typically 298 K, which is 25°C)

T = temperature (K)

β = a constant (K) provided by the manufacturer

A plot of this equation was given previously in Figure 7–60. You can confirm that your thermistor resistance calculations in this circuit application are in reasonable agreement with this plot.

Temperature Measuring Circuit

The Wheatstone bridge is designed to be balanced at 20°C. The resistance of the thermistor is approximately 33 kΩ at this temperature. You can confirm this value by substituting the temperature (in Kelvin) into the equation for R_T. Remember that the temperature in K is °C + 273.

◆ Substituting into the equation for R_T, calculate the resistance of the thermistor at a temperature of 50°C (full-scale deflection of the meter).

◆ Thevenize the bridge between terminals A and B by keeping the ground reference and forming two facing Thevenin circuits as was illustrated in Figure 8–42. Assume the thermistor temperature is 50°C and its resistance is the value calculated previously. Draw the Thevenin circuit for this temperature but do not show a load.

◆ Show the load resistor for the Thevenin circuit you drew. The load is a resistor in series with the ammeter, which will have a current of 50 μA at full scale (50°C). You can find the value of the required load resistor by applying the

◀ FIGURE 8–66

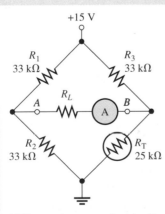

(a) Temperature-measuring circuit with thermistor

(b) Meter panel

superposition theorem to the two sources and calculating the total resistance from Ohm's law (using the full-scale deflection as the current). Subtract the Thevenin resistance of each arm from the total resistance to obtain the required load resistance. Neglect the meter resistance. Show the value calculated on the Thevenin circuit.

◆ Calculate the thermistor resistance for the lower and upper limits of temperature (30°C and 40°C). Draw Thevenin circuits for each temperature and calculate the current through the load resistor.

The Meter Scale

A requirement for the temperature monitor is to mark three color bands on the meter to indicate that the temperature is within the desired range. The desired range is between a low of 30°C and 40°C. The meter should indicate a too-cold range from 20°C to 30°C, a proper operating range from 30°C to 40°C, and a too hot range from 40°C and 50°C. Full-scale deflection of the meter should be set for 50°C.

◆ Indicate how you would mark the meter to have a quick visual indication of the temperature in the tank.

Review

1. At 35°C, what is the current in the meter?
2. What change is needed if a 100 μA meter is used instead of a 50 μA meter?

SUMMARY

◆ An ideal voltage source has zero internal resistance. It provides a constant voltage across its terminals regardless of the load resistance.
◆ A practical voltage source has a nonzero internal resistance.
◆ An ideal current source has infinite internal resistance. It provides a constant current regardless of the load resistance.
◆ A practical current source has a finite internal resistance.
◆ The superposition theorem is useful for multiple-source circuits.
◆ Thevenin's theorem provides for the reduction of any two-terminal linear resistive circuit to an equivalent form consisting of an equivalent voltage source in series with an equivalent resistance.
◆ The term *equivalency,* as used in Thevenin's and Norton's theorems, means that when a given load resistance is connected to the equivalent circuit, it will have the same voltage across it and the same current through it as when it was connected to the original circuit.
◆ Norton's theorem provides for the reduction of any two-terminal linear resistive circuit to an equivalent form consisting of an equivalent current source in parallel with an equivalent resistance.
◆ Maximum power is transferred to a load from a source when the load resistance equals the internal source resistance.

KEY TERMS

These key terms are also defined in the end-of-book glossary.

Maximum power transfer For a given source voltage, a transfer of maximum power from a source to a load occurs when the load resistance equals the internal source resistance.

Norton's theorem A method for simplifying a two-terminal linear circuit to an equivalent circuit with only a current source in parallel with a resistance.

Superposition theorem A method for the analysis of circuits with more than one source.

Terminal equivalency The concept that when any given load resistance is connected to two sources, the same load voltage and load current are produced by both sources.

Thevenin's theorem A method for simplifying a two-terminal linear circuit to an equivalent circuit with only a voltage source in series with a resistance.

FORMULAS

Δ-to-Y Conversions

8–1 $R_1 = \dfrac{R_A R_C}{R_A + R_B + R_C}$

8–2 $R_2 = \dfrac{R_B R_C}{R_A + R_B + R_C}$

8–3 $R_3 = \dfrac{R_A R_B}{R_A + R_B + R_C}$

Y-to-Δ Conversions

8–4 $R_A = \dfrac{R_1 R_2 + R_1 R_3 + R_2 R_3}{R_2}$

8–5 $R_B = \dfrac{R_1 R_2 + R_1 R_3 + R_2 R_3}{R_1}$

8–6 $R_C = \dfrac{R_1 R_2 + R_1 R_3 + R_2 R_3}{R_3}$

SELF-TEST

Answers are at the end of the chapter.

1. A 100 Ω load is connected across an ideal voltage source with $V_S = 10$ V. The voltage across the load is

   (a) 0 V   (b) 10 V   (c) 100 V

2. A 100 Ω load is connected across a voltage source with $V_S = 10$ V and $R_S = 10$ Ω. The voltage across the load is

   (a) 10 V   (b) 0 V   (c) 9.09 V   (d) 0.909 V

3. A certain voltage source has the values $V_S = 25$ V and $R_S = 5$ Ω. The values for an equivalent current source are

   (a) 5 A, 5 Ω   (b) 25 A, 5 Ω   (c) 5 A, 125 Ω

4. A certain current source has the values $I_S = 3 \mu A$ and $R_S = 1.0 M\Omega$. The values for an equivalent voltage source are

   (a) 3 μV, 1.0 MΩ   (b) 3 V, 1.0 MΩ   (c) 1 V, 3.0 MΩ

5. In a two-source circuit, one source acting alone produces 10 mA through a given branch. The other source acting alone produces 8 mA in the opposite direction through the same branch. The actual current through the branch is

   (a) 10 mA   (b) 18 mA   (c) 8 mA   (d) 2 mA

6. Thevenin's theorem converts a circuit to an equivalent form consisting of

   (a) a current source and a series resistance

   (b) a voltage source and a parallel resistance

   (c) a voltage source and a series resistance

   (d) a current source and a parallel resistance

7. The Thevenin equivalent voltage for a given circuit is found by

   (a) shorting the output terminals

   (b) opening the output terminals

   (c) shorting the voltage source

   (d) removing the voltage source and replacing it with a short

8. A certain circuit produces 15 V across its open output terminals, and when a 10 kΩ load is connected across its output terminals, it produces 12 V. The Thevenin equivalent for this circuit is

   (a) 15 V in series with 10 kΩ   (b) 12 V in series with 10 kΩ

   (c) 12 V in series with 2.5 kΩ   (d) 15 V in series with 2.5 kΩ

9. Maximum power is transferred from a source to a load when
   (a) the load resistance is very large
   (b) the load resistance is very small
   (c) the load resistance is twice the source resistance
   (d) the load resistance equals the source resistance

10. For the circuit described in Question 8, maximum power is transferred to a
    (a) 10 kΩ load (b) 2.5 kΩ load (c) an infinitely large resistance load

CIRCUIT DYNAMICS QUIZ

Answers are at the end of the chapter.

Refer to Figure 8–69.

1. If a short develops across R_4, the voltage across R_5
   (a) increases (b) decreases (c) stays the same

2. If the 2 V source opens, the voltage across R_1
   (a) increases (b) decreases (c) stays the same

3. If R_2 opens, the current through R_1
   (a) increases (b) decreases (c) stays the same

Refer to Figure 8–77.

4. If R_L opens, the voltage at the output terminal with respect to ground
   (a) increases (b) decreases (c) stays the same

5. If either of the 5.6 kΩ resistors are shorted, the current through the load resistor
   (a) increases (b) decreases (c) stays the same

6. If either of the 5.6 kΩ resistors are shorted, current from the source
   (a) increases (b) decreases (c) stays the same

Refer to Figure 8–79.

7. If the input to the amplifier becomes shorted to ground, the current drawn from both voltage sources
   (a) increases (b) decreases (c) stays the same

Refer to Figure 8–82.

8. If R_1 is actually 1.0 kΩ instead of 10 kΩ, the expected voltage between A and B
   (a) increases (b) decreases (c) stays the same

9. If a 10 MΩ load resistor is connected from A to B, the voltage between A and B
   (a) increases (b) decreases (c) stays the same

10. If a short develops across R_4, the magnitude of the voltage between A and B
    (a) increases (b) decreases (c) stays the same

Refer to Figure 8–84.

11. If the 220 Ω resistor opens, V_{AB}
    (a) increases (b) decreases (c) stays the same

12. If a short develops across the 330 Ω resistor, V_{AB}
    (a) increases (b) decreases (c) stays the same

Refer to Figure 8–85(d).

13. If the 680 Ω resistor opens, the current through R_L
    (a) increases (b) decreases (c) stays the same

14. If the 47 Ω resistor becomes shorted, the voltage across R_L
    (a) increases (b) decreases (c) stays the same

PROBLEMS

More difficult problems are indicated by an asterisk (*).
Answers to odd-numbered problems are at the end of the book.

SECTION 8–3 Source Conversions

1. A voltage source has the values $V_S = 300$ V and $R_S = 50$ Ω. Convert it to an equivalent current source.

2. Convert the practical voltage sources in Figure 8–67 to equivalent current sources.

▶ FIGURE 8–67

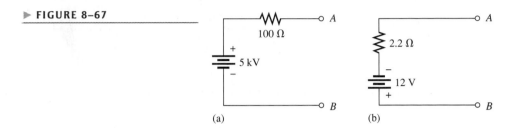

(a)   (b)

3. A fresh D cell battery has a terminal voltage of 1.6 V and can supply up to 8.0 A into a short for a very short time. What is the internal resistance of the battery?

4. Draw the voltage and current source equivalent circuits for the D cell in Problem 3.

5. A current source has an I_S of 600 mA and an R_S of 1.2 kΩ. Convert it to an equivalent voltage source.

6. Convert the practical current sources in Figure 8–68 to equivalent voltage sources.

▶ FIGURE 8–68

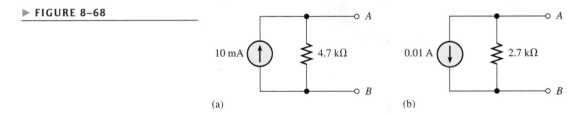

(a)   (b)

SECTION 8–4 The Superposition Theorem

7. Using the superposition method, calculate the current through R_5 in Figure 8–69.

8. Use the superposition theorem to find the current in and the voltage across the R_2 branch of Figure 8–69.

▶ FIGURE 8–69

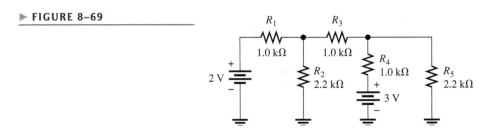

9. Using the superposition theorem, solve for the current through R_3 in Figure 8–70.

▶ FIGURE 8–70

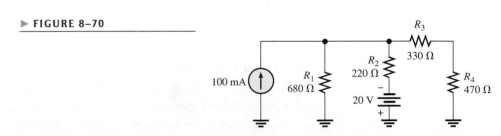

10. Using the superposition theorem, find the load current in each circuit of Figure 8–71.

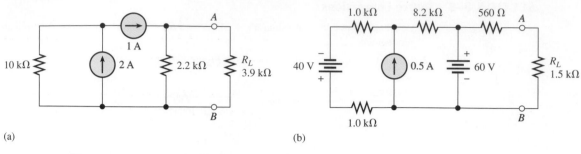

(a)

(b)

▲ FIGURE 8–71

11. A comparator circuit is shown in Figure 8–72. The input voltage, V_{IN}, is compared to the reference voltage, V_{REF}, and an output is generated that is negative if $V_{REF} > V_{IN}$; otherwise it is positive. The comparator does not load either input. If R_2 is $1.0\,k\Omega$, what is the range of the reference voltage?

12. Repeat Problem 11 if R_2 is $10\,k\Omega$.

▶ FIGURE 8–72

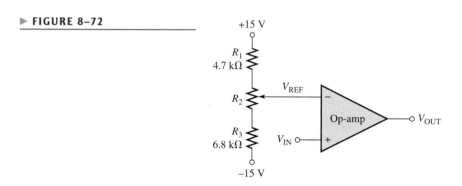

*13. Determine the voltage from point A to point B in Figure 8–73.

▶ FIGURE 8–73

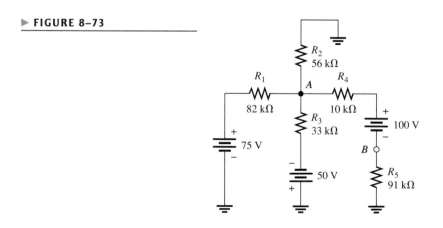

14. The switches in Figure 8–74 are closed in sequence, SW1 first. Find the current through R_4 after each switch closure.

▶ FIGURE 8–74

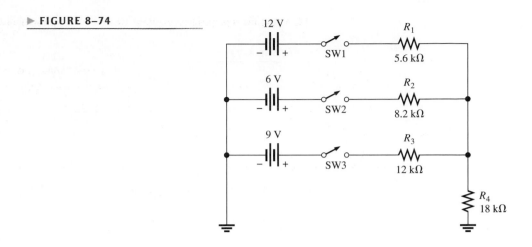

*15. Figure 8–75 shows two ladder networks. Determine the current provided by each of the batteries when terminals A are connected (A to A) and terminals B are connected (B to B).

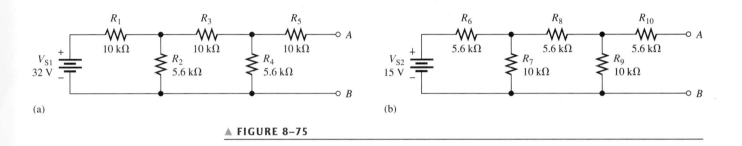

(a)

(b)

▲ FIGURE 8–75

SECTION 8–5 Thevenin's Theorem

16. For each circuit in Figure 8–76, determine the Thevenin equivalent as seen from terminals A and B.

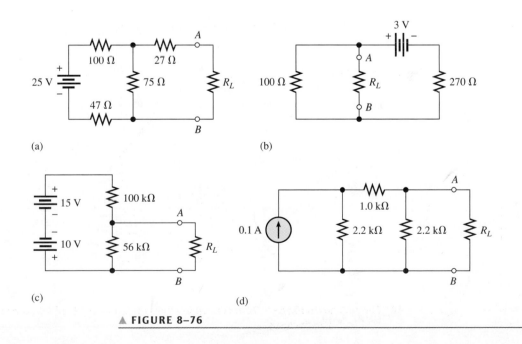

(a)

(b)

(c)

(d)

▲ FIGURE 8–76

17. Using Thevenin's theorem, determine the current through the load R_L in Figure 8–77.

▶ **FIGURE 8–77**

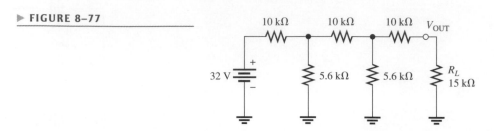

*18.** Using Thevenin's theorem, find the voltage across R_4 in Figure 8–78.

▶ **FIGURE 8–78**

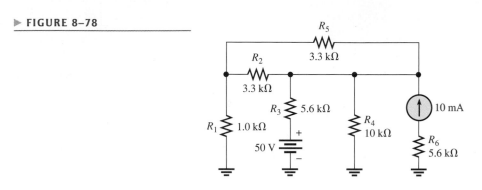

19. Find the Thevenin equivalent for the circuit external to the amplifier in Figure 8–79.

▶ **FIGURE 8–79**

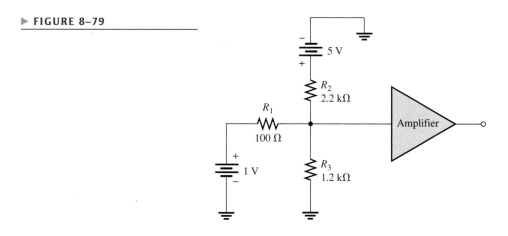

20. Determine the current into point A when R_8 is $1.0\,\text{k}\Omega$, $5\,\text{k}\Omega$, and $10\,\text{k}\Omega$ in Figure 8–80.

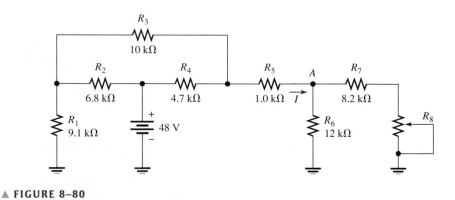

▲ **FIGURE 8–80**

*21. Find the current through the load resistor in the bridge circuit of Figure 8–81.

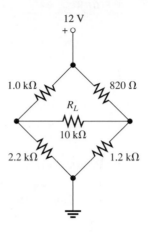

▲ FIGURE 8–81

22. Determine the Thevenin equivalent looking from terminals A and B for the circuit in Figure 8–82.

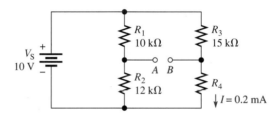

▲ FIGURE 8–82

SECTION 8–6 Norton's Theorem

23. For each circuit in Figure 8–76, determine the Norton equivalent as seen by R_L.

24. Using Norton's theorem, find the current through the load resistor R_L in Figure 8–77.

*25. Using Norton's theorem, find the voltage across R_5 in Figure 8–78.

26. Using Norton's theorem, find the current through R_1 in Figure 8–80 when $R_8 = 8\,k\Omega$.

27. Determine the Norton equivalent circuit for the bridge in Figure 8–81 with R_L removed.

28. Reduce the circuit between terminals A and B in Figure 8–83 to its Norton equivalent.

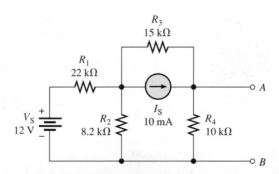

▲ FIGURE 8–83

29. Apply Norton's theorem to the circuit of Figure 8–84.

▶ FIGURE 8–84

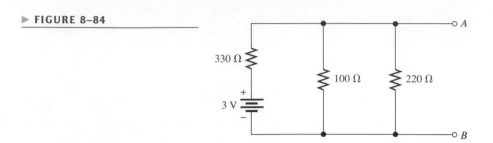

SECTION 8–7   Maximum Power Transfer Theorem

30. For each circuit in Figure 8–85, maximum power is to be transferred to the load R_L. Determine the appropriate value for R_L in each case.

▶ FIGURE 8–85

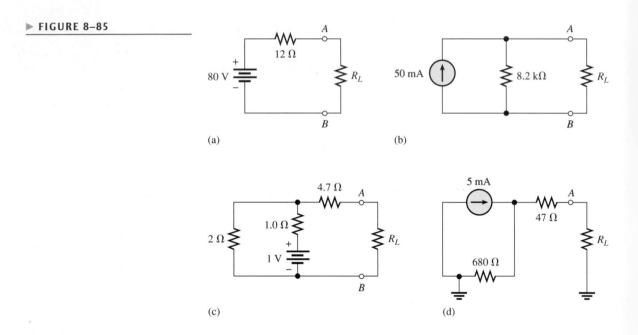

(a)   (b)

(c)   (d)

31. Determine the value of R_L for maximum power in Figure 8–86.

▶ FIGURE 8–86

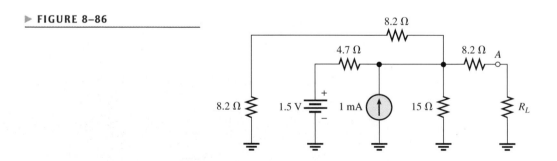

*32. How much power is delivered to the load when R_L is 10% higher than its value for maximum power in Figure 8–86?

*33. What are the values of R_4 and R_{TH} when the maximum power is transferred from the thevenized source to the ladder network in Figure 8–87?

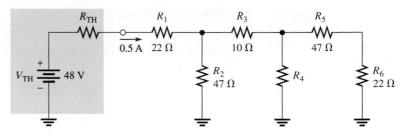

▲ FIGURE 8–87

SECTION 8–8 Delta-to-Wye (Δ-to-Y) and Wye-to-Delta (Y-to-Δ) Conversions

34. In Figure 8–88, convert each delta network to a wye network.

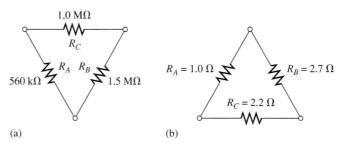

(a)   (b)

▲ FIGURE 8–88

35. In Figure 8–89, convert each wye network to a delta network.

► FIGURE 8–89

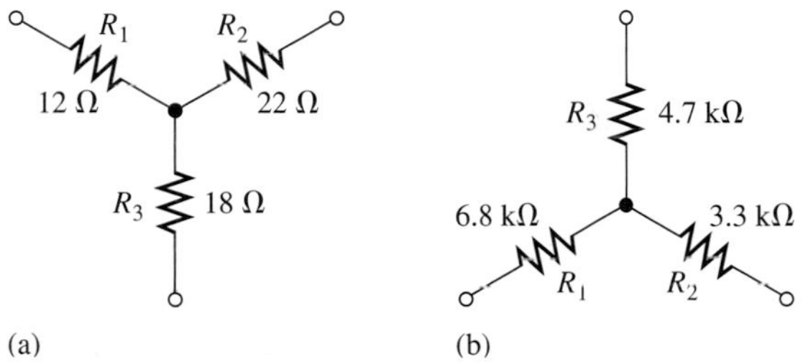

(a)   (b)

*36. Find all currents in the circuit of Figure 8–90.

► FIGURE 8–90

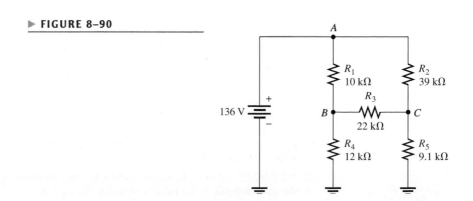

Multisim Troubleshooting and Analysis

These problems require your Multisim CD-ROM.

37. Open file P08-37 and verify that the current through each resistor is correct and, if not, determine the fault.

38. Open file P08-38 and determine by measurement the Thevenin equivalent for the circuit between terminal *A* and ground.

39. Open file P08-39 and determine by measurement the Norton equivalent for the circuit between terminal *A* and ground.

40. Open file P08-40 and determine the fault, if any.

41. Open file P08-41 and determine the value of a load resistor to be connected between terminals *A* and *B* to achieve maximum power transfer.

ANSWERS

SECTION REVIEWS

SECTION 8–1   The DC Voltage Source

1. For ideal voltage source, see Figure 8–91.
2. For practical voltage source, see Figure 8–92.
3. The internal resistance of an ideal voltage source is zero ohms.
4. Output voltage of a voltage source varies directly with load resistance.

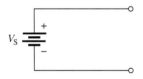

▲ FIGURE 8–91

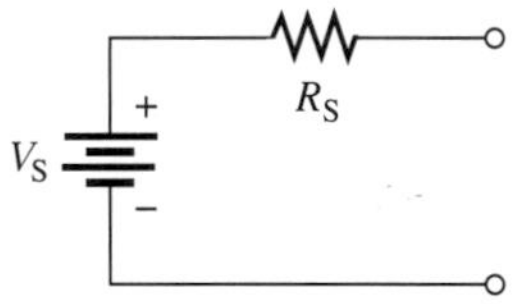

▲ FIGURE 8–92

SECTION 8–2   The Current Source

1. For ideal current source, see Figure 8–93.
2. For practical current source, see Figure 8–94.
3. An ideal current source has infinite internal resistance.
4. Load current from a current source varies inversely with load resistance.

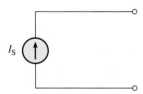

▲ FIGURE 8–93

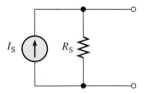

▲ FIGURE 8–94

SECTION 8–3   Source Conversions

1. $I_S = V_S/R_S$
2. $V_S = I_S R_S$
3. See Figure 8–95.
4. See Figure 8–96.

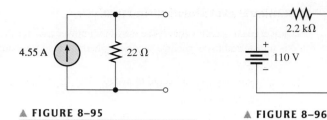

▲ FIGURE 8–95   ▲ FIGURE 8–96

SECTION 8–4   The Superposition Theorem

1. The superposition theorem states that the total current in any branch of a multiple-source linear circuit is equal to the algebraic sum of the currents due to the individual sources acting alone, with the other sources replaced by their internal resistances.

2. The superposition theorem allows each source to be treated independently.

3. A short simulates the internal resistance of an ideal voltage source; an open simulates the internal resistance of an ideal current source.

4. $I_{R1} = 6.67\,\text{mA}$

5. The net current is in the direction of the larger current.

SECTION 8–5   Thevenin's Theorem

1. A Thevenin equivalent circuit consists of V_{TH} and R_{TH}.

2. See Figure 8–97 for the general form of a Thevenin equivalent circuit.

3. V_{TH} is the open circuit voltage between two terminals in a circuit.

4. R_{TH} is the resistance as viewed from two terminals in a circuit, with all sources replaced by their internal resistances.

5. See Figure 8–98.

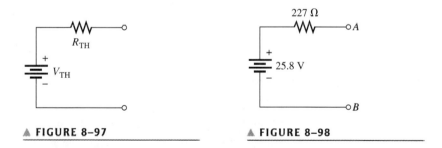

▲ FIGURE 8–97   ▲ FIGURE 8–98

SECTION 8–6   Norton's Theorem

1. A Norton equivalent circuit consists of I_N and R_N.

2. See Figure 8–99 for the general form of a Norton equivalent circuit.

3. I_N is the short circuit current between two terminals in a circuit.

4. R_N is the resistance as viewed from the two open terminals in a circuit.

5. See Figure 8–100.

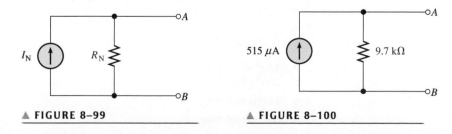

▲ FIGURE 8–99   ▲ FIGURE 8–100

SECTION 8–7   **Maximum Power Transfer Theorem**

1. For a given source voltage, the maximum power transfer theorem states that maximum power is transferred from a source to a load when the load resistance is equal to the internal source resistance.

2. Maximum power is delivered to a load when $R_L = R_S$.

3. $R_L = R_S = 50 \, \Omega$

SECTION 8–8   **Delta-to-Wye (Δ-to-Y) and Wye-to-Delta (Y-to-Δ) Conversions**

1. For a delta circuit see Figure 8–101.

2. For a wye circuit, see Figure 8–102.

▲ FIGURE 8–101

▲ FIGURE 8–102

3. The delta-to-wye conversion equations are

$$R_1 = \frac{R_A R_C}{R_A + R_B + R_C}$$

$$R_2 = \frac{R_B R_C}{R_A + R_B + R_C}$$

$$R_3 = \frac{R_A R_B}{R_A + R_B + R_C}$$

4. The wye-to-delta conversion equations are

$$R_A = \frac{R_1 R_2 + R_1 R_3 + R_2 R_3}{R_2}$$

$$R_B = \frac{R_1 R_2 + R_1 R_3 + R_2 R_3}{R_1}$$

$$R_C = \frac{R_1 R_2 + R_1 R_3 + R_2 R_3}{R_3}$$

A Circuit Application

1. 27.8 μA

2. The total series resistance at 50°C needs to be 47.1 kΩ (bridge arms, thermistor, and limiting resistor). The series resistor needs to be reduced to 26.2 kΩ.(47.1 kΩ − (16.5 kΩ + 4.38 kΩ))

RELATED PROBLEMS FOR EXAMPLES

8–1   99.5 V

8–2   100 V

8–3   3.33 kΩ

8–4   1.2 A; 10 Ω

8–5   300 V; 600 Ω

8–6   16.6 mA

8–7 I_S is not affected.

8–8 7 mA

8–9 5 mA

8–10 2.36 V; 1240 Ω

8–11 1.17 mA

8–12 0.7 V; V_{OUT} is 17.7 greater than V_{IN} at the base (B).

8–13 25.4 mA

8–14 131 Ω

8–15 $R_N = 248$ Ω

8–16 6.93 V

8–17 41.7 mW

8–18 $R_1 = 792$ Ω, $R_2 = 360$ Ω, $R_3 = 440$ Ω

8–19 $R_A = 712$ Ω, $R_B = 2.35$ kΩ, $R_C = 500$ Ω

8–20 0.3 μA

SELF-TEST

1. (b) **2.** (c) **3.** (a) **4.** (b) **5.** (d) **6.** (c) **7.** (b) **8.** (d)

9. (d) **10.** (b)

CIRCUIT DYNAMICS QUIZ

1. (a) **2.** (b) **3.** (b) **4.** (a) **5.** (b) **6.** (a) **7.** (a) **8.** (a)

9. (b) **10.** (a) **11.** (a) **12.** (a) **13.** (a) **14.** (a)

9

BRANCH, LOOP, AND NODE ANALYSES

CHAPTER OBJECTIVES

◆ Discuss three methods to solve simultaneous equations
◆ Use the branch current method to find unknown quantities in a circuit
◆ Use loop analysis to find unknown quantities in a circuit
◆ Use node analysis to find unknown quantities in a circuit

KEY TERMS

◆ Simultaneous equations
◆ Matrix
◆ Determinant
◆ Loop
◆ Node
◆ Branch

A CIRCUIT APPLICATION PREVIEW

In the application, you will analyze a model of an amplifier using the methods covered in this chapter.

VISIT THE COMPANION WEBSITE

Study aids for this chapter are available at
http://www.prenhall.com/floyd

INTRODUCTION

In the last chapter, you learned about the superposition theorem, Thevenin's theorem, Norton's theorem, maximum power transfer theorem, and several types of conversion methods. These theorems and conversion methods are useful in solving some types of circuit problems for both dc and ac.

In this chapter, three more circuit analysis methods are introduced. These methods, based on Ohm's law and Kirchhoff's laws, are particularly useful in the analysis of multiple-loop circuits having two or more voltage or current sources. The methods presented here can be used alone or in conjunction with the techniques covered in the previous chapters. With experience, you will learn which method is best for a particular problem or you may develop a preference for one of them.

In the branch current method, you will apply Kirchhoff's laws to solve for current in various branches of a multiple-loop circuit. A loop is a complete current path within a circuit. In the loop current method, you will solve for loop currents rather than branch currents. In the node voltage method, you will find the voltages at the independent nodes in a circuit. As you know, a node is the junction of two or more components.

9–1 SIMULTANEOUS EQUATIONS IN CIRCUIT ANALYSIS

The circuit analysis methods covered in this chapter allow you to find two or more unknown currents or voltages by solving simultaneous equations. These analysis methods, which include branch current, loop current, and node voltage methods, result in a number of equations equal to the number of unknowns. Coverage is limited to equations with two unknowns (2nd order) and equations with three unknowns (3rd order). These equations can then be solved simultaneously for the unknown quantities using one of the methods covered in this section.

After completing this section, you should be able to

◆ **Discuss three methods to solve simultaneous equations**

  ◆ Write a set of simultaneous equations in standard form

  ◆ Solve simultaneous equations using algebraic substitution

  ◆ Solve simultaneous equations using determinants

  ◆ Solve simultaneous equations using a calculator

Simultaneous equations consist of a set of n equations containing n unknowns, where n is a number with a value of 2 or more. The number of equations in the set must equal the number of unknowns. For example, to solve for two unknown variables, you must have two equations; to solve for three unknowns you must have three equations, and so on.

Second-Order Standard Form Equations

An equation with two variables is called a *second-order equation*. In circuit analysis, the variables can represent unknowns such as current or voltage. In order to solve for variables x_1 and x_2, there must be a set of two equations containing those variables expressed in standard form.

In standard form, the x_1 variables are in the first position in each equation, and the x_2 variables are in the second position in each equation. The variables with their coefficients are on the left side of the equation, and the constants are on the right side.

The set of two simultaneous second-order equations written in standard form is

$$a_{1,1}x_1 + a_{1,2}x_2 = b_1$$
$$a_{2,1}x_1 + a_{2,2}x_2 = b_2$$

In these simultaneous equations, the "a" is the coefficient of the variables x_1 and x_2 and can represent values of circuit components. Notice that the subscripts of the coefficients contain two numbers. For example, $a_{1,1}$ appears in the *first* equation as the coefficient of x_1 and $a_{2,1}$ appears in the *second* equation as the coefficient of x_1. The "b" is the constant and can represent a voltage source. This notation will be useful when you use a calculator to solve the equations.

EXAMPLE 9–1

Assume the following two equations describe a particular circuit with two unknown currents I_1 and I_2. The coefficients are resistance values and the constants are voltages in the circuit. Write the equations in standard form.

$$2I_1 = 8 - 5I_2$$
$$4I_2 - 5I_1 + 6 = 0$$

Solution   Rearrange the equations in standard form as follows:

$$2I_1 + 5I_2 = 8$$
$$5I_1 + 4I_2 = -6$$

*Related Problem**   Convert these two equations to standard form:

$$20x_1 + 15 = 11x_2$$
$$10 = 25x_2 + 18x_1$$

---

*Answers are at the end of the chapter.

A third-order equation contains three variables and a constant term. Just as in second-order equations, each variable has a coefficient. In order to solve for the variables x_1, x_2, and x_3, there must be a set of three simultaneous equations containing those variables. The general form for three simultaneous third-order equations written in standard form is

$$a_{1,1}x_1 + a_{1,2}x_2 + a_{1,3}x_3 = b_1$$
$$a_{2,1}x_1 + a_{2,2}x_2 + a_{2,3}x_3 = b_2$$
$$a_{3,1}x_1 + a_{3,2}x_2 + a_{3,3}x_3 = b_3$$

EXAMPLE 9–2   Assume the following three equations describe a particular circuit with three unknown currents I_1, I_2, and I_3. The coefficients are resistance values and the constants are known voltages in the circuit. Write the equations in standard form.

$$4I_3 + 2I_2 + 7I_1 = 0$$
$$5I_1 + 6I_2 + 9I_3 - 7 = 0$$
$$8 = 1I_1 + 2I_2 + 5I_3$$

Solution   The equations are rearranged to put them in standard form as follows:

$$7I_1 + 2I_2 + 4I_3 = 0$$
$$5I_1 + 6I_2 + 9I_3 = 7$$
$$1I_1 + 2I_2 + 5I_3 = 8$$

Related Problem   Convert these three equations to standard form:

$$10V_1 + 15 = 21V_2 + 50V_3$$
$$10 + 12V_3 = 25V_2 + 18V_1$$
$$12V_3 - 25V_2 + 18V_1 = 9$$

Solutions of Simultaneous Equations

Three ways for solving simultaneous equations are algebraic substitution, the determinant method, and using a calculator.

Solving by Substitution   You can solve two or three simultaneous equations in standard form using algebraic substitution by first solving one of the variables in terms of the others.

However, because the process can become quite lengthy, we will restrict this method to second-order equations. Consider the following set of simultaneous equations:

$$2x_1 + 6x_2 = 8 \quad \text{(Eq. 1)}$$
$$3x_1 + 6x_2 = 2 \quad \text{(Eq. 2)}$$

Step 1. Solve for x_1 in terms of x_2 in Eq. 1.

$$2x_1 = 8 - 6x_2$$
$$x_1 = 4 - 3x_2$$

Step 2. Substitute the result for x_1 into Eq. 2 and solve for x_2.

$$3x_1 + 6x_2 = 2$$
$$3(4 - 3x_2) + 6x_2 = 2$$
$$12 - 9x_2 + 6x_2 = 2$$
$$-3x_2 = -10$$
$$x_2 = \frac{-10}{-3} = 3.33$$

Step 3. Substitute the value for x_2 into the equation for x_1 in Step 1.

$$x_1 = 4 - 3x_2 = 4 - 3(3.33) = 4 - 9.99 = -5.99$$

Solving by Determinants The determinant method is a part of matrix algebra and provides a "cookbook" approach for solving simultaneous equations with two or three variables. A **matrix** is an array of numbers, and a **determinant** is effectively the solution to a matrix, resulting in a specific value. Second-order determinants are used for two variables and third-order determinants are used for three variables. The equations must be in standard form for a solution.

To illustrate the determinant method for second-order equations, let's find the values of I_1 and I_2 in the following two equations expressed in standard form:

$$10I_1 + 5I_2 = 15$$
$$2I_1 + 4I_2 = 8$$

First, form the characteristic determinant from the matrix of the coefficients of the unknown currents. The first column in the determinant consists of the coefficients of I_1, and the second column consists of the coefficients of I_2. The resulting determinant is

$$\begin{vmatrix} 10 & 5 \\ 2 & 4 \end{vmatrix}$$

An evaluation of this characteristic determinant requires three steps.

Step 1. Multiply the first number in the left column by the second number in the right column.

$$\begin{vmatrix} 10 & 5 \\ 2 & 4 \end{vmatrix} = 10 \times 4 = 40$$

Step 2. Multiply the second number in the left column by the first number in the right column.

$$\begin{vmatrix} 10 & 5 \\ 2 & 4 \end{vmatrix} = 2 \times 5 = 10$$

Step 3. Subtract the product in Step 2 from the product in Step 1.

$$40 - 10 = 30$$

This difference is the value of the characteristic determinant (30 in this case).

Next, replace the coefficients of I_1 in the first column of the characteristic determinant with the constants (fixed numbers) on the right side of the equations to form another determinant:

$$\begin{vmatrix} 15 & 5 \\ 8 & 4 \end{vmatrix}$$

Replace coefficients of I_1 with constants from right sides of equations.

Evaluate this I_1 determinant as follows:

$$\begin{vmatrix} 15 & 5 \\ 8 & 4 \end{vmatrix} = 15 \times 4 = 60$$

$$\begin{vmatrix} 15 & 5 \\ 8 & 4 \end{vmatrix} = 60 - (8 \times 5) = 60 - 40 = 20$$

The value of this determinant is 20.

Now solve for I_1 by dividing the I_1 determinant by the characteristic determinant as follows:

$$I_1 = \frac{\begin{vmatrix} 15 & 5 \\ 8 & 4 \end{vmatrix}}{\begin{vmatrix} 10 & 5 \\ 2 & 4 \end{vmatrix}} = \frac{20}{30} = 0.667 \text{ A}$$

To find I_2, form another determinant by substituting the constants on the right side of the given equations for the coefficients of I_2 in the second column of the characteristic determinant.

$$\begin{vmatrix} 10 & 15 \\ 2 & 8 \end{vmatrix}$$

Replace coefficients of I_2 with constants from right sides of equations.

Solve for I_2 by dividing this determinant by the characteristic determinant previously found.

$$I_2 = \frac{\begin{vmatrix} 10 & 15 \\ 2 & 8 \end{vmatrix}}{30} = \frac{(10 \times 8) - (2 \times 15)}{30} = \frac{80 - 30}{30} = \frac{50}{30} = 1.67 \text{ A}$$

EXAMPLE 9–3

Solve the following set of equations for the unknown currents:

$$2I_1 - 5I_2 = 10$$
$$6I_1 + 10I_2 = 20$$

Solution Evaluate the characteristic determinant as follows:

$$\begin{vmatrix} 2 & -5 \\ 6 & 10 \end{vmatrix} = (2)(10) - (-5)(6) = 20 - (-30) = 20 + 30 = 50$$

Solving for I_1 yields

$$I_1 = \frac{\begin{vmatrix} 10 & -5 \\ 20 & 10 \end{vmatrix}}{50} = \frac{(10)(10) - (-5)(20)}{50} = \frac{100 - (-100)}{50} = \frac{200}{50} = \textbf{4 A}$$

Solving for I_2 yields

$$I_2 = \frac{\begin{vmatrix} 2 & 10 \\ 6 & 20 \end{vmatrix}}{50} = \frac{(2)(20) - (6)(10)}{50} = \frac{40 - 60}{50} = -0.4 \text{ A}$$

In a circuit problem, a result with a negative sign indicates that the direction of actual current is opposite to the assigned direction.

Note that multiplication can be expressed either by the multiplication sign such as 2×10 or by parentheses such as $(2)(10)$.

Related Problem Solve the following set of equations for I_1:

$$5I_1 + 3I_2 = 4$$
$$I_1 + 2I_2 = -6$$

Third-order determinants can be evaluated by the expansion method. We will illustrate this method by finding the unknown current values in the following three equations expressed in standard form:

$$1I_1 + 3I_2 - 2I_3 = 7$$
$$0I_1 + 4I_2 + 1I_3 = 8$$
$$-5I_1 + 1I_2 + 6I_3 = 9$$

The characteristic determinant for the matrix of coefficients for this set of equations is formed in a similar way to that used earlier for the second-order determinant. The first column consists of the coefficients of I_1, the second column consists of the coefficients of I_2, and the third column consists of the coefficients of I_3, as shown below.

$$\begin{vmatrix} 1 & 3 & -2 \\ 0 & 4 & 1 \\ -5 & 1 & 6 \end{vmatrix}$$

This third-order determinant is evaluated by the expansion method as follows:

Step 1. Rewrite the first two columns immediately to the right of the determinant.

$$\begin{vmatrix} 1 & 3 & -2 \\ 0 & 4 & 1 \\ -5 & 1 & 6 \end{vmatrix} \begin{matrix} 1 & 3 \\ 0 & 4 \\ -5 & 1 \end{matrix}$$

Step 2. Identify the three downward diagonal groups of three coefficients each.

Step 3. Multiply the numbers in each diagonal and add the products.

$$(1)(4)(6) + (3)(1)(-5) + (-2)(0)(1) = 24 + (-15) + 0 = 9$$

Step 4. Repeat Steps 2 and 3 for the three upward diagonal groups of three coefficients.

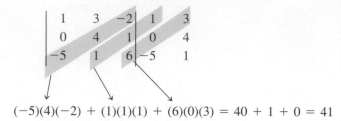

$$(-5)(4)(-2) + (1)(1)(1) + (6)(0)(3) = 40 + 1 + 0 = 41$$

Step 5. Subtract the result in Step 4 from the result in Step 3 to get the value of the characteristic determinant.

$$9 - 41 = -32$$

Next, form another determinant by substituting the constants on the right of the equations for the coefficients of I_1 in the characteristic determinant.

$$\begin{vmatrix} 7 & 3 & -2 \\ 8 & 4 & 1 \\ 9 & 1 & 6 \end{vmatrix}$$

Evaluate this determinant using the method described in the previous steps.

$$= [(7)(4)(6) + (3)(1)(9) + (-2)(8)(1)] - [(9)(4)(-2) + (1)(1)(7) + (6)(8)(3)]$$
$$= (168 + 27 - 16) - (-72 + 7 + 144) = 179 - 79 = 100$$

Solve for I_1 by dividing this determinant by the characteristic determinant. The negative result indicates that the actual current is in a direction opposite to the original assumption.

$$I_1 = \frac{\begin{vmatrix} 7 & 3 & -2 \\ 8 & 4 & 1 \\ 9 & 1 & 6 \end{vmatrix}}{\begin{vmatrix} 1 & 3 & -2 \\ 0 & 4 & 1 \\ -5 & 1 & 6 \end{vmatrix}} = \frac{100}{-32} = -3.125 \text{ A}$$

You can find I_2 and I_3 in a similar way.

EXAMPLE 9–4

Determine the value of I_2 from the following set of equations:

$$2I_1 + 0.5I_2 + 1I_3 = 0$$
$$0.75I_1 + 0I_2 + 2I_3 = 1.5$$
$$3I_1 + 0.2I_2 + 0I_3 = -1$$

Solution Evaluate the characteristic determinant as follows:

$$\begin{vmatrix} 2 & 0.5 & 1 \\ 0.75 & 0 & 2 \\ 3 & 0.2 & 0 \end{vmatrix}\begin{matrix} 2 & 0.5 \\ 0.75 & 0 \\ 3 & 0.2 \end{matrix}$$
$$= [(2)(0)(0) + (0.5)(2)(3) + (1)(0.75)(0.2)] - [(3)(0)(1) + (0.2)(2)(2) + (0)(0.75)(0.5)]$$
$$= (0 + 3 + 0.15) - (0 + 0.8 + 0) = 3.15 - 0.8 = 2.35$$

Evaluate the determinant for I_2 is as follows:

$$\begin{vmatrix} 2 & 0 & 1 \\ 0.75 & 1.5 & 2 \\ 3 & -1 & 0 \end{vmatrix} \begin{matrix} 2 & 0 \\ 0.75 & 1.5 \\ 3 & -1 \end{matrix}$$

$$= [(2)(1.5)(0) + (0)(2)(3) + (1)(0.75)(-1)] - [(3)(1.5)(1) + (-1)(2)(2) + (0)(0.75)(0)]$$

$$= [0 + 0 + (-0.75)] - [4.5 + (-4) + 0] = -0.75 - 0.5 = -1.25$$

Finally, divide the two determinants.

$$I_2 = \frac{-1.25}{2.35} = -0.532 \text{ A} = -532 \text{ mA}$$

Related Problem   Determine the value of I_1 in the set of equations used in this example.

Solving by Calculator   Calculators generally employ matrix algorithms for simultaneous equation solution and make the results much easier to obtain. As with the two "manual" methods, it is important to first get the equations into the standard form before entering data on a calculator. Calculators that offer simultaneous equation solutions generally employ the notation discussed earlier in relation to the general form of equations. Variables are labeled as x_1, x_2, etc, coefficients are designated as $a_{1,1}$, $a_{1,2}$, $a_{2,1}$, $a_{2,2}$, etc., and constants are designated as b_1, b_2, etc.

A typical sequence for entering the data for a specific set of equations into a calculator is illustrated in a generic way for three simultaneous equations in Figure 9–1.

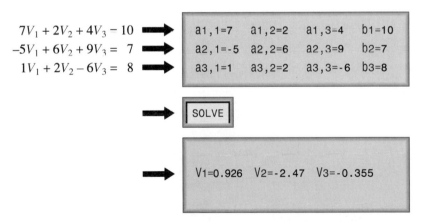

◀ **FIGURE 9–1**

We have selected the TI-86 and the TI-89 calculators to illustrate the procedure in the following two examples, although other scientific calculators can be used. If your calculator has simultaneous equation capability, consult the user's manual for the correct procedure.

EXAMPLE 9–5

Use the TI-86 calculator to solve the following three simultaneous equations for the three unknowns.

$$8I_1 + 4I_2 + 1I_3 = 7$$
$$2I_1 - 5I_2 + 6I_3 = 3$$
$$3I_1 + 3I_2 - 2I_3 = -5$$

Solution   Press the 2nd key, then SIMULT to enter the number of equations as shown in Figure 9–2.

▶ **FIGURE 9–2**

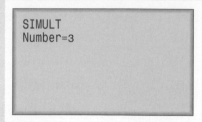

After you enter 3 and press ENTER, the first equation screen comes up. Enter the coefficients 8, 4, 1 and the constant 7 by pressing each number key followed by the ENTER key, which results in the screen shown in Figure 9–3(a). After you enter the last number and press ENTER, the second equation screen appears. Enter the coefficients 2, −5, 6 and the constant 3 as shown in Figure 9–3(b) (A negative value is entered by first pressing the (−) key.). Finally, enter the coefficients of the third equation (3, 3, −2) and the constant −5 as shown in Figure 9–3(c).

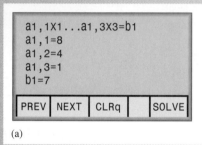

(a)

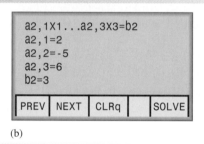

(b)

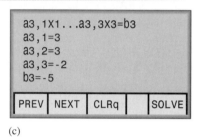

(c)

▲ **FIGURE 9–3**

Selecting SOLVE, which is the F5 key, produces the results displayed in Figure 9–4. X1 is I_1, X2 is I_2, and X3 is I_3.

▶ **FIGURE 9–4**

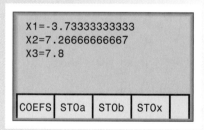

Related Problem   Edit the equations to change a1,2 from 4 to −3, a2,3 from 6 to 2.5, and b3 from −5 to 8 and solve the modified equations.

EXAMPLE 9–6

Use the TI-89 Titanium calculator to solve the same three simultaneous equations that were given in Example 9–5.

$$8I_1 + 4I_2 + 1I_3 = 7$$
$$2I_1 - 5I_2 + 6I_3 = 3$$
$$3I_1 + 3I_2 - 2I_3 = -5$$

Solution From the Home screen select the simultaneous equation icon

Press ENTER. Select 3:New, then press ENTER again. Next specify the number of equations and the number of unknowns and press ENTER. On the simultaneous equation screen enter the coefficients and the constants as shown in Figure 9–5(a). Press ENTER after each number.

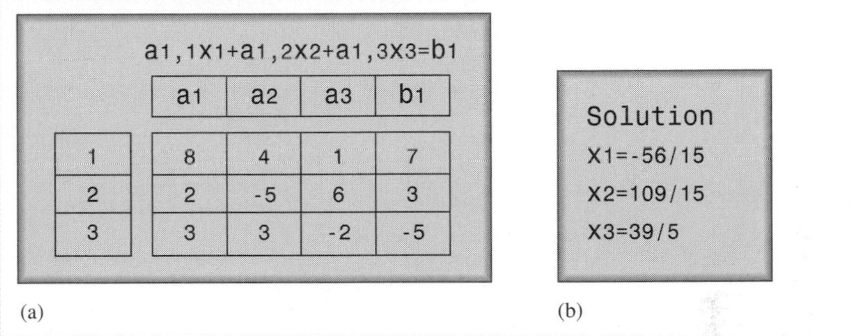

a1,1X1+a1,2X2+a1,3X3=b1

	a1	a2	a3	b1
1	8	4	1	7
2	2	-5	6	3
3	3	3	-2	-5

Solution
X1=-56/15
X2=109/15
X3=39/5

(a)   (b)

▲ **FIGURE 9–5**

After you have entered the coefficients and constants, press the F5 key to solve. As indicated by the screen in part (b), the results are returned as fractions. They agree with the TI-86 results, which were returned as decimal numbers.

Related Problem Repeat the Related Problem from Example 9–5 using the TI-89.

SECTION 9–1 REVIEW

1. Evaluate the following determinants:

   (a) $\begin{vmatrix} 0 & -1 \\ 4 & 8 \end{vmatrix}$ (b) $\begin{vmatrix} 0.25 & 0.33 \\ -0.5 & 1 \end{vmatrix}$ (c) $\begin{vmatrix} 1 & 3 & 7 \\ 2 & -1 & 7 \\ -4 & 0 & -2 \end{vmatrix}$

2. Set up the characteristic determinant for the following set of simultaneous equations:

$$2I_1 + 3I_2 = 0$$
$$5I_1 + 4I_2 = 1$$

3. Find I_2 in Question 2.

4. Use your calculator to solve the following set of simultaneous equations for I_1, I_2, I_3, and I_4.

$$100I_1 + 220I_2 + 180I_3 + 330I_4 = 0$$
$$470I_1 + 390I_2 + 100I_3 + 100I_4 = 12$$
$$120I_1 - 270I_2 + 150I_3 - 180I_4 = -9$$
$$560I_1 + 680I_2 - 220I_3 + 390I_4 = 0$$

5. Modify the equations in Question 4 by changing the constant in the first equation to 8.5, the coefficient of I_3 in the second equation to 220, and the coefficient of I_1 in the fourth equation to 330. Solve the new set of equations for the currents.

9–2   BRANCH CURRENT METHOD

The branch current method is a circuit analysis method using Kirchhoff's voltage and current laws to find the current in each branch of a circuit by generating simultaneous equations. Once you know the branch currents, you can determine voltages.

After completing this section, you should be able to

◆ **Use the branch current method to find unknown quantities in a circuit**

  ◆ Identify loops and nodes in a circuit

  ◆ Develop a set of branch current equations

  ◆ Solve the branch current equations

Figure 9–6 shows a circuit that will be used as the basic model throughout the chapter to illustrate each of the three circuit analysis methods. In this circuit, there are only two nonredundant closed loops. A **loop** is a complete current path within a circuit, and you can view a set of nonredundant closed loops as a set of "windowpanes," where each windowpane represents one nonredundant loop. Also, there are four nodes as indicated by the letters A, B, C, and D. A **node** is a point where two or more components are connected. A **branch** is a path that connects two nodes, and there are three branches in this circuit: one containing R_1, one containing R_2, and one containing R_3.

◆ **FIGURE 9–6**

Circuit showing loops, nodes, and branches.

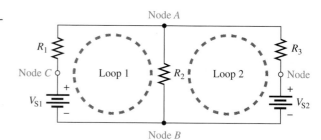

The following are the general steps used in applying the branch current method.

Step 1. Assign a current in each circuit branch in an arbitrary direction.

Step 2. Show the polarities of the resistor voltages according to the assigned branch current directions.

Step 3. Apply Kirchhoff's voltage law around each closed loop (algebraic sum of voltages is equal to zero).

Step 4. Apply Kirchhoff's current law at the minimum number of nodes so that all branch currents are included (algebraic sum of currents at a node equals zero).

Step 5. Solve the equations resulting from Steps 3 and 4 for the branch current values.

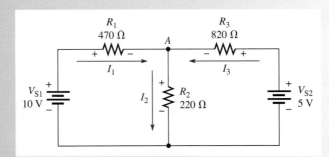

Circuit for demonstrating branch current analysis.

These steps are demonstrated with the aid of Figure 9–7. First, the **branch currents** I_1, I_2, and I_3 are assigned in the direction shown. Don't worry about the actual current directions at this point. Second, the polarities of the voltage drops across R_1, R_2, and R_3 are indicated in the figure according to the assigned current directions. Third, Kirchhoff's voltage law applied to the two loops gives the following equations where the resistance values are the coefficients for the unknown currents:

Equation 1: $R_1I_1 + R_2I_2 - V_{S1} = 0$ for loop 1

Equation 2: $R_2I_2 + R_3I_3 - V_{S2} = 0$ for loop 2

Fourth, Kirchhoff's current law is applied to node A, including all branch currents as follows:

Equation 3: $I_1 - I_2 + I_3 = 0$

The negative sign indicates that I_2 is out of the node. Fifth and last, the three equations must be solved for the three unknown currents, I_1, I_2, and I_3. Example 9–7 shows how to solve equations by the *substitution* method.

EXAMPLE 9–7

Use the branch current method to find each branch current in Figure 9–8.

[circuit diagram: R_1 470 Ω, R_3 820 Ω, node A, I_1, I_3, V_{S1} 10 V, I_2, R_2 220 Ω, V_{S2} 5 V]

▲ **FIGURE 9–8**

Solution   **Step 1:** Assign branch currents as shown in Figure 9–8. Keep in mind that you can assume any current direction at this point and that the final solution will have a negative sign if the actual current is opposite to the assigned current.

Step 2: Mark the polarities of the resistor voltage drops in accordance with the assigned current directions as shown in the figure.

Step 3: Applying Kirchhoff's voltage law around the left loop gives

$$470I_1 + 220I_2 - 10 = 0$$

Around the right loop gives

$$220I_2 + 820I_3 - 5 = 0$$

where all resistance values are in ohms and voltage values are in volts. For simplicity, the units are not shown.

Step 4: At node A, the current equation is

$$I_1 - I_2 + I_3 = 0$$

Step 5: The equations are solved by substitution as follows. First, find I_1 in terms of I_2 and I_3.

$$I_1 = I_2 - I_3$$

Now, substitute $I_2 - I_3$ for I_1 in the left loop equation.

$$470(I_2 - I_3) + 220I_2 = 10$$
$$470I_2 - 470I_3 + 220I_2 = 10$$
$$690I_2 - 470I_3 = 10$$

Next, take the right loop equation and solve for I_2 in terms of I_3.

$$220I_2 = 5 - 820I_3$$
$$I_2 = \frac{5 - 820I_3}{220}$$

Substituting this expression for I_2 into $820I_2 - 470I_3 = 10$ yields

$$690\left(\frac{5 - 820I_3}{220}\right) - 470I_3 = 10$$
$$\frac{3450 - 565800I_3}{220} - 470I_3 = 10$$
$$15.68 - 2571.8I_3 - 470I_3 = 10$$
$$-3041.8I_3 = -5.68$$
$$I_3 = \frac{5.68}{3041.8} = 0.00187 \text{ A} = \mathbf{1.87\,mA}$$

Now, substitute the value of I_3 in amps into the right loop equation.

$$220I_2 + 820(0.00187) = 5$$

Solve for I_2.

$$I_2 = \frac{5 - 820(0.00187)}{220} = \frac{3.47}{220} = 0.0158 = \mathbf{15.8\,mA}$$

Substituting I_2 and I_3 values into the current equation at node A yields

$$I_1 - 0.0158 + 0.00187 = 0$$
$$I_1 = 0.0158 - 0.00187 = 0.0139 \text{ A} = \mathbf{13.9\,mA}$$

Related Problem   Determine the branch currents in Figure 9–8 with the polarity of the 5 V source reversed.

 Use Multisim file E09-07 to verify the calculated results in this example and to confirm your calculations for the related problem.

SECTION 9–2 REVIEW

1. What basic circuit laws are used in the branch current method?
2. When assigning branch currents, you should be careful that the assigned directions match the actual directions.   (T or F)
3. What is a loop?
4. What is a node?

9–3 LOOP CURRENT METHOD

In the loop current method (also known as the mesh current method), you will work with loop currents instead of branch currents. An ammeter placed in a given branch will measure the branch current. Unlike branch currents, loop currents are mathematical quantities, rather than actual physical currents, that are used to make circuit analysis somewhat easier than with the branch current method.

After completing this section, you should be able to

♦ **Use loop analysis to find unknown quantities in a circuit**

   ♦ Assign loop currents

   ♦ Apply Kirchhoff's voltage law around each loop

   ♦ Develop the loop equations

   ♦ Solve the loop equations

A systematic method of loop analysis is given in the following steps and is illustrated in Figure 9–9, which is the same circuit configuration used in the branch current analysis. It demonstrates the basic principles well.

Step 1. Although direction of an assigned loop current is arbitrary, we will assign a current in the clockwise (CW) direction around each nonredundant closed loop, for consistency. This may not be the actual current direction, but it does not matter. The number of loop-current assignments must be sufficient to include current through all components in the circuit.

Step 2. Indicate the voltage drop polarities in each loop based on the assigned current directions.

Step 3. Apply Kirchhoff's voltage law around each closed loop. When more than one loop current passes through a component, include its voltage drop. This results in one equation for each loop.

Step 4. Using substitution or determinants, solve the resulting equations for the loop currents.

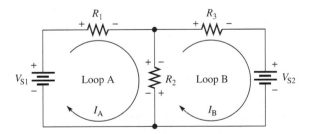

▲ **FIGURE 9–9**

First, the **loop currents** I_A and I_B are assigned in the CW direction as shown in Figure 9–9. A loop current could be assigned around the outer perimeter of the circuit, but this would be redundant since I_A and I_B already pass through all of the components.

Second, the polarities of the voltage drops across R_1, R_2, and R_3 are shown based on the loop-current directions. Notice that I_A and I_B are in opposite directions through R_2 because R_2 is common to both loops. Therefore, two voltage polarities are indicated. In reality, the R_2 current cannot be separated into two parts, but remember that the loop currents are basically mathematical quantities used for analysis purposes. The polarities of the voltage sources are fixed and are not affected by the current assignments.

Third, Kirchhoff's voltage law applied to the two loops results in the following two equations:

$$R_1I_A + R_2(I_A - I_B) = V_{S1} \qquad \text{for loop A}$$
$$R_3I_B + R_2(I_B - I_A) = -V_{S2} \qquad \text{for loop B}$$

Notice that I_A is positive in loop A and I_B is positive in loop B.

Fourth, the like terms in the equations are combined and rearranged into standard form for convenient solution so that they have the same position in each equation, that is, the I_A term is first and the I_B term is second. The equations are rearranged into the following form. Once the loop currents are evaluated, all of the branch currents can be determined.

$$(R_1 + R_2)I_A - R_2I_B = V_{S1} \qquad \text{for loop A}$$
$$-R_2I_A + (R_2 + R_3)I_B = -V_{S2} \qquad \text{for loop B}$$

Notice that in the loop current method only two equations are required for the same circuit that required three equations in the branch current method. The last two equations (developed in the fourth step) follow a form to make loop analysis easier. Referring to these last two equations, notice that for loop A, the total resistance in the loop, $R_1 + R_2$, is multiplied by I_A (its loop current). Also in the loop A equation, the resistance common to both loops, R_2, is multiplied by the other loop current, I_B, and subtracted from the first term. The same form is seen in the loop B equation except that the terms have been rearranged. From these observations, a concise rule for applying steps 1 to 4 is as follows:

(Sum of resistors in loop) times (loop current) minus (each resistor common to both loops) times (associated adjacent loop current) equals (source voltage in the loop).

Example 9–8 illustrates the application of this rule to the loop current analysis of a circuit.

EXAMPLE 9–8

Using the loop current method, find the branch currents in Figure 9–10.

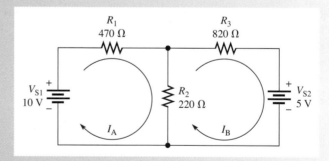

▲ **FIGURE 9–10**

Solution Assign the loop currents (I_A and I_B) as shown in Figure 9–10; resistance values are in ohms and voltage values are in volts. Use the rule described to set up the two loop equations.

$$(470 + 220)I_A - 220I_B = 10$$
$$690I_A - 220I_B = 10 \qquad \text{for loop A}$$

$$-220I_A + (220 + 820)I_B = -5$$
$$-220I_A + 1040I_B = -5 \qquad \text{for loop B}$$

Use determinants to find I_A.

$$I_A = \frac{\begin{vmatrix} 10 & -220 \\ -5 & 1040 \end{vmatrix}}{\begin{vmatrix} 690 & -220 \\ -220 & 1040 \end{vmatrix}} = \frac{(10)(1040) - (-5)(-220)}{(690)(1040) - (-220)(-220)} = \frac{104000 - 1100}{717600 - 48400} = \frac{102900}{669200} = 13.9\,\text{mA}$$

Solving for I_B yields

$$I_B = \frac{\begin{vmatrix} 690 & 10 \\ -220 & -5 \end{vmatrix}}{669200} = \frac{(690)(-5) - (-220)(10)}{669200} = \frac{-3450 - (-2200)}{669200} = -1.87\,\text{mA}$$

The negative sign on I_B means that its assigned direction is opposite to the actual current.

Now find the actual branch currents. Since I_A is the only current through R_1, it is also the branch current I_1.

$$I_1 = I_A = \textbf{13.9\,mA}$$

Since I_B is the only current through R_3, it is also the branch current I_3.

$$I_3 = I_B = \textbf{-1.87\,mA}$$

The negative sign indicates opposite direction of that originally assigned to I_B.

As originally assigned, both loop currents I_A and I_B are through R_2 in opposite directions. The branch current I_2 is the difference between I_A and I_B.

$$I_2 = I_A - I_B = 13.9\,\text{mA} - (-1.87\,\text{mA}) = \textbf{15.8\,mA}$$

Keep in mind that once you know the branch currents, you can find the voltages by using Ohm's law. Notice that these results are the same as in Example 9–7 where the branch current method was used.

Related Problem   Solve for the two loop currents using your calculator.

Use Multisim file E09-08 to verify the calculated results in this example and to verify your calculations for the related problem.

Circuits with More Than Two Loops

The loop current method can be systematically applied to circuits with any number of loops. Of course, the more loops there are, the more difficult is the solution, but calculators have greatly simplified solving simultaneous equations. Most circuits that you will encounter will not have more than three loops. Keep in mind that the loop currents are not the actual physical currents but are mathematical quantities assigned for analysis purposes.

A widely used circuit that you have already encountered is the Wheatstone bridge. The Wheatstone bridge was originally designed as a stand-alone measuring instrument but has largely been replaced with other instruments. However, the Wheatstone bridge circuit is incorporated in automated measuring instruments, and as explained previously, is widely used in the scale industry and in other measurement applications.

One method for solving the bridge parameters, which directly leads to finding the current in each arm of the bridge and the load current, is to write loop equations for the bridge. Figure 9–11 shows a Wheatstone bridge with three loops. Example 9–9 illustrates how to solve for all of the currents in the bridge.

Wheatstone bridge with three loops.

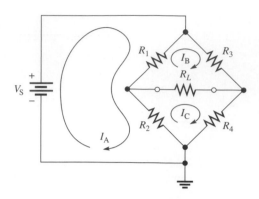

EXAMPLE 9–9

For the circuit in Figure 9–12, find the loop currents. Use the loop currents to solve for the current in each resistor (branch current).

▶ FIGURE 9–12

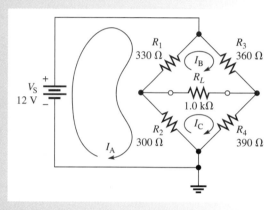

Solution　Assign three clockwise loop currents (I_A, I_B, and I_C) as shown in Figure 9–12. Then write the loop equations. The equations for the loops are

Loop A:　$-12 + 330(I_A - I_B) + 330(I_A - I_C) = 0$

Loop B:　$330(I_B - I_A) + 360I_B + 1000(I_B - I_C) = 0$

Loop C:　$300(I_C - I_A) + 1000(I_C - I_B) + 390I_C = 0$

Rearrange the equations into standard form:

Loop A:　　$630I_A - 330I_B - 300I_C = 12$ V

Loop B:　$-330I_A + 1690I_B - 1000I_C = 0$

Loop C:　$-300I_A - 1000I_B + 1690I_C = 0$

You can solve these equations with substitution, but this is tedious with three unknowns. The determinant method or directly solving with your calculator are simpler ways. Units are not shown until the end of the problem.

Evaluating the characteristic determinant using the expansion method,

$$
\begin{vmatrix}
630 & -330 & -300 \\
-330 & 1690 & -1000 \\
-300 & -1000 & 1690
\end{vmatrix}
\begin{matrix}
630 & -330 \\
-330 & 1690 \\
-300 & -1000
\end{matrix}
$$

$= [(630)(1690)(1690) + (-330)(-1000)(-300) + (-300)(-330)(-1000)]$

　$- [(-300)(1690)(-300) + (-1000)(-1000)(630) + (1690)(-330)(-330)]$

$= 635202000$

Solving for I_A:

$$\frac{\begin{vmatrix} 12 & -330 & -300 \\ 0 & 1690 & -1000 \\ 0 & -1000 & 1690 \end{vmatrix}}{635202000} = \frac{(12)(1690)(1690) - (12)(-1000)(-1000)}{635202000} = 0.0351 \text{ A} = 35.1 \text{ mA}$$

Solving for I_B:

$$\frac{\begin{vmatrix} 630 & 12 & -300 \\ -330 & 0 & 1000 \\ -300 & 0 & 1690 \end{vmatrix}}{635202000} = \frac{(12)(-1000)(-300) - (-330)(12)(1690)}{635202000} = 0.0162 \text{ A} = 16.2 \text{ mA}$$

Solving for I_C:

$$\frac{\begin{vmatrix} 630 & -330 & 12 \\ -330 & 1690 & 0 \\ -300 & -1000 & 0 \end{vmatrix}}{635202000} = \frac{(12)(-330)(-1000) - (-300)(1690)(12)}{635202000} = 0.0158 \text{ A} = 15.8 \text{ mA}$$

The current in R_1 is the difference between I_A and I_B:

$$I_1 = (I_A - I_B) = 35.1 \text{ mA} - 16.2 \text{ mA} = \mathbf{18.9 \text{ mA}}$$

The current in R_2 is the difference between I_A and I_C:

$$I_2 = (I_A - I_C) = 35.1 \text{ mA} - 15.8 \text{ mA} = \mathbf{19.3 \text{ mA}}$$

The current in R_3 is I_B:

$$I_3 - I_B = \mathbf{16.2 \text{ mA}}$$

The current in R_4 is I_C:

$$I_4 = I_C = \mathbf{15.8 \text{ mA}}$$

The current in R_L is the difference between I_B and I_C:

$$I_L = (I_B - I_C) = 16.2 \text{ mA} - 15.8 \text{ mA} = \mathbf{0.4 \text{ mA}}$$

Related Problem   Use a calculator to verify the loop currents in this example.

 Use Multisim file E09-09 to verify the calculated results in this example and to confirm your calculations for the related problem.

Another useful three-loop circuit is the bridged-T circuit. While the circuit is primarily applied in ac filter circuits using reactive components, it is introduced here to illustrate the three-loop circuit solution. A loaded resistive bridged-T is shown in Figure 9–13.

Resistors will often be in kΩ (or even MΩ) so the coefficients for simultaneous equations will become quite large if they are shown explicitly in solving equations. To simplify entering and solving equations with kΩ, it is common practice to drop the kΩ in the equations and recognize that the unit for current is the mA if the voltage is volts. The following example of a bridged-T circuit illustrates this idea.

▶ FIGURE 9–13

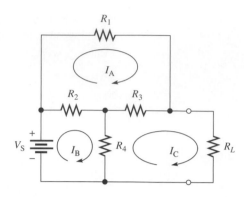

EXAMPLE 9–10

Figure 9–14 shows a bridged-T circuit with three loops. Set up the standard form equations for the loop currents. Solve the equations with a calculator and find the current in each resistor.

▶ FIGURE 9–14

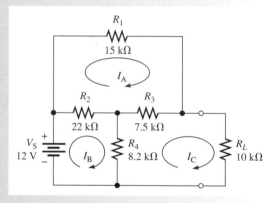

Solution   Assign three clockwise loop currents(I_A, I_B, and I_C) as shown in Figure 9–14. Write the loop equations, but drop the k prefix from the resistances. Current will be in mA.

Loop A: $22(I_A - I_B) + 15I_A + 7.5(I_A - I_C) = 0$
Loop B: $-12 + 22(I_B - I_A) + 8.2(I_B - I_C) = 0$
Loop C: $8.2(I_C - I_B) + 7.5(I_C - I_A) + 10I_C = 0$

Rearrange the equations into standard form:

Loop A: $44.5I_A - 22I_B - 7.5I_C = 0$
Loop B: $-22I_A + 30.2I_B - 8.2I_C = 12$
Loop C: $-7.5I_A - 8.2I_B + 25.7I_C = 0$

Calculator Solution: A calculator solution requires entry of the number of equations (3), the coefficients, and the constants. The calculator SOLVE function produces the results as shown in Figure 9–15. Because the resistors were in kΩ, the unit for the loop currents is mA. Solve for the current in each resistor. The current in $R_1 = I_A$.

$$I_1 = 0.512 \text{ mA}$$

FIGURE 9–15

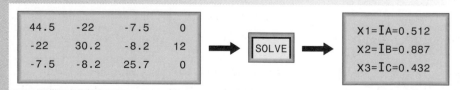

The current in R_2 is the difference between I_A and I_B.

$$I_2 = (I_A - I_B) = 0.512\,\text{mA} - 0.887\,\text{mA} = -0.375\,\text{mA}$$

The negative sign indicates that current is the opposite direction to I_A; the positive side of the resistor is the right side.

The current in R_3 is $I_A - I_C$.

$$I_3 = 0.512\,\text{mA} - 0.432\,\text{mA} = \mathbf{0.08\,mA}$$

The current in R_4 is $I_B - I_C$.

$$I_4 = I_B - I_C = 0.887\,\text{mA} - 0.432\,\text{mA} = \mathbf{0.455\,mA}$$

The current in R_L is I_C.

$$I_L = \mathbf{0.432\,mA}$$

Related Problem Find the voltage across each resistor.

Use Multisim file E09-10 to verify the calculated results in this example and to confirm your calculations for the related problem.

SECTION 9–3 REVIEW

1. Do the loop currents necessarily represent the actual currents in the branches?
2. When you solve for a current using the loop method and get a negative value, what does it mean?
3. What circuit law is used in the loop current method?

9–4 NODE VOLTAGE METHOD

Another method of analysis of multiple-loop circuits is called the node voltage method. It is based on finding the voltages at each node in the circuit using Kirchhoff's current law. Recall that a node is the junction of two or more components.

After completing this section, you should be able to

◆ **Use node analysis to find unknown quantities in a circuit**

  ◆ Select the nodes at which the voltage is unknown and assign currents

  ◆ Apply Kirchhoff's current law at each node

  ◆ Develop and solve the node equations

The general steps for the node voltage method of circuit analysis are as follows:

Step 1. Determine the number of nodes.

Step 2. Select one node as a reference. All voltages will be relative to the reference node. Assign voltage designations to each node where the voltage is unknown.

Step 3. Assign currents at each node where the voltage is unknown, except at the reference node. The directions are arbitrary.

Step 4. Apply Kirchhoff's current law to each node where currents are assigned.

Step 5. Express the current equations in terms of voltages, and solve the equations for the unknown node voltages using Ohm's law.

▶ FIGURE 9–16

Circuit for node voltage analysis.

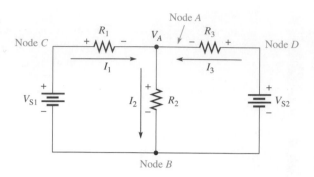

▶ FIGURE 9–16

Circuit for node voltage analysis.

We will use Figure 9–16 to illustrate the general approach to node voltage analysis. First, establish the nodes. In this case, there are four nodes, as indicated in the figure. Second, let's use node B as the reference. Think of it as the circuit's reference ground. Node voltages C and D are already known to be the source voltages. The voltage at node A is the only unknown; it is designated as V_A. Third, arbitrarily assign the branch currents at node A as indicated in the figure. Fourth, the Kirchhoff current equation at node A is

$$I_1 - I_2 + I_3 = 0$$

Fifth, express the currents in terms of circuit voltages using Ohm's law.

$$I_1 = \frac{V_1}{R_1} = \frac{V_{S1} - V_A}{R_1}$$

$$I_2 = \frac{V_2}{R_2} = \frac{V_A}{R_2}$$

$$I_3 = \frac{V_3}{R_3} = \frac{V_{S3} - V_A}{R_3}$$

Substituting these terms into the current equation yields

$$\frac{V_{S1} - V_A}{R_1} - \frac{V_A}{R_2} + \frac{V_{S2} - V_A}{R_3} = 0$$

The only unknown is V_A; so solve the single equation by combining and rearranging terms. Once you know the voltage, you can calculate all branch currents. Example 9–11 illustrates this method further.

EXAMPLE 9–11

Find the node voltage V_A in Figure 9–17 and determine the branch currents.

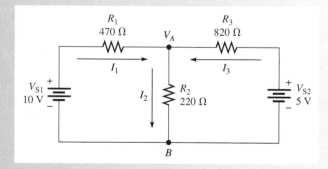

▲ FIGURE 9–17

Solution   The reference node is chosen at *B*. The unknown node voltage is V_A, as indicated in Figure 9–17. This is the only unknown voltage. Branch currents are assigned at node *A* as shown. The current equation is

$$I_1 - I_2 + I_3 = 0$$

Substitution for currents using Ohm's law gives the equation in terms of voltages.

$$\frac{10 - V_A}{470} - \frac{V_A}{220} + \frac{5 - V_A}{820} = 0$$

Rearranging the terms yields

$$\frac{10}{470} - \frac{V_A}{470} - \frac{V_A}{220} + \frac{5}{820} - \frac{V_A}{820} = 0$$

$$-\frac{V_A}{470} - \frac{V_A}{220} - \frac{V_A}{820} = -\frac{10}{470} - \frac{5}{820}$$

To solve for V_A, combine the terms on each side of the equation and find the common denominator.

$$\frac{1804V_A + 3854V_A + 1034V_A}{847880} = \frac{820 + 235}{38540}$$

$$\frac{6692V_A}{847880} = \frac{1055}{38540}$$

$$V_A = \frac{(1055)(847880)}{(6692)(38540)} = \textbf{3.47 V}$$

You can now determine the branch currents.

$$I_1 = \frac{10\text{ V} - 3.47\text{ V}}{470\text{ }\Omega} = \textbf{13.9 mA}$$

$$I_2 = \frac{3.47\text{ V}}{220\text{ }\Omega} = \textbf{15.8 mA}$$

$$I_3 = \frac{5\text{ V} - 3.47\text{ V}}{820\text{ }\Omega} = \textbf{1.87 mA}$$

These results agree with those for the same circuit in Example 9–7 and in Example 9 8 using the branch and loop current methods.

Related Problem   Find V_A in Figure 9–17 if the 5 V source is reversed.

Use Multisim file E09-11 to verify the calculated results in this example and to confirm your calculation for the related problem.

Example 9–11 illustrated an obvious advantage to the node method. The branch current method required three equations for the three unknown currents. The loop current method reduced the number of simultaneous equations but required the extra step of converting the fictitious loop currents to the actual currents in the resistors. The node method for the circuit in Figure 9–17 reduced the equations to one, in which all of the currents were written in terms of one unknown node voltage. The node voltage method also has the advantage of finding unknown voltages, which are easier to directly measure than current.

Node Voltage Method for a Wheatstone Bridge

The node voltage method can be applied to a Wheatstone bridge. The Wheatstone bridge is shown with nodes identified in Figure 9–18 with currents shown. Node *D* is usually selected as the reference node, and node A has the same potential as the source voltage.

Wheatstone bridge with node assignments.

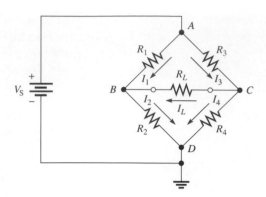

When setting up the equations for the two unknown node voltages (B and C), it is necessary to specify a current direction as described in the general steps. The direction of current in R_L is dependent on the bridge resistances; if the assigned direction is incorrect, it will show up as a negative current in the solution.

Kirchhoff's current law is then written for each of the unknown nodes. Each current is then expressed in terms of node voltages using Ohm's law as follows:

Node B:

$$I_1 + I_L = I_2$$

$$\frac{V_A - V_B}{R_1} + \frac{V_C - V_B}{R_L} = \frac{V_B}{R_2}$$

Node C:

$$I_3 = I_L + I_4$$

$$\frac{V_A - V_C}{R_3} = \frac{V_C - V_B}{R_L} + \frac{V_C}{R_4}$$

The equations are put in standard form and can be solved with any of the methods you have learned. The following example illustrates this for the Wheatstone bridge that was solved by loop equations in Example 9–9.

EXAMPLE 9–12

For the circuit in Figure 9–19, find the node voltages at node B and node C. Node D is the reference, and node A has the same voltage as the source. Use the results to calculate the current in each resistor. Compare the result to the loop current method in Example 9–9.

▶ FIGURE 9–19

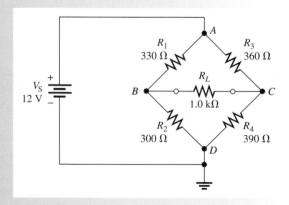

Solution   Apply Kirchhoff's current law at node B and node C in terms of node voltages. To keep coefficients more manageable, all resistances are shown in kΩ; current will be in mA.

Node B:

$$I_1 + I_L = I_2$$

$$\frac{V_A - V_B}{R_1} + \frac{V_C - V_B}{R_L} = \frac{V_B}{R_2}$$

$$\frac{12 - V_B}{0.330 \,\text{k}\Omega} + \frac{V_C - V_B}{1.0 \,\text{k}\Omega} = \frac{V_B}{0.300 \,\text{k}\Omega}$$

Node C:

$$I_3 = I_L + I_4$$

$$\frac{V_A - V_C}{R_3} = \frac{V_C - V_B}{R_L} + \frac{V_C}{R_4}$$

$$\frac{12\,\text{V} - V_C}{0.360 \,\text{k}\Omega} = \frac{V_C - V_B}{1.0 \,\text{k}\Omega} + \frac{V_C}{0.390 \,\text{k}\Omega}$$

Rearrange the equations for each node into standard form. For convenience, units are not shown until the end of the problem.

Node B: Multiply each term in the expression for node B by $R_1 R_2 R_L$ and combine like terms to get the standard form.

$$R_2 R_L (V_A - V_B) + R_1 R_2 (V_C - V_B) = R_1 R_L V_B$$

$$(1.0)(0.30)(12 - V_B) + (0.33)(0.30)(V_C - V_B) = (0.33)(1.0)V_B$$

$$0.729 V_B - 0.099 V_C = 3.6$$

Node C: Multiply each term in the expression for node C by $R_3 R_4 R_L$ and combine like terms to get the standard form.

$$R_4 R_L (V_A - V_C) = R_3 R_4 (V_C - V_B) + R_3 R_L V_C$$

$$(1.0)(0.39)(12 - V_C) = (0.36)(0.39)(V_C - V_B) + (0.36)(1.0)V_C$$

$$0.1404 V_B - 0.8904 V_C = -4.68$$

You can solve the two simultaneous equations using substitution, determinants, or the calculator. Solving by determinants,

$$0.729 V_B - 0.099 V_C = 3.6$$
$$0.1404 V_B - 0.8904 V_C = -4.68$$

$$V_B = \frac{\begin{vmatrix} 3.6 & -0.099 \\ -4.68 & -0.8904 \end{vmatrix}}{\begin{vmatrix} 0.729 & -0.099 \\ 0.1404 & -0.8904 \end{vmatrix}} = \frac{(3.6)(-0.8904) - (-0.099)(-4.68)}{(0.729)(-0.8904) - (0.1404)(-0.099)} = \textbf{5.78 V}$$

$$V_C = \frac{\begin{vmatrix} 0.729 & 3.6 \\ 0.1404 & -4.68 \end{vmatrix}}{\begin{vmatrix} 0.729 & -0.099 \\ 0.1404 & -0.8904 \end{vmatrix}} = \frac{(0.729)(-4.68) - (0.1404)(3.6)}{(0.729)(-0.8904) - (0.1404)(-0.099)} = \textbf{6.17 V}$$

Related Problem   Using Ohm's law, determine the current in each resistor.

Use Multisim file E09-12 to verify the calculated results in this example and to confirm your calculations for the related problem.

Node Voltage Method for the Bridged-T Circuit

Applying the node voltage method to the bridged-T circuit also results in two equations with two unknowns. As in the case of the Wheatstone bridge, there are four nodes as shown in Figure 9–20. Node D is the reference and node A is the source voltage, so the two unknown voltages are at nodes C and D. The effect of a load resistor on the circuit is usually the most important question, so the voltage at node C is the focus. A calculator solution of the simultaneous equations is simplified for analyzing the effect of various loads because only the equation for node C is affected when the load changes. Example 9–13 illustrates this idea.

▶ **FIGURE 9–20**

The bridged-T circuit with node assignments.

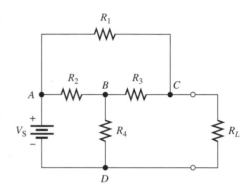

EXAMPLE 9–13

The circuit in Figure 9–21 is the same as in Example 9–10.

(a) Solve for the voltage across R_L using node analysis and a calculator.

(b) Find the effect on the load voltage when the load resistor is changed to 15 kΩ.

▶ **FIGURE 9–21**

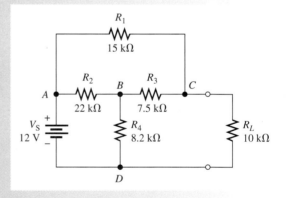

Solution (a) Apply Kirchhoff's current law at nodes B and C in terms of node voltages. To keep coefficients more manageable, all resistances are shown in kΩ. The current will be in mA.

Node B:

$$I_2 = I_3 + I_4$$

$$\frac{V_A - V_B}{R_2} = \frac{V_B - V_C}{R_3} + \frac{V_B}{R_4}$$

$$\frac{12 - V_B}{22 \text{ k}\Omega} = \frac{V_B - V_C}{7.5 \text{ k}\Omega} + \frac{V_B}{8.2 \text{ k}\Omega}$$

Node C:

$$I_1 + I_3 = I_L$$

$$\frac{V_A - V_C}{R_1} + \frac{V_B - V_C}{R_3} = \frac{V_C}{R_L}$$

$$\frac{12\,V - V_C}{15\,k\Omega} + \frac{V_B - V_C}{7.5\,k\Omega} = \frac{V_C}{10\,k\Omega}$$

Rearrange the equations for each node into standard form. For convenience, units are not shown until the end of the problem.

Node B: Multiply each term in the equation for node B by $R_2 R_3 R_4$ to cancel the denominator. Combine like terms to get the standard form.

$$R_3 R_4 (V_A - V_B) = R_2 R_4 (V_B - V_C) + R_2 R_3 V_B$$

$$(7.5)(8.2)(12 - V_B) = (22)(8.2)(V_B - V_C) + (22)(7.5)V_B$$

$$406.9V_B - 180.4V_C = 738$$

Node C: Multiply each term in the equation for node C by $R_1 R_3 R_L$ and combine like terms to get the standard form.

$$R_3 R_L (V_A - V_C) + R_1 R_L (V_B - V_C) = R_1 R_3 V_C$$

$$(7.5)(10)(12 - V_C) + (15)(10)(V_B - V_C) = (15)(7.5)V_C$$

$$150V_B - 337.5V_C = -900$$

Calculator Solution: The two equations in standard form are

$$406.9V_B - 180.4V_C = 738$$

$$150V_B - 337.5V_C = -900$$

Enter the number of equations (2), the coefficients, and the constants into a calculator to solve for V_B and V_C as shown in Figure 9–22. As a check, notice that this voltage implies that the load current is 0.432 mA, which is in agreement with the result found by the loop current method in Example 9–10.

FIGURE 9–22

```
406.9    -180.4    738
150      -337.5   -900
```
SOLVE
```
X1=VB=3.73
X2=VC=4.32
```

(b) To calculate the load voltage with a 15 kΩ load resistor, notice that the equation for node B is unaffected. The node C equation is modified as follows:

$$\frac{12\,V - V_C}{15\,k\Omega} + \frac{V_B - V_C}{7.5\,k\Omega} = \frac{V_C}{15\,k\Omega}$$

$$(7.5)(15)(12 - V_C) + (15)(15)(V_B - V_C) = (15)(7.5)V_C$$

$$225V_B - 450V_C = -1350$$

Change the parameters for the node C equation and press solve. The result is

$$V_C = V_L = \mathbf{5.02\ V}$$

Related Problem   For the 15 kΩ load, what is the voltage at node B?

Use Multisim file E09-13 to verify the calculated results in this example and to confirm your calculation for the related problem.

A Circuit Application

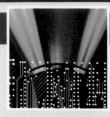

Dependent sources were covered in Chapter 8 and applied to the modeling of transistors and amplifiers. In this circuit application, you will see how a particular type of amplifier can be modeled and analyzed using the methods introduced in this chapter. The point is not to learn how amplifiers work because that is beyond the scope of this text and will be covered in a later course. The focus is on the application of circuit analysis methods to circuit models. The amplifier is simply used as an example to illustrate how you can apply analysis methods to a practical circuit.

Operational amplifiers are integrated circuit devices that are widely used in analog applications for signal processing. An operational amplifier symbol is shown in Figure 9–23(a). The equivalent dependent source model is shown in part (b). The gain (A) of the dependent source can be positive or negative, depending on how it is configured.

Assume you need to calculate in detail the effect of an operational amplifier circuit on a transducer that serves as an input. Some transducers, such as pH meters, appear as a small source voltage with a high series resistance. The transducer that is shown here is modeled as a small Thevenin dc voltage source in series with a Thevenin resistance of $10\,\text{k}\Omega$.

Practical amplifiers are created using an operational amplifier with external components. Figure 9–24(a) shows one type of amplifier configuration that includes the Thevenin resistance of the source along with two other external resistors. R_S represents the Thevenin source resistance. R_L is connected from the operational amplifier output to ground as a load, and R_F is connected from the output to the input as a feedback resistor. Feedback is used in most operational amplifier circuits and is simply a path from the output back to the input. It has many advantages as you will learn in a later course.

▶ FIGURE 9–23

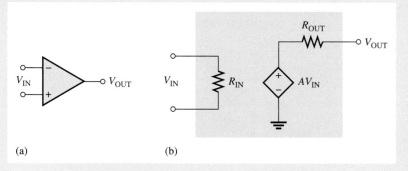

(a) (b)

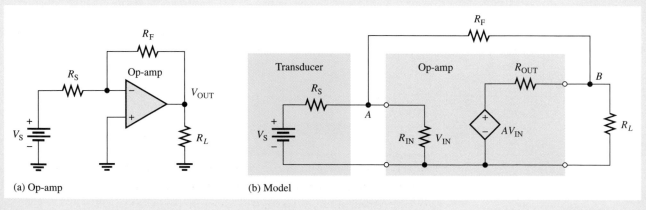

(a) Op-amp (b) Model

▲ FIGURE 9–24

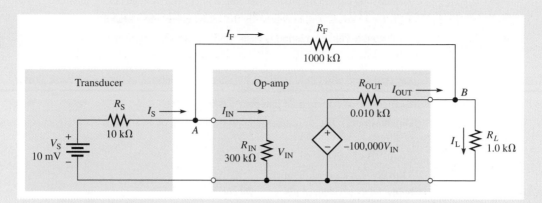

▲ **FIGURE 9–25**

Figure 9–24(b) is the equivalent circuit model of the source, the op-amp, and the load that we will use for analysis purposes. The internal gain of the dependent source, shown with the letter A on the op-amp block, is negative because it is an inverting amplifier (the output has the opposite sign of the input). This internal gain is generally extremely high. Although it is a very large number, the actual gain of the circuit with the external components is much lower because it is controlled by the external components rather than the internal gain.

The specific values for the circuit in this application are given in Figure 9–25 along with assigned currents. All values are shown in kΩ to simplify entering the coefficients in the equations that you will use. Although op-amp circuits have excellent simple approximations for determining the output voltage, there are times you may want to know the exact output. You can apply your knowledge from this chapter to the circuit to find the exact output voltage.

The amplifier model in Figure 9–25 can be analyzed easiest with the node voltage method because there are only two nodes, A and B, that have unknown voltages. At node A, the voltage is designated V_A and is the same as the input to the op-amp (V_{IN}). At node B, the voltage is designated V_B and also represents the output (or load voltage), V_L. Current names and directions are assigned as shown.

Analysis

Apply Kirchhoff's current law at each of the unknown nodes to write the node equations.

Node A: $I_S = I_F + I_{IN}$

Node B: $I_{OUT} + I_F = I_L$

Next, apply Ohm's law and let $V_{IN} = V_A$. The internal source voltage for the op-amp is AV_{IN}, so this is written as AV_A to express the unknowns in terms of V_A and V_B.

Node A: $\dfrac{V_S - V_A}{R_S} = \dfrac{V_A}{R_{IN}} + \dfrac{V_A - V_B}{R_F}$

Node B: $\dfrac{AV_A - V_B}{R_{OUT}} + \dfrac{V_A - V_B}{R_F} = \dfrac{V_B}{R_L}$

Expressing the equations in standard form,

Node A: $-\left(\dfrac{1}{R_S} + \dfrac{1}{R_{IN}} + \dfrac{1}{R_F}\right)V_A + \left(\dfrac{1}{R_F}\right)V_B = -\left(\dfrac{1}{R_S}\right)V_S$

Node B: $-\left(\dfrac{A}{R_{OUT}} + \dfrac{1}{R_F}\right)V_A + \left(\dfrac{1}{R_L} + \dfrac{1}{R_{OUT}} + \dfrac{1}{R_F}\right)V_B = 0$

◆ Substitute the values that were given in Figure 9–25 into the standard form equations. Solve the equations to find V_{IN} and V_L. (Resistance can be entered in kΩ.)

◆ Calculate the input current, I_{IN}, and the current in the feedback resistor, I_F.

Review

1. Does the output voltage change if the load resistor R_L is doubled?

2. Does the output voltage change if the feedback resistor R_F is doubled?

SUMMARY

◆ Simultaneous equations can be solved by substitution, by determinants, or by a graphics calculator.

◆ The number of equations must be equal to the number of unknowns.

◆ Second-order determinants are evaluated by adding the signed cross-products.

◆ Third-order determinants are evaluated by the expansion method.

◆ The branch current method is based on Kirchhoff's voltage law and Kirchhoff's current law.

◆ The loop current method is based on Kirchhoff's voltage law.

◆ A loop current is not necessarily the actual current in a branch.

◆ The node voltage method is based on Kirchhoff's current law.

KEY TERMS

Key terms and other bold terms in the chapter are defined in the end-of-book glossary.

Branch   One current path that connects two nodes.

Determinant   The solution of a matrix consisting of an array of coefficients and constants for a set of simultaneous equations.

Loop   A closed current path in a circuit.

Matrix   An array of numbers.

Node   The junction of two or more components.

Simultaneous equations   A set of n equations containing n unknowns, where n is a number with a value of 2 or more.

SELF-TEST

Answers are at the end of the chapter.

1. Assuming the voltage source values in Figure 9–6 are known, there is/are
   (a) 3 nonredundant loops   (b) 1 unknown node   (c) 2 nonredundant loops
   (d) 2 unknown nodes   (e) both answers (b) and (c)

2. In assigning the direction of branch currents,
   (a) the directions are critical   (b) they must all be in the same direction
   (c) they must all point into a node   (d) the directions are not critical

3. The branch current method uses
   (a) Ohm's law and Kirchhoff's voltage law
   (b) Kirchhoff's voltage and current laws
   (c) the superposition theorem and Kirchhoff's current law
   (d) Thevenin's theorem and Kirchhoff's voltage law

4. A characteristic determinant for two simultaneous equations will have
   (a) 2 rows and 1 column   (b) 1 row and 2 columns
   (c) 2 rows and 2 columns

5. The first row of a certain determinant has the numbers 2 and 4. The second row has the numbers 6 and 1. The value of this determinant is
   (a) 22   (b) 2   (c) −22   (d) 8

6. The expansion method for evaluating determinants is
   (a) good only for second-order determinants
   (b) good only for both second and third-order determinants
   (c) good for any determinant
   (d) easier than using a calculator

7. The loop current method is based on
   (a) Kirchhoff's current law   (b) Ohm's law
   (c) the superposition theorem   (d) Kirchhoff's voltage law

8. The node voltage method is based on
   (a) Kirchhoff's current law   (b) Ohm's law
   (c) the superposition theorem   (d) Kirchhoff's voltage law

9. In the node voltage method,
   (a) currents are assigned at each node
   (b) currents are assigned at the reference node
   (c) the current directions are arbitrary

(d) currents are assigned only at the nodes where the voltage is unknown

(e) both answers (c) and (d)

10. Generally, the node voltage method results in

(a) more equations than the loop current method

(b) fewer equations than the loop current method

(c) the same number of equations as the loop current method

CIRCUIT DYNAMICS QUIZ

Answers are at the end of the chapter.

Refer to Figure 9–26.

1. If R_2 opens, the current through R_3

(a) increases   (b) decreases   (c) stays the same

2. If the 6 V source shorts out, the voltage at point A with respect to ground

(a) increases   (b) decreases   (c) stays the same

3. If R_2 becomes disconnected from ground, the voltage at point A with respect to ground

(a) increases   (b) decreases   (c) stays the same

Refer to Figure 9–27.

4. If the current source fails open, the current through R_2

(a) increases   (b) decreases   (c) stays the same

5. If R_2 opens, the current through R_3

(a) increases   (b) decreases   (c) stays the same

Refer to Figure 9–30.

6. If R_1 opens, the magnitude of the voltage between the A and B terminals

(a) increases   (b) decreases   (c) stays the same

7. If R_3 is replaced by a 10 Ω resistor, V_{AB}

(a) increases   (b) decreases   (c) stays the same

8. If point B shorts to the negative side of the source, V_{AB}

(a) increases   (b) decreases   (c) stays the same

9. If the negative side of the source is grounded, V_{AB}

(a) increases   (b) decreases   (c) stays the same

Refer to Figure 9–32.

10. If a voltage source V_{S2} fails open, the voltage at A with respect to ground

(a) increases   (b) decreases   (c) stays the same

11. If a short develops from point A to ground, the current through R_3

(a) increases   (b) decreases   (c) stays the same

12. If R_2 opens, the voltage across R_3

(a) increases   (b) decreases   (c) stays the same

PROBLEMS

More difficult problems are indicated by an asterisk (*).
Answers to odd-numbered problems are at the end of the book.

SECTION 9–1   Simultaneous Equations in Circuit Analysis

1. Using the substitution method, solve the following set of equations for I_{R1} and I_{R2}.

$$100I_1 + 50I_2 = 30$$
$$75I_1 + 90I_2 = 15$$

2. Evaluate each determinant:

(a) $\begin{vmatrix} 4 & 6 \\ 2 & 3 \end{vmatrix}$ (b) $\begin{vmatrix} 9 & -1 \\ 0 & 5 \end{vmatrix}$ (c) $\begin{vmatrix} 12 & 15 \\ -2 & -1 \end{vmatrix}$ (d) $\begin{vmatrix} 100 & 50 \\ 30 & -20 \end{vmatrix}$

3. Using determinants, solve the following set of equations for both currents:

$$-I_1 + 2I_2 = 4$$
$$7I_1 + 3I_2 = 6$$

4. Evaluate each of the determinants:

(a) $\begin{vmatrix} 1 & 0 & -2 \\ 5 & 4 & 1 \\ 2 & 10 & 0 \end{vmatrix}$ (b) $\begin{vmatrix} 0.5 & 1 & -0.8 \\ 0.1 & 1.2 & 1.5 \\ -0.1 & -0.3 & 5 \end{vmatrix}$

5. Evaluate each of the determinants:

(a) $\begin{vmatrix} 25 & 0 & -20 \\ 10 & 12 & 5 \\ -8 & 30 & -16 \end{vmatrix}$ (b) $\begin{vmatrix} 1.08 & 1.75 & 0.55 \\ 0 & 2.12 & -0.98 \\ 1 & 3.49 & -1.05 \end{vmatrix}$

6. Find I_3 in Example 9–4.

7. Solve for I_1, I_2, I_3 in the following set of equations using determinants:

$$2I_1 - 6I_2 + 10I_3 = 9$$
$$3I_1 + 7I_2 - 8I_3 = 3$$
$$10I_1 + 5I_2 - 12I_3 = 0$$

*8. Find V_1, V_2, V_3, and V_4 from the following set of equations using the calculator:

$$16V_1 + 10V_2 - 8V_3 - 3V_4 = 15$$
$$2V_1 + 0V_2 + 5V_3 + 2V_4 = 0$$
$$-7V_1 - 12V_2 + 0V_3 + 0V_4 = 9$$
$$-1V_1 + 20V_2 - 18V_3 + 0V_4 = 10$$

9. Solve the two simultaneous equations in Problem 1 using your calculator.

10. Solve the three simultaneous equations in Problem 7 using your calculator.

SECTION 9–2   **Branch Current Method**

11. Write the Kirchhoff current equation for the current assignment shown at node A in Figure 9–26.

12. Solve for each of the branch currents in Figure 9–26.

13. Find the voltage drop across each resistor in Figure 9–26 and indicate its actual polarity.

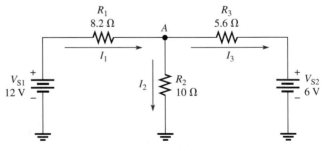

▲ FIGURE 9–26

*14. Find the current through each resistor in Figure 9–27.

15. In Figure 9–27, determine the voltage across the current source (points A and B).

▶ FIGURE 9–27

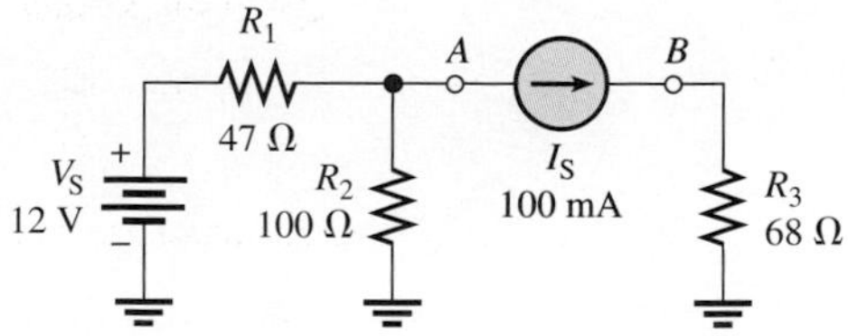

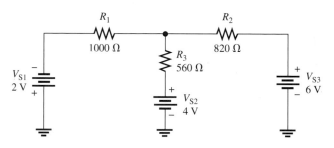

SECTION 9–3 Loop Current Method

16. Write the characteristic determinant for the equations:

$$0.045I_A + 0.130I_B + 0.066I_C = 0$$
$$0.177I_A + 0.0420I_B + 0.109I_C = 12$$
$$0.078I_A + 0.196I_B + 0.029I_C = 3.0$$

17. Using the loop current method, find the loop currents in Figure 9–28.

18. Find the branch currents in Figure 9–28.

19. Determine the voltages and their proper polarities for each resistor in Figure 9–28.

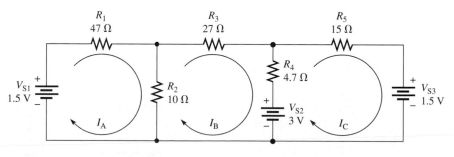

▲ FIGURE 9–28

20. Write the loop equations for the circuit in Figure 9–29.

21. Solve for the loop currents in Figure 9–29 using your calculator.

22. Find the current through each resistor in Figure 9–29.

▲ FIGURE 9–29

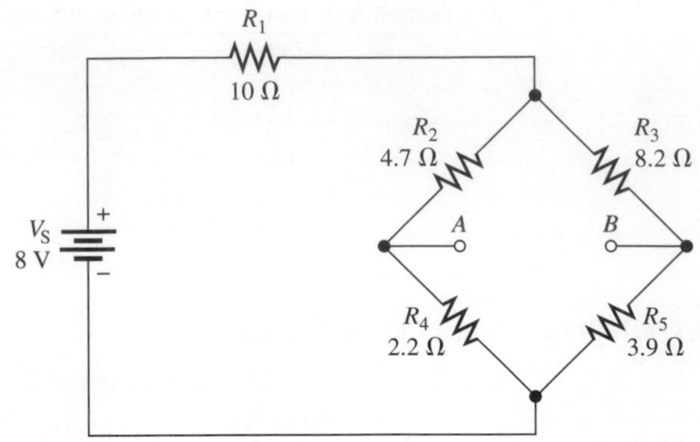

▲ FIGURE 9–30

23. Determine the voltage across the open bridge terminals, A and B, in Figure 9–30.

24. When a 10 Ω resistor is connected from terminal A to terminal B in Figure 9–30, what is the current through it?

25. Write the loop equations in standard form for the bridged-T circuit in Figure 9–31.

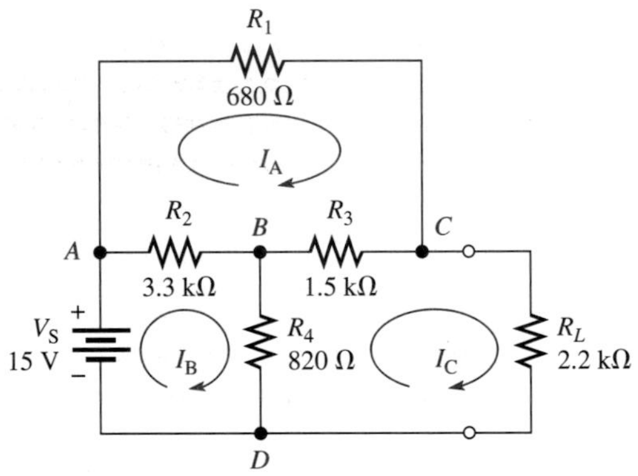

▲ FIGURE 9–31

SECTION 9–4 Node Voltage Method

26. In Figure 9–32, use the node voltage method to find the voltage at point A with respect to ground.

27. What are the branch current values in Figure 9–32? Show the actual direction of current in each branch.

28. Write the node voltage equations for Figure 9–29. Use your calculator to find the node voltages.

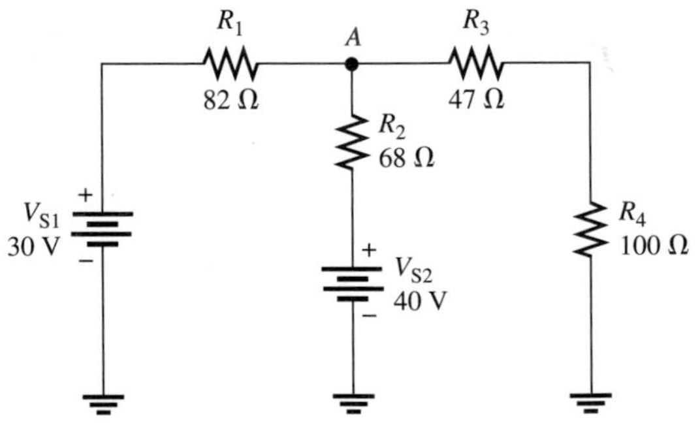

▲ FIGURE 9–32

29. Use node analysis to determine the voltage at points *A* and *B* with respect to ground in Figure 9–33.

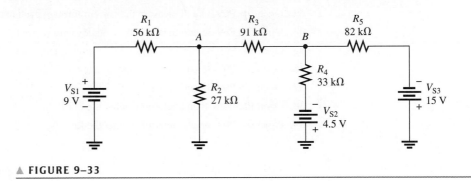

▲ FIGURE 9–33

*30. Find the voltage at points *A*, *B*, and *C* in Figure 9–34.

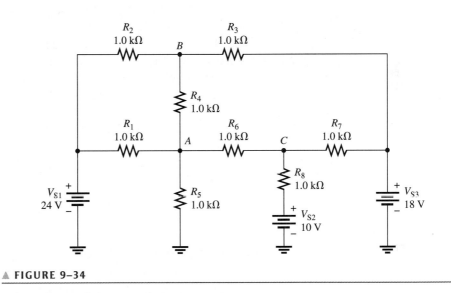

▲ FIGURE 9–34

*31. Use node analysis, loop analysis, or any other procedure to find all currents and the voltages at each unknown node in Figure 9–35.

► FIGURE 9–35

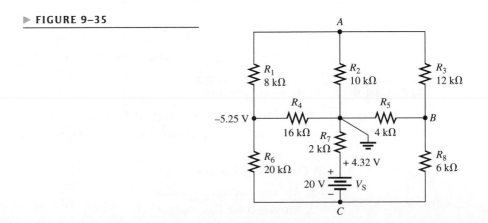

Multisim Troubleshooting and Analysis

These problems require your Multisim CD-ROM.

32. Open file P09-32 and measure the current through each resistor.

33. Open file P09-33 and measure the current through each resistor.

34. Open file P09-34 and measure the voltages with respect to ground at nodes *A* and *B*.

35. Open file P09-35. Determine if there is a fault and, if so, specify the fault.

36. Open file P09-36 and measure the voltages with respect to ground at output terminals 1 and 2.

37. Open file P09-37 and determine what the fault is.

38. Open file P09-38 and determine what the fault is.

39. Open file P09-39 and determine what the fault is.

ANSWERS

SECTION REVIEWS

SECTION 9–1   Simultaneous Equations in Circuit Analysis

1. (a) 4   **(b)** 0.415   **(c)** −98

2. $\begin{vmatrix} 2 & 3 \\ 5 & 4 \end{vmatrix}$

3. −0.286 A = −286 mA

4. $I_1 = -.038893513289$
   $I_2 = .084110232475$
   $I_3 = .041925798204$
   $I_4 = -.067156192401$

5. $I_1 = -.056363148617$
   $I_2 = .07218287729$
   $I_3 = .065684612774$
   $I_4 = -.041112571034$

SECTION 9–2   Branch Current Method

1. Kirchhoff's voltage law and Kirchhoff's current law are used in the branch current method.

2. False, but write the equations so that they are consistent with your assigned directions.

3. A loop is a closed path within a circuit.

4. A node is a junction of two or more components.

SECTION 9–3   Loop Current Method

1. No, loop currents are not necessarily the same as branch currents.

2. A negative value means the direction should be reversed.

3. Kirchhoff's voltage law is used in loop analysis.

SECTION 9–4   Node Voltage Method

1. Kirchhoff's current law is the basis for node analysis.

2. A reference node is the junction to which all circuit voltages are referenced.

A Circuit Application

1. The output voltage is unaffected.

2. The output voltage is doubled.

RELATED PROBLEMS FOR EXAMPLES

9–1 $20x_1 - 11x_2 = -15$
$18x_1 + 25x_2 = 10$

9–2 $10V_1 - 21V_2 - 50V_3 = -15$
$18V_1 + 25V_2 - 12V_3 = 10$
$18V_1 - 25V_2 + 12V_3 = 9$

9–3   3.71 A

9–4   −298 mA

9–5 $X_1 = -1.76923076923$; $X_2 = -18.5384615385$; $X_3 = -34.4615384615$

9–6   Same results as answer for 9–5

9–7 $I_1 = 17.2$ mA; $I_2 = 8.74$ mA; $I_3 = -8.44$ mA

9–8 $I_1 = X_1 = .013897190675$ (≈ 13.9 mA); $I_2 = X_2 = -.001867901972$ (≈ -1.87 mA)

9–9   They are correct.

9–10 $V_1 = 7.68$ V, $V_2 = 8.25$ V, $V_3 = 0.6$ V, $V_4 = 3.73$ V, $V_L = 4.32$ V

9–11   1.92 V

9–12 $I_1 = 18.8$ mA, $I_2 = 19.3$ mA, $I_3 = 16.2$ mA, $I_4 = 15.8$ mA, $I_L = 0.39$ mA

9–13 $V_B = 4.04$ V

SELF-TEST

1. (e)   **2.** (d)   **3.** (b)   **4.** (c)   **5.** (c)

6. (b)   **7.** (d)   **8.** (a)   **9.** (e)   **10.** (b)

CIRCUIT DYNAMICS QUIZ

1. (a)   **2.** (b)   **3.** (a)   **4.** (a)   **5.** (c)   **6.** (b)

7. (a)   **8.** (a)   **9.** (c)   **10.** (b)   **11.** (b)   **12.** (b)

10

MAGNETISM AND ELECTROMAGNETISM

CHAPTER OBJECTIVES

◆ Explain the principles of a magnetic field
◆ Explain the principles of electromagnetism
◆ Describe the principle of operation for several types of electromagnetic devices
◆ Explain magnetic hysteresis
◆ Discuss the principle of electromagnetic induction
◆ Describe some applications of electromagnetic induction

KEY TERMS

◆ Magnetic field
◆ Lines of force
◆ Magnetic flux
◆ Weber (Wb)
◆ Tesla
◆ Electromagnetism
◆ Electromagnetic field
◆ Permeability
◆ Reluctance
◆ Magnetomotive force (mmf)
◆ Ampere-turn (At)

◆ Solenoid
◆ Relay
◆ Speaker
◆ Hysteresis
◆ Retentivity
◆ Induced voltage (v_{ind})
◆ Induced current (i_{ind})
◆ Electromagnetic induction
◆ Faraday's law
◆ Lenz's law

A CIRCUIT APPLICATION PREVIEW

In the circuit application, you will learn how electromagnetic relays can be used in burglar alarm systems, and you will develop a procedure to check out a basic alarm system.

VISIT THE COMPANION WEBSITE

Study aids for this chapter are available at
http://www.prenhall.com/floyd

INTRODUCTION

This chapter departs from the coverage of dc circuits and introduces the concepts of magnetism and electromagnetism. The operation of devices such as the relay, the solenoid, and the speaker is based partially on magnetic or electromagnetic principles. Electromagnetic induction is important in an electrical component called an inductor or coil, which is the topic in Chapter 13.

Two types of magnets are the permanent magnet and the electromagnet. The permanent magnet maintains a constant magnetic field between its two poles with no external excitation. The electromagnet produces a magnetic field only when there is current through it. The electromagnet is basically a coil of wire wound around a magnetic core material.

10–1 THE MAGNETIC FIELD

A permanent magnet has a magnetic field surrounding it. The **magnetic field** consists of **lines of force** that radiate from the north pole (N) to the south pole (S) and back to the north pole through the magnetic material.

After completing this section, you should be able to

- ◆ **Explain the principles of a magnetic field**
  - ◆ Define *magnetic flux*
  - ◆ Define *magnetic flux density*
  - ◆ Discuss how materials are magnetized
  - ◆ Explain how a magnetic switch works

A permanent magnet, such as the bar magnet shown in Figure 10–1, has a magnetic field surrounding it that consists of lines of force, or flux lines. For clarity, only a few lines of force are shown in the figure. Imagine, however, that many lines surround the magnet in three dimensions. The lines shrink to the smallest possible size and blend together, although they do not touch. This effectively forms a continuous magnetic field surrounding the magnet.

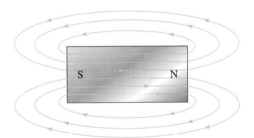

◀ **FIGURE 10–1**

Magnetic lines of force around a bar magnet.

Blue lines represent only a few of the many magnetic lines of force in the magnetic field.

When unlike poles of two permanent magnets are placed close together, their magnetic fields produce an attractive force, as indicated in Figure 10–2(a). When two like poles are brought close together, they repel each other, as shown in part (b).

When a nonmagnetic material such as paper, glass, wood, or plastic is placed in a magnetic field, the lines of force are unaltered, as shown in Figure 10–3(a). However, when a magnetic material such as iron is placed in the magnetic field, the lines of force tend to change course and pass through the iron rather than through the surrounding air. They do so because the iron provides a magnetic path that is more easily established than that of air. Figure 10–3(b) illustrates this principle. The fact that magnetic lines of force follow a path through iron or other materials is a consideration in the design of shields that prevent stray magnetic fields from affecting sensitive circuits.

Magnetic Flux (ϕ)

The group of force lines going from the north pole to the south pole of a magnet is called the **magnetic flux**, symbolized by ϕ (the Greek letter phi). The number of lines of force in a magnetic field determines the value of the flux. The more lines of force, the greater the flux and the stronger the magnetic field.

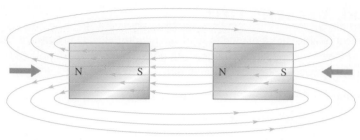

(a) Unlike poles attract.

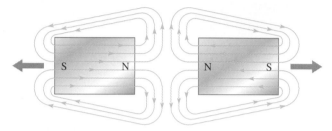

(b) Like poles repel.

▲ **FIGURE 10–2**

Magnetic attraction and repulsion.

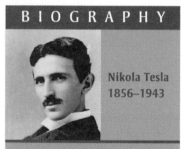
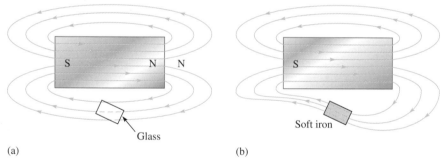

(a)   (b)

▲ **FIGURE 10–3**

Effect of (a) nonmagnetic and (b) magnetic materials on a magnetic field.

The unit of magnetic flux is the **weber (Wb)**. One weber equals 10^8 lines. The weber is a very large unit; thus, in most practical situations, the microweber (μWb) is used. One microweber equals 100 lines of magnetic flux.

Magnetic Flux Density (*B*)

The **magnetic flux density** is the amount of flux per unit area perpendicular to the magnetic field. Its symbol is B, and its SI unit is the **tesla (T)**. One tesla equals one weber per square meter (Wb/m^2). The following formula expresses the flux density:

Equation 10–1

$$B = \frac{\phi}{A}$$

where ϕ is the flux (Wb) and A is the cross-sectional area in square meters (m^2) of the magnetic field.

EXAMPLE 10–1

Compare the flux and the flux density in the two magnetic cores shown in Figure 10–4. The diagram represents the cross section of a magnetized material. Assume that each dot represents 100 lines or 1 μWb.

FIGURE 10–4

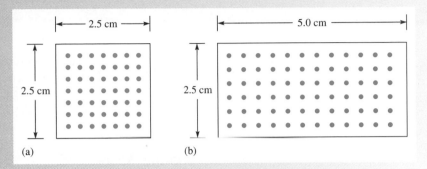

(a)   (b)

Solution   The flux is simply the number of lines. In Figure 10–4(a) there are 49 dots. Each represents 1 μWb, so the flux is 49 μWb. In Figure 10–4(b) there are 72 dots, so the flux is 72 μWb.

To calculate the flux density in Figure 10–4(a), first calculate the area in m^2.

$$A = l \times w = 0.025 \,\text{m} \times 0.025 \,\text{m} = 6.25 \times 10^{-4} \,\text{m}^2$$

For Figure 10–4(b) the area is

$$A = l \times w = 0.025 \,\text{m} \times 0.050 \,\text{m} = 1.25 \times 10^{-3} \,\text{m}^2$$

Use Equation 10–1 to calculate the flux density. For Figure 10–4(a) the flux density is

$$B = \frac{\phi}{A} = \frac{49 \,\mu\text{Wb}}{6.25 \times 10^{-4} \,\text{m}^2} = 78.4 \times 10^{-3} \,\text{Wb/m}^2 = 78.4 \times 10^{-3} \,\text{T}$$

For Figure 10–4(b) the flux density is

$$B = \frac{\phi}{A} = \frac{72 \,\mu\text{Wb}}{1.25 \times 10^{-3} \,\text{m}^2} = 57.6 \times 10^{-3} \,\text{Wb/m}^2 = 57.6 \times 10^{-3} \,\text{T}$$

The data in Table 10–1 compares the two cores. Note that the core with the largest flux does not necessarily have the highest flux density.

TABLE 10–1

	FLUX (Wb)	AREA (m^2)	FLUX DENSITY (T)
Figure 10–4(a)	49 μWb	$6.25 \times 10^{-4} \,\text{m}^2$	$78.4 \times 10^{-3} \,\text{T}$
Figure 10–4(b)	72 μWb	$1.25 \times 10^{-3} \,\text{m}^2$	$57.6 \times 10^{-3} \,\text{T}$

*Related Problem**   What happens to the flux density if the same flux shown in Figure 10–4(a) is in a core that is 5.0 cm × 5.0 cm?

*Answers are at the end of the chapter.

EXAMPLE 10–2

If the flux density in a certain magnetic material is 0.23 T and the area of the material is 0.38 in.^{2}, what is the flux through the material?

Solution   First, 0.38 in.^{2} must be converted to square meters. 39.37 in. = 1 m; therefore,

$$A = 0.38 \,\text{in.}^2[1 \,\text{m}^2/(39.37 \,\text{in.})^2] = 245 \times 10^{-6} \,\text{m}^2$$

The flux through the material is

$$\phi = BA = (0.23 \text{ T})(245 \times 10^{-6} \text{ m}^2) = \mathbf{56.4 \ \mu Wb}$$

Related Problem   Calculate B if $A = 0.05$ in.^{2} and $\phi = 1000 \ \mu$Wb.

The Gauss   Although the tesla (T) is the SI unit for flux density, another unit called the **gauss**, from the CGS (centimeter-gram-second) system, is occasionally used (10^4 gauss $= 1$ T). In fact, the instrument used to measure flux density is the gaussmeter. The gauss is a convenient unit for small magnetic fields such as the earth's field, which ranges from 0.3–0.6 gauss, depending on location.

How Materials Become Magnetized

Ferromagnetic materials such as iron, nickel, and cobalt become magnetized when placed in the magnetic field of a magnet. We have all seen a permanent magnet pick up things like paper clips, nails, and iron filings. In these cases, the object becomes magnetized (that is, it actually becomes a magnet itself) under the influence of the permanent magnetic field and becomes attracted to the magnet. When removed from the magnetic field, the object tends to lose its magnetism.

Ferromagnetic materials have minute magnetic domains created within their atomic structure. These domains can be viewed as very small bar magnets with north and south poles. When the material is not exposed to an external magnetic field, the magnetic domains are randomly oriented, as shown in Figure 10–5(a). When the material is placed in a magnetic field, the domains align themselves as shown in part (b). Thus, the object itself effectively becomes a magnet.

(a) The magnetic domains (N ◄▬ S) are randomly oriented in the unmagnetized material.

(b) The magnetic domains become aligned when the material is magnetized.

▲ **FIGURE 10–5**

Magnetic domains in (a) an unmagnetized and in (b) a magnetized material.

Applications   Permanent magnets are used in switches, such as the normally closed (NC) magnetic switch. When a magnet is near the switch mechanism as in Figure 10–6(a), the

▶ **FIGURE 10–6**

Operation of a magnetic switch.

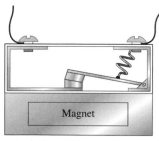

(a) Contact is closed when magnet is near.

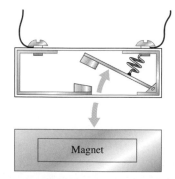

(b) Contact opens when magnet is moved away.

metallic arm is held in its NC position. When the magnet is moved away, the spring pulls the arm up, which breaks the contact, as shown in part (b).

Switches of this type are commonly used in perimeter alarm systems to detect entry into a building through windows or doors. As Figure 10–7 shows, several openings can be protected by magnetic switches wired to a common transmitter. When any one of the switches opens, the transmitter is activated and sends a signal to a central receiver and alarm unit.

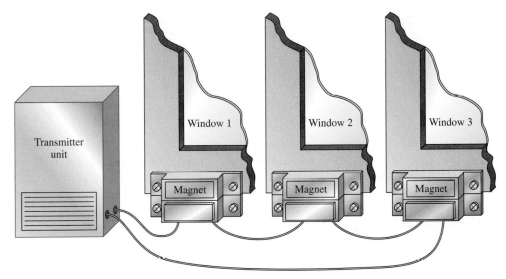

◀ **FIGURE 10–7**

Connection of a typical perimeter alarm system.

SECTION 10–1 REVIEW Answers are at the end of the chapter.	1. When the north poles of two magnets are placed close together, do they repel or attract each other?
	2. What is magnetic flux?
	3. What is the flux density when $\phi = 4.5 \ \mu$Wb and $A = 5 \times 10^{-3} \ \text{m}^2$?

10–2 ELECTROMAGNETISM

Electromagnetism is the production of a magnetic field by current in a conductor.

After completing this section, you should be able to

♦ **Explain the principles of electromagnetism**

  ♦ Determine the direction of the magnetic lines of force

  ♦ Define *permeability*

  ♦ Define *reluctance*

  ♦ Define *magnetomotive force*

  ♦ Describe a basic electromagnet

Current produces a magnetic field, called an **electromagnetic field**, around a conductor, as illustrated in Figure 10–8. The invisible lines of force of the magnetic field form a concentric circular pattern around the conductor and are continuous along its length. Unlike the bar magnet, the magnetic field surrounding a wire does not have a north or south

► FIGURE 10–8

Magnetic field around a current-carrying conductor. The red arrows indicate the direction of conventional (+ to −) current.

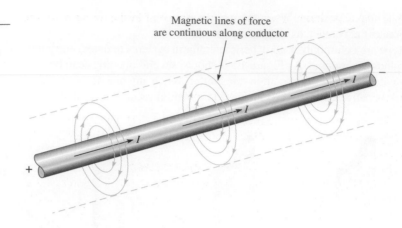

Magnetic lines of force are continuous along conductor

pole. The direction of the lines of force surrounding the conductor shown in the figure is for conventional current. The lines are in a clockwise direction. When current is reversed, the magnetic field lines are in a counterclockwise direction.

Although the magnetic field cannot be seen, it is capable of producing visible effects. For example, if a current-carrying wire is inserted through a sheet of paper in a perpendicular direction, iron filings placed on the surface of the paper arrange themselves along the magnetic lines of force in concentric rings, as illustrated in Figure 10–9(a). Part (b) of the figure illustrates that the north pole of a compass placed in the electromagnetic field will point in the direction of the lines of force. The field is stronger closer to the conductor and becomes weaker with increasing distance from the conductor.

► FIGURE 10–9

Visible effects of an electromagnetic field.

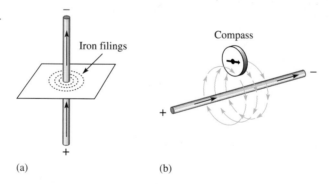

Iron filings

Compass

(a)   (b)

Right-Hand Rule An aid to remembering the direction of the lines of force is illustrated in Figure 10–10. Imagine that you are grasping the conductor with your right hand, with

► FIGURE 10–10

Illustration of right-hand rule. The right-hand rule is used for conventional current (+ to −).

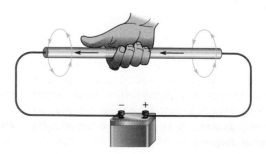

your thumb pointing in the direction of current. Your fingers point in the direction of the magnetic lines of force.

Electromagnetic Properties

Several important properties are related to electromagnetic fields.

Permeability (μ)   The ease with which a magnetic field can be established in a given material is measured by the **permeability** of that material. The higher the permeability, the more easily a magnetic field can be established.

The symbol of permeability is μ (the Greek letter mu), and its value varies depending on the type of material. The permeability of a vacuum (μ_0) is $4\pi \times 10^{-7}$ Wb/At·m (webers/ampere-turn·meter) and is used as a reference. Ferromagnetic materials typically have permeabilities hundreds of times larger than that of a vacuum, indicating that a magnetic field can be set up with relative ease in these materials. Ferromagnetic materials include iron, steel, nickel, cobalt, and their alloys.

The *relative permeability* (μ_r) of a material is the ratio of its absolute permeability to the permeability of a vacuum.

$$\mu_r = \frac{\mu}{\mu_0}$$

Equation 10–2

Because it is a ratio of permeabilities, μ_r is dimensionless. Typical magnetic materials, such as iron, have a relative permeability of a few hundred. Highly permeable materials can have a relative permeability as high as 100,000.

Reluctance ($\mathcal{R}$)   The opposition to the establishment of a magnetic field in a material is called **reluctance**. The value of reluctance is directly proportional to the length (l) of the magnetic path and inversely proportional to the permeability (μ) and to the cross-sectional area (A) of the material, as expressed by the following equation:

$$\mathcal{R} = \frac{l}{\mu A}$$

Equation 10–3

Reluctance in magnetic circuits is analogous to resistance in electric circuits. The unit of reluctance can be derived using l in meters, A (area) in square meters, and μ in Wb/At·m as follows:

$$\mathcal{R} = \frac{l}{\mu A} = \frac{\not{m}}{(\text{Wb/At}\cdot\not{m})(\not{m}^2)} = \frac{\text{At}}{\text{Wb}}$$

At/Wb is ampere-turns/weber.

Equation 10–3 is similar to Equation 2–6 for determining wire resistance. Recall that Equation 2–6 is

$$R = \frac{\rho l}{A}$$

The reciprocal of resistivity (ρ) is conductivity (σ). By substituting $1/\sigma$ for ρ, Equation 2–6 can be written as

$$R = \frac{l}{\sigma A}$$

Compare this last equation for wire resistance with Equation 10–3. The length (l) and the area (A) have the same meaning in both equations. The conductivity (σ) in electrical circuits is analogous to permeability (μ) in magnetic circuits. Also, resistance (R) in electric circuits is analogous to reluctance ($\mathcal{R}$) in magnetic circuits; both are oppositions. Typically, the reluctance of a magnetic circuit is 50,000 At/Wb or more, depending on the size and type of material.

EXAMPLE 10–3

Calculate the reluctance of a torus (a doughnut-shaped core) made of low-carbon steel. The inner radius of the torus is 1.75 cm and the outer radius of the torus is 2.25 cm. Assume the permeability of low-carbon steel is 2×10^{-4} Wb/At·m.

Solution

You must convert centimeters to meters before you calculate the area and length. From the dimensions given, the thickness (diameter) is 0.5 cm = 0.005 m. Thus, the cross-sectional area is

$$A = \pi r^2 = \pi(0.0025)^2 = 1.96 \times 10^{-5}\, m^2$$

The length is equal to the circumference of the torus measured at the average radius of 2.0 cm or 0.020 m.

$$l = C = 2\pi r = 2\pi(0.020\,m) = 0.125\,m$$

Substituting values into Equation 10–3,

$$\mathcal{R} = \frac{l}{\mu A} = \frac{0.125\,m}{(2 \times 10^{-4}\,Wb/At\cdot m)(1.96 \times 10^{-5}\,m^2)} = \textbf{31.9} \times \textbf{10}^{\textbf{6}}\ \textbf{At/Wb}$$

Related Problem

What happens to the reluctance if cast steel with a permeability of 5×10^{-4} Wb/At·m is substituted for the cast iron core?

EXAMPLE 10–4

Mild steel has a relative permeability of 800. Calculate the reluctance of a mild steel core that has a length of 10 cm and has a cross section of 1.0 cm × 1.2 cm.

Solution

First, determine the permeability of mild steel.

$$\mu = \mu_0 \mu_r = (4\pi \times 10^{-7}\,Wb/At\cdot m)(800) = 1.00 \times 10^{-3}\,Wb/At\cdot m$$

Next, convert the length to meters and the area to square meters.

$$l = 10\,cm = 0.10\,m$$
$$A = 0.010\,m \times 0.012\,m = 1.2 \times 10^{-4}\,m^2$$

Substituting values into Equation 10–3,

$$\mathcal{R} = \frac{l}{\mu A} = \frac{0.10\,m}{(1.00 \times 10^{-3}\,Wb/At\cdot m)(1.2 \times 10^{-4}\,m^2)} = \textbf{8.33} \times \textbf{10}^{\textbf{5}}\ \textbf{At/Wb}$$

Related Problem

What happens to the reluctance if the core is made from 78 Permalloy with a relative permeability of 4000?

Magnetomotive Force (mmf)   As you have learned, current in a conductor produces a magnetic field. The cause of a magnetic field is called the **magnetomotive force (mmf)**. Magnetomotive force is something of a misnomer because in a physics sense it is not really a force, rather a direct result of the movement of charge (current). The unit of mmf, the **ampere-turn (At)**, is established on the basis of the current in a single loop (turn) of wire. The formula for mmf is

Equation 10–4

$$F_m = NI$$

where F_m is the magnetomotive force, N is the number of turns of wire, and I is the current in amperes.

Figure 10–11 illustrates that a number of turns of wire carrying a current around a magnetic material creates a force that sets up flux lines through the magnetic path. The amount

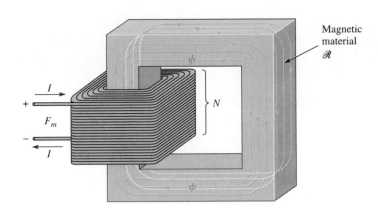

◀ FIGURE 10–11

A basic magnetic circuit.

of flux depends on the magnitude of the mmf and on the reluctance of the material, as expressed by the following equation:

$$\phi = \frac{F_m}{\mathcal{R}}$$

Equation 10–5

Equation 10–5 is known as the *Ohm's law for magnetic circuits* because the flux (ϕ) is analogous to current, the mmf (F_m) is analogous to voltage, and the reluctance ($\mathcal{R}$) is analogous to resistance.

EXAMPLE 10–5

How much flux is established in the magnetic path of Figure 10–12 if the reluctance of the material is 2.8×10^5 At/Wb?

 ▶ **FIGURE 10–12**

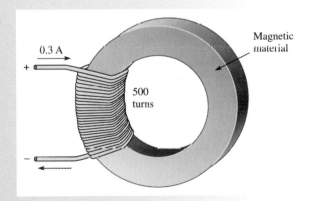

Magnetic material

0.3 A

500 turns

Solution

$$\phi = \frac{F_m}{\mathcal{R}} = \frac{NI}{\mathcal{R}} = \frac{(500 \text{ t})(0.300 \text{ A})}{2.8 \times 10^5 \text{ At/Wb}} = \mathbf{536 \ \mu Wb}$$

Related Problem

How much flux is established in the magnetic path of Figure 10–12 if the reluctance is 7.5×10^3 At/Wb, the number of turns is 300, and the current is 0.18 A?

EXAMPLE 10–6

There is 0.1 A of current through a coil with 400 turns.

(a) What is the mmf?

(b) What is the reluctance of the circuit if the flux is 250 μWb?

Solution **(a)** $N = 400$ and $I = 0.1$ A
$$F_m = NI = (400\,\text{t})(0.1\,\text{A}) = \mathbf{40\,At}$$

(b) $\mathcal{R} = \dfrac{F_m}{\phi} = \dfrac{40\,\text{At}}{250\,\mu\text{Wb}} = \mathbf{1.60 \times 10^5\,At/Wb}$

Related Problem Rework the example for $I = 85$ mA and $N = 500$. The flux is 500 μWb.

The Electromagnet

An electromagnet is based on the properties that you have just learned. A basic electromagnet is simply a coil of wire wound around a core material that can be easily magnetized.

The shape of an electromagnet can be designed for various applications. For example, Figure 10–13 shows a U-shaped magnetic core. When the coil of wire is connected to a battery and there is current, as shown in part (a), a magnetic field is established as indicated. If the current is reversed, as shown in part (b), the direction of the magnetic field is also reversed. The closer the north and south poles are brought together, the smaller the air gap between them becomes, and the easier it becomes to establish a magnetic field because the reluctance is lessened.

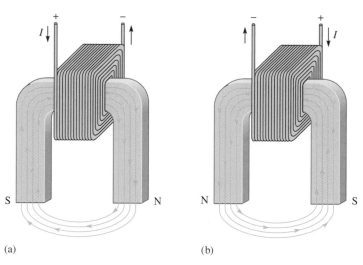

(a)   (b)

▲ **FIGURE 10–13**

Reversing the current in the coil causes the electromagnetic field to reverse.

SECTION 10–2 REVIEW

1. Explain the difference between magnetism and electromagnetism.
2. What happens to the magnetic field in an electromagnet when the current through the coil is reversed?
3. State Ohm's law for a magnetic circuit.
4. Compare each quantity in Question 3 to its electrical counterpart.

10–3 ELECTROMAGNETIC DEVICES

Many types of useful devices such as tape recorders, electric motors, speakers, solenoids, and relays are based on electromagnetism.

After completing this section, you should be able to

- ◆ **Describe the principle of operation for several types of electromagnetic devices**
  - ◆ Discuss how a solenoid and a solenoid valve work
  - ◆ Discuss how a relay works
  - ◆ Discuss how a speaker works
  - ◆ Discuss the basic analog meter movement
  - ◆ Explain a magnetic disk and tape Read/Write operation
  - ◆ Explain the concept of the magneto-optical disk

The Solenoid

The **solenoid** is a type of electromagnetic device that has a movable iron core called a *plunger*. The movement of this iron core depends on both an electromagnetic field and a mechanical spring force. The basic structure of a solenoid is shown in Figure 10–14. It consists of a cylindrical coil of wire wound around a nonmagnetic hollow form. A stationary iron core is fixed in position at the end of the shaft and a sliding iron core (plunger) is attached to the stationary core with a spring.

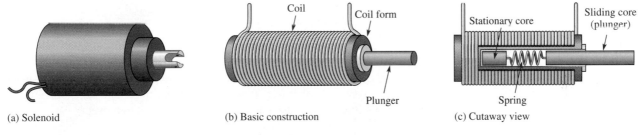

(a) Solenoid   (b) Basic construction   (c) Cutaway view

▲ **FIGURE 10–14**

Basic solenoid structure.

The basic solenoid operation is illustrated in Figure 10–15 for the unenergized and the energized conditions. In the at-rest (or unenergized) state, the plunger is extended. The solenoid is energized by current through the coil, which sets up an electromagnetic field that magnetizes both iron cores. The south pole of the stationary core attracts the north pole of the movable core causing it to slide inward, thus retracting the plunger and compressing the spring. As long as there is coil current, the plunger remains retracted by the attractive force of the magnetic fields. When the current is cut off, the magnetic fields collapse and the force of the compressed spring pushes the plunger back out. The solenoid is used for applications such as opening and closing valves and automobile door locks.

The Solenoid Valve In industrial controls, **solenoid valves** are widely used to control the flow of air, water, steam, oils, refrigerants, and other fluids. Solenoid valves are used in both pneumatic (air) and hydraulic (oil) systems, common in machine controls. Solenoid valves are also common in the aerospace and medical fields. Solenoid valves can either move a plunger to open or close a port or can rotate a blocking flap a fixed amount.

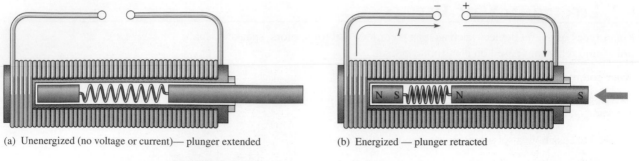

(a) Unenergized (no voltage or current)— plunger extended   (b) Energized — plunger retracted

▲ **FIGURE 10–15**

Basic solenoid operation.

A solenoid valve consists of two functional units: a solenoid coil that provides the magnetic field to provide the required movement to open or close the valve and a valve body, which is isolated from the coil assembly via a leakproof seal and includes a pipe and butterfly valve. Figure 10–16 shows a cutaway of one type of solenoid valve. When the solenoid is energized, the butterfly valve is turned to open a normally closed (NC) valve or to close a normally open (NO) valve.

▶ **FIGURE 10–16**

A basic solenoid valve structure.

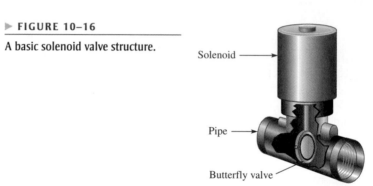

Solenoid

Pipe

Butterfly valve

Solenoid valves are available with a wide variety of configurations including normally open or normally closed valves. They are rated for different types of fluids (for example, gas or water), pressures, number of pathways, sizes, and more. The same valve may control more than one line and may have more than one solenoid to move.

The Relay

The **relay** differs from the solenoid in that the electromagnetic action is used to open or close electrical contacts rather than to provide mechanical movement. Figure 10–17 shows the basic operation of an *armature-type relay* with one normally open (NO) contact and one normally closed (NC) contact (single pole–double throw). When there is no coil current, the armature is held against the upper contact by the spring, thus providing continuity from terminal 1 to terminal 2, as shown in part (a) of the figure. When energized with coil current, the armature is pulled down by the attractive force of the electromagnetic field and makes connection with the lower contact to provide continuity from terminal 1 to terminal 3, as shown in Figure 10–17(b). A typical armature relay is shown in part (c) and the schematic symbol is shown in part (d).

Another widely used type of relay is the *reed relay,* which is shown in Figure 10–18. The reed relay, like the armature relay, uses an electromagnetic coil. The contacts are thin reeds of magnetic material and are usually located inside the coil. When there is no coil current, the reeds are in the open position as shown in part (b). When there is current through the coil, the reeds make contact because they are magnetized and attract each other as shown in part (c).

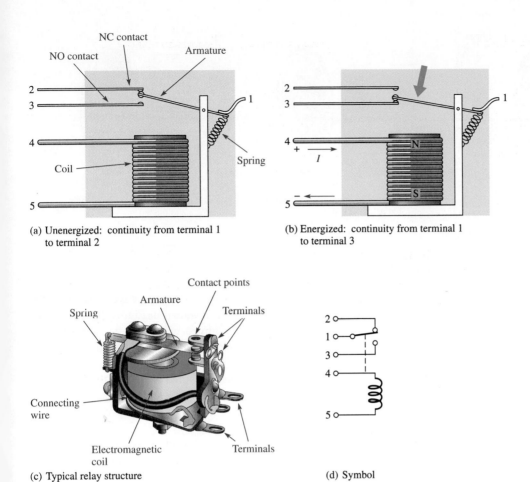

(a) Unenergized: continuity from terminal 1 to terminal 2

(b) Energized: continuity from terminal 1 to terminal 3

(c) Typical relay structure

(d) Symbol

▲ FIGURE 10–17

Basic structure of a single-pole–double-throw armature relay.

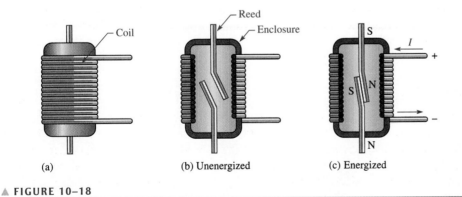

(a)

(b) Unenergized

(c) Energized

▲ FIGURE 10–18

Basic structure of a reed relay.

Reed relays are superior to armature relays in that they are faster, more reliable, and produce less contact arcing. However, they have less current-handling capability than armature relays and are more susceptible to mechanical shock.

The Speaker

A **speaker** is an electromagnetic device that converts electrical signals to sound waves. Permanent-magnet speakers are commonly used in stereos, radios, and TVs, and their

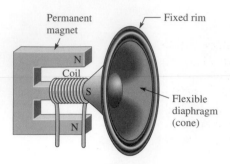
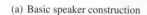

Permanent magnet

Fixed rim

Coil

S

N

N

Flexible diaphragm (cone)

(a) Basic speaker construction

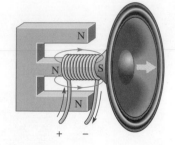

+ −

(b) Coil current producing movement of cone to the right

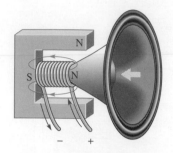

− +

(c) Coil current producing movement of cone to the left

▲ **FIGURE 10–19**

Basic speaker operation.

operation is based on the principle of electromagnetism. A typical speaker is constructed with both a permanent magnet and an electromagnet, as shown in Figure 10–19(a). The cone of the speaker consists of a paper-like diaphragm to which is attached a hollow cylinder with a coil around it, forming an electromagnet. One of the poles of the permanent magnet is positioned within the cylindrical coil. When there is current through the coil in one direction, the interaction of the permanent magnetic field with the electromagnetic field causes the cylinder to move to the right, as indicated in Figure 10–19(b). Current through the coil in the other direction causes the cylinder to move to the left, as shown in part (c).

The movement of the coil cylinder causes the flexible diaphragm also to move in or out, depending on the direction of the coil current. The amount of coil current determines the intensity of the magnetic field, which controls the amount that the diaphragm moves.

As shown in Figure 10–20, when an audio signal (voice or music) is applied to the coil, the current varies in both direction and amount. In response, the diaphragm will vibrate in and out by varying amounts and at varying rates corresponding to the audio signal. Vibration in the diaphragm causes the air that is in contact with it to vibrate in the same manner. These air vibrations move through the air as sound waves.

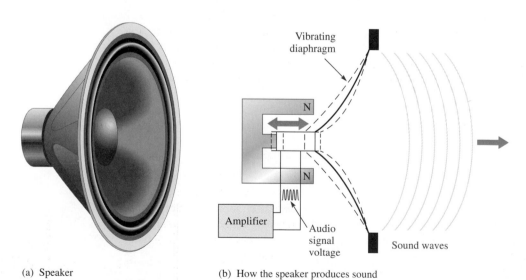

(a) Speaker

(b) How the speaker produces sound

▲ **FIGURE 10–20**

The speaker converts audio signal voltages into sound waves.

Meter Movement

The d'Arsonval meter movement is the most common type used in analog multimeters. In this type of meter movement, the pointer is deflected in proportion to the amount of current through a coil. Figure 10–21 shows a basic d'Arsonval meter movement. It consists of a coil of wire wound on a bearing-mounted assembly that is placed between the poles of a permanent magnet. A pointer is attached to the moving assembly. With no current through the coil, a spring mechanism keeps the pointer at its left-most (zero) position. When there is current through the coil, electromagnetic forces act on the coil, causing a rotation to the right. The amount of rotation depends on the amount of current.

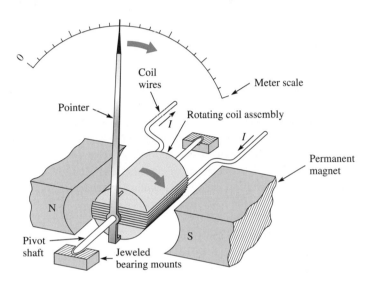

Figure 10–22 illustrates how the interaction of magnetic fields produces rotation of the coil assembly. Current is inward at the "cross" and outward at the "dot" in the single winding shown. The inward current produces a clockwise electromagnetic field that reinforces the permanent magnetic field above it. The result is a downward force on the right side of the coil as shown. The outward current produces a counterclockwise electromagnetic field that reinforces the permanent magnetic field below it. The result is an upward force on the left side of the coil as shown. These forces produce a clockwise rotation of the coil assembly and are opposed by a spring mechanism. The indicated forces and the spring force are balanced at the value of the current. When current is removed, the spring force returns the pointer to its zero position.

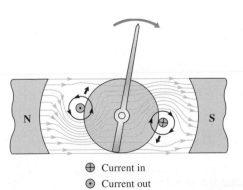

◀ FIGURE 10–22

When the electromagnetic field interacts with the permanent magnetic field, forces are exerted on the rotating coil assembly, causing it to move clockwise and thus deflecting the pointer.

Magnetic Disk and Tape Read/Write Head

A simplified diagram of a magnetic disk or tape surface read/write operation is shown in Figure 10–23. A data bit (1 or 0) is written on the magnetic surface by the magnetization of a small segment of the surface as it moves by the write head. The direction of the magnetic flux lines is controlled by the direction of the current pulse in the winding, as shown in Figure 10–23(a). At the air gap in the write head, the magnetic flux takes a path through the surface of the storage device. This magnetizes a small spot on the surface in the direction of the field. A magnetized spot of one polarity represents a binary 1, and one of the opposite polarity represents a binary 0. Once a spot on the surface is magnetized, it remains until written over with an opposite magnetic field.

When the magnetic surface passes a read head, the magnetized spots produce magnetic fields in the read head, which induce voltage pulses in the winding. The polarity of these pulses depends on the direction of the magnetized spot and indicates whether the stored bit is a 1 or a 0. This process is illustrated in Figure 10–23(b). Often the read and write heads are combined into a single unit.

The Magneto-Optical Disk

The magneto-optical disk uses an electromagnet and laser beams to read and write (record) data on a magnetic surface. Magneto-optical disks are formatted in tracks and sectors similar to magnetic floppy disks and hard disks. However, because of the ability of a laser beam to be precisely directed to an extremely small spot, magneto-optical disks are capable of storing much more data than standard magnetic hard disks.

Figure 10–24(a) illustrates a small cross-sectional area of a disk before recording, with an electromagnet positioned below it. Tiny magnetic particles, represented by the arrows, are all magnetized in the same direction.

Writing (recording) on the disk is accomplished by applying an external magnetic field opposite to the direction of the magnetic particles, as indicated in Figure 10–24(b), and then directing a high-power laser beam to heat the disk at a precise point where a binary 1 is to be stored. The disk material, a magneto-optic alloy, is highly resistant to magnetization at room temperature; but at the spot where the laser beam heats the material, the inherent direction of magnetism is reversed by the external magnetic field produced by the electromagnet. At points where binary 0s are to be stored, the laser beam is not applied and the inherent upward direction of the magnetic particle remains.

As illustrated in Figure 10–24(c), reading data from the disk is accomplished by turning off the external magnetic field and directing a low-power laser beam at a spot

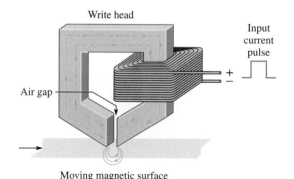

(a) The magnetic flux from the write head follows the low reluctance path through the moving magnetic surface.

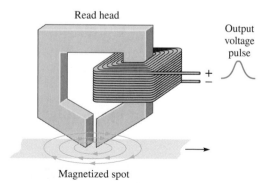

(b) When read head passes over magnetized spot, an induced voltage appears at the output.

▲ FIGURE 10–23

Read/write function on a magnetic surface.

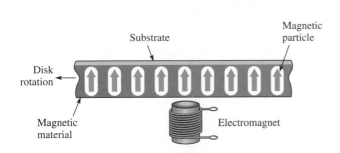

(a) Small cross-section of unrecorded disk

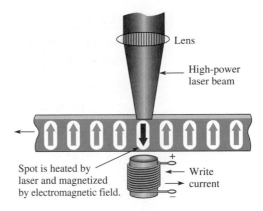

(b) Writing: A high-power laser beam heats the spot, causing the magnetic particle to align with the electromagnetic field.

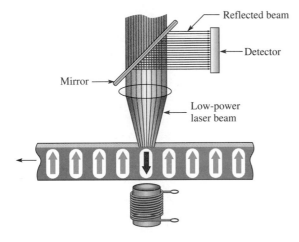

(c) Reading: A low-power laser beam reflects off of the reversed-polarity magnetic particle and its polarization shifts. If the particle is not reversed, the polarization of the reflected beam is unchanged.

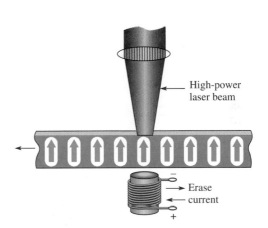

(d) Erasing: The electromagnetic field is reversed as the high-power laser beam heats the spot, causing the magnetic particle to be restored to the original polarity.

▲ FIGURE 10–24

Basic concept of the magneto-optical disk.

where a bit is to be read. Basically, if a binary 1 is stored at the spot (reversed magnetization), the laser beam is reflected and its polarization is shifted; but if a binary 0 is stored, the polarization of the reflected laser beam is unchanged. A detector senses the difference in the polarity of the reflected laser beam to determine if the bit being read is a 1 or a 0.

Figure 10–24(d) shows that the disk is erased by restoring the original magnetic direction of each particle by reversing the external magnetic field and applying the high power laser beam.

SECTION 10–3 REVIEW	1. Explain the difference between a solenoid and a relay. 2. What is the movable part of a solenoid called? 3. What is the movable part of a relay called? 4. Upon what basic principle is the d'Arsonval meter movement based?

10–4 MAGNETIC HYSTERESIS

When a magnetizing force is applied to a material, the flux density in the material changes in a certain way.

After completing this section, you should be able to

◆ **Explain magnetic hysteresis**

  ◆ State the formula for magnetic field intensity

  ◆ Discuss a hysteresis curve

  ◆ Define *retentivity*

Magnetic Field Intensity (*H*)

The **magnetic field intensity** (also called *magnetizing force*) in a material is defined to be the magnetomotive force (F_m) per unit length (l) of the material, as expressed by the following equation. The unit of magnetic field intensity (*H*) is ampere-turns per meter (At/m).

Equation 10–6

$$H = \frac{F_m}{l}$$

where $F_m = NI$. Note that the magnetic field intensity depends on the number of turns (*N*) of the coil of wire, the current (*I*) through the coil, and the length (*l*) of the material. It does not depend on the type of material.

Since $\phi = F_m/\mathcal{R}$, as F_m increases, the flux increases. Also, the magnetic field intensity (*H*) increases. Recall that the flux density (*B*) is the flux per unit cross-sectional area ($B = \phi/A$), so *B* is also proportional to *H*. The curve showing how these two quantities (*B* and *H*) are related is called the *B-H* curve or the hysteresis curve. The parameters that influence both *B* and *H* are illustrated in Figure 10–25.

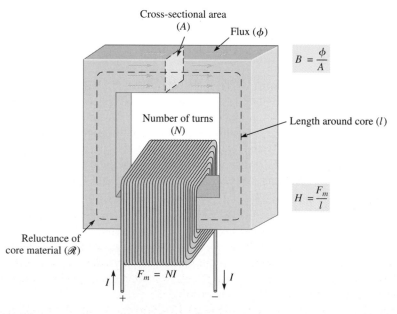

▲ FIGURE 10–25

Parameters that determine the magnetic field intensity (*H*) and the flux density (*B*).

The Hysteresis Curve and Retentivity

Hysteresis is a characteristic of a magnetic material whereby a change in magnetization lags the application of the magnetic field intensity. The magnetic field intensity (*H*) can be readily increased or decreased by varying the current through the coil of wire, and it can be reversed by reversing the voltage polarity across the coil.

Figure 10–26 illustrates the development of the hysteresis curve. Let's start by assuming a magnetic core is unmagnetized so that $B = 0$. As the magnetic field intensity (*H*) is increased from zero, the flux density (*B*) increases proportionally, as indicated by the curve in Figure 10–26(a). When *H* reaches a certain value, *B* begins to level off. As *H* continues to increase, *B* reaches a saturation value (B_{sat}) when *H* reaches a value (H_{sat}), as illustrated in Figure 10–26(b). Once saturation is reached, a further increase in *H* will not increase *B*.

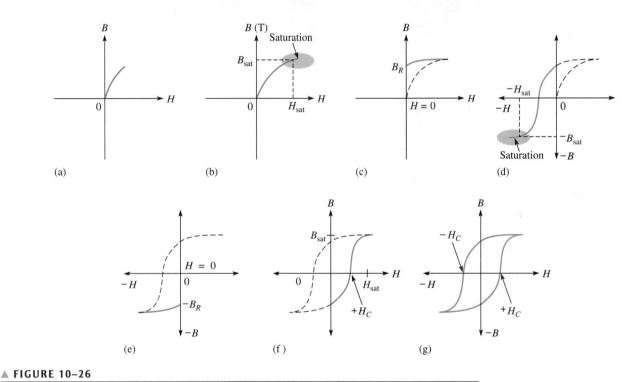

(a)   (b)   (c)   (d)

(e)   (f)   (g)

▲ **FIGURE 10–26**

Development of a magnetic hysteresis curve.

Now, if *H* is decreased to zero, *B* will fall back along a different path to a residual value (B_R), as shown in Figure 10–26(c). This indicates that the material continues to be magnetized even when the magnetic field intensity is zero ($H = 0$). The ability of a material to maintain a magnetized state without magnetic field intensity is called **retentivity**. The retentivity of a material represents the maximum flux that can be retained after the material has been magnetized to saturation and is indicated by the ratio of B_R to B_{sat}.

Reversal of the magnetic field intensity is represented by negative values of *H* on the curve and is achieved by reversing the current in the coil of wire. An increase in *H* in the negative direction causes saturation to occur at a value ($-H_{sat}$) where the flux density is at its maximum negative value, as indicated in Figure 10–26(d).

When the magnetic field intensity is removed ($H = 0$), the flux density goes to its negative residual value ($-B_R$), as shown in Figure 10–26(e). From the $-B_R$ value, the flux density follows the curve indicated in part (f) back to its maximum positive value when the magnetic field intensity equals H_{sat} in the positive direction.

The complete *B-H* curve is shown in Figure 10–26(g) and is called the *hysteresis curve*. The magnetic field intensity required to make the flux density zero is called the *coercive force*, H_C.

Materials with a low retentivity do not retain a magnetic field very well, while those with high retentivities exhibit values of B_R very close to the saturation value of B. Depending on the application, retentivity in a magnetic material can be an advantage or a disadvantage. In permanent magnets and magnetic tape, for example, high retentivity is required; while in tape recorder read/write heads, low retentivity is necessary. In ac motors, retentivity is undesirable because the residual magnetic field must be overcome each time the current reverses, thus wasting energy.

SECTION 10–4 REVIEW	1. For a given wirewound core, how does an increase in current through the coil affect the flux density?
	2. Define *retentivity*.
	3. Why is low retentivity required for tape recorder read/write heads but high retentivity is required for magnetic tape?

10–5   ELECTROMAGNETIC INDUCTION

When a conductor is moved through a magnetic field, a voltage is produced across the conductor. This principle is known as electromagnetic induction and the resulting voltage is an induced voltage. The principle of electromagnetic induction is what makes transformers, electrical generators, and many other devices possible.

After completing this section, you should be able to

◆ **Discuss the principle of electromagnetic induction**

    ◆ Explain how voltage is induced in a conductor in a magnetic field

    ◆ Determine polarity of an induced voltage

    ◆ Discuss forces on a conductor in a magnetic field

    ◆ State Faraday's law

    ◆ State Lenz's law

Relative Motion

When a conductor is moved across a magnetic field, there is a relative motion between the conductor and the magnetic field. Likewise, when a magnetic field is moved past a stationary conductor, there is also relative motion. In either case, this relative motion results in an **induced voltage (v_{ind})** in the conductor, as Figure 10–27 indicates. The lowercase v stands for instantaneous voltage. Voltage is only induced when the conductor "cuts" magnetic lines as shown.

The amount of the induced voltage (v_{ind}) depends on the flux density, B, the length of the conductor, l, which is exposed to the magnetic field, and the rate at which the conductor and the magnetic field move with respect to each other. The faster the relative speed, the greater the induced voltage. The equation for induced voltage in a conductor is

Equation 10–7

$$v_{ind} = Blv$$

where v_{ind} is the induced voltage, B is the flux density in tesla, l is the length of the conductor exposed to the magnetic field in meters, and v is the relative velocity in meters per second.

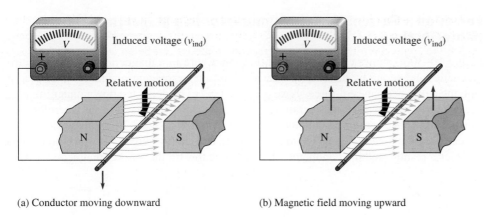

(a) Conductor moving downward   (b) Magnetic field moving upward

▲ **FIGURE 10–27**

Relative motion between a conductor and a magnetic field.

Polarity of the Induced Voltage

If the conductor in Figure 10–27 is moved first one way and then another in the magnetic field, a reversal of the polarity of the induced voltage will be observed. As the conductor is moved downward, a voltage is induced with the polarity indicated in Figure 10–28(a). As the conductor is moved upward, the polarity is as indicated in part (b) of the figure.

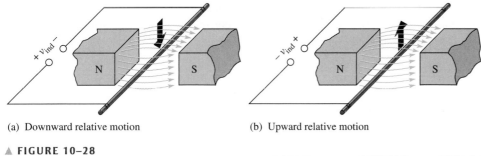

(a) Downward relative motion   (b) Upward relative motion

▲ **FIGURE 10–28**

Polarity of induced voltage depends on direction of motion.

Induced Current

When a load resistor is connected to the conductor in Figure 10–28, the voltage induced by the relative motion in the magnetic field will cause a current in the load, as shown in Figure 10–29. This current is called the **induced current (i_{ind})**. The lowercase i stands for instantaneous current.

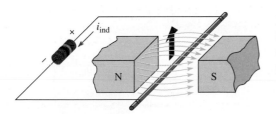

◀ **FIGURE 10–29**

Induced current in a load as the conductor moves through the magnetic field.

The action of producing a voltage and a resulting current in a load by moving a conductor across a magnetic field is the basis for electrical generators. A single conductor will have a small induced current, so practical generators use coils with many turns. The concept of a conductor in a moving magnetic field is the basis for inductance in an electric circuit.

Forces on a Current-Carrying Conductor in a Magnetic Field (Motor Action)

Figure 10–30(a) shows current outward through a wire in a magnetic field. The electromagnetic field set up by the current interacts with the permanent magnetic field; as a result, the permanent lines of force above the wire tend to be deflected down under the wire, because they are opposite in direction to the electromagnetic lines of force. Therefore, the flux density above is reduced, and the magnetic field is weakened. The flux density below the conductor is increased, and the magnetic field is strengthened. An upward force on the conductor results, and the conductor tends to move toward the weaker magnetic field.

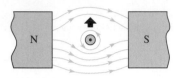

(a) Upward force: weak field above, strong field below.

(b) Downward force: strong field above, weak field below

⊙ Current out
⊕ Current in

▲ FIGURE 10–30

Forces on a current-carrying conductor in a magnetic field (motor action).

Figure 10–30(b) shows the current inward, resulting in a force on the conductor in the downward direction. This force is the basis for electrical motors. The discovery of it was one of the factors that led to the Industrial Revolution.

Faraday's Law

Michael Faraday discovered the principle of **electromagnetic induction** in 1831. He found that moving a magnet through a coil of wire induced a voltage across the coil, and that when a complete path was provided, the induced voltage caused an induced current, as you have learned. Faraday's two observations are stated as follows:

1. The amount of voltage induced in a coil is directly proportional to the rate of change of the magnetic field with respect to the coil ($d\phi/dt$).

2. The amount of voltage induced in a coil is directly proportional to the number of turns of wire in the coil (N).

Faraday's first observation is demonstrated in Figure 10–31, where a bar magnet is moved through a coil, thus creating a changing magnetic field. In part (a) of the figure, the

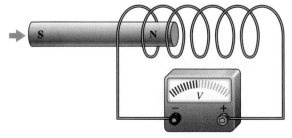

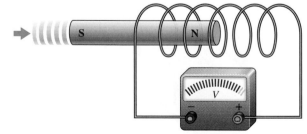

(a) As the magnet moves slowly to the right, its magnetic field is changing with respect to coil, and a voltage is induced.

(b) As the magnet moves more rapidly to the right, its magnetic field is changing more rapidly with respect to coil, and a greater voltage is induced.

▲ FIGURE 10–31

A demonstration of Faraday's first observation: The amount of induced voltage is directly proportional to the rate of change of the magnetic field with respect to the coil.

magnet is moved at a certain rate, and a certain induced voltage is produced as indicated. In part (b), the magnet is moved at a faster rate through the coil, creating a greater induced voltage.

Faraday's second observation is demonstrated in Figure 10–32. In part (a), the magnet is moved through the coil and a voltage is induced as shown. In part (b), the magnet is moved at the same speed through a coil that has a greater number of turns. The greater number of turns creates a greater induced voltage.

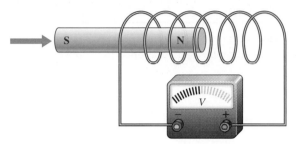

(a) Magnet moves through a coil and induces a voltage.

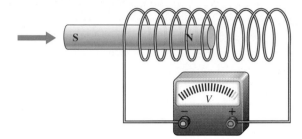

(b) Magnet moves at same rate through a coil with more turns (loops) and induces a greater voltage.

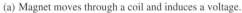

▲ FIGURE 10–32

A demonstration of Faraday's second observation: The amount of induced voltage is directly proportional to the number of turns in the coil.

Faraday's law is stated as follows:

The voltage induced across a coil of wire equals the number of turns in the coil times the rate of change of the magnetic flux.

Faraday's law is expressed in equation form as

$$v_{ind} = N\left(\frac{d\phi}{dt}\right)$$

Equation 10–8

EXAMPLE 10–7   Apply Faraday's law to find the induced voltage across a coil with 500 turns that is located in a magnetic field that is changing at a rate of 8000 μWb/s.

Solution $$v_{ind} = N\left(\frac{d\phi}{dt}\right) = (500 \text{ t})(8000 \ \mu\text{Wb/s}) = \textbf{4.0 V}$$

Related Problem   Find the induced voltage across a 250 turn coil in a magnetic field that is changing at 50 μWb/s.

Lenz's Law

Faraday's law states that a changing magnetic field induces a voltage in a coil that is directly proportional to the rate of change of the magnetic field and the number of turns in the coil. **Lenz's law** defines the polarity or direction of the induced voltage.

When the current through a coil changes, an induced voltage is created as a result of the changing electromagnetic field and the polarity of the induced voltage is such that it always opposes the change in current.

**SECTION 10–5
REVIEW**
1. What is the induced voltage across a stationary conductor in a stationary magnetic field?
2. When the speed at which a conductor is moved through a magnetic field is increased, does the induced voltage increase, decrease, or remain the same?
3. When there is current through a conductor in a magnetic field, what happens?

10–6 APPLICATIONS OF ELECTROMAGNETIC INDUCTION

Two applications of electromagnetic induction are an automotive crankshaft position sensor and a dc generator. Although there are many varied applications, these two are representative.

After completing this section, you should be able to

◆ **Describe some applications of electromagnetic induction**

  ◆ Explain how a crankshaft position sensor works

  ◆ Explain how a dc generator works

Automotive Crankshaft Position Sensor

An automotive application of electromagnetic induction involves a type of engine sensor that detects the crankshaft position. The electronic engine controller in many automobiles uses the position of the crankshaft to set ignition timing and, sometimes, to adjust the fuel control system. Figure 10–33 shows the basic concept. A steel disk is attached to the engine's crankshaft by an extension rod; the protruding tabs on the disk represent specific crankshaft positions.

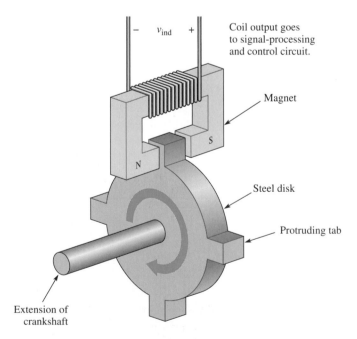

▲ **FIGURE 10–33**

A crankshaft position sensor that produces a voltage when a tab passes through the air gap of the magnet.

As the steel disk rotates with the crankshaft, the tabs periodically pass through the air gap of the permanent magnet. Since steel has a much lower reluctance than does air (a magnetic field can be established in steel much more easily than in air), the magnetic flux suddenly increases as a tab comes into the air gap, causing a voltage to be induced across the coil. This process is illustrated in Figure 10–34. The electronic engine control circuit uses the induced voltage as an indicator of the crankshaft position.

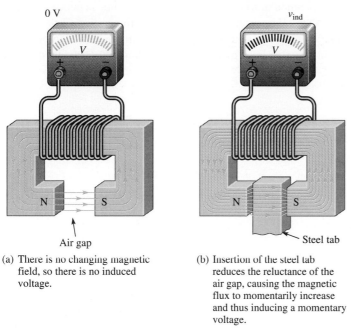

0 V

v_{ind}

Air gap

Steel tab

(a) There is no changing magnetic field, so there is no induced voltage.

(b) Insertion of the steel tab reduces the reluctance of the air gap, causing the magnetic flux to momentarily increase and thus inducing a momentary voltage.

▲ FIGURE 10–34

As the tab passes through the air gap of the magnet, the coil senses a change in the magnetic field, and a voltage is induced.

DC Generator

Figure 10–35 shows a simplified dc generator consisting of a single loop of wire in a permanent magnetic field. Notice that each end of the wire loop is connected to a split-ring arrangement. This conductive metal ring is called a *commutator*. As the wire loop is rotated in the magnetic field, the split commutator ring also rotates. Each half of the split

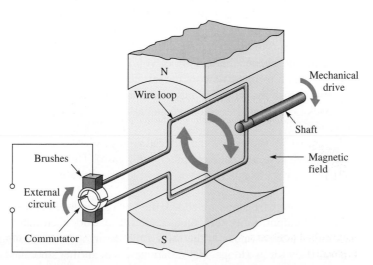

◄ FIGURE 10–35

A simplified dc generator.

N

Wire loop

Mechanical drive

Shaft

Brushes

External circuit

Magnetic field

Commutator

S

ring rubs against the fixed contacts, called *brushes,* and connects the wire loop to an external circuit.

As the wire loop rotates through the magnetic field, it cuts through the flux lines at varying angles, as illustrated in Figure 10–36. At position *A* in its rotation, the loop of wire is effectively moving parallel with the magnetic field. Therefore, at this instant, the rate at which it is cutting through the magnetic flux lines is zero. As the loop moves from position *A* to position *B*, it cuts through the flux lines at an increasing rate. At position *B*, it is moving effectively perpendicular to the magnetic field and thus is cutting through a maximum number of lines. As the loop rotates from position *B* to position *C*, the rate at which it cuts the flux lines decreases to minimum (zero) at *C*. From position *C* to position *D*, the rate at which the loop cuts the flux lines increases to a maximum at *D* and then back to a minimum again at *A*.

▶ **FIGURE 10–36**

End view of wire loop cutting through the magnetic field.

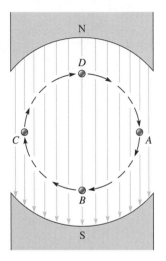

As you have learned, when a wire moves through a magnetic field, a voltage is induced, and by Faraday's law, the amount of induced voltage is proportional to the number of loops (turns) in the wire and the rate at which it is moving with respect to the magnetic field. Now you know that the angle at which the wire moves with respect to the magnetic flux lines determines the amount of induced voltage because the rate at which the wire cuts through the flux lines depends on the angle of motion.

Figure 10–37 illustrates how a voltage is induced in the external circuit as the single loop rotates in the magnetic field. Assume that the loop is in its instantaneous horizontal position, so the induced voltage is zero. As the loop continues in its rotation, the induced voltage builds up to a maximum at position *B*, as shown in part (a) of the figure. Then, as the loop continues from *B* to *C*, the voltage decreases to zero at position *C*, as shown in part (b).

During the second half of the revolution, shown in Figure 10–37(c) and (d), the brushes switch to opposite commutator sections, so the polarity of the voltage remains the same across the output. Thus, as the loop rotates from position *C* to position *D* and then back to position *A*, the voltage increases from zero at *C* to a maximum at *D* and back to zero at *A*.

Figure 10–38 shows how the induced voltage varies as the wire loop in the dc generator goes through several rotations (three in this case). This voltage is a dc voltage because its polarities do not change. However, the voltage is pulsating between zero and its maximum value.

When more wire loops are added, the voltages induced across each loop are combined across the output. Since the voltages are offset from each other, they do not reach their maximum or zero values at the same time. A smoother dc voltage results, as shown in Figure 10–39 for two loops. The variations can be further smoothed out by filters to achieve a nearly constant dc voltage. (Filters are covered in Chapter 18.)

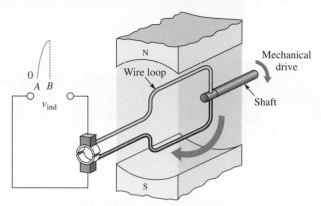

(a) Position *B*: Loop is moving perpendicular to flux lines, and voltage is maximum.

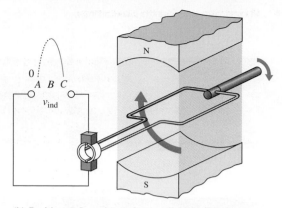

(b) Position *C*: Loop is moving parallel with flux lines, and voltage is zero.

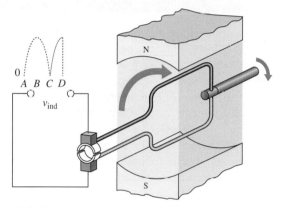

(c) Position *D*: Loop is moving perpendicular to flux lines, and voltage is maximum.

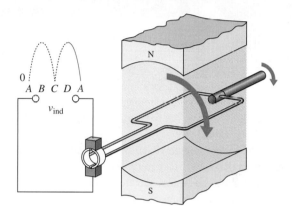

(d) Position *A*: Loop is moving parallel with flux lines, and voltage is zero.

▲ FIGURE 10–37

Operation of a basic dc generator.

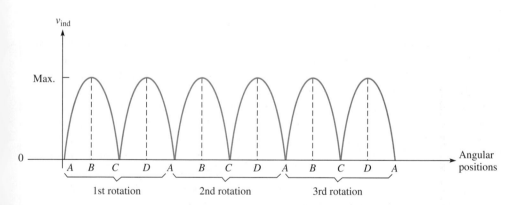

◄ FIGURE 10–38

Induced voltage over three rotations of the wire loop in the dc generator.

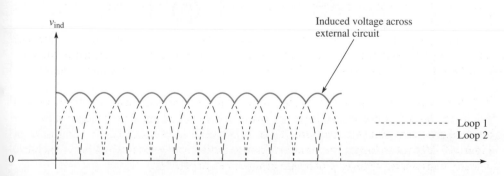

◄ FIGURE 10–39

The induced voltage for a two-loop generator. There is much less variation in the induced voltage.

SECTION 10–6
REVIEW

1. If the steel disk in the crankshaft position sensor has stopped, with a tab in the magnet's air gap, what is the induced voltage?

2. What happens to the induced voltage if the wire loop in the basic dc generator suddenly begins rotating at a faster speed?

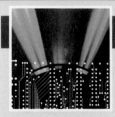

A Circuit Application

The relay is a common type of electromagnetic device that is used in many types of control applications. With a relay, a lower voltage, such as from a battery, can be used to switch a much higher voltage, such as the 110 V from an ac outlet. You will see how a relay can be used in a basic burglar alarm system.

The schematic in Figure 10–40 shows a simplified intrusion alarm system that uses a relay to turn on an audible alarm (siren) and lights. The system operates from a 9 V battery so that even if power to the house is off, the audible alarm will still work.

The detection switches are normally open (NO) magnetic switches that are parallel connected and located in the windows and doors. The relay is a triple-pole–double-throw device that

operates with a coil voltage of 9 V dc and draws approximately 50 mA. When an intrusion occurs, one of the switches closes and allows current from the battery to the relay coil, which energizes the relay and causes the three sets of normally open contacts to close. Closure of contact *A* turns on the alarm, which draws 2 A from the battery. Closure of contact *C* turns on a light circuit in the house. Closure of contact *B* latches the relay and keeps it energized even if the intruder closes the door or window through which entry was made. If not for contact *B* in parallel with the detection switches, the alarm and lights would go off as soon as the window or door was shut behind the intruder.

The relay contacts are not physically remote in relation to the coil as the schematic indicates. The schematic is drawn this way for functional clarity. The entire relay is housed in the package

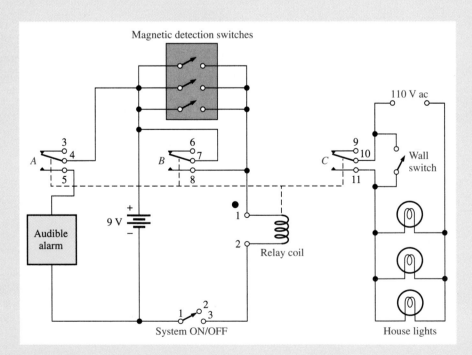

▲ **FIGURE 10–40**

Simplified burglar alarm system.

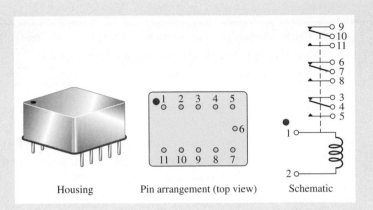

▲ FIGURE 10–41

Triple-pole–double-throw relay.

shown in Figure 10–41. Also shown are the pin diagram and internal schematic for the relay.

System Interconnections

◆ Develop a connection block diagram and point-to-point wire list for interconnecting the components in Figure 10–42 to create the alarm system shown in the schematic of Figure 10–40. The connection points on the components are indicated by letters.

A Test Procedure

◆ Develop a detailed step-by-step procedure to check out the completely wired burglar alarm system.

Review

1. What is the purpose of the detection switches?
2. What is the purpose of contact *B* in the relay in Figure 10–40?

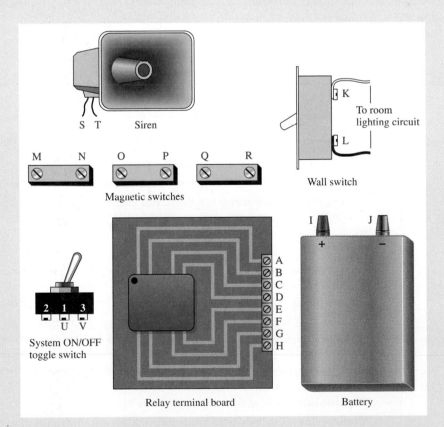

▲ FIGURE 10–42

Array of burglar alarm components.

SUMMARY

- Unlike magnetic poles attract each other, and like poles repel each other.
- Materials that can be magnetized are called *ferromagnetic*.
- When there is current through a conductor, it produces an electromagnetic field around the conductor.
- You can use the right-hand rule to establish the direction of the electromagnetic lines of force around a conductor.
- An electromagnet is basically a coil of wire around a magnetic core.
- When a conductor moves within a magnetic field, or when a magnetic field moves relative to a conductor, a voltage is induced across the conductor.
- The faster the relative motion between a conductor and a magnetic field, the greater the induced voltage.
- Table 10–2 summarizes the magnetic quantities and units.

▶ TABLE 10–2

SYMBOL	QUANTITY	SI UNIT
B	Magnetic flux density	Tesla (T)
ϕ	Magnetic flux	Weber (Wb)
μ	Permeability	Webers/ampere-turn · meter (Wb/At · m)
$\mathcal{R}$	Reluctance	Ampere-turns/weber (At/Wb)
F_m	Magnetomotive force (mmf)	Ampere-turn (At)
H	Magnetic field intensity	Ampere-turns/meter (At/m)

KEY TERMS

Key terms and other bold terms in the chapter are defined in the end-of-book glossary.

Ampere-turn (At)   The current in a single loop (turn) of wire.

Electromagnetic field   A formation of a group of magnetic lines of force surrounding a conductor created by electrical current in the conductor.

Electromagnetic induction   The phenomenon or process by which a voltage is produced in a conductor when there is relative motion between the conductor and a magnetic or electromagnetic field.

Electromagnetism   The production of a magnetic field by current in a conductor.

Faraday's law   A law stating that the voltage induced across a coil of wire equals the number of turns in the coil times the rate of change of the magnetic flux.

Hysteresis   A characteristic of a magnetic material whereby a change in magnetization lags the application of the magnetic field intensity.

Induced current (i_{ind})   A current induced in a conductor when the conductor moves through a magnetic field.

Induced voltage (v_{ind})   Voltage produced as a result of a changing magnetic field.

Lenz's law   A law that states when the current through a coil changes, the polarity of the induced voltage created by the changing magnetic field is such that it always opposes the change in current that caused it. The current cannot change instantaneously.

Lines of force   Magnetic flux lines in a magnetic field radiating from the north pole to the south pole.

Magnetic field   A force field radiating from the north pole to the south pole of a magnet.

Magnetic flux   The lines of force between the north and south poles of a permanent magnet or an electromagnet.

Magnetomotive force (mmf)   The cause of a magnetic field, measured in ampere-turns.

Permeability   The measure of ease with which a magnetic field can be established in a material.

Relay An electromagnetically controlled mechanical device in which electrical contacts are opened or closed by a magnetizing current.

Reluctance The opposition to the establishment of a magnetic field in a material.

Retentivity The ability of a material, once magnetized, to maintain a magnetized state without the presence of a magnetizing force.

Solenoid An electromagnetically controlled device in which the mechanical movement of a shaft or plunger is activated by a magnetizing current.

Speaker An electromagnetic device that converts electrical signals to sound waves.

Tesla (T) The SI unit for magnetic flux density.

Weber (Wb) The SI unit of magnetic flux, which represents 10^8 lines.

FORMULAS

10–1	$B = \dfrac{\phi}{A}$	Magnetic flux density
10–2	$\mu_r = \dfrac{\mu}{\mu_0}$	Relative permeability
10–3	$\mathcal{R} = \dfrac{l}{\mu A}$	Reluctance
10–4	$F_m = NI$	Magnetomotive force
10–5	$\phi = \dfrac{F_m}{\mathcal{R}}$	Magnetic flux
10–6	$H = \dfrac{F_m}{l}$	Magnetic field intensity
10–7	$v_{ind} = Blv$	Induced voltage across a moving conductor
10–8	$v_{ind} = N\left(\dfrac{d\phi}{dt}\right)$	Faraday's law

SELF-TEST

Answers are at the end of the chapter.

1. When the south poles of two bar magnets are brought close together, there will be
   (a) a force of attraction   (b) a force of repulsion
   (c) an upward force   (d) no force

2. A magnetic field is made up of
   (a) positive and negative charges   (b) magnetic domains
   (c) flux lines   (d) magnetic poles

3. The direction of a magnetic field is from
   (a) north pole to south pole   (b) south pole to north pole
   (c) inside to outside the magnet   (d) front to back

4. Reluctance in a magnetic circuit is analogous to
   (a) voltage in an electric circuit   (b) current in an electric circuit
   (c) power in an electric circuit   (d) resistance in an electric circuit

5. The unit of magnetic flux is the
   (a) tesla   (b) weber   (c) ampere-turn   (d) ampere-turns/weber

6. The unit of magnetomotive force is the
   (a) tesla   (b) weber   (c) ampere-turn   (d) ampere-turns/weber

7. The unit of flux density is the
   (a) tesla   (b) weber   (c) ampere-turn   (d) electron-volt

8. The electromagnetic activation of a movable shaft is the basis for

(a) relays   (b) circuit breakers   (c) magnetic switches   (d) solenoids

9. When there is current through a wire placed in a magnetic field,

(a) the wire will overheat   (b) the wire will become magnetized

(c) a force is exerted on the wire   (d) the magnetic field will be cancelled

10. A coil of wire is placed in a changing magnetic field. If the number of turns in the coil is increased, the voltage induced across the coil will

(a) remain unchanged   (b) decrease   (c) increase   (d) be excessive

11. If a conductor is moved back and forth at a constant rate in a constant magnetic field, the voltage induced in the conductor will

(a) remain constant   (b) reverse polarity   (c) be reduced   (d) be increased

12. In the crankshaft position sensor in Figure 10–33, the induced voltage across the coil is caused by

(a) current in the coil

(b) rotation of the steel disk

(c) a tab passing through the magnetic field

(d) acceleration of the steel disk's rotational speed

PROBLEMS

More difficult problems are indicated by an asterisk (*).
Answers to odd-numbered problems are at the end of the book.

SECTION 10–1 The Magnetic Field

1. The cross-sectional area of a magnetic field is increased, but the flux remains the same. Does the flux density increase or decrease?

2. In a certain magnetic field the cross-sectional area is 0.5 m^2 and the flux is 1500 μWb. What is the flux density?

3. What is the flux in a magnetic material when the flux density is 2500 $\times$ 10^{-6} T and the cross-sectional area is 150 cm^2?

4. At a given location, assume the earth's magnetic field is 0.6 gauss. Express this flux density in tesla.

5. A very strong permanent magnet has a magnetic field of 100,000 μT. Express this flux density in gauss.

SECTION 10–2 Electromagnetism

6. What happens to the compass needle in Figure 10–9 when the current through the conductor is reversed?

7. What is the relative permeability of a ferromagnetic material whose absolute permeability is 750 $\times$ 10^{-6} Wb/At $\cdot$ m?

8. Determine the reluctance of a material with a length of 0.28 m and a cross-sectional area of 0.08 m^2 if the absolute permeability is 150 $\times$ 10^{-7} Wb/At $\cdot$ m.

9. What is the magnetomotive force in a 50 turn coil of wire when there are 3 A of current through it?

SECTION 10–3 Electromagnetic Devices

10. Typically, when a solenoid is activated, is the plunger extended or retracted?

11. (a) What force moves the plunger when a solenoid is activated?

(b) What force causes the plunger to return to its at-rest position?

12. Explain the sequence of events in the circuit of Figure 10–43 starting when switch 1 (SW1) is closed.

13. What causes the pointer in a d'Arsonval movement to deflect when there is current through the coil?

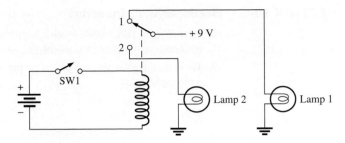

▲ FIGURE 10–43

SECTION 10–4   Magnetic Hysteresis

14. What is the magnetizing force in Problem 9 if the length of the core is 0.2 m?

15. How can the flux density in Figure 10–44 be changed without altering the physical characteristics of the core?

16. In Figure 10–44, there are 500 turns. Determine

    (a) H **(b)** ϕ **(c)** B

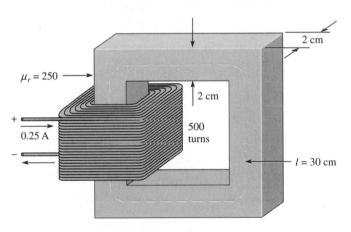

▲ FIGURE 10–44

17. Determine from the hysteresis curves in Figure 10–45 which material has the most retentivity.

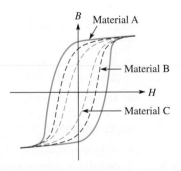

▲ FIGURE 10–45

SECTION 10–5   **Electromagnetic Induction**

    18. According to Faraday's law, what happens to the induced voltage across a given coil if the rate of change of magnetic flux doubles?

    19. What are three factors that determine the voltage in a conductor that is moving perpendicular to a magnetic field?

    20. A magnetic field is changing at a rate of 3500×10^{-3} Wb/s. How much voltage is induced across a 50 turn coil that is placed in the magnetic field?

    21. How does Lenz's law complement Faraday's law?

SECTION 10–6   **Applications of Electromagnetic Induction**

    22. In Figure 10–33, why is there no induced voltage when the steel disk is not rotating?

    23. Explain the purpose of the commutator and brushes in Figure 10–35.

  *24.** A basic one-loop dc generator is rotated at 60 rev/s. How many times each second does the dc output voltage peak (reach a maximum)?

  *25.** Assume that another loop, 90 degrees from the first loop, is added to the dc generator in Problem 24. Make a graph of voltage versus time to show how the output voltage appears. Let the maximum voltage be 10 V.

ANSWERS

SECTION REVIEWS

SECTION 10–1   **The Magnetic Field**

    1. North poles repel.

    2. Magnetic flux is the group of lines of force that make up a magnetic field.

    3. $B = \phi/A = 900\ \mu\text{T}$

SECTION 10–2   **Electromagnetism**

    1. Electromagnetism is produced by current through a conductor. An electromagnetic field exists only when there is current. A magnetic field exists independently of current.

    2. When current reverses, the direction of the magnetic field also reverses.

    3. Flux (ϕ) equals magnetomotive force (F_m) divided by reluctance ($\mathcal{R}$).

    4. Flux: current, mmf: voltage, reluctance: resistance.

SECTION 10–3   **Electromagnetic Devices**

    1. A solenoid produces a movement only. A relay provides an electrical contact closure.

    2. The movable part of a solenoid is the plunger.

    3. The movable part of a relay is the armature.

    4. The d'Arsonval movement is based on the interaction of magnetic fields.

SECTION 10–4   **Magnetic Hysteresis**

    1. An increase in current increases the flux density.

    2. Retentivity is the ability of a material to remain magnetized after removal of the magnetizing force.

    3. Heads should not remain magnetized after magnetic force is removed, but tape should.

SECTION 10–5   **Electromagnetic Induction**

    1. Zero voltage is induced.

    2. Induced voltage increases.

    3. A force is exerted on the conductor when there is current.

SECTION 10–6   Applications of Electromagnetic Induction

1. Zero voltage is induced in the air gap.

2. A faster rotation increases induced voltage.

A Circuit Application

1. The detection switches when closed indicate an intrusion through a window or door.

2. Contact B latches the relay and keeps it energized when intrusion is detected.

RELATED PROBLEMS FOR EXAMPLES

10–1 Flux density will decrease.

10–2 31.0 T

10–3 The reluctance is reduced to 12.8×10^6 At/Wb.

10–4 1.66×10^5 At/Wb

10–5 7.2 mWb

10–6 (a) $F_m = 42.5$ At

    (b) $\mathcal{R} = 85 \times 10^3$ At/Wb

10–7 12.5 mV

SELF-TEST

1. (b)   2. (c)   3. (a)   4. (d)   5. (b)   6. (c)   7. (a)   8. (d)

9. (c)   10. (c)   11. (b)   12. (c)

11

INTRODUCTION TO ALTERNATING CURRENT AND VOLTAGE

CHAPTER OBJECTIVES

◆ Identify a sinusoidal waveform and measure its characteristics

◆ Describe how sine waves are generated

◆ Determine the various voltage and current values of a sine wave

◆ Describe angular relationships of sine waves

◆ Mathematically analyze a sinusoidal waveform

◆ Use a phasor to represent a sine wave

◆ Apply the basic circuit laws to resistive ac circuits

◆ Determine total voltages that have both ac and dc components

◆ Identify the characteristics of basic nonsinusoidal waveforms

◆ Use an oscilloscope to measure waveforms

KEY TERMS

◆ Waveform
◆ Sine wave
◆ Cycle
◆ Period (T)
◆ Frequency (f)
◆ Hertz (Hz)
◆ Amplitude
◆ Oscillator
◆ Function generator
◆ Instantaneous value
◆ Peak value
◆ Peak-to-peak value
◆ rms value
◆ Average value
◆ Degree
◆ Radian
◆ Phase
◆ Phasor
◆ Angular velocity
◆ Pulse
◆ Rise time (t_r)
◆ Fall time (t_f)
◆ Pulse width (t_W)
◆ Periodic
◆ Duty cycle
◆ Ramp
◆ Fundamental frequency
◆ Harmonics
◆ Oscilloscope

A CIRCUIT APPLICATION PREVIEW

In the circuit application, you will learn to measure voltage signals in an AM receiver using an oscilloscope.

VISIT THE COMPANION WEBSITE

Study aids for this chapter are available at
http://www.prenhall.com/floyd

INTRODUCTION

In the preceding chapters, you have studied resistive circuits with dc currents and voltages. This chapter provides an introduction to ac circuit analysis in which time-varying electrical signals, particularly the sine wave, are studied. An electrical signal is a voltage or current that changes in some consistent manner with time. In other words, the voltage or current fluctuates according to a certain pattern called a waveform.

An alternating voltage is one that changes polarity at a certain rate, and an alternating current is one that changes direction at a certain rate. The sinusoidal waveform (sine wave) is the most common and fundamental type because all other types of repetitive waveforms can be broken down into composite sine waves. The sine wave is a periodic type of waveform that repeats at fixed intervals.

Special emphasis is given to the sinusoidal waveform (sine wave) because of its fundamental importance in ac circuit analysis. Other types of waveforms are also introduced, including pulse, triangular, and sawtooth. The use of the oscilloscope for displaying and measuring waveforms is introduced. The use of phasors to represent sine waves is discussed.

11–1 THE SINUSOIDAL WAVEFORM

The sinusoidal waveform or *sine wave* is the fundamental type of alternating current (ac) and alternating voltage. It is also referred to as a sinusoidal wave or, simply, sinusoid. The electrical service provided by the power company is in the form of sinusoidal voltage and current. In addition, other types of repetitive **waveforms** are composites of many individual sine waves called harmonics.

After completing this section, you should be able to

- ◆ **Identify a sinusoidal waveform and measure its characteristics**
  - ◆ Determine the period
  - ◆ Determine the frequency
  - ◆ Relate the period and the frequency

Sinusoidal voltages are produced by two types of sources: rotating electrical machines (ac generators) or electronic oscillator circuits, which are used in instruments commonly known as electronic signal generators. Figure 11–1 shows the symbol used to represent either source of sinusoidal voltage.

Figure 11–2 is a graph showing the general shape of a **sine wave**, which can be either an alternating current or an alternating voltage. Voltage (or current) is displayed on the vertical axis and time (*t*) is displayed on the horizontal axis. Notice how the voltage (or current) varies with time. Starting at zero, the voltage (or current) increases to a positive maximum (peak), returns to zero, and then increases to a negative maximum (peak) before returning again to zero, thus completing one full cycle.

▲ **FIGURE 11–1**

Symbol for a sinusoidal voltage source.

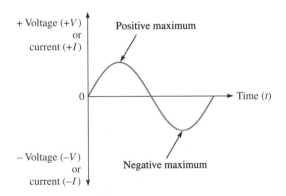

▲ **FIGURE 11–2**

Graph of one cycle of a sine wave.

Polarity of a Sine Wave

As mentioned, a sine wave changes polarity at its zero value; that is, it alternates between positive and negative values. When a sinusoidal voltage source (V_s) is applied to a resistive circuit, as in Figure 11–3, an alternating sinusoidal current results. When the voltage changes polarity, the current correspondingly changes direction as indicated.

During the positive alternation of the applied voltage V_s, the current is in the direction shown in Figure 11–3(a). During a negative alternation of the applied voltage, the current is in the opposite direction, as shown in Figure 11–3(b). The combined positive and negative alternations make up one **cycle** of a sine wave.

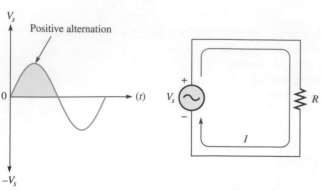

(a) During a positive alternation of voltage, current is in the direction shown.

(b) During a negative alternation of voltage, current reverses direction, as shown.

▲ FIGURE 11–3

Alternating current and voltage.

Period of a Sine Wave

A sine wave varies with time (t) in a definable manner.

The time required for a sine wave to complete one full cycle is called the period (T).

Figure 11–4(a) illustrates the period of a sine wave. Typically, a sine wave continues to repeat itself in identical cycles, as shown in Figure 11–4(b). Since all cycles of a repetitive sine wave are the same, the period is always a fixed value for a given sine wave. The period of a sine wave can be measured from a zero crossing to the next corresponding zero crossing, as indicated in Figure 11–4(a). The period can also be measured from any peak in a given cycle to the corresponding peak in the next cycle.

▶ FIGURE 11–4

The period of a sine wave is the same for each cycle.

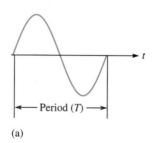

(a)

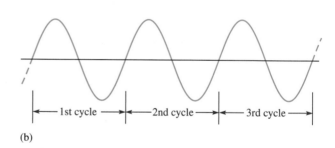

(b)

EXAMPLE 11–1

What is the period of the sine wave in Figure 11–5?

FIGURE 11–5

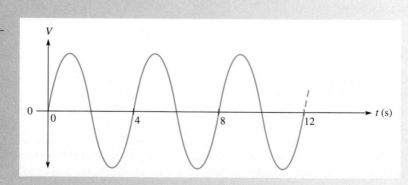

Solution   As shown in Figure 11–5, it takes four seconds (4 s) to complete each cycle. Therefore, the period is 4 s.

$$T = 4\,\text{s}$$

Related Problem*   What is the period if the sine wave goes through five cycles in 12 s?

*Answers are at the end of the chapter.

EXAMPLE 11–2

Show three possible ways to measure the period of the sine wave in Figure 11–6. How many cycles are shown?

▶ **FIGURE 11–6**

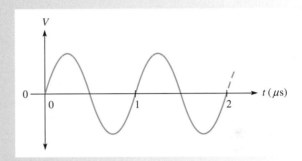

Solution   **Method 1:** The period can be measured from one zero crossing to the corresponding zero crossing in the next cycle (the slope must be the same at the corresponding zero crossings).

Method 2: The period can be measured from the positive peak in one cycle to the positive peak in the next cycle.

Method 3: The period can be measured from the negative peak in one cycle to the negative peak in the next cycle.

These measurements are indicated in Figure 11–7, where **two cycles of the sine wave** are shown. Keep in mind that you obtain the same value for the period no matter which corresponding points on the waveform you use.

▶ **FIGURE 11–7**

Measurement of the period of a sine wave.

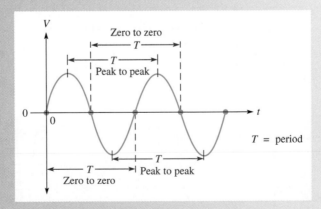

Related Problem   If a positive peak occurs at 1 ms and the next positive peak occurs at 2.5 ms, what is the period?

Frequency of a Sine Wave

Frequency (f) is the number of cycles that a sine wave completes in one second.

The more cycles completed in one second, the higher the frequency. Frequency (f) is measured in units of hertz. One **hertz (Hz)** is equivalent to one cycle per second; 60 Hz is 60 cycles per second, for example. Figure 11–8 shows two sine waves. The sine wave in part (a) completes two full cycles in one second. The one in part (b) completes four cycles in one second. Therefore, the sine wave in part (b) has twice the frequency of the one in part (a).

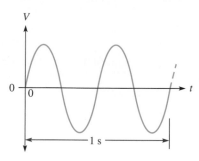
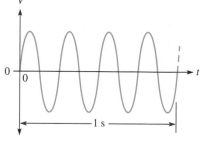

(a) Lower frequency: fewer cycles per second   (b) Higher frequency: more cycles per second

▲ **FIGURE 11–8**

Illustration of frequency.

Relationship of Frequency and Period

The formulas for the relationship between frequency (f) and period (T) are as follows:

Equation 11–1
$$f = \frac{1}{T}$$

Equation 11–2
$$T = \frac{1}{f}$$

There is a reciprocal relationship between f and T. Knowing one, you can calculate the other with the x^{-1} or $1/x$ key on your calculator. This inverse relationship makes sense because a sine wave with a longer period goes through fewer cycles in one second than one with a shorter period.

EXAMPLE 11–3

Which sine wave in Figure 11–9 has a higher frequency? Determine the frequency and the period of both waveforms.

FIGURE 11–9

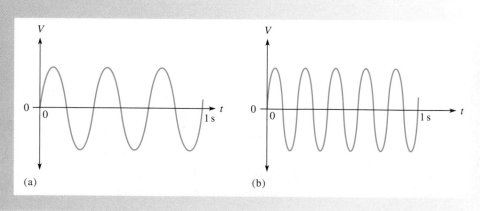

Solution   The sine wave in Figure 11–9(b) has the higher frequency because it completes more cycles in 1 s than does the sine wave in part (a).

In Figure 11–9(a), three cycles are completed in 1 s; therefore,

$$f = \textbf{3 Hz}$$

One cycle takes 0.333 s (one-third second), so the period is

$$T = 0.333 \text{ s} = \textbf{333 ms}$$

In Figure 11–9(b), five cycles are completed in 1 s; therefore,

$$f = \textbf{5 Hz}$$

One cycle takes 0.2 s (one-fifth second), so the period is

$$T = 0.2 \text{ s} = \textbf{200 ms}$$

Related Problem   If the time between negative peaks of a given sine wave is 50 μs, what is the frequency?

EXAMPLE 11–4   The period of a certain sine wave is 10 ms. What is the frequency?

Solution   Use Equation 11–1.

$$f = \frac{1}{T} = \frac{1}{10 \text{ ms}} = \frac{1}{10 \times 10^{-3} \text{ s}} = \textbf{100 Hz}$$

Related Problem   A certain sine wave goes through four cycles in 20 ms. What is the frequency?

EXAMPLE 11–5   The frequency of a sine wave is 60 Hz. What is the period?

Solution   Use Equation 11–2.

$$T = \frac{1}{f} = \frac{1}{60 \text{ Hz}} = \textbf{16.7 ms}$$

Related Problem   If $T = 15 \ \mu$s, what is f?

**SECTION 11–1
REVIEW**
Answers are at the end of the chapter.

1. Describe one cycle of a sine wave.
2. At what point does a sine wave change polarity?
3. How many maximum points does a sine wave have during one cycle?
4. How is the period of a sine wave measured?
5. Define *frequency*, and state its unit.
6. Determine f when $T = 5 \ \mu$s.
7. Determine T when $f = 120$ Hz.

11–2 SINUSOIDAL VOLTAGE SOURCES

Two basic methods of generating sinusoidal voltages are electromagnetic and electronic. Sine waves are produced electromagnetically by ac generators and electronically by oscillator circuits.

After completing this section, you should be able to

◆ **Describe how sine waves are generated**

  ◆ Discuss the basic operation of an ac generator

  ◆ Discuss factors that affect frequency in ac generators

  ◆ Discuss factors that affect voltage in ac generators

An AC Generator

Figure 11–10 shows a greatly simplified ac **generator** consisting of a single loop of wire in a permanent magnetic field. Notice that each end of the wire loop is connected to a separate solid conductive ring called a *slip ring*. A mechanical drive, such as a motor, turns the shaft to which the wire loop is connected. As the wire loop rotates in the magnetic field between the north and south poles, the slip rings also rotate and rub against the brushes that connect the loop to an external load. Compare this generator to the basic dc generator in Figure 10–35, and note the difference in the ring and brush arrangements.

◆ **FIGURE 11–10**

A simplified ac generator.

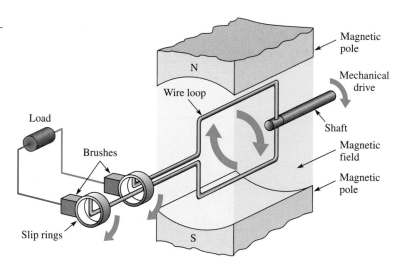

As you learned in Chapter 10, when a conductor moves through a magnetic field, a voltage is induced. Figure 11–11 illustrates how a sinusoidal voltage is produced by the basic ac generator as the wire loop rotates. An oscilloscope is used to display the voltage waveform.

To begin, Figure 11–11(a) shows the wire loop rotating through the first quarter of a revolution. It goes from an instantaneous horizontal position, where the induced voltage is zero, to an instantaneous vertical position, where the induced voltage is maximum. At the horizontal position, the loop is instantaneously moving parallel with the flux lines, which exist between the north (N) and south (S) poles of the magnet. Thus, no lines are being cut and the voltage is zero. As the wire loop rotates through the first quarter-cycle, it cuts through the flux lines at an increasing rate until it is instantaneously moving perpendicular to the flux lines at the vertical position and cutting through them at a maximum rate. Thus, the induced voltage increases from zero to a peak during the quarter-cycle. As shown on the

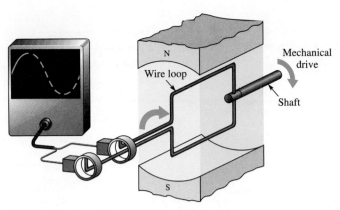

(a) First quarter-cycle (positive alternation)

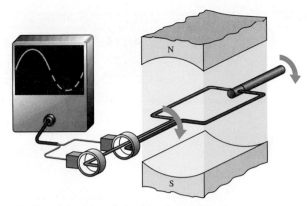

(b) Second quarter-cycle (positive alternation)

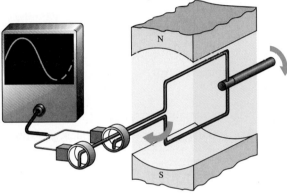

(c) Third quarter-cycle (negative alternation)

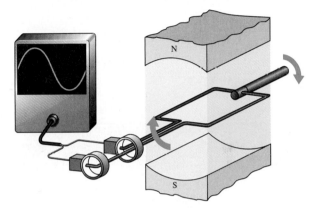

(d) Fourth quarter-cycle (negative alternation)

▲ FIGURE 11–11

One revolution of the wire loop generates one cycle of the sinusoidal voltage.

display in part (a), this part of the rotation produces the first quarter of the sine wave cycle as the voltage builds up from zero to its positive maximum.

Figure 11–11(b) shows the loop completing the first half of a revolution. During this part of the rotation, the voltage decreases from its positive maximum back to zero as the rate at which the loop cuts through the flux lines decreases.

During the second half of the revolution, illustrated in Figures 11–11(c) and 11–11(d), the loop is cutting through the magnetic field in the opposite direction, so the voltage produced has a polarity opposite to that produced during the first half of the revolution. After one complete revolution of the loop, one full cycle of the sinusoidal voltage has been produced. As the wire loop continues to rotate, repetitive cycles of the sine wave are generated.

Frequency   You have seen that one revolution of the conductor through the magnetic field in the basic ac generator (also called an *alternator*) produces one cycle of induced sinusoidal voltage. It is obvious that the rate at which the conductor is rotated determines the time for completion of one cycle. For example, if the conductor completes 60 revolutions in one second (rps), the period of the resulting sine wave is 1/60 s, corresponding to a frequency of 60 Hz. Thus, the faster the conductor rotates, the higher the resulting frequency of the induced voltage, as illustrated in Figure 11–12.

Another way of achieving a higher frequency is to increase the number of magnetic poles. In the previous discussion, two magnetic poles were used to illustrate the ac generator principle. During one revolution, the conductor passes under a north pole and a south pole, thus producing one cycle of a sine wave. When four magnetic poles are used instead of two, as shown in Figure 11–13, one cycle is generated during one-half a revolution. This doubles the frequency for the same rate of rotation.

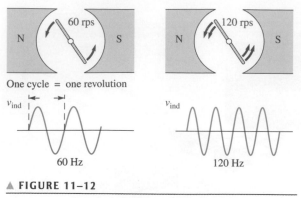

▲ FIGURE 11–12

Frequency is directly proportional to the rate of rotation of the wire loop in an ac generator.

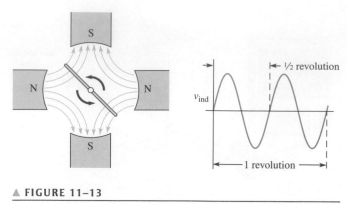

▲ FIGURE 11–13

Four poles achieve a higher frequency than two for the same rps.

An expression for frequency in terms of the number of pole pairs and the number of revolutions per second (rps) is as follows:

Equation 11–3

$$f = (\text{number of pole pairs})(\text{rps})$$

EXAMPLE 11–6

A four-pole generator has a rotation speed of 100 rps. Determine the frequency of the output voltage.

Solution

$$f = (\text{number of pole pairs})(\text{rps}) = 2(100\,\text{rps}) = \textbf{200 Hz}$$

Related Problem   If the frequency of the output of a four-pole generator is 60 Hz, what is the rps?

Voltage Amplitude   Recall from Faraday's law that the voltage induced in a conductor depends on the number of turns (N), and the rate of change with respect to the magnetic field. Therefore, when the speed of rotation of the conductor is increased, not only the frequency of the induced voltage increases—so also does the **amplitude**, which is its maximum value. Since the frequency value normally is fixed, the most practical method of increasing the amount of induced voltage is to increase the number of wire loops.

Electronic Signal Generators

The signal generator is an instrument that electronically produces sine waves for use in testing or controlling electronic circuits and systems. There are a variety of signal generators, ranging from special-purpose instruments that produce only one type of waveform in a limited frequency range, to programmable instruments that produce a wide range of frequencies and a variety of waveforms. All signal generators consist basically of an **oscillator**, which is an electronic circuit that produces repetitive waves. All generators have controls for adjusting the amplitude and frequency.

Function Generators and Arbitrary Waveform Generators   A **function generator** is an instrument that produces more than one type of waveform. It provides pulse waveforms as well as sine waves and triangular waves. A typical function generator is shown in Figure 11–14(a).

An arbitrary waveform generator can be used to generate standard signals like sine waves, triangular waves, and pulses as well as signals with various shapes and characteristics. Waveforms can be defined by mathematical or graphical input. A typical arbitrary waveform generator is shown in Figure 11–14(b).

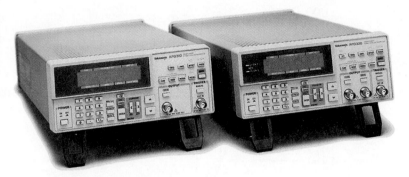

(a) Examples of function generators

(b) A typical arbitrary waveform generator

▲ FIGURE 11–14

Typical signal generators. Copyright © Tektronix, Inc. Reproduced by permission.

SECTION 11–2 REVIEW	1. What two basic methods are used to generate sinusoidal voltages?
	2. How are the speed of rotation and the frequency in an ac generator related?
	3. What is an oscillator?

11–3 SINUSOIDAL VOLTAGE AND CURRENT VALUES

Five ways to express the value of a sine wave in terms of its voltage or its current magnitude are instantaneous, peak, peak-to-peak, rms, and average values.

After completing this section, you should be able to

◆ **Determine the various voltage and current values of a sine wave**

 ◆ Find the instantaneous value at any point

 ◆ Find the peak value

 ◆ Find the peak-to-peak value

 ◆ Define *rms*

 ◆ Explain why the average value is always zero over a complete cycle

 ◆ Find the half-cycle average value

Instantaneous Value

Figure 11–15 illustrates that at any point in time on a sine wave, the voltage (or current) has an **instantaneous value**. This instantaneous value is different at different points along the curve. Instantaneous values are positive during the positive alternation and negative during the negative alternation. Instantaneous values of voltage and current are symbolized by lowercase v and i, respectively. The curve in part (a) shows voltage only, but it applies equally for current when the v's are replaced with i's. An example of instantaneous values is shown in part (b) where the instantaneous voltage is 3.1 V at 1 μs, 7.07 V at 2.5 μs, 10 V at 5 μs, 0 V at 10 μs, −3.1 V at 11 μs, and so on.

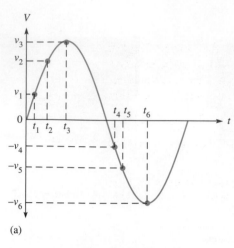

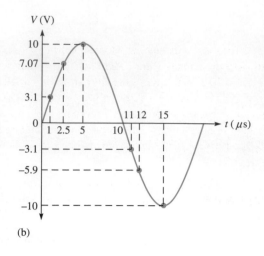

(a)   (b)

▲ FIGURE 11–15

Instantaneous values.

Peak Value

The **peak value** of a sine wave is the value of voltage (or current) at the positive or the negative maximum (peak) with respect to zero. Since the positive and negative peak values are equal in **magnitude**, a sine wave is characterized by a single peak value. This is illustrated in Figure 11–16. For a given sine wave, the peak value is constant and is represented by V_p or I_p.

▶ FIGURE 11–16

Peak values.

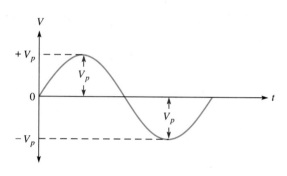

Peak-to-Peak Value

The **peak-to-peak value** of a sine wave, as shown in Figure 11–17, is the voltage or current from the positive peak to the negative peak. It is always twice the peak value as expressed in the following equations. Peak-to-peak voltage or current values are represented by V_{pp} or I_{pp}.

Equation 11–4

$$V_{pp} = 2V_p$$

Equation 11–5

$$I_{pp} = 2I_p$$

▶ FIGURE 11–17

Peak-to-peak value.

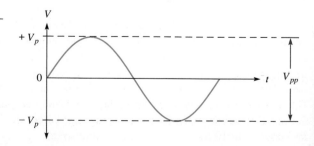

RMS Value

The term *rms* stands for *root mean square*. Most ac voltmeters display rms voltage. The 110 volts at your wall outlet is an rms value. The **rms value**, also referred to as the **effective value**, of a sinusoidal voltage is actually a measure of the heating effect of the sine wave. For example, when a resistor is connected across an ac (sinusoidal) voltage source, as shown in Figure 11–18(a), a certain amount of heat is generated by the power in the resistor. Figure 11–18(b) shows the same resistor connected across a dc voltage source. The value of the dc voltage can be adjusted so that the resistor gives off the same amount of heat as it does when connected to the ac source.

The rms value of a sinusoidal voltage is equal to the dc voltage that produces the same amount of heat in a resistance as does the sinusoidal voltage.

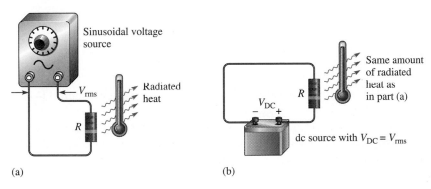

(a)   (b)

▲ **FIGURE 11–18**

When the same amount of heat is produced in both setups, the sinusoidal voltage has an rms value equal to the dc voltage.

The peak value of a sine wave can be converted to the corresponding rms value using the following relationships, derived in Appendix B, for either voltage or current:

$$V_{rms} = 0.707V_p$$

Equation 11–6

$$I_{rms} = 0.707I_p$$

Equation 11–7

Using these formulas, you can also determine the peak value if you know the rms value.

$$V_p = \frac{V_{rms}}{0.707}$$

$$V_p = 1.414V_{rms}$$

Equation 11–8

Similarly,

$$I_p = 1.414I_{rms}$$

Equation 11–9

To get the peak-to-peak value, simply double the peak value.

$$V_{pp} = 2.828V_{rms}$$

Equation 11–10

and

$$I_{pp} = 2.828I_{rms}$$

Equation 11–11

Average Value

The average value of a sine wave taken over one complete cycle is always zero because the positive values (above the zero crossing) offset the negative values (below the zero crossing).

To be useful for certain purposes such as measuring types of voltages found in power supplies, the average value of a sine wave is defined over a half-cycle rather than over a full cycle. The **average value** is the total area under the half-cycle curve divided by the distance in radians of the curve along the horizontal axis. The result is derived in Appendix B and is expressed in terms of the peak value as follows for both voltage and current sine waves:

Equation 11–12

$$V_{avg} = \left(\frac{2}{\pi}\right)V_p$$

$$V_{avg} = 0.637V_p$$

$$I_{avg} = \left(\frac{2}{\pi}\right)I_p$$

Equation 11–13

$$I_{avg} = 0.637I_p$$

EXAMPLE 11–7

Determine V_p, V_{pp}, V_{rms}, and the half-cycle V_{avg} for the sine wave in Figure 11–19.

▶ **FIGURE 11–19**

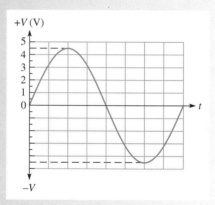

Solution $V_p = \textbf{4.5 V}$ is read directly from the graph. From this, calculate the other values.

$$V_{pp} = 2V_p = 2(4.5\ \text{V}) = \textbf{9 V}$$
$$V_{rms} = 0.707V_p = 0.707(4.5\ \text{V}) = \textbf{3.18 V}$$
$$V_{avg} = 0.637V_p = 0.637(4.5\ \text{V}) = \textbf{2.87 V}$$

Related Problem   If $V_p = 25$ V, determine V_{pp}, V_{rms}, and V_{avg} for a voltage sine wave.

SECTION 11–3 REVIEW

1. Determine V_{pp} in each case when
   (a) $V_p = 1$ V   (b) $V_{rms} = 1.414$ V   (c) $V_{avg} = 3$ V
2. Determine V_{rms} in each case when
   (a) $V_p = 2.5$ V   (b) $V_{pp} = 10$ V   (c) $V_{avg} = 1.5$ V
3. Determine the half-cycle V_{avg} in each case when
   (a) $V_p = 10$ V   (b) $V_{rms} = 2.3$ V   (c) $V_{pp} = 60$ V

11–4 ANGULAR MEASUREMENT OF A SINE WAVE

As you have seen, sine waves can be measured along the horizontal axis on a time basis; however, since the time for completion of one full cycle or any portion of a cycle is frequency-dependent, it is often useful to specify points on the sine wave in terms of an angular measurement expressed in degrees or radians.

After completing this section, you should be able to

◆ **Describe angular relationships of sine waves**

  ◆ Show how to measure a sine wave in terms of angles

  ◆ Define *radian*

  ◆ Convert radians to degrees

  ◆ Determine the phase angle of a sine wave

A sinusoidal voltage can be produced by an ac generator. As the windings on the rotor of the ac generator go through a full 360° of rotation, the resulting voltage output is one full cycle of a sine wave. Thus, the angular measurement of a sine wave can be related to the angular rotation of a generator, as shown in Figure 11–20.

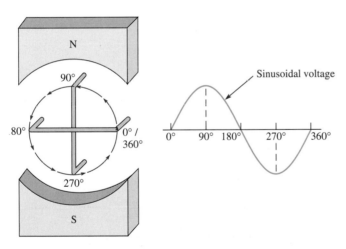

▲ FIGURE 11–20

Relationship of a sine wave to the rotational motion in an ac generator.

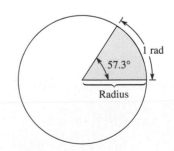

▲ FIGURE 11–21

Angular measurement showing relationship of radian (rad) to degrees (°).

Angular Measurement

A **degree** is an angular measurement corresponding to 1/360 of a circle or a complete revolution. A **radian** is the angular measurement along the circumference of a circle that is equal to the radius of the circle. One radian (rad) is equivalent to 57.3°, as illustrated in Figure 11–21. In a 360° revolution, there are 2π radians.

> The Greek letter π (pi) represents the ratio of the circumference of any circle to its diameter and has a constant value of approximately 3.1416.

Scientific calculators have a π function so that the actual numerical value does not have to be entered.

► TABLE 11–1

DEGREES (°)	RADIANS (RAD)
0	0
45	$\pi/4$
90	$\pi/2$
135	$3\pi/4$
180	π
225	$5\pi/4$
270	$3\pi/2$
315	$7\pi/4$
360	2π

Table 11–1 lists several values of degrees and the corresponding radian values. These angular measurements are illustrated in Figure 11–22.

► FIGURE 11–22

Angular measurements starting at 0° and going counterclockwise.

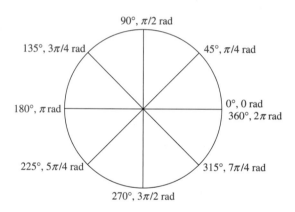

Radian/Degree Conversion

Degrees can be converted to radians.

Equation 11–14
$$\text{rad} = \left(\frac{\pi \text{ rad}}{180°}\right) \times \text{degrees}$$

Similarly, radians can be converted to degrees.

Equation 11–15
$$\text{degrees} = \left(\frac{180°}{\pi \text{ rad}}\right) \times \text{rad}$$

EXAMPLE 11–8   (a) Convert 60° to radians.   (b) Convert $\pi/6$ rad to degrees.

Solution   (a) Rad $= \left(\dfrac{\pi \text{ rad}}{180°}\right)60° = \dfrac{\pi}{3}$ **rad**   (b) Degrees $= \left(\dfrac{180°}{\pi \text{ rad}}\right)\left(\dfrac{\pi}{6}\text{ rad}\right) = \textbf{30°}$

Related Problem   (a) Convert 15° to radians.   (b) Convert $5\pi/8$ rad to degrees.

Sine Wave Angles

The angular measurement of a sine wave is based on 360° or 2π rad for a complete cycle. A half-cycle is 180° or π rad; a quarter-cycle is 90° or $\pi/2$ rad; and so on. Figure 11–23(a) shows angles in degrees for a full cycle of a sine wave; part (b) shows the same points in radians.

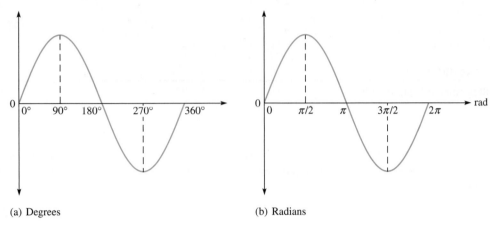

(a) Degrees   (b) Radians

▲ FIGURE 11–23

Sine wave angles.

Phase of a Sine Wave

The **phase** of a sine wave is an angular measurement that specifies the position of that sine wave relative to a reference. Figure 11–24 shows one cycle of a sine wave to be used as the reference. Note that the first positive-going crossing of the horizontal axis (zero crossing) is at 0° (0 rad), and the positive peak is at 90° ($\pi/2$ rad). The negative-going zero crossing is at 180° (π rad), and the negative peak is at 270° ($3\pi/2$ rad). The cycle is completed at 360° (2π rad). When the sine wave is shifted left or right with respect to this reference, there is a phase shift.

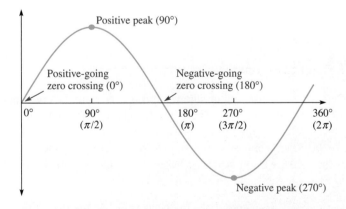

▲ FIGURE 11–24

Phase reference.

Figure 11–25 illustrates phase shifts of a sine wave. In part (a), sine wave B is shifted to the right by 90° ($\pi/2$ rad) with respect to sine wave A. Thus, there is a phase angle of 90° between sine wave A and sine wave B. In terms of time, the positive peak of sine wave B occurs later than the positive peak of sine wave A because time increases to the right along the horizontal axis. In this case, sine wave B is said to **lag** sine wave A by 90° or $\pi/2$ radians. Stated another way, sine wave A leads sine wave B by 90°.

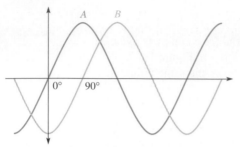

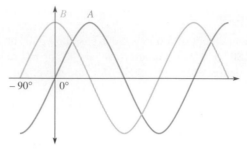

a) *A* leads *B* by 90°, or *B* lags *A* by 90°.

(b) *B* leads *A* by 90°, or *A* lags *B* by 90°.

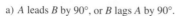

▲ FIGURE 11–25

Illustration of a phase shift.

In Figure 11–25(b), sine wave *B* is shown shifted left by 90° with respect to sine wave *A*. Thus, again there is a phase angle of 90° between sine wave *A* and sine wave *B*. In this case, the positive peak of sine wave *B* occurs earlier in time than that of sine wave *A*; therefore, sine wave *B* is said to **lead** sine wave *A* by 90°.

EXAMPLE 11–9

What are the phase angles between the two sine waves in parts (a) and (b) of Figure 11–26?

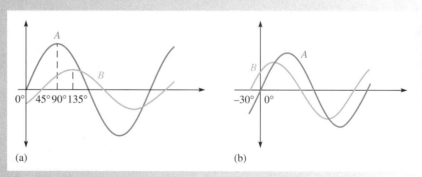

(a)

(b)

▲ FIGURE 11–26

Solution   In Figure 11–26(a) the zero crossing of sine wave *A* is at 0°, and the corresponding zero crossing of sine wave *B* is at 45°. There is a 45° phase angle between the two waveforms with sine wave *B* lagging sine wave *A*.

In Figure 11–26(b) the zero crossing of sine wave *B* is at −30°, and the corresponding zero crossing of sine wave *A* is at 0°. There is a 30° phase angle between the two waveforms with sine wave *B* leading sine wave *A*.

Related Problem   If the positive-going zero crossing of one sine wave is at 15° and that of the second sine wave is at 23°, what is the phase angle between them?

As a practical matter, when you measure the phase shift between two waveforms on an oscilloscope, you should make them appear to have the same amplitude. This is done by taking one of the oscilloscope channels out of vertical calibration and adjusting the corresponding waveform until its apparent amplitude equals that of the other waveform. This procedure eliminates the error caused if both waveforms are not measured at their exact center.

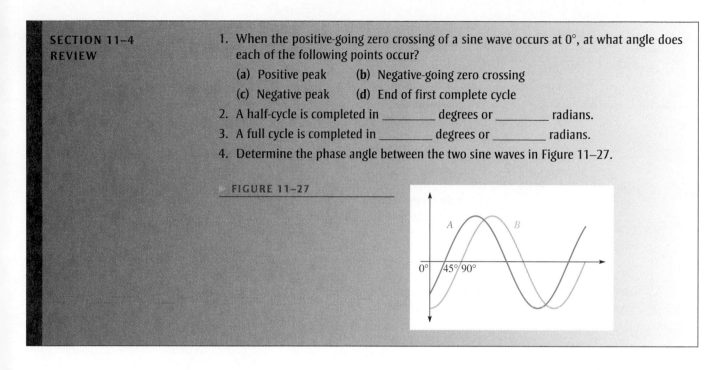

SECTION 11–4 REVIEW

1. When the positive-going zero crossing of a sine wave occurs at 0°, at what angle does each of the following points occur?
   (a) Positive peak   (b) Negative-going zero crossing
   (c) Negative peak   (d) End of first complete cycle
2. A half-cycle is completed in _____ degrees or _____ radians.
3. A full cycle is completed in _____ degrees or _____ radians.
4. Determine the phase angle between the two sine waves in Figure 11–27.

◢ FIGURE 11–27

11–5 THE SINE WAVE FORMULA

A sine wave can be graphically represented by voltage or current values on the vertical axis and by angular measurement (degrees or radians) along the horizontal axis. This graph can be expressed mathematically, as you will see.

After completing this section, you should be able to

- ◆ **Mathematically analyze a sinusoidal waveform**
  - ◆ State the sine wave formula
  - ◆ Find instantaneous values using the sine wave formula

A generalized graph of one cycle of a sine wave is shown in Figure 11–28. The sine wave amplitude (A) is the maximum value of the voltage or current on the vertical axis; angular values run along the horizontal axis. The variable y is an instantaneous value that represents either voltage or current at a given angle, θ. The symbol θ is the Greek letter *theta*.

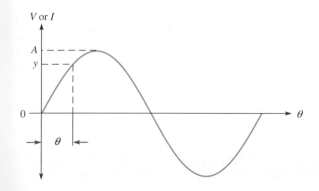

◀ **FIGURE 11–28**

One cycle of a generic sine wave showing amplitude and phase.

All electrical sine waves follow a specific mathematical formula. The general expression for the sine wave curve in Figure 11–28 is

Equation 11–16

$$y = A \sin \theta$$

This formula states that any point on the sine wave, represented by an instantaneous value (y), is equal to the maximum value A times the sine (sin) of the angle θ at that point. For example, a certain voltage sine wave has a peak value of 10 V. You can calculate the instantaneous voltage at a point 60° along the horizontal axis as follows, where $y = v$ and $A = V_p$:

$$v = V_p\sin \theta = (10 \text{ V})\sin 60° = (10 \text{ V})(0.866) = 8.66 \text{ V}$$

Figure 11–29 shows this particular instantaneous value of the curve. You can find the sine of any angle on most calculators by first entering the value of the angle and then pressing the sin key. Verify that your calculator is in the degree mode.

▶ **FIGURE 11–29**

Illustration of the instantaneous value of a voltage sine wave at $\theta = 60°$.

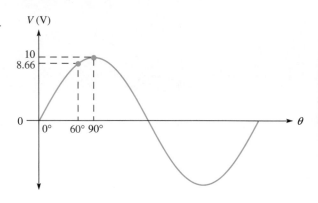

Expressions for Phase-Shifted Sine Waves

When a sine wave is shifted to the right of the reference (lagging) by a certain angle, ϕ (Greek letter phi), as illustrated in Figure 11–30(a) where the reference is the vertical axis, the general expression is

Equation 11–17

$$y = A \sin(\theta - \phi)$$

where y represents instantaneous voltage or current, and A represents the peak value (amplitude). When a sine wave is shifted to the left of the reference (leading) by a certain angle, ϕ, as shown in Figure 11–30(b), the general expression is

Equation 11–18

$$y = A \sin(\theta + \phi)$$

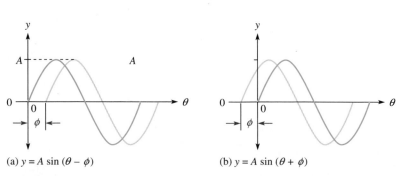

(a) $y = A \sin (\theta - \phi)$ (b) $y = A \sin (\theta + \phi)$

▲ **FIGURE 11–30**

Shifted sine waves.

EXAMPLE 11–10

Determine the instantaneous value at the 90° reference point on the horizontal axis for each voltage sine wave in Figure 11–31.

▶ **FIGURE 11–31**

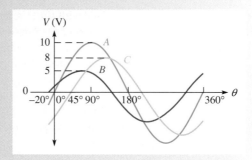

Solution Sine wave *A* is the reference. Sine wave *B* is shifted left by 20° with respect to *A*, so it leads. Sine wave *C* is shifted right by 45° with respect to *A*, so it lags.

$$v_A = V_p \sin \theta$$
$$= (10\ \text{V})\sin(90°) = (10\ \text{V})(1) = \textbf{10 V}$$

$$v_B = V_p \sin(\theta + \phi_B)$$
$$= (5\ \text{V})\sin(90° + 20°) = (5\ \text{V})\sin(110°) = (5\ \text{V})(0.9397) = \textbf{4.70 V}$$

$$v_C = V_p \sin(\theta - \phi_C)$$
$$= (8\ \text{V})\sin(90° - 45°) = (8\ \text{V})\sin(45°) = (8\ \text{V})(0.7071) = \textbf{5.66 V}$$

Related Problem A voltage sine wave has a peak value of 20 V. What is its instantaneous value at 65° from its zero crossing?

SECTION 11–5 REVIEW

1. Calculate the instantaneous value at 120° for the voltage sine wave in Figure 11–29.
2. Determine the instantaneous value at 45° of a voltage sine wave that leads the reference by 10° ($V_p = 10$ V).
3. Find the instantaneous value of 90° of a voltage sine wave that leads the reference by 25° ($V_p = 5$ V).

11–6 INTRODUCTION TO PHASORS

Phasors provide a graphic means for representing quantities that have both magnitude and direction (angular position). Phasors are especially useful for representing sine waves in terms of their magnitude and phase angle and also for analysis of reactive circuits discussed in later chapters.

After completing this section, you should be able to

◆ **Use a phasor to represent a sine wave**

   ◆ Define *phasor*

   ◆ Explain how phasors are related to the sine wave formula

   ◆ Draw a phasor diagram

   ◆ Discuss angular velocity

You may already be familiar with vectors. In math and science, a vector is any quantity with both magnitude and direction. Examples of vectors are force, velocity, and acceleration. The simplest way to describe a vector is to assign a magnitude and an angle to a quantity.

In electronics, a **phasor** is a type of vector but the term generally refers to quantities that vary with time, such as sine waves. Examples of phasors are shown in Figure 11–32. The length of the phasor "arrow" represents the magnitude of a quantity. The angle, θ (relative to 0°), represents the angular position, as shown in part (a) for a positive angle. The specific phasor example in part (b) has a magnitude of 2 and a phase angle of 45°. The phasor in part (c) has a magnitude of 3 and a phase angle of 180°. The phasor in part (d) has a magnitude of 1 and a phase angle of −45° (or +315°). Notice that positive angles are measured counterclockwise (CCW) from the reference (0°) and negative angles are measured clockwise (CW) from the reference.

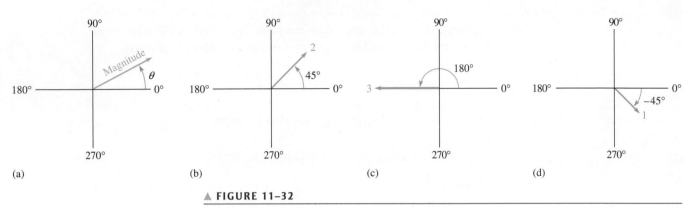

(a)   (b)   (c)   (d)

▲ **FIGURE 11–32**

Examples of phasors.

Phasor Representation of a Sine Wave

A full cycle of a sine wave can be represented by rotation of a phasor through 360 degrees.

> **The instantaneous value of the sine wave at any point is equal to the vertical distance from the tip of the phasor to the horizontal axis.**

Figure 11–33 shows how the phasor traces out the sine wave as it goes from 0° to 360°. You can relate this concept to the rotation in an ac generator. Notice that the length of the phasor is equal to the peak value of the sine wave (observe the 90° and the 270° points). The angle of the phasor measured from 0° is the corresponding angular point on the sine wave.

▶ **FIGURE 11–33**

Sine wave represented by rotational phasor motion.

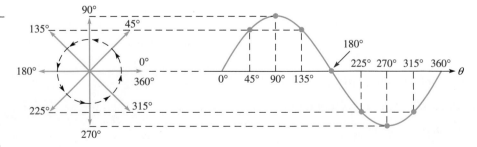

Phasors and the Sine Wave Formula

Let's examine a phasor representation at one specific angle. Figure 11–34 shows a voltage phasor at an angular position of 45° and the corresponding point on the sine wave. The instantaneous value of the sine wave at this point is related to both the position and the length

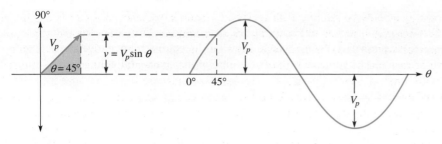

▲ FIGURE 11–34

Right triangle derivation of sine wave formula.

of the phasor. As previously mentioned, the vertical distance from the phasor tip down to the horizontal axis represents the instantaneous value of the sine wave at that point.

Notice that when a vertical line is drawn from the phasor tip down to the horizontal axis, a right triangle is formed, as shown shaded in Figure 11–34. The length of the phasor is the hypotenuse of the triangle, and the vertical projection is the opposite side. From trigonometry,

The opposite side of a right triangle is equal to the hypotenuse times the sine of the angle θ.

The length of the phasor is the peak value of the sinusoidal voltage, V_p. Thus, the opposite side of the triangle, which is the instantaneous value, can be expressed as

$$v = V_p \sin \theta$$

Recall that this formula is the one stated earlier for calculating instantaneous sinusoidal voltage. A similar formula applies to a sinusoidal current.

$$i = I_p \sin \theta$$

Positive and Negative Phasor Angles

The position of a phasor at any instant can be expressed as a positive angle, as you have seen, or as an equivalent negative angle. Positive angles are measured counterclockwise from 0°. Negative angles are measured clockwise from 0°. For a given positive angle θ, the corresponding negative angle is $\theta - 360°$, as illustrated in Figure 11–35(a). In part (b), a specific example is shown. The angle of the phasor in this case can be expressed as +225° or −135°.

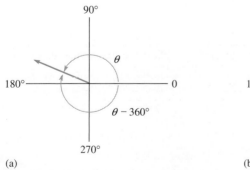

(a)

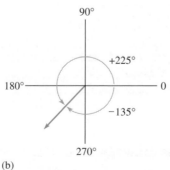

(b)

▲ FIGURE 11–35

Positive and negative phasor angles.

EXAMPLE 11-11

For the phasor in each part of Figure 11–36, determine the instantaneous voltage value. Also express each positive angle shown as an equivalent negative angle. The length of each phasor represents the peak value of the sinusoidal voltage.

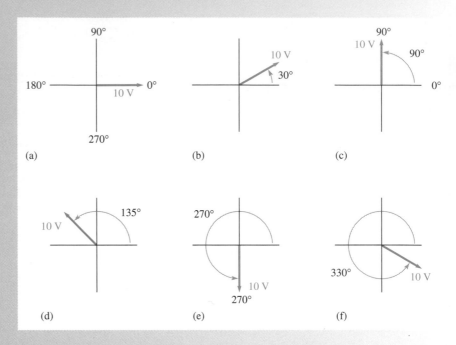

▲ FIGURE 11–36

Solution

(a) $v = (10\,\text{V})\sin 0° = (10\,\text{V})(0) = \mathbf{0\,V}$
$0° - 360° = \mathbf{-360°}$

(b) $v = (10\,\text{V})\sin 30° = (10\,\text{V})(0.5) = \mathbf{5\,V}$
$30° - 360° = \mathbf{-330°}$

(c) $v = (10\,\text{V})\sin 90° = (10\,\text{V})(1) = \mathbf{10\,V}$
$90° - 360° = \mathbf{-270°}$

(d) $v = (10\,\text{V})\sin 135° = (10\,\text{V})(0.707) = \mathbf{7.07\,V}$
$135° - 360° = \mathbf{-225°}$

(e) $v = (10\,\text{V})\sin 270° = (10\,\text{V})(-1) = \mathbf{-10\,V}$
$270° - 360° = \mathbf{-90°}$

(f) $v = (10\,\text{V})\sin 330° = (10\,\text{V})(-0.5) = \mathbf{-5\,V}$
$330° - 360° = \mathbf{-30°}$

Related Problem

If a phasor is at 45° and its length represents 15 V, what is the instantaneous sine wave value?

Phasor Diagrams

A phasor diagram can be used to show the relative relationship of two or more sine waves of the same frequency. A phasor in a *fixed position* is used to represent a complete sine wave because once the phase angle between two or more sine waves of the same frequency or between the sine wave and a reference is established, the phase angle remains constant throughout the cycles. For example, the two sine waves in Figure 11–37(a) can be represented by a phasor diagram, as shown in part (b). As you can see, sine wave *B* leads sine wave *A* by 30° and has less amplitude than sine wave *A*, as indicated by the lengths of the phasors.

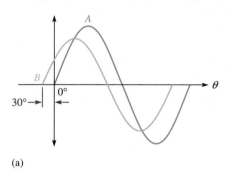 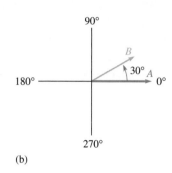

(a)   (b)

▲ FIGURE 11–37

Example of a phasor diagram representing sinusoidal waveforms.

EXAMPLE 11–12   Use a phasor diagram to represent the sine waves in Figure 11–38.

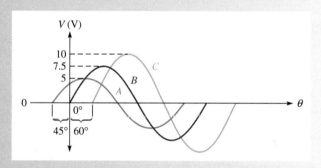

▲ FIGURE 11–38

Solution   The phasor diagram representing the sine waves is shown in Figure 11–39. The length of each phasor represents the peak value of the sine wave.

▶ FIGURE 11–39

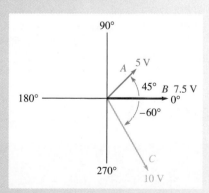

Related Problem   Describe a phasor to represent a 5 V peak sine wave that lags sine wave *C* in Figure 11–38 by 25°.

Angular Velocity of a Phasor

As you have seen, one cycle of a sine wave is traced out when a phasor is rotated through 360 degrees or 2π radians. The faster it is rotated, the faster the sine wave cycle is traced out. Thus, the period and frequency are related to the velocity of rotation of the phasor. The velocity of rotation is called the **angular velocity** and is designated ω (the small Greek letter omega).

When a phasor rotates through 2π radians, one complete cycle is traced out. Therefore, the time required for the phasor to go through 2π radians is the period of the sine wave. Because the phasor rotates through 2π radians in a time equal to the period, T, the angular velocity can be expressed as

$$\omega = \frac{2\pi}{T}$$

Since $f = 1/T$,

Equation 11–19
$$\omega = 2\pi f$$

When a phasor is rotated at an angular velocity ω, then ωt is the angle through which the phasor has passed at any instant. Therefore, the following relationship can be stated:

Equation 11–20
$$\theta = \omega t$$

Substituting $2\pi f$ for ω results in $\theta = 2\pi ft$. With this relationship between angle and time, the equation for the instantaneous value of a sinusoidal voltage, $v = V_p \sin\theta$, can be written as

Equation 11–21
$$v = V_p \sin 2\pi ft$$

You can calculate the instantaneous value at any point in time along the sine wave curve if you know the frequency and peak value. The unit of $2\pi ft$ is the radian so your calculator must be in the radian mode.

EXAMPLE 11–13

What is the value of a sinusoidal voltage at 3 μs from the positive-going zero crossing when $V_p = 10\,\text{V}$ and $f = 50\,\text{kHz}$?

Solution
$$v = V_p \sin 2\pi ft$$
$$= (10\,\text{V})\sin[2\pi(50\,\text{kHz})(3 \times 10^{-6}\,\text{s})] = \textbf{8.09 V}$$

Related Problem
What is the value of a sinusoidal voltage at 12 μs from the positive-going zero crossing when $V_p = 50\,\text{V}$ and $f = 10\,\text{kHz}$?

SECTION 11–6 REVIEW

1. What is a phasor?
2. What is the angular velocity of a phasor representing a sine wave with a frequency of 1500 Hz?
3. A certain phasor has an angular velocity of 628 rad/s. To what frequency does this correspond?

4. Draw a phasor diagram to represent the two sine waves in Figure 11–40. Use peak values.

◢ **FIGURE 11–40**

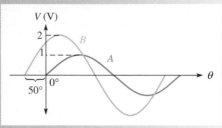

11–7 ANALYSIS OF AC CIRCUITS

When a time-varying ac voltage such as a sinusoidal voltage is applied to a circuit, the circuit laws and power formulas that you studied earlier still apply. Ohm's law, Kirchhoff's laws, and the power formulas apply to ac circuits in the same way that they apply to dc circuits.

After completing this section, you should be able to

* **Apply the basic circuit laws to resistive ac circuits**
  * Apply Ohm's law to resistive circuits with ac sources
  * Apply Kirchhoff's voltage law and current law to resistive circuits with ac sources
  * Determine power in resistive ac circuits

If a sinusoidal voltage is applied across a resistor as shown in Figure 11–41, there is a sinusoidal current. The current is zero when the voltage is zero and is maximum when the voltage is maximum. When the voltage changes polarity, the current reverses direction. As a result, the voltage and current are said to be in phase with each other.

Sine wave generator

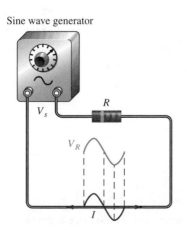

◀ **FIGURE 11–41**

A sinusoidal voltage produces a sinusoidal current.

When you use Ohm's law in ac circuits, remember that both the voltage and the current must be expressed consistently, that is, both as peak values, both as rms values, both as average values, and so on. Kirchhoff's voltage and current laws apply to ac circuits as well as to dc circuits. Figure 11–42 illustrates Kirchhoff's voltage law in a resistive circuit that has a sinusoidal voltage source. The source voltage is the sum of all the voltage drops across the resistors, just as in a dc circuit.

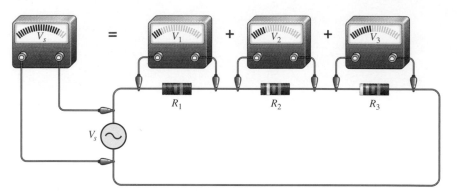

▲ FIGURE 11–42

Illustration of Kirchhoff's voltage law in an ac circuit.

Power in resistive ac circuits is determined the same as for dc circuits except that you must use rms values of current and voltage. Recall that the rms value of a sine wave voltage is equivalent to a dc voltage of the same value in terms of its heating effect. The general power formulas are restated for a resistive ac circuit as

$$P = V_{rms}I_{rms}$$

$$P = \frac{V_{rms}^2}{R}$$

$$P = I_{rms}^2 R$$

EXAMPLE 11–14

Determine the rms voltage across each resistor and the rms current in Figure 11–43. The source voltage is given as an rms value. Also determine the total power.

▶ FIGURE 11–43

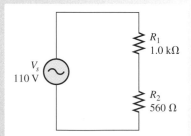

Solution   The total resistance of the circuit is

$$R_{tot} = R_1 + R_2 = 1.0\,k\Omega + 560\,\Omega = 1.56\,k\Omega$$

Use Ohm's law to find the rms current.

$$I_{rms} = \frac{V_{s(rms)}}{R_{tot}} = \frac{110\,V}{1.56\,k\Omega} = \textbf{70.5 mA}$$

The rms voltage drop across each resistor is

$$V_{1(rms)} = I_{rms}R_1 = (70.5 \, \text{mA})(1.0 \, \text{k}\Omega) = \textbf{70.5 V}$$
$$V_{2(rms)} = I_{rms}R_2 = (70.5 \, \text{mA})(560 \, \Omega) = \textbf{39.5 V}$$

The total power is

$$P_{tot} = I_{rms}^2 R_{tot} = (70.5 \, \text{mA})^2(1.56 \, \text{k}\Omega) = 7.75 \, \text{W}$$

Related Problem   Repeat this example for a source voltage of 10 V peak.

Use Multisim file E11-14 to verify the calculated results in this example and to confirm your calculations for the related problem.

EXAMPLE 11–15   All values in Figure 11–44 are given in rms.

(a) Find the unknown peak voltage drop in Figure 11–44(a).

(b) Find the total rms current in Figure 11–44(b).

(c) Find the total power in Figure 11–44(b) if $V_{rms} = 24 \, \text{V}$.

FIGURE 11–44

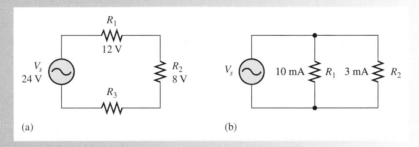

(a)   (b)

Solution   (a) Use Kirchhoff's voltage law to find V_3.

$$V_s = V_1 + V_2 + V_3$$
$$V_{3(rms)} = V_{s(rms)} - V_{1(rms)} - V_{2(rms)} = 24 \, \text{V} - 12 \, \text{V} - 8 \, \text{V} = 4 \, \text{V}$$

Convert rms to peak.

$$V_{3(p)} = 1.414V_{3(rms)} = 1.414(4 \, \text{V}) = \textbf{5.66 V}$$

(b) Use Kirchhoff's current law to find I_{tot}.

$$I_{tot(rms)} = I_{1(rms)} + I_{2(rms)} = 10 \, \text{mA} + 3 \, \text{mA} = \textbf{13 mA}$$

(c) $P_{tot} = V_{rms}I_{rms} = (24 \, \text{V})(13 \, \text{mA}) = \textbf{312 mW}$

Related Problem   A series circuit has the following voltage drops: $V_{1(rms)} = 3.50 \, \text{V}$, $V_{2(p)} = 4.25 \, \text{V}$, $V_{3(avg)} = 1.70 \, \text{V}$. Determine the peak-to-peak source voltage.

SECTION 11–7 REVIEW

1. A sinusoidal voltage with a half-cycle average value of 12.5 V is applied to a circuit with a resistance of 330 Ω. What is the peak current in the circuit?

2. The peak voltage drops in a series resistive circuit are 6.2 V, 11.3 V, and 7.8 V. What is the rms value of the source voltage?

11–8 SUPERIMPOSED DC AND AC VOLTAGES

In many practical circuits, you will find both dc and ac voltages combined. An example of this is in amplifier circuits where ac signal voltages are superimposed on dc operating voltages. This is a common application of the superposition theorem studied in Chapter 8.

After completing this section, you should be able to

◆ **Determine total voltages that have both ac and dc components**

Figure 11–45 shows a dc source and an ac source in series. These two voltages will add algebraically to produce an ac voltage "riding" on a dc level, as measured across the resistor.

▶ FIGURE 11–45

Superimposed dc and ac voltages.

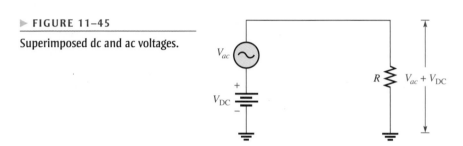

If V_{DC} is greater than the peak value of the sinusoidal voltage, the combined voltage is a sine wave that never reverses polarity and is therefore nonalternating. That is, the sine wave is riding on a dc level, as shown in Figure 11–46(a). If V_{DC} is less than the peak value of the sine wave, the sine wave will be negative during a portion of its lower half-cycle, as illustrated in Figure 11–46(b), and is therefore alternating. In either case, the sine wave will reach a maximum voltage equal to $V_{DC} + V_p$, and it will reach a minimum voltage equal to $V_{DC} - V_p$.

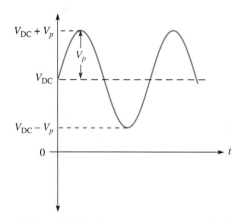

(a) $V_{DC} > V_p$. The sine wave never goes negative.

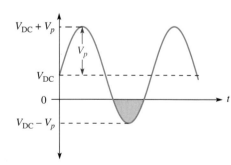

(b) $V_{DC} < V_p$. The sine wave reverses polarity during a portion of its cycle, as indicated by the gray area.

▲ FIGURE 11–46

Sine waves with dc levels.

EXAMPLE 11–16

Determine the maximum and minimum voltage across the resistor in each circuit of Figure 11–47.

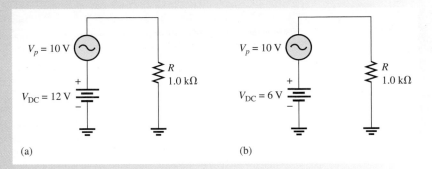

▲ FIGURE 11–47

Solution   In Figure 11–47(a), the maximum voltage across R is

$$V_{max} = V_{DC} + V_p = 12\,V + 10\,V = \mathbf{22\,V}$$

The minimum voltage across R is

$$V_{min} = V_{DC} - V_p = 12\,V - 10\,V = \mathbf{2\,V}$$

Therefore, $V_{R(tot)}$ is a nonalternating sine wave that varies from +22 V to +2 V, as shown in Figure 11–48(a).

In Figure 11–47(b), the maximum voltage across R is

$$V_{max} = V_{DC} + V_p = 6\,V + 10\,V = \mathbf{16\,V}$$

The minimum voltage across R is

$$V_{min} - V_{DC} - V_p = \mathbf{-4\,V}$$

Therefore, $V_{R(tot)}$ is an alternating sine wave that varies from +16 V to −4 V, as shown in Figure 11–48(b).

FIGURE 11–48

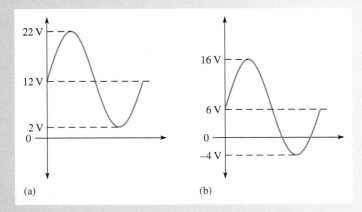

Related Problem   Explain why the waveform in Figure 11–48(a) is nonalternating but the waveform in part (b) is considered to be alternating.

 Use Multisim files E11-16A and E11-16B to verify the calculated results in this example.

11–9 NONSINUSOIDAL WAVEFORMS

Sine waves are important in electronics, but they are by no means the only type of ac or time-varying waveform. Two other major types of waveforms are the pulse waveform and the triangular waveform.

After completing this section, you should be able to

◆ **Identify the characteristics of basic nonsinusoidal waveforms**

  ◆ Discuss the properties of a pulse waveform

  ◆ Define *duty cycle*

  ◆ Discuss the properties of triangular and sawtooth waveforms

  ◆ Discuss the harmonic content of a waveform

Pulse Waveforms

Basically, a **pulse** can be described as a very rapid transition (**leading edge**) from one voltage or current level (**baseline**) to an amplitude level, and then, after an interval of time, a very rapid transition (**trailing edge**) back to the original baseline level. The transitions in level are also called *steps*. An ideal pulse consists of two opposite-going steps of equal amplitude. When the leading or trailing edge is positive-going, it is called a **rising edge**. When the leading or trailing edge is negative-going, it is called a **falling edge**.

Figure 11–49(a) shows an ideal positive-going pulse consisting of two equal but opposite instantaneous steps separated by an interval of time called the *pulse width*. Part (b) of Figure 11–49 shows an ideal negative-going pulse. The height of the pulse measured from the baseline is its voltage (or current) amplitude.

▶ **FIGURE 11–49**

Ideal pulses.

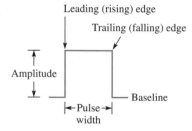

(a) Positive-going pulse

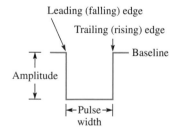

(b) Negative-going pulse

In many applications, analysis is simplified by treating all pulses as ideal (composed of instantaneous steps and perfectly rectangular in shape). Actual pulses, however, are never ideal. All pulses possess certain characteristics that cause them to be different from the ideal.

In practice, pulses cannot change from one level to another instantaneously. Time is always required for a transition (step), as illustrated in Figure 11–50(a). As you can see, there is an interval of time during the rising edge in which the pulse is going from its lower value to its higher value. This interval is called the *rise time, t_r*.

Rise time is the time required for the pulse to go from 10% of its amplitude to 90% of its amplitude.

The interval of time during the falling edge in which the pulse is going from its higher value to its lower value is called the *fall time, t_f*.

Fall time is the time required for the pulse to go from 90% of its amplitude to 10% of its amplitude.

Pulse width, t_W, also requires a precise definition for the nonideal pulse because the rising and falling edges are not vertical.

Pulse width is the time between the point on the rising edge, where the value is 50% of amplitude, to the point on the falling edge, where the value is 50% of amplitude.

Pulse width is shown in Figure 11–50(b).

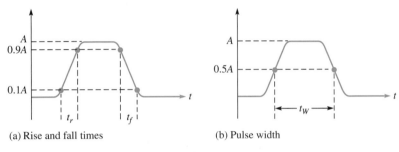

(a) Rise and fall times   (b) Pulse width

▲ FIGURE 11–50

Nonideal pulse.

Repetitive Pulses   Any waveform that repeats itself at fixed intervals is **periodic**. Some examples of periodic pulse waveforms are shown in Figure 11–51. Notice that, in each case, the pulses repeat at regular intervals. The rate at which the pulses repeat is the **pulse repetition frequency**, which is the fundamental frequency of the waveform. The frequency can be expressed in hertz or in pulses per second. The time from one pulse to the corresponding point on the next pulse is the period, T. The relationship between frequency and period is the same as with the sine wave, $f = 1/T$.

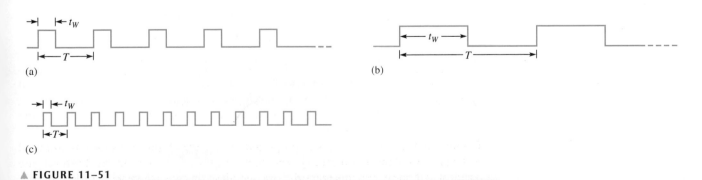

▲ FIGURE 11–51

Repetitive pulse waveforms.

An important characteristic of repetitive pulse waveforms is the duty cycle.

The duty cycle is the ratio of the pulse width (t_W) to the period (T) and is usually expressed as a percentage.

Equation 11–22

$$\text{percent duty cycle} = \left(\frac{t_W}{T}\right)100\%$$

EXAMPLE 11–17

Determine the period, frequency, and duty cycle for the pulse waveform in Figure 11–52.

▲ FIGURE 11–52

Solution

$$T = 10\ \mu s$$

$$f = \frac{1}{T} = \frac{1}{10\ \mu s} = 100\ \text{kHz}$$

$$\text{percent duty cycle} = \left(\frac{1\ \mu s}{10\ \mu s}\right)100\% = 10\%$$

Related Problem

A certain pulse waveform has a frequency of 200 kHz and a pulse width of 0.25 μs. Determine the duty cycle.

Square Waves A square wave is a pulse waveform with a duty cycle of 50%. Thus, the pulse width is equal to one-half of the period. A square wave is shown in Figure 11–53.

▲ FIGURE 11–53

Square wave.

The Average Value of a Pulse Waveform The average value (V_{avg}) of a pulse waveform is equal to its baseline value plus its duty cycle times its amplitude. The lower level of a positive-going waveform or the upper level of a negative-going waveform is taken as the baseline. The formula is as follows:

Equation 11–23

$$V_{avg} = \text{baseline} + (\text{duty cycle})(\text{amplitude})$$

The following example illustrates the calculation of the average value.

EXAMPLE 11–18

Determine the average value of each of the waveforms in Figure 11–54.

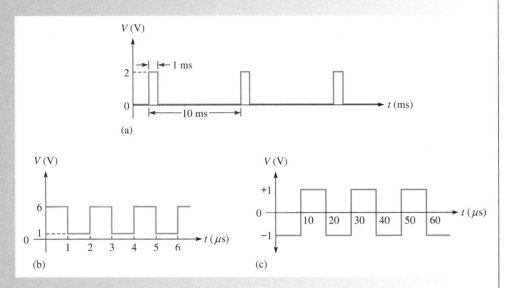

▲ **FIGURE 11–54**

Solution

In Figure 11–54(a), the baseline is at 0 V, the amplitude is 2 V, and the duty cycle is 10%. The average value is

$$V_{avg} = \text{baseline} + (\text{duty cycle})(\text{amplitude})$$
$$= 0\ V + (0.1)(2\ V) = \mathbf{0.2\ V}$$

The waveform in Figure 11–54(b) has a baseline of +1 V, an amplitude of 5 V, and a duty cycle of 50%. The average value is

$$V_{avg} = \text{baseline} + (\text{duty cycle})(\text{amplitude})$$
$$= 1\ V + (0.5)(5\ V) = 1\ V + 2.5\ V = \mathbf{3.5\ V}$$

Figure 11–54(c) shows a square wave with a baseline of −1 V and an amplitude of 2 V. The average value is

$$V_{avg} = \text{baseline} + (\text{duty cycle})(\text{amplitude})$$
$$= -1\ V + (0.5)(2\ V) = -1\ V + 1\ V = \mathbf{0\ V}$$

This is an alternating square wave, and, as with an alternating sine wave, it has an average of zero.

Related Problem

If the baseline of the waveform in Figure 11–54(a) is shifted to 1 V, what is the average value?

Triangular and Sawtooth Waveforms

Triangular and sawtooth waveforms are formed by voltage or current ramps. A **ramp** is a linear increase or decrease in the voltage or current. Figure 11–55 shows both positive- and negative-going ramps. In part (a), the ramp has a positive slope; in part (b), the ramp has a negative slope. The slope of a voltage ramp is $\pm V/t$ and is expressed in units of V/s. The slope of a current ramp is $\pm I/t$ and is expressed in units of A/s.

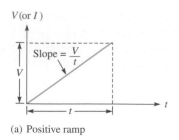

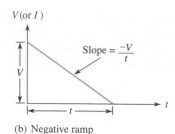

(a) Positive ramp   (b) Negative ramp

▲ FIGURE 11–55

Ramps.

EXAMPLE 11–19

What are the slopes of the voltage ramps in Figure 11–56?

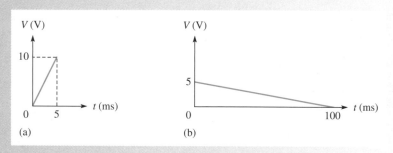

(a)   (b)

▲ FIGURE 11–56

Solution   In Figure 11–56(a), the voltage increases from 0 V to +10 V in 5 ms. Thus, $V = 10$ V and $t = 5$ ms. The slope is

$$\frac{V}{t} = \frac{10 \text{ V}}{5 \text{ ms}} = \textbf{2 V/ms}$$

In Figure 11–56(b), the voltage decreases from +5 V to 0 V in 100 ms. Thus, $V = -5$ V and $t = 100$ ms. The slope is

$$\frac{V}{t} = \frac{-5 \text{ V}}{100 \text{ ms}} = \textbf{-0.05 V/ms}$$

Related Problem   A certain voltage ramp has a slope of $+12$ V/μs. If the ramp starts at zero, what is the voltage at 0.01 ms?

Triangular Waveforms   Figure 11–57 shows that a **triangular waveform** is composed of positive-going and negative-going ramps having equal slopes. The period of this waveform

▶ **FIGURE 11–57**

Alternating triangular waveform with a zero average value.

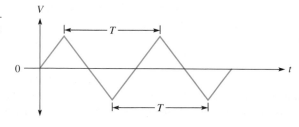

is measured from one peak to the next corresponding peak, as illustrated. This particular triangular waveform is alternating and has an average value of zero.

Figure 11–58 depicts a triangular waveform with a nonzero average value. The frequency for triangular waves is determined in the same way as for sine waves, that is, $f = 1/T$.

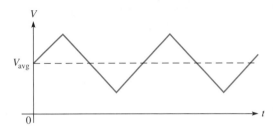

◀ FIGURE 11–58

Nonalternating triangular waveform with a nonzero average value.

Sawtooth Waveforms   The **sawtooth waveform** is actually a special case of the triangular wave consisting of two ramps, one of much longer duration than the other. Sawtooth waveforms are used in many electronic systems. For example, the electron beam that sweeps across the screen of your TV receiver, creating the picture, is controlled by sawtooth voltages and currents. One sawtooth wave produces the horizontal beam movement, and the other produces the vertical beam movement. A sawtooth voltage is sometimes called a *sweep voltage*.

Figure 11–59 is an example of a sawtooth wave. Notice that it consists of a positive-going ramp of relatively long duration, followed by a negative-going ramp of relatively short duration.

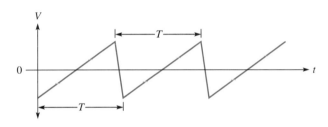

▲ FIGURE 11–59

Alternating sawtooth waveform.

Harmonics

A repetitive nonsinusoidal waveform is composed of a fundamental frequency and harmonic frequencies. The **fundamental frequency** is the repetition rate of the waveform, and the **harmonics** are higher frequency sine waves that are multiples of the fundamental.

Odd Harmonics   *Odd harmonics* are frequencies that are odd multiples of the fundamental frequency of a waveform. For example, a 1 kHz square wave consists of a fundamental of 1 kHz and odd harmonics of 3 kHz, 5 kHz, 7 kHz, and so on. The 3 kHz frequency in this case is called the third harmonic, the 5 kHz frequency is the fifth harmonic, and so on.

Even Harmonics   *Even harmonics* are frequencies that are even multiples of the fundamental frequency. For example, if a certain wave has a fundamental of 200 Hz, the second harmonic is 400 Hz, the fourth harmonic is 800 Hz, the sixth harmonic is 1200 Hz, and so on. These are even harmonics.

Composite Waveform   Any variation from a pure sine wave produces harmonics. A nonsinusoidal wave is a composite of the fundamental and the harmonics. Some types of waveforms have only odd harmonics, some have only even harmonics, and some contain both. The shape of the wave is determined by its harmonic content. Generally, only the fundamental and the first few harmonics are of significant importance in determining the wave shape.

A square wave is an example of a waveform that consists of a fundamental and only odd harmonics. When the instantaneous values of the fundamental and each odd harmonic are added algebraically at each point, the resulting curve will have the shape of a square wave, as illustrated in Figure 11–60. In part (a) of the figure, the fundamental and the third harmonic produce a wave shape that begins to resemble a square wave. In part (b), the fundamental, third, and fifth harmonics produce a closer resemblance. When the seventh harmonic is included, as in part (c), the resulting wave shape becomes even more like a square wave. As more harmonics are included, a periodic square wave is approached.

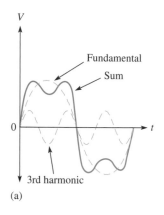
(a)

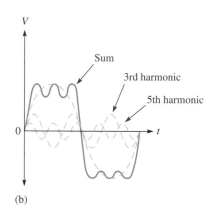
(b)

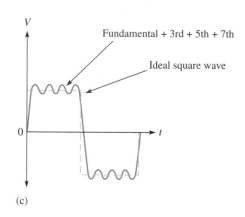
(c)

▲ FIGURE 11–60

Odd harmonics produce a square wave.

SECTION 11–9 REVIEW

1. Define the following parameters:
   (a) rise time   **(b)** fall time   **(c)** pulse width
2. In a certain repetitive pulse waveform, the pulses occur once every millisecond. What is the frequency of this waveform?
3. Determine the duty cycle, amplitude, and average value of the waveform in Figure 11–61(a).
4. What is the period of the triangular wave in Figure 11–61(b)?
5. What is the frequency of the sawtooth wave in Figure 11–61(c)?

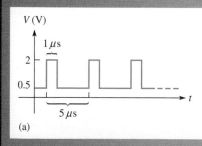
(a)

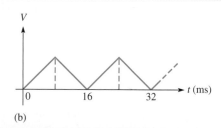
(b)

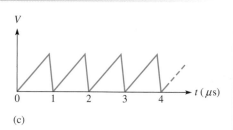
(c)

▲ FIGURE 11–61

6. Define *fundamental frequency*.
7. What is the second harmonic of a fundamental frequency of 1 kHz?
8. What is the fundamental frequency of a square wave having a period of 10 μs?

11–10 THE OSCILLOSCOPE

The oscilloscope (scope for short) is a widely used and versatile test instrument for observing and measuring waveforms.

After completing this section, you should be able to

* **Use an oscilloscope to measure waveforms**
  * Recognize common oscilloscope controls
  * Measure the amplitude of a waveform
  * Measure the period and frequency of a waveform

The **oscilloscope** is basically a graph-displaying device that traces a graph of a measured electrical signal on its screen. In most applications, the graph shows how signals change over time. The vertical axis of the display screen represents voltage, and the horizontal axis represents time. You can measure amplitude, period, and frequency of a signal using an oscilloscope. Also, you can determine the pulse width, duty cycle, rise time, and fall time of a pulse waveform. Most scopes can display at least two signals on the screen at one time, enabling you to observe their time relationship. Typical oscilloscopes are shown in Figure 11–62.

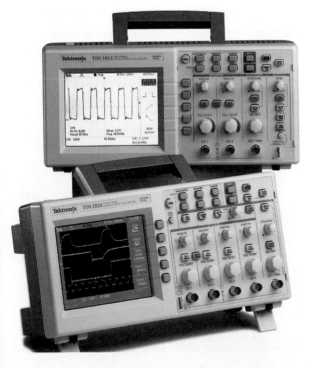

◀ FIGURE 11–62

Typical oscilloscopes. Copyright © Tektronix. Reproduced by permission.

Two basic types of oscilloscopes, analog and digital, can be used to view digital waveforms. As shown in Figure 11–63(a), the analog scope works by applying the measured waveform directly to control the up and down motion of the electron beam in the cathode-ray tube (CRT) as it sweeps across the screen. As a result, the beam traces out the waveform pattern on the screen. As shown in Figure 11–63(b), the digital scope converts the measured waveform to digital information by a sampling process in an analog-to-digital converter (ADC). The digital information is then used to reconstruct the waveform on the screen.

▶ FIGURE 11–63

Comparison of analog and digital
oscilloscopes.

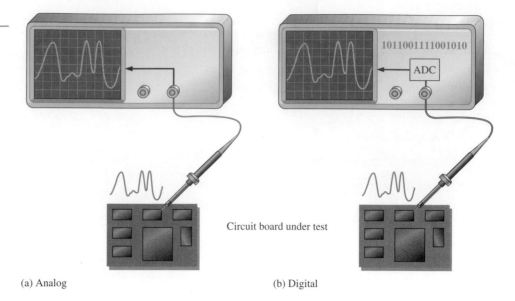

(a) Analog   (b) Digital

The digital scope is more widely used than the analog scope. However, either type can be
used in many applications; each has characteristics that make it more suitable for certain situations. An analog scope displays waveforms as they occur in "real time." Digital scopes are
useful for measuring transient pulses that may occur randomly or only once. Also, because
information about the measured waveform can be stored in a digital scope, it may be viewed
at some later time, printed out, or thoroughly analyzed by a computer or other means.

Basic Operation of Analog Oscilloscopes   To measure a voltage, a probe must be connected from the scope to the point in a circuit at which the voltage is present. Generally, a
$\times 10$ probe is used that reduces (attenuates) the signal amplitude by ten. The signal goes
through the probe into the vertical circuits where it is either further attenuated or amplified,
depending on the actual amplitude and on where you set the vertical control of the scope.
The vertical circuits then drive the vertical deflection plates of the CRT. Also, the signal
goes to the trigger circuits that trigger the horizontal circuits to initiate repetitive horizontal sweeps of the electron beam across the screen using a sawtooth waveform. There are
many sweeps per second so that the beam appears to form a solid line across the screen in
the shape of the waveform. This basic operation is illustrated in Figure 11–64.

▶ FIGURE 11–64

Block diagram of an analog
oscilloscope.

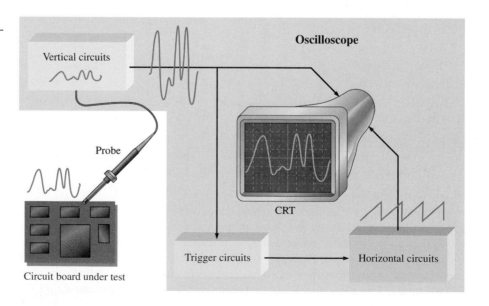

Basic Operation of Digital Oscilloscopes   Some parts of a digital scope are similar to the analog scope. However, the digital scope is more complex than an analog scope and typically has an LCD screen rather than a CRT. Rather than displaying a waveform as it occurs, the digital scope first acquires the measured analog waveform and converts it to a digital format using an analog-to-digital converter (ADC). The digital data is stored and processed. The data then goes to the reconstruction and display circuits for display in its original analog form. Figure 11–65 show a basic block diagram for a digital oscilloscope.

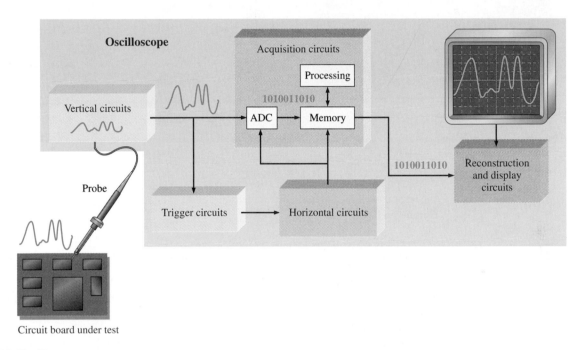

▲ FIGURE 11–65

Block diagram of a digital oscilloscope.

Oscilloscope Controls

A front panel view of a typical dual-channel oscilloscope is shown in Figure 11–66. Instruments vary depending on model and manufacturer, but most have certain common features. For example, the two vertical sections contain a Position control, a channel menu button, and a Volts/Div control. The horizontal section contains a Sec/Div control.

Some of the main controls are now discussed. Refer to the user manual for complete details of your particular scope.

Vertical Controls   In the vertical section of the scope in Figure 11–66, there are identical controls for each of the two channels (CH1 and CH2). The Position control lets you move a displayed waveform up or down vertically on the screen. The Menu button provides for the selection of several items that appear on the screen, such as the coupling modes (ac, dc, or ground) and coarse or fine adjustment for the Volts/Div, as indicated in Figure 11–67(a). The Volts/Div control adjusts the number of volts represented by each vertical division on the screen. The Volts/Div setting for each channel is displayed on the bottom of the screen. The Math Menu button provides a selection of operations that can be performed on the input waveforms, such as subtraction and addition of signals, as indicated in Figure 11–67(b).

Horizontal Controls   In the horizontal section, the controls apply to both channels. The Position control lets you move a displayed waveform left or right horizontally on the

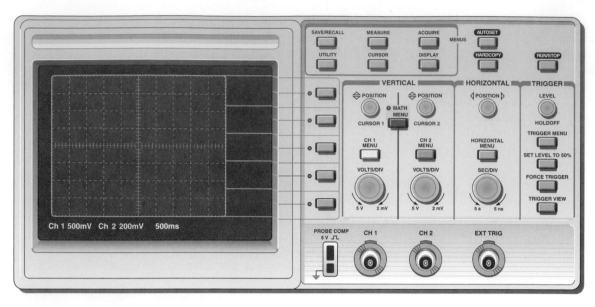

▲ FIGURE 11–66

A typical dual-channel oscilloscope. Numbers below screen indicate the values for each division on the vertical (voltage) and horizontal (time) scales and can be varied using the vertical and horizontal controls on the scope.

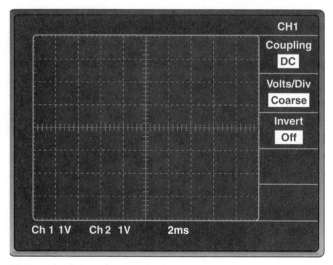

(a) Example of channel menu selection

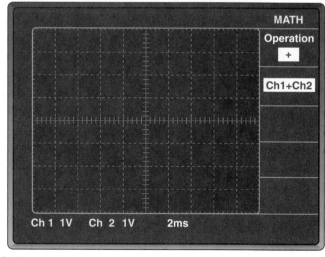

(b) Example of math menu selection

▲ FIGURE 11–67

Scope screens showing examples of menu selections.

screen. The Menu button provides for the selection of several items that appear on the screen such as the main time base, expanded view of a portion of a waveform, and other parameters. The Sec/Div control adjusts the time represented by each horizontal division or main time base. The Sec/Div setting is displayed at the bottom of the screen.

Trigger Controls In the Trigger section, the Level control determines the point on the triggering waveform where triggering occurs to initiate the sweep to display input waveforms. The Menu button provides for the selection of several items that appear on the screen including edge or slope triggering, trigger source, trigger mode, and other parameters, as shown in Figure 11–68. There is also an input for an external trigger signal.

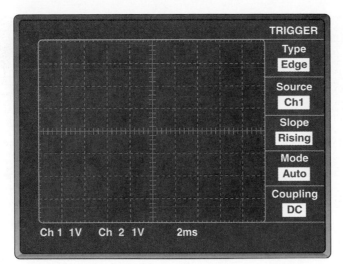

Triggering stabilizes a waveform on the screen and properly triggers on a pulse that occurs only one time or randomly. Also, it allows you to observe time delays between two waveforms. Figure 11–69 compares a triggered to an untriggered signal. The untriggered signal tends to drift across the screen producing what appears to be multiple waveforms.

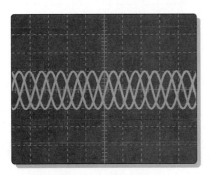

(a) Untriggered waveform display

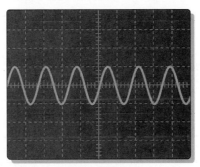

(b) Triggered waveform display

◀ FIGURE 11–69

Comparison of an untriggered and a triggered waveform on an oscilloscope.

Coupling a Signal into the Scope   Coupling is the method used to connect a signal voltage to be measured into the oscilloscope. The DC and AC coupling modes are selected from the Vertical menu. DC coupling allows a waveform including its dc component to be displayed. AC coupling blocks the dc component of a signal so that you see the waveform centered at 0 V. The Ground mode allows you to connect the channel input to ground to see where the 0 V reference is on the screen. Figure 11–70 illustrates the result of DC and AC coupling using a sinusoidal waveform that has a dc component.

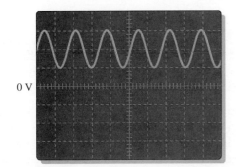

(a) DC coupled waveform

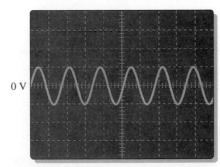

(b) AC coupled waveform

▶ FIGURE 11–71

An oscilloscope voltage probe. Copyright © Tektronix, Inc. Reproduced by permission.

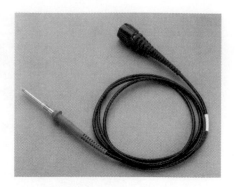

The voltage probe, shown in Figure 11–71, is used for connecting a signal to the scope. Since all instruments tend to affect the circuit being measured due to loading, most scope probes provide a high series resistance to minimize loading effects. Probes that have a series resistance ten times larger than the input resistance of the scope are called ×10 (times ten) probes. Probes with no series resistance are called ×1 (times one) probes. The oscilloscope adjusts its calibration for the attenuation of the type of probe being used. For most measurements, the ×10 probe should be used. However, if you are measuring very small signals, a ×1 may be the best choice.

The probe has an adjustment that allows you to compensate for the input capacitance of the scope. Most scopes have a probe compensation output that provides a calibrated square wave for probe compensation. Before making a measurement, you should make sure that the probe is properly compensated to eliminate any distortion introduced. Typically, there is a screw or other means of adjusting compensation on a probe. Figure 11–72 shows scope waveforms for three probe conditions: properly compensated, undercompensated, and overcompensated. If the waveform appears either over- or under-compensated, adjust the probe until the properly compensated square wave is achieved.

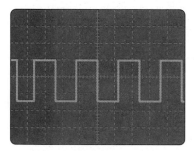

Properly compensated

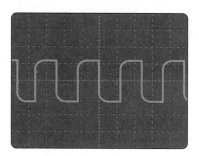

Undercompensated

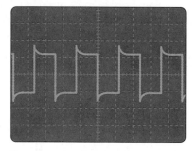

Overcompensated

▲ FIGURE 11–72

Probe compensation conditions.

EXAMPLE 11–20

Determine the peak-to-peak value and period of each sine wave in Figure 11–73 from the digital scope screen displays and the settings for Volts/Div and Sec/Div, which are indicated under the screens. Sine waves are centered vertically on the screens.

Solution　Looking at the vertical scale in Figure 11–73(a),

$$V_{pp} = 6 \text{ divisions} \times 0.5 \text{ V/division} = \textbf{3.0 V}$$

From the horizontal scale (one cycle covers ten divisions),

$$T = 10 \text{ divisions} \times 2 \text{ ms/division} = \textbf{20 ms}$$

Looking at the vertical scale in Figure 11–73(b),

$$V_{pp} = 5 \text{ divisions} \times 50 \text{ mV/division} = \mathbf{250\ mV}$$

From the horizontal scale (one cycle covers six divisions),

$$T = 6 \text{ divisions} \times 0.1 \text{ ms/division} = 0.6 \text{ ms} = \mathbf{600\ \mu s}$$

Looking at the vertical scale in Figure 11–73(c),

$$V_{pp} = 6.8 \text{ divisions} \times 2 \text{ V/division} = \mathbf{13.6\ V}$$

From the horizontal scale (one-half cycle covers ten divisions),

$$T = 20 \text{ divisions} \times 10 \text{ } \mu s/\text{division} = \mathbf{200\ \mu s}$$

Looking at the vertical scale in Figure 11–73(d),

$$V_{pp} = 4 \text{ divisions} \times 5 \text{ V/division} = \mathbf{20\ V}$$

From the horizontal scale (one cycle covers two divisions),

$$T = 2 \text{ divisions} \times 2 \text{ } \mu s/\text{division} = \mathbf{4\ \mu s}$$

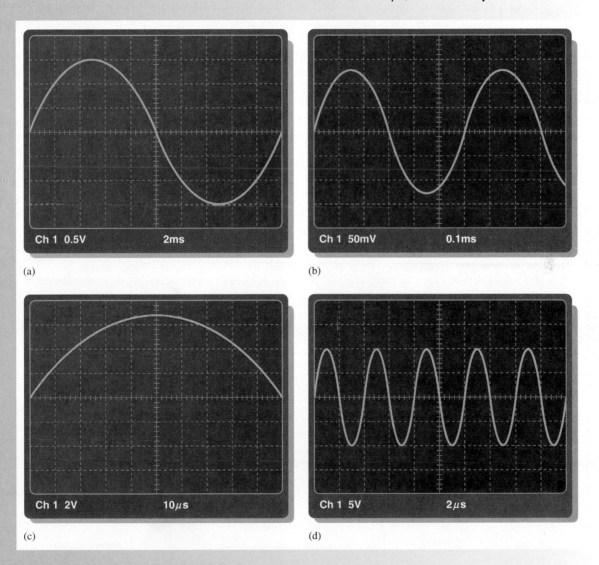

Ch 1 0.5V   2ms

(a)

Ch 1 50mV   0.1ms

(b)

Ch 1 2V   10 μs

(c)

Ch 1 5V   2 μs

(d)

FIGURE 11–73

Related Problem   Determine the rms value and the frequency for each waveform displayed in Figure 11–73.

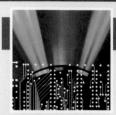

A Circuit Application

As you learned in this chapter, nonsinusoidal waveforms contain a combination of various harmonic frequencies.

Each of these harmonics is a sinusoidal waveform with a certain frequency. Certain sinusoidal frequencies are audible; that is, they can be heard by the human ear. A single audible frequency, or pure sine wave, is called a tone and generally falls in the frequency range from about 300 Hz to about 15 kHz. When you hear a tone reproduced through a speaker, its loudness, or volume, depends on its voltage amplitude. You will use your knowledge of sine wave characteristics and the operation of an oscilloscope to measure the frequency and amplitude of signals at various points in a basic radio receiver.

Actual voice or music signals that are picked up by a radio receiver contain many harmonic frequencies with different voltage values. A voice or music signal is continuously changing, so its harmonic content is also changing. However, if a single sinusoidal frequency is transmitted and picked up by the receiver, you will hear a constant tone from the speaker.

Although, at this point you do not have the background to study amplifiers and receiver systems in detail, you can observe the signals at various points in the receiver. A block diagram of a typical AM receiver is shown in Figure 11–74. AM stands for

amplitude modulation, a topic that will be covered in another course. Figure 11–75 shows what a basic AM signal looks like, and for now that's all you need to know. As you can see, the amplitude of the sinusoidal waveform is changing. The higher radio frequency (RF) signal is called the *carrier;* and its amplitude is varied or modulated by a lower frequency signal, which is the audio (a tone in this case). Normally, however, the audio signal is a complex voice or music waveform.

Oscilloscope Measurements

Signals that are indicated by circled numbers at several test points on the receiver block diagram in Figure 11–74 are displayed on the oscilloscope screen in Figure 11–76 as indicated by the same corresponding circled numbers. In all cases, the upper waveform on the screen is channel 1 and the lower waveform is channel 2. The readings on the bottom of the screen show readings for both channels.

The signal at point 1 is an AM signal, but you can't see the amplitude variation because of the short time base. The waveform is spread out too much to see the modulating audio signal, which causes amplitude variations; so what you see is just one cycle of the carrier. At point 3, the higher carrier frequency is difficult to determine because the time base was selected to

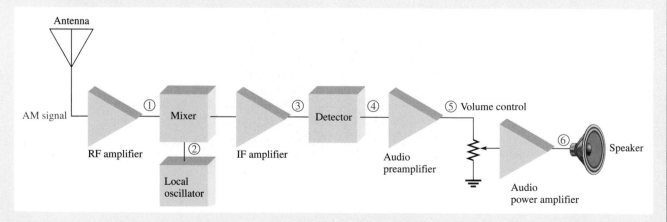

▲ FIGURE 11–74

Simplified block diagram of a basic radio receiver. Circled numbers represent test points.

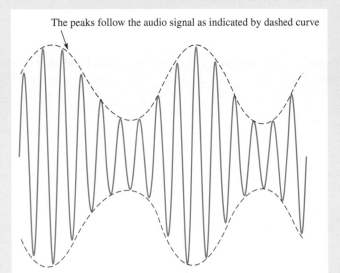

The peaks follow the audio signal as indicated by dashed curve

Example of an amplitude modulated (AM) signal.

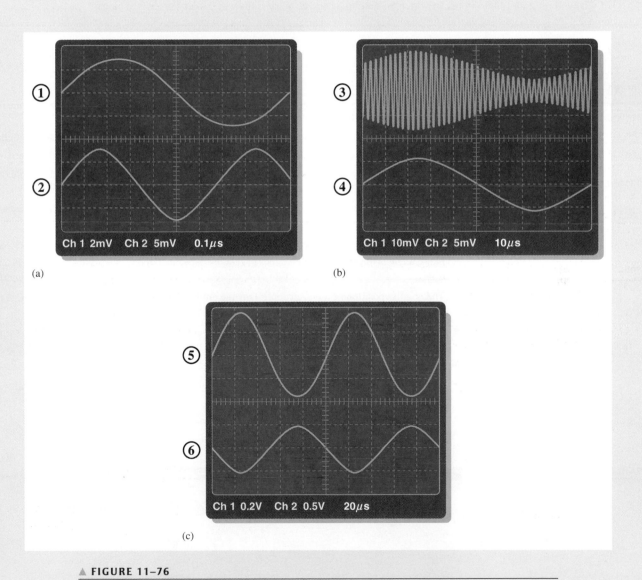

(a)

Ch 1 2mV   Ch 2 5mV   0.1μs

(b)

Ch 1 10mV   Ch 2 5mV   10μs

(c)

Ch 1 0.2V   Ch 2 0.5V   20μs

▲ FIGURE 11–76

Circled numbers correspond to the numbered test points in Figure 11–74.

allow viewing of one full cycle of the modulating signal. In an AM receiver, this intermediate frequency is 455 kHz. In actual practice, the modulated carrier signal at point 3 cannot easily be viewed on the scope because it contains two frequencies that make it difficult to synchronize in order to obtain a stable pattern. Sometimes external triggering or TV field is used to obtain a stable display. A stable pattern is shown in this case to illustrate what the modulated waveform looks like.

◆ For each waveform in Figure 11–76, except point 3, determine the frequency and rms value. The signal at point 4 is the modulating tone extracted by the detector from the higher intermediate frequency (455 kHz).

Amplifier Analysis

◆ All voltage amplifiers have a characteristic known as voltage gain. The voltage gain is the amount by which the amplitude of the output signal is greater than the amplitude of the input signal. Using this definition and the appropriate scope measurements, determine the gain of the audio preamplifier in this particular receiver.

◆ When an electrical signal is converted to sound by a speaker, the loudness of the sound depends on the amplitude of the signal applied to the speaker. Based on this, explain how the volume control potentiometer is used to adjust the loudness (volume) of the sound and determine the rms amplitude at the speaker.

Review

1. What does RF stand for?
2. What does IF stand for?
3. Which frequency is higher, the carrier or the audio?
4. What is the variable in a given AM signal?

SUMMARY

◆ The sine wave is a time-varying, periodic waveform.

◆ Alternating current changes direction in response to changes in the polarity of the source voltage.

◆ One cycle of an alternating sine wave consists of a positive alternation and a negative alternation.

◆ Two common sources of sine waves are the electromagnetic ac generator and the electronic oscillator circuit.

◆ A full cycle of a sine wave is 360°, or 2π radians. A half-cycle is 180°, or π radian. A quarter-cycle is 90°, or $\pi/2$ radians.

◆ A sinusoidal voltage can be generated by a conductor rotating in a magnetic field.

◆ Phase angle is the difference in degrees or radians between a given sine wave and a reference sine wave.

◆ The angular position of a phasor represents the angle of the sine wave with respect to a 0° reference, and the length or magnitude of a phasor represents the amplitude.

◆ A pulse consists of a transition from a baseline level to an amplitude level, followed by a transition back to the baseline level.

◆ A triangle or sawtooth wave consists of positive-going and negative-going ramps.

◆ Harmonic frequencies are odd or even multiples of the repetition rate of a nonsinusoidal waveform.

◆ Conversions of sine wave values are summarized in Table 11–2.

▶ TABLE 11–2

TO CHANGE FROM	TO	MULTIPLY BY
Peak	rms	0.707
Peak	Peak-to-peak	2
Peak	Average	0.637
rms	Peak	1.414
Peak-to-peak	Peak	0.5
Average	Peak	1.57

KEY TERMS

Key terms and other bold terms in the chapter are defined in the end-of-book glossary.

Amplitude (A)   The maximum value of a voltage or current.

Angular velocity   The rotational rate of a phasor that is related to the frequency of the sine wave that it represents.

Average value   The average of a sine wave over one half-cycle. It is 0.637 times the peak value.

Cycle   One repetition of a periodic waveform.

Degree   The unit of angular measure corresponding to 1/360 of a complete revolution.

Duty cycle   A characteristic of a pulse waveform that indicates the percentage of time that a pulse is present during a cycle; the ratio of pulse width to period expressed as either a fraction or as a percentage.

Fall time (t_f)   The time interval required for a pulse to change from 90% to 10% of its amplitude.

Frequency (f)   A measure of the rate of change of a periodic function; the number of cycles completed in 1 s. The unit of frequency is the hertz.

Function generator   An instrument that produces more than one type of waveform.

Fundamental frequency   The repetition rate of a waveform.

Harmonics   The frequencies contained in a composite waveform, which are integer multiples of the pulse repetition frequency (fundamental).

Hertz (Hz)   The unit of frequency. One hertz equals one cycle per second.

Instantaneous value   The voltage or current value of a waveform at a given instant in time.

Oscillator   An electronic circuit that produces a time-varying signal without an external input signal using positive feedback.

Oscilloscope   A measurement instrument that displays signal waveforms on a screen.

Peak-to-peak value   The voltage or current value of a waveform measured from its minimum to its maximum points.

Peak value   The voltage or current value of a waveform at its maximum positive or negative points.

Period (T)   The time interval of one complete cycle of a periodic waveform.

Periodic   Characterized by a repetition at fixed-time intervals.

Phase   The relative angular displacement of a time-varying waveform in terms of its occurrence with respect to a reference.

Phasor   A representation of a sine wave in terms of its magnitude (amplitude) and direction (phase angle).

Pulse   A type of waveform that consists of two equal and opposite steps in voltage or current separated by a time interval.

Pulse width (t_W)   For a nonideal pulse, the time between the 50% points on the leading and trailing edges; the time interval between the opposite steps of an ideal pulse.

Radian   A unit of angular measurement. There are 2π radians in one complete 360° revolution. One radian equals 57.3°.

Ramp   A type of waveform characterized by a linear increase or decrease in voltage or current.

Rise time (t_r)   The time interval required for a pulse to change from 10% to 90% of its amplitude.

rms value   The value of a sinusoidal voltage that indicates its heating effect, also known as the effective value. It is equal to 0.707 times the peak value. *rms* stands for root mean square.

Sine wave   A type of waveform that follows a cyclic sinusoidal pattern defined by the formula $y = A \sin\theta$.

Waveform   The pattern of variations of a voltage or current showing how the quantity changes with time.

FORMULAS

11–1 $f = \dfrac{1}{T}$ Frequency

11–2 $T = \dfrac{1}{f}$ Period

11–3	$f = $ (number of pole pairs)(rps)	Output frequency of a generator
11–4	$V_{pp} = 2V_p$	Peak-to-peak voltage (sine wave)
11–5	$I_{pp} = 2I_p$	Peak-to-peak current (sine wave)
11–6	$V_{rms} = 0.707V_p$	Root-mean-square voltage (sine wave)
11–7	$I_{rms} = 0.707I_p$	Root-mean-square current (sine wave)
11–8	$V_p = 1.414V_{rms}$	Peak voltage (sine wave)
11–9	$I_p = 1.414I_{rms}$	Peak current (sine wave)
11–10	$V_{pp} = 2.828V_{rms}$	Peak-to-peak voltage (sine wave)
11–11	$I_{pp} = 2.828I_{rms}$	Peak to peak current (sine wave)
11–12	$V_{avg} = 0.637V_p$	Half-cycle average voltage (sine wave)
11–13	$I_{avg} = 0.637I_p$	Half-cycle average current (sine wave)
11–14	$rad = \left(\dfrac{\pi \text{ rad}}{180°}\right) \times degrees$	Degrees to radian conversion
11–15	$degrees = \left(\dfrac{180°}{\pi \text{ rad}}\right) \times rad$	Radian to degrees conversion
11–16	$y = A \sin\theta$	General formula for a sine wave
11–17	$y = A \sin(\theta - \phi)$	Sine wave lagging the reference
11–18	$y = A \sin(\theta + \phi)$	Sine wave leading the reference
11–19	$\omega = 2\pi f$	Angular velocity
11–20	$\theta = \omega t$	Phase angle
11–21	$v = V_p \sin 2\pi f t$	Sine wave voltage
11–22	percent duty cycle $= \left(\dfrac{t_W}{T}\right)100\%$	Duty cycle
11–23	$V_{avg} = $ baseline + (duty cycle)(amplitude)	Average value of a pulse waveform

SELF-TEST
Answers are at the end of the chapter.

1. The difference between alternating current (ac) and direct current (dc) is
   (a) ac changes value and dc does not   (b) ac changes direction and dc does not
   (c) both answers (a) and (b)   (d) neither answer (a) nor (b)
2. During each cycle, a sine wave reaches a peak value
   (a) one time   (b) two times
   (c) four times   (d) a number of times depending on the frequency
3. A sine wave with a frequency of 12 kHz is changing at a faster rate than a sine wave with a frequency of
   (a) 20 kHz   (b) 15,000 Hz   (c) 10,000 Hz   (d) 1.25 MHz
4. A sine wave with a period of 2 ms is changing at a faster rate than a sine wave with a period of
   (a) 1 ms   (b) 0.0025 s   (c) 1.5 ms   (d) 1200 ms
5. When a sine wave has a frequency of 60 Hz, in 10 s it goes through
   (a) 6 cycles   (b) 10 cycles   (c) 1/16 cycle   (d) 600 cycles
6. If the peak value of a sine wave is 10 V, the peak-to-peak value is
   (a) 20 V   (b) 5 V   (c) 100 V   (d) none of these
7. If the peak value of a sine wave is 20 V, the rms value is
   (a) 14.14 V   (b) 6.37 V   (c) 7.07 V   (d) 0.707 V
8. The average value of a 10 V peak sine wave over one complete cycle is
   (a) 0 V   (b) 6.37 V   (c) 7.07 V   (d) 5 V
9. The average half-cycle value of a sine wave with a 20 V peak is
   (a) 0 V   (b) 6.37 V   (c) 12.74 V   (d) 14.14 V

10. One sine wave has a positive-going zero crossing at 10° and another sine wave has a positive-going zero crossing at 45°. The phase angle between the two waveforms is

    (a) 55°   (b) 35°   (c) 0°   (d) none of these

11. The instantaneous value of a 15 A peak sine wave at a point 32° from its positive-going zero crossing is

    (a) 7.95 A   (b) 7.5 A   (c) 2.13 A   (d) 7.95 V

12. A phasor represents

    (a) the magnitude of a quantity   (b) the magnitude and direction of a quantity

    (c) the phase angle   (d) the length of a quantity

13. If the rms current through a 10 kΩ resistor is 5 mA, the rms voltage drop across the resistor is

    (a) 70.7 V   (b) 7.07 V   (c) 5 V   (d) 50 V

14. Two series resistors are connected to an ac source. If there are 6.5 V rms across one resistor and 3.2 V rms across the other, the peak source voltage is

    (a) 9.7 V   (b) 9.19 V   (c) 13.72 V   (d) 4.53 V

15. A 10 kHz pulse waveform consists of pulses that are 10 μs wide. Its duty cycle is

    (a) 100%   (b) 10%   (c) 1%   (d) not determinable

16. The duty cycle of a square wave

    (a) varies with the frequency   (b) varies with the pulse width

    (c) both answers (a) and (b)   (d) is 50%

CIRCUIT DYNAMICS QUIZ

Answers are at the end of the chapter.

Refer to Figure 11–81.

1. If the source voltage increases, the voltage across R_3

   (a) increases   (b) decreases   (c) stays the same

2. If R_4 opens, the voltage across R_3

   (a) increases   (b) decreases   (c) stays the same

3. If the half-cycle average value of the source voltage is decreased, the rms voltage across R_2

   (a) increases   (b) decreases   (c) stays the same

Refer to Figure 11–83.

4. If the dc voltage is reduced, the average current through R_L

   (a) increases   (b) decreases   (c) stays the same

5. If the dc voltage source is reversed, the rms current through R_L

   (a) increases   (b) decreases   (c) stays the same

Refer to Figure 11–90.

6. If the resistor in the upper left of the protoboard has a color code of blue, gray, brown, gold instead of the color bands shown, the CH2 voltage measured by the oscilloscope

   (a) increases   (b) decreases   (c) stays the same

7. If the CH2 probe shown connected to the right side of the resistor is moved to the left side of the resistor, the amplitude of the measured voltage

   (a) increases   (b) decreases   (c) stays the same

8. If the bottom lead of the right-most resistor becomes disconnected, the CH2 voltage

   (a) increases   (b) decreases   (c) stays the same

9. If the wire connecting the two upper resistors becomes disconnected, altering the loading effect on the input signal source, the CH1 voltage

   (a) increases   (b) decreases   (c) stays the same

Refer to Figure 11–91.

10. If the right-most resistor has a third band that is orange instead of red, the CH1 voltage

    (a) increases   (b) decreases   (c) stays the same

11. If the resistor at the upper left opens, the CH1 voltage

    (a) increases   (b) decreases   (c) stays the same

12. If the resistor at the lower left opens, the CH1 voltage

    (a) increases   (b) decreases   (c) stays the same

PROBLEMS

More difficult problems are indicated by an asterisk (*).
Answers to odd-numbered problems are at the end of the book.

SECTION 11–1   The Sinusoidal Waveform

1. Calculate the frequency for each of the following values of period:

   (a) 1 s   (b) 0.2 s   (c) 50 ms   (d) 1 ms   (e) 500 μs   (f) 10 μs

2. Calculate the period of each of the following values of frequency:

   (a) 1 Hz   (b) 60 Hz   (c) 500 Hz   (d) 1 kHz   (e) 200 kHz   (f) 5 MHz

3. A sine wave goes through 5 cycles in 10 μs. What is its period?

4. A sine wave has a frequency of 50 kHz. How many cycles does it complete in 10 ms?

SECTION 11–2   Sinusoidal Voltage Sources

5. The conductive loop on the rotor of a simple two-pole, single-phase generator rotates at a rate of 250 rps. What is the frequency of the induced output voltage?

6. A certain four-pole generator has a speed of rotation of 3600 rpm. What is the frequency of the voltage produced by this generator?

7. At what speed of rotation must a four-pole generator be operated to produce a 400 Hz sinusoidal voltage?

SECTION 11–3   Sinusoidal Voltage and Current Values

8. A sine wave has a peak value of 12 V. Determine the following values:

   (a) rms   (b) peak-to-peak   (c) average

9. A sinusoidal current has an rms value of 5 mA. Determine the following values:

   (a) peak   (b) average   (c) peak-to-peak

10. For the sine wave in Figure 11–77, determine the peak, peak-to-peak, rms, and average values.

▶ **FIGURE 11–77**

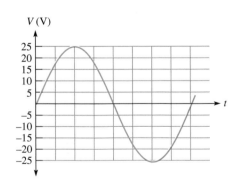

SECTION 11–4   Angular Measurement of a Sine Wave

11. Convert the following angular values from degrees to radians:

    (a) 30°   (b) 45°   (c) 78°   (d) 135°   (e) 200°   (f) 300°

12. Convert the following angular values from radians to degrees:

    (a) $\pi/8$ rad   **(b)** $\pi/3$ rad   **(c)** $\pi/2$ rad

    (d) $3\pi/5$ rad   **(e)** $6\pi/5$ rad   **(f)** 1.8π rad

13. Sine wave A has a positive-going zero crossing at $30°$. Sine wave B has a positive-going zero crossing at $45°$. Determine the phase angle between the two signals. Which signal leads?

14. One sine wave has a positive peak at $75°$, and another has a positive peak at $100°$. How much is each sine wave shifted in phase from the $0°$ reference? What is the phase angle between them?

15. Make a sketch of two sine waves as follows: Sine wave A is the reference, and sine wave B lags A by $90°$. Both have equal amplitudes.

SECTION 11–5   The Sine Wave Formula

16. A certain sine wave has a positive-going zero crossing at $0°$ and an rms value of 20 V. Calculate its instantaneous value at each of the following angles:

    (a) $15°$ **(b)** $33°$ **(c)** $50°$ **(d)** $110°$

    (e) $70°$ **(f)** $145°$ **(g)** $250°$ **(h)** $325°$

17. For a particular $0°$ reference sinusoidal current, the peak value is 100 mA. Determine the instantaneous value at each of the following points:

    (a) $35°$ **(b)** $95°$ **(c)** $190°$ **(d)** $215°$ **(e)** $275°$ **(f)** $360°$

18. For a $0°$ reference sine wave with an rms value of 6.37 V, determine its instantaneous value at each of the following points:

    (a) $\pi/8$ rad   **(b)** $\pi/4$ rad   **(c)** $\pi/2$ rad   **(d)** $3\pi/4$ rad

    (e) π rad   **(f)** $3\pi/2$ rad   **(g)** 2π rad

19. Sine wave A lags sine wave B by $30°$. Both have peak values of 15 V. Sine wave A is the reference with a positive-going crossing at $0°$. Determine the instantaneous value of sine wave B at $30°$, $45°$, $90°$, $180°$, $200°$, and $300°$.

20. Repeat Problem 19 for the case when sine wave A leads sine wave B by $30°$.

*21. A certain sine wave has a frequency of 2.2 kHz and an rms value of 25 V. Assuming a given cycle begins (zero crossing) at $t = 0$ s, what is the change in voltage from 0.12 ms to 0.2 ms?

SECTION 11–6   Introduction to Phasors

22. Draw a phasor diagram to represent the sine waves in Figure 11–78 with respect to a $0°$ reference.

  ▶ **FIGURE 11–78**

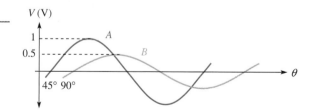

23. Draw the sine waves represented by the phasor diagram in Figure 11–79. The phasor lengths represent peak values.

  ▶ **FIGURE 11–79**

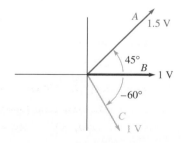

24. Determine the frequency for each angular velocity:

    (a) 60 rad/s   **(b)** 360 rad/s   **(c)** 2 rad/s   **(d)** 1256 rad/s

25. Determine the value of sine wave A in Figure 11–78 at each of the following times, measured from the positive-going zero crossing. Assume the frequency is 5 kHz.

    (a) 30 μs   **(b)** 75 μs   **(c)** 125 μs

SECTION 11–7   Analysis of AC Circuits

26. A sinusoidal voltage is applied to the resistive circuit in Figure 11–80. Determine the following:

    (a) I_{rms} **(b)** I_{avg} **(c)** I_p **(d)** I_{pp} **(e)** i at the positive peak

▲ **FIGURE 11–80**

27. Find the half-cycle average values of the voltages across R_1 and R_2 in Figure 11–81. All values shown are rms.

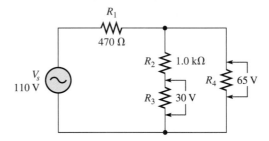

▲ **FIGURE 11–81**

28. Determine the rms voltage across R_3 in Figure 11–82.

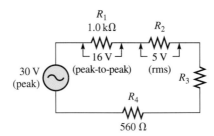

▲ **FIGURE 11–82**

SECTION 11–8   Superimposed DC and AC Voltages

29. A sine wave with an rms value of 10.6 V is riding on a dc level of 24 V. What are the maximum and minimum values of the resulting waveform?

30. How much dc voltage must be added to a 3 V rms sine wave in order to make the resulting voltage nonalternating (no negative values)?

31. A 6 V peak sine wave is riding on a dc voltage of 8 V. If the dc voltage is lowered to 5 V, how far negative will the sine wave go?

*32.** Figure 11–83 shows a sinusoidal voltage source in series with a dc source. Effectively, the two voltages are superimposed. Determine the power dissipation in the load resistor.

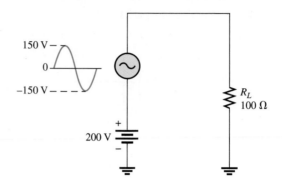

▲ FIGURE 11–83

SECTION 11–9   Nonsinusoidal Waveforms

33. From the graph in Figure 11–84, determine the approximate values of t_r, t_f, t_W, and amplitude.

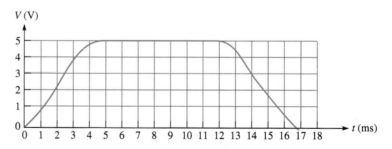

▲ FIGURE 11–84

34. The repetition frequency of a pulse waveform is 2 kHz, and the pulse width is 1 μs. What is the percent duty cycle?

35. Calculate the average value of the pulse waveform in Figure 11–85.

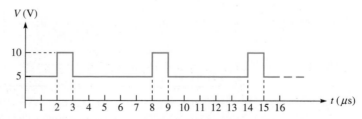

▲ FIGURE 11–85

36. Determine the duty cycle for each waveform in Figure 11–86.

37. Find the average value of each pulse waveform in Figure 11–86.

38. What is the frequency of each waveform in Figure 11–86?

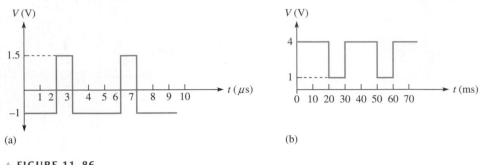

(a)   (b)

▲ FIGURE 11–86

39. What is the frequency of each sawtooth waveform in Figure 11–87?

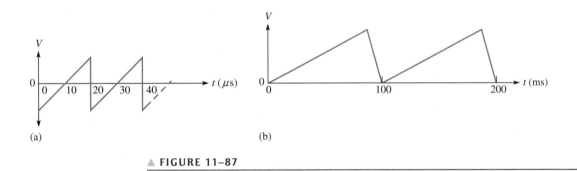

(a)   (b)

▲ FIGURE 11–87

*****40.** A nonsinusoidal waveform called a *stairstep* is shown in Figure 11–88. Determine its average value.

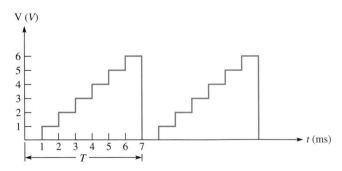

▲ FIGURE 11–88

41. A square wave has a period of 40 μs. List the first six odd harmonics.

42. What is the fundamental frequency of the square wave mentioned in Problem 41?

SECTION 11–10 The Oscilloscope

43. Determine the peak value and the period of the sine wave displayed on the scope screen in Figure 11–89.

◄ **FIGURE 11–89**

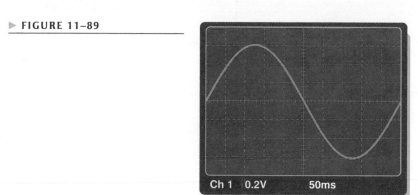

*44. Based on the instrument settings and an examination of the scope display and the protoboard in Figure 11–90, determine the frequency and peak value of the input signal and output signal. The waveform shown is channel 1. Draw the channel 2 waveform as it would appear on the scope with the indicated settings.

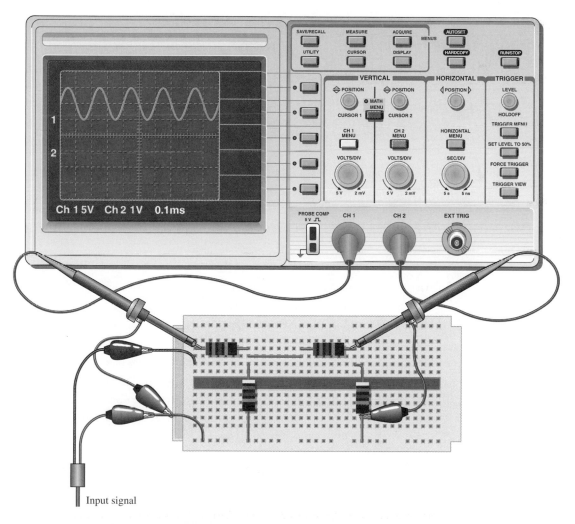

▲ **FIGURE 11–90**

*45. Examine the protoboard and the oscilloscope display in Figure 11–91 and determine the peak value and the frequency of the unknown input signal.

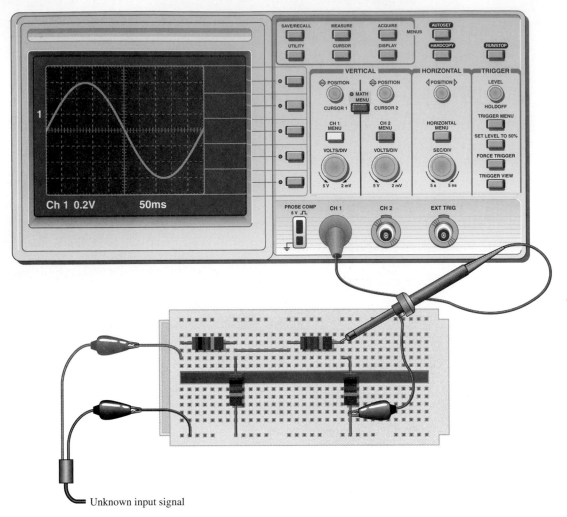

Unknown input signal

▲ FIGURE 11–91

Multisim Troubleshooting and Analysis

These problems require your Multisim CD-ROM.

46. Open file P11-46 and measure the peak and rms voltage across each of the resistors.

47. Open file P11-47 and measure the peak and rms voltage across each of the resistors.

48. Open file P11-48. Determine if there is a fault and, if so, identify the fault.

49. Open file P11-49 and measure the rms current in each branch of the circuit.

50. Open file P11-50. Determine if there is a fault and, if so, identify the fault.

51. Open file P11-51 and measure the total voltage across the resistor using the oscilloscope.

52. Open file P11-52 and measure the total voltage across the resistor using the oscilloscope.

ANSWERS

SECTION REVIEWS

SECTION 11–1   The Sinusoidal Waveform

1. One cycle of a sine wave is from the zero crossing through a positive peak, then through zero to a negative peak and back to the zero crossing.

2. A sine wave changes polarity at the zero crossings.

3. A sine wave has two maximum points (peaks) per cycle.

4. The period is from one zero crossing to the next corresponding zero crossing, or from one peak to the next corresponding peak.

5. Frequency is the number of cycles completed in one second; the unit of frequency is the hertz.

6. $f = 1/T = 200$ kHz

7. $T = 1/f = 8.33$ ms

SECTION 11–2 Sinusoidal Voltage Sources

1. Sine waves are generated by electromagnetic and electronic methods.

2. Speed and frequency are directly proportional.

3. An oscillator is an electronic circuit that produces repetitive waveforms.

SECTION 11–3 Sinusoidal Voltage and Current Values

1. (a) $V_{pp} = 2(1\text{ V}) = 2$ V   (b) $V_{pp} = 2(1.414)(1.414\text{ V}) = 4$ V

   (c) $V_{pp} = 2(1.57)(3\text{ V}) = 9.42$ V

2. (a) $V_{rms} = (0.707)(2.5\text{ V}) = 1.77$ V   (b) $V_{rms} = (0.5)(0.707)(10\text{ V}) = 3.54$ V

   (c) $V_{rms} = (0.707)(1.57)(1.5\text{ V}) = 1.66$ V

3. (a) $V_{avg} = (0.637)(10\text{ V}) = 6.37$ V   (b) $V_{avg} = (0.637)(1.414)(2.3\text{ V}) = 2.07$ V

   (c) $V_{avg} = (0.637)(0.5)(60\text{ V}) = 19.1$ V

SECTION 11–4 Angular Measurement of a Sine Wave

1. (a) Positive peak at 90°   (b) Negative-going zero crossing at 180°

   (c) Negative peak at 270°   (d) End of cycle at 360°

2. Half-cycle: 180°; π

3. Full cycle: 360°; 2π

4. $90° - 45° = 45°$

SECTION 11–5 The Sine Wave Formula

1. $v = (10\text{ V})\sin(120°) = 8.66$ V

2. $v = (10\text{ V})\sin(45° + 10°) = 8.19$ V

3. $v = (5\text{ V})\sin(90° - 25°) = 4.53$ V

SECTION 11–6 Introduction to Phasors

1. A graphic representation of the magnitude and angular position of a time-varying quantity

2. 9425 rad/s

3. 100 Hz

4. See Figure 11–92.

▶ **FIGURE 11–92**

SECTION 11–7 Analysis of AC Circuits

1. $I_p = V_p/R = (1.57)(12.5\text{ V})/330\ \Omega = 59.5$ mA

2. $V_{s(rms)} = (0.707)(25.3\text{ V}) = 17.9$ V

SECTION 11–8 Superimposed DC and AC Voltages

1. $+V_{max} = 5\text{ V} + 2.5\text{ V} = 7.5$ V

2. Yes, it will alternate.

3. $+V_{max} = 5\text{ V} - 2.5\text{ V} = 2.5$ V

SECTION 11–9   Nonsinusoidal Waveforms

1. (a) Rise time is the time interval from 10% to 90% of the rising pulse edge;

   (b) Fall time is the time interval from 90% to 10% of the falling pulse edge;

   (c) Pulse width is the time interval from 50% of the leading pulse edge to 50% of the trailing pulse edge.

2. $f = 1/1$ ms $= 1$ kHz

3. d.c. $= (1/5)100\% = 20\%$; Ampl. 1.5 V; $V_{avg} = 0.5$ V $+ 0.2(1.5$ V$) = 0.8$ V

4. $T = 16$ ms

5. $f = 1/T = 1/1\,\mu$s $= 1$ MHz

6. Fundamental frequency is the repetition rate of the waveform.

7. 2nd harm.: 2 kHz

8. $f = 1/10\,\mu$s $= 100$ kHz

SECTION 11–10   The Oscilloscope

1. Analog : Signal drives display directly.

   Digital : Signal is converted to digital for processing and then reconstructed for display.

2. Voltage is measured vertically; time is measured horizontally.

3. The Volts/Div control adjusts the voltage scale.

4. The Sec/Div control adjusts the time scale.

5. Always, unless you are trying to measure a very small, low-frequency signal.

A Circuit Application

1. RF is radio frequency.

2. IF is intermediate frequency.

3. Carrier frequency is higher than audio.

4. The amplitude varies in an AM signal.

RELATED PROBLEMS FOR EXAMPLES

11–1   2.4 s

11–2   1.5 ms

11–3   20 kHz

11–4   200 Hz

11–5   66.7 kHz

11–6   30 rps

11–7 $V_{pp} = 50$ V; $V_{rms} = 17.7$ V; $V_{avg} = 15.9$ V

11–8   (a) $\pi/12$ rad   (b) 112.5°

11–9   8°

11–10 18.1 V

11–11 10.6 V

11–12 5 V at $-85°$

11–13 34.2 V

11–14 $I_{rms} = 4.53$ mA; $V_{1(rms)} = 4.53$ V; $V_{2(rms)} = 2.54$ V; $P_{tot} = 32.0$ mW

11–15 23.7 V

11–16 The waveform in part (a) never goes negative. The waveform in part (b) goes negative for a portion of its cycle.

11–17 5%

11–18 1.2 V

11–19 120 V

11–20 Part (a) 1.06 V, 50 Hz;

part (b) 88.4 mV, 1.67 kHz

part (c) 4.81 V, 5 kHz;

part (d) 7.07 V, 250 kHz

SELF-TEST

1. (b)   **2.** (b)   **3.** (c)   **4.** (b)   **5.** (d)   **6.** (a)   **7.** (a)   **8.** (a)

9. (c)   **10.** (b)   **11.** (a)   **12.** (b)   **13.** (d)   **14.** (c)   **15.** (b)   **16.** (d)

CIRCUIT DYNAMICS QUIZ

1. (a)   **2.** (a)   **3.** (b)   **4.** (b)   **5.** (c)   **6.** (b)

7. (a)   **8.** (a)   **9.** (a)   **10.** (a)   **11.** (b)   **12.** (a)

12

CAPACITORS

CHAPTER OBJECTIVES

◆ Describe the basic construction and characteristics of a capacitor
◆ Discuss various types of capacitors
◆ Analyze series capacitors
◆ Analyze parallel capacitors
◆ Analyze capacitive dc switching circuits
◆ Analyze capacitive ac circuits
◆ Discuss some capacitor applications
◆ Describe the operation of switched-capacitor circuits

KEY TERMS

◆ Capacitor
◆ Dielectric
◆ Farad (F)
◆ Coulomb's law
◆ *RC* time constant
◆ Capacitive reactance
◆ Instantaneous power
◆ True power
◆ Reactive power
◆ VAR (volt-ampere reactive)
◆ Ripple voltage

A CIRCUIT APPLICATION PREVIEW

In the circuit application, you will see how a capacitor is used to couple signal voltages to and from an amplifier. You will also troubleshoot the amplifier using oscilloscope waveforms.

VISIT THE COMPANION WEBSITE

Study aids for this chapter are available at http://www.prenhall.com/floyd

INTRODUCTION

In previous chapters, the resistor has been the only passive electrical component that you have studied. Capacitors and inductors are other types of basic passive electrical components. You will study inductors in Chapter 13.

In this chapter, you will learn about the capacitor and its characteristics. The physical construction and electrical properties are examined, and the effects of connecting capacitors in series and in parallel are analyzed. How a capacitor works in both dc and ac circuits is an important part of this coverage and forms the basis for the study of reactive circuits in terms of both frequency response and time response.

The *capacitor* is an electrical device that can store electrical charge, thereby creating an electric field that, in turn, stores energy. The measure of the energy-storing ability of a capacitor is its *capacitance*. When a sinusoidal signal is applied to a capacitor, it reacts in a certain way and produces an opposition to current, which depends on the frequency of the applied signal. This opposition to current is called *capacitive reactance*.

12–1 THE BASIC CAPACITOR

A **capacitor** is a passive electrical component that stores electrical charge and has the property of capacitance.

After completing this section, you should be able to

- ◆ **Describe the basic construction and characteristics of a capacitor**
  - ◆ Explain how a capacitor stores charge
  - ◆ Define *capacitance* and state its unit
  - ◆ State Coulomb's law
  - ◆ Explain how a capacitor stores energy
  - ◆ Discuss voltage rating and temperature coefficient
  - ◆ Explain capacitor leakage
  - ◆ Specify how the physical characteristics affect the capacitance

Basic Construction

In its simplest form, a capacitor is an electrical device that stores electrical charge and is constructed of two parallel conductive plates separated by an insulating material called the **dielectric**. Connecting leads are attached to the parallel plates. A basic capacitor is shown in Figure 12–1(a), and a schematic symbol is shown in part (b).

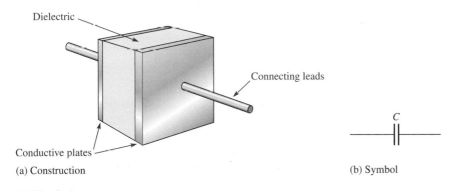

Dielectric

Connecting leads

Conductive plates

(a) Construction

C

(b) Symbol

▲ **FIGURE 12–1**

The basic capacitor.

How a Capacitor Stores Charge

In the neutral state, both plates of a capacitor have an equal number of free electrons, as indicated in Figure 12–2(a). When the capacitor is connected to a voltage source through a resistor, as shown in part (b), electrons (negative charge) are removed from plate A, and an equal number are deposited on plate B. As plate A loses electrons and plate B gains electrons, plate A becomes positive with respect to plate B. During this charging process, electrons flow only through the connecting leads. No electrons flow through the dielectric of the capacitor because it is an insulator. The movement of electrons ceases when the voltage across the capacitor equals the source voltage, as indicated in Figure 12–2(c). If the capacitor is disconnected from the source, it retains the stored charge for a long period of time (the length of time depends on the type of capacitor) and still has the voltage across it, as shown in Figure 12–2(d). A charged capacitor can act as a temporary battery.

SAFETY NOTE

Capacitors are capable of storing electrical charge for a long time after power has been turned off in a circuit. Be careful when touching or handling capacitors in or out of a circuit. If you touch the leads, you may be in for a shock as the capacitor discharges through you! It is usually good practice to discharge a capacitor using a shorting tool with an insulated grip of some sort before handling the capacitor.

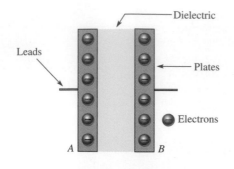

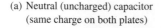

(a) Neutral (uncharged) capacitor
(same charge on both plates)

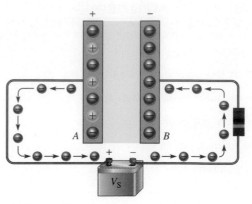

(b) Electrons flow from plate A to plate B as the capacitor
charges when connected to a voltage source.

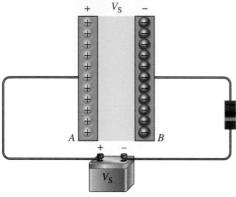

(c) After the capacitor charges to V_S, no electrons flow
while connected to the voltage source.

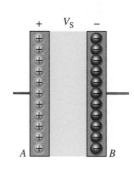

(d) Ideally, the capacitor retains charge when
disconnected from the voltage source.

▲ FIGURE 12–2

Illustration of a capacitor storing charge.

Capacitance

The amount of charge that a capacitor can store per unit of voltage across its plates is its capacitance, designated C. That is, **capacitance** is a measure of a capacitor's ability to store charge. The more charge per unit of voltage that a capacitor can store, the greater its capacitance, as expressed by the following formula:

Equation 12–1

$$C = \frac{Q}{V}$$

where C is capacitance, Q is charge, and V is voltage.

By rearranging the terms in Equation 12–1, you can obtain two other formulas.

Equation 12–2

$$Q = CV$$

Equation 12–3

$$V = \frac{Q}{C}$$

The Unit of Capacitance   The farad (F) is the basic unit of capacitance. Recall that the coulomb (C) is the unit of electrical charge.

One farad is the amount of capacitance when one coulomb (C) of charge is stored with one volt across the plates.

Most capacitors that are used in electronics work have capacitance values that are specified in microfarads (μF) and picofarads (pF). A microfarad is one-millionth of a farad ($1\,\mu\text{F} = 1 \times 10^{-6}\,\text{F}$), and a picofarad is one-trillionth of a farad ($1\,\text{pF} = 1 \times 10^{-12}\,\text{F}$). Conversions for farads, microfarads, and picofarads are given in Table 12–1.

◀ TABLE 12–1

TO CONVERT FROM	TO	MOVE THE DECIMAL POINT
Farads	Microfarads	6 places to right ($\times 10^{6}$)
Farads	Picofarads	12 places to right ($\times 10^{12}$)
Microfarads	Farads	6 places to left ($\times 10^{-6}$)
Microfarads	Picofarads	6 places to right ($\times 10^{6}$)
Picofarads	Farads	12 places to left ($\times 10^{-12}$)
Picofarads	Microfarads	6 places to left ($\times 10^{-6}$)

EXAMPLE 12–1

(a) A certain capacitor stores 50 microcoulombs ($50\,\mu\text{C}$) with 10 V across its plates. What is its capacitance in units of microfarads?

(b) A $2.2\,\mu\text{F}$ capacitor has 100 V across its plates. How much charge does it store?

(c) Determine the voltage across a 1000 pF capacitor that is storing 20 microcoulombs ($20\,\mu\text{C}$) of charge.

Solution

(a) $C = \dfrac{Q}{V} = \dfrac{50\,\mu\text{C}}{10\,\text{V}} = \mathbf{5\,\mu F}$

(b) $Q = CV = (2.2\,\mu\text{F})(100\,\text{V}) = \mathbf{220\,\mu C}$

(c) $V = \dfrac{Q}{C} = \dfrac{20\,\mu\text{C}}{1000\,\text{pF}} = \mathbf{20\,kV}$

*Related Problem**   Determine V if $C = 1000\,\text{pF}$ and $Q = 100\,\mu\text{C}$.

*Answers are at the end of the chapter.

EXAMPLE 12–2

Convert the following values to microfarads:

(a) 0.00001 F   (b) 0.0047 F   (c) 1000 pF   (d) 220 pF

Solution

(a) $0.00001\,\text{F} \times 10^{6}\,\mu\text{F/F} = \mathbf{10\,\mu F}$ (b) $0.0047\,\text{F} \times 10^{6}\,\mu\text{F/F} = \mathbf{4700\,\mu F}$

(c) $1000\,\text{pF} \times 10^{-6}\,\mu\text{F/pF} = \mathbf{0.001\,\mu F}$ (d) $220\,\text{pF} \times 10^{-6}\,\mu\text{F/pF} = \mathbf{0.00022\,\mu F}$

Related Problem   Convert 47,000 pF to microfarads.

EXAMPLE 12–3

Convert the following values to picofarads:

(a) $0.1 \times 10^{-8}\,\text{F}$ (b) 0.000022 F   (c) $0.01\,\mu\text{F}$ (d) $0.0047\,\mu\text{F}$

Solution

(a) $0.1 \times 10^{-8}\,\text{F} \times 10^{12}\,\text{pF/F} = \mathbf{1000\,pF}$

(b) $0.000022\,\text{F} \times 10^{12}\,\text{pF/F} = \mathbf{22 \times 10^{6}\,pF}$

(c) $0.01\,\mu\text{F} \times 10^{6}\,\text{pF/}\mu\text{F} = \mathbf{10{,}000\,pF}$

(d) $0.0047\,\mu\text{F} \times 10^{6}\,\text{pF/}\mu\text{F} = \mathbf{4700\,pF}$

Related Problem   Convert $100\,\mu\text{F}$ to picofarads.

How a Capacitor Stores Energy

A capacitor stores energy in the form of an electric field that is established by the opposite charges stored on the two plates. The electric field is represented by lines of force between the positive and negative charges and is concentrated within the dielectric, as shown in Figure 12–3.

▶ FIGURE 12–3

The electric field stores energy in a capacitor.

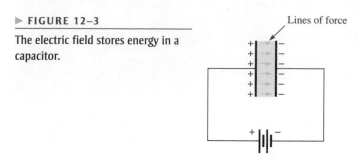

Lines of force

Coulomb's law states

A force (F) exists between two point-source charges (Q_1, Q_2) that is directly proportional to the product of the two charges and inversely proportional to the square of the distance (d) between the charges.

Figure 12–4(a) illustrates a line of force between a positive and a negative charge. Figure 12–4(b) shows that many opposite charges distributed on the plates of a capacitor create lines of force, which form an electric field that stores energy within the dielectric. Although the distributed charges no longer act as point-source charges and do not exactly obey Coulomb's law, the force is still dependent on the amount of charge and the distance between the plates.

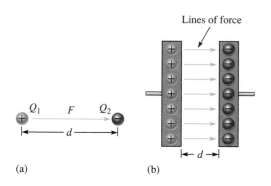

Lines of force

(a)   (b)

▲ FIGURE 12–4

Lines of force are created by opposite charges.

The greater the forces between the charges on the plates of a capacitor, the more energy is stored. Therefore, the amount of energy stored is directly proportional to the capacitance because the more charge stored, the greater the force.

Also, from Equation 12–2, the amount of charge stored is directly related to the voltage as well as to the capacitance. Therefore, the amount of energy stored is also dependent on the square of the voltage across the plates of the capacitor. The formula for the energy stored by a capacitor is

Equation 12–4

$$W = \frac{1}{2}CV^2$$

When capacitance (C) is in farads and voltage (V) is in volts, energy (W) is in joules.

Voltage Rating

Every capacitor has a limit on the amount of voltage that it can withstand across its plates. The voltage rating specifies the maximum dc voltage that can be applied without risk of damage to the device. If this maximum voltage, commonly called the *breakdown voltage* or *working voltage,* is exceeded, permanent damage to the capacitor can result.

You must consider both the capacitance and the voltage rating before you use a capacitor in a circuit application. The choice of capacitance value is based on particular circuit requirements. The voltage rating should always be above the maximum voltage expected in a particular application.

Dielectric Strength The breakdown voltage of a capacitor is determined by the **dielectric strength** of the dielectric material used. The dielectric strength is expressed in V/mil (1 mil = 0.001 in. = 2.54×10^{-5} m). Table 12–2 lists typical values for several materials. Exact values vary depending on the specific composition of the material.

MATERIAL	DIELECTRIC STRENGTH (V/MIL)
Air	80
Oil	375
Ceramic	1000
Paper (paraffined)	1200
Teflon®	1500
Mica	1500
Glass	2000

◀ TABLE 12–2

Some common dielectric materials and their dielectric strengths.

A capacitor's dielectric strength can best be explained by an example. Assume that a certain capacitor has a plate separation of 1 mil and that the dielectric material is ceramic. This particular capacitor can withstand a maximum voltage of 1000 V because its dielectric strength is 1000 V/mil. If the maximum voltage is exceeded, the dielectric may break down and conduct current, causing permanent damage to the capacitor. Similarly, if the ceramic capacitor has a plate separation of 2 mils, its breakdown voltage is 2000 V.

Temperature Coefficient

The **temperature coefficient** indicates the amount and direction of a change in capacitance value with temperature. A positive temperature coefficient means that the capacitance increases with an increase in temperature or decreases with a decrease in temperature. A negative coefficient means that the capacitance decreases with an increase in temperature or increases with a decrease in temperature.

Temperature coefficients are typically specified in parts per million per Celsius degree (ppm/°C). For example, a negative temperature coefficient of 150 ppm/°C for a 1 μF capacitor means that for every degree rise in temperature, the capacitance decreases by 150 pF (there are one million picofarads in one microfarad).

Leakage

No insulating material is perfect. The dielectric of any capacitor will conduct some very small amount of current. Thus, the charge on a capacitor will eventually leak off. Some types of capacitors, such as large electrolytic types, have higher leakages than others. An equivalent circuit for a nonideal capacitor is shown in Figure 12–5. The parallel resistor R_{leak} represents the extremely high resistance (several hundred kilohms or more) of the dielectric material through which there is leakage current.

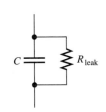

▲ FIGURE 12–5

Equivalent circuit for a nonideal capacitor.

Physical Characteristics of a Capacitor

The following parameters are important in establishing the capacitance and the voltage rating of a capacitor: plate area, plate separation, and dielectric constant.

Plate Area *Capacitance is directly proportional to the physical size of the plates as determined by the plate area, A.* A larger plate area produces more capacitance, and a smaller plate area produces less capacitance. Figure 12–6(a) shows that the plate area of a parallel plate capacitor is the area of one of the plates. If the plates are moved in relation to each other, as shown in Figure 12–6(b), the overlapping area determines the effective plate area. This variation in effective plate area is the basis for a certain type of variable capacitor.

Capacitance is directly proportional
to plate area (*A*).

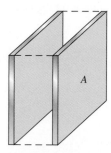 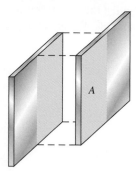

(a) Full plate area:
more capacitance

(b) Reduced plate area:
less capacitance

Plate Separation *Capacitance is inversely proportional to the distance between the plates.* The plate separation is designated *d*, as shown in Figure 12–7. A greater separation of the plates produces a smaller capacitance, as illustrated in the figure. As previously discussed, the breakdown voltage is directly proportional to the plate separation. The further the plates are separated, the greater the breakdown voltage.

Capacitance is inversely proportional
to the distance between the plates.

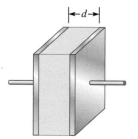 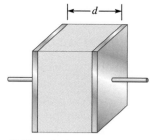

(a) Plates closer together:
more capacitance

(b) Plates further apart:
less capacitance

Dielectric Constant As you know, the insulating material between the plates of a capacitor is called the *dielectric*. Dielectric materials tend to reduce the voltage between plates for a given charge and thus increase the capacitance. If the voltage is fixed, more charge can be stored due to the presence of a dielectric than can be stored without a dielectric. The measure of a material's ability to establish an electric field is called the **dielectric constant** or *relative permittivity,* symbolized by ε_r. (ε is the Greek letter epsilon.)

Capacitance is directly proportional to the dielectric constant. The dielectric constant of a vacuum is defined as 1 and that of air is very close to 1. These values are used as a reference, and all other materials have values of ε_r specified with respect to that of a vacuum or air. For example, a material with $\varepsilon_r = 8$ can result in a capacitance eight times greater than that of air with all other factors being equal.

Table 12–3 lists several common dielectric materials and a typical dielectric constant for each. The values can vary because it depends on the specific composition of the material.

▼ TABLE 12–3

Some common dielectric materials and their dielectric constants.

MATERIAL	TYPICAL ε_r VALUE
Air (vacuum)	1.0
Teflon®	2.0
Paper (paraffined)	2.5
Oil	4.0
Mica	5.0
Glass	7.5
Ceramic	1200

The dielectric constant (relative permittivity) is dimensionless because it is a relative measure. It is a ratio of the absolute permittivity of a material, ε, to the absolute permittivity of a vacuum, ε_0, as expressed by the following formula:

$$\varepsilon_r = \frac{\varepsilon}{\varepsilon_0}$$

Equation 12–5

The value of ε_0 is 8.85×10^{-12} F/m (farads per meter).

Formula You have seen how capacitance is directly related to plate area, A, and the dielectric constant, ε_r, and inversely related to plate separation, d. An exact formula for calculating the capacitance in terms of these three quantities is

$$C = \frac{A\varepsilon_r(8.85 \times 10^{-12}\,\text{F/m})}{d}$$

Equation 12–6

where A is in square meters (m^2), d is in meters (m), and C is in farads (F). Recall that the absolute permittivity of a vacuum, ε_0, is 8.85×10^{-12} F/m and that the absolute permittivity of a dielectric (ε), as derived from Equation (12–5), is

$$\varepsilon = \varepsilon_r(8.85 \times 10^{-12}\,\text{F/m})$$

EXAMPLE 12–4

Determine the capacitance of a parallel plate capacitor having a plate area of $0.01\,\text{m}^2$ and a plate separation of 1 mil (2.54×10^{-5} m). The dielectric is mica, which has a dielectric constant of 5.0.

Solution Use Equation 12–6.

$$C = \frac{A\varepsilon_r(8.85 \times 10^{-12}\,\text{F/m})}{d} = \frac{(0.01\,\text{m}^2)(5.0)(8.85 \times 10^{-12}\,\text{F/m})}{2.54 \times 10^{-5}\,\text{m}} = \textbf{0.017}\,\boldsymbol{\mu}\textbf{F}$$

Related Problem Determine C where $A = 0.005\,\text{m}^2$, $d = 3$ mil (7.62×10^{-5} m), and ceramic is the dielectric.

**SECTION 12–1
REVIEW**
Answers are at the end of the chapter.

1. Define *capacitance*.
2. (a) How many microfarads are in one farad?
   (b) How many picofarads are in one farad?
   (c) How many picofarads are in one microfarad?
3. Convert 0.0015 μF to picofarads. To farads.
4. How much energy in joules is stored by a 0.01 μF capacitor with 15 V across its plates?
5. (a) When the plate area of a capacitor is increased, does the capacitance increase or decrease?
   (b) When the distance between the plates is increased, does the capacitance increase or decrease?
6. The plates of a ceramic capacitor are separated by 2 mils. What is the typical break-down voltage?
7. A capacitor with a value of 2 μF at 25°C has a positive temperature coefficient of 50 ppm/°C. What is the capacitance value when the temperature increases to 125°C?

12–2 TYPES OF CAPACITORS

Capacitors normally are classified according to the type of dielectric material and whether they are polarized or nonpolarized. The most common types of dielectric materials are mica, ceramic, plastic-film, and electrolytic (aluminum oxide and tantalum oxide).

After completing this section, you should be able to

- ◆ **Discuss various types of capacitors**
  - ◆ Describe the characteristics of mica, ceramic, plastic-film, and electrolytic capacitors
  - ◆ Describe types of variable capacitors
  - ◆ Identify capacitor labeling
  - ◆ Discuss capacitance measurement

Fixed Capacitors

Mica Capacitors Two types of mica capacitors are stacked-foil and silver-mica. The basic construction of the stacked-foil type is shown in Figure 12–8. It consists of alternate layers

▶ **FIGURE 12–8**

Construction of a typical radial-lead mica capacitor.

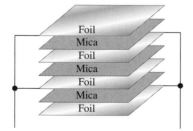

(a) Stacked layer arrangement

(b) Layers are pressed together and encapsulated.

of metal foil and thin sheets of mica. The metal foil forms the plate, with alternate foil sheets connected together to increase the plate area. More layers are used to increase the plate area, thus increasing the capacitance. The mica/foil stack is encapsulated in an insulating material such as Bakelite®, as shown in Figure 12–8(b). A silver-mica capacitor is formed in a similar way by stacking mica sheets with silver electrode material screened on them.

Mica capacitors are available with capacitance values ranging from 1 pF to 0.1 μF and voltage ratings from 100 V dc to 2500 V dc. Common temperature coefficients range from -20 ppm/°C to $+100$ ppm/°C. Mica has a typical dielectric constant of 5.

Ceramic Capacitors Ceramic dielectrics provide very high dielectric constants (1200 is typical). As a result, comparatively high capacitance values can be achieved in a small physical size. Ceramic capacitors are commonly available in a ceramic disk form, as shown in Figure 12–9, in a multilayer radial-lead configuration, as shown in Figure 12–10, or in a leadless ceramic chip, as shown in Figure 12–11, for surface mounting on printed circuit boards.

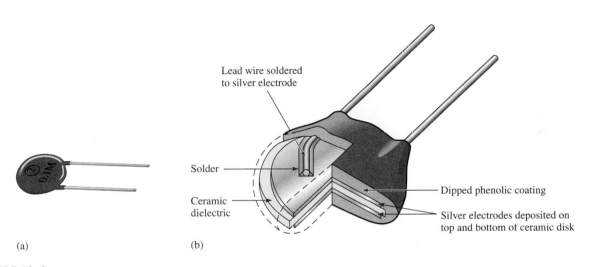

(a) (b)

▲ **FIGURE 12–9**

A ceramic disk capacitor and its basic construction.

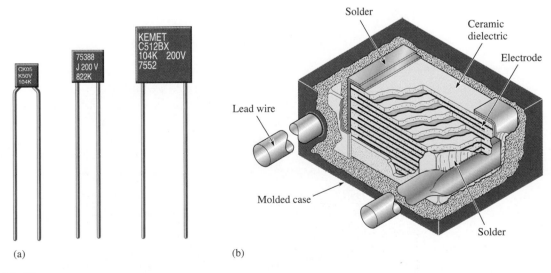

(a) (b)

▲ **FIGURE 12–10**

(a) Typical ceramic capacitors. (b) Construction view.

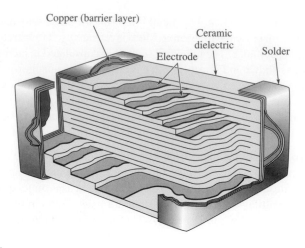

Copper (barrier layer)

Ceramic dielectric

Electrode

Solder

▲ FIGURE 12–11

Construction view of a typical ceramic chip capacitor used for surface mounting on printed circuit boards.

Ceramic capacitors typically are available in capacitance values ranging from 1 pF to 2.2 μF with voltage ratings up to 6 kV. A typical temperature coefficient for ceramic capacitors is 200,000 ppm/°C. A special type of disk ceramic has a zero temperature coefficient.

Plastic-Film Capacitors Common dielectric materials used in plastic-film capacitors include polycarbonate, propylene, polyester, polystyrene, polypropylene, and mylar. Some of these types have capacitance values up to 100 μF but most are less than 1 μF.

Figure 12–12 shows a common basic construction used in many plastic-film capacitors. A thin strip of plastic-film dielectric is sandwiched between two thin metal strips that act as plates. One lead is connected to the inner plate and one is connected to the outer plate as indicated. The strips are then rolled in a spiral configuration and encapsulated in a molded case. Thus, a large plate area can be packaged in a relatively small physical size, thereby achieving large capacitance values. Another method uses metal deposited directly on the film dielectric to form the plates.

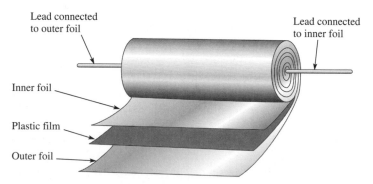

Lead connected to outer foil

Lead connected to inner foil

Inner foil

Plastic film

Outer foil

▲ FIGURE 12–12

Basic construction of axial-lead tubular plastic-film dielectric capacitors.

Figure 12–13(a) shows typical plastic-film capacitors. Figure 12–13(b) shows a construction view for one type of plastic-film capacitor.

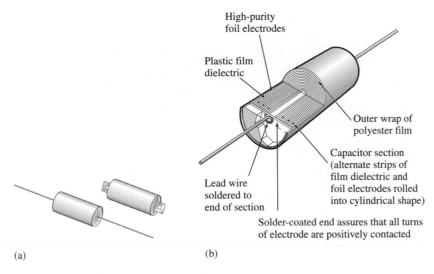

High-purity
foil electrodes

Plastic film
dielectric

Outer wrap of
polyester film

Capacitor section
(alternate strips of
film dielectric and
foil electrodes rolled
into cylindrical shape)

Lead wire
soldered to
end of section

Solder-coated end assures that all turns
of electrode are positively contacted

(a)   (b)

▲ FIGURE 12–13

(a) Typical capacitors. (b) Construction view of plastic-film capacitor.

Electrolytic Capacitors   Electrolytic capacitors are polarized so that one plate is positive and the other negative. These capacitors are used for capacitance values from 1 μF up to over 200,000 μF, but they have relatively low breakdown voltages (350 V is a typical maximum) and high amounts of leakage. In this text, capacitors with values of 1 μF or greater are considered to be polarized.

Electrolytic capacitors offer much higher capacitance values than mica or ceramic capacitors, but their voltage ratings are typically lower. Aluminum electrolytics are probably the most commonly used type. While other capacitors use two similar plates, the electrolytic consists of one plate of aluminum foil and another plate made of a conducting electrolyte applied to a material such as plastic film. These two "plates" are separated by a layer of aluminum oxide that forms on the surface of the aluminum plate. Figure 12–14(a) illustrates the basic construction of a typical aluminum electrolytic capacitor with axial leads. Other electrolytics with radial leads are shown in Figure 12–14(b); the symbol for an electrolytic capacitor is shown in part (c).

Tantalum electrolytics can be in either a tubular configuration similar to Figure 12–14 or "tear drop" shape as shown in Figure 12–15. In the tear drop configuration, the positive plate is actually a pellet of tantalum powder rather than a sheet of foil. Tantalum pentoxide forms the dielectric, and manganese dioxide forms the negative plate.

Because of the process used for the insulating oxide dielectric, the metallic (aluminum or tantalum) plate must be connected so that it is always positive with respect to the electrolyte plate, and, thus all electrolytic capacitors are polarized. The metal plate (positive lead) is usually indicated by a plus sign or some other obvious marking and must always be connected in a dc circuit where the voltage across the capacitor does not change polarity regardless of any ac present. Reversal of the polarity of the voltage will usually result in complete destruction of the capacitor.

The problem of dielectric absorption occurs mostly in electrolytic capacitors when they do not completely discharge during use and retain a residual charge. Approximately 25% of defective capacitors exhibit this condition.

Variable Capacitors

Variable capacitors are used in a circuit when there is a need to adjust the capacitance value either manually or automatically. These capacitors are generally less than 300 pF but are

SAFETY NOTE

Be extremely careful with electrolytic capacitors because it does make a difference which way an electrolytic capacitor is connected. Always observe the proper polarity. If a polarized capacitor is connected backwards, it may explode and cause injury.

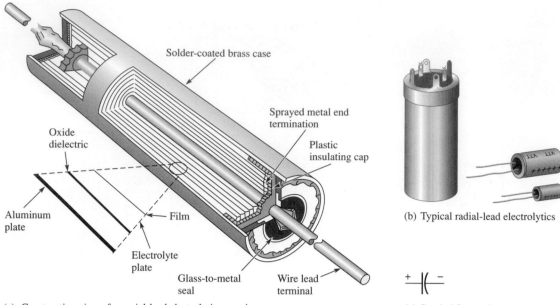

(a) Construction view of an axial-lead electrolytic capacitor

Solder-coated brass case

Sprayed metal end termination

Plastic insulating cap

Oxide dielectric

Aluminum plate

Film

Electrolyte plate

Glass-to-metal seal

Wire lead terminal

(b) Typical radial-lead electrolytics

(c) Symbol for an electrolytic capacitor. The straight plate is positive and the curved plate is negative, as indicated.

▲ FIGURE 12–14

Examples of electrolytic capacitors.

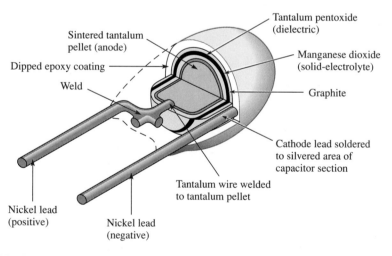

Tantalum pentoxide (dielectric)

Sintered tantalum pellet (anode)

Manganese dioxide (solid-electrolyte)

Dipped epoxy coating

Weld

Graphite

Cathode lead soldered to silvered area of capacitor section

Tantalum wire welded to tantalum pellet

Nickel lead (positive)

Nickel lead (negative)

▲ FIGURE 12–15

Construction view of a typical "tear drop" shaped tantalum electrolytic capacitor.

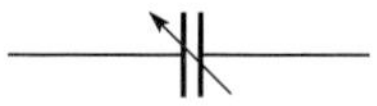

▲ FIGURE 12–16

Schematic symbol for a variable capacitor.

available in larger values for specialized applications. The schematic symbol for a variable capacitor is shown in Figure 12–16.

Adjustable capacitors that normally have slotted screw-type adjustments and are used for very fine adjustments in a circuit are called **trimmers**. Ceramic or mica is a common dielectric in these types of capacitors, and the capacitance usually is changed by adjusting the plate separation. Generally, trimmer capacitors have values less than 100 pF. Figure 12–17 shows some typical devices.

◄ **FIGURE 12–17**

Examples of trimmer capacitors.

The **varactor** is a semiconductor device that exhibits a capacitance characteristic that is varied by changing the voltage across its terminals. This device usually is covered in detail in a course on electronic devices.

Capacitor Labeling

Capacitor values are indicated on the body of the capacitor either by typographical labels or by color codes. Typographical labels consist of letters and numbers that indicate various parameters such as capacitance, voltage rating, and tolerance.

Some capacitors carry no unit designation for capacitance. In these cases, the units are implied by the value indicated and are recognized by experience. For example, a ceramic capacitor marked .001 or .01 has units of microfarads because picofarad values that small are not available. As another example, a ceramic capacitor labeled 50 or 330 has units of picofarads because microfarad units that large normally are not available in this type. In some cases, a 3-digit designation is used. The first two digits are the first two digits of the capacitance value. The third digit is the number of zeros after the second digit. For example, 103 means 10,000 pF. In some instances, the units are labeled as pF or μF; sometimes the microfarad unit is labeled as MF or MFD.

A voltage rating appears on some types of capacitors with WV or WVDC and is omitted on others. When it is omitted, the voltage rating can be determined from information supplied by the manufacturer. The tolerance of the capacitor is usually labeled as a percentage, such as $\pm$10%. The temperature coefficient is indicated by a *parts per million* marking. This type of label consists of a P or N followed by a number. For example, N750 means a negative temperature coefficient of 750 ppm/°C, and P30 means a positive temperature coefficient of 30 ppm/°C. An NP0 designation means that the positive and negative coefficients are zero; thus the capacitance does not change with temperature. Certain types of capacitors are color coded. Refer to Appendix C for additional capacitor labeling and color code information.

Capacitance Measurement

An *LCR* meter such as the one shown in Figure 12–18 can be used to check the value of a capacitor. Also, many DMMs provide a capacitance measurement feature. Most capacitors change value over a period of time, some more than others. Ceramic capacitors, for example, often exhibit a 10% to 15% change in value during the first year. Electrolytic capacitors are particularly subject to value change due to drying of the electrolytic solution. In other cases, capacitors may be labeled incorrectly or the wrong value may have been installed in the circuit. Although a value change represents less than 25% of defective capacitors, a value check can quickly eliminate this as a source of trouble when troubleshooting a circuit.

Typically, values from 200 pF to 20 mF can be measured on an *LCR* meter by simply connecting the capacitor, setting the switch, and reading the value on the display. Some *LCR* meters can also be used to check for leakage current in capacitors. In order to check for leakage, a sufficient voltage must be applied across the capacitor to simulate operating conditions. This is automatically done by the test instrument. Over 40% of all defective capacitors have excessive leakage current and electrolytics are particularly susceptible to this problem.

▲ **FIGURE 12–18**

A typical *LCR* meter. (Courtesy of B+K Precision)

12–3 SERIES CAPACITORS

The total capacitance of a series connection of capacitors is less than the individual capacitance of any of the capacitors. Capacitors in series divide voltage across them in proportion to their capacitance.

After completing this section, you should be able to

◆ **Analyze series capacitors**

   ◆ Determine total capacitance

   ◆ Determine capacitor voltages

Total Capacitance

When capacitors are connected in series, the total capacitance is less than the smallest capacitance value because the effective plate separation increases. The calculation of total series capacitance is analogous to the calculation of total resistance of parallel resistors (Chapter 6).

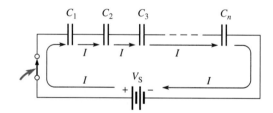

(a) Charging current is same for each capacitor, $I = Q/t$.

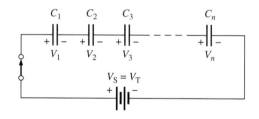

(b) All capacitors store same amount of charge and $V = Q/C$.

▲ **FIGURE 12–19**

A series capacitive circuit.

Consider the generalized circuit in Figure 12–19(a), which has n capacitors in series with a voltage source and a switch. When the switch is closed, the capacitors charge as current is established through the circuit. Since this is a series circuit, the current must be the same at all points, as illustrated. Since current is the rate of flow of charge, the amount of charge stored by each capacitor is equal to the total charge, expressed as

Equation 12–7
$$Q_T = Q_1 = Q_2 = Q_3 = \cdots = Q_n$$

Next, according to Kirchhoff's voltage law, the sum of the voltages across the charged capacitors must equal the total voltage, V_T, as shown in Figure 12–19(b). This is expressed in equation form as

$$V_T = V_1 + V_2 + V_3 + \cdots + V_n$$

From Equation 12–3, $V = Q/C$. When this relationship is substituted into each term of the voltage equation, the following result is obtained:

$$\frac{Q_T}{C_T} = \frac{Q_1}{C_1} + \frac{Q_2}{C_2} + \frac{Q_3}{C_3} + \cdots + \frac{Q_n}{C_n}$$

Since the charges on all the capacitors are equal, the Q terms can be factored and canceled, resulting in

$$\frac{1}{C_T} = \frac{1}{C_1} + \frac{1}{C_2} + \frac{1}{C_3} + \cdots + \frac{1}{C_n}$$

Equation 12–8

Taking the reciprocal of both sides of Equation 12–8 yields the following general formula for total series capacitance:

$$C_T = \frac{1}{\dfrac{1}{C_1} + \dfrac{1}{C_2} + \dfrac{1}{C_3} + \cdots + \dfrac{1}{C_n}}$$

Equation 12–9

Remember,

The total series capacitance is always less than the smallest capacitance.

Two Capacitors in Series   When only two capacitors are in series, a special form of Equation 12–8 can be used.

$$\frac{1}{C_T} = \frac{1}{C_1} + \frac{1}{C_2} = \frac{C_1 + C_2}{C_1 C_2}$$

Taking the reciprocal of the left and right terms gives the formula for total capacitance of two capacitors in series.

$$C_T = \frac{C_1 C_2}{C_1 + C_2}$$

Equation 12–10

Capacitors of Equal Value in Series   This special case is another in which a formula can be developed from Equation 12–8. When all capacitor values are the same and equal to C, the formula is

$$\frac{1}{C_T} = \frac{1}{C} + \frac{1}{C} + \frac{1}{C} + \cdots + \frac{1}{C}$$

Adding all the terms on the right yields

$$\frac{1}{C_T} = \frac{n}{C}$$

where n is the number of equal-value capacitors. Taking the reciprocal of both sides yields

$$C_T = \frac{C}{n}$$

Equation 12–11

The capacitance value of the equal capacitors divided by the number of equal series capacitors gives the total capacitance.

EXAMPLE 12–5

Determine the total capacitance between points A and B in Figure 12–20.

▶ **FIGURE 12–20**

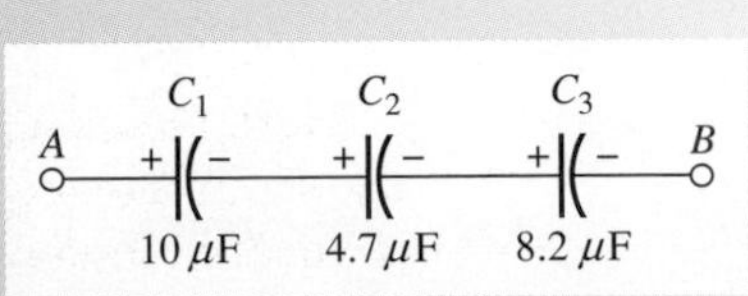

Solution   Use Equation 12–9.

$$C_T = \frac{1}{\dfrac{1}{C_1} + \dfrac{1}{C_2} + \dfrac{1}{C_3}} = \frac{1}{\dfrac{1}{10\ \mu F} + \dfrac{1}{4.7\ \mu F} + \dfrac{1}{8.2\ \mu F}} = \textbf{2.30}\ \boldsymbol{\mu F}$$

Related Problem   If a 4.7 μF capacitor is connected in series with the three existing capacitors in Figure 12–20, what is C_T?

EXAMPLE 12–6

Find the total capacitance, C_T, in Figure 12–21.

▶ **FIGURE 12–21**

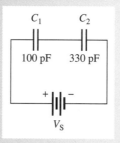

Solution   From Equation 12–10,

$$C_T = \frac{C_1 C_2}{C_1 + C_2} = \frac{(100\ \text{pF})(330\ \text{pF})}{430\ \text{pF}} = \textbf{76.7 pF}$$

You can also use Equation 12–9.

$$C_T = \frac{1}{\dfrac{1}{100\ \text{pF}} + \dfrac{1}{330\ \text{pF}}} = \textbf{76.7 pF}$$

Related Problem   Determine C_T if $C_1 = 470$ pF and $C_2 = 680$ pF in Figure 12–21.

EXAMPLE 12–7

Determine C_T for the series capacitors in Figure 12–22.

▶ **FIGURE 12–22**

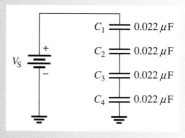

Solution   Since $C_1 = C_2 = C_3 = C_4 = C$, use Equation 12–11,

$$C_T = \frac{C}{n} = \frac{0.022\ \mu F}{4} = \textbf{0.0055}\ \boldsymbol{\mu F}$$

Related Problem   Determine C_T if the capacitor values in Figure 12–22 are doubled.

Capacitor Voltages

A series connection of charged capacitors acts as a voltage divider. The voltage across each capacitor in series is inversely proportional to its capacitance value, as shown by the formula $V = Q/C$. You can determine the voltage across any individual capacitor in series with the following formula:

$$V_x = \left(\frac{C_T}{C_x}\right)V_T$$

Equation 12–12

where C_x is any capacitor in series (such as C_1, C_2, or C_3), V_x is the voltage across C_x, and V_T is the total voltage across the capacitors. The derivation is as follows: Since the charge on any capacitor in series is the same as the total charge ($Q_x = Q_T$), and since $Q_x = V_xC_x$ and $Q_T = V_TC_T$, then

$$V_xC_x = V_TC_T$$

Solving for V_x yields

$$V_x = \frac{C_TV_T}{C_x}$$

The largest-value capacitor in a series connection will have the smallest voltage across it. The smallest-value capacitor will have the largest voltage across it.

EXAMPLE 12–8

Find the voltage across each capacitor in Figure 12–23.

▶ **FIGURE 12–23**

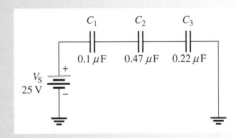

Solution Calculate the total capacitance.

$$\frac{1}{C_T} = \frac{1}{C_1} + \frac{1}{C_2} + \frac{1}{C_3} = \frac{1}{0.1\,\mu F} + \frac{1}{0.47\,\mu F} + \frac{1}{0.22\,\mu F}$$

$$C_T = 0.06\,\mu F$$

From Figure 12–24, $V_S = V_T = 25$ V. Therefore, use Equation 12–12 to calculate the voltage across each capacitor.

$$V_1 = \left(\frac{C_T}{C_1}\right)V_T = \left(\frac{0.06\,\mu F}{0.1\,\mu F}\right)25\text{ V} = \textbf{15.0 V}$$

$$V_2 = \left(\frac{C_T}{C_2}\right)V_T = \left(\frac{0.06\,\mu F}{0.47\,\mu F}\right)25\text{ V} = \textbf{3.19 V}$$

$$V_3 = \left(\frac{C_T}{C_3}\right)V_T = \left(\frac{0.06\,\mu F}{0.22\,\mu F}\right)25\text{ V} = \textbf{6.82 V}$$

Related Problem Another 0.47 μF capacitor is connected in series with the existing capacitor in Figure 12–23. Determine the voltage across the new capacitor, assuming all the capacitors are initially uncharged.

SECTION 12–3 REVIEW

1. Is the total capacitance of a series connection less than or greater than the value of the smallest capacitor?
2. The following capacitors are in series: 100 pF, 220 pF, and 560 pF. What is the total capacitance?
3. A 0.01 μF and a 0.015 μF capacitor are in series. Determine the total capacitance.
4. Five 100 pF capacitors are connected in series. What is C_T?
5. Determine the voltage across C_1 in Figure 12–24.

▶ FIGURE 12–24

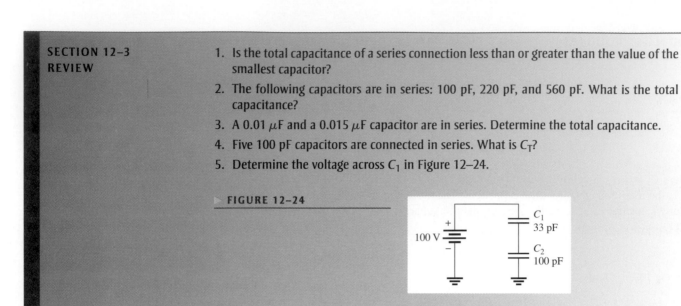

12–4 PARALLEL CAPACITORS

Capacitances add when capacitors are connected in parallel.

After completing this section, you should be able to

◆ **Analyze parallel capacitors**

  ◆ Determine total capacitance

When capacitors are connected in parallel, the total capacitance is the sum of the individual capacitances because the effective plate area increases. The calculation of total parallel capacitance is analogous to the calculation of total series resistance (Chapter 5).

Consider what happens when the switch in Figure 12–25 is closed. The total charging current from the source divides at the junction of the parallel branches. There is a separate charging current through each branch so that a different charge can be stored by each capacitor. By Kirchhoff's current law, the sum of all of the charging currents is equal to the total current. Therefore, the sum of the charges on the capacitors is equal to the total charge. Also, the voltages across all of the parallel branches are equal. These observations are used to develop a formula for total parallel capacitance as follows for the general case of n capacitors in parallel.

Equation 12–13

$$Q_T = Q_1 + Q_2 + Q_3 + \cdots + Q_n$$

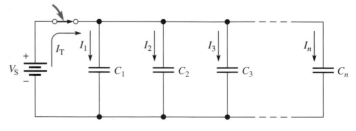

▲ FIGURE 12–25

Capacitors in parallel.

From Equation 12–2, $Q = CV$. When this relationship is substituted into each term of Equation 12–13, the following result is obtained:

$$C_T V_T = C_1 V_1 + C_2 V_2 + C_3 V_3 + \cdots + C_n V_n$$

Since $V_T = V_1 = V_2 = V_3 = \cdots = V_n$, the voltages can be factored and canceled, giving

$$C_T = C_1 + C_2 + C_3 + \cdots + C_n \qquad\qquad \textbf{Equation 12–14}$$

Equation 12–14 is the general formula for total parallel capacitance where n is the number of capacitors. Remember,

The total parallel capacitance is the sum of all the capacitors in parallel.

For the special case when all of the capacitors have the same value, C, multiply the value by the number (n) of capacitors in parallel.

$$C_T = nC \qquad\qquad \textbf{Equation 12–15}$$

EXAMPLE 12–9

What is the total capacitance in Figure 12–26? What is the voltage across each capacitor?

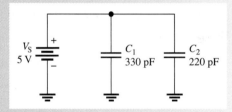

▲ FIGURE 12–26

Solution The total capacitance is

$$C_T = C_1 + C_2 = 330\,\text{pF} + 220\,\text{pF} = \textbf{550 pF}$$

The voltage across each capacitor in parallel is equal to the source voltage.

$$V_S = V_1 = V_2 = \textbf{5 V}$$

Related Problem What is C_T if a 100 pF capacitor is connected in parallel with C_2 in Figure 12–26?

EXAMPLE 12–10

Determine C_T in Figure 12–27.

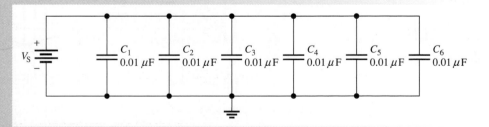

▲ FIGURE 12–27

Solution   There are six equal-value capacitors in parallel, so $n = 6$.

$$C_T = nC = (6)(0.01 \ \mu\text{F}) = \mathbf{0.06 \ \mu F}$$

Related Problem   If three more 0.01 μF capacitors are connected in parallel in Figure 12–27, what is the total capacitance?

**SECTION 12–4
REVIEW**

1. How is total parallel capacitance determined?
2. In a certain application, you need 0.05 μF. The only values available are 0.01 μF, which are available in large quantities. How can you get the total capacitance that you need?
3. The following capacitors are in parallel: 10 pF, 56 pF, 33 pF, and 68 pF. What is C_T?

12–5 CAPACITORS IN DC CIRCUITS

A capacitor will charge up when it is connected to a dc voltage source. The buildup of charge across the plates occurs in a predictable manner that is dependent on the capacitance and the resistance in a circuit.

After completing this section, you should be able to

◆ **Analyze capacitive dc switching circuits**

   ◆ Describe the charging and discharging of a capacitor

   ◆ Define *RC time constant*

   ◆ Relate the time constant to charging and discharging of a capacitor

   ◆ Write equations for the charging and discharging curves

   ◆ Explain why a capacitor blocks dc

Charging a Capacitor

A capacitor will charge when it is connected to a dc voltage source, as shown in Figure 12–28. The capacitor in part (a) of the figure is uncharged; that is, plate *A* and plate *B* have equal numbers of free electrons. When the switch is closed, as shown in part (b), the source moves electrons away from plate *A* through the circuit to plate *B* as the arrows indicate. As

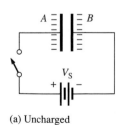

(a) Uncharged

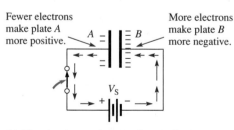

(b) Charging (arrows indicate electron flow)

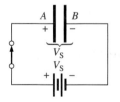

(c) Fully charged.
$I = 0$

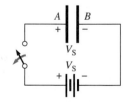

(d) Retains charge

▲ FIGURE 12–28

Charging a capacitor.

plate A loses electrons and plate B gains electrons, plate A becomes positive with respect to plate B. As this charging process continues, the voltage across the plates builds up rapidly until it is equal to the applied voltage, V_S, but opposite in polarity, as shown in part (c). When the capacitor is fully charged, there is no current.

A capacitor blocks constant dc.

When the charged capacitor is disconnected from the source, as shown in Figure 12–28(d), it remains charged for long periods of time, depending on its leakage resistance, and can cause severe electrical shock. The charge on an electrolytic capacitor generally leaks off more rapidly than in other types of capacitors.

Discharging a Capacitor

When a wire is connected across a charged capacitor, as shown in Figure 12–29, the capacitor will discharge. In this particular case, a very low resistance path (the wire) is connected across the capacitor with a switch. Before the switch is closed, the capacitor is charged to 50 V, as indicated in part (a). When the switch is closed, as shown in part (b), the excess electrons on plate B move through the circuit to plate A (indicated by the arrows); as a result of the electrons moving through the low resistance of the wire, the energy stored by the capacitor is dissipated in the wire. The charge is neutralized when the numbers of free electrons on both plates are again equal. At this time, the voltage across the capacitor is zero, and the capacitor is completely discharged, as shown in part (c).

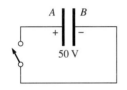

(a) Retains charge

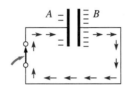

(b) Discharging (arrows indicate electron flow)

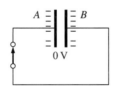

(c) Uncharged

◀ **FIGURE 12–29**

Discharging a charged capacitor.

Current and Voltage During Charging and Discharging

Notice in Figures 12–28 and 12–29 that the direction of electron flow during discharge is opposite to that during charging. It is important to understand that ideally *there is no current through the dielectric of the capacitor during charging or discharging because the dielectric is an insulating material.* There is current from one plate to the other only through the external circuit.

Figure 12–30(a) shows a capacitor connected in series with a resistor and a switch to a dc voltage source. Initially, the switch is open and the capacitor is uncharged with zero volts across its plates. At the instant the switch is closed, the current jumps to its maximum value and the capacitor begins to charge. The current is maximum initially because the capacitor has zero volts across it and, therefore, effectively acts as a short; thus, the current is limited only by the resistance. As time passes and the capacitor charges, the current decreases and the voltage across the capacitor (V_C) increases. The resistor voltage is proportional to the current during this charging period.

After a certain period of time, the capacitor reaches full charge. At this point, the current is zero and the capacitor voltage is equal to the dc source voltage, as shown in Figure 12–30(b). If the switch were opened now, the capacitor would retain its full charge (neglecting any leakage).

In Figure 12–30(c), the voltage source has been removed. When the switch is closed, the capacitor begins to discharge. Initially, the current jumps to a maximum but in a direction

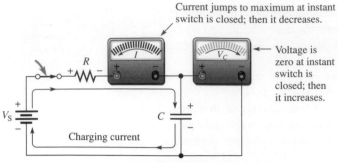

Current jumps to maximum at instant switch is closed; then it decreases.

Voltage is zero at instant switch is closed; then it increases.

Charging current

(a) Charging: Capacitor voltage increases as the current and resistor voltage decrease.

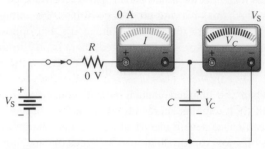

(b) Fully charged: Capacitor voltage equals source voltage. The current is zero.

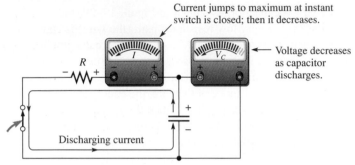

Current jumps to maximum at instant switch is closed; then it decreases.

Voltage decreases as capacitor discharges.

Discharging current

(c) Discharging: Capacitor voltage, resistor voltage, and the current decrease from their initial maximum values. Note that the discharge current is opposite to the charge current.

▲ FIGURE 12–30

Current and voltage in a charging and discharging capacitor.

opposite to its direction during charging. As time passes, the current and capacitor voltage decrease. The resistor voltage is always proportional to the current. When the capacitor has fully discharged, the current and the capacitor voltage are zero.

Remember the following rules about capacitors in dc circuits:

1. A capacitor appears as an *open* to constant voltage.

2. A capacitor appears as a *short* to an instantaneous change in voltage.

Now let's examine in more detail how the voltage and current change with time in a capacitive circuit.

The *RC* Time Constant

In a practical situation, there cannot be capacitance without some resistance in a circuit. It may simply be the small resistance of a wire, a Thevenin source resistance, or it may be a physical resistor. Because of this, the charging and discharging characteristics of a capacitor must always be considered with the associated resistance. The resistance introduces the element of *time* in the charging and discharging of a capacitor.

When a capacitor charges or discharges through a resistance, a certain time is required for the capacitor to charge fully or discharge fully. The voltage across a capacitor cannot change instantaneously because a finite time is required to move charge from one point to another. The time constant of a series *RC* circuit determines the rate at which the capacitor charges or discharges.

The *RC* time constant is a fixed time interval that equals the product of the resistance and the capacitance in a series *RC* circuit.

The time constant is expressed in seconds when resistance is in ohms and capacitance is in farads. It is symbolized by τ (Greek letter tau), and the formula is

$$\tau = RC$$

Equation 12–16

Recall that $I = Q/t$. The current depends on the amount of charge moved in a given time. When the resistance is increased, the charging current is reduced, thus increasing the charging time of the capacitor. When the capacitance is increased, the amount of charge increases; thus, for the same current, more time is required to charge the capacitor.

EXAMPLE 12–11

A series RC circuit has a resistance of $1.0\,M\Omega$ and a capacitance of $4.7\,\mu F$. What is the time constant?

Solution

$$\tau = RC = (1.0 \times 10^6\,\Omega)(4.7 \times 10^{-6}\,F) = \textbf{4.7 s}$$

Related Problem

A series RC circuit has a $270\,k\Omega$ resistor and a $3300\,pF$ capacitor. What is the time constant?

When a capacitor is charging or discharging between two voltage levels, the charge on the capacitor changes by approximately 63% of the difference in the levels in one time constant. An uncharged capacitor charges to 63% of its fully charged voltage in one time constant. When a capacitor is discharging, its voltage drops to approximately 100% − 63% = 37% of its initial value in one time constant, which is a 63% change.

The Charging and Discharging Curves

A capacitor charges and discharges following a nonlinear curve, as shown in Figure 12–31. In these graphs, the approximate percentage of full charge is shown at each time-constant interval. This type of curve follows a precise mathematical formula and is called an *exponential curve*. The charging curve is an increasing exponential, and the discharging curve is a decreasing exponential. It takes five time constants to change the voltage by 99% (considered 100%). This five time-constant interval is generally accepted as the time to fully charge or discharge a capacitor and is called the *transient time*.

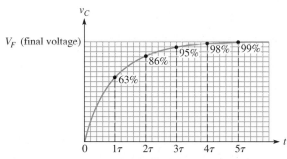
(a) Charging curve with percentages of the final voltage

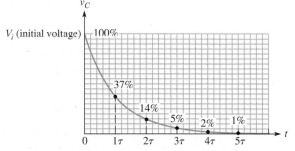
(b) Discharging curve with percentages of the initial voltage

▲ **FIGURE 12–31**

Exponential voltage curves for the charging and discharging of an RC circuit.

General Formula   The general expressions for either increasing or decreasing exponential curves are given in the following equations for both instantaneous voltage and instantaneous current.

$$v = V_F + (V_i - V_F)e^{-t/\tau}$$

Equation 12–17

$$i = I_F + (I_i - I_F)e^{-t/\tau}$$

Equation 12–18

where V_F and I_F are the final values of voltage and current, and V_i and I_i are the initial values of voltage and current. The lowercase italic letters v and i are the instantaneous values of the capacitor voltage and current at time t, and e is the base of natural logarithms. The e^x key on a calculator makes it easy to work with this exponential term.

Charging from Zero   The formula for the special case in which an increasing exponential voltage curve begins at zero ($V_i = 0$), as shown in Figure 12–31(a), is given in Equation 12–19. It is developed as follows, starting with the general formula, Equation 12–17.

$$v = V_F + (V_i - V_F)e^{-t/\tau} = V_F + (0 - V_F)e^{-t/RC} = V_F - V_F e^{-t/RC}$$

Factoring out V_F, you have

Equation 12–19
$$v = V_F(1 - e^{-t/RC})$$

Using Equation 12–19, you can calculate the value of the charging voltage of a capacitor at any instant of time if it is initially uncharged. You can calculate an increasing current by substituting i for v and I_F for V_F in Equation 12–19.

EXAMPLE 12–12

In Figure 12–32, determine the capacitor voltage 50 μs after the switch is closed if the capacitor is initially uncharged. Draw the charging curve.

▶ **FIGURE 12–32**

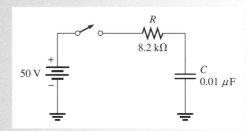

Solution   The time constant is $RC = (8.2\text{ k}\Omega)(0.01\ \mu\text{F}) = 82\ \mu\text{s}$. The voltage to which the capacitor will fully charge is 50 V (this is V_F). The initial voltage is zero. Notice that 50 μs is less than one time constant; so the capacitor will charge less than 63% of the full voltage in that time.

$$v_C = V_F(1 - e^{-t/RC}) = (50\text{ V})(1 - e^{-50\mu s/82\mu s})$$
$$= (50\text{ V})(1 - e^{-0.61}) = (50\text{ V})(1 - 0.543) = \textbf{22.8 V}$$

The charging curve for the capacitor is shown in Figure 12–33.

▶ **FIGURE 12–33**

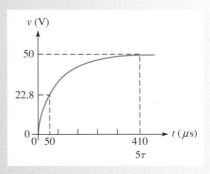

You can determine an exponential function on your calculator by using the e^x key and entering the value of the exponent of e.

Related Problem Determine the capacitor voltage 15 μs after switch closure in Figure 12–32.

 Use Multisim file E12-12 to verify the calculated results in this example and to confirm your calculation for the related problem.

Discharging to Zero The formula for the special case in which a decreasing exponential voltage curve ends at zero ($V_F = 0$), as shown in Figure 12–31(b), is derived from the general formula as follows:

$$v = V_F + (V_i - V_F)e^{-t/\tau} = 0 + (V_i - 0)e^{-t/RC}$$

This reduces to

$$v = V_i e^{-t/RC}$$

Equation 12–20

where V_i is the voltage at the beginning of the discharge. You can use this formula to calculate the discharging voltage at any instant, as Example 12–13 illustrates.

EXAMPLE 12–13

Determine the capacitor voltage in Figure 12–34 at a point in time 6 ms after the switch is closed. Draw the discharging curve.

▶ **FIGURE 12–34**

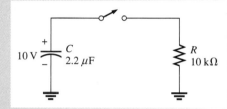

Solution The discharge time constant is $RC = (10\,k\Omega)(2.2\,\mu F) = 22\,ms$. The initial capacitor voltage is 10 V. Notice that 6 ms is less than one time constant, so the capacitor will discharge less than 63%. Therefore, it will have a voltage greater than 37% of the initial voltage at 6 ms.

$$v_C = V_i e^{-t/RC} = (10\,V)e^{-6ms/22ms} = (10\,V)e^{-0.27} = (10\,V)(0.761) = \textbf{7.61 V}$$

The discharging curve for the capacitor is shown in Figure 12–35.

▶ **FIGURE 12–35**

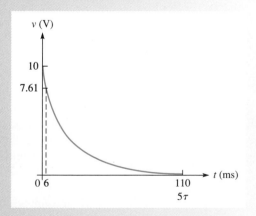

Related Problem In Figure 12–34, change R to 2.2 kΩ and determine the capacitor voltage 1 ms after the switch is closed.

Graphical Method Using Universal Exponential Curves The universal curves in Figure 12–36 provide a graphic solution of the charge and discharge of capacitors. Example 12–14 illustrates this graphical method.

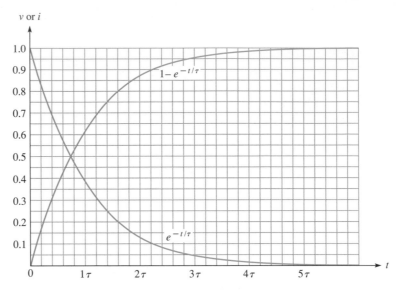

▲ **FIGURE 12–36**

Normalized universal exponential curves.

EXAMPLE 12–14

How long will it take the initially uncharged capacitor in Figure 12–37 to charge to 75 V? What is the capacitor voltage 2 ms after the switch is closed? Use the normalized universal exponential curves in Figure 12–36 to determine the answers.

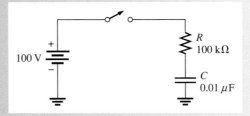

▲ **FIGURE 12–37**

Solution The full charge voltage is 100 V, which is at the 100% level (1.0) on the normalized vertical scale of the graph. The value 75 V is 75% of maximum, or 0.75 on the graph. You can see that this value occurs at 1.4 time constants. In this circuit, one time constant is $RC = (100 \text{ k}\Omega)(0.01 \ \mu\text{F}) = 1 \text{ ms}$. Therefore, the capacitor voltage reaches 75 V at 1.4 ms after the switch is closed.

On the universal exponential curve, you see that the capacitor is at approximately 86 V (0.86 on the vertical axis) in 2 ms, which is 2 time constants. These graphic solutions are shown in Figure 12–38.

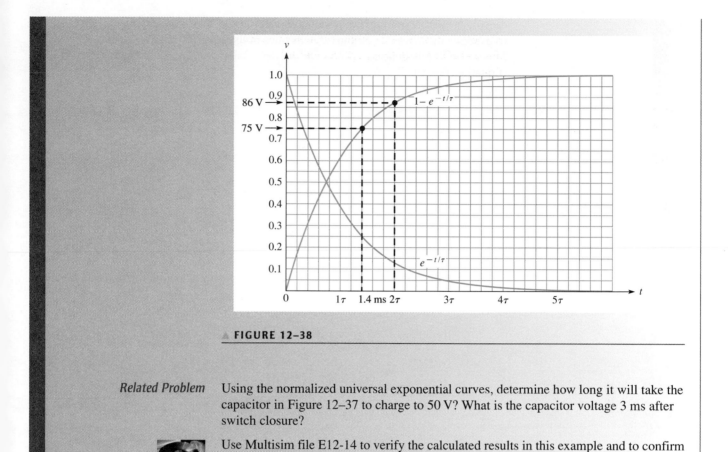

▲ FIGURE 12–38

Related Problem   Using the normalized universal exponential curves, determine how long it will take the capacitor in Figure 12–37 to charge to 50 V? What is the capacitor voltage 3 ms after switch closure?

Use Multisim file E12-14 to verify the calculated results in this example and to confirm your calculation for the related problem. Use a square wave to replace the dc voltage source and the switch.

Time-Constant Percentage Tables   The percentages of full charge or discharge at each time-constant interval can be calculated using the exponential formulas, or they can be extracted from the universal exponential curves. The results are summarized in Tables 12–4 and 12–5.

▼ TABLE 12–4

Percentage of final charge after each charging time-constant interval.

NUMBER OF TIME CONSTANTS	APPROXIMATE % OF FINAL CHARGE
1	63
2	86
3	95
4	98
5	99 (considered 100%)

▼ TABLE 12–5

Percentage of initial charge after each discharging time-constant interval.

NUMBER OF TIME CONSTANTS	APPROXIMATE % OF INITIAL CHARGE
1	37
2	14
3	5
4	2
5	1 (considered 0)

Solving for Time

Occasionally, it is necessary to determine how long it will take a capacitor to charge or discharge to a specified voltage. Equations 12–17 and 12–19 can be solved for t if v is specified. The natural logarithm (abbreviated ln) of $e^{-t/RC}$ is the exponent $-t/RC$. Therefore, taking the natural logarithm of both sides of the equation allows you to solve for time. This procedure is done as follows for the decreasing exponential formula when $V_F = 0$ (Equation 12–20).

$$v = V_i e^{-t/RC}$$

$$\frac{v}{V_i} = e^{-t/RC}$$

$$\ln\left(\frac{v}{V_i}\right) = \ln e^{-t/RC}$$

$$\ln\left(\frac{v}{V_i}\right) = \frac{-t}{RC}$$

Equation 12–21
$$t = -RC \ln\left(\frac{v}{V_i}\right)$$

The same procedure can be used for the increasing exponential formula in Equation 12–19 as follows:

$$v = V_F(1 - e^{-t/RC})$$

$$\frac{v}{V_F} = 1 - e^{-t/RC}$$

$$1 - \frac{v}{V_F} = e^{-t/RC}$$

$$\ln\left(1 - \frac{v}{V_F}\right) = \ln e^{-t/RC}$$

$$\ln\left(1 - \frac{v}{V_F}\right) = \frac{-t}{RC}$$

Equation 12–22
$$t = -RC \ln\left(1 - \frac{v}{V_F}\right)$$

EXAMPLE 12–15

In Figure 12–39, how long will it take the capacitor to discharge to 25 V when the switch is closed?

▶ **FIGURE 12–39**

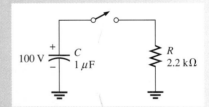

Solution   Use Equation 12–21 to find the discharge time.

$$t = -RC \ln\left(\frac{v}{V_i}\right) = -(2.2\,k\Omega)(1\,\mu F)\ln\left(\frac{25\,V}{100\,V}\right)$$

$$= -(2.2\,ms)\ln(0.25) = -(2.2\,ms)(-1.39) = \textbf{3.05 ms}$$

You can determine ln(0.25) with your calculator by using the LN key.

Related Problem   How long will it take the capacitor in Figure 12–39 to discharge to 50 V?

RC Response to a Square Wave

A common case that illustrates the rising and falling exponential occurs when an *RC* circuit is driven with a square wave that has a long period compared to the time constant. The square wave provides on and off action but, unlike a single switch, it provides a discharge path back through the generator when the wave drops back to zero.

When the square wave rises, the voltage across the capacitor rises exponentially toward the maximum value of the square wave in a time that depends on the time constant. When the square wave returns to the zero level, the capacitor voltage decreases exponentially, again depending on the time constant. The Thevenin resistance of the generator is part of the *RC* time constant; however, it can be ignored if it is small compared to *R*. Example 12–16 shows the waveforms for the case where the period is long compared to the time constant; other cases will be covered in detail in Chapter 20.

EXAMPLE 12–16

In Figure 12–40, calculate the voltage across the capacitor every 0.1 ms for one complete period of the input. Then sketch the capacitor waveform. Assume the Thevenin resistance of the generator is negligible.

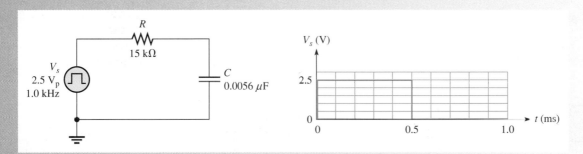

▲ FIGURE 12–40

Solution

$$\tau = RC = (15 \text{ k}\Omega)(0.0056 \,\mu\text{F}) = 0.084 \text{ ms}$$

The period of the square wave is 1 ms, which is approximately 12τ. This means that 6τ will elapse after each change of the pulse, allowing the capacitor to fully charge and fully discharge.

For the rising exponential,

$$v = V_\text{F}(1 - e^{-t/RC}) = V_\text{F}(1 - e^{-t/\tau})$$

At 0.1 ms: $v = 2.5 \text{ V}(1 - e^{-0.1\text{ms}/0.084\text{ms}}) = 1.74 \text{ V}$

At 0.2 ms: $v = 2.5 \text{ V}(1 - e^{-0.2\text{ms}/0.084\text{ms}}) = 2.27 \text{ V}$

At 0.3 ms: $v = 2.5 \text{ V}(1 - e^{-0.3\text{ms}/0.084\text{ms}}) = 2.43 \text{ V}$

At 0.4 ms: $v = 2.5 \text{ V}(1 - e^{-0.4\text{ms}/0.084\text{ms}}) = 2.48 \text{ V}$

At 0.5 ms: $v = 2.5 \text{ V}(1 - e^{-0.5\text{ms}/0.084\text{ms}}) = 2.49 \text{ V}$

For the falling exponential,

$$v = V_i(e^{-t/RC}) = V_i(e^{-t/\tau})$$

In the equation, time is shown from the point when the change occurs (subtracting 0.5 ms from the actual time). For example, at 0.6 ms, $t = 0.6\,\text{ms} - 0.5\,\text{ms} = 0.1\,\text{ms}$.

At 0.6 ms: $v = 2.5\,\text{V}(e^{-0.1\text{ms}/0.084\text{ms}}) = 0.76\,\text{V}$

At 0.7 ms: $v = 2.5\,\text{V}(e^{-0.2\text{ms}/0.084\text{ms}}) = 0.23\,\text{V}$

At 0.8 ms: $v = 2.5\,\text{V}(e^{-0.3\text{ms}/0.084\text{ms}}) = 0.07\,\text{V}$

At 0.9 ms: $v = 2.5\,\text{V}(e^{-0.4\text{ms}/0.084\text{ms}}) = 0.02\,\text{V}$

At 1.0 ms: $v = 2.5\,\text{V}(e^{-0.5\text{ms}/0.084\text{ms}}) = 0.01\,\text{V}$

Figure 12–41 is a plot of these results.

FIGURE 12–41

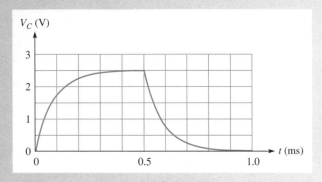

Related Problem   What is the capacitor voltage at 0.65 ms?

SECTION 12–5 REVIEW

1. Determine the time constant when $R = 1.2\,\text{k}\Omega$ and $C = 1000\,\text{pF}$.
2. If the circuit mentioned in Question 1 is charged with a 5 V source, how long will it take the capacitor to reach full charge? At full charge, what is the capacitor voltage?
3. For a certain circuit, $\tau = 1\,\text{ms}$. If it is charged with a 10 V battery, what will the capacitor voltage be at each of the following times: 2 ms, 3 ms, 4 ms, and 5 ms?
4. A capacitor is charged to 100 V. If it is discharged through a resistor, what is the capacitor voltage at one time constant?

12–6 CAPACITORS IN AC CIRCUITS

As you know, a capacitor blocks dc. A capacitor passes ac but with an amount of opposition, called capacitive reactance, that depends on the frequency of the ac.

After completing this section, you should be able to

♦ **Analyze capacitive ac circuits**

   ♦ Explain why a capacitor causes a phase shift between voltage and current

   ♦ Define *capacitive reactance*

   ♦ Determine the value of capacitive reactance in a given circuit

   ♦ Discuss instantaneous, true, and reactive power in a capacitor

To explain fully how capacitors work in ac circuits, the concept of the derivative must be introduced. *The derivative of a time-varying quantity is the instantaneous rate of change of that quantity.*

Recall that current is the rate of flow of charge (electrons). Therefore, instantaneous current, i, can be expressed as the instantaneous rate of change of charge, q, with respect to time, t.

$$i = \frac{dq}{dt}$$

Equation 12–23

The term dq/dt is the derivative of q with respect to time and represents the instantaneous rate of change of q. Also, in terms of instantaneous quantities, $q = Cv$. Therefore, from a basic rule of differential calculus, the derivative of q with respect to time is $dq/dt = C(dv/dt)$. Since $i = dq/dt$, we get the following relationship:

$$i = C\left(\frac{dv}{dt}\right)$$

Equation 12–24

This formula states

The instantaneous capacitor current is equal to the capacitance times the instantaneous rate of change of the voltage across the capacitor.

The faster the voltage across a capacitor changes, the greater the current.

Phase Relationship of Current and Voltage in a Capacitor

Consider what happens when a sinusoidal voltage is applied across a capacitor, as shown in Figure 12–42(a). The voltage waveform has a maximum rate of change (dv/dt = max) at the zero crossings and a zero rate of change (dv/dt = 0) at the peaks, as indicated in Figure 12–42(b).

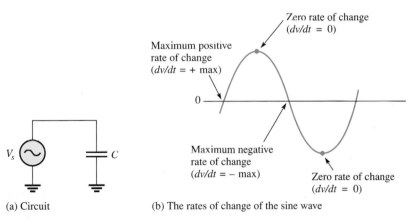

▲ FIGURE 12–42

A sine wave applied to a capacitor.

The phase relationship between the current and the voltage for the capacitor can be established from Equation 12–24. When $dv/dt = 0$, i is also zero because $i = C(dv/dt) = C(0) = 0$. When dv/dt is a positive-going maximum, i is a positive maximum; when dv/dt is a negative-going maximum, i is a negative maximum.

A sinusoidal voltage always produces a sinusoidal current in a capacitive circuit. Therefore, you can plot the current with respect to the voltage if you know the points on the voltage curve at which the current is zero and those at which it is maximum. This relationship is shown in Figure 12–43(a). Notice that the current leads the voltage in phase by 90°. This is always true in a purely capacitive circuit. The relationship between the voltage and current phasors is shown in Figure 12–43(b).

Phase relation of V_C and I_C in a capacitor. Current always leads the capacitor voltage by 90°.

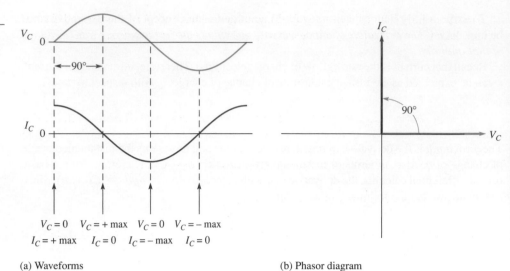

(a) Waveforms

(b) Phasor diagram

Capacitive Reactance, X_C

Capacitive reactance is the opposition to sinusoidal current, expressed in ohms. The symbol for capacitive reactance is X_C.

To develop a formula for X_C, we use the relationship $i = C(dv/dt)$ and the curves in Figure 12–44. The rate of change of voltage is directly related to frequency. The faster the voltage changes, the higher the frequency. For example, you can see that in Figure 12–44 the slope of sine wave A at the zero crossings is steeper than that of sine wave B. The slope of a curve at a point indicates the rate of change at that point. Sine wave A has a higher frequency than sine wave B, as indicated by a greater maximum rate of change (dv/dt is greater at the zero crossings).

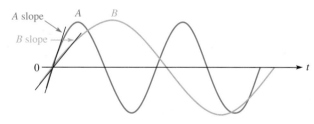

The higher frequency waveform (A) has a greater slope at its zero crossings, corresponding to a higher rate of change.

When frequency increases, dv/dt increases, and thus i increases. When frequency decreases, dv/dt decreases, and thus i decreases.

$$i = C(dv/dt) \qquad \text{and} \qquad i = C(dv/dt)$$

An increase in i means that there is less opposition to current (X_C is less), and a decrease in i means a greater opposition to current (X_C is greater). Therefore, X_C is inversely proportional to i and thus inversely proportional to frequency.

X_C **is inversely proportional to** f, **shown as** $\dfrac{1}{f}$.

From the same relationship $i = C(dv/dt)$, you can see that if dv/dt is constant and C is varied, an increase in C produces an increase in i, and a decrease in C produces a decrease in i.

$$\overset{\uparrow}{i} = C(\overset{\uparrow}{dv/dt}) \quad \text{and} \quad \underset{\downarrow}{i} = C(\underset{\downarrow}{dv/dt})$$

Again, an increase in i means less opposition (X_C is less), and a decrease in i means greater opposition (X_C is greater). Therefore, X_C is inversely proportional to i and thus inversely proportional to capacitance.

The capacitive reactance is inversely proportional to both f and C.

X_C is inversely proportional to fC, shown as $\dfrac{1}{fC}$.

Thus far, we have determined a proportional relationship between X_C and $1/fC$. Equation 12–25 is the complete formula for calculating X_C. The derivation is given in Appendix B.

$$X_C = \frac{1}{2\pi fC}$$

Equation 12–25

Capacitive reactance, X_C, is in ohms when f is in hertz and C is in farads. Notice that 2π appears in the denominator as a constant of proportionality. This term is derived from the relationship of a sine wave to rotational motion.

EXAMPLE 12–17

A sinusoidal voltage is applied to a capacitor, as shown in Figure 12–45. The frequency of the sine wave is 1 kHz. Determine the capacitive reactance.

▶ **FIGURE 12–45**

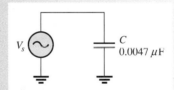

Solution

$$X_C = \frac{1}{2\pi fC} = \frac{1}{2\pi(1 \times 10^3 \text{ Hz})(0.0047 \times 10^{-6} \text{ F})} = \mathbf{33.9\,k\Omega}$$

Related Problem Determine the frequency required to make the capacitive reactance in Figure 12–45 equal to $10\,k\Omega$.

 Use Multisim file E12-17 to verify the calculated results in this example and to confirm your calculation for the related problem.

Ohm's Law The reactance of a capacitor is analogous to the resistance of a resistor, as shown in Figure 12–46. In fact, both are expressed in ohms. Since both R and X_C are forms of opposition to current, Ohm's law applies to capacitive circuits as well as to resistive circuits.

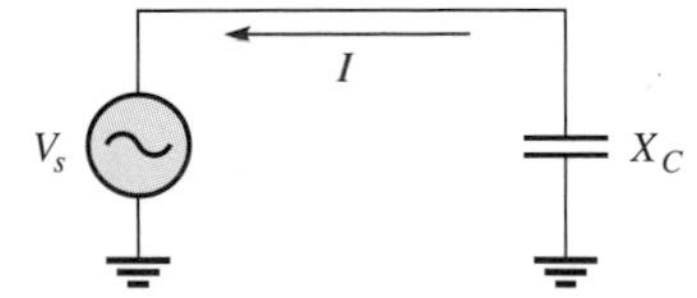

$$I = \frac{V}{X_C}$$

▲ **FIGURE 12–46**

When applying Ohm's law in ac circuits, you must express both the current and the voltage in the same way, that is, both in rms, both in peak, and so on.

EXAMPLE 12–18

Determine the rms current in Figure 12–47.

▶ FIGURE 12–47

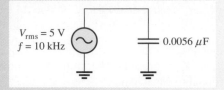

$V_{rms} = 5$ V
$f = 10$ kHz
0.0056 μF

Solution

First, determine the capacitive reactance.

$$X_C = \frac{1}{2\pi fC} = \frac{1}{2\pi(10 \times 10^3 \text{ Hz})(0.0056 \times 10^{-6} \text{ F})} = 2.84 \text{ k}\Omega$$

Then apply Ohm's law.

$$I_{rms} = \frac{V_{rms}}{X_C} = \frac{5 \text{ V}}{2.84 \text{ k}\Omega} = \mathbf{1.76 \text{ mA}}$$

Related Problem

Change the frequency in Figure 12–47 to 25 kHz and determine the rms current.

Use Multisim file E12-18 to verify the calculated results in this example and to confirm your calculation for the related problem.

Power in a Capacitor

As discussed earlier in this chapter, a charged capacitor stores energy in the electric field within the dielectric. An ideal capacitor does not dissipate energy; it only stores it temporarily. When an ac voltage is applied to a capacitor, energy is stored by the capacitor during a portion of the voltage cycle; then the stored energy is returned to the source during another portion of the cycle. There is no net energy loss. Figure 12–48 shows the power curve that results from one cycle of capacitor voltage and current.

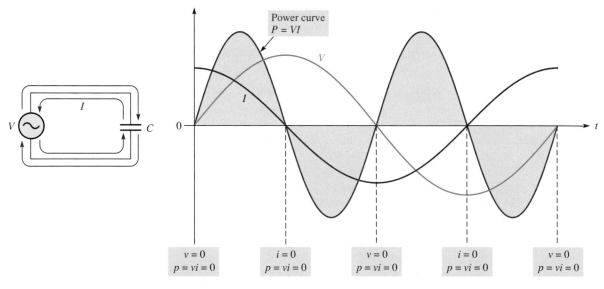

▲ FIGURE 12–48

Power curve.

Instantaneous Power (p)   The product of v and i gives **instantaneous power**. At points where v or i is zero, p is also zero. When both v and i are positive, p is also positive. When either v or i is positive and the other is negative, p is negative. When both v and i are negative, p is positive. As you can see, the power follows a sinusoidal-shaped curve. Positive values of power indicate that energy is stored by the capacitor. Negative values of power indicate that energy is returned from the capacitor to the source. Note that the power fluctuates at a frequency twice that of the voltage or current as energy is alternately stored and returned to the source.

True Power (P$_{true}$)   Ideally, all of the energy stored by a capacitor during the positive portion of the power cycle is returned to the source during the negative portion. No net energy is lost due to conversion to heat in the capacitor, so the **true power** is zero. Actually, because of leakage and foil resistance in a practical capacitor, a small percentage of the total power is dissipated in the form of true power.

Reactive Power (P$_r$)   The rate at which a capacitor stores or returns energy is called its **reactive power**. The reactive power is a nonzero quantity, because at any instant in time, the capacitor is actually taking energy from the source or returning energy to it. Reactive power does not represent an energy loss. The following formulas apply:

$$P_r = V_{rms}I_{rms}$$

Equation 12–26

$$P_r = \frac{V_{rms}^2}{X_C}$$

Equation 12–27

$$P_r = I_{rms}^2 X_C$$

Equation 12–28

Notice that these equations are of the same form as those introduced in Chapter 4 for power in a resistor. The voltage and current are expressed in rms. The unit of reactive power is **VAR (volt-ampere reactive)**.

EXAMPLE 12–19   Determine the true power and the reactive power in Figure 12–49.

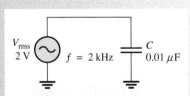

▲ **FIGURE 12–49**

Solution   The true power, P_{true}, is *always* **zero** for an ideal capacitor. The reactive power is determined by first finding the value for the capacitive reactance and then using Equation 12–27.

$$X_C = \frac{1}{2\pi fC} = \frac{1}{2\pi(2 \times 10^3 \text{ Hz})(0.01 \times 10^{-6}\text{ F})} = 7.96 \text{ k}\Omega$$

$$P_r = \frac{V_{rms}^2}{X_C} = \frac{(2\text{ V})^2}{7.96 \text{ k}\Omega} = 503 \times 10^{-6}\text{ VAR} = \textbf{503 }\boldsymbol{\mu}\textbf{VAR}$$

Related Problem   If the frequency is doubled in Figure 12–49, what are the true power and the reactive power?

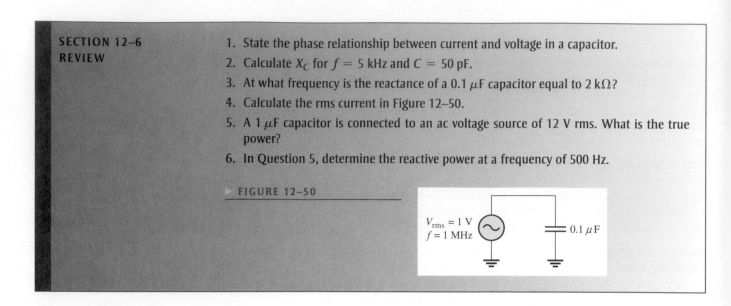

SECTION 12–6
REVIEW

1. State the phase relationship between current and voltage in a capacitor.
2. Calculate X_C for $f = 5$ kHz and $C = 50$ pF.
3. At what frequency is the reactance of a 0.1 μF capacitor equal to 2 kΩ?
4. Calculate the rms current in Figure 12–50.
5. A 1 μF capacitor is connected to an ac voltage source of 12 V rms. What is the true power?
6. In Question 5, determine the reactive power at a frequency of 500 Hz.

▶ FIGURE 12–50

$V_{rms} = 1$ V
$f = 1$ MHz
0.1 μF

12–7 CAPACITOR APPLICATIONS

Capacitors are widely used in many electrical and electronic applications.

After completing this section, you should be able to

♦ **Discuss some capacitor applications**

   ♦ Describe a power supply filter

   ♦ Explain the purpose of coupling and bypass capacitors

   ♦ Discuss the basics of capacitors applied to tuned circuits, timing circuits, and computer memories

If you pick up any circuit board, open any power supply, or look inside any piece of electronic equipment, chances are you will find capacitors of one type or another. These components are used for a variety of reasons in both dc and ac applications.

Electrical Storage

One of the most basic applications of a capacitor is as a backup voltage source for low-power circuits such as certain types of semiconductor memories in computers. This particular application requires a very high capacitance value and negligible leakage.

The storage capacitor is connected between the dc power supply input to the circuit and ground. When the circuit is operating from its normal power supply, the capacitor remains fully charged to the dc power supply voltage. If the normal power source is disrupted, effectively removing the power supply from the circuit, the storage capacitor temporarily becomes the power source for the circuit.

A capacitor provides voltage and current to a circuit as long as its charge remains sufficient. As current is drawn by the circuit, charge is removed from the capacitor and the voltage decreases. For this reason, the storage capacitor can only be used as a temporary power source. The length of time that a capacitor can provide sufficient power to the circuit depends on the capacitance and the amount of current drawn by the circuit. The smaller the current and the higher the capacitance, the longer the time a capacitor can provide power to a circuit.

Power Supply Filtering

A basic dc power supply consists of a circuit known as a **rectifier** followed by a **filter**. The rectifier converts the 110 V, 60 Hz sinusoidal voltage available at a standard outlet to a pulsating dc voltage that can be either a half-wave rectified voltage or a full-wave rectified voltage, depending on the type of rectifier circuit. As shown in Figure 12–51(a), a half-wave rectifier removes each negative half-cycle of the sinusoidal voltage. As shown in Figure 12–51(b), a full-wave rectifier actually reverses the polarity of the negative portion of each cycle. Both half-wave and full-wave rectified voltages are dc because, even though they are changing, they do not alternate polarity.

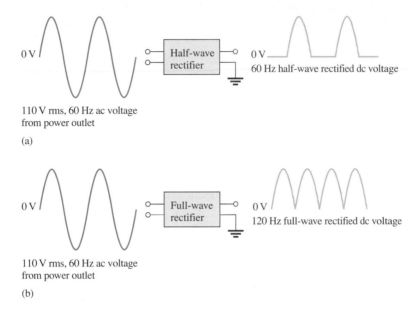

◀ **FIGURE 12–51**

Half-wave and full-wave rectifier operation.

To be useful for powering electronic circuits, the rectified voltage must be changed to constant dc voltage because all circuits require constant power. The filter nearly eliminates the fluctuations in the rectified voltage and ideally provides a smooth constant-value dc voltage to the load that is the electronic circuit, as indicated in Figure 12–52.

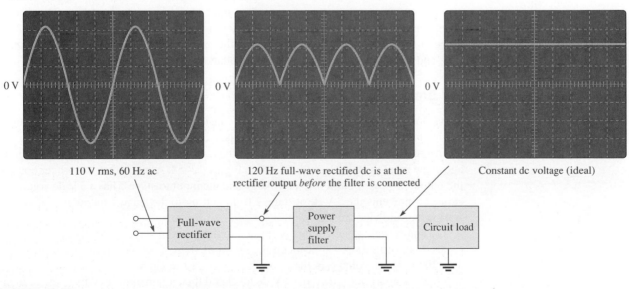

▲ **FIGURE 12–52**

Basic waveforms showing the operation of a dc power supply.

The Capacitor as a Power Supply Filter   Capacitors are used as filters in dc power supplies because of their ability to store electrical charge. Figure 12–53(a) shows a dc power supply with a full-wave rectifier and a capacitor filter. The operation can be described from a charging and discharging point of view as follows. Assume the capacitor is initially uncharged. When the power supply is first turned on and the first cycle of the rectified voltage occurs, the capacitor will quickly charge through the low forward resistance of the rectifier. The capacitor voltage will follow the rectified voltage curve up to the peak of the rectified voltage. As the rectified voltage passes the peak and begins to decrease, the capacitor will begin to discharge very slowly through the high resistance of the load circuit, as indicated in Figure 12–53(b). The amount of discharge is typically very small and is exaggerated in the figure for purposes of illustration. The next cycle of the rectified voltage will recharge the capacitor back to the peak value by replenishing the small amount of charge lost since the previous peak. This pattern of a small amount of charging and discharging continues as long as the power is on.

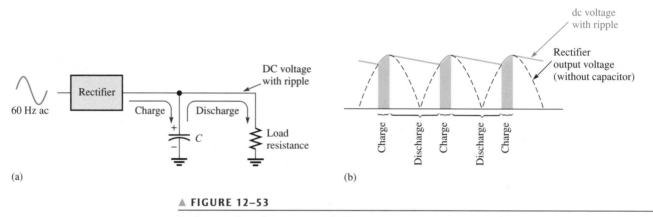

▲ FIGURE 12–53

Basic operation of a power supply filter capacitor.

A rectifier is designed so that it allows current only in the direction to charge the capacitor. The capacitor will not discharge back through the rectifier but will only discharge a small amount through the relatively high resistance of the load. The small fluctuation in voltage due to the charging and discharging of the capacitor is called the **ripple voltage**. A good dc power supply has a very small amount of ripple on its dc output. The discharge time constant of a power supply filter capacitor depends on its capacitance and the resistance of the load; consequently, the higher the capacitance value, the longer the discharge time and therefore, the smaller the ripple voltage.

DC Blocking and AC Coupling

Capacitors are commonly used to block the constant dc voltage in one part of a circuit from getting to another part. As an example of this, a capacitor is connected between two stages of an amplifier to prevent the dc voltage at the output of stage 1 from affecting the dc voltage at the input of stage 2, as illustrated in Figure 12–54. Assume that, for proper operation, the output of stage 1 has a zero dc voltage and the input to stage 2 has a 3 V dc voltage. The capacitor prevents the 3 V dc at stage 2 from getting to the stage 1 output and affecting its zero value, and vice versa.

If a sinusoidal signal voltage is applied to the input to stage 1, the signal voltage is increased (amplified) and appears on the ouput of stage 1, as shown in Figure 12–54. The amplified signal voltage is then coupled through the capacitor to the input of stage 2 where it is superimposed on the 3 V dc level and then again amplified by stage 2. In order for the signal voltage to be passed through the capacitor without being reduced, the capacitor must be large enough so that its reactance at the frequency of the signal voltage

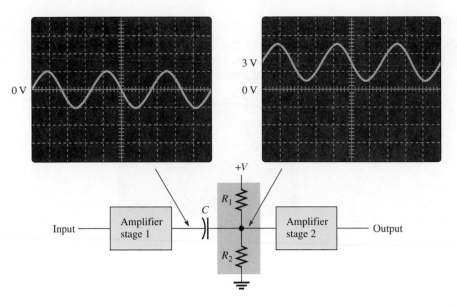

▲ FIGURE 12–54

An application of a capacitor to block dc and couple ac in an amplifier.

is negligible. In this type of application, the capacitor is known as a *coupling capacitor*, which ideally appears as an open to dc and as a short to ac. As the signal frequency is reduced, the capacitive reactance increases and, at some point, the capacitive reactance becomes large enough to cause a significant reduction in ac voltage between stage 1 and stage 2.

Power Line Decoupling

Capacitors connected from the dc supply voltage line to ground are used on circuit boards to decouple unwanted voltage transients or spikes that occur on the dc supply voltage because of fast switching digital circuits. A voltage transient contains high frequencies that may affect the operation of the circuits. These transients are shorted to ground through the very low reactance of the decoupling capacitors. Several decoupling capacitors are often used at various points along the supply voltage line on a circuit board.

Bypassing

Another capacitor application is to bypass an ac voltage around a resistor in a circuit without affecting the dc voltage across the resistor. In amplifier circuits, for example, dc voltages called *bias voltages* are required at various points. For the amplifier to operate properly, certain bias voltages must remain constant and, therefore, any ac voltages must be removed. A sufficiently large capacitor connected from a bias point to ground provides a low reactance path to ground for ac voltages, leaving the constant dc bias voltage at the given point. At lower frequencies, the bypass capacitor becomes less effective because of its increased reactance. This bypass application is illustrated in Figure 12–55.

Signal Filters

Capacitors are essential to the operation of a class of circuits called *filters* that are used for selecting one ac signal with a certain specified frequency from a wide range of signals with

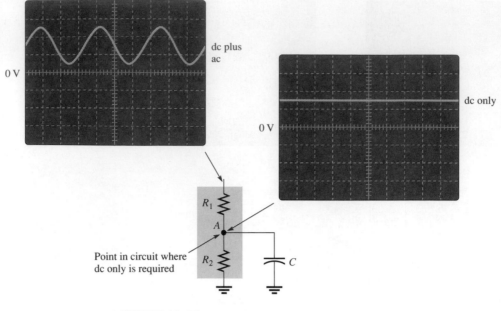

▲ **FIGURE 12–55**

Example of the operation of a bypass capacitor. Point *A* is at ac ground due to the low reactance path through the capacitor.

many different frequencies or for selecting a certain band of frequencies and eliminating all others. A common example of this application is in radio and television receivers where it is necessary to select the signal transmitted from a given station and eliminate or filter out the signals transmitted from all the other stations in the area.

When you tune your radio or TV, you are actually changing the capacitance in the tuner circuit (which is a type of filter) so that only the signal from the station or channel you want passes through to the receiver circuitry. Capacitors are used in conjunction with resistors, inductors (covered in the next chapter), and other components in these types of filters. The topic of filters will be covered in Chapter 18.

The main characteristic of a filter is its frequency selectivity, which is based on the fact that the reactance of a capacitor depends on frequency ($X_C = 1/2\pi fC$).

Timing Circuits

Another important area in which capacitors are used is in timing circuits that generate specified time delays or produce waveforms with specific characteristics. Recall that the time constant of a circuit with resistance and capacitance can be controlled by selecting appropriate values for *R* and *C*. The charging time of a capacitor can be used as a basic time delay in various types of circuits. An example is the circuit that controls the turn indicators on your car where the light flashes on and off at regular intervals.

Computer Memories

Dynamic memories in computers use very tiny capacitors as the basic storage element for binary information, which consists of two binary digits, 1 and 0. A charged capacitor can represent a stored 1 and a discharged capacitor can represent a stored 0. Patterns of 1s and 0s that make up binary data are stored in a memory that consists of an array of capacitors with associated circuitry. You will study this topic in a computer or digital fundamentals course.

SECTION 12-7
REVIEW
1. Explain how half-wave or full-wave rectified dc voltages are smoothed out by a filter capacitor.

2. Explain the purpose of a coupling capacitor.

3. How large must a coupling capacitor be?

4. Explain the purpose of a decoupling capacitor.

5. Discuss how the relationship of frequency and capacitive reactance is important in frequency-selective circuits such as signal filters.

6. What characteristic of a capacitor is most important in time-delay applications?

12–8 SWITCHED-CAPACITOR CIRCUITS

Capacitors are applied in programmable analog arrays, which are implemented in integrated circuit (IC) form. Switched-capacitors are used to implement various types of programmable analog circuits in which capacitors take the place of resistors. Capacitors can be implemented on an IC chip more easily than a resistor can, and they offer other advantages such as zero power dissipation. When a resistance is required in a circuit, the switched capacitor can be made to emulate a resistor. Using switched-capacitor emulation, resistor values can be readily changed by reprogramming and accurate and stable resistance values can be achieved.

After completing this section, you should be able to

♦ **Describe the basic operation of switched-capacitor circuits**

  ♦ Explain how switched-capacitor circuits emulate resistors

Recall that current is defined in terms of charge Q and time t as

$$I = \frac{Q}{t}$$

This formula states that current is the rate at which charge flows through a circuit. Also recall that the basic definition of charge in terms of capacitance and voltage is

$$Q = CV$$

Substituting CV for Q, the current can be expressed as

$$I = \frac{CV}{t}$$

Basic Operation

A general model of a switched-capacitor circuit is shown in Figure 12–56. It consists of a capacitor, two arbitrary voltage sources (V_1 and V_2), and a two-pole switch. Let's examine this circuit for a specified period of time, T, which is then repeated. Assume that V_1 and V_2 are constant during the time period T. Of particular interest is the average current I_1 from the source V_1 during the period of time, T.

During the first half of the time period T, the switch is in position 1, as indicated in Figure 12–56. Therefore, there is a current I_1 due to V_1 that is charging the capacitor during the interval from $t = 0$ to $t = T/2$. During the second half of the time period, the

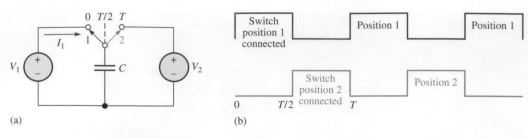

▲ FIGURE 12–56

Basic operation of a switched-capacitor circuit. The voltage source symbol represents a time-varying symbol.

switch is in position 2, as indicated, and there is no current from V_1; therefore, the average current from the source V_1 over the time period T is

$$I_{1(\text{avg})} = \frac{Q_{1(T/2)} - Q_{1(0)}}{T}$$

$Q_{1(0)}$ is the charge at $t = 0$ and $Q_{1(T/2)}$ is the charge at $t = T/2$. So, $Q_{1(T/2)} - Q_{1(0)}$ is the net charge transferred while the switch is in position 1.

The capacitor voltage at $T/2$ is equal to V_1, and the capacitor voltage at 0 or T is equal to V_2. Using the formula $Q = CV$ and substituting into the previous equation, you obtain

$$I_{1(\text{avg})} = \frac{CV_{1(T/2)} - CV_{2(0)}}{T} = \frac{C(V_{1(T/2)} - V_{2(0)})}{T}$$

Since V_1 and V_2 are assumed to be constant during T, the average current can be expressed as

Equation 12–29

$$I_{1(\text{avg})} = \frac{C(V_1 - V_2)}{T}$$

Figure 12–57 shows an equivalent circuit with a resistor instead of the capacitor and switches.

▶ FIGURE 12–57

Resistive circuit.

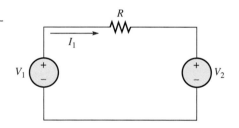

Applying Ohm's law to the resistive circuit, the current is

$$I_1 = \frac{V_1 - V_2}{R}$$

Setting $I_{1(\text{avg})}$ in the switched-capacitor circuit equal to current in the resistive circuit, you have

$$\frac{C(V_1 - V_2)}{T} = \frac{V_1 - V_2}{R}$$

Canceling the $V_1 - V_2$ terms and solving for R gives the equivalent resistance.

Equation 12–30

$$R = \frac{T}{C}$$

This important result shows that a switched-capacitor circuit can emulate a resistor with a value determined by the time T and the capacitance C. Remember that the switch is in each position for one-half of the time period T and that T can be varied by varying the frequency at which the switches are operated. In a programmable analog device, the switching frequency is a programmable parameter for each emulated resistor and can be set to achieve a precise resistor value. Since $T = 1/f$, the resistance in terms of frequency is

$$R = \frac{1}{fC}$$

Equation 12–31

Practical Switches

The two-pole switch that has been used to illustrate the basic concept of a switched-capacitor circuit is an impractical form for implementation in a programmable amplifier or other analog circuit. Figure 12–58 shows how the simple two-pole switch can be replaced by two single-pole switches in a mechanical analogy. You can see that when SW1 is closed and SW2 is open, it is equivalent to the two-pole switch being in position 1. When SW1 is open and SW2 is closed, it is equivalent to the two-pole switch being in position 2.

◀ **FIGURE 12–58**

Mechanical switch analogy.

Switches in electronic circuits are implemented with transistors. A switched-capacitor circuit is shown in Figure 12–59 with two transistors (Q_1 and Q_2) acting as the switches. Their *on* and *off* times are controlled by pulse waveform voltages that are programmable. The two pulse waveforms that turn the transistors on and off are 180° out of phase so that when one transistor is *on,* the other is *off* and vice versa with no overlap.

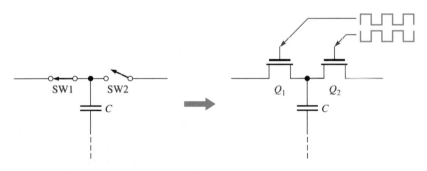

◀ **FIGURE 12–59**

A switched-capacitor with transistor switches.

EXAMPLE 12–20

Replace the input resistor, R, in the amplifier circuit of Figure 12–60 with a switched-capacitor circuit. The triangle symbol represents an operational amplifier, which is covered in a later course. For now, you need only be concerned with the input resistor.

▶ **FIGURE 12–60**

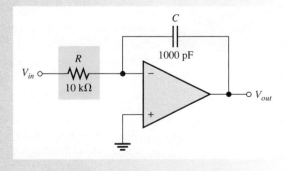

Solution   Assume that the switched capacitor value is 1000 pF. You want the switched-capacitor to emulate a 10 kΩ resistor by effectively providing the same average current as the actual resistor would. Using the formula $R = T/C$,

$$T = RC = (10\,\text{k}\Omega)(1000\,\text{pF}) = 10\,\mu\text{s}$$

This means that each switch must be operated at a frequency of

$$f = \frac{1}{T} = \frac{1}{10\,\mu\text{s}} = \mathbf{100\,kHz}$$

The duty cycle is 50% so that the switch is in each position half of the period. Two non-overlapping 100 kHz, 50% duty cycle voltages that are 180° out-of-phase with each other are applied to the transistor switches in Figure 12–61.

FIGURE 12–61

Switched-capacitor equivalent of the circuit in Figure 12–60.

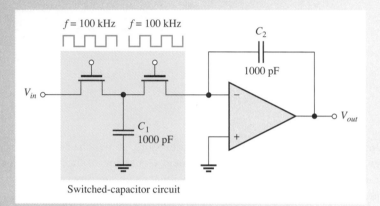

Switched-capacitor circuit

Related Problem   At what frequency must the transistor switches be operated in Figure 12–61 to emulate a 5.6 kΩ resistor?

**SECTION 12–8
REVIEW**

1. How does a switched capacitor emulate a resistor?
2. What factors determine the resistance value that a given switched-capacitor circuit can emulate?
3. In a practical implementation, what devices are used for switches?

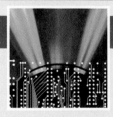

A Circuit Application

Capacitors are used in certain types of amplifiers to couple the ac signal while blocking the dc voltage. Capacitors are used in many other applications, but in this application, you will focus on the coupling capacitors in an amplifier circuit. This topic was introduced in Section 12–7. A knowledge of amplifier circuits is not necessary for this assignment.

All amplifier circuits contain transistors that require dc voltages to establish proper operating conditions for amplify-

ing ac signals. These dc voltages are referred to as bias voltages. As indicated in Figure 12–62(a), a common type of dc bias circuit used in amplifiers is the voltage divider formed by R_1 and R_2, which sets up the proper dc voltage at the input to the amplifier.

When an ac signal voltage is applied to the amplifier, the input coupling capacitor, C_1, prevents the internal resistance of the ac source from changing the dc bias voltage. Without the capacitor, the internal source resistance would appear in parallel with R_2 and drastically change the value of the dc voltage.

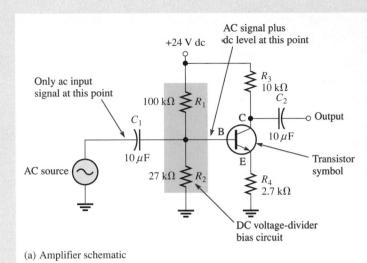

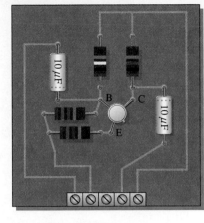

(a) Amplifier schematic

(b) Amplifier board

▲ **FIGURE 12–62**

A capacitively coupled amplifier.

The coupling capacitance is chosen so that its reactance (X_C) at the frequency of the ac signal is very small compared to the bias resistor values. The coupling capacitance therefore efficiently couples the ac signal from the source to the input of the amplifier. On the source side of the input coupling capacitor there is only ac but on the amplifier side there is ac plus dc (the signal voltage is riding on the dc bias voltage set by the voltage divider), as indicated in Figure 12–62(a). Capacitor C_2 is the output coupling capacitor, which couples the amplified ac signal to another amplifier stage that would be connected to the output.

You will check three amplifier boards like the one in Figure 12–62(b) for the proper input voltages using an oscilloscope. If the voltages are incorrect, you will determine the most likely fault. For all measurements, assume the amplifier has no dc loading effect on the voltage-divider bias circuit.

The Printed Circuit Board and the Schematic

◆ Check the printed circuit board in Figure 12–62(b) to make sure it agrees with the amplifier schematic in part (a).

Testing Board 1

The oscilloscope probe is connected from channel 1 to the board as shown in Figure 12–63. The input signal from a sinusoidal voltage source is connected to the board and set to a frequency of 5 kHz with an amplitude of 1 V rms.

◆ Determine if the voltage and frequency displayed on the scope are correct. If the scope measurement is incorrect, specify the most likely fault in the circuit.

Testing Board 2

The oscilloscope probe is connected from channel 1 to board 2 the same as was shown in Figure 12–63 for board 1. The input signal from the sinusoidal voltage source is the same as it was for board 1.

◆ Determine if the scope display in Figure 12–64 is correct. If the scope measurement is incorrect, specify the most likely fault in the circuit.

Testing Board 3

The oscilloscope probe is connected from channel 1 to board 3 the same as was shown in Figure 12–63 for board 1. The input signal from the sinusoidal voltage source is the same as before.

◆ Determine if the scope display in Figure 12–65 is correct. If the scope measurement is incorrect, specify the most likely fault in the circuit.

Review

1. Explain why the input coupling capacitor is necessary when connecting an ac source to the amplifier.

2. Capacitor C_2 in Figure 12–62 is an output coupling capacitor. Generally, what would you expect to measure at the point in the circuit labelled C and at the output of the circuit when an ac input signal is applied to the amplifier?

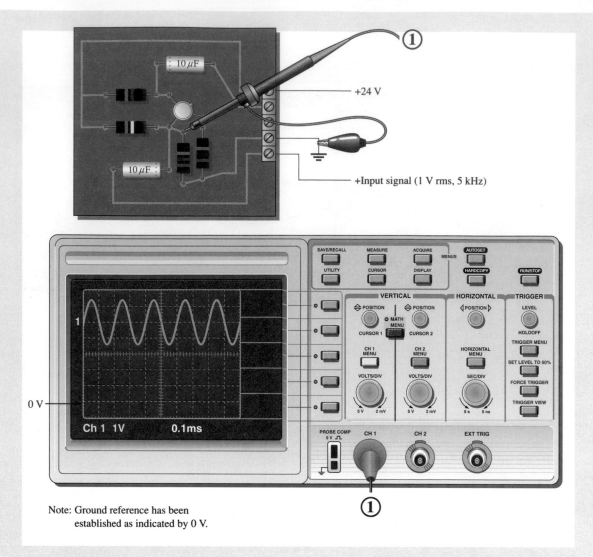

▲ FIGURE 12–63

Testing board 1.

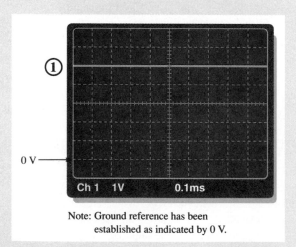

▲ FIGURE 12–64

Testing board 2.

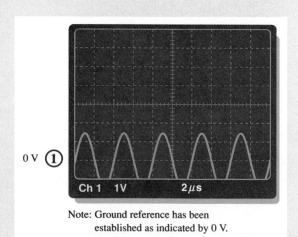

▲ FIGURE 12–65

Testing board 3.

SUMMARY

- A capacitor is composed of two parallel conducting plates separated by an insulating material called the *dielectric*.
- Energy is stored by a capacitor in the electric field between the plates.
- One farad is the amount of capacitance when one coulomb of charge is stored with one volt across the plates.
- Capacitance is directly proportional to the plate area and inversely proportional to the plate separation.
- The dielectric constant is an indication of the ability of a material to establish an electric field.
- The dielectric strength is one factor that determines the breakdown voltage of a capacitor.
- A capacitor blocks constant dc.
- The time constant for a series *RC* circuit is the resistance times the capacitance.
- In an *RC* circuit, the voltage and current in a charging or discharging capacitor make a 63% change during each time-constant interval.
- Five time constants are required for a capacitor to charge fully or to discharge fully. This is called the *transient time.*
- Charging and discharging follow exponential curves.
- Total series capacitance is less than that of the smallest capacitor in series.
- Capacitance adds in parallel.
- Current leads voltage by 90° in a capacitor.
- Capacitive reactance, X_C, is inversely proportional to frequency and capacitance.
- The true power in a capacitor is zero; that is, no energy is lost in an ideal capacitor due to conversion to heat.

KEY TERMS

Key terms and other bold terms in the chapter are defined in the end-of-book glossary.

Capacitive reactance The opposition of a capacitor to sinusoidal current. The unit is the ohm.

Capacitor An electrical device consisting of two conductive plates separated by an insulating material and possessing the property of capacitance.

Coulomb's law A law that states a force exists between two charged bodies that is directly proportional to the product of the two charges and inversely proportional to the square of the distance between them.

Dielectric The insulating material between the plates of a capacitor.

Farad (F) The unit of capacitance.

Instantaneous power (p) The value of power in a circuit at any given instant of time.

***RC* time constant** A fixed time interval set by R and C values that determines the time response of a series *RC* circuit. It equals the product of the resistance and the capacitance.

Reactive power (P_r) The rate at which energy is alternately stored and returned to the source by a capacitor. The unit is the VAR.

Ripple voltage The small fluctuation in voltage due to the charging and discharging of a capacitor.

True power (P_{true}) The power that is dissipated in a circuit, usually in the form of heat.

VAR (volt-ampere reactive) The unit of reactive power.

FORMULAS

12–1 $C = \dfrac{Q}{V}$ Capacitance in terms of charge and voltage

12–2 $Q = CV$ Charge in terms of capacitance and voltage

12–3 $V = \dfrac{Q}{C}$ Voltage in terms of charge and capacitance

12–4 $W = \dfrac{1}{2}\,CV^2$ Energy stored by a capacitor

12–5 $\varepsilon_r = \dfrac{\varepsilon}{\varepsilon_0}$ Dielectric constant (relative permittivity)

12–6 $C = \dfrac{A\varepsilon_r(8.85 \times 10^{-12}\ \text{F/m})}{d}$ Capacitance in terms of physical parameters

12–7 $Q_T = Q_1 = Q_2 = Q_3 = \cdots = Q_n$ Total charge of series capacitors (general)

12–8 $\dfrac{1}{C_T} = \dfrac{1}{C_1} + \dfrac{1}{C_2} + \dfrac{1}{C_3} + \cdots + \dfrac{1}{C_n}$ Reciprocal of total series capacitance (general)

12–9 $C_T = \dfrac{1}{\dfrac{1}{C_1} + \dfrac{1}{C_2} + \dfrac{1}{C_3} + \cdots + \dfrac{1}{C_n}}$ Total series capacitance (general)

12–10 $C_T = \dfrac{C_1 C_2}{C_1 + C_2}$ Total capacitance of two capacitors in series

12–11 $C_T = \dfrac{C}{n}$ Total capacitance of equal-value capacitors in series

12–12 $V_x = \left(\dfrac{C_T}{C_x}\right)V_T$ Capacitor voltage for capacitors in series

12–13 $Q_T = Q_1 + Q_2 + Q_3 + \cdots + Q_n$ Total charge of parallel capacitors (general)

12–14 $C_T = C_1 + C_2 + C_3 + \cdots + C_n$ Total parallel capacitance (general)

12–15 $C_T = nC$ Total capacitance of equal-value capacitors in parallel

12–16 $\tau = RC$ Time constant

12–17 $v = V_F + (V_i - V_F)e^{-t/\tau}$ Exponential voltage (general)

12–18 $i = I_F + (I_i - I_F)e^{-t/\tau}$ Exponential current (general)

12–19 $v = V_F(1 - e^{-t/RC})$ Increasing exponential voltage beginning at zero

12–20 $v = V_i e^{-t/RC}$ Decreasing exponential voltage ending at zero

12–21 $t = -RC\ln\left(\dfrac{v}{V_i}\right)$ Time on decreasing exponential ($V_F = 0$)

12–22 $t = -RC\ln\left(1 - \dfrac{v}{V_F}\right)$ Time on increasing exponential ($V_i = 0$)

12–23 $i = \dfrac{dq}{dt}$ Instantaneous current using charge derivative

12–24 $i = C\left(\dfrac{dv}{dt}\right)$ Instantaneous capacitor current using voltage derivative

12–25 $X_C = \dfrac{1}{2\pi f C}$ Capacitive reactance

12–26 $P_r = V_{\text{rms}}I_{\text{rms}}$ Reactive power in a capacitor

12–27 $P_r = \dfrac{V_{\text{rms}}^2}{X_C}$ Reactive power in a capacitor

12–28 $P_r = I_{\text{rms}}^2 X_C$ Reactive power in a capacitor

$$12\text{-}29 \quad I_{1(avg)} = \frac{C(V_1 - V_2)}{T} \qquad \text{Switched-capacitor average current}$$

$$12\text{-}30 \quad R = \frac{T}{C} \qquad \text{Switched-capacitor equivalent resistance}$$

$$12\text{-}31 \quad R = \frac{1}{fC} \qquad \text{Switched-capacitor equivalent resistance}$$

SELF-TEST

Answers are at the end of the chapter.

1. The following statement(s) accurately describes a capacitor:
   (a) The plates are conductive.
   (b) The dielectric is an insulator between the plates.
   (c) There is constant direct current (dc) through a fully charged capacitor.
   (d) A practical capacitor stores charge indefinitely when disconnected from the source.
   (e) none of the above answers
   (f) all of the above answers
   (g) only answers (a) and (b)

2. Which one of the following statements is true?
   (a) There is current through the dielectric of a charging capacitor.
   (b) When a capacitor is connected to a dc voltage source, it will charge to the value of the source.
   (c) An ideal capacitor can be discharged by disconnecting it from the voltage source.

3. A capacitance of 0.01 mF is larger than
   (a) 0.00001 F   (b) 100,000 pF   (c) 1000 pF   (d) all of these answers

4. A capacitance of 1000 pF is smaller than
   (a) 0.01 μF   (b) 0.001 μF   (c) 0.00000001 F   (d) both (a) and (c)

5. When the voltage across a capacitor is increased, the stored charge
   (a) increases   (b) decreases   (c) remains constant   (d) fluctuates

6. When the voltage across a capacitor is doubled, the stored charge
   (a) stays the same   (b) is halved   (c) increases by four   (d) doubles

7. The voltage rating of a capacitor is increased by
   (a) increasing the plate separation   (b) decreasing the plate separation
   (c) increasing the plate area   (d) answers (b) and (c)

8. The capacitance value is increased by
   (a) decreasing the plate area   (b) increasing the plate separation
   (c) decreasing the plate separation   (d) increasing the plate area
   (e) answers (a) and (b)   (f) answers (c) and (d)

9. A 1 μF, a 2.2 μF, and a 0.047 μF capacitor are connected in series. The total capacitance is less than
   (a) 1 μF   (b) 2.2 μF   (c) 0.047 μF   (d) 0.001 μF

10. Four 0.022 μF capacitors are in parallel. The total capacitance is
    (a) 0.022 μF   (b) 0.088 μF   (c) 0.011 μF   (d) 0.044 μF

11. An uncharged capacitor and a resistor are connected in series with a switch and a 12 V battery. At the instant the switch is closed, the voltage across the capacitor is
    (a) 12 V   (b) 6 V   (c) 24 V   (d) 0 V

12. In Question 11, the voltage across the capacitor when it is fully charged is
    (a) 12 V   (b) 6 V   (c) 24 V   (d) −6 V

13. In Question 11, the capacitor will reach full charge in a time equal to approximately

(a) RC (b) $5RC$ (c) $12RC$ (d) cannot be predicted

14. A sinusoidal voltage is applied across a capacitor. When the frequency of the voltage is increased, the current

(a) increases   (b) decreases   (c) remains constant   (d) ceases

15. A capacitor and a resistor are connected in series to a sine wave generator. The frequency is set so that the capacitive reactance is equal to the resistance and, thus, an equal amount of voltage appears across each component. If the frequency is decreased,

(a) $V_R > V_C$ (b) $V_C > V_R$ (c) $V_R = V_C$

16. Switched-capacitor circuits are used to

(a) increase capacitance   (b) emulate inductance

(c) emulate resistance   (d) generate sine wave voltages

CIRCUIT DYNAMICS QUIZ

Answers are at the end of the chapter.

Refer to Figure 12–73.

1. If the capacitors are initially uncharged and the switch is thrown to the closed position, the charge on C_1

(a) increases   (b) decreases   (c) stays the same

2. If C_4 is shorted with the switch closed, the charge on C_1

(a) increases   (b) decreases   (c) stays the same

3. If the switch is closed and C_2 fails to open, the charge on C_1

(a) increases   (b) decreases   (c) stays the same

Refer to Figure 12–74.

4. Assume the switch is closed and C is allowed to fully charge. When the switch is opened, the voltage across C

(a) increases   (b) decreases   (c) stays the same

5. If C has failed open when the switch is closed, the voltage across C

(a) increases   (b) decreases   (c) stays the same

Refer to Figure 12–77.

6. If the switch is closed allowing the capacitor to charge and then the switch is opened, the voltage across the capacitor

(a) increases   (b) decreases   (c) stays the same

7. If R_2 opens, the time it takes the capacitor to fully charge

(a) increases   (b) decreases   (c) stays the same

8. If R_4 opens, the maximum voltage to which the capacitor can charge

(a) increases   (b) decreases   (c) stays the same

9. If V_S is reduced, the time required for the capacitor to fully change

(a) increases   (b) decreases   (c) stays the same

Refer to Figure 12–80(b).

10. If the frequency of the ac source is increased, the total current

(a) increases   (b) decreases   (c) stays the same

11. If C_1 opens, the current through C_2

(a) increases   (b) decreases   (c) stays the same

12. If the value of C_2 is changed to 1 μF, the current through it

(a) increases   (b) decreases   (c) stays the same

PROBLEMS

More difficult problems are indicated by an asterisk (*).
Answers to odd-numbered problems are at the end of the book.

SECTION 12–1   The Basic Capacitor

1. (a) Find the capacitance when $Q = 50\,\mu C$ and $V = 10\,V$.
   (b) Find the charge when $C = 0.001\,\mu F$ and $V = 1\,kV$.
   (c) Find the voltage when $Q = 2\,mC$ and $C = 200\,\mu F$.

2. Convert the following values from microfarads to picofarads:
   (a) $0.1\,\mu F$ (b) $0.0025\,\mu F$ (c) $4.7\,\mu F$

3. Convert the following values from picofarads to microfarads:
   (a) $1000\,pF$ (b) $3500\,pF$ (c) $250\,pF$

4. Convert the following values from farads to microfarads:
   (a) $0.0000001\,F$ (b) $0.0022\,F$ (c) $0.0000000015\,F$

5. How much energy is stored in a $1000\,\mu F$ capacitor that is charged to $500\,V$?

6. What size capacitor is capable of storing $10\,mJ$ of energy with $100\,V$ across its plates?

7. Calculate the absolute permittivity, ε, for each of the following materials. Refer to Table 12–3 for ε_r values.
   (a) air   (b) oil   (c) glass   (d) Teflon®

8. A mica capacitor has square plates that are 3.8 cm on a side and separated by 2.5 mils. What is the capacitance?

9. An air capacitor has a total plate area of $0.05\,m^2$. The plates are separated by $4.5 \times 10^{-4}\,m$. Calculate the capacitance.

*10. A student wants to construct a 1 F capacitor out of two square plates for a science fair project. He plans to use a paper dielectric ($\varepsilon_r = 2.5$) that is $8 \times 10^{-5}\,m$ thick. The science fair is to be held in the Astrodome. Will his capacitor fit in the Astrodome? What would be the size of the plates if it could be constructed?

11. A student decides to construct a capacitor using two conducting plates 30 cm on a side. He separates the plates with a paper dielectric ($\varepsilon_r = 2.5$) that is $8 \times 10^{-5}\,m$ thick. What is the capacitance of his capacitor?

12. At ambient temperature (25°C), a certain capacitor is specified to be 1000 pF. It has a negative temperature coefficient of 200 ppm/°C. What is its capacitance at 75°C?

13. A $0.001\,\mu F$ capacitor has a positive temperature coefficient of 500 ppm/°C. How much change in capacitance will a 25°C increase in temperature cause?

SECTION 12–2   Types of Capacitors

14. In the construction of a stacked-foil mica capacitor, how is the plate area increased?

15. Of mica or ceramic, which type of capacitor has the highest dielectric constant?

16. Show how to connect an electrolytic capacitor across R_2 between points A and B in Figure 12–66.

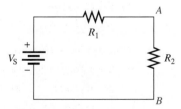

▲ FIGURE 12–66

17. Name two types of electrolytic capacitors. How do electrolytics differ from other capacitors?

▶ FIGURE 12-67

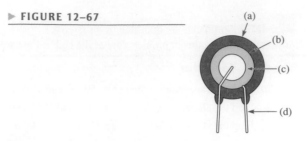

18. Identify the parts of the ceramic disk capacitor shown in the cutaway view of Figure 12–67.

19. Determine the value of the ceramic disk capacitors in Figure 12–68.

▶ FIGURE 12-68

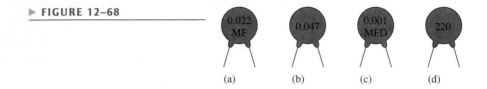

(a) (b) (c) (d)

SECTION 12-3   Series Capacitors

20. Five 1000 pF capacitors are in series. What is the total capacitance?

21. Find the total capacitance for each circuit in Figure 12–69.

22. For each circuit in Figure 12–69, determine the voltage across each capacitor.

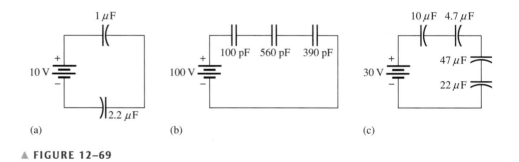

▲ FIGURE 12-69

23. Two series capacitors (one 1 μF, the other of unknown value) are charged from a 12 V source. The 1 μF capacitor is charged to 8 V and the other to 4 V. What is the value of the unknown capacitor?

24. The total charge stored by the series capacitors in Figure 12–70 is 10 μC. Determine the voltage across each of the capacitors.

▶ FIGURE 12-70

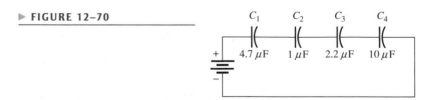

SECTION 12-4   Parallel Capacitors

25. Determine C_T for each circuit in Figure 12–71.

26. What is the charge on each capacitor in Figure 12–71?

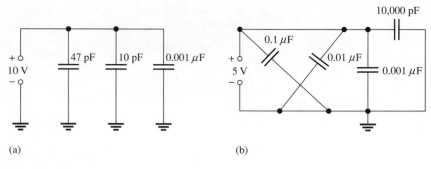

(a)   (b)

▲ FIGURE 12–71

27. Determine C_T for each circuit in Figure 12–72.

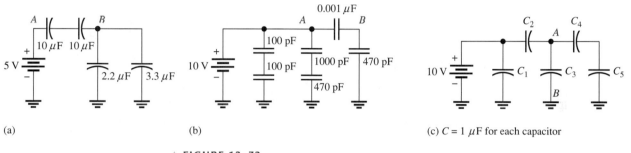

(a)   (b)   (c) $C = 1\ \mu F$ for each capacitor

▲ FIGURE 12–72

28. What is the voltage between nodes A and B in each circuit in Figure 12–72?

*29. Initially, the capacitors in the circuit in Figure 12–73 are uncharged.

    (a) After the switch is closed, what total charge is supplied by the source?

    (b) What is the voltage across each capacitor?

▶ FIGURE 12–73

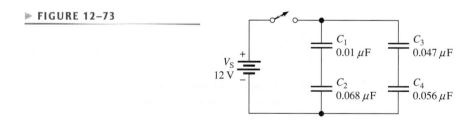

SECTION 12–5   Capacitors in DC Circuits

30. Determine the time constant for each of the following series RC combinations:

    (a) $R = 100\ \Omega, C = 1\ \mu F$ **(b)** $R = 10\ M\Omega, C = 47\ pF$

    (c) $R = 4.7\ k\Omega, C = 0.0047\ \mu F$ **(d)** $R = 1.5\ M\Omega, C = 0.01\ \mu F$

31. Determine how long it takes the capacitor to reach full charge for each of the following combinations:

    (a) $R = 56\ \Omega, C = 47\ \mu F$ **(b)** $R = 3300\ \Omega, C = 0.015\ \mu F$

    (c) $R = 22\ k\Omega, C = 100\ pF$ **(d)** $R = 5.6\ M\Omega, C = 10\ pF$

32. In the circuit of Figure 12–74, the capacitor is initially uncharged. Determine the capacitor voltage at the following times after the switch is closed:

    (a) 10 μs   **(b)** 20 μs   **(c)** 30 μs   **(d)** 40 μs   **(e)** 50 μs

▶ FIGURE 12–74

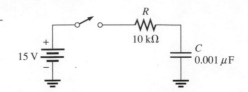

33. In Figure 12–75, the capacitor is charged to 25 V. When the switch is closed, what is the capacitor voltage after the following times?

    (a) 1.5 ms   **(b)** 4.5 ms   **(c)** 6 ms   **(d)** 7.5 ms

▶ FIGURE 12–75

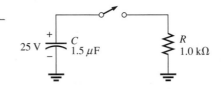

34. Repeat Problem 32 for the following time intervals:

    (a) 2 μs   **(b)** 5 μs   **(c)** 15 μs

35. Repeat Problem 33 for the following times:

    (a) 0.5 ms   **(b)** 1 ms   **(c)** 2 ms

*36. Derive the formula for finding the time at any point on an increasing exponential voltage curve. Use this formula to find the time at which the voltage in Figure 12–76 reaches 6 V after switch closure.

▶ FIGURE 12–76

37. How long does it take C to charge to 8 V in Figure 12–74?

38. How long does it take C to discharge to 3 V in Figure 12–75?

39. Determine the time constant for the circuit in Figure 12–77.

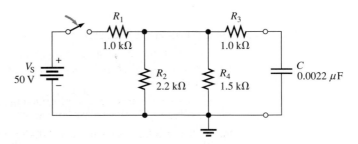

▲ FIGURE 12–77

*40. In Figure 12–78, the capacitor is initially uncharged. At $t = 10 \, \mu s$ after the switch is closed, the instantaneous capacitor voltage is 7.2 V. Determine the value of R.

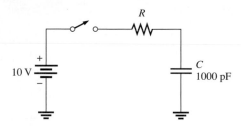

▲ FIGURE 12–78

*41. (a) The capacitor in Figure 12–79 is uncharged when the switch is thrown into position 1. The switch remains in position 1 for 10 ms and is then thrown into position 2, where it remains indefinitely. Draw the complete waveform for the capacitor voltage.

(b) If the switch is thrown back to position 1 after 5 ms in position 2, and then left in position 1, how would the waveform appear?

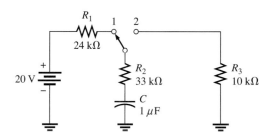

▲ FIGURE 12–79

SECTION 12–6 Capacitors in AC Circuits

42. What is the value of the total capacitive reactance in each circuit in Figure 12–80?

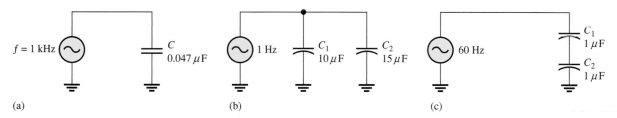

(a)   (b)   (c)

▲ FIGURE 12–80

43. In Figure 12–72, each dc voltage source is replaced by a 10 V rms, 2 kHz ac source. Determine the total reactance in each case.

44. In each circuit of Figure 12–80, what frequency is required to produce an X_C of 100 Ω? An X_C of 1 kΩ?

45. A sinusoidal voltage of 20 V rms produces an rms current of 100 mA when connected to a certain capacitor. What is the reactance?

46. A 10 kHz voltage is applied to a 0.0047 μF capacitor, and 1 mA of rms current is measured. What is the value of the voltage?

47. Determine the true power and the reactive power in Problem 46.

*48. Determine the ac voltage across each capacitor and the current in each branch of the circuit in Figure 12–81.

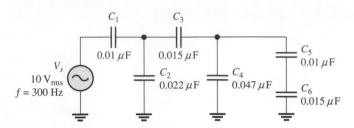

▲ FIGURE 12–81

49. Find the value of C_1 in Figure 12–82.

*50. If C_4 in Figure 12–81 opened, determine the voltages that would be measured across the other capacitors.

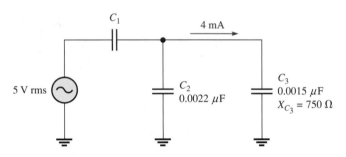

▲ FIGURE 12–82

SECTION 12–7   Capacitor Applications

51. If another capacitor is connected in parallel with the existing capacitor in the power supply filter of Figure 12–53, how is the ripple voltage affected?

52. Ideally, what should the reactance of a bypass capacitor be in order to eliminate a 10 kHz ac voltage at a given point in an amplifier circuit?

SECTION 12–8   Switched-Capacitor Circuits

53. The capacitor in a switched-capacitor circuit has a value of 2200 pF and is switched with a waveform having a period of 10 μs. Determine the value of the resistor that it emulates.

54. In a switched-capacitor circuit, the 100 pF capacitor is switched at a frequency of 8 kHz. What resistor value is emulated?

Multisim Troubleshooting and Analysis

These problems require your Multisim CD-ROM.

55. Open file P12-55 and measure the voltage across each capacitor.

56. Open file P12-56 and measure the voltage across each capacitor.

57. Open file P12-57 and measure the current. Decrease the frequency by one-half and measure the current again. Double the original frequency and measure the current again. Explain your observations.

58. Open file P12-58 and find the open capacitor if there is one.

59. Open file P12-59 and find the shorted capacitor if there is one.

ANSWERS

SECTION REVIEWS

SECTION 12–1 The Basic Capacitor
1. Capacitance is the ability (capacity) to store electrical charge.
2. (a) 1,000,000 μF in 1 F (b) 1×10^{12} pF in 1 F (c) 1,000,000 pF in 1 μF
3. 0.0015 μF = 1500 pF; 0.0015 μF = 0.0000000015 F
4. $W = \frac{1}{2}CV^2 = 1.125\ \mu$J
5. (a) C increases. (b) C decreases.
6. (1000 V/mil) (2 mil) = 2 kV
7. $C = 2.01\ \mu$F

SECTION 12–2 Types of Capacitors
1. Capacitors can be classified by the dielectric material.
2. The capacitance value of a fixed capacitor cannot be changed; the capacitance value of a variable capacitor can be changed.
3. Electrolytic capacitors are polarized.
4. When connecting a polarized capacitor, make sure the voltage rating is sufficient. Connect the positive end to the positive side of the circuit.

SECTION 12–3 Series Capacitors
1. Series C_T is less than smallest C.
2. $C_T = 61.2$ pF
3. $C_T = 0.006\ \mu$F
4. $C_T = 20$ pF
5. $V_{C1} = 75.2$ V

SECTION 12–4 Parallel Capacitors
1. The values of the individual capacitors are added in parallel.
2. Achieve C_T by using five 0.01 μF capacitors in parallel.
3. $C_T = 167$ pF

SECTION 12–5 Capacitors in DC Circuits
1. $\tau - RC - 1.2\ \mu$s
2. $5\tau = 6\ \mu$s; $V_C = 4.97$ V
3. $v_{2ms} = 8.65$ V; $v_{3ms} = 9.50$ V; $v_{4ms} = 9.82$ V; $v_{5ms} = 9.93$ V
4. $v_C = 36.8$ V

SECTION 12–6 Capacitors in AC Circuits
1. Current leads voltage by 90° in a capacitor.
2. $X_C = 1/2\pi fC = 637$ kΩ
3. $f = 1/2\pi X_C C = 796$ Hz
4. $I_{rms} = 629$ mA
5. $P_{true} = 0$ W
6. $P_r = 0.453$ VAR

SECTION 12–7 Capacitor Applications
1. Once the capacitor charges to the peak voltage, it discharges very little before the next peak, thus smoothing the rectified voltage.
2. A coupling capacitor allows ac to pass from one point to another, but blocks constant dc.

3. A coupling capacitor must be large enough to have a negligible reactance at the frequency that is to be passed without opposition.

4. A decoupling capacitor shorts power line voltage transients to ground.

5. X_C is inversely proportional to frequency and so is the filter's ability to pass ac signals.

6. Capacitance

SECTION 12–8 Switched-Capacitor Circuits

1. By moving the same amount of charge corresponding to the current in the equivalent resistance

2. Switching frequency and capacitance value

3. Transistors

A Circuit Application

1. The coupling capacitor prevents the source from affecting the dc voltage but passes the input signal.

2. An ac voltage riding on a dc voltage is at point C. An ac voltage only at the output.

RELATED PROBLEMS FOR EXAMPLES

12–1   100 kV

12–2   0.047 μF

12–3   100 × 10^6 pF

12–4   0.697 μF

12–5   1.54 μF

12–6   278 pF

12–7   0.011 μF

12–8   2.83 V

12–9   650 pF

12–10 0.09 μF

12–11 891 μs

12–12 8.36 V

12–13 8.13 V

12–14 ≈0.74 ms; 95 V

12–15 1.52 ms

12–16 0.42 V

12–17 3.39 kHz

12–18 4.40∠90° mA

12–19 0 W; 1.01 mVAR

12–20 178.57 kHz

SELF-TEST

1. (g)	2. (b)	3. (c)	4. (d)	5. (a)	6. (d)	7. (a)	8. (f)
9. (c)	10. (b)	11. (d)	12. (a)	13. (b)	14. (a)	15. (b)	16. (c)

CIRCUIT DYNAMICS QUIZ

1. (a)	2. (c)	3. (c)	4. (c)	5. (a)	6. (b)	7. (a)	8. (a)
9. (c)	10. (a)	11. (c)	12. (b)				

INDUCTORS

13

CHAPTER OBJECTIVES

◆ Describe the basic construction and characteristics of an inductor

◆ Discuss various types of inductors

◆ Analyze series and parallel inductors

◆ Analyze inductive dc switching circuits

◆ Analyze inductive ac circuits

◆ Discuss some inductor applications

KEY TERMS

◆ Inductor   ◆ Henry (H)
◆ Winding   ◆ RL time constant
◆ Induced voltage   ◆ Inductive reactance
◆ Inductance   ◆ Quality factor (Q)

A CIRCUIT APPLICATION PREVIEW

In the circuit application, you will determine the inductance of coils by measuring the time constant of a test circuit using oscilloscope waveforms.

VISIT THE COMPANION WEBSITE

Study aids for this chapter are available at
http://www.prenhall.com/floyd

INTRODUCTION

You have learned about the resistor and the capacitor. In this chapter, you will learn about a third type of basic passive component, the *inductor,* and study its characteristics.

The basic construction and electrical properties of inductors are discussed, and the effects of connecting them in series and in parallel are analyzed. How an inductor works in both dc and ac circuits is an important part of this coverage and forms the basis for the study of reactive circuits in terms of both frequency response and time response. You will also learn how to check for a faulty inductor.

The inductor, which is basically a coil of wire, is based on the principle of electromagnetic induction, which you studied in Chapter 10. Inductance is the property of a coil of wire that opposes a change in current. The basis for inductance is the electromagnetic field that surrounds any conductor when there is current through it. The electrical component designed to have the property of inductance is called an *inductor, coil,* or in certain applications a *choke.* All of these terms refer to the same type of device.

13–1 THE BASIC INDUCTOR

An **inductor** is a passive electrical component formed by a wire wound around a core and which exhibits the property of inductance.

After completing this section, you should be able to

◆ **Describe the basic construction and characteristics of an inductor**

    ◆ Define *inductance* and state its unit

    ◆ Discuss induced voltage

    ◆ Explain how an inductor stores energy

    ◆ Specify how the physical characteristics affect inductance

    ◆ Discuss winding resistance and winding capacitance

    ◆ State Faraday's law

    ◆ State Lenz's law

When a length of wire is formed into a coil, as shown in Figure 13–1, it becomes an inductor. The terms *coil* and *inductor* are used interchangeably. Current through the coil produces an electromagnetic field, as illustrated. The magnetic lines of force around each loop (turn) in the **winding** of the coil effectively add to the lines of force around the adjoining loops, forming a strong electromagnetic field within and around the coil. The net direction of the total electromagnetic field creates a north and a south pole.

▶ FIGURE 13–1

A coil of wire forms an inductor. When there is current through it, a three-dimensional electromagnetic field is created, surrounding the coil in all directions.

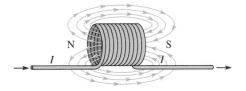

To understand the formation of the total electromagnetic field in a coil, consider the interaction of the electromagnetic fields around two adjacent loops. The magnetic lines of force around adjacent loops are each deflected into a single outer path when the loops are brought close together. This effect occurs because the magnetic lines of force are in opposing directions between adjacent loops and therefore cancel out when the loops are close together, as illustrated in Figure 13–2. The total electromagnetic field for the two loops is depicted in part (b). This effect is additive for many closely adjacent loops in a coil; that is,

Opposing fields
between loops

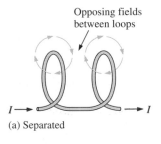

(a) Separated

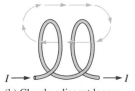

(b) Closely adjacent loops; opposing fields between loops cancel.

▲ FIGURE 13–2

Interaction of magnetic lines of force in two adjacent loops of a coil.

each additional loop adds to the strength of the electromagnetic field. For simplicity, only single lines of force are shown, although there are many. Figure 13–3 shows a schematic symbol for an inductor.

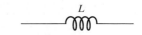

Inductance

When there is current through an inductor, an electromagnetic field is established. When the current changes, the electromagnetic field also changes. An increase in current expands the electromagnetic field, and a decrease in current reduces it. Therefore, a changing current produces a changing electromagnetic field around the inductor. In turn, the changing electromagnetic field causes an **induced voltage** across the coil in a direction to oppose the change in current. This property is called *self-inductance* but is usually referred to as simply *inductance,* symbolized by *L*.

Inductance is a measure of a coil's ability to establish an induced voltage as a result of a change in its current, and that induced voltage is in a direction to oppose the change in current.

The inductance (*L*) of a coil and the time rate of change of the current (*di/dt*) determine the induced voltage (v_{ind}). A change in current causes a change in the electromagnetic field, which, in turn, induces a voltage across the coil, as you know. The induced voltage is directly proportional to *L* and *di/dt*, as stated by the following formula:

$$v_{ind} = L\left(\frac{di}{dt}\right)$$

Equation 13–1

This formula indicates that the greater the inductance, the greater the induced voltage. Also, it shows that the faster the coil current changes (greater *di/dt*), the greater the induced voltage. Notice the similarity of Equation 13–1 to Equation 12–24: *i* = *C(dv/dt)*.

The Unit of Inductance The **henry (H)** is the basic unit of inductance. By definition, the inductance of a coil is one henry when current through the coil, changing at the rate of one ampere per second, induces one volt across the coil. The henry is a large unit, so in practical applications, millihenries (mH) and microhenries (μH) are the more common units.

EXAMPLE 13–1   Determine the induced voltage across a 1 henry (1 H) inductor when the current is changing at a rate of 2 A/s.

Solution $$v_{ind} = L\left(\frac{di}{dt}\right) = (1\ \text{H})(2\ \text{A/s}) = \textbf{2 V}$$

*Related Problem**   Determine the inductance when a current changing at a rate of 10 A/s causes 50 V to be induced.

---

*Answers are at the end of the chapter.

Energy Storage An inductor stores energy in the electromagnetic field created by the current. The energy stored is expressed as follows:

$$W = \frac{1}{2}LI^2$$

Equation 13–2

As you can see, the energy stored is proportional to the inductance and the square of the current. When current (*I*) is in amperes and inductance (*L*) is in henries, energy (*W*) is in joules.

Physical Characteristics of an Inductor

The following parameters are important in establishing the inductance of a coil: permeability of the core material, number of turns of wire, core length, and cross-sectional area of the core.

Core Material As discussed earlier, an inductor is basically a coil of wire that surrounds a magnetic or nonmagnetic material called the **core**. Examples of magnetic materials are iron, nickel, steel, cobalt, or alloys. These materials have permeabilities that are hundreds or thousands of times greater than that of a vacuum and are classified as *ferromagnetic*. A ferromagnetic core provides a better path for the magnetic lines of force and thus permits a stronger magnetic field. Examples of nonmagnetic materials are air, copper, plastic, and glass. The permeabilities of these materials are the same as for a vacuum.

As you learned in Chapter 10, the permeability (μ) of the core material determines how easily a magnetic field can be established, and is measured in Wb/At·m, which is the same as H/m. The inductance is directly proportional to the permeability of the core material.

Physical Parameters As indicated in Figure 13–4, the number of turns of wire, the length, and the cross-sectional area of the core are factors in setting the value of inductance. The inductance is inversely proportional to the length of the core and directly proportional to the cross-sectional area. Also, the inductance is directly related to the number of turns squared. This relationship is as follows:

Equation 13–3

$$L = \frac{N^2 \mu A}{l}$$

where L is the inductance in henries (H), N is the number of turns of wire, μ is the permeability in henries per meter (H/m), A is the cross-sectional area in meters squared, and l is the core length in meters (m).

▶ **FIGURE 13–4**

Physical parameters of an inductor.

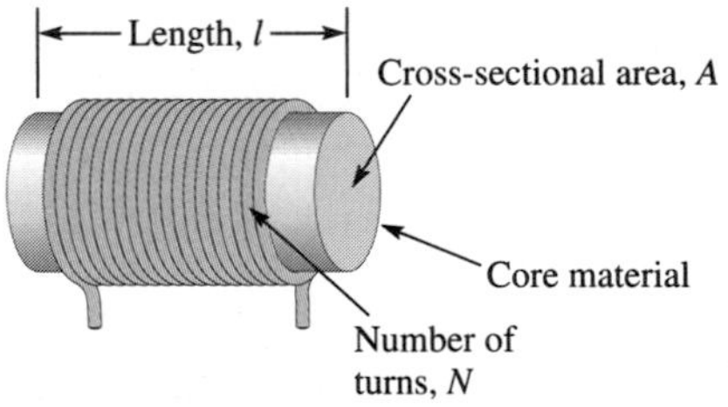

EXAMPLE 13–2

Determine the inductance of the coil in Figure 13–5. The permeability of the core is 0.25×10^{-3} H/m.

▶ **FIGURE 13–5**

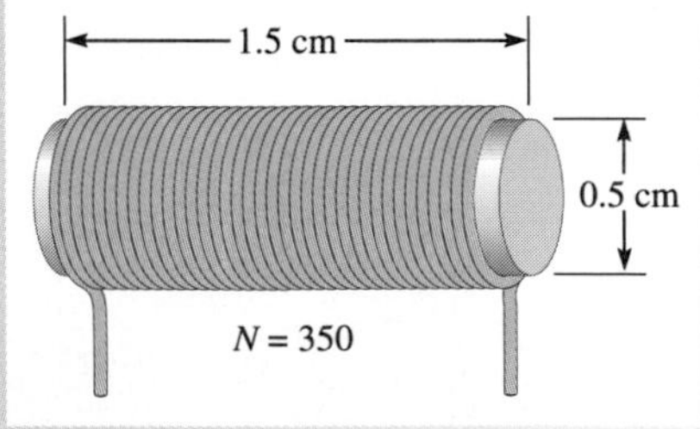

Solution First determine the length and area in meters.

$$l = 1.5 \text{ cm} = 0.015 \text{ m}$$
$$A = \pi r^2 = \pi (0.25 \times 10^{-2} \text{ m})^2 = 1.96 \times 10^{-5} \text{ m}^2$$

The inductance of the coil is

$$L = \frac{N^2 \mu A}{l} = \frac{(350)^2 (0.25 \times 10^{-3}\,\text{H/m})(1.96 \times 10^{-5}\,\text{m}^2)}{0.015\,\text{m}} = \textbf{40 mH}$$

Related Problem   Determine the inductance of a coil with 90 turns around a core that is 1.0 cm long and has a diameter of 0.8 cm. The permeability is 0.25×10^{-3} H/m.

Winding Resistance

When a coil is made of a certain material, for example, insulated copper wire, that wire has a certain resistance per unit of length. When many turns of wire are used to construct a coil, the total resistance may be significant. This inherent resistance is called the *dc resistance* or the *winding resistance* (R_W).

Although this resistance is distributed along the length of the wire, it effectively appears in series with the inductance of the coil, as shown in Figure 13–6. In many applications, the winding resistance may be small enough to be ignored and the coil can be considered an ideal inductor. In other cases, the resistance must be considered.

(a) The wire has resistance distributed along its length.

(b) Equivalent circuit

▲ FIGURE 13–6

Winding resistance of a coil.

Winding Capacitance

When two conductors are placed side by side, there is always some capacitance between them. Thus, when many turns of wire are placed close together in a coil, a certain amount of stray capacitance, called *winding capacitance* (C_W), is a natural side effect. In many applications, this winding capacitance is very small and has no significant effect. In other cases, particularly at high frequencies, it may become quite important.

The equivalent circuit for an inductor with both its winding resistance (R_W) and its winding capacitance (C_W) is shown in Figure 13–7. The capacitance effectively acts in parallel. The total of the stray capacitances between each loop of the winding is indicated in a schematic as a capacitance appearing in parallel with the coil and its winding resistance, as shown in Figure 13–7(b).

SAFETY NOTE

Be careful when working with inductors because high induced voltages can be developed due to a rapidly changing magnetic field. This occurs when the current is interrupted or its value abruptly changed.

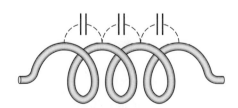

(a) Stray capacitance between each loop appears as a total parallel capacitance (C_W).

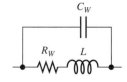

(b) Equivalent circuit

▲ FIGURE 13–7

Winding capacitance of a coil.

Review of Faraday's Law

Faraday's law was introduced in Chapter 10 and is reviewed here because of its importance in the study of inductors. Michael Faraday discovered the principle of electromagnetic induction in 1831. He found that by moving a magnet through a coil of wire, a voltage was induced across the coil and that when a complete path was provided, the induced voltage caused an induced current. Faraday observed that

The amount of voltage induced in a coil is directly proportional to the rate of change of the magnetic field with respect to the coil.

This principle is illustrated in Figure 13–8, where a bar magnet is moved through a coil of wire. An induced voltage is indicated by the voltmeter connected across the coil. The faster the magnet is moved, the greater the induced voltage.

▶ FIGURE 13–8

Induced voltage created by a changing magnetic field.

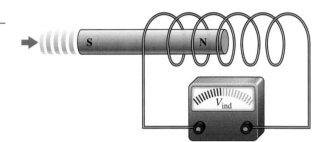

When a wire is formed into a certain number of loops or turns and is exposed to a changing magnetic field, a voltage is induced across the coil. The induced voltage is proportional to the number of turns of wire in the coil, N, and to the rate at which the magnetic field changes. The rate of change of the magnetic field is designated $d\phi/dt$, where ϕ is the magnetic flux. The ratio $d\phi/dt$ is expressed in webers/second (Wb/s). Faraday's law states that the induced voltage across a coil is equal to the number of turns (loops) times the rate of flux change and is expressed in concise form as follows:

Equation 13–4

$$v_{ind} = N\left(\frac{d\phi}{dt}\right)$$

EXAMPLE 13–3

Apply Faraday's law to find the induced voltage across a coil with 500 turns located in a magnetic field that is changing at a rate of 5 Wb/s.

Solution

$$v_{ind} = N\left(\frac{d\phi}{dt}\right) = (500\,t)(5\,\text{Wb/s}) = \textbf{2.5 kV}$$

Related Problem

A 1000 turn coil has an induced voltage of 500 V across it. What is the rate of change of the magnetic field?

Lenz's Law

Lenz's law was introduced in Chapter 10 and is restated here.

When the current through a coil changes, an induced voltage is created as a result of the changing electromagnetic field and the polarity of the induced voltage is such that it always opposes the change in current.

Figure 13–9 illustrates Lenz's law. In part (a), the current is constant and is limited by R_1. There is no induced voltage because the electromagnetic field is unchanging. In part (b),

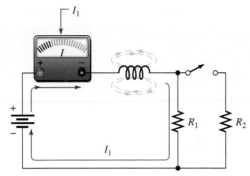

(a) Switch open: Constant current and constant magnetic field; no induced voltage.

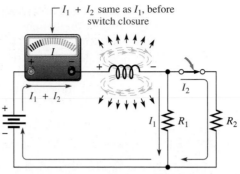

(b) At instant of switch closure: Expanding magnetic field induces voltage, which opposes an increase in total current. The total current remains the same at this instant.

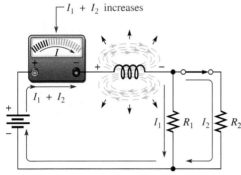

(c) Right after switch closure: The rate of expansion of the magnetic field decreases, allowing the current to increase exponentially as induced voltage decreases.

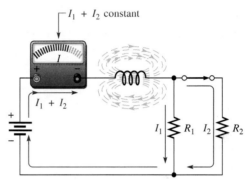

(d) Switch remains closed: Current and magnetic field reach constant value.

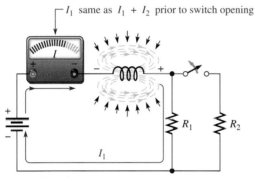

(e) At instant of switch opening: Magnetic field begins to collapse, creating an induced voltage, which opposes a decrease in current.

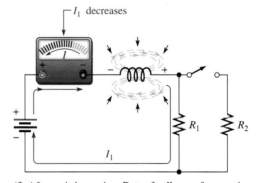

(f) After switch opening: Rate of collapse of magnetic field decreases, allowing current to decrease exponentially back to original value.

▲ FIGURE 13–9

Demonstration of Lenz's law in an inductive circuit: When the current tries to change suddenly, the electromagnetic field changes and induces a voltage in a direction that opposes that change in current.

the switch suddenly is closed, placing R_2 in parallel with R_1 and thus reducing the resistance. Naturally, the current tries to increase and the electromagnetic field begins to expand, but the induced voltage opposes this attempted increase in current for an instant.

In Figure 13–9(c), the induced voltage gradually decreases, allowing the current to increase. In part (d), the current has reached a constant value as determined by the parallel resistors, and the induced voltage is zero. In part (e), the switch has been suddenly opened, and, for an instant, the induced voltage prevents any decrease in current, and arcing between the switch contacts results. In part (f), the induced voltage gradually decreases, allowing the current to decrease back to a value determined by R_1. Notice that the induced voltage has a polarity that opposes any current change. The polarity of the induced voltage is opposite that of the battery voltage for an increase in current and aids the battery voltage for a decrease in current.

**SECTION 13–1
REVIEW**
Answers are at the end of the chapter.

1. List the parameters that contribute to the inductance of a coil.
2. The current through a 15 mH inductor is changing at the rate of 500 mA/s. What is the induced voltage?
3. Describe what happens to L when
   (a) N is increased
   (b) The core length is increased
   (c) The cross-sectional area of the core is decreased
   (d) A ferromagnetic core is replaced by an air core
4. Explain why inductors have some winding resistance.
5. Explain why inductors have some winding capacitance.

13–2 TYPES OF INDUCTORS

Inductors normally are classified according to the type of core material.

After completing this section, you should be able to

◆ **Discuss various types of inductors**

   ◆ Describe the basic types of fixed inductors

   ◆ Distinguish between fixed and variable inductors

Inductors are made in a variety of shapes and sizes. Basically, they fall into two general categories: fixed and variable. The standard schematic symbols are shown in Figure 13–10.

Both fixed and variable inductors can be classified according to the type of core material. Three common types are the air core, the iron core, and the ferrite core. Each has a unique symbol, as shown in Figure 13–11.

Adjustable (variable) inductors usually have a screw-type adjustment that moves a sliding core in and out, thus changing the inductance. A wide variety of inductors exist, and some are shown in Figure 13–12. Small fixed inductors are frequently encapsulated in an insulating material that protects the fine wire in the coil. Encapsulated inductors have an appearance similar to a resistor.

▶ **FIGURE 13–10**

Symbols for fixed and variable inductors.

(a) Fixed   (b) Variable

▶ **FIGURE 13–11**

Inductor symbols.

(a) Air core   (b) Iron core   (c) Ferrite core

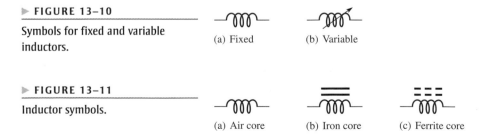

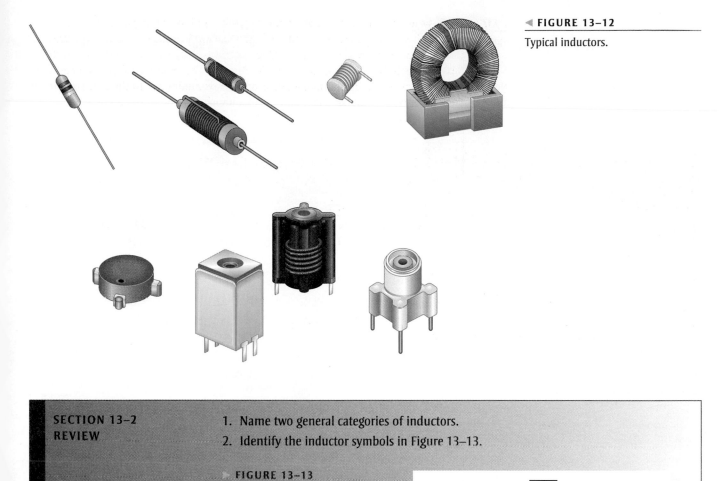

◀ **FIGURE 13–12**

Typical inductors.

---

1. Name two general categories of inductors.
2. Identify the inductor symbols in Figure 13–13.

▶ **FIGURE 13–13**

(a)   (b)   (c)

13–3 SERIES AND PARALLEL INDUCTORS

When inductors are connected in series, the total inductance increases. When inductors are connected in parallel, the total inductance decreases.

After completing this section, you should be able to

- ◆ **Analyze series and parallel inductors**
  - ◆ Determine total series inductance
  - ◆ Determine total parallel inductance

Total Series Inductance

When inductors are connected in series, as in Figure 13–14, the total inductance, L_T, is the sum of the individual inductances. The formula for L_T is expressed in the following equation for the general case of n inductors in series:

$$L_T = L_1 + L_2 + L_3 + \cdots + L_n$$

Equation 13–5

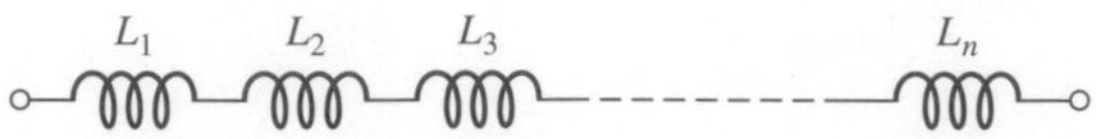

▶ FIGURE 13–14

Inductors in series.

L_1 L_2 L_3 L_n

Notice that the calculation of total inductance in series is analogous to the calculations of total resistance in series (Chapter 5) and total capacitance in parallel (Chapter 12).

EXAMPLE 13–4

Determine the total inductance for each of the series connections in Figure 13–15.

FIGURE 13–15

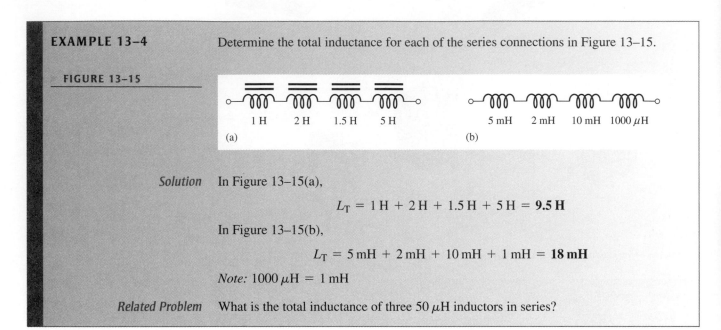

1 H   2 H   1.5 H   5 H   5 mH   2 mH   10 mH   1000 μH

(a)   (b)

Solution   In Figure 13–15(a),

$$L_T = 1\,H + 2\,H + 1.5\,H + 5\,H = \mathbf{9.5\,H}$$

In Figure 13–15(b),

$$L_T = 5\,mH + 2\,mH + 10\,mH + 1\,mH = \mathbf{18\,mH}$$

Note: 1000 μH = 1 mH

Related Problem   What is the total inductance of three 50 μH inductors in series?

Total Parallel Inductance

When inductors are connected in parallel, as in Figure 13–16, the total inductance is less than the smallest inductance. The general formula states that the reciprocal of the total inductance is equal to the sum of the reciprocals of the individual inductances.

Equation 13–6

$$\frac{1}{L_T} = \frac{1}{L_1} + \frac{1}{L_2} + \frac{1}{L_3} + \cdots + \frac{1}{L_n}$$

You can calculate total inductance, L_T, by taking the reciprocal of both sides of Equation 13–6.

Equation 13–7

$$L_T = \frac{1}{\left(\dfrac{1}{L_1}\right) + \left(\dfrac{1}{L_2}\right) + \left(\dfrac{1}{L_3}\right) + \cdots + \left(\dfrac{1}{L_n}\right)}$$

The calculation for total inductance in parallel is analogous to the calculations for total parallel resistance (Chapter 6) and total series capacitance (Chapter 12). For series-parallel combination of inductors, determine, the total inductance is in the same way as total resistance in resistive circuits.

▶ FIGURE 13–16

Inductors in parallel.

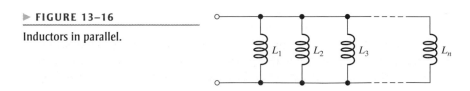

L_1 L_2 L_3 L_n

EXAMPLE 13–5

Determine L_T in Figure 13–17.

▶ **FIGURE 13–17**

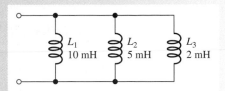

Solution Use Equation 13–7 to determine the total inductance.

$$L_T = \cfrac{1}{\left(\cfrac{1}{L_1}\right) + \left(\cfrac{1}{L_2}\right) + \left(\cfrac{1}{L_3}\right)} = \cfrac{1}{\cfrac{1}{10\,\text{mH}} + \cfrac{1}{5\,\text{mH}} + \cfrac{1}{2\,\text{mH}}} = \mathbf{1.25\,mH}$$

Related Problem Determine L_T for a parallel connection of 50 μH, 80 μH, 100 μH, and 150 μH.

**SECTION 13–3
REVIEW**

1. State the rule for combining inductors in series.
2. What is L_T for a series connection of 100 μH, 500 μH, and 2 mH?
3. Five 100 mH coils are connected in series. What is the total inductance?
4. Compare the total inductance in parallel with the smallest-value individual inductor.
5. The calculation of total parallel inductance is analogous to that for total parallel resistance. (T or F)
6. Determine L_T for each parallel combination:
   (a) 40 μH and 60 μH
   (b) 100 mH, 50 mH, and 10 mH

13–4 INDUCTORS IN DC CIRCUITS

Energy is stored in the electromagnetic field of an inductor when it is connected to a dc voltage source. The buildup of current through the inductor occurs in a predictable manner, which is dependent on the time constant determined by the inductance and the resistance in a circuit.

After completing this section, you should be able to

◆ **Analyze inductive dc switching circuits**

  ◆ Describe the increase and decrease of current an inductor

  ◆ Define *RL time constant*

  ◆ Describe induced voltage

  ◆ Write the exponential equations for current in an inductor

When there is constant direct current in an inductor, there is no induced voltage. There is, however, a voltage drop due to the winding resistance of the coil. The inductance itself appears as a short to dc. Energy is stored in the electromagnetic field according to the formula previously stated in Equation 13–2, $W = \frac{1}{2}LI^2$. The only energy conversion to heat occurs in the winding resistance ($P = I^2R_W$). This condition is illustrated in Figure 13–18.

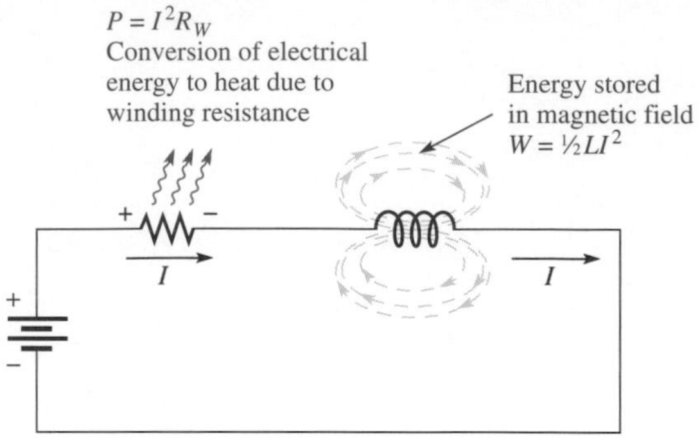

▶ FIGURE 13–18

Energy storage and conversion to heat in an inductor in a dc circuit.

$P = I^2R_W$
Conversion of electrical energy to heat due to winding resistance

Energy stored in magnetic field $W = \frac{1}{2}LI^2$

The *RL* Time Constant

Because the inductor's basic action is to develop a voltage that opposes a change in its current, it follows that current cannot change instantaneously in an inductor. A certain time is required for the current to make a change from one value to another. The rate at which the current changes is determined by the *RL* time constant.

> **The *RL* time constant is a fixed time interval that equals the ratio of the inductance to the resistance.**

The formula is

Equation 13–8

$$\tau = \frac{L}{R}$$

where τ is in seconds when inductance (L) is in henries and resistance (R) is in ohms.

EXAMPLE 13–6

A series *RL* circuit has a resistance of $1.0\,k\Omega$ and an inductance of $1\,mH$. What is the time constant?

Solution

$$\tau = \frac{L}{R} = \frac{1\,\text{mH}}{1.0\,k\Omega} = \frac{1 \times 10^{-3}\,\text{H}}{1 \times 10^{3}\,\Omega} = 1 \times 10^{-6}\,\text{s} = \mathbf{1\,\mu s}$$

Related Problem

Find the time constant for $R = 2.2\,k\Omega$ and $L = 500\,\mu H$.

Current in an Inductor

Increasing Current In a series *RL* circuit, the current will increase to approximately 63% of its full value in one time-constant interval after voltage is applied. This buildup of current is analogous to the buildup of capacitor voltage during the charging in an *RC* circuit; they both follow an exponential curve and reach the approximate percentages of the final current as indicated in Table 13–1 and as illustrated in Figure 13–19.

The change in current over five time-constant intervals is illustrated in Figure 13–20. When the current reaches its final value at 5τ, it ceases to change. At this time, the inductor acts as a short (except for winding resistance) to the constant current. The final value of the current is

$$I_F = \frac{V_S}{R} = \frac{10\,\text{V}}{1.0\,k\Omega} = 10\,\text{mA}$$

NUMBER OF TIME CONSTANTS	APPROXIMATE % OF FINAL CURRENT
1	63
2	86
3	95
4	98
5	99 (considered 100%)

◀ **TABLE 13–1**

Percentage of the final current after each time-constant interval during current buildup.

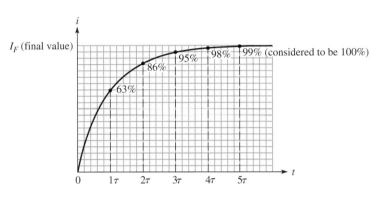

▲ **FIGURE 13–19**

Increasing current in an inductor.

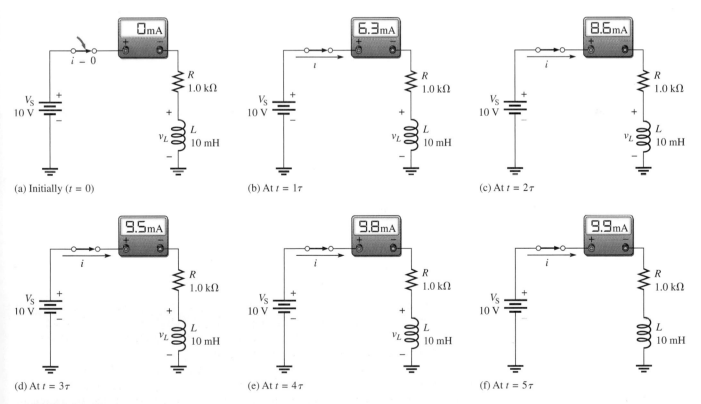

(a) Initially (t = 0)   (b) At t = 1τ   (c) At t = 2τ

(d) At t = 3τ   (e) At t = 4τ   (f) At t = 5τ

▲ **FIGURE 13–20**

Illustration of the exponential buildup of current in an inductor. The current increases approximately 63% during each time-constant interval after the switch is closed. A voltage (v_L) is induced in the coil that tends to oppose the increase in current.

EXAMPLE 13–7

Calculate the time constant for Figure 13–21. Then determine the current and the time at each time-constant interval, measured from the instant the switch is closed.

▶ **FIGURE 13–21**

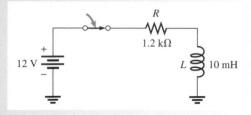

Solution   The time constant is

$$\tau = \frac{L}{R} = \frac{10\,\text{mH}}{1.2\,\text{k}\Omega} = \textbf{8.33}\,\boldsymbol{\mu}\textbf{s}$$

The current at each time-constant interval is a certain percentage of the final current. The final current is

$$I_F = \frac{V_S}{R} = \frac{12\,\text{V}}{1.2\,\text{k}\Omega} = 10\,\text{mA}$$

Using the time-constant percentage values from Table 13–1,

At 1τ: $i = 0.63(10\,\text{mA}) = \textbf{6.3\,mA}$; $t = \textbf{8.33}\,\boldsymbol{\mu}\textbf{s}$

At 2τ: $i = 0.86(10\,\text{mA}) = \textbf{8.6\,mA}$; $t = \textbf{16.7}\,\boldsymbol{\mu}\textbf{s}$

At 3τ: $i = 0.95(10\,\text{mA}) = \textbf{9.5\,mA}$; $t = \textbf{25.0}\,\boldsymbol{\mu}\textbf{s}$

At 4τ: $i = 0.98(10\,\text{mA}) = \textbf{9.8\,mA}$; $t = \textbf{33.3}\,\boldsymbol{\mu}\textbf{s}$

At 5τ: $i = 0.99(10\,\text{mA}) = 9.9\,\text{mA} \cong \textbf{10\,mA}$; $t = \textbf{41.7}\,\boldsymbol{\mu}\textbf{s}$

Related Problem   Repeat the calculations if R is 680 Ω and L is 100 μH.

Use Multisim file E13-07 to verify the calculated results in this example and to confirm your calculation for the related problem. Use a square wave to replace the dc voltage source and the switch.

Decreasing Current   Current in an inductor decreases exponentially according to the approximate percentage values in Table 13–2 and in Figure 13–22.

The change in current over five time-constant intervals is illustrated in Figure 13–23. When the current reaches its final value of approximately 0 A, it ceases to change. Before

▶ **TABLE 13–2**

Percentage of initial current after each time-constant interval while current is decreasing.

NUMBER OF TIME CONSTANTS	APPROXIMATE % OF INITIAL CURRENT
1	37
2	14
3	5
4	2
5	1 (considered 0)

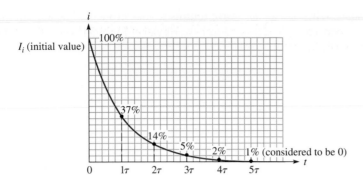

▲ FIGURE 13–22

Decreasing current in an inductor.

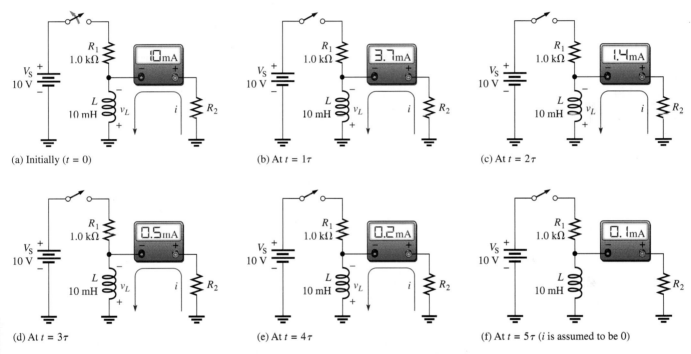

(a) Initially ($t = 0$)

(b) At $t = 1\tau$

(c) At $t = 2\tau$

(d) At $t = 3\tau$

(e) At $t = 4\tau$

(f) At $t = 5\tau$ (i is assumed to be 0)

▲ FIGURE 13–23

Illustration of the exponential decrease of current in an inductor. The current decreases approximately 63% during each time-constant interval after the switch is opened. A voltage (v_L) is induced in the coil that tends to oppose the decrease in current.

the switch is opened, the current through L is at a constant value of 10 mA, which is determined by R_1 because L acts ideally as a short. When the switch is opened, the induced inductor voltage initially provides 10 mA through R_2. The current then decreases by 63% during each time constant interval.

A good way to demonstrate both increasing and decreasing current in an RL circuit is to use a square wave voltage as the input. The square wave is a useful signal for observing the dc response of a circuit because it automatically provides on and off action similar to a switch. (Time response will be covered further in Chapter 20.) When the square wave goes from its low level to its high level, the current in the circuit responds by exponentially rising to its final value. When the square wave returns to the zero level, the current in the circuit responds by exponentially decreasing to its zero value. Figure 13–24 shows input voltage and current waveforms.

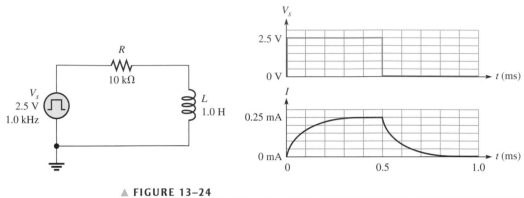

▲ FIGURE 13–24

EXAMPLE 13–8

For the circuit in Figure 13–24, what is the current at 0.1 ms and 0.6 ms?

Solution   The RL time constant for the circuit is

$$\tau = \frac{L}{R} = \frac{1.0\,\text{H}}{10\,\text{k}\Omega} = 0.1\,\text{ms}$$

If the square-wave generator period is long enough for the current to reach its maximum value in 5τ, the current will increase exponentially and during each time constant interval will have a value equal to the percentage of the final current given in Table 13–1. The final current is

$$I_F = \frac{V_S}{R} = \frac{2.5\,\text{V}}{10\,\text{k}\Omega} = 0.25\,\text{mA}$$

The current at 0.1 ms is

$$i = 0.63(0.25\,\text{mA}) = \textbf{0.158\,mA}$$

At 0.6 ms, the square-wave input has been at the 0 V level for 0.1 ms, or 1τ; and the current decreases from the maximum value toward its final value of 0 mA by 63%. Therefore,

$$i = 0.25\,\text{mA} - 0.63(0.25\,\text{mA}) = \textbf{0.092\,mA}$$

Related Problem   What is the current at 0.2 ms and 0.8 ms?

Voltages in a Series *RL* Circuit

As you know, when current changes in an inductor, a voltage is induced. Let's examine what happens to the induced voltage across the inductor in the series circuit in Figure 13–25 during one complete cycle of a square wave input. Keep in mind that the generator produces a level that is like switching a dc source on and then puts an "automatic" low resistance (ideally zero) path across the source when it returns to its zero level.

An ammeter placed in the circuit shows the current in the circuit at any instant in time. The V_L waveform is the voltage across the inductor. In Figure 13–25(a), the square wave has just

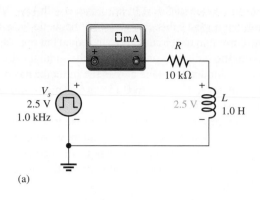

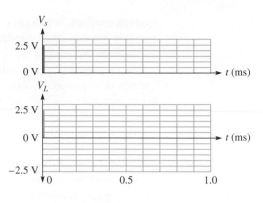

(a)

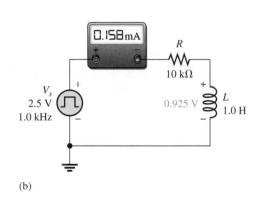

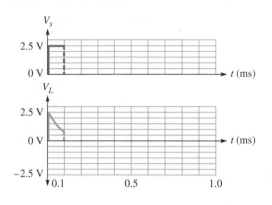

(b)

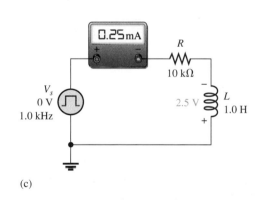

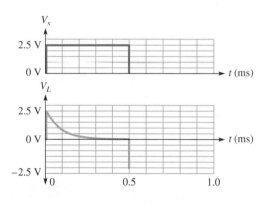

(c)

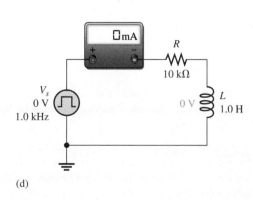

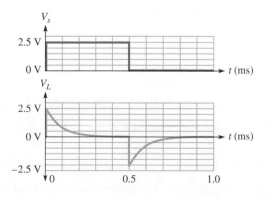

(d)

▲ FIGURE 13–25

transitioned from zero to its maximum value of 2.5 V. In accordance with Lenz's law, a voltage is induced across the inductor that opposes this *change* as the magnetic field surrounding the inductor builds up. There is no current in the circuit due to the equal but opposing voltages.

As the magnetic field builds up, the induced voltage across the inductor decreases, and current is in the circuit. After 1τ, the induced voltage across the inductor has decreased by 63%, which causes the current to increase by 63% to 0.158 mA. This is shown in Figure 13–25(b) at the end of one time constant (0.1 ms).

The voltage on the inductor continues to exponentially decrease to zero, at which point the current is limited only by the circuit resistance. Then the square wave goes back to zero (at $t = 0.5$ ms) as shown in Figure 13–25(c). Again a voltage is induced across the inductor opposing this change. This time, the polarity of the inductor voltage is reversed due to the collapsing magnetic field. Although the source voltage is 0, the collapsing magnetic field maintains current in the same direction until the current decreases to zero, as shown in Figure 13–25(d).

EXAMPLE 13–9

(a) The circuit in Figure 13–26 has a square wave input. What is the highest frequency that can be used and still observe the complete waveform across the inductor?

(b) Assume the generator is set to the frequency determined in (a). Describe the voltage waveform across the resistor?

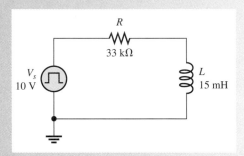

▲ **FIGURE 13–26**

Solution **(a)** $\tau = \dfrac{L}{R} = \dfrac{15 \text{ mH}}{33 \text{ k}\Omega} = 0.454 \text{ } \mu s$

The period needs to be 10 times longer than τ to observe the entire wave.

$$T = 10\tau = 4.54 \text{ } \mu s$$

$$f = \frac{1}{T} = \frac{1}{4.54 \text{ } \mu s} = \textbf{220 kHz}$$

(b) The voltage across the resistor has the same shape as the current waveform. The general shape was shown in Figure 13–24 and has a maximum value of 10 V (the same V_s assuming no winding resistance).

Related Problem What is the maximum voltage across the resistor for $f = 220$ kHz?

Use Multisim file E13-09 to verify the calculated results in this example and to confirm your calculation for the related problem.

The Exponential Formulas

The formulas for the exponential current and voltage in an RL circuit are similar to those used in Chapter 12 for the RC circuit, and the universal exponential curves in Figure 12–36 apply to inductors as well as capacitors. The general formulas for RL circuits are stated as follows:

$$v = V_F + (V_i - V_F)e^{-Rt/L}$$ **Equation 13–9**

$$i = I_F + (I_i - I_F)e^{-Rt/L}$$ **Equation 13–10**

where V_F and I_F are the final values of voltage and current, V_i and I_i are the initial values of voltage and current. The lowercase italic letters v and i are the instantaneous values of the inductor voltage and current at time t.

Increasing Current The formula for the special case in which an increasing exponential current curve begins at zero is derived by setting $I_i = 0$ in Equaton 13–10.

$$i = I_F(1 - e^{-Rt/L})$$ **Equation 13–11**

Using Equation 13–11, you can calculate the value of the increasing inductor current at any instant of time. You can calculate voltage by substituting v for i and V_F for I_F in Equation 13–11. Notice that the exponent Rt/L can also be written as $t/(L/R) = t/\tau$.

EXAMPLE 13–10

In Figure 13–27, determine the inductor current 30 μs after the switch is closed.

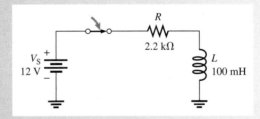

FIGURE 13–27

Solution The time constant is

$$\tau = \frac{L}{R} = \frac{100\,\text{mH}}{2.2\,\text{k}\Omega} = 45.5\,\mu\text{s}$$

The final current is

$$I_F = \frac{V_S}{R} = \frac{12\,\text{V}}{2.2\,\text{k}\Omega} = 5.45\,\text{mA}$$

The initial current is zero. Notice that 30 μs is less than one time constant, so the current will reach less than 63% of its final value in that time.

$$i_L = I_F(1 - e^{-Rt/L}) = 5.45\,\text{mA}(1 - e^{-0.66}) = 5.45\,\text{mA}(1 - 0.517) = \textbf{2.63\,mA}$$

Related Problem In Figure 13–27, determine the inductor current 55 μs after the switch is closed.

Decreasing Current The formula for the special case in which a decreasing exponential current has a final value of zero is derived by setting $I_F = 0$ in Equation 13–10.

Equation 13–12

$$i = I_i e^{-Rt/L}$$

This formula can be used to calculate the decreasing inductor current at any instant, as the next example shows.

EXAMPLE 13–11

In Figure 13–28, what is the current at each microsecond interval for one complete cycle of the input square wave, V_s After calculating the current at each time, sketch the current waveform.

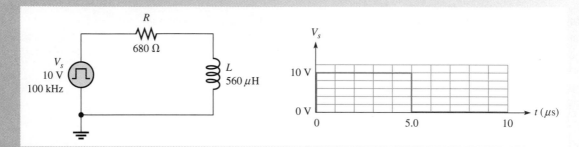

▲ **FIGURE 13–28**

Solution

$$\tau = \frac{L}{R} = \frac{560 \ \mu H}{680 \ \Omega} = 0.824 \ \mu s$$

When the pulse goes from 0 V to 10 V at $t = 0$, the final current is

$$I_F = \frac{V_s}{R} = \frac{10 \ V}{680 \ \Omega} = 14.7 \ mA$$

For the increasing current,

$$i = I_F(1 - e^{-Rt/L}) = I_F(1 - e^{-t/\tau})$$

At 1 μs: $i = 14.7 \ mA(1 - e^{-1\mu s/0.824\mu s}) = \textbf{10.3 mA}$

At 2 μs: $i = 14.7 \ mA(1 - e^{-2\mu s/0.824\mu s}) = \textbf{13.4 mA}$

At 3 μs: $i = 14.7 \ mA(1 - e^{-3\mu s/0.824\mu s}) = \textbf{14.3 mA}$

At 4 μs: $i = 14.7 \ mA(1 - e^{-4\mu s/0.824\mu s}) = \textbf{14.6 mA}$

At 5 μs: $i = 14.7 \ mA(1 - e^{-5\mu s/0.824\mu s}) = \textbf{14.7 mA}$

When the pulse goes from 10 V to 0 V at $t = 5 \ \mu$s, the current decreases exponentially. For the decreasing current,

$$i = I_i(e^{-Rt/L}) = I_i(e^{-t/\tau})$$

The initial current is the value at 5 μs, which is 14.7 mA.

At 6 μs: $i = 14.7 \ mA(e^{-1\mu s/0.824\mu s}) = \textbf{4.37 mA}$

At 7 μs: $i = 14.7 \ mA(e^{-2\mu s/0.824\mu s}) = \textbf{1.30 mA}$

At 8 μs: $i = 14.7 \ mA(e^{-3\mu s/0.824\mu s}) = \textbf{0.38 mA}$

At 9 μs: $i = 14.7 \ mA(e^{-4\mu s/0.824\mu s}) = \textbf{0.11 mA}$

At 10 μs: $i = 14.7 \ mA(e^{-5 \ \mu s/0.824 \ \mu s}) = \textbf{0.03 mA}$

Figure 13–29 is a graph of these results.

▶ **FIGURE 13–29**

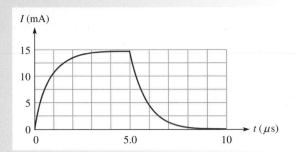

Related Problem What is the current at 0.5 µs?

Use Multisim file E13-11 to verify the calculated results in this example and to confirm your calculation for the related problem.

SECTION 13–4 REVIEW

1. A 15 mH inductor with a winding resistance of 10 Ω has a constant direct current of 10 mA through it. What is the voltage drop across the inductor?
2. A 20 V dc source is connected to a series *RL* circuit with a switch. At the instant of switch closure, what are the values of i and v_L?
3. In the same circuit as in Question 2, after a time interval equal to 5τ from switch closure, what is v_L?
4. In a series *RL* circuit where $R = 1.0\,k\Omega$ and $L = 500\,\mu H$, what is the time constant? Determine the current 0.25 µs after a switch connects 10 V across the circuit.

13–5 INDUCTORS IN AC CIRCUITS

An inductor passes ac with an amount of opposition called inductive reactance that depends on the frequency of the ac. The concept of the derivative was introduced in Chapter 12, and the expression for induced voltage in an inductor was stated earlier in Equation 13–1. These are used again in this section.

After completing this section, you should be able to

- ◆ **Analyze inductive ac circuits**
  - ◆ Explain why an inductor causes a phase shift between voltage and current
  - ◆ Define *inductive reactance*
  - ◆ Determine the value of inductive reactance in a given circuit
  - ◆ Discuss instantaneous, true, and reactive power in an inductor

Phase Relationship of Current and Voltage in an Inductor

From Equation 13–1, the formula for induced voltage, you can see that the faster the current through an inductor changes, the greater the induced voltage will be. For example, if the rate of change of current is zero, the voltage is zero $[v_{ind} = L(di/dt) = L(0) = 0 \text{ V}]$. When di/dt is a positive-going maximum, v_{ind} is a positive maximum; when di/dt is a negative-going maximum, v_{ind} is a negative maximum.

A sinusoidal current always induces a sinusoidal voltage in inductive circuits. Therefore, you can plot the voltage with respect to the current if you know the points on the current curve at which the voltage is zero and those at which it is maximum. This phase relationship is shown in Figure 13–30(a). Notice that the voltage leads the current by 90°. This is always true in a purely inductive circuit. The current and voltage of this relationship is shown by the phasors in Figure 13–30(b).

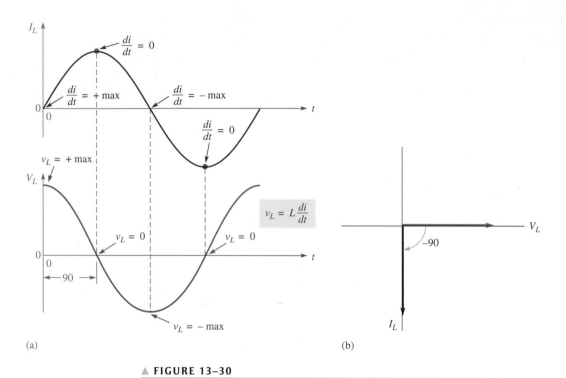

(a) (b)

▲ **FIGURE 13–30**

Phase relation of V_L and I_L in an inductor. Current always lags the inductor voltage by 90°.

Inductive Reactance, X_L

Inductive reactance is the opposition to sinusoidal current, expressed in ohms. The symbol for inductive reactance is X_L.

To develop a formula for X_L, we use the relationship $v_{ind} = L(di/dt)$ and the curves in Figure 13–31. The rate of change of current is directly related to frequency. The faster the current changes, the higher the frequency. For example, you can see that in Figure 13–31, the slope of sine wave A at the zero crossings is steeper than that of sine wave B. Recall that the slope of a curve at a point indicates the rate of change at that point. Sine wave A has a higher frequency than sine wave B, as indicated by a greater maximum rate of change (di/dt is greater at the zero crossings).

▶ **FIGURE 13–31**

Slope indicates rate of change. Sine wave A has a greater rate of change at the zero crossing than B, and thus A has a higher frequency.

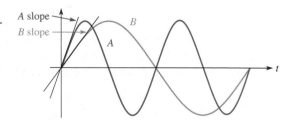

When frequency increases, di/dt increases, and thus v_{ind} increases. When frequency decreases, di/dt decreases, and thus v_{ind} decreases. The induced voltage is directly dependent on frequency.

$$\uparrow \qquad \uparrow$$
$$v_{ind} = L(di/dt) \qquad \text{and} \qquad v_{ind} = L(di/dt)$$
$$\downarrow \qquad \downarrow$$

An increase in induced voltage means more opposition (X_L is greater). Therefore, X_L is directly proportional to induced voltage and thus directly proportional to frequency.

X_L is proportional to f.

Now, if di/dt is constant and the inductance is varied, an increase in L produces an increase in v_{ind}, and a decrease in L produces a decrease in v_{ind}.

$$\uparrow \qquad \uparrow$$
$$v_{ind} = L(di/dt) \qquad \text{and} \qquad v_{ind} = L(di/dt)$$
$$\downarrow \qquad \downarrow$$

Again, an increase in v_{ind} means more opposition (greater X_L). Therefore, X_L is directly proportional to induced voltage and thus directly proportional to inductance. The inductive reactance is directly proportional to both f and L.

X_L is proportional to fL.

The formula (derived in Appendix B) for inductive reactance, X_L, is

$$X_L = 2\pi fL$$

Equation 13–13

Inductive reactance, X_L, is in ohms when f is in hertz and L is in henries. As with capacitive reactance, the 2π term is a constant factor in the equation, which comes from the relationship of a sine wave to rotational motion.

EXAMPLE 13–12

A sinusoidal voltage is applied to the circuit in Figure 13–32. The frequency is 10 kHz. Determine the inductive reactance.

▶ **FIGURE 13–32**

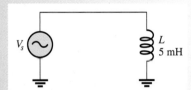

Solution Convert 10 kHz to 10×10^3 Hz and 5 mH to 5×10^{-3} H. Therefore, the inductive reactance is

$$X_L = 2\pi fL = 2\pi(10 \times 10^3 \text{ Hz})(5 \times 10^{-3} \text{ H}) = \textbf{314 } \Omega$$

Related Problem What is X_L in Figure 13–32 if the frequency is increased to 35 kHz?

Ohm's Law The reactance of an inductor is analogous to the resistance of a resistor as shown in Figure 13–33. In fact, X_L, just like X_C and R, is expressed in ohms. Since inductive reactance is a form of opposition to current, Ohm's law applies to inductive circuits as well as to resistive circuits and capacitive circuits; and it is stated as follows:

$$I = \frac{V}{X_L}$$

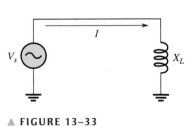

▲ **FIGURE 13–33**

When applying Ohm's law in ac circuits, you must express both the current and the voltage in the same way, that is, both in rms, both in peak, and so on.

EXAMPLE 13–13

Determine the rms current in Figure 13–34.

▶ **FIGURE 13–34**

$V_{rms} = 5$ V
$f = 10$ kHz

L
100 mH

Solution   Convert 10 kHz to 10×10^3 Hz and 100 mH to 100×10^{-3} H. Then calculate X_L.

$$X_L = 2\pi fL = 2\pi(10 \times 10^3 \text{ Hz})(100 \times 10^{-3}\text{ H}) = 6283 \; \Omega$$

Apply Ohm's law to determine the rms current.

$$I_{rms} = \frac{V_{rms}}{X_L} = \frac{5 \text{ V}}{6283 \; \Omega} = \textbf{796 } \boldsymbol{\mu}\textbf{A}$$

Related Problem   Determine the rms current in Figure 13–34 for the following values: $V_{rms} = 12$ V, $f = 4.9$ kHz, and $L = 680$ mH.

Use Multisim file E13-13 to verify the calculated results in this example and to confirm your calculation for the related problem.

Power in an Inductor

As discussed earlier, an inductor stores energy in its magnetic field when there is current through it. An ideal inductor (assuming no winding resistance) does not dissipate energy; it only stores it. When an ac voltage is applied to an ideal inductor, energy is stored by the inductor during a portion of the cycle; then the stored energy is returned to the source during another portion of the cycle. No net energy is lost in an ideal inductor due to conversion to heat. Figure 13–35 shows the power curve that results from one cycle of inductor current and voltage.

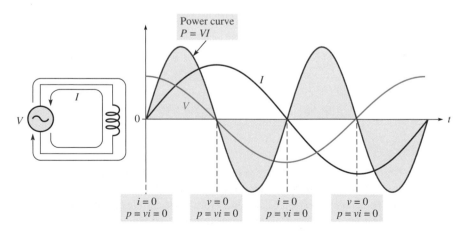

▲ **FIGURE 13–35**

Power curve.

Instantaneous Power (p)   The product of v and i gives instantaneous power. At points where v or i is zero, p is also zero. When both v and i are positive, p is also positive. When either v or i is positive and the other negative, p is negative. When both v and i are negative, p is positive. As you can see in Figure 13–35, the power follows a sinusoidal-shaped curve. Positive values of power indicate that energy is stored by the inductor. Negative values of power indicate that energy is returned from the inductor to the source. Note that the power fluctuates at a frequency twice that of the voltage or current as energy is alternately stored and returned to the source.

True Power (P_{true})   Ideally, all of the energy stored by an inductor during the positive portion of the power cycle is returned to the source during the negative portion. No net energy is lost due to conversion to heat in the inductor, so the true power is zero. Actually, because of winding resistance in a practical inductor, some power is always dissipated; and there is a very small amount of true power, which can normally be neglected.

$$P_{\text{true}} = (I_{\text{rms}})^2 R_W \qquad\qquad \textbf{Equation 13–14}$$

Reactive Power (P_r)   The rate at which an inductor stores or returns energy is called its **reactive power**, with the unit of VAR (volt-ampere reactive). The reactive power is a nonzero quantity because at any instant in time the inductor is actually taking energy from the source or returning energy to it. Reactive power does not represent an energy loss due to conversion to heat. The following formulas apply:

$$P_r = V_{\text{rms}}I_{\text{rms}} \qquad\qquad \textbf{Equation 13–15}$$

$$P_r = \frac{V_{\text{rms}}^2}{X_L} \qquad\qquad \textbf{Equation 13–16}$$

$$P_r = I_{\text{rms}}^2 X_L \qquad\qquad \textbf{Equation 13–17}$$

EXAMPLE 13–14   A 10 V rms signal with a frequency of 1 kHz is applied to a 10 mH coil with a negligible winding resistance. Determine the reactive power (P_r).

Solution   First, calculate the inductive reactance and current values.

$$X_L = 2\pi fL = 2\pi(1\,\text{kHz})(10\,\text{mH}) = 62.8\,\Omega$$

$$I = \frac{V_s}{X_L} = \frac{10\,\text{V}}{62.8\,\Omega} = 159\,\text{mA}$$

Then, use Equation 13–17.

$$P_r = I^2 X_L = (159\,\text{mA})^2(62.8\,\Omega) = \textbf{1.59 VAR}$$

Related Problem   What happens to the reactive power if the frequency increases?

The Quality Factor (Q) of a Coil

The **quality factor (Q)** is the ratio of the reactive power in an inductor to the true power in the winding resistance of the coil or the resistance in series with the coil. It is a ratio of the power in L to the power in R_W. The quality factor is important in resonant circuits, which are studied in Chapter 17. A formula for Q is developed as follows:

$$Q = \frac{\text{reactive power}}{\text{true power}} = \frac{I^2 X_L}{I^2 R_W}$$

The current is the same in L and R_W; thus, the I^2 terms cancel, leaving

Equation 13–18

$$Q = \frac{X_L}{R_W}$$

When the resistance is just the winding resistance of the coil, the circuit Q and the coil Q are the same. Note that Q is a ratio of like units and, therefore, has no unit itself. The quality factor is also known as unloaded Q because it is defined with no load across the coil.

SECTION 13–5 REVIEW

1. State the phase relationship between current and voltage in an inductor.
2. Calculate X_L for $f = 5$ kHz and $L = 100$ mH.
3. At what frequency is the reactance of a 50 μH inductor equal to 800 Ω?
4. Calculate the rms current in Figure 13–36.
5. An ideal 50 mH inductor is connected to a 12 V rms source. What is the true power? What is the reactive power at a frequency of 1 kHz?

▶ **FIGURE 13–36**

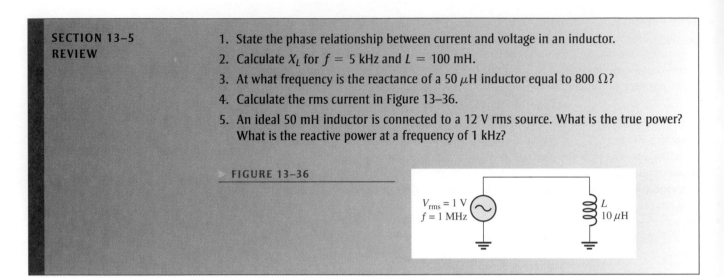

$V_{rms} = 1$ V
$f = 1$ MHz

L
$10\,\mu$H

13–6 INDUCTOR APPLICATIONS

Inductors are not as versatile as capacitors and tend to be more limited in their application due, in part, to size, cost factors, and nonideal behavior (internal resistance, etc.). One of the most common applications for inductors is noise reduction applications.

After completing this section, you should be able to

♦ **Discuss some inductor applications**

  ♦ Discuss two ways in which noise enters a circuit

  ♦ Describe the suppression of electromagnetic interference (EMI)

  ♦ Explain how a ferrite bead is used

  ♦ Discuss the basics of tuned circuits

Noise Suppression

One of the most important applications of inductors has to do with suppressing unwanted electrical noise. The inductors used in these applications are generally wound on a closed core to avoid having the inductor become a source of radiated noise itself. Two types of noise are conductive noise and radiated noise.

Conductive Noise Many systems have common conductive paths connecting different parts of the system, which can conduct high frequency noise from one part of the system to another. Consider the case of two circuits connected with common lines as shown in Figure 13–37(a). A path for high frequency noise exists though the common grounds, creating a condition known as a *ground loop*. Ground loops are particularly a problem in instrumentation

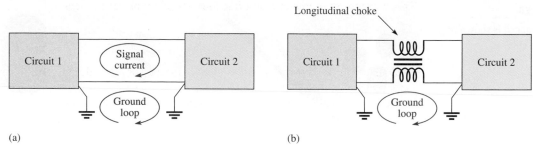

▲ FIGURE 13–37

systems, where a transducer may be located a distance from the recording system and noise current in the ground can affect the signal.

If the signal of interest changes slowly, a special inductor, called a *longitudinal choke,* can be installed in the signal line as shown in Figure 13–37(b). The longitudinal choke is a form of transformer (covered in Chapter 14) that acts as inductors in each signal line. The ground loop sees a high impedance path, thus reducing the noise, while the low-frequency signal is coupled through the low impedance of the choke.

Switching circuits also tend to generate high frequency noise (above 10 MHz) by virtue of the high frequency components present. (Recall from Section 11–9 that a pulse waveform contains many high frequency harmonics). Certain types of power supplies use high-speed switching circuits that are a source of conductive noise themselves.

Because an inductor's impedance increases with frequency, inductors are good for blocking electrical noise from these supplies, which should carry only dc. Inductors are frequently installed in the power supply lines to suppress this conductive noise, so that one circuit does not adversely affect another circuit. One or more capacitors may also be used in conjunction with the inductor to improve filtering action.

Radiated Noise   Noise can also enter a circuit by way of the electric field. The noise source can be an adjacent circuit or a nearby power supply. There are several approaches to reducing the effects of radiated noise. Usually, the first step is to determine the cause of the noise and isolate it using shielding or filtering.

Inductors are widely employed in filters that are used to suppress radio-frequency noise. The inductor used for noise suppression must be carefully selected so as not to become a source of radiated noise itself. For high frequencies(>20 MHz), inductors wound on highly permeable toroidal cores are widely used, as they tend to keep the magnetic flux restricted to the core.

RF Chokes

Inductors used for the purpose of blocking very high frequencies are called *Radio Frequency (RF) chokes*. RF chokes are used for conductive or radiated noise. They are special inductors designed to block high frequencies from getting into or leaving parts of a system by providing a high impedance path for high frequencies. In general, the choke is placed in series with the line for which RF suppression is required. Depending on the frequency of the interference, different types of chokes are required. A common type of electromagnetic interference (EMI) filter wraps the signal line on a toroidal core several times. The toroidal configuration is desired because it contains the magnetic field so that the choke does not become a source of noise itself.

Another common type of RF choke is a ferrite bead, such as the ones shown in Figure 13–38. All wires have inductance, and the ferrite bead is a small ferromagnetic material that is strung onto the wire to increase its inductance. The impedance presented by the bead is a function of both the material and the frequency, as well as the size of the bead. It is an effective and inexpensive "choke" for high frequencies. Ferrite beads are common in high-frequency communication systems. Sometimes several are strung together in series to increase the effective inductance.

▶ FIGURE 13–38

▶ FIGURE 13–38

Ferrite beads. A pencil is shown to illustrate relative size.

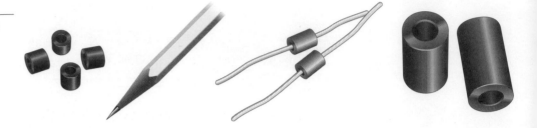

Tuned Circuits

Inductors are used in conjunction with capacitors to provide frequency selection in communications systems. These tuned circuits allow a narrow band of frequencies to be selected while all other frequencies are rejected. The tuners in your TV and radio receivers are based on this principle and permit you to select one channel or station out of the many that are available.

Frequency selectivity is based on the fact that the reactances of both capacitors and inductors depend on the frequency and on the interaction of these two components when connected in series or parallel. Since the capacitor and the inductor produce opposite phase shifts, their combined opposition to current can be used to obtain a desired response at a selected frequency. Tuned *RLC* circuits are covered in Chapter 17.

SECTION 13–6 REVIEW

1. Name two types of unwanted noise.
2. What do the letters *EMI* stand for?
3. How is a ferrite bead used?

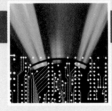

A Circuit Application

In this application, you will see how you can test coils for their unknown inductance values using a test setup consisting of a square wave generator and an oscilloscope. You are given two coils for which the inductance values are not known. You are to test the coils using simple laboratory instruments to determine the inductance values. The method is to place the coil in series with a resistor with a known value and measure the time constant. Knowing the time constant and the resistance value, you can calculate the value of *L*.

The method of determining the time constant is to apply a square wave to the circuit and measure the resulting voltage across the resistor. Each time the square wave input voltage goes high, the inductor is energized and each time the square wave goes back to zero, the inductor is deenergized. The time it takes for the exponential resistor voltage to increase to approximately its final value equals five time constants. This operation is illustrated in Figure 13–39. To make sure that the winding resistance of the coil can be neglected, it must be measured and the value of the resistor used in the circuit must

▶ FIGURE 13–39

Circuit for time-constant measurement.

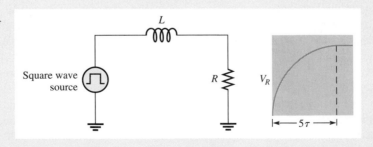

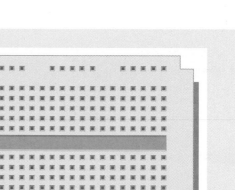

▲ FIGURE 13–40

Breadboard setup for measuring the time constant.

be selected to be considerably larger than the winding and source resistances.

The Winding Resistance

Assume that the winding resistance of the coil in Figure 13–40 has been measured with an ohmmeter and found to be 85 Ω. To make the winding and source resistances negligible for time constant measurement, a 10 kΩ series resistor is used in the circuit.

◆ If 10 V dc is connected with the clip leads as shown, how much current is in the circuit after $t = 5\tau$?

Inductance of Coil 1

Refer to Figure 13–41. To measure the inductance of coil 1, a square wave voltage is applied to the circuit. The amplitude of the square wave is adjusted to 10 V. The frequency is adjusted so

that the inductor has time to fully energize during each square wave pulse; the scope is set to view a complete energizing curve as shown.

◆ Determine the approximate circuit time constant.
◆ Calculate the inductance of coil 1.

The Inductance of Coil 2

Refer to Figure 13–42 in which coil 2 replaces coil 1. To determine the inductance, a 10 V square wave is applied to the breadboarded circuit. The frequency of the square wave is adjusted so that the inductor has time to fully energize during each square wave pulse; the scope is set to view a complete energizing curve as shown.

◆ Determine the approximate circuit time constant.
◆ Calculate the inductance of coil 2.

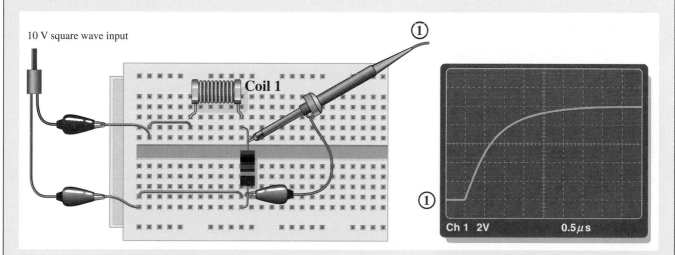

▲ FIGURE 13–41

Testing coil 1.

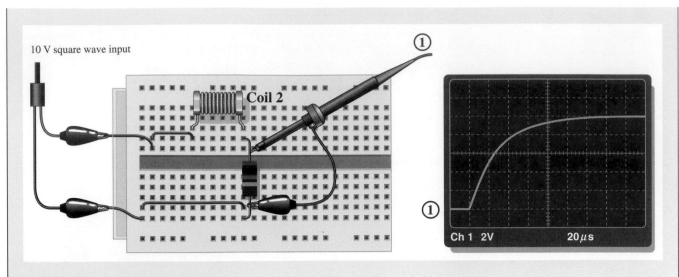

▲ FIGURE 13–42

Testing coil 2.

◆ Discuss any difficulty in using this method.

◆ Specify how you can use a sinusoidal input voltage instead of a square wave to determine inductance.

Review

1. What is the maximum square wave frequency that can be used in Figure 13–41?

2. What is the maximum square wave frequency that can be used in Figure 13–42?

3. What happens if the frequency exceeds the maximum you determined in Questions 1 and 2? Explain how your measurements would be affected.

SUMMARY

◆ Inductance is a measure of a coil's ability to establish an induced voltage as a result of a change in its current.

◆ An inductor opposes a change in its own current.

◆ Faraday's law states that relative motion between a magnetic field and a coil induces a voltage across the coil.

◆ The amount of induced voltage is directly proportional to the inductance and to the rate of change in current.

◆ Lenz's law states that the polarity of induced voltage is such that the resulting induced current is in a direction that opposes the change in the magnetic field that produced it.

◆ Energy is stored by an inductor in its magnetic field.

◆ One henry is the amount of inductance when current, changing at the rate of one ampere per second, induces one volt across the inductor.

◆ Inductance is directly proportional to the square of the number of turns, the permeability, and the cross-sectional area of the core. It is inversely proportional to the length of the core.

◆ The permeability of a core material is an indication of the ability of the material to establish a magnetic field.

◆ The time constant for a series RL circuit is the inductance divided by the resistance.

◆ In an RL circuit, the increasing or decreasing voltage and current in an inductor make a 63% change during each time-constant interval.

◆ Increasing and decreasing voltages and currents follow exponential curves.

◆ Inductors add in series.

◆ Total parallel inductance is less than that of the smallest inductor in parallel.

◆ Voltage leads current by 90° in an inductor.

◆ Inductive reactance, X_L, is directly proportional to frequency and inductance.

◆ The true power in an inductor is zero; that is, no energy is lost in an ideal inductor due to conversion to heat, only in its winding resistance.

KEY TERMS

Key terms and other bold terms in the chapter are defined in the end-of-book glossary.

Henry (H) The unit of inductance.

Induced voltage Voltage produced as a result of a changing magnetic field.

Inductance The property of an inductor whereby a change in current causes the inductor to produce a voltage that opposes the change in current.

Inductive reactance The opposition of an inductor to sinusoidal current. The unit is the ohm.

Inductor An electrical device formed by a wire wound around a core having the property of inductance; also known as *coil*.

Quality factor (Q) The ratio of reactive power to true power in an inductor.

RL time constant A fixed time interval, set by the L and R values, that determines the time response of a circuit and is equal to L/R.

Winding The loops or turns of wire in an inductor.

FORMULAS

13–1	$v_{\text{ind}} = L\left(\dfrac{di}{dt}\right)$	Induced voltage
13–2	$W = \dfrac{1}{2}LI^2$	Energy stored by an inductor
13–3	$L = \dfrac{N^2 \mu A}{l}$	Inductance in terms of physical parameters
13–4	$v_{\text{ind}} = N\left(\dfrac{d\phi}{dt}\right)$	Faraday's law
13–5	$L_T = L_1 + L_2 + L_3 + \cdots + L_n$	Series inductance
13–6	$\dfrac{1}{L_T} = \dfrac{1}{L_1} + \dfrac{1}{L_2} + \dfrac{1}{L_3} + \cdots + \dfrac{1}{L_n}$	Reciprocal of total parallel inductance
13–7	$L_T = \dfrac{1}{\left(\dfrac{1}{L_1}\right) + \left(\dfrac{1}{L_2}\right) + \left(\dfrac{1}{L_3}\right) + \cdots + \left(\dfrac{1}{L_n}\right)}$	Total parallel inductance
13–8	$\tau = \dfrac{L}{R}$	Time constant
13–9	$v = V_F + (V_i - V_F)e^{-Rt/L}$	Exponential voltage (general)
13–10	$i = I_F + (I_i - I_F)e^{-Rt/L}$	Exponential current (general)
13–11	$i = I_F(1 - e^{-Rt/L})$	Increasing exponential current beginning at zero
13–12	$i = I_i e^{-Rt/L}$	Decreasing exponential current ending at zero
13–13	$X_L = 2\pi fL$	Inductive reactance
13–14	$P_{\text{true}} = (I_{\text{rms}})^2 R_W$	True power
13–15	$P_r = V_{\text{rms}} I_{\text{rms}}$	Reactive power
13–16	$P_r = \dfrac{V_{\text{rms}}^2}{X_L}$	Reactive power
13–17	$P_r = I_{\text{rms}}^2 X_L$	Reactive power
13–18	$Q = \dfrac{X_L}{R_W}$	Quality factor

SELF-TEST

Answers are at the end of the chapter.

1. An inductance of 0.05 μH is larger than

   (a) 0.0000005 H   (b) 0.000005 H   (c) 0.000000008 H   (d) 0.00005 mH

2. An inductance of 0.33 mH is smaller than

   (a) 33 μH   (b) 330 μH   (c) 0.05 mH   (d) 0.0005 H

3. When the current through an inductor increases, the amount of energy stored in the electromagnetic field

   (a) decreases   (b) remains constant   (c) increases   (d) doubles

4. When the current through an inductor doubles, the stored energy

   (a) doubles   (b) quadruples   (c) is halved   (d) does not change

5. The winding resistance of a coil can be decreased by

   (a) reducing the number of turns   (b) using a larger wire

   (c) changing the core material   (d) either answer (a) or (b)

6. The inductance of an iron-core coil increases if

   (a) the number of turns is increased   (b) the iron core is removed

   (c) the length of the core is increased   (d) larger wire is used

7. Four 10 mH inductors are in series. The total inductance is

   (a) 40 mH   (b) 2.5 mH   (c) 40,000 μH   (d) answers (a) and (c)

8. A 1 mH, a 3.3 mH, and a 0.1 mH inductor are connected in parallel. The total inductance is

   (a) 4.4 mH   (b) greater than 3.3 mH

   (c) less than 0.1 mH   (d) answers (a) and (b)

9. An inductor, a resistor, and a switch are connected in series to a 12 V battery. At the instant the switch is closed, the inductor voltage is

   (a) 0 V   (b) 12 V   (c) 6 V   (d) 4 V

10. A sinusoidal voltage is applied across an inductor. When the frequency of the voltage is increased, the current

    (a) decreases   (b) increases
    (c) does not change   (d) momentarily goes to zero

11. An inductor and a resistor are in series with a sinusoidal voltage source. The frequency is set so that the inductive reactance is equal to the resistance. If the frequency is increased, then

    (a) $V_R > V_L$ (b) $V_L < V_R$ (c) $V_L = V_R$ (d) $V_L > V_R$

12. An ohmmeter is connected across an inductor and the pointer indicates an infinite value. The inductor is

    (a) good   (b) open   (c) shorted   (d) resistive

CIRCUIT DYNAMICS QUIZ

Answers are at the end of the chapter.

Refer to Figure 13–45.

1. The switch is in position 1. When it is thrown into position 2, the inductance between A and B

   (a) increases   (b) decreases   (c) stays the same

2. If the switch is moved from position 3 to position 4, the inductance between A and B

   (a) increases   (b) decreases   (c) stays the same

Refer to Figure 13–48.

3. If R were 10 kΩ instead of 1.0 kΩ and the switch is closed, the time it takes for the current to reach its maximum value

   (a) increases   (b) decreases   (c) stays the same

4. If L is decreased from 10 mH to 1 mH and the swtich is closed, the time constant
   (a) increases   (b) decreases   (c) stays the same

5. If the source voltage drops from +15 V to +10 V, the time constant
   (a) increases   (b) decreases   (c) stays the same

Refer to Figure 13–51.

6. If the frequency of the source voltage is increased, the total current
   (a) increases   (b) decreases   (c) stays the same

7. If L_2 opens, the current through L_1
   (a) increases   (b) decreases   (c) stays the same

8. If the frequency of the source voltage is decreased, the ratio of the values of the currents through L_2 and L_3
   (a) increases   (b) decreases   (c) stays the same

Refer to Figure 13–52.

9. If the frequency of the source voltage increases, the voltage across L_1
   (a) increases   (b) decreases   (c) stays the same

10. If L_3 opens, the voltage across L_2
    (a) increases   (b) decreases   (c) stays the same

PROBLEMS

More difficult problems are indicated by an asterisk (*).
Answers to odd-numbered problems are at the end of the book.

SECTION 13–1   The Basic Inductor

1. Convert the following to millihenries:
   (a) 1 H   (b) 250 μH   (c) 10 μH   (d) 0.0005 H

2. Convert the following to microhenries:
   (a) 300 mH   (b) 0.08 H   (c) 5 mH   (d) 0.00045 mH

3. What is the voltage across a coil when $di/dt = 10\,\text{mA}/\mu\text{s}$ and $L = 5\,\mu\text{H}$?

4. Fifty volts are induced across a 25 mH coil. At what rate is the current changing?

5. The current through a 100 mH coil is changing at a rate of 200 mA/s. How much voltage is induced across the coil?

6. How many turns are required to produce 30 mH with a coil wound on a cylindrical core having a cross-sectional area of $10 \times 10^{-5}\,\text{m}^2$ and a length of 0.05 m? The core has a permeability of $1.2 \times 10^{-6}\,\text{H/m}$.

7. What amount of energy is stored in a 4.7 mH inductor when the current is 20 mA?

8. Compare the inductance of two inductors that are identical except that inductor 2 has twice the number of turns as inductor 1.

9. Compare the inductance of two inductors that are identical except that inductor 2 is wound on an iron coil (relative permeability = 150) and inductor 1 is wound on a low carbon steel core (relative permeability = 200).

10. A student wraps 100 turns of wire on a pencil that is 7 mm in diameter as shown in Figure 13–43. The pencil is a nonmagnetic core so has the same permeability as a vacuum ($4\pi \times 10^{-6}\,\text{H/m}$). Determine the inductance of the coil that is formed.

▶ FIGURE 13–43

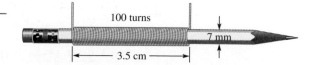

100 turns

7 mm

3.5 cm

SECTION 13–3 **Series and Parallel Inductors**

11. Five inductors are connected in series. The lowest value is 5 μH. If the value of each inductor is twice that of the preceding one, and if the inductors are connected in order of ascending values, what is the total inductance?

12. Suppose that you require a total inductance of 50 mH. You have available a 10 mH coil and a 22 mH coil. How much additional inductance do you need?

13. Determine the total inductance in Figure 13–44.

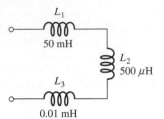

▲ FIGURE 13–44

14. What is the total inductance between points A and B for each switch position in Figure 13–45?

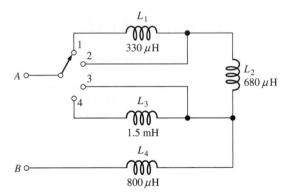

▲ FIGURE 13–45

15. Determine the total parallel inductance for the following coils in parallel: 75 μH, 50 μH, 25 μH, and 15 μH.

16. You have a 12 mH inductor, and it is your smallest value. You need an inductance of 8 mH. What value can you use in parallel with the 12 mH to obtain 8 mH?

17. Determine the total inductance of each circuit in Figure 13–46.

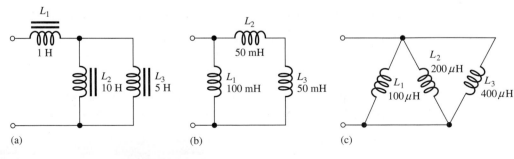

(a)   (b)   (c)

▲ FIGURE 13–46

18. Determine the total inductance of each circuit in Figure 13–47.

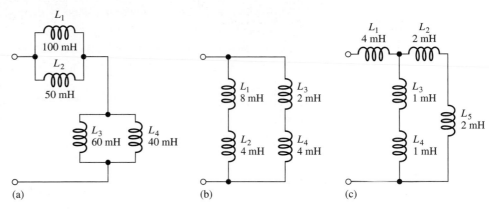

(a)   (b)   (c)

▲ FIGURE 13–47

SECTION 13–4 Inductors in DC Circuits

19. Determine the time constant for each of the following series RL combinations:

   (a) $R = 100 \, \Omega, L = 100 \, \mu H$ (b) $R = 4.7 \, k\Omega, L = 10 \, mH$

   (c) $R = 1.5 \, M\Omega, L = 3 \, H$

20. In a series RL circuit, determine how long it takes the current to build up to its full value for each of the following:

   (a) $R = 56 \, \Omega, L = 50 \, \mu H$ (b) $R = 3300 \, \Omega, L = 15 \, mH$

   (c) $R = 22 \, k\Omega, L = 100 \, mH$

21. In the circuit of Figure 13–48, there is initially no current. Determine the inductor voltage at the following times after the switch is closed:

   (a) $10 \, \mu s$ (b) $20 \, \mu s$ (c) $30 \, \mu s$ (d) $40 \, \mu s$ (e) $50 \, \mu s$

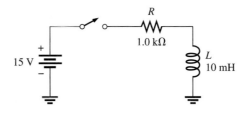

▲ FIGURE 13–48

*22. For the ideal inductor in Figure 13–49, calculate the current at each of the following times:

   (a) $10 \, \mu s$ (b) $20 \, \mu s$ (c) $30 \, \mu s$

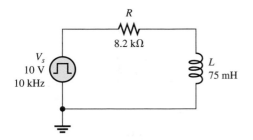
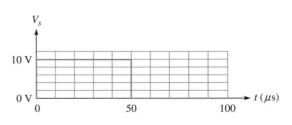

▲ FIGURE 13–49

23. Repeat Problem 21 for the following times:

(a) 2 μs (b) 5 μs (c) 15 μs

*24. Repeat Problem 22 for the following times:

(a) 65 μs (b) 75 μs (c) 85 μs

25. In Figure 13–48, at what time after switch closure does the inductor voltage reach 5 V?

26. (a) What is the polarity of the induced voltage across the inductor in Figure 13–49 when the square wave is rising?

(b) What is the current just before the square wave drops to zero?

27. Determine the time constant for the circuit in Figure 13–50.

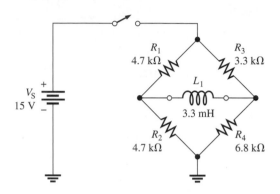

▲ FIGURE 13–50

*28. (a) What is the current in the inductor 1.0 μs after the switch closes in Figure 13–50?

(b) What is the current after 5τ have elapsed?

*29. For the circuit in Figure 13–50, assume the switch has been closed for more than 5τ and is opened. What is the current in the inductor 1.0 μs after the switch is opened?

SECTION 13–5   Inductors in AC Circuits

30. Find the total reactance for each circuit in Figure 13–46 when a voltage with a frequency of 5 kHz is applied across the terminals.

31. Find the total reactance for each circuit in Figure 13–47 when a 400 Hz voltage is applied.

32. Determine the total rms current in Figure 13–51. What are the currents through L_2 and L_3?

33. What frequency will produce 500 mA total rms current in each circuit of Figure 13–47 with an rms input voltage of 10 V?

34. Determine the reactive power in Figure 13–51.

35. Determine I_{L2} in Figure 13–52.

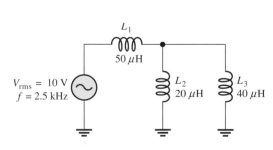

▲ FIGURE 13–51

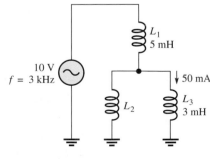

▲ FIGURE 13–52

Multisim Troubleshooting and Analysis

These problems require your Multisim CD-ROM.

36. Open file P13-36 and measure the voltage across each inductor.

37. Open file P13-37 and measure the voltage across each inductor.

38. Open file P13-38 and measure the current. Double the frequency and measure the current again. Reduce the original frequency by one-half and measure the current. Explain your observations.

39. Open file P13-39 and determine the fault if there is one.

40. Open file P13-40 and find the fault if there is one.

ANSWERS

SECTION REVIEWS

SECTION 13–1 The Basic Inductor

1. Inductance depends on number of turns of wire, permeability, cross-sectional area, and core length.

2. $v_{\text{ind}} = 7.5 \, \text{mV}$

3. (a) L increases when N increases.

   (b) L decreases when core length increases.

   (c) L decreases when core cross-sectional area decreases.

   (d) L decreases when ferromagnetic core is replaced by air core.

4. All wire has some resistance, and because inductors are made from turns of wire, there is always resistance.

5. Adjacent turns in a coil act as plates of a capacitor.

SECTION 13–2 Types of Inductors

1. Two categories of inductors are fixed and variable.

2. (a) air core   (b) iron core   (c) variable

SECTION 13–3 Series and Parallel Inductors

1. Inductances are added in series.

2. $L_T = 2.60 \, \text{mH}$

3. $L_T = 5(100 \, \text{mH}) = 500 \, \text{mH}$

4. The total parallel inductance is smaller than that of the smallest-value inductor in parallel.

5. True, calculation of parallel inductance is similar to parallel resistance.

6. (a) $L_T = 24 \, \mu\text{H}$ (b) $L_T = 7.69 \, \text{mH}$

SECTION 13–4 Inductors in DC Circuits

1. $V_L = I R_W = 100 \, \text{mV}$

2. $i = 0 \, \text{V}, v_L = 20 \, \text{V}$

3. $v_L = 0 \, \text{V}$

4. $\tau = 500 \, \text{ns}, i_L = 3.93 \, \text{mA}$

SECTION 13–5 Inductors in AC Circuits

1. Voltage leads current by 90 degrees in an inductor.

2. $X_L = 2\pi f L = 3.14 \, \text{k}\Omega$

3. $f = X_L/2\pi L = 2.55 \, \text{MHz}$

4. $I_{\text{rms}} = 15.9 \, \text{mA}$

5. $P_{\text{true}} = 0 \, \text{W}; P_r = 458 \, \text{mVAR}$

SECTION 13–6 **Inductor Applications**

1. Conductive and radiated

2. Electromagnetic interference

3. A ferrite bead is placed on a wire to increase its inductance, creating an RF choke.

A Circuit Application

1. $f_{max} = 125\,kHz\ (5\tau = 4\,ms)$

2. $f_{max} = 3.13\,kHz\ (5\tau = 160\,\mu s)$

3. If $f > f_{max}$, the inductor will not fully energize because $T/2 < 5\tau$.

RELATED PROBLEMS FOR EXAMPLES

13–1   5 H

13–2   10.1 mH

13–3   0.5 Wb/s

13–4   150 μH

13–5   20.3 μH

13–6   227 ns

13–7 $I_F = 17.6\,mA, \tau = 147\,ns$

at 1τ: $i = 11.1\,mA$; $t = 147\,ns$

at 2τ: $i = 15.1\,mA$; $t = 294\,ns$

at 3τ: $i = 16.7\,mA$; $t = 441\,ns$

at 4τ: $i = 17.2\,mA$; $t = 588\,ns$

at 5τ: $i = 17.4\,mA$; $t = 735\,ns$

13–8   at 0.2 ms, $i = 0.215\,mA$

at 0.8 ms, $i = 0.035\,mA$

13–9   10 V

13–10   3.83 mA

13–11   6.7 mA

13–12   1.1 kΩ

13–13   573 mA

13–14 P_r decreases.

SELF-TEST

1. (c)   **2.** (d)   **3.** (c)   **4.** (b)   **5.** (d)   **6.** (a)   **7.** (d)   **8.**(c)

9. (b)   **10.** (a)   **11.** (d)   **12.** (b)

CIRCUIT DYNAMICS QUIZ

1. (a)   **2.** (b)   **3.** (b)   **4.** (b)   **5.** (b)

6. (b)   **7.** (b)   **8.** (c)   **9.** (c)   **10.** (a)

TRANSFORMERS

14

CHAPTER OBJECTIVES

◆ Explain mutual inductance
◆ Describe how a transformer is constructed and how it operates
◆ Describe how transformers increase and decrease voltage
◆ Discuss the effect of a resistive load across the secondary winding
◆ Discuss the concept of a reflected load in a transformer
◆ Discuss impedance matching with transformers
◆ Describe a nonideal transformer
◆ Describe several types of transformers
◆ Troubleshoot transformers

KEY TERMS

◆ Mutual inductance (L_M)
◆ Transformer
◆ Primary winding
◆ Secondary winding
◆ Magnetic coupling
◆ Turns ratio (n)
◆ Reflected resistance
◆ Impedance matching
◆ Apparent power rating
◆ Center tap (CT)

A CIRCUIT APPLICATION PREVIEW

In the circuit application you will learn to troubleshoot a type of dc power supply that uses a transformer to couple the ac voltage from a standard electrical outlet. By making voltage measurements at various points, you can determine if there is a fault and be able to specify the part of the power supply that is faulty.

VISIT THE COMPANION WEBSITE

Study aids for this chapter are available at
http://www.prenhall.com/floyd

INTRODUCTION

In Chapter 13, you learned about self-inductance. In this chapter, you will study mutual inductance, which is the basis for the operation of transformers. Transformers are used in all types of applications such as power supplies, electrical power distribution, and signal coupling in communications systems.

The operation of the transformer is based on the principle of mutual inductance, which occurs when two or more coils are in close proximity. A simple transformer is actually two coils that are electromagnetically coupled by their mutual inductance. Because there is no electrical contact between two magnetically coupled coils, the transfer of energy from one coil to the other can be achieved in a situation of complete electrical isolation. In relation to transformers, the term *winding* or *coil* is commonly used to describe the primary and secondary.

14–1 MUTUAL INCTANCE

When two coils are placed close to each other, a changing electromagnetic field produced by the current in one coil will cause an induced voltage in the second coil because of the mutual inductance between the two coils.

After completing this section, you should be able to

- ◆ **Explain mutual inductance**
    - ◆ Discuss magnetic coupling
    - ◆ Define *electrical isolation*
    - ◆ Define *coefficient of coupling*
    - ◆ Identify the factors that affect mutual inductance and state the formula

Recall from Chapter 10 that the electromagnetic field surrounding a coil of wire expands, collapses, and reverses as the current increases, decreases, and reverses.

When a second coil is placed very close to the first coil so that the changing magnetic lines of force cut through the second coil, the coils are magnetically coupled and a voltage is induced, as indicated in Figure 14–1. When two coils are magnetically coupled, they provide **electrical isolation** because there is no electrical connection between them, only a magnetic link. If the current in the first coil is sinusoidal, the voltage induced in the second coil is also sinusoidal. The amount of voltage induced in the second coil as a result of the current in the first coil is dependent on the **mutual inductance, (L_M)**, which is the inductance between the two coils. The mutual inductance is established by the inductance of each coil and by the amount of coupling (k) between the two coils. To maximize coupling, the two coils are wound on a common core.

▶ **FIGURE 14–1**

A voltage is induced in the second coil as a result of the changing current in the first coil, producing a changing electromagnetic field that links the second coil.

Lines of force cutting second coil as the electromagnetic field expands, collapses, and reverses in response to the sinusoidal input.

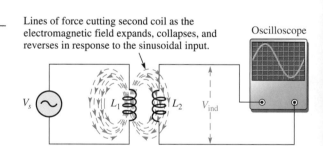

Coefficient of Coupling

The **coefficient of coupling, k**, between two coils is the ratio of the magnetic lines of force (flux) produced by coil 1 linking coil 2 (ϕ_{1-2}) to the total flux produced by coil 1 (ϕ_1).

Equation 14–1

$$k = \frac{\phi_{1-2}}{\phi_1}$$

For example, if half of the total flux produced by coil 1 links coil 2, then $k = 0.5$. A greater value of k means that more voltage is induced in coil 2 for a certain rate of change of current in coil 1. Note that k has no units. Recall that the unit of magnetic lines of force (flux) is the weber, abbreviated Wb.

The coefficient of coupling, k, depends on the physical closeness of the coils and the type of core material on which they are wound. Also, the construction and shape of the cores are factors.

Formula for Mutual Inductance

The three factors influencing mutual inductance (k, L_1, and L_2) are shown in Figure 14–2. The formula for mutual inductance is

$$L_M = k\sqrt{L_1 L_2}$$

Equation 14–2

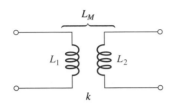

◀ FIGURE 14–2

The mutual inductance of two coils.

EXAMPLE 14–1

One coil produces a total magnetic flux of 50 μWb, and 20 μWb link coil 2. What is the coefficient of coupling, k?

Solution

$$k = \frac{\phi_{1\text{-}2}}{\phi_1} = \frac{20\,\mu\text{Wb}}{50\,\mu\text{Wb}} = \mathbf{0.4}$$

*Related Problem** Determine k when $\phi_1 = 500\,\mu$Wb and $\phi_{1\text{-}2} = 375\,\mu$Wb.

*Answers are at the end of the chapter.

EXAMPLE 14–2

Two coils are wound on a single core, and the coefficient of coupling is 0.3. The inductance of coil 1 is 10 μH, and the inductance of coil 2 is 15 μH. What is L_M?

Solution

$$L_M = k\sqrt{L_1 L_2} = 0.3\sqrt{(10\,\mu\text{H})(15\,\mu\text{H})} = \mathbf{3.67\,\mu H}$$

Related Problem Determine the mutual inductance when $k = 0.5$, $L_1 = 1$ mH, and $L_2 = 600\,\mu$H.

SECTION 14–1 REVIEW

Answers are at the end of the chapter.

1. Define *mutual inductance*.
2. Two 50 mH coils have $k = 0.9$. What is L_M?
3. If k is increased, what happens to the voltage induced in one coil as a result of a current change in the other coil?

14–2 THE BASIC TRANSFORMER

A basic **transformer** is an electrical device constructed of two coils of wire (windings) magnetically coupled to each other so that there is a mutual inductance for the transfer of power from one winding to the other.

After completing this section, you should be able to

◆ **Describe how a transformer is constructed and how it operates**

  ◆ Identify the parts of a basic transformer

  ◆ Discuss the importance of the core material

- Define *primary winding* and *secondary winding*
- Define *turns ratio*
- Discuss how the direction of windings affects voltage polarities

A schematic of a transformer is shown in Figure 14–3(a). As shown, one coil is called the **primary winding**, and the other is called the **secondary winding**. The source voltage is applied to the primary winding, and the load is connected to the secondary winding, as shown in Figure 14–3(b). The primary winding is the input winding, and the secondary winding is the output winding. It is common to refer to the side of the transformer that has the source voltage as the *primary,* and the side that has the induced voltage as the *secondary.*

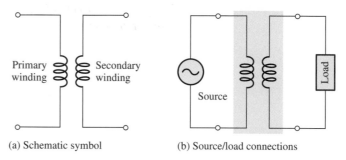

(a) Schematic symbol   (b) Source/load connections

▲ **FIGURE 14–3**

The basic transformer.

The windings of a transformer are formed around the core. The core provides both a physical structure for placement of the windings and a magnetic path so that the magnetic flux is concentrated close to the coils. There are three general categories of core material: air, ferrite, and iron. The schematic symbol for each type is shown in Figure 14–4.

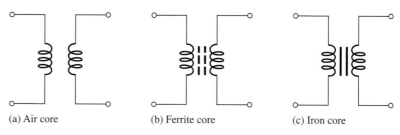

(a) Air core   (b) Ferrite core   (c) Iron core

▲ **FIGURE 14–4**

Schematic symbols based on type of core.

Air-core and ferrite-core transformers generally are used for high-frequency applications and consist of windings on an insulating shell which is hollow (air) or constructed of ferrite, such as depicted in Figure 14–5. The wire is typically covered by a varnish-type coating to prevent the windings from shorting together. The amount of **magnetic coupling** between the primary winding and the secondary winding is set by the type of core material and by the relative positions of the windings. In Figure 14–5(a), the windings are loosely coupled because they are separated, and in part (b) they are tightly coupled because they are overlapping. The tighter the coupling, the greater the induced voltage in the secondary for a given current in the primary.

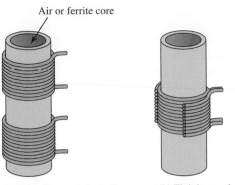

Air or ferrite core

(a) Loosely coupled windings

(b) Tightly coupled windings.
Cutaway view shows both
windings.

◀ FIGURE 14–5

Transformers with cylindrical-shaped
cores.

Iron-core transformers generally are used for audio frequency (AF) and power applications. These transformers consist of windings on a core constructed from laminated sheets of ferromagnetic material insulated from each other, as shown in Figure 14–6. This construction provides an easy path for the magnetic flux and increases the amount of coupling between the windings. Figure 14–6 shows the basic construction of two major configurations of iron-core transformers. In the core-type construction, shown in part (a), the windings are on separate legs of the laminated core. In the shell-type construction, shown in part (b), both windings are on the same leg. Each type has certain advantages. Generally, the core type has more room for insulation and can handle higher voltages. The shell type can produce higher magnetic fluxes in the core, resulting in the need for fewer turns. A variety of transformers is shown in Figure 14–7.

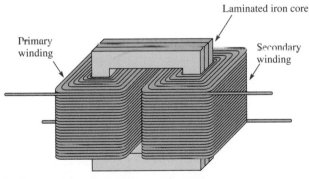
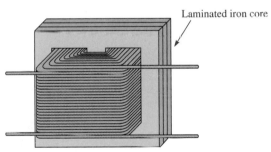

Laminated iron core

Primary
winding

Secondary
winding

Laminated iron core

(a) Core type has each winding on a separate leg.

(b) Shell type has both windings on the same leg.

▲ FIGURE 14–6

Iron-core transformer construction with multilayer windings.

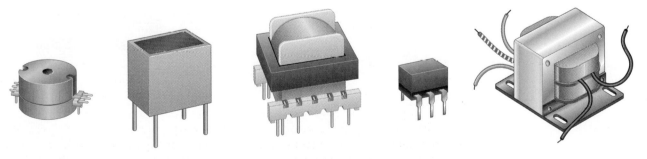

▲ FIGURE 14–7

Some common types of transformers.

Turns Ratio

A transformer parameter that is useful in understanding how a transformer operates is the turns ratio. In this text, the **turns ratio (n)** is defined as the ratio of the number of turns in the secondary winding (N_{sec}) to the number of turns in the primary winding (N_{pri}).

Equation 14–3

$$n = \frac{N_{sec}}{N_{pri}}$$

This definition of turns ratio is based on the IEEE standard for electronics power transformers as specified in the IEEE dictionary. Other categories of transformer may have a different definition, so some sources define the turns ratio as N_{pri}/N_{sec}. Either definition is correct as long as it is clearly stated and used consistently. The turns ratio of a transformer is rarely if ever given as a transformer specification. Generally, the input and output voltages and the power rating are the key specifications. However, the turns ratio is useful in studying the operating principle of a transformer.

EXAMPLE 14–3

A transformer primary winding has 100 turns, and the secondary winding has 400 turns. What is the turns ratio?

Solution N_{sec} = 400 and N_{pri} = 100; therefore, the turns ratio is

$$n = \frac{N_{sec}}{N_{pri}} = \frac{400}{100} = \textbf{4}$$

Related Problem   A certain transformer has a turns ratio of 10. If N_{pri} = 500, what is N_{sec}?

Direction of Windings

Another important transformer parameter is the direction in which the windings are placed around the core. As illustrated in Figure 14–8, the direction of the windings determines the polarity of the voltage across the secondary winding (secondary voltage) with respect to the voltage across the primary winding (primary voltage). Phase dots are sometimes used on the schematic symbols to indicate polarities, as shown in Figure 14–9.

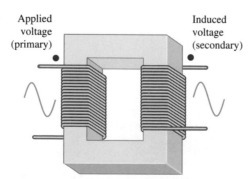
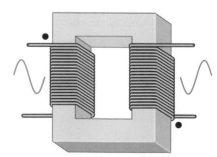

Applied voltage (primary)   Induced voltage (secondary)

(a) The primary and secondary voltages are in phase when the windings are in the same effective direction around the magnetic path.

(b) The primary and secondary voltages are 180° out of phase when the windings are in the opposite direction.

▲ **FIGURE 14–8**

The direction of the windings determines the relative polarities of the voltages.

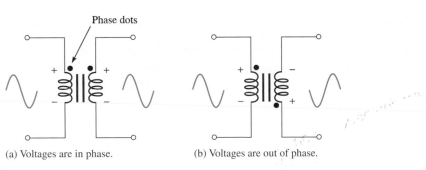

(a) Voltages are in phase.   (b) Voltages are out of phase.

▲ FIGURE 14–9

Phase dots indicate relative polarities of primary and secondary voltages.

SECTION 14–2 REVIEW	1. Upon what principle is the operation of a transformer based? 2. Define *turns ratio*. 3. Why are the directions of the windings of a transformer important? 4. A certain transformer has a primary winding with 500 turns and a secondary winding with 250 turns. What is the turns ratio?

14–3 STEP-UP AND STEP-DOWN TRANSFORMERS

A step-up transformer has more turns in its secondary winding than in its primary winding and is used to increase ac voltage. A step-down transformer has more turns in its primary winding than in its secondary winding and is used to decrease ac voltage.

After completing this section, you should be able to

◆ **Describe how transformers increase and decrease voltage**

  ◆ Explain how a step-up transformer works

  ◆ Identify a step-up transformer by its turns ratio

  ◆ State the relationship between primary and secondary voltages and the turns ratio

  ◆ Explain how a step-down transformer works

  ◆ Identify a step-down transformer by its turns ratio

  ◆ Describe dc isolation

The Step-Up Transformer

A transformer in which the secondary voltage is greater than the primary voltage is called a **step-up transformer**. The amount that the voltage is stepped up depends on the turns ratio.

The ratio of secondary voltage (V_{sec}) to primary voltage (V_{pri}) is equal to the ratio of the number of turns in the secondary winding (N_{sec}) to the number of turns in the primary winding (N_{pri}).

$$\frac{V_{sec}}{V_{pri}} = \frac{N_{sec}}{N_{pri}}$$

Equation 14–4

Recall that N_{sec}/N_{pri} defines the turns ratio, n. Therefore, from this relationship,

Equation 14–5

$$V_{sec} = nV_{pri}$$

Equation 14–5 shows that the secondary voltage is equal to the turns ratio times the primary voltage. This condition assumes that the coefficient of coupling is 1, and a good iron-core transformer approaches this value.

The turns ratio for a step-up transformer is always greater than 1 because the number of turns in the secondary winding (N_{sec}) is always greater than the number of turns in the primary winding (N_{pri}).

EXAMPLE 14–4

The transformer in Figure 14–10 has a turns ratio of 3. What is the voltage across the secondary winding?

▶ **FIGURE 14–10**

Solution   The secondary voltage is

$$V_{sec} = nV_{pri} = (3)120 \text{ V} = \mathbf{360 \text{ V}}$$

Note that the turns ratio of 3 is indicated on the schematic as 1:3, meaning that there are three secondary turns for each primary turn.

Related Problem   The transformer in Figure 14–10 is changed to one with a turns ratio of 4. Determine V_{sec}.

Use Multisim file E14-04 to verify the calculated results in this example and to confirm your calculation for the related problem.

The Step-Down Transformer

A transformer in which the secondary voltage is less than the primary voltage is called a **step-down transformer**. The amount by which the voltage is stepped down depends on the turns ratio. Equation 14–5 applies also to a step-down transformer.

The turns ratio of a step-down transformer is always less than 1 because the number of turns in the secondary winding is always fewer than the number of turns in the primary winding.

EXAMPLE 14–5

The transformer in Figure 14–11 has a turns ratio of 0.2. What is the secondary voltage?

▶ **FIGURE 14–11**

Solution   The secondary voltage is

$$V_{sec} = nV_{pri} = (0.2)120 \text{ V} = \mathbf{24 \text{ V}}$$

Related Problem   The transformer in Figure 14–11 is changed to one with a turns ratio of 0.48. Determine the secondary voltage.

Use Multisim file E14-05 to verify the calculated results in this example and to confirm your calculation for the related problem.

DC Isolation

As illustrated in Figure 14–12(a), if there is a direct current in the primary of a transformer, nothing happens in the secondary. The reason is that a changing current in the primary winding is necessary to induce a voltage in the secondary winding, as shown in part (b). Therefore, the transformer isolates the secondary circuit from any dc voltage in the primary circuit. A transformer that is used strictly for isolation has a turns ratio of 1.

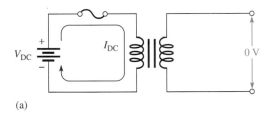

(a)

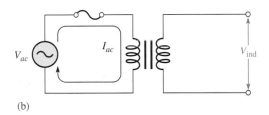

(b)

▲ FIGURE 14–12

DC isolation and ac coupling.

In a typical high frequency application, a small transformer can be used to keep the dc voltage on the output of an amplifier stage from affecting the dc bias of the next amplifier stage. Only the ac signal is coupled through the transformer from one stage to the next, as Figure 14–13 illustrates.

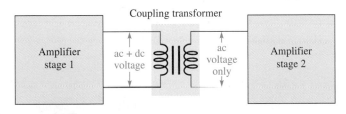

▲ FIGURE 14–13

Amplifier stages with transformer coupling for dc isolation.

SECTION 14–3 REVIEW

1. What does a step-up transformer do?
2. If the turns ratio is 5, how much greater is the secondary voltage than the primary voltage?
3. When 240 V ac are applied to the primary winding of a transformer with a turns ratio of 10, what is the secondary voltage?
4. What does a step-down transformer do?
5. A voltage of 120 V ac is applied to the primary winding of a transformer with a turns ratio of 0.5. What is the secondary voltage?
6. A primary voltage of 120 V ac is reduced to 12 V ac. What is the turns ratio?
7. What does the term *electrical isolation* mean?

14–4 LOADING THE SECONDARY WINDING

When a resistive load is connected to the secondary winding of a transformer, the relationship of the load (secondary) current and the current in the primary circuit is determined by the turns ratio.

After completing this section, you should be able to

- ◆ **Discuss the effect of a resistive load across the secondary winding**
  - ◆ Determine the current delivered by the secondary when a step-up transformer is loaded
  - ◆ Determine the current delivered by the secondary when a step-down transformer is loaded
  - ◆ Discuss power in a transformer

When a load resistor is connected to the secondary winding, as shown in Figure 14–14, there is current through the resulting secondary circuit because of the voltage induced in the secondary coil. It can be shown that the ratio of the primary current, I_{pri}, to the secondary current, I_{sec}, is equal to the turns ratio, as expressed in the following equation:

Equation 14–6

$$\frac{I_{pri}}{I_{sec}} = n$$

▶ **FIGURE 14–14**

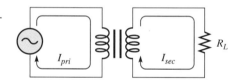

A manipulation of the terms in Equation 14–6 gives Equation 14–7, which shows that I_{sec} is equal to I_{pri} times the reciprocal of the turns ratio.

Equation 14–7

$$I_{sec} = \left(\frac{1}{n}\right)I_{pri}$$

Thus, for a step-up transformer, in which n is greater than 1, the secondary current is less than the primary current. For a step-down transformer, n is less than 1, and I_{sec} is greater than I_{pri}. When the secondary voltage is greater than the primary voltage, the secondary current is lower than the primary current and vice versa.

EXAMPLE 14–6

The two transformers in Figure 14–15 have loaded secondary windings. If the primary current is 100 mA in each case, what is the load current?

FIGURE 14–15

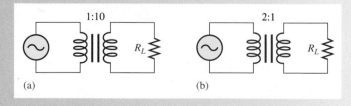

(a)   (b)

Solution   In Figure 14–15(a), the turns ratio is 10. The current through the load is

$$I_{sec} = \left(\frac{1}{n}\right)I_{pri} = (0.1)100 \text{ mA} = \textbf{10 mA}$$

In Figure 14–15(b), the turns ratio is 0.5. The current through the load is

$$I_{sec} = \left(\frac{1}{n}\right)I_{pri} = (2)100 \text{ mA} = \textbf{200 mA}$$

Related Problem   What is the secondary current in Figure 14–15(a) if the turns ratio is doubled? What is the secondary current in Figure 14–15(b) if the turns ratio is halved? Assume I_{pri} remains the same in both circuits.

Primary Power Equals Load Power

When a load is connected to the secondary winding of a transformer, the power transferred to the load can never be greater than the power in the primary winding. For an ideal transformer, the power delivered to the primary equals the power delivered by the secondary to the load. When losses are considered, some of the power is dissipated in the transformer rather than the load; therefore, the load power is always less than the power in the primary.

Power is dependent on voltage and current, and there can be no increase in power in a transformer. Therefore, if the voltage is stepped up, the current is stepped down and vice versa. In an ideal transformer, the secondary power is equal to the primary power regardless of the turns ratio, as the following equations show. The power delivered to the primary is

$$P_{pri} = V_{pri}I_{pri}$$

and the power delivered to the load is

$$P_{sec} = V_{sec}I_{sec}$$

From Equations 14–7 and 14–5,

$$I_{sec} = \left(\frac{1}{n}\right)I_{pri} \quad \text{and} \quad V_{sec} = nV_{pri}$$

By substitution,

$$P_{sec} = \left(\frac{1}{\cancel{n}}\right)\cancel{n}V_{pri}I_{pri}$$

Canceling terms yields

$$P_{sec} = V_{pri}I_{pri} = P_{pri}$$

This result is closely approached in practice by power transformers because of the very high efficiencies.

SECTION 14–4
REVIEW

1. If the turns ratio of a transformer is 2, is the secondary current greater than or less than the primary current? By how much?

2. A transformer has 1000 turns in its primary winding and 250 turns in its secondary winding, and I_{pri} is 0.5 A. What is the value of I_{sec}?

3. In Problem 2, how much primary current is necessary to produce a secondary load current of 10 A?

14–5 REFLECTED LOAD

From the viewpoint of the primary, a load connected across the secondary winding of a transformer appears to have a resistance that is not necessarily equal to the actual resistance of the load. The actual load is "reflected" into the primary as determined by the turns ratio. This reflected load is what the primary source effectively sees, and it determines the amount of primary current.

After completing this section, you should be able to

♦ **Discuss the concept of a reflected load in a transformer**

♦ Define *reflected resistance*

♦ Explain how the turns ratio affects the reflected resistance

♦ Calculate reflected resistance

The concept of the **reflected load** is illustrated in Figure 14–16. The load (R_L) in the secondary of a transformer is reflected into the primary by transformer action. The load appears to the source in the primary to be a resistance (R_{pri}) with a value determined by the turns ratio and the actual value of the load resistance. The resistance R_{pri} is called the **reflected resistance**.

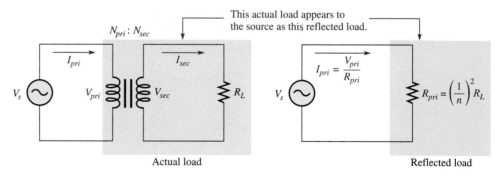

▲ **FIGURE 14–16**

Reflected load in a transformer circuit.

The resistance in the primary of Figure 14–16 is $R_{pri} = V_{pri}/I_{pri}$. The resistance in the secondary is $R_L = V_{sec}/I_{sec}$. From Equations 14–4 and 14–6, you know that $V_{sec}/V_{pri} = n$ and $I_{pri}/I_{sec} = n$. Using these relationships, a formula for R_{pri} in terms of R_L is determined as follows:

$$\frac{R_{pri}}{R_L} = \frac{V_{pri}/I_{pri}}{V_{sec}/I_{sec}} = \left(\frac{V_{pri}}{V_{sec}}\right)\left(\frac{I_{sec}}{I_{pri}}\right) = \left(\frac{1}{n}\right)\left(\frac{1}{n}\right) = \left(\frac{1}{n}\right)^2$$

Solving for R_{pri} yields

Equation 14–8

$$R_{pri} = \left(\frac{1}{n}\right)^2 R_L$$

Equation 14–8 shows that the resistance reflected into the primary circuit is the square of the reciprocal of the turns ratio times the load resistance.

EXAMPLE 14–7

Figure 14–17 shows a source that is transformer-coupled to a load resistor of 100 Ω. The transformer has a turns ratio of 4. What is the reflected resistance seen by the source?

▶ **FIGURE 14–17**

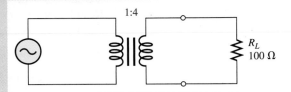

Solution   The reflected resistance is determined by Equation 14–8.

$$R_{pri} = \left(\frac{1}{n}\right)^2 R_L = \left(\frac{1}{4}\right)^2 R_L = \left(\frac{1}{16}\right)100\ \Omega = \mathbf{6.25\ \Omega}$$

The source sees a resistance of 6.25 Ω just as if it were connected directly, as shown in the equivalent circuit of Figure 14–18.

▶ **FIGURE 14–18**

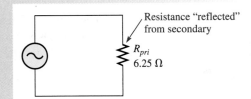

Related Problem   If the turns ratio in Figure 14–17 is 10 and R_L is 600 Ω, what is the reflected resistance?

EXAMPLE 14–8

In Figure 14–17, if a transformer is used having a turns ratio of 0.25, what is the reflected resistance?

Solution   The reflected resistance is

$$R_{pri} = \left(\frac{1}{n}\right)^2 R_L = \left(\frac{1}{0.25}\right)^2 100\ \Omega = (4)^2 100\ \Omega = \mathbf{1600\ \Omega}$$

This result illustrates the difference that the turns ratio makes.

Related Problem   To achieve a reflected resistance of 800 Ω, what turns ratio is required in Figure 14–17?

In a step-up transformer ($n > 1$), the reflected resistance is less than the actual load resistance; in a step-down transformer ($n < 1$), the reflected resistance is greater than the load resistance. This was illustrated in Examples 14–7 and 14–8, respectively.

SECTION 14–5 REVIEW

1. Define *reflected resistance*.
2. What transformer characteristic determines the reflected resistance?
3. A given transformer has a turns ratio of 10, and the load is 50 Ω. How much resistance is reflected into the primary?
4. What is the turns ratio required to reflect a 4 Ω load resistance into the primary as 400 Ω?

14–6 IMPEDANCE MATCHING

One application of transformers is in the matching of a load resistance to a source resistance to achieve maximum transfer of power. This technique is called impedance matching. Recall that the maximum power transfer theorem was studied in Chapter 8. In audio systems, special impedance matching transformers are often used to get the maximum amount of available power from the amplifier to the speaker.

After completing this section, you should be able to

◆ **Discuss impedance matching with transformers**

  ◆ Give a general definition of impedance

  ◆ Define *impedance matching*

  ◆ Explain the purpose of impedance matching

  ◆ Describe a practical application

Impedance is the opposition to current, including the effects of both resistance and reactance combined. We will confine our usage in this chapter to resistance only.

The concept of power transfer is illustrated in the basic circuit of Figure 14–19. Part (a) shows an ac voltage source with a series resistor representing its internal resistance. Some fixed internal resistance is inherent in all sources due to their internal circuitry or physical makeup. When the source is connected directly to a load, as shown in part (b), generally the objective is to transfer as much of the power produced by the source to the load as possible. However, a certain amount of the power produced by the source is dissipated in its internal resistance, and the remaining power goes to the load.

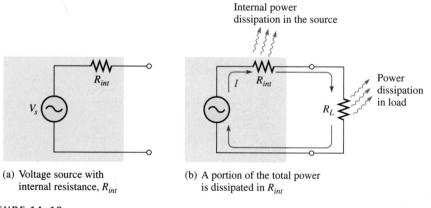

(a) Voltage source with internal resistance, R_{int}

(b) A portion of the total power is dissipated in R_{int}

▲ **FIGURE 14–19**

Power transfer from a nonideal voltage source to a load.

In most practical situations, the internal source resistance of various types of sources is fixed. Also, in many cases, the resistance of a device that acts as a load is fixed and cannot be altered. If you need to connect a given source to a given load, remember that only by chance will their resistances match. In this situation, a special type of wide band transformer comes in handy. You can use the reflected-resistance characteristic provided by a transformer to make the load resistance appear to have the same value as the source resistance. This technique is called **impedance matching**, and the transformer is called an impedance-matching transformer because it also transforms reactances as well as resistances.

Let's take a practical, everyday situation to illustrate the concept of impedance matching. Assume that the input resistance of a TV receiver is 300 Ω. An antenna must be connected to this input by a lead-in cable in order to receive TV signals. In this situation, the antenna and the lead-in act as the source, and the input resistance of the TV receiver is the load, as illustrated in Figure 14–20.

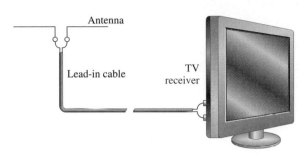
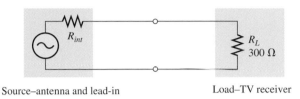

(a) The antenna/lead-in is the source; the TV input is the load.　　(b) Circuit equivalent of antenna and TV receiver system

▲ FIGURE 14–20

An antenna directly coupled to a TV receiver.

It is common for an antenna system to have a characteristic impedance of 75 Ω. This means that the antenna and the lead-in wire appear as a 75 Ω source. Thus, if the 75 Ω source (antenna and lead-in) is connected directly to the 300 Ω TV input, maximum power will not be delivered to the input of the TV, and you will have poor signal reception. The solution is to use a matching transformer, connected as indicated in Figure 14–21, in order to match the 300 Ω load resistance to the 75 Ω source resistance.

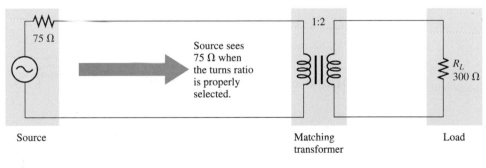

▲ FIGURE 14–21

Example of a load matched to a source by transformer coupling for maximum power transfer.

To match the resistances, that is, to reflect the load resistance (R_L) into the primary circuit so that it appears to have a value equal to the internal source resistance (R_{int}), you must select a proper value of turns ratio (n). You want the 300 Ω load to look like 75 Ω to the source. You can use Equation 14–8 to determine a formula for the turns ratio, n, when you know the values for R_L and R_{pri}, the reflected resistance, as follows:

$$R_{pri} = \left(\frac{1}{n}\right)^2 R_L$$

Transpose terms and divide both sides by R_L.

$$\left(\frac{1}{n}\right)^2 = \frac{R_{pri}}{R_L}$$

Take the square root of both sides.

$$\frac{1}{n} = \sqrt{\frac{R_{pri}}{R_L}}$$

Invert both sides to get the following formula for the turns ratio:

Equation 14–9

$$n = \sqrt{\frac{R_L}{R_{pri}}}$$

Finally, solve for this particular turns ratio.

$$n = \sqrt{\frac{R_L}{R_{pri}}} = \sqrt{\frac{300\ \Omega}{75\ \Omega}} = \sqrt{4} = 2$$

Therefore, a matching transformer with a turns ratio of 2 must be used in this application.

EXAMPLE 14–9

A certain amplifier has an 800 Ω internal resistance looking from its output. In order to provide maximum power to an 8 Ω speaker, what turns ratio must be used in the coupling transformer?

Solution The reflected resistance must equal 800 Ω. Thus, from Equation 14–9, you can determine the turns ratio.

$$n = \sqrt{\frac{R_L}{R_{pri}}} = \sqrt{\frac{8\ \Omega}{800\ \Omega}} = \sqrt{0.01} = \mathbf{0.1}$$

There must be ten primary turns for each secondary turn. The diagram and its equivalent reflected circuit are shown in Figure 14–22.

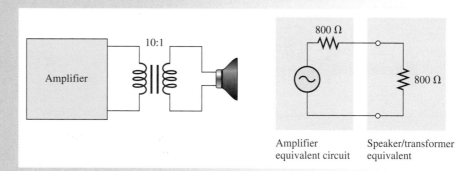

▲ FIGURE 14–22

Related Problem What must be the turns ratio in Figure 14–22 to provide maximum power to two 8 Ω speakers in parallel?

SECTION 14–6 REVIEW

1. What does impedance matching mean?
2. What is the advantage of matching the load resistance to the resistance of a source?
3. A transformer has a turns ratio of 0.5. What is the reflected resistance with 100 Ω across the secondary winding?

14–7 NONIDEAL TRANSFORMER CHARACTERISTICS

Transformer operation has been discussed from an ideal point of view. That is, the winding resistance, the winding capacitance, and nonideal core characteristics were all neglected and the transformer was treated as if it had an efficiency of 100%. For studying the basic concepts and in many applications, the ideal model is valid. However, the practical transformer has several nonideal characteristics.

After completing this section, you should be able to

◆ **Describe a nonideal transformer**

    ◆ List and describe the nonideal characteristics

    ◆ Explain power rating of a transformer

    ◆ Define *efficiency* of a transformer

Winding Resistance

Both the primary and the secondary windings of a practical transformer have winding resistance. You learned about the winding resistance of inductors in Chapter 13. The winding resistances of a practical transformer are represented as resistors in series with the windings as shown in Figure 14–23.

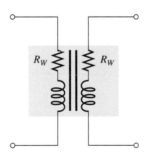

◀ **FIGURE 14–23**

Winding resistances in a practical transformer.

Winding resistance in a practical transformer results in less voltage across a secondary load. Voltage drops due to the winding resistance effectively subtract from the primary and secondary voltages and result in load voltage that is less than that predicted by the relationship $V_{sec} = nV_{pri}$. In many cases, the effect is relatively small and can be neglected.

Losses in the Core

There is always some energy conversion in the core material of a practical transformer. This conversion is seen as a heating of ferrite and iron cores, but it does not occur in air cores. Part of this energy conversion is because of the continuous reversal of the magnetic field due to the changing direction of the primary current; this component of the energy conversion is called *hysteresis loss*. The rest of the energy conversion to heat is caused by eddy currents produced when voltage is induced in the core material by the changing magnetic flux, according to Faraday's law. The eddy currents occur in circular patterns in the core resistance, thus producing heat. This conversion to heat is greatly reduced by the use of laminated construction of iron cores. The thin layers of ferromagnetic material are insulated from each other to minimize the buildup of eddy currents by confining them to a small area and to keep core losses to a minimum.

Magnetic Flux Leakage

In an ideal transformer, all of the magnetic flux produced by the primary current is assumed to pass through the core to the secondary winding, and vice versa. In a practical transformer, some of the magnetic flux lines break out of the core and pass through the surrounding air back to the other end of the winding, as illustrated in Figure 14–24 for the magnetic field produced by the primary current. Magnetic flux leakage results in a reduced secondary voltage.

▶ FIGURE 14–24

Flux leakage in a practical transformer.

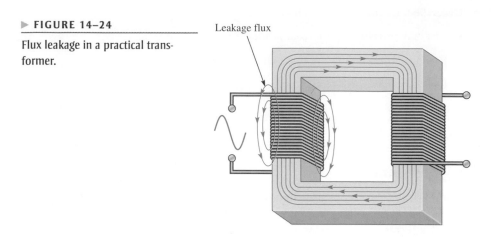
Leakage flux

The percentage of magnetic flux that actually reaches the secondary winding determines the coefficient of coupling of the transformer. For example, if nine out of ten flux lines remain inside the core, the coefficient of coupling is 0.90 or 90%. Most iron-core transformers have very high coefficients of coupling (greater than 0.99), while ferrite-core and air-core devices have lower values.

Winding Capacitance

As you learned in Chapter 13, there is always some stray capacitance between adjacent turns of a winding. These stray capacitances result in an effective capacitance in parallel with each winding of a transformer, as indicated in Figure 14–25.

▶ FIGURE 14–25

Winding capacitance in a practical transformer.

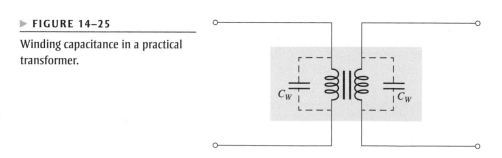

These stray capacitances have very little effect on the transformer's operation at low frequencies (such as power line frequencies) because the reactances (X_C) are very high. However, at higher frequencies, the reactances decrease and begin to produce a bypassing effect across the primary winding and across the secondary load. As a result, less of the total primary current is through the primary winding, and less of the total secondary current is through the load. This effect reduces the load voltage as the frequency goes up.

Transformer Power Rating

A power transformer is typically rated in volt-amperes (VA), primary/secondary voltage, and operating frequency. For example, a given transformer rating may be specified as 2 kVA, 500/50, 60 Hz. The 2 kVA value is the **apparent power rating**. The 500 and the 50 can be either secondary or primary voltages. The 60 Hz is the operating frequency.

The transformer rating can be helpful in selecting the proper transformer for a given application. Let's assume, for example, that 50 V is the secondary voltage. In this case the load current is

$$I_L = \frac{P_{sec}}{V_{sec}} = \frac{2\,kVA}{50\,V} = 40\,A$$

On the other hand, if 500 V is the secondary voltage, then

$$I_L = \frac{P_{sec}}{V_{sec}} = \frac{2\,kVA}{500\,V} = 4\,A$$

These are the maximum currents that the secondary can handle in either case.

The reason that the power rating is in volt-amperes (apparent power) rather than in watts (true power) is as follows: If the transformer load is purely capacitive or purely inductive, the true power (watts) delivered to the load is zero. However, the current for $V_{sec} = 500\,V$ and $X_C = 100\,\Omega$ at 60 Hz, for example, is 5 A. This current exceeds the maximum that the 2 kVA secondary can handle, and the transformer may be damaged. So it is meaningless to specify power in watts.

Transformer Efficiency

Recall that the power delivered to the load is equal to the power delivered to the primary in an ideal transformer. Because the nonideal characteristics just discussed result in a power loss in the transformer, the secondary (output) power is always less than the primary (input) power. The **efficiency (η)** of a transformer is a measure of the percentage of the input power that is delivered to the output.

$$\eta = \left(\frac{P_{out}}{P_{in}}\right)100\%$$

Equation 14–10

Most power transformers have efficiencies in excess of 95% under load.

EXAMPLE 14–10

A certain type of transformer has a primary current of 5 A and a primary voltage of 4800 V. The secondary current is 90 A, and the secondary voltage is 240 V. Determine the efficiency of this transformer.

Solution The input power is

$$P_{in} = V_{pri}I_{pri} = (4800\,V)(5\,A) = 24\,kVA$$

The output power is

$$P_{out} = V_{sec}I_{sec} = (240\,V)(90\,A) = 21.6\,kVA$$

The efficiency is

$$\eta = \left(\frac{P_{out}}{P_{in}}\right)100\% = \left(\frac{21.6\,kVA}{24\,kVA}\right)100\% = \mathbf{90\%}$$

Related Problem A transformer has a primary current of 8 A with a primary voltage of 440 V. The secondary current is 30 A and the secondary voltage is 100 V. What is the efficiency?

**SECTION 14–7
REVIEW**

1. Explain how a practical transformer differs from the ideal model.
2. The coefficient of coupling of a certain transformer is 0.85. What does this mean?
3. A certain transformer has a rating of 10 kVA. If the secondary voltage is 250 V, how much load current can the transformer handle?

14–8 TAPPED AND MULTIPLE-WINDING TRANSFORMERS

The basic transformer has several important variations. They include tapped transformers, multiple-winding transformers, and autotransformers.

After completing this section, you should be able to

◆ **Describe several types of transformers**

    ◆ Describe center-tapped transformers

    ◆ Describe multiple-winding transformers

    ◆ Describe autotransformers

Tapped Transformers

A schematic of a transformer with a center-tapped secondary winding is shown in Figure 14–26(a). The **center tap (CT)** is equivalent to two secondary windings with half the total voltage across each.

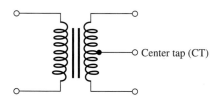

(a) Center-tapped transformer

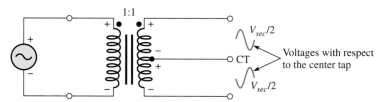

(b) Output voltages with respect to the center tap are 180° out of phase with each other and are one-half the magnitude of the secondary voltage.

▲ **FIGURE 14–26**

Operation of a center-tapped transformer.

The voltages between either end of the secondary winding and the center tap are, at any instant, equal in magnitude but opposite in polarity, as illustrated in Figure 14–26(b). Here, for example, at some instant on the sinusoidal voltage, the polarity across the entire secondary winding is as shown (top end +, bottom end −). At the center tap, the voltage is less positive than the top end but more positive than the bottom end of the secondary. Therefore, measured with respect to the center tap, the top end of the secondary is positive, and the bottom end is negative. This center-tapped feature is used in many power supply rectifiers in which the ac voltage is converted to dc, as illustrated in Figure 14–27.

Some tapped transformers have taps on the secondary winding at points other than the electrical center. Also, single and multiple primary and secondary taps are sometimes used in certain applications, such as impedance-matching transformers that normally have a center-tapped primary. Examples of these types of transformers are shown in Figure 14–28.

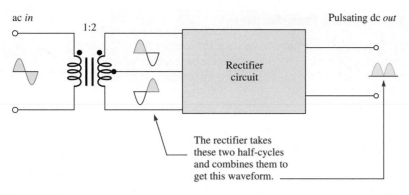

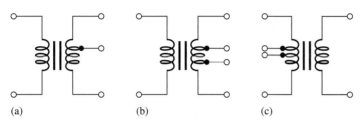

▲ FIGURE 14–27

Application of a center-tapped transformer in ac-to-dc conversion.

(a)   (b)   (c)

▲ FIGURE 14–28

Tapped transformers.

One example of a transformer with a multiple-tap primary winding and a center-tapped secondary winding is the utility-pole transformer used by power companies to step down the high voltage from the power line to 110 V/220 V service for residential and commercial customers, as shown in Figure 14–29. The multiple taps on the primary winding are used for minor adjustments in the turns ratio in order to overcome line voltages that are slightly too high or too low.

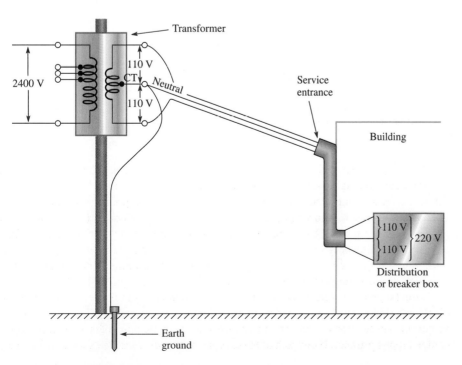

◀ FIGURE 14–29

Utility-pole transformer in a typical power distribution system.

Multiple-Winding Transformers

Some transformers are designed to operate from either 110 V ac or 220 V ac lines. These transformers usually have two primary windings, each of which is designed for 110 V ac. When the two are connected in series, the transformer can be used for 220 V ac operations, as illustrated in Figure 14–30.

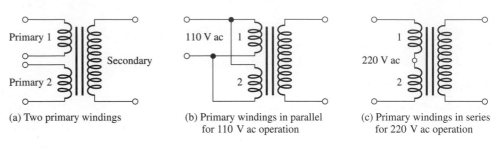

(a) Two primary windings   (b) Primary windings in parallel for 110 V ac operation   (c) Primary windings in series for 220 V ac operation

▲ **FIGURE 14–30**

Multiple-primary transformers.

More than one secondary can be wound on a common core. Transformers with several secondary windings are often used to achieve several voltages by either stepping up or stepping down the primary voltage. These types are commonly used in power supply applications in which several voltage levels are required for the operation of an electronic instrument.

A typical schematic of a multiple-secondary transformer is shown in Figure 14–31; this transformer has three secondaries. Sometimes you will find combinations of multiple-primary, multiple-secondary, and tapped transformers all in one unit.

▶ **FIGURE 14–31**

A multiple-secondary transformer.

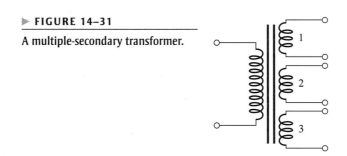

EXAMPLE 14–11

The transformer shown in Figure 14–32 has the numbers of turns indicated. One of the secondaries is also center tapped. If 120 V ac are connected to the primary, determine each secondary voltage and the voltages with respect to the center tap (CT) on the middle secondary.

▶ **FIGURE 14–32**

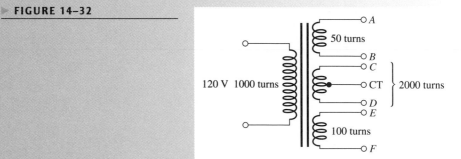

Solution

$$V_{AB} = n_{AB}V_{pri} = (0.05)120 \text{ V} = \textbf{6 V}$$

$$V_{CD} = n_{CD}V_{pri} = (2)120 \text{ V} = \textbf{240 V}$$

$$V_{(CT)C} = V_{(CT)D} = \frac{240 \text{ V}}{2} = \textbf{120 V}$$

$$V_{EF} = n_{EF}V_{pri} = (0.1)120 \text{ V} = \textbf{12 V}$$

Related Problem   Repeat the calculations for a primary with 500 turns.

Autotransformers

In an **autotransformer**, one winding serves as both the primary and the secondary. The winding is tapped at the proper points to achieve the desired turns ratio for stepping up or stepping down the voltage.

Autotransformers differ from conventional transformers in that there is no electrical isolation between the primary and the secondary because both are on one winding. Autotransformers normally are smaller and lighter than equivalent conventional transformers because they require a much lower kVA rating for a given load. Many autotransformers provide an adjustable tap using a sliding contact mechanism so that the output voltage can be varied (these are often called *variacs*). Figure 14–33 shows schematic symbols for various types of autotransformers. Autotransformers are used in starting industrial induction motors and regulating transmission line voltages.

Example 14–12 illustrates why an autotransformer has a kVA requirement that is less than the input or output kVA.

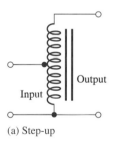
(a) Step-up

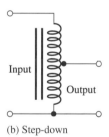
(b) Step-down

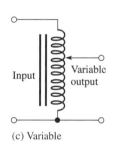
(c) Variable

◀ **FIGURE 14–33**

Types of autotransformers.

EXAMPLE 14–12

A certain autotransformer is used to change a source voltage of 220 V to a load voltage of 160 V across an 8 Ω load resistance. Determine the input and output power in kilovolt-amperes, and show that the actual kVA requirement is less than this value. Assume that this transformer is ideal.

Solution   The circuit is shown in Figure 14–34 with voltages and currents indicated.

▶ **FIGURE 14–34**

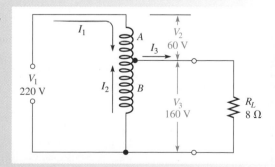

The load current, I_3, is determined as

$$I_3 = \frac{V_3}{R_L} = \frac{160\ \text{V}}{8\ \Omega} = 20\ \text{A}$$

The input power is the total source voltage (V_1) times the total current from the source (I_1).

$$P_{in} = V_1 I_1$$

The output power is the load voltage, V_3, times the load current, I_3.

$$P_{out} = V_3 I_3$$

For an ideal transformer, $P_{in} = P_{out}$; thus,

$$V_1 I_1 = V_3 I_3$$

Solving for I_1 yields

$$I_1 = \frac{V_3 I_3}{V_1} = \frac{(160\ \text{V})(20\ \text{A})}{220\ \text{V}} = 14.55\ \text{A}$$

Applying Kirchhoff's current law at the tap junction,

$$I_1 = I_2 + I_3$$

Solving for I_2, the current through winding B, yields

$$I_2 = I_1 - I_3 = 14.55\ \text{A} - 20\ \text{A} = -5.45\ \text{A}$$

The minus sign indicates I_2 is out of phase with I_1.

The input and output power are

$$P_{in} = P_{out} = V_3 I_3 = (160\ \text{V})(20\ \text{A}) = \textbf{3.2 kVA}$$

The power in winding A is

$$P_A = V_2 I_1 = (60\ \text{V})(14.55\ \text{A}) = 873\ \text{VA} = 0.873\ \text{kVA}$$

The power in winding B is

$$P_B = V_3 I_2 = (160\ \text{V})(5.45\ \text{A}) = 872\ \text{VA} = 0.872\ \text{kVA}$$

Thus, the power rating required for each winding is less than the power that is delivered to the load. The slight difference in the calculated powers in windings A and B is due to rounding.

Related Problem   What happens to the kVA requirement if the load is changed to 4 Ω?

SECTION 14–8 REVIEW

1. A certain transformer has two secondary windings. The turns ratio from the primary winding to the first secondary is 10. The turns ratio from the primary to the other secondary is 0.2. If 240 V ac are applied to the primary, what are the secondary voltages?

2. Name one advantage and one disadvantage of an autotransformer over a conventional transformer.

14–9 TROUBLESHOOTING

Transformers are reliable devices when they are operated within their specified range. Common failures in transformers are opens in either the primary or the secondary windings. One cause of such failures is the operation of the device under conditions that exceed its ratings. Normally, when a transformer fails, it is difficult to repair and therefore the simplest procedure is to replace it. A few transformer failures and the associated symptoms are covered in this section.

After completing this section, you should be able to

◆ **Troubleshoot transformers**

   ◆ Find an open primary or secondary winding

   ◆ Find a shorted or partially shorted primary or secondary winding

Open Primary Winding

When there is an open primary winding, there is no primary current and, therefore, no induced voltage or current in the secondary. This condition is illustrated in Figure 14–35(a), and the method of checking with an ohmmeter is shown in part (b).

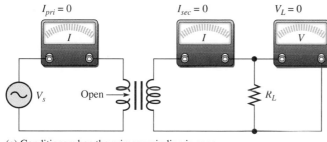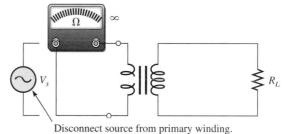

(a) Conditions when the primary winding is open

Disconnect source from primary winding.

(b) Checking the primary winding with an ohmmeter

▲ **FIGURE 14–35**

Open primary winding.

Open Secondary Winding

When there is an open secondary winding, there is no current in the secondary circuit and, as a result, no voltage across the load. Also, an open secondary causes the primary current to be very small (there is only a small magnetizing current). In fact, the primary current may be practically zero. This condition is illustrated in Figure 14–36(a), and the ohmmeter check is shown in part (b).

Shorted or Partially Shorted Windings

Shorted windings are very rare and if they do occur are very difficult to find unless there is a visual indication or a large number of windings are shorted. A completely shorted primary winding will draw excessive current from the source; and unless there is a breaker or a fuse in the circuit, either the source or the transformer or both will burn out. A partial short in the primary winding can cause higher than normal or even excessive primary current.

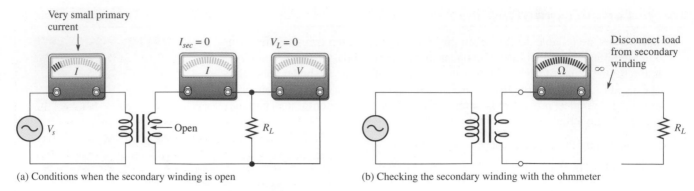

(a) Conditions when the secondary winding is open

(b) Checking the secondary winding with the ohmmeter

▲ **FIGURE 14–36**

Open secondary winding.

In the case of a shorted or partially shorted secondary winding, there is an excessive primary current because of the low reflected resistance due to the short. Often, this excessive current will burn out the primary winding and result in an open. The short-circuit current in the secondary winding causes the load current to be zero (full short) or smaller than normal (partial short), as demonstrated in Figure 14–37(a) and (b). The ohmmeter check for this condition is shown in part (c).

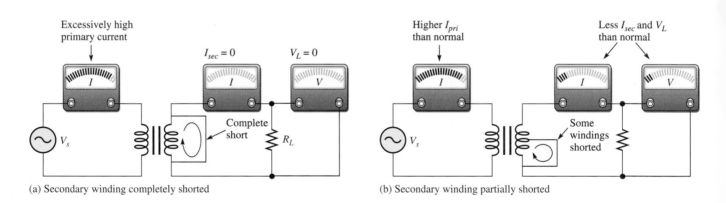

(a) Secondary winding completely shorted

(b) Secondary winding partially shorted

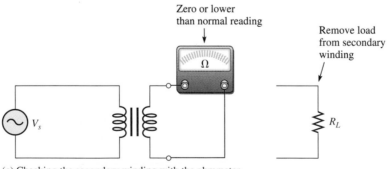

(c) Checking the secondary winding with the ohmmeter

▲ **FIGURE 14–37**

Shorted secondary winding.

A Circuit Application

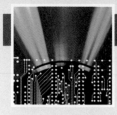

A common application of the transformer is in dc power supplies. The transformer is used to change and couple the ac line voltage into the power supply circuitry where it is converted to a dc voltage. You will troubleshoot four identical transformer-coupled dc power supplies and, based on a series of measurements, determine the fault, if any, in each.

The transformer (T_1) in the power supply schematic of Figure 14–38 steps the 110 V rms at the ac outlet down to a level that can be converted by the diode bridge rectifier, filtered, and regulated to obtain a 6 V dc output. The diode rectifier changes the ac to a pulsating full-wave dc voltage that is smoothed by the capacitor filter C_1. The voltage regulator is an integrated circuit that takes the filtered voltage and provides a constant 6 V dc over a range of load values and line voltage variations. Additional filtering is provided by capacitor C_2. You will learn about these circuits in a later course. The circled numbers in Figure 14–38 correspond to measurement points on the power supply board.

The Power Supply

You have four identical power supply boards to troubleshoot like the one shown in Figure 14–39. The power line to the primary winding of transformer T_1 is protected by the fuse. The secondary winding is connected to the circuit board containing the rectifier, filter, and regulator. Measurement points are indicated by the circled numbers.

Measuring Voltages on Power Supply Board 1

After plugging the power supply into a standard wall outlet, an autoranging portable multimeter is used to measure the voltages. In an autoranging meter, the appropriate measurement range is automatically selected instead of being manually selected as in a standard multimeter.

◆ Determine from the meter readings in Figure 14–40 whether or not the power supply is operating properly. If it is not, isolate the problem to one of the following: the circuit board con-

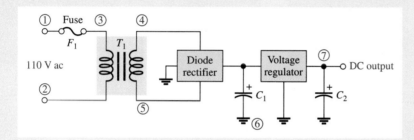

◀ FIGURE 14–38

Basic transformer-coupled dc power supply.

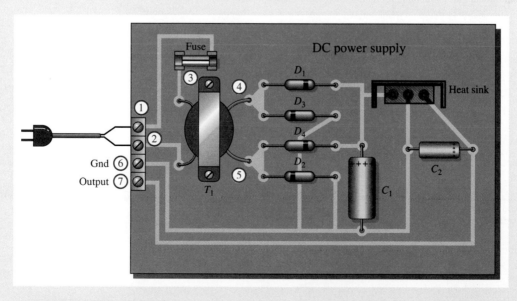

▲ FIGURE 14–39

Power supply board (top view).

taining the rectifier, filter, and regulator; the transformer; the fuse; or the power source. The circled numbers on the meter inputs correspond to the numbered points on the power supply in Figure 14–39.

Measuring Voltages on Power Supply Boards 2, 3, and 4

◆ Determine from the meter readings for boards 2, 3, and 4 in Figure 14–41 whether or not each power supply is operating properly. If it is not, isolate the problem to one of the following:

the circuit board containing the rectifier, filter, and regulator; the transformer; the fuse; or the power source. Only the meter displays and corresponding measurement points are shown.

Review

1. In the case where the transformer was found to be faulty, how can you determine the specific fault (open windings or shorted windings)?

2. What type of fault could cause the fuse to blow?

▲ FIGURE 14–40

Voltage measurements on power supply board 1.

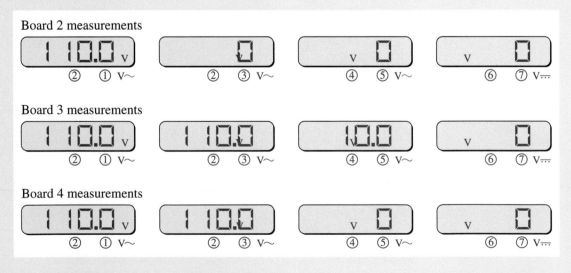

▲ FIGURE 14–41

Measurements for power supply boards 2, 3, and 4.

SUMMARY

- A normal transformer consists of two or more coils that are magnetically coupled on a common core.
- There is mutual inductance between two magnetically coupled coils.
- When current in one coil changes, voltage is induced in the other coil.
- The primary is the winding connected to the source, and the secondary is the winding connected to the load.
- The number of turns in the primary and the number of turns in the secondary determine the turns ratio.
- The relative polarities of the primary and secondary voltages are determined by the direction of the windings around the core.
- A step-up transformer has a turns ratio greater than 1.
- A step-down transformer has a turns ratio less than 1.
- A transformer cannot increase power.
- In an ideal transformer, the power from the source (input power) is equal to the power delivered to the load (output power).
- If the voltage is stepped up, the current is stepped down, and vice versa.
- A load across the secondary winding of a transformer appears to the source as a reflected load having a value dependent on the reciprocal of the turns ratio squared.
- Certain transformers can match a load resistance to a source resistance to achieve maximum power transfer to the load by selecting the proper turns ratio.
- A typical transformer does not respond to dc.
- Conversion of electrical energy to heat in an actual transformer results from winding resistances, hysteresis loss in the core, eddy currents in the core, and flux leakage.

KEY TERMS

Key terms and other bold terms in the chapter are defined in the end-of-book glossary.

Apparent power rating   The method of rating transformers in which the power capability is expressed in volt-amperes (VA).

Center tap (CT)   A connection at the midpoint of a winding in a transformer.

Impedance matching   A technique used to match a load resistance to a source resistance in order to achieve maximum transfer of power.

Magnetic coupling   The magnetic connection between two coils as a result of the changing magnetic flux lines of one coil cutting through the second coil.

Mutual inductance (L_M)   The inductance between two separate coils, such as in a transformer.

Primary winding   The input winding of a transformer; also called *primary*.

Reflected resistance   The resistance in the secondary circuit reflected into the primary circuit.

Secondary winding   The output winding of a transformer; also called *secondary*.

Transformer   An electrical device constructed of two or more coils (windings) that are electromagnetically coupled to each other to provide a transfer of power from one coil to another.

Turns ratio (n)   The ratio of turns in the secondary winding to turns in the primary winding.

FORMULAS

14–1 $k = \dfrac{\phi_{1-2}}{\phi_1}$ 　　　　　　　　　Coefficient of coupling

14–2 $L_M = k\sqrt{L_1 L_2}$ 　　　　　　　　Mutual inductance

14–3 $n = \dfrac{N_{sec}}{N_{pri}}$ 　　　　　　　　　Turns ratio

14–4 $\dfrac{V_{sec}}{V_{pri}} = \dfrac{N_{sec}}{N_{pri}}$ Voltage ratio

14–5 $V_{sec} = nV_{pri}$ Secondary voltage

14–6 $\dfrac{I_{pri}}{I_{sec}} = n$ Current ratio

14–7 $I_{sec} = \left(\dfrac{1}{n}\right)I_{pri}$ Secondary current

14–8 $R_{pri} = \left(\dfrac{1}{n}\right)^{2} R_{L}$ Reflected resistance

14–9 $n = \sqrt{\dfrac{R_{L}}{R_{pri}}}$ Turns ratio for impedance matching

14–10 $\eta = \left(\dfrac{P_{out}}{P_{in}}\right)100\%$ Transformer efficiency

SELF-TEST

Answers are at the end of the chapter.

1. A transformer is used for
   (a) dc voltages   (b) ac voltages   (c) both dc and ac

2. Which one of the following is affected by the turns ratio of a transformer?
   (a) primary voltage   (b) dc voltage
   (c) secondary voltage   (d) none of these

3. If the windings of a certain transformer with a turns ratio of 1 are in opposite directions around the core, the secondary voltage is
   (a) in phase with the primary voltage   (b) less than the primary voltage
   (c) greater than the primary voltage   (d) out of phase with the primary voltage

4. When the turns ratio of a transformer is 10 and the primary ac voltage is 6 V, the secondary voltage is
   (a) 60 V   (b) 0.6 V   (c) 6 V   (d) 36 V

5. When the turns ratio of a transformer is 0.5 and the primary ac voltage is 100 V, the secondary voltage is
   (a) 200 V   (b) 50 V   (c) 10 V   (d) 100 V

6. A certain transformer has 500 turns in the primary winding and 2500 turns in the secondary winding. The turns ratio is
   (a) 0.2   (b) 2.5   (c) 5   (d) 0.5

7. If 10 W of power are applied to the primary of an ideal transformer with a turns ratio of 5, the power delivered to the secondary load is
   (a) 50 W   (b) 0.5 W   (c) 0 W   (d) 10 W

8. In a certain loaded transformer, the secondary voltage is one-third the primary voltage. The secondary current is
   (a) one-third the primary current   (b) three times the primary current
   (c) equal to the primary current   (d) less than the primary current

9. When a 1.0 kΩ load resistor is connected across the secondary winding of a transformer with a turns ratio of 2, the source "sees" a reflected load of
   (a) 250 Ω   (b) 2 kΩ   (c) 4 kΩ   (d) 1.0 kΩ

10. In Question 9, if the turns ratio is 0.5, the source "sees" a reflected load of
    (a) 1.0 kΩ   (b) 2 kΩ   (c) 4 kΩ   (d) 500 Ω

11. The turns required to match a 50 Ω source to a 200 Ω load is
    (a) 0.25   (b) 0.5   (c) 4   (d) 2

12. Maximum power is transferred from a source to a load in a transformer coupled circuit when
    (a) $R_L > R_{int}$ (b) $R_L < R_{int}$ (c) $(1/n)^2 R_L = R_{int}$ (d) $R_L = nR_{int}$

13. When a 12 V battery is connected across the primary of a transformer with a turns ratio of 4, the secondary voltage is
    (a) 0 V   (b) 12 V   (c) 48 V   (d) 3 V

14. A certain transformer has a turns ratio of 1 and a 0.95 coefficient of coupling. When 1 V ac is applied to the primary, the secondary voltage is
    (a) 1 V   (b) 1.95 V   (c) 0.95 V

CIRCUIT DYNAMICS QUIZ

Answers are at the end of the chapter.

Refer to Figure 14–43(c).

1. If the ac source shorts out, the voltage across R_L
   (a) increases   (b) decreases   (c) stays the same

2. If the dc source shorts out, the voltage across R_L
   (a) increases   (b) decreases   (c) stays the same

3. If R_L opens, the voltage across it
   (a) increases   (b) decreases   (c) stays the same

Refer to Figure 14–45.

4. If the fuse opens, the voltage across R_L
   (a) increases   (b) decreases   (c) stays the same

5. If the turns ratio is changed to 2, the current through R_L
   (a) increases   (b) decreases   (c) stays the same

6. If the frequency of the source voltage is increased, the voltage across R_L
   (a) increases   (b) decreases   (c) stays the same

Refer to Figure 14–49.

7. If the source voltage is increased, the loudness of the sound from the speaker
   (a) increases   (b) decreases   (c) stays the same

8. If the turns ratio is increased, the loudness of the sound from the speaker
   (a) increases   (b) decreases   (c) stays the same

Refer to Figure 14–50.

9. 10 V rms are applied across the primary. If the left switch is moved form position 1 to position 2, the voltage from the top of R_1 to ground
   (a) increases   (b) decreases   (c) stays the same

10. Again, 10 V rms are applied across the primary. With both switches in position 1 as shown and if R_1 opens, the voltage across R_1
    (a) increases   (b) decreases   (c) stays the same

PROBLEMS

More difficult problems are indicated by an asterisk (*).
Answers to odd-numbered problems are at the end of the book.

SECTION 14–1   Mutual Inductance

1. What is the mutual inductance when $k = 0.75$, $L_1 = 1\,\mu H$, and $L_2 = 4\,\mu H$?

2. Determine the coefficient of coupling when $L_M = 1\,\mu H$, $L_1 = 8\,\mu H$, and $L_2 = 2\,\mu H$.

SECTION 14–2   The Basic Transformer

3. What is the turns ratio of a transformer having 250 primary turns and 1000 secondary turns? What is the turns ratio when the primary winding has 400 turns and the secondary winding has 100 turns?

4. A certain transformer has 250 turns in its primary winding. In order to double the voltage, how many turns must be in the secondary winding?

5. For each transformer in Figure 14–42, draw the secondary voltage showing its relationship to the primary voltage. Also indicate the amplitude.

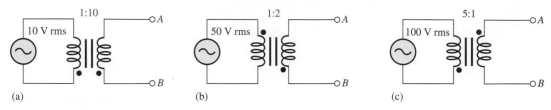

▲ FIGURE 14–42

SECTION 14–3   Step-Up and Step-Down Transformers

6. To step 240 V ac up to 720 V, what must the turns ratio be?

7. The primary winding of a transformer has 120 V ac across it. What is the secondary voltage if the turns ratio is 5?

8. How many primary volts must be applied to a transformer with a turns ratio of 10 to obtain a secondary voltage of 60 V ac?

9. To step 120 V down to 30 V, what must the turns ratio be?

10. The primary winding of a transformer has 1200 V across it. What is the secondary voltage if the turns ratio is 0.2?

11. How many primary volts must be applied to a transformer with a turns ratio of 0.1 to obtain a secondary voltage of 6 V ac?

12. What is the voltage across the load in each circuit of Figure 14–43?

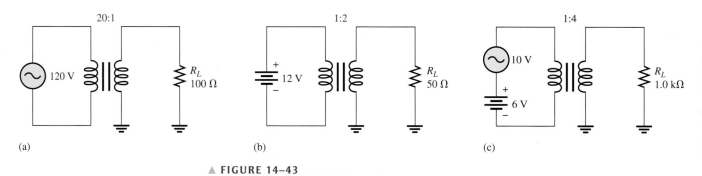

▲ FIGURE 14–43

13. Determine the unspecified meter readings in Figure 14–44.

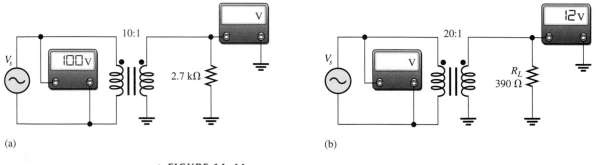

▲ FIGURE 14–44

SECTION 14–4   Loading the Secondary Winding

14. Determine I_s in Figure 14–45. What is the value of R_L?

▶ **FIGURE 14–45**

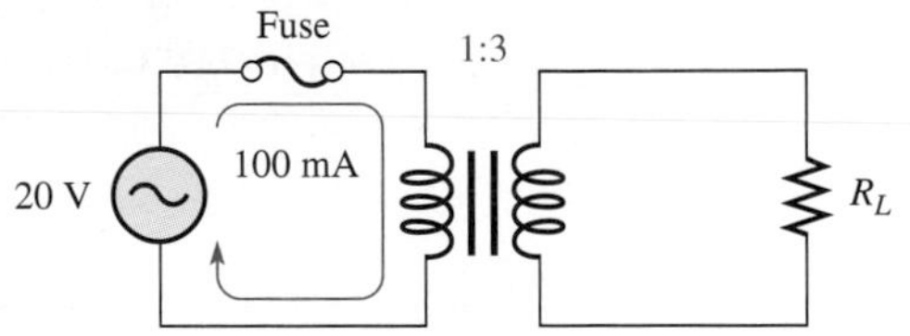

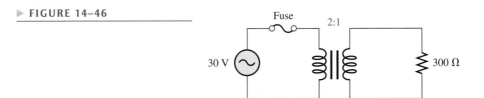

15. Determine the following quantities in Figure 14–46.

(a) Primary current   (b) Secondary current

(c) Secondary voltage   (d) Power in the load

▶ **FIGURE 14–46**

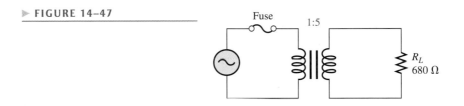

SECTION 14–5   Reflected Load

16. What is the load resistance as seen by the source in Figure 14–47?

▶ **FIGURE 14–47**

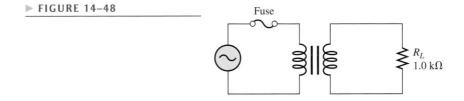

17. What must the turns ratio be in Figure 14–48 in order to reflect 300 Ω into the primary circuit?

▶ **FIGURE 14–48**

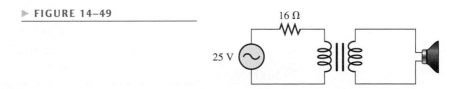

SECTION 14–6   Matching the Load and Source Resistances

18. For the circuit in Figure 14–49, find the turns ratio required to deliver maximum power to the 4 Ω speaker.

19. In Figure 14–49, what is the maximum power that can be delivered to the 4 Ω speaker?

▶ **FIGURE 14–49**

*20. Find the appropriate turns ratio for each switch position in Figure 14–50 in order to transfer the maximum power to each load when the source resistance is 10 Ω. Specify the number of turns for the secondary winding if the primary winding has 1000 turns.

► FIGURE 14–50

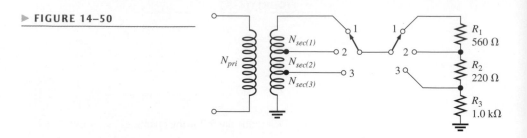

SECTION 14–7   Nonideal Transformer Characteristics

21. In a certain transformer, the input power to the primary is 100 W. If 5.5 W are lost in the winding resistances, what is the output power to the load, neglecting any other losses?

22. What is the efficiency of the transformer in Problem 21?

23. Determine the coefficient of coupling for a transformer in which 2% of the total flux generated in the primary does not pass through the secondary.

*24. A certain transformer is rated at 1 kVA. It operates on 60 Hz, 120 V ac. The secondary voltage is 600 V.

   (a) What is the maximum load current?

   (b) What is the smallest R_L that you can drive?

   (c) What is the largest capacitor that can be connected as a load?

25. What kVA rating is required for a transformer that must handle a maximum load current of 10 A with a secondary voltage of 2.5 kV?

*26. A certain transformer is rated at 5 kVA, 2400/120 V, at 60 Hz.

   (a) What is the turns ratio if the 120 V is the secondary voltage?

   (b) What is the current rating of the secondary if 2400 V is the primary voltage?

   (c) What is the current rating of the primary winding if 2400 V is the primary voltage?

SECTION 14–8   Tapped and Multiple-Winding Transformers

27. Determine each unknown voltage indicated in Figure 14–51.

► FIGURE 14–51

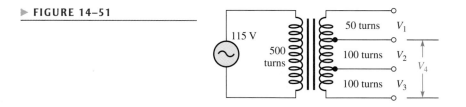

28. Using the indicated secondary voltages in Figure 14–52, determine the turns ratio of each tapped section of the secondary winding to the primary winding.

► FIGURE 14–52

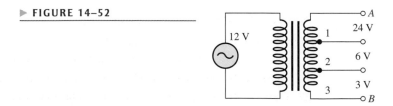

29. Find the secondary voltage for each autotransformer in Figure 14–53.

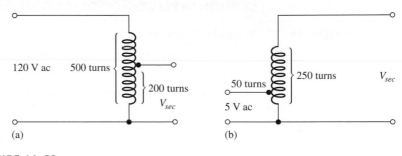

▲ FIGURE 14–53

30. In Figure 14–54, each primary can accommodate 120 V ac. How should the primaries be connected for 240 V ac operation? Determine each secondary voltage for 240 V operation.

▶ FIGURE 14–54

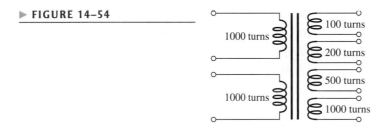

*31. For the loaded, tapped-secondary transformer in Figure 14–55, determine the following:

(a) All load voltages and currents

(b) The resistance reflected into the primary

▶ FIGURE 14–55

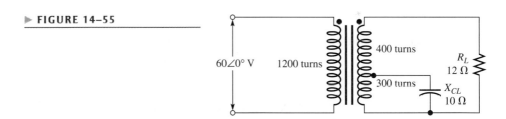

SECTION 14–9 Troubleshooting

32. When you apply 120 V ac across the primary winding of a transformer and check the voltage across the secondary winding, you get 0 V. Further investigation shows no primary or secondary currents. List the possible faults. What is your next step in investigating the problem?

33. What is likely to happen if the primary winding of a transformer shorts?

34. While checking out a transformer, you find that the secondary voltage is less than it should be although it is not zero. What is the most likely fault?

Multisim Troubleshooting and Analysis

These problems require your Multisim CD-ROM.

35. Open file P14-35 and measure the secondary voltage. Determine the turns ratio.

36. Open file P14-36 and determine by measurement if there is an open winding.

37. Open file P14-37 and determine if there is a fault in the circuit.

ANSWERS

SECTION REVIEWS

SECTION 14-1 Mutual Inductance

1. Mutual inductance is the inductance between two coils.
2. $L_M = k\sqrt{L_1 L_2} = 45\,\text{mH}$
3. The induced voltage increases if k increases.

SECTION 14-2 The Basic Transformer

1. Transformer operation is based on mutual inductance.
2. The turns ratio is the ratio of turns in the secondary winding to turns in the primary winding.
3. The directions of the windings determine the relative polarities of the voltages.
4. $n = 250/500 = 0.5$

SECTION 14-3 Step-Up and Step-Down Transformers

1. A step-up transformer produces a secondary voltage that is greater than the primary voltage.
2. V_{sec} is five times greater than V_{pri}.
3. $V_{sec} = nV_{pri} = 10(240\,\text{V}) = 2400\,\text{V}$
4. A step-down transformer produces a secondary voltage that is less than the primary voltage.
5. $V_{sec} = (0.5)120\,\text{V} = 60\,\text{V}$
6. $n = 12\,\text{V}/120\,\text{V} = 0.1$
7. Electrical isolation is the condition that exists when two coils are magnetically linked but have no electrical connection between them.

SECTION 14-4 Loading the Secondary Winding

1. I_{sec} is less than I_{pri} by half.
2. $I_{sec} = (1000/250)0.5\,\text{A} = 2\,\text{A}$
3. $I_{pri} = (250/1000)10\,\text{A} = 2.5\,\text{A}$

SECTION 14-5 Reflected Load

1. Reflected resistance is the resistance in the secondary, altered by the reciprocal of the turns ratio squared, as it appears to the primary.
2. The turns ratio determines reflected resistance.
3. $R_{pri} = (0.1)^2 50\,\Omega = 0.5\,\Omega$
4. $n = 0.1$

SECTION 14-6 Matching the Load and Source Resistances

1. Impedance matching makes the load resistance equal the source resistance.
2. Maximum power is delivered to the load when $R_L = R_s$.
3. $R_{pri} = (100/50)^2 100\,\Omega = 400\,\Omega$

SECTION 14-7 Nonideal Transformer Characteristics

1. In a practical transformer, conversion of electrical energy to heat reduces the efficiency. An ideal transformer has an efficiency of 100%.
2. When $k = 0.85$, 85% of the magnetic flux generated in the primary winding passes through the secondary winding.
3. $I_L = 10\,\text{kVA}/250\,\text{V} = 40\,\text{A}$

SECTION 14–8 **Tapped and Multiple-Winding Transformers**

1. $V_{sec} = (10)240\,V = 2400\,V$, $V_{sec} = (0.2)240\,V = 48\,V$

2. An autotransformer is smaller and lighter for the same rating than a conventional one. An autotransformer has no electrical isolation.

SECTION 14–9 **Troubleshooting**

1. Transformer faults: open windings are the most common, shorted windings are much less common.

2. Operating above rated values will cause a failure.

A Circuit Application

1. Use an ohmmeter to check for open windings. Shorted windings are indicated by an incorrect secondary voltage.

2. A short will cause the fuse to blow.

RELATED PROBLEMS FOR EXAMPLES

14–1 0.75

14–2 $387\,\mu H$

14–3 5000 turns

14–4 480 V

14–5 57.6 V

14–6 5 mA; 400 mA

14–7 6 Ω

14–8 0.354

14–9 0.0707 or 14.14:1

14–10 85.2%

14–11 $V_{AB} = 12\,V$, $V_{CD} = 480\,V$, $V_{(CT)C} = V_{(CT)D} = 240\,V$, $V_{EF} = 24\,V$

14–12 Increases to 1.75 kVA

SELF-TEST

1. (b)   **2.** (c)   **3.** (d)   **4.** (a)   **5.** (b)   **6.** (c)   **7.** (d)   **8.** (b)

9. (a)   **10.** (c)   **11.** (d)   **12.** (c)   **13.** (a)   **14.** (c)

CIRCUIT DYNAMICS QUIZ

1. (b)   **2.** (c)   **3.** (c)   **4.** (b)   **5.** (a)

6. (c)   **7.** (a)   **8.** (a)   **9.** (b)   **10.** (c)

15

RC CIRCUITS

CHAPTER OBJECTIVES

PART 1: SERIES CIRCUITS

◆ Use complex numbers to express phasor quantities
◆ Describe the relationship between current and voltage in a series *RC* circuit
◆ Determine the impedance of a series *RC* circuit
◆ Analyze a series *RC* circuit

PART 2: PARALLEL CIRCUITS

◆ Determine impedance and admittance in a parallel *RC* circuit
◆ Analyze a parallel *RC* circuit

PART 3: SERIES-PARALLEL CIRCUITS

◆ Analyze series-parallel *RC* circuits

PART 4: SPECIAL TOPICS

◆ Determine power in *RC* circuits
◆ Discuss some basic *RC* applications
◆ Troubleshoot *RC* circuits

KEY TERMS

◆ Complex plane   ◆ Polar form
◆ Real number   ◆ Impedance
◆ Imaginary number   ◆ Capacitive susceptance (B_C)
◆ Rectangular form   ◆ Admittance (Y)

◆ Apparent power (P_a)   ◆ Frequency response
◆ Power factor   ◆ Cutoff frequency
◆ Filter   ◆ Bandwidth

A CIRCUIT APPLICATION PREVIEW

The frequency response of the *RC* input circuit in an amplifier is similar to the one you worked with in Chapter 12 and is the subject of this chapter's circuit application.

VISIT THE COMPANION WEBSITE

Study aids for this chapter are available at
http://www.prenhall.com/floyd

INTRODUCTION

An *RC* circuit contains both resistance and capacitance. In this chapter, basic series and parallel *RC* circuits and their responses to sinusoidal ac voltages are presented. Series-parallel combinations are also analyzed. True, reactive, and apparent power in *RC* circuits are discussed and some basic *RC* circuit applications are introduced. Applications of *RC* circuits include filters, amplifier coupling, oscillators, and wave-shaping circuits. Troubleshooting is also covered in this chapter.

The first section of this chapter provides an introduction to complex numbers, an important tool for the analysis of ac circuits. The complex number system is a way to mathematically express a phasor quantity and allows phasor quantities to be added, subtracted, multiplied, and divided. You will use complex numbers in Chapters 15, 16, and 17.

COVERAGE OPTIONS

This chapter and Chapters 16 and 17 are each divided into four parts: Series Circuits, Parallel Circuits, Series-Parallel Circuits, and Special Topics. This organization facilitates either of two options to the coverage of reactive circuits in Chapters 15, 16, and 17.

Option 1   Study all *RC* circuit topics (Chapter 15) first, followed by all *RL* circuit topics (Chapter 16), and then all *RLC* circuit topics (Chapter 17). Using this approach, you simply cover Chapters 15, 16, and 17 in sequence.

Option 2   Study *series* reactive circuits first. Then study *parallel* reactive circuits, followed by *series-parallel* reactive circuits and finally *special topics*. Using this approach, you cover Part 1: Series Circuits in Chapters 15, 16, and 17; then Part 2: Parallel Circuits in Chapters 15, 16, and 17; then Part 3: Series-Parallel Circuits in Chapters 15, 16, and 17. Finally, Part 4: Special Topics can be covered in each of the chapters.

SERIES CIRCUITS
With an Introduction to Complex Numbers

15–1 THE COMPLEX NUMBER SYSTEM

Complex numbers allow mathematical operations with phasor quantities and are useful in the analysis of ac circuits. With the complex number system, you can add, subtract, multiply, and divide quantities that have both magnitude and angle, such as sine waves and other ac circuit quantities. Most scientific calculators can perform operations with complex numbers. Consult your user's manual for the exact procedure.

After completing this section, you should be able to

- **Use complex numbers to express phasor quantities**
    - Describe the complex plane
    - Represent a point on the complex plane
    - Discuss real and imaginary numbers
    - Express phasor quantities in both rectangular and polar forms
    - Convert between rectangular and polar forms
    - Do arithmetic operations with complex numbers

Positive and Negative Numbers

Positive numbers are represented by points to the right of the origin on the horizontal axis of a graph, and negative numbers are represented by points to the left of the origin, as illustrated in Figure 15–1(a). Also, positive numbers are represented by points on the vertical axis above

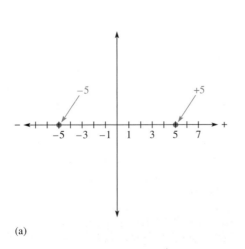

(a)

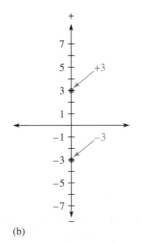

(b)

◀ **FIGURE 15–1**

Graphic representation of positive and negative numbers.

the origin, and negative numbers are represented by points below the origin, as shown in Figure 15–1(b).

The Complex Plane

To distinguish between values on the horizontal axis and values on the vertical axis, a **complex plane** is used. In the complex plane, the horizontal axis is called the *real axis,* and the vertical axis is called the *imaginary axis,* as shown in Figure 15–2. In electrical circuit work, a $\pm j$ prefix is used to designate numbers that lie on the imaginary axis in order to distinguish them from numbers lying on the real axis. This prefix is known as the *j operator.* In mathematics, an *i* is used instead of a *j*, but in electric circuits, the *i* can be confused with instantaneous current, so *j* is used.

▶ **FIGURE 15–2**

The complex plane.

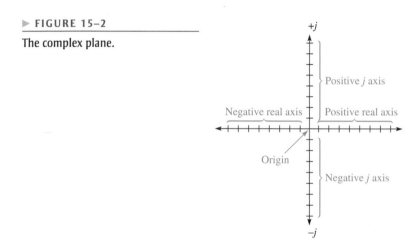

Angular Position on the Complex Plane   Angular positions are represented on the complex plane, as shown in Figure 15–3. The positive real axis represents zero degrees. Proceeding counterclockwise, the $+j$ axis represents 90°, the negative real axis represents 180°, the $-j$ axis is the 270° point, and, after a full rotation of 360°, you are back to the positive real axis. Notice that the plane is divided into four quadrants.

▶ **FIGURE 15–3**

Angles on the complex plane.

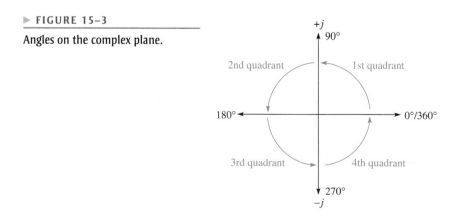

Representing a Point on the Complex Plane   A point located on the complex plane is classified as real, imaginary ($\pm j$), or a combination of the two. For example, a point located 4 units from the origin on the positive real axis is the positive **real number,**

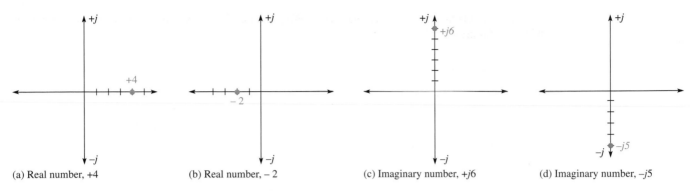

(a) Real number, +4 (b) Real number, −2 (c) Imaginary number, +j6 (d) Imaginary number, −j5

▲ **FIGURE 15–4**

Real and imaginary (*j*) numbers on the complex plane.

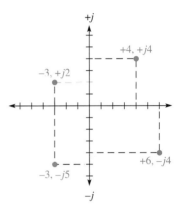

+4, as shown in Figure 15–4(a). A point 2 units from the origin on the negative real axis is the negative real number, −2, as shown in part (b). A point on the +*j* axis 6 units from the origin, as shown in part (c), is the positive **imaginary number**, +j6. Finally, a point 5 units along the −*j* axis is the negative imaginary number, −j5, as shown in part (d).

When a point lies not on any axis but somewhere in one of the four quadrants, it is a complex number and is defined by its coordinates. For example, in Figure 15–5, the point located in the first quadrant has a real value of +4 and a *j* value of +j4 and is expressed as +4, +j4. The point located in the second quadrant has coordinates −3 and +j2. The point located in the third quadrant has coordinates −3 and −j5. The point located in the fourth quadrant has coordinates of +6 and −j4.

▲ **FIGURE 15–5**

Coordinate points on the complex plane.

EXAMPLE 15–1

(a) Locate the following points on the complex plane: 7, j5; 5, −j2; −3.5, j1; and −5.5, −j6.5.

(b) Determine the coordinates for each point in Figure 15–6.

▶ **FIGURE 15–6**

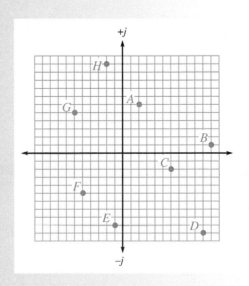

Solution   (a) See Figure 15–7.

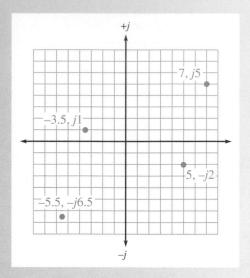

▲ **FIGURE 15–7**

(b) *A*: **2, *j*6**   *B*: **11, *j*1**   *C*: **6, −*j*2**   *D*: **10, −*j*10**
     E: **−1, −*j*9**   *F*: **−5, −*j*5**   *G*: **−6, *j*5**   *H*: **−2, *j*11**

Related Problem*   In what quadrant is each of the following points located?

(a) +2.5, +*j*1   (b) 7, −*j*5   (c) −10, −*j*5   (d) −11, +*j*6.8

*Answers are at the end of the chapter.

Value of *j*

If you multiply the positive real value of +2 by *j*, the result is +*j*2. This multiplication has effectively moved the +2 through a 90° angle to the +*j* axis. Similarly, multiplying +2 by −*j* rotates it −90° to the −*j* axis. Thus, *j* is considered a rotational operator.

Mathematically, the *j* operator has a value of $\sqrt{-1}$. If +*j*2 is multiplied by *j*, you get

$$j^2 2 = (\sqrt{-1})(\sqrt{-1})(2) = (-1)(2) = -2$$

This calculation effectively places the value on the negative real axis. Therefore, multiplying a positive real number by j^2 converts it to a negative real number, which, in effect, is a rotation of 180° on the complex plane. This operation is illustrated in Figure 15–8.

▶ **FIGURE 15–8**

Effect of the *j* operator on location of a number on the complex plane.

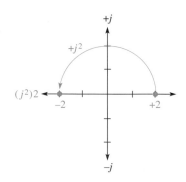

Rectangular and Polar Forms

Rectangular and polar are two forms of complex numbers that are used to represent phasor quantities. Each has certain advantages when used in circuit analysis, depending on the particular application. A phasor quantity contains both *magnitude* and angular position or *phase*. In this text, italic letters such as *V* and *I* are used to represent magnitude only, and boldfaced nonitalic letters such as **V** and **I** are used to represent complete phasor quantities.

Rectangular Form   A phasor quantity is represented in **rectangular form** by the algebraic sum of the real value (*A*) of the coordinate and the *j* value (*B*) of the coordinate, expressed in the following general form:

$$A + jB$$

Examples of phasor quantities are $1 + j2$, $5 - j3$, $-4 + j4$, and $-2 - j6$, which are shown on the complex plane in Figure 15–9. As you can see, the rectangular coordinates describe the phasor in terms of its values projected onto the real axis and the *j* axis. An "arrow" drawn from the origin to the coordinate point in the complex plane represents graphically the phasor quantity.

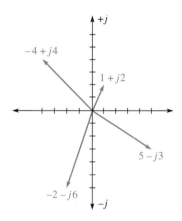

◀ **FIGURE 15–9**

Examples of phasors specified by rectangular coordinates.

Polar Form   Phasor quantities can also be expressed in **polar form**, which consists of the phasor magnitude (*C*) and the angular position relative to the positive real axis (*θ*), expressed in the following general form:

$$C\angle \pm \theta$$

Examples are $2\angle 45°$, $5\angle 120°$, $4\angle -110°$, and $8\angle -30°$. The first number is the magnitude, and the symbol $\angle$ precedes the value of the angle. Figure 15–10 shows these phasors on the complex plane. The length of the phasor, of course, represents the magnitude of the

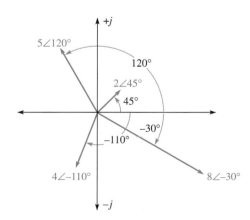

◀ **FIGURE 15–10**

Examples of phasors specified by polar values.

quantity. Keep in mind that for every phasor expressed in polar form, there is also an equivalent expression in rectangular form.

Conversion from Rectangular to Polar Form A phasor can exist in any of the four quadrants of the complex plane, as indicated in Figure 15–11. The phase angle θ in each case is measured relative to the positive real axis (0°), and ϕ (phi) is the angle in the 2nd and 3rd quadrants relative to the negative real axis, as shown.

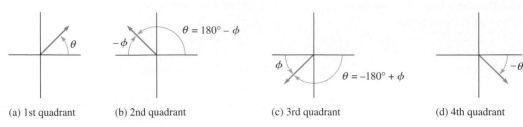

(a) 1st quadrant   (b) 2nd quadrant   (c) 3rd quadrant   (d) 4th quadrant

▲ **FIGURE 15–11**

All possible phasor quadrant locations.

The first step to convert from rectangular form to polar form is to determine the magnitude of the phasor. A phasor can be visualized as forming a right triangle in the complex plane, as indicated in Figure 15–12, for each quadrant location. The horizontal side of the triangle is the real value, A, and the vertical side is the j value, B. The hypotenuse of the triangle is the length of the phasor, C, representing the magnitude, and can be expressed, using the Pythagorean theorem, as

Equation 15–1

$$C = \sqrt{A^2 + B^2}$$

Next, the angle θ indicated in parts (a) and (d) of Figure 15–12 is expressed as an inverse tangent function.

Equation 15–2

$$\theta = \tan^{-1}\left(\frac{\pm B}{A}\right)$$

The angle θ indicated in parts (b) and (c) of Figure 15–12 is

$$\theta = \pm 180° \mp \phi$$

which includes both conditions as indicated by the dual signs.

$$\theta = \pm 180° \mp \tan^{-1}\left(\frac{B}{A}\right)$$

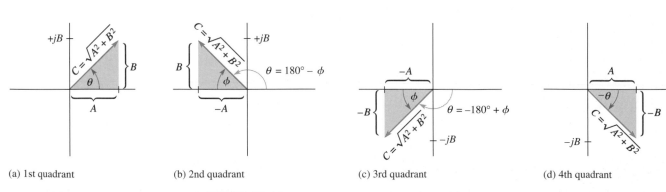

(a) 1st quadrant   (b) 2nd quadrant   (c) 3rd quadrant   (d) 4th quadrant

▲ **FIGURE 15–12**

Right angle relationships in the complex plane.

In each case the appropriate signs must be used in the calculation.

The general formula for converting from rectangular to polar is

$$\pm A \pm jB = C\angle\pm\theta$$

Equation 15–3

Example 15–2 illustrates the conversion procedure.

EXAMPLE 15–2

Convert the following complex numbers from rectangular form to polar form by determining the magnitude and angle:

(a) $8 + j6$ (b) $10 - j5$

Solution (a) The magnitude of the phasor represented by $8 + j6$ is

$$C = \sqrt{A^2 + B^2} = \sqrt{8^2 + 6^2} = \sqrt{100} = 10$$

Since the phasor is in the first quadrant, use Equation 15–2. The angle is

$$\theta = \tan^{-1}\left(\frac{\pm B}{A}\right) = \tan^{-1}\left(\frac{6}{8}\right) = 36.9°$$

θ is the angle relative to the positive real axis. The polar form of $8 + j6$ is

$$C\angle\theta = \mathbf{10\angle 36.9°}$$

(b) The magnitude of the phasor represented by $10 - j5$ is

$$C = \sqrt{10^2 + (-5)^2} = \sqrt{125} = 11.2$$

Since the phasor is in the fourth quadrant, use Equation 15–2. The angle is

$$\theta = \tan^{-1}\left(\frac{-5}{10}\right) = -26.6°$$

θ is the angle relative to the positive real axis. The polar form of $10 - j5$ is

$$C\angle\theta = \mathbf{11.2\angle -26.6°}$$

Related Problem Convert $18 + j23$ to polar form.

Conversion from Polar to Rectangular Form The polar form gives the magnitude and angle of a phasor quantity, as indicated in Figure 15–13.

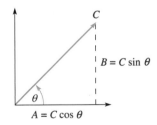

◀ **FIGURE 15–13**

Polar components of a phasor.

To get the rectangular form, you must find sides A and B of the triangle, using the rules from trigonometry stated below:

$$A = C\cos\theta$$ **Equation 15–4**

$$B = C\sin\theta$$ **Equation 15–5**

The polar-to-rectangular conversion formula is

$$C\angle\theta = A + jB$$ **Equation 15–6**

The following example demonstrates this conversion.

EXAMPLE 15–3

Convert the following polar quantities to rectangular form:

(a) $10\angle 30°$ (b) $200\angle -45°$

Solution   (a) The real part of the phasor represented by $10\angle 30°$ is

$$A = C\cos\theta = 10\cos 30° = 10(0.866) = 8.66$$

The j part of this phasor is

$$jB = jC\sin\theta = j10\sin 30° = j10(0.5) = j5$$

The rectangular form of $10\angle 30°$ is

$$A + jB = \mathbf{8.66 + j5}$$

(b) The real part of the phasor represented by $200\angle -45°$ is

$$A = 200\cos(-45°) = 200(0.707) = 141$$

The j part is

$$jB = j200\sin(-45°) = j200(-0.707) = -j141$$

The rectangular form of $200\angle -45°$ is

$$A + jB = \mathbf{141 - j141}$$

Related Problem   Convert $78\angle -26°$ to rectangular form.

Mathematical Operations

Addition   Complex numbers must be in rectangular form in order to add them. The rule is

Add the real parts of each complex number to get the real part of the sum. Then add the j parts of each complex number to get the j part of the sum.

EXAMPLE 15–4

Add the following sets of complex numbers:

(a) $8 + j5$ and $2 + j1$ (b) $20 - j10$ and $12 + j6$

Solution   (a) $(8 + j5) + (2 + j1) = (8 + 2) + j(5 + 1) = \mathbf{10 + j6}$

(b) $(20 - j10) + (12 + j6) = (20 + 12) + j(-10 + 6) = 32 + j(-4) = \mathbf{32 - j4}$

Related Problem   Add $5 - j11$ and $-6 + j3$.

Subtraction   As in addition, the numbers must be in rectangular form to be subtracted. The rule is

Subtract the real parts of the numbers to get the real part of the difference. Then subtract the j parts of the numbers to get the j part of the difference.

EXAMPLE 15–5

Perform the following subtractions:

(a) Subtract $1 + j2$ from $3 + j4$.

(b) Subtract $10 - j8$ from $15 + j15$.

Solution **(a)** $(3 + j4) - (1 + j2) = (3 - 1) + j(4 - 2) = \mathbf{2 + j2}$

(b) $(15 + j15) - (10 - j8) = (15 - 10) + j[15 - (-8)] = \mathbf{5 + j23}$

Related Problem Subtract $3.5 - j4.5$ from $-10 - j9$.

Multiplication Multiplication of two complex numbers in rectangular form is accomplished by multiplying, in turn, each term in one number by both terms in the other number and then combining the resulting real terms and the resulting j terms (recall that $j \times j = -1$). As an example,

$$(5 + j3)(2 - j4) = 10 - j20 + j6 + 12 = 22 - j14$$

Multiplication of two complex numbers is easier when both numbers are in polar form, so it is best to convert to polar form before multiplying. The rule is

Multiply the magnitudes, and add the angles algebraically.

EXAMPLE 15–6

Perform the following multiplications:

(a) $10\angle45°$ times $5\angle20°$ **(b)** $2\angle60°$ times $4\angle-30°$

Solution **(a)** $(10\angle45°)(5\angle20°) = (10)(5)\angle(45° + 20°) = \mathbf{50\angle65°}$

(b) $(2\angle60°)(4\angle-30°) = (2)(4)\angle[60° + (-30°)] = \mathbf{8\angle30°}$

Related Problem Multiply $50\angle10°$ times $30\angle-60°$.

Division Division of two complex numbers in rectangular form is accomplished by multiplying both the numerator and the denominator by the **complex conjugate** of the denominator and then combining terms and simplifying. The complex conjugate of a number is found by changing the sign of the j term. As an example,

$$\frac{10 + j5}{2 + j4} = \frac{(10 + j5)(2 - j4)}{(2 + j4)(2 - j4)} = \frac{20 - j30 + 20}{4 + 16} = \frac{40 - j30}{20} = 2 - j1.5$$

Like multiplication, division is easier when the numbers are in polar form, so it is best to convert to polar form before dividing. The rule is

Divide the magnitude of the numerator by the magnitude of the denominator to get the magnitude of the quotient. Then subtract the denominator angle from the numerator angle to get the angle of the quotient.

EXAMPLE 15–7

Perform the following divisions:

(a) Divide $100\angle50°$ by $25\angle20°$.   **(b)** Divide $15\angle10°$ by $3\angle-30°$.

Solution **(a)** $\dfrac{100\angle50°}{25\angle20°} = \left(\dfrac{100}{25}\right)\angle(50° - 20°) = \mathbf{4\angle30°}$

(b) $\dfrac{15\angle10°}{3\angle-30°} = \left(\dfrac{15}{3}\right)\angle[10° - (-30°)] = \mathbf{5\angle40°}$

Related Problem Divide $24\angle-30°$ by $6\angle12°$.

1. Convert $2 + j2$ to polar form. In which quadrant does this phasor lie?
2. Convert $5\angle-45°$ to rectangular form. In which quadrant does this phasor lie?
3. Add $1 + j2$ and $3 - j1$.
4. Subtract $12 + j18$ from $15 + j25$.
5. Multiply $8\angle45°$ times $2\angle65°$.
6. Divide $30\angle75°$ by $6\angle60°$.

15–2 SINUSOIDAL RESPONSE OF SERIES *RC* CIRCUITS

When a sinusoidal voltage is applied to a series *RC* circuit, each resulting voltage drop and the current in the circuit are also sinusoidal and have the same frequency as the applied voltage. The capacitance causes a phase shift between the voltage and current that depends on the relative values of the resistance and the capacitive reactance.

After completing this section, you should be able to

◆ **Describe the relationship between current and voltage in a series *RC* circuit**

   ◆ Discuss voltage and current waveforms

   ◆ Discuss phase shift

As shown in Figure 15–14, the resistor voltage (V_R), the capacitor voltage (V_C), and the current (I) are all sine waves with the frequency of the source. Phase shifts are introduced because of the capacitance. The resistor voltage and current *lead* the source voltage, and the capacitor voltage *lags* the source voltage. The phase angle between the current and the capacitor voltage is always 90°. These generalized phase relationships are indicated in Figure 15–14.

▶ **FIGURE 15–14**

Illustration of sinusoidal response with general phase relationships of V_R, V_C, and I relative to the source voltage. V_R and I are in phase while V_R and V_C are 90° out of phase.

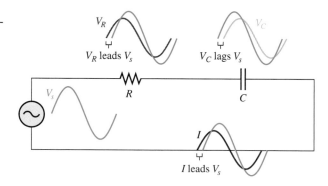

The amplitudes and the phase relationships of the voltages and current depend on the values of the resistance and the **capacitive reactance**. When a circuit is purely resistive, the phase angle between the applied (source) voltage and the total current is zero. When a circuit is purely capacitive, the phase angle between the applied voltage and the total current is 90°, with the current leading the voltage. When there is a combination of both resistance and capacitive reactance in a circuit, the phase angle between the applied voltage and the total current is somewhere between 0° and 90°, depending on the relative values of the resistance and the capacitive reactance.

15–3 IMPEDANCE OF SERIES *RC* CIRCUITS

The **impedance** of a series *RC* circuit consists of resistance and capacitive reactance and is the total opposition to sinusoidal current. Its unit is the ohm. The impedance also causes a phase difference between the total current and the source voltage. Therefore, the impedance consists of a magnitude component and a phase angle component.

After completing this section, you should be able to

◆ **Determine the impedance of a series *RC* circuit**

  ◆ Define *impedance*

  ◆ Express capacitive reactance in complex form

  ◆ Express total impedance in complex form

  ◆ Draw an impedance triangle

  ◆ Calculate impedance magnitude and the phase angle

In a purely resistive circuit, the impedance is simply equal to the total resistance. In a purely capacitive circuit, the impedance is equal to the total capacitive reactance. The impedance of a series *RC* circuit is determined by both the resistance and the capacitive reactance. These cases are illustrated in Figure 15–15. The magnitude of the impedance is symbolized by *Z*.

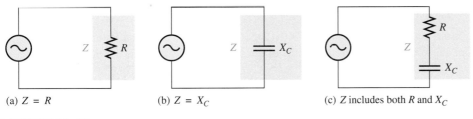

(a) $Z = R$ (b) $Z = X_C$ (c) *Z* includes both *R* and X_C

▲ **FIGURE 15–15**

Three cases of impedance.

Capacitive reactance is a phasor quantity and is expressed as a complex number in rectangular form as

$$\mathbf{X}_C = -jX_C$$

where boldface $\mathbf{X}_C$ designates a phasor quantity (representing both magnitude and angle) and X_C is just the magnitude.

In the series *RC* circuit of Figure 15–16, the total impedance is the phasor sum of *R* and $-jX_C$ and is expressed as

Equation 15–7

$$\mathbf{Z} = R - jX_C$$

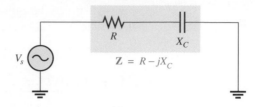

▲ FIGURE 15–16

Impedance in a series *RC* circuit.

In ac analysis, both *R* and X_C are as shown in the phasor diagram of Figure 15–17(a), with X_C appearing at a $-90°$ angle with respect to *R*. This relationship comes from the fact that the capacitor voltage in a series *RC* circuit lags the current, and thus the resistor voltage, by 90°. Since **Z** is the phasor sum of *R* and $-jX_C$, its phasor representation is as shown in Figure 15–17(b). A repositioning of the phasors, as shown in part (c), forms a right triangle called the *impedance triangle*. The length of each phasor represents the magnitude in ohms, and the angle θ is the phase angle of the *RC* circuit and represents the phase difference between the applied voltage and the current.

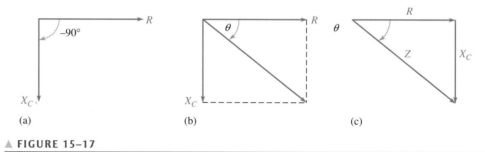

▲ FIGURE 15–17

Development of the impedance triangle for a series *RC* circuit.

From right-angle trigonometry (Pythagorean theorem), the magnitude (length) of the impedance can be expressed in terms of the resistance and reactance as

$$Z = \sqrt{R^2 + X_C^2}$$

The italic letter *Z* represents the magnitude of the phasor quantity **Z** and is expressed in ohms.

The phase angle, θ, is expressed as

$$\theta = -\tan^{-1}\left(\frac{X_C}{R}\right)$$

The symbol $\tan^{-1}$ stands for inverse tangent. You can find the $\tan^{-1}$ value on your calculator. Combining the magnitude and angle, the phasor expression for impedance in polar form is

Equation 15–8

$$\mathbf{Z} = \sqrt{R^2 + X_C^2}\angle -\tan^{-1}\left(\frac{X_C}{R}\right)$$

EXAMPLE 15–8

For each circuit in Figure 15–18, write the phasor expression for the impedance in both rectangular form and polar form.

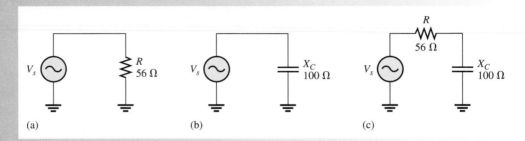

(a)   (b)   (c)

▲ **FIGURE 15–18**

Solution   For the circuit in Figure 15–18(a), the impedance is

$$\mathbf{Z} = R - j0 = R = \mathbf{56\ \Omega} \qquad \text{in rectangular form } (X_C = 0)$$
$$\mathbf{Z} = R\angle 0° = \mathbf{56\angle 0°\ \Omega} \qquad \text{in polar form}$$

The impedance is simply the resistance, and the phase angle is zero because pure resistance does not cause a phase shift between the voltage and current.

For the circuit in Figure 15–18(b), the impedance is

$$\mathbf{Z} = 0 - jX_C = \mathbf{-j100\ \Omega} \qquad \text{in rectangular form } (R = 0)$$
$$\mathbf{Z} = X_C\angle -90° = \mathbf{100\angle -90°\ \Omega} \qquad \text{in polar form}$$

The impedance is simply the capacitive reactance, and the phase angle is −90° because the capacitance causes the current to lead the voltage by 90°.

For the circuit in Figure 15–18(c), the impedance in rectangular form is

$$\mathbf{Z} = R - jX_C = \mathbf{56\ \Omega - j100\ \Omega}$$

The impedance in polar form is

$$\mathbf{Z} = \sqrt{R^2 + X_C^2}\angle -\tan^{-1}\left(\frac{X_C}{R}\right)$$

$$= \sqrt{(56\ \Omega)^2 + (100\ \Omega)^2}\angle -\tan^{-1}\left(\frac{100\ \Omega}{56\ \Omega}\right) = \mathbf{115\angle -60.8°\ \Omega}$$

In this case, the impedance is the phasor sum of the resistance and the capacitive reactance. The phase angle is fixed by the relative values of X_C and R. Rectangular to polar conversion can be done on a calculator (refer to your user's manual).

Related Problem   Use your calculator to convert the impedance in Figure 15–18(c) from rectangular to polar form. Draw the impedance phasor diagram.

SECTION 15–3
REVIEW

1. The impedance of a certain *RC* circuit is 150 Ω − *j*220 Ω. What is the value of the resistance? The capacitive reactance?

2. A series *RC* circuit has a total resistance of 33 kΩ and a capacitive reactance of 50 kΩ. Write the phasor expression for the impedance in rectangular form.

3. For the circuit in Question 2, what is the magnitude of the impedance? What is the phase angle?

15–4 ANALYSIS OF SERIES *RC* CIRCUITS

In this section, Ohm's law and Kirchhoff's voltage law are used in the analysis of series *RC* circuits to determine voltage, currents and impedance. Also, *RC* lead and lag circuits are examined.

After completing this section, you should be able to

♦ **Analyze a series *RC* circuit**

  ♦ Apply Ohm's law and Kirchhoff's voltage law to series *RC* circuits

  ♦ Express the voltages and current as phasor quantities

  ♦ Show how impedance and phase angle vary with frequency

  ♦ Discuss and analyze the *RC* lag circuit

  ♦ Discuss and analyze the *RC* lead circuit

Ohm's Law

The application of Ohm's law to series *RC* circuits involves the use of the phasor quantities of **Z**, **V**, and **I**. Keep in mind that the use of boldface nonitalic letters indicates phasor quantities where both magnitude and angle are included. The three equivalent forms of Ohm's law are as follows:

Equation 15–9
$$\mathbf{V} = \mathbf{IZ}$$

Equation 15–10
$$\mathbf{I} = \frac{\mathbf{V}}{\mathbf{Z}}$$

Equation 15–11
$$\mathbf{Z} = \frac{\mathbf{V}}{\mathbf{I}}$$

Recall that multiplication and division are most easily accomplished with the polar forms. Since Ohm's law calculations involve multiplications and divisions, you should express the voltage, current, and impedance in polar form. The following two examples show the relationship between the source voltage and source current. In Example 15–9, the current is the reference and in Example 15–10, the voltage is the reference. Notice that the reference is drawn along the x-axis in both cases.

EXAMPLE 15–9

The current in Figure 15–19 is expressed in polar form as $\mathbf{I} = 0.2\angle 0°$ mA. Determine the source voltage expressed in polar form, and draw a phasor diagram showing the relation between source voltage and current.

▶ **FIGURE 15–19**

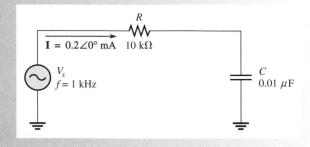

Solution   The magnitude of the capacitive reactance is

$$X_C = \frac{1}{2\pi f C} = \frac{1}{2\pi(1000\ \text{Hz})(0.01\ \mu\text{F})} = 15.9\ \text{k}\Omega$$

The total impedance in rectangular form is

$$\mathbf{Z} = R - jX_C = 10\ \text{k}\Omega - j15.9\ \text{k}\Omega$$

Converting to polar form yields

$$\mathbf{Z} = \sqrt{R^2 + X_C^2} \angle -\tan^{-1}\!\left(\frac{X_C}{R}\right)$$

$$= \sqrt{(10\ \text{k}\Omega)^2 + (15.9\ \text{k}\Omega)^2} \angle -\tan^{-1}\!\left(\frac{15.9\ \text{k}\Omega}{10\ \text{k}\Omega}\right) = 18.8 \angle -57.8^\circ\ \text{k}\Omega$$

Use Ohm's law to determine the source voltage.

$$\mathbf{V}_s = \mathbf{IZ} = (0.2 \angle 0^\circ\ \text{mA})(18.8 \angle -57.8^\circ\ \text{k}\Omega) = \mathbf{3.76 \angle -57.8^\circ\ V}$$

The magnitude of the source voltage is 3.76 V at an angle of -57.8° with respect to the current; that is, the voltage lags the current by 57.8°, as shown in the phasor diagram of Figure 15–20.

▶ **FIGURE 15–20**

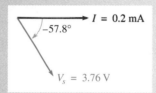

Related Problem   Determine $\mathbf{V}_s$ in Figure 15–19 if $f = 2\ \text{kHz}$ and $\mathbf{I} = 0.2 \angle 0^\circ\ \text{A}$.

EXAMPLE 15–10   Determine the current in the circuit of Figure 15–21, and draw a phasor diagram showing the relation between source voltage and current.

▶ **FIGURE 15–21**

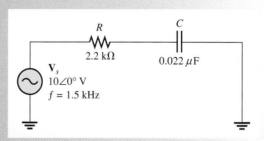

Solution   The magnitude of the capacitive reactance is

$$X_C = \frac{1}{2\pi f C} = \frac{1}{2\pi(1.5\ \text{kHz})(0.022\ \mu\text{F})} = 4.82\ \text{k}\Omega$$

The total impedance in rectangular form is

$$\mathbf{Z} = R - jX_C = 2.2\ \text{k}\Omega - j4.82\ \text{k}\Omega$$

Converting to polar form yields

$$\mathbf{Z} = \sqrt{R^2 + X_C^2} \angle -\tan^{-1}\left(\frac{X_C}{R}\right)$$

$$= \sqrt{(2.2\,\text{k}\Omega)^2 + (4.82\,\text{k}\Omega)^2} \angle -\tan^{-1}\left(\frac{4.82\,\text{k}\Omega}{2.2\,\text{k}\Omega}\right) = 5.30 \angle -65.5^\circ\,\text{k}\Omega$$

Use Ohm's law to determine the current.

$$\mathbf{I} = \frac{\mathbf{V}}{\mathbf{Z}} = \frac{10 \angle 0^\circ\,\text{V}}{5.30 \angle -65.5^\circ\,\text{k}\Omega} = \mathbf{1.89 \angle 65.5^\circ\,\text{mA}}$$

The magnitude of the current is 1.89 mA. The positive phase angle of 65.5° indicates that the current leads the voltage by that amount, as shown in the phasor diagram of Figure 15–22.

▶ FIGURE 15–22

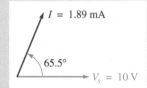

$I = 1.89\,\text{mA}$

65.5°

$V_s = 10\,\text{V}$

Related Problem   Determine **I** in Figure 15–21 if the frequency is increased to 5 kHz.

Use Multisim file E15-10 to verify the calculated results in this example and to confirm your calculation for the related problem.

Phase Relationships of Current and Voltages

In a series *RC* circuit, the current is the same through both the resistor and the capacitor. Thus, the resistor voltage is in phase with the current, and the capacitor voltage lags the current by 90°. Therefore, there is a phase difference of 90° between the resistor voltage, V_R, and the capacitor voltage, V_C, as shown in the waveform diagram of Figure 15–23.

▶ FIGURE 15–23

Phase relation of voltages and current in a series *RC* circuit.

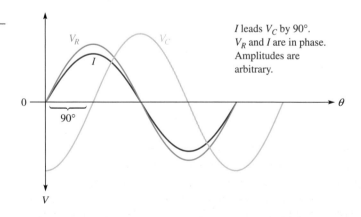

I leads V_C by 90°.
V_R and I are in phase.
Amplitudes are arbitrary.

From Kirchhoff's voltage law, the sum of the voltage drops must equal the applied voltage. However, since V_R and V_C are not in phase with each other, they must be added as phasor quantities, with V_C lagging V_R by 90°, as shown in Figure 15–24(a). As shown in

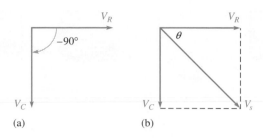

▲ FIGURE 15–24

Voltage phasor diagram for a series *RC* circuit.

Figure 15–24(b), $\mathbf{V}_s$ is the phasor sum of V_R and V_C, as expressed in rectangular form in the following equation:

$$\mathbf{V}_s = V_R - jV_C$$

Equation 15–12

This equation can be expressed in polar form as

$$\mathbf{V}_s = \sqrt{V_R^2 + V_C^2}\angle -\tan^{-1}\left(\frac{V_C}{V_R}\right)$$

Equation 15–13

where the magnitude of the source voltage is

$$V_s = \sqrt{V_R^2 + V_C^2}$$

and the phase angle between the resistor voltage and the source voltage is

$$\theta = -\tan^{-1}\left(\frac{V_C}{V_R}\right)$$

Since the resistor voltage and the current are in phase, θ also represents the phase angle between the source voltage and the current. Figure 15–25 shows a complete voltage and current phasor diagram that represents the waveform diagram of Figure 15–23.

Variation of Impedance and Phase Angle with Frequency

As you know, capacitive reactance varies inversely with frequency. Since $Z = \sqrt{R^2 + X_C^2}$, you can see that when X_C increases, the entire term under the square root sign increases and thus the magnitude of the total impedance also increases; and when X_C decreases, the magnitude of the total impedance also decreases. Therefore, *in an RC circuit, Z is inversely dependent on frequency.*

Figure 15–26 illustrates how the voltages and current in a series *RC* circuit vary as the frequency increases or decreases, with the source voltage held at a constant value. In part (a), as the frequency is increased, X_C decreases; so less voltage is dropped across the capacitor. Also, *Z* decreases as X_C decreases, causing the current to increase. An increase in the current causes more voltage across *R*.

In Figure 15–26(b), as the frequency is decreased, X_C increases; so more voltage is dropped across the capacitor. Also, *Z* increases as X_C increases, causing the current to decrease. A decrease in the current causes less voltage across *R*.

The effect of changes in *Z* and X_C can be observed as shown in Figure 15–27. As the frequency increases, the voltage across *Z* remains constant because V_s is constant. Also, the voltage across *C* decreases. The increasing current indicates that *Z* is decreasing. It does so because of the inverse relationship stated in Ohm's law ($Z = V_Z/I$). The increasing current also indicates that X_C is decreasing ($X_C = V_C/I$). The decrease in V_C corresponds to the decrease in X_C.

Since X_C is the factor that introduces the phase angle in a series *RC* circuit, a change in X_C produces a change in the phase angle. As the frequency is increased, X_C becomes

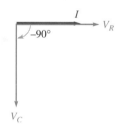

▲ FIGURE 15–25

Voltage and current phasor diagram for the waveforms in Figure 15–23.

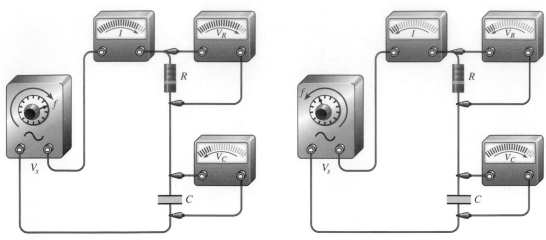

(a) As frequency is increased, Z decreases as X_C decreases, causing I and V_R to increase and V_C to decrease.

(b) As frequency is decreased, Z increases as X_C increases, causing I and V_R to decrease and V_C to increase.

▲ FIGURE 15–26

An illustration of how the variation of impedance affects the voltages and current as the source frequency is varied. The source voltage is held at a constant amplitude.

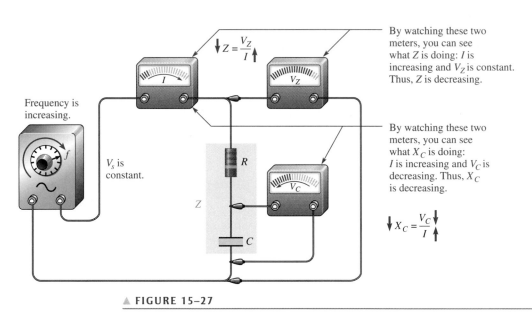

▲ FIGURE 15–27

An illustration of how Z and X_C change with frequency.

smaller, and thus the phase angle decreases. As the frequency is decreased, X_C becomes larger, and thus the phase angle increases. The angle between V_s and V_R is the phase angle of the circuit because I is in phase with V_R. By measuring the phase of V_R, you are effectively measuring the phase of I. An oscilloscope is normally used to observe the phase angle by measuring the phase angle between V_s and one of the component voltages.

Figure 15–28 uses the impedance triangle to illustrate the variations in X_C, Z, and θ as the frequency changes. Of course, R remains constant. The main point is that because X_C varies inversely with the frequency, so also do the magnitude of the total impedance and the phase angle. Example 15–11 illustrates this.

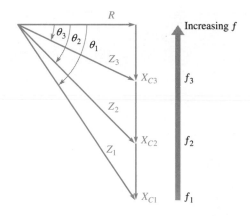

▲ FIGURE 15–28

As the frequency increases, X_C decreases, Z decreases, and θ decreases. Each value of frequency can be visualized as forming a different impedance triangle.

EXAMPLE 15–11

For the series *RC* circuit in Figure 15–29, determine the magnitude of the total impedance and the phase angle for each of the following values of input frequency:

(a) 10 kHz (b) 20 kHz (c) 30 kHz

▶ **FIGURE 15–29**

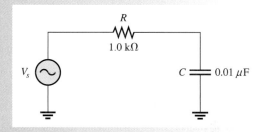

Solution (a) For $f = 10\,\text{kHz}$,

$$X_C = \frac{1}{2\pi fC} = \frac{1}{2\pi(10\,\text{kHz})(0.01\,\mu\text{F})} = 1.59\,\text{k}\Omega$$

$$\mathbf{Z} = \sqrt{R^2 + X_C^2}\angle -\tan^{-1}\left(\frac{X_C}{R}\right)$$

$$= \sqrt{(1.0\,\text{k}\Omega)^2 + (1.59\,\text{k}\Omega)^2}\angle -\tan^{-1}\left(\frac{1.59\,\text{k}\Omega}{1.0\,\text{k}\Omega}\right) = 1.88\angle -57.8°\,\text{k}\Omega$$

Thus, $Z = \mathbf{1.88\,k\Omega}$ and $\theta = \mathbf{-57.8°}$.

(b) For $f = 20\,\text{kHz}$,

$$X_C = \frac{1}{2\pi(20\,\text{kHz})(0.01\,\mu\text{F})} = 796\,\Omega$$

$$\mathbf{Z} = \sqrt{(1.0\,\text{k}\Omega)^2 + (796\,\Omega)^2}\angle -\tan^{-1}\left(\frac{796\,\Omega}{1.0\,\text{k}\Omega}\right) = 1.28\angle -38.5°\,\text{k}\Omega$$

Thus, $Z = \mathbf{1.28\,k\Omega}$ and $\theta = \mathbf{-38.5°}$.

(c) For $f = 30\,\text{kHz}$,

$$X_C = \frac{1}{2\pi(30\,\text{kHz})(0.01\,\mu\text{F})} = 531\,\Omega$$

$$Z = \sqrt{(1.0\,\text{k}\Omega)^2 + (531\,\Omega)^2}\angle-\tan^{-1}\!\left(\frac{531\,\Omega}{1.0\,\text{k}\Omega}\right) = 1.13\angle-28.0°\,\text{k}\Omega$$

Thus, $Z = \mathbf{1.13\,k\Omega}$ and $\theta = \mathbf{-28.0°}$.

Notice that as the frequency increases, X_C, Z, and θ decrease.

Related Problem Find the magnitude of the total impedance and the phase angle in Figure 15–29 for $f = 1\,\text{kHz}$.

The *RC* Lag Circuit

An *RC* lag circuit is a phase shift circuit in which the output voltage lags the input voltage by a specified amount. Figure 15–30(a) shows a series *RC* circuit with the output voltage taken across the capacitor. The source voltage is the input, V_{in}. As you know, θ, the phase angle between the current and the input voltage, is also the phase angle between the resistor voltage and the input voltage because V_R and I are in phase with each other.

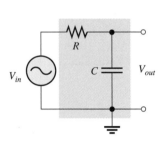

(a) A basic *RC* lag circuit

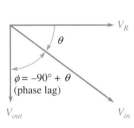

(b) Phasor voltage diagram showing the phase lag between V_{in} and V_{out}

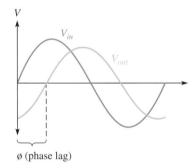

(c) Input and output voltage waveforms

▲ **FIGURE 15–30**

RC lag circuit ($V_{out} = V_C$).

Since V_C lags V_R by 90°, the phase angle between the capacitor voltage and the input voltage is the difference between −90° and θ, as shown in Figure 15–30(b). The capacitor voltage is the output, and it lags the input, thus creating a basic lag circuit.

The input and output voltage waveforms of the lag circuit are shown in Figure 15–30(c). The amount of phase difference, designated ϕ, between the input and the output is dependent on the relative sizes of the capacitive reactance and the resistance, as is the magnitude of the output voltage.

Phase Difference Between Input and Output As already established, θ is the phase angle between I and V_{in}. The angle between V_{out} and V_{in} is designated ϕ (phi) and is developed as follows.

The polar expressions for the input voltage and the current are $V_{in}\angle0°$ and $I\angle\theta$, respectively. The output voltage in polar form is

$$\mathbf{V}_{out} = (I\angle\theta)(X_C\angle-90°) = IX_C\angle(-90° + \theta)$$

The preceding equation states that the output voltage is at an angle of $-90° + \theta$ with respect to the input voltage. Since $\theta = -\tan^{-1}(X_C/R)$, the angle ϕ between the input and output is

$$\phi = -90° + \tan^{-1}\left(\frac{X_C}{R}\right)$$

Equivalently, this angle can be expressed as

$$\phi = -\tan^{-1}\left(\frac{R}{X_C}\right)$$

Equation 15–14

This angle is always negative, indicating that the output voltage lags the input voltage, as shown in Figure 15–31.

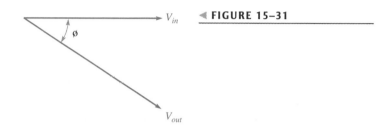

◀ FIGURE 15–31

EXAMPLE 15–12

Determine the amount of phase lag from input to output in each lag circuit in Figure 15–32.

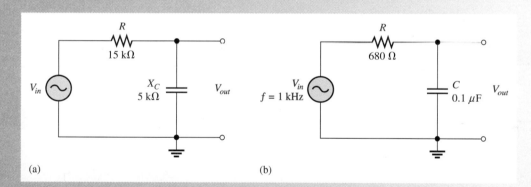

(a) (b)

▲ FIGURE 15–32

Solution For the lag circuit in Figure 15–32(a),

$$\phi = -\tan^{-1}\left(\frac{R}{X_C}\right) = -\tan^{-1}\left(\frac{15\,k\Omega}{5\,k\Omega}\right) = -71.6°$$

The output lags the input by 71.6°.
For the lag circuit in Figure 15–32(b), first determine the capacitive reactance.

$$X_C = \frac{1}{2\pi fC} = \frac{1}{2\pi(1\,kHz)(0.1\,\mu F)} = 1.59\,k\Omega$$

$$\phi = -\tan^{-1}\left(\frac{R}{X_C}\right) = -\tan^{-1}\left(\frac{680\,\Omega}{1.59\,k\Omega}\right) = -23.2°$$

The output lags the input by 23.2°.

Magnitude of the Output Voltage To evaluate the output voltage in terms of its magnitude, visualize the *RC* lag circuit as a voltage divider. A portion of the total input voltage is dropped across the resistor and a portion across the capacitor. Because the output voltage is the voltage across the capacitor, it can be calculated using either Ohm's law ($V_{out} = IX_C$) or the voltage divider formula.

Equation 15–15

$$V_{out} = \left(\frac{X_C}{\sqrt{R^2 + X_C^2}}\right)V_{in}$$

The phasor expression for the output voltage of an *RC* lag circuit is

$$\mathbf{V}_{out} = V_{out}\angle\phi$$

EXAMPLE 15–13

For the lag circuit in Figure 15–32(b) (Example 15–12), determine the output voltage in phasor form when the input voltage has an rms value of 10 V. Draw the input and output voltage waveforms showing the proper phase relationship. The capacitive reactance X_C (1.59 kΩ) and ϕ (−23.2°) were found in Example 15–12.

Solution The output voltage in phasor form is

$$\mathbf{V}_{out} = V_{out}\angle\phi = \left(\frac{X_C}{\sqrt{R^2 + X_C^2}}\right)V_{in}\angle\phi$$

$$= \left(\frac{1.59\text{ k}\Omega}{\sqrt{(680\text{ }\Omega)^2 + (1.59\text{ k}\Omega)^2}}\right)10\angle-23.2°\text{ V} = \mathbf{9.20\angle-23.2°\text{ V rms}}$$

The waveforms are shown in Figure 15–33. Notice that the output voltage lags the input voltage by 23.2°.

FIGURE 15–33

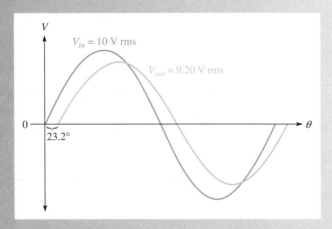

Related Problem In a lag circuit, what happens to the output voltage if the frequency increases?

Use Multisim file E15-13 to verify the calculated results in this example and to confirm your calculation for the related problem.

The *RC* Lead Circuit

An *RC* lead circuit is a phase shift circuit in which the output voltage leads the input voltage by a specified amount. When the output of a series *RC* circuit is taken across the resistor rather than across the capacitor, as shown in Figure 15–34(a), it becomes a lead circuit.

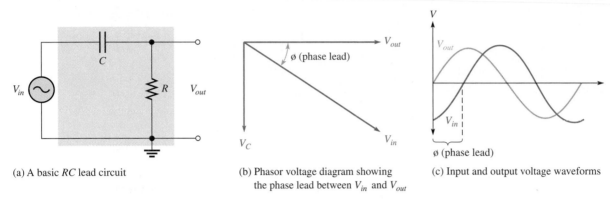

(a) A basic *RC* lead circuit

(b) Phasor voltage diagram showing the phase lead between V_{in} and V_{out}

(c) Input and output voltage waveforms

▲ **FIGURE 15–34**

RC lead circuit ($V_{out} = V_R$).

Phase Difference Between Input and Output In a series *RC* circuit, the current leads the input voltage. Also, as you know, the resistor voltage is in phase with the current. Since the output voltage is taken across the resistor, the output leads the input, as indicated by the phasor diagram in Figure 15–34(b). The waveforms are shown in Figure 15–34(c).

As in the lag circuit, the amount of phase difference between the input and output and the magnitude of the output voltage in the lead circuit are dependent on the relative values of the resistance and the capacitive reactance. When the input voltage is assigned a reference angle of 0°, the angle of the output voltage is the same as θ (the angle between total current and applied voltage) because the resistor voltage (output) and the current are in phase with each other. Therefore, since $\phi = \theta$ in this case, the expression is

$$\phi = \tan^{-1}\left(\frac{X_C}{R}\right)$$

Equation 15–16

This angle is positive because the output leads the input.

EXAMPLE 15–14 Calculate the output phase angle for each circuit in Figure 15–35.

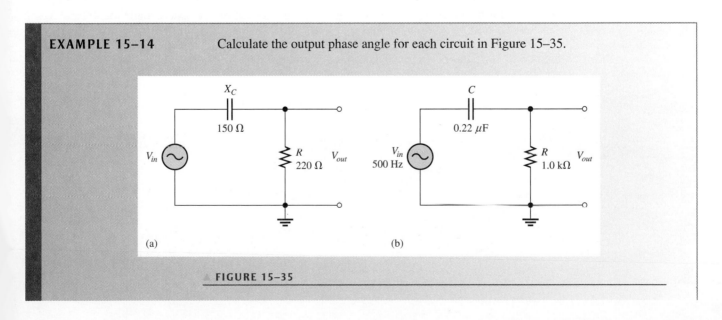

(a)

(b)

▲ **FIGURE 15–35**

Solution For the lead circuit in Figure 15–35(a),

$$\phi = \tan^{-1}\left(\frac{X_C}{R}\right) = \tan^{-1}\left(\frac{150\ \Omega}{220\ \Omega}\right) = \mathbf{34.3°}$$

The output leads the input by 34.3°.

For the lead circuit in Figure 15–35(b), first determine the capacitive reactance.

$$X_C = \frac{1}{2\pi f C} = \frac{1}{2\pi(500\ \text{Hz})(0.22\ \mu\text{F})} = 1.45\ \text{k}\Omega$$

$$\phi = \tan^{-1}\left(\frac{X_C}{R}\right) = \tan^{-1}\left(\frac{1.45\ \text{k}\Omega}{1.0\ \text{k}\Omega}\right) = \mathbf{55.4°}$$

The output leads the input by 55.4°.

Related Problem In a lead circuit, what happens to the phase lead if the frequency increases?

 Use Multisim files E15-14A and E15-14B to verify the calculated results in this example and to confirm your calculation for the related problem.

Magnitude of the Output Voltage Since the output voltage of an *RC* lead circuit is taken across the resistor, the magnitude can be calculated using either Ohm's law ($V_{out} = IR$) or the voltage-divider formula.

Equation 15–17

$$V_{out} = \left(\frac{R}{\sqrt{R^2 + X_C^2}}\right)V_{in}$$

The expression for the output voltage in phasor form is

$$\mathbf{V}_{out} = V_{out}\angle\phi$$

EXAMPLE 15–15

The input voltage in Figure 15–35(b) (Example 15–14) has an rms value of 10 V. Determine the phasor expression for the output voltage. Draw the waveform relationships for the input and output voltages showing peak values. The phase angle (55.4°) and X_C (1.45 kΩ) were found in Example 15–14.

Solution The phasor expression for the output voltage is

$$\mathbf{V}_{out} = V_{out}\angle\phi = \left(\frac{R}{\sqrt{R^2 + X_C^2}}\right)V_{in}\angle\phi$$

$$= \left(\frac{1.0\ \text{k}\Omega}{1.76\ \text{k}\Omega}\right)10\angle 55.4°\ \text{V} = \mathbf{5.68\angle 55.4°\ V\ rms}$$

The peak value of the input voltage is

$$V_{in(p)} = 1.414 V_{in(rms)} = 1.414(10\ \text{V}) = 14.14\ \text{V}$$

The peak value of the output voltage is

$$V_{out(p)} = 1.414 V_{out(rms)} = 1.414(5.68\ \text{V}) = 8.03\ \text{V}$$

The waveforms are shown in Figure 15–36.

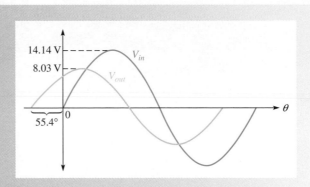

▲ FIGURE 15–36

Related Problem   In a lead circuit, what happens to the output voltage if the frequency is reduced?

  Use Multisim file E15-15 to verify the calculated results in this example and to confirm your calculation for the related problem.

SECTION 15–4
REVIEW

1. In a certain series *RC* circuit, $V_R = 4$ V, and $V_C = 6$ V. What is the magnitude of the source voltage?
2. In Question 1, what is the phase angle between the source voltage and the current?
3. What is the phase difference between the capacitor voltage and the resistor voltage in a series *RC* circuit?
4. When the frequency of the applied voltage in a series *RC* circuit is increased, what happens to the capacitive reactance? What happens to the magnitude of the total impedance? What happens to the phase angle?
5. A certain *RC* lag circuit consists of a 4.7 kΩ resistor and a 0.022 μF capacitor. Determine the phase shift between input and output at a frequency of 3 kHz.
6. An *RC* lead circuit has the same component values as the lag circuit in Question 5. What is the magnitude of the output voltage at 3 kHz when the input is 10 V rms?

OPTION 2 NOTE

Coverage of series reactive circuits continues in Chapter 16, Part 1, on page 678.

15–5 IMPEDANCE AND ADMITTANCE OF PARALLEL *RC* CIRCUITS

In this section, you will learn how to determine the impedance and phase angle of a parallel *RC* circuit. The impedance consists of a magnitude component and a phase angle component. Also, capacitive susceptance and admittance of a parallel *RC* circuit are introduced.

After completing this section, you should be able to

+ **Determine impedance and admittance in a parallel *RC* circuit**

    + Express total impedance in complex form

    + Define and calculate *conductance, capacitive susceptance,* and *admittance*

Figure 15–37 shows a basic parallel *RC* circuit connected to an ac voltage source.

▶ FIGURE 15–37

Basic parallel *RC* circuit.

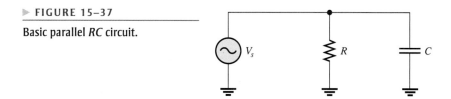

The expression for the total impedance is developed as follows, using complex numbers. Since there are only two circuit components, *R* and *C*, the total impedance can be found from the product-over-sum rule.

$$\mathbf{Z} = \frac{(R\angle 0°)(X_C\angle -90°)}{R - jX_C}$$

By multiplying the magnitudes, adding the angles in the numerator, and converting the denominator to polar form, you get

$$\mathbf{Z} = \frac{RX_C\angle(0° - 90°)}{\sqrt{R^2 + X_C^2}\angle -\tan^{-1}\left(\frac{X_C}{R}\right)}$$

Now, by dividing the magnitude expression in the numerator by that in the denominator, and by subtracting the angle in the denominator from that in the numerator, you get

$$\mathbf{Z} = \left(\frac{RX_C}{\sqrt{R^2 + X_C^2}}\right)\angle\left(-90° + \tan^{-1}\left(\frac{X_C}{R}\right)\right)$$

Equivalently, this expression can be written as

$$\mathbf{Z} = \frac{RX_C}{\sqrt{R^2 + X_C^2}} \angle -\tan^{-1}\left(\frac{R}{X_C}\right)$$

Equation 15–18

EXAMPLE 15–16

For each circuit in Figure 15–38, determine the magnitude and phase angle of the total impedance.

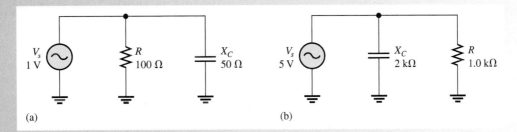

(a)   (b)

▲ **FIGURE 15–38**

Solution For the circuit in Figure 15–38(a), the total impedance is

$$\mathbf{Z} = \left(\frac{RX_C}{\sqrt{R^2 + X_C^2}}\right)\angle -\tan^{-1}\left(\frac{R}{X_C}\right)$$

$$= \left(\frac{(100\ \Omega)(50\ \Omega)}{\sqrt{(100\ \Omega)^2 + (50\ \Omega)^2}}\right)\angle -\tan^{-1}\left(\frac{100\ \Omega}{50\ \Omega}\right) = \mathbf{44.7}\angle\mathbf{-63.4°\ \Omega}$$

Thus, $Z = 44.7\ \Omega$ and $\theta = -63.4°$.

For the circuit in Figure 15–38(b), the total impedance is

$$\mathbf{Z} = \left(\frac{(1.0\ k\Omega)(2\ k\Omega)}{\sqrt{(1.0\ k\Omega)^2 + (2\ k\Omega)^2}}\right)\angle -\tan^{-1}\left(\frac{1.0\ k\Omega}{2\ k\Omega}\right) = \mathbf{894}\angle\mathbf{-26.6°\ \Omega}$$

Thus, $Z = 894\ \Omega$ and $\theta = -26.6°$.

Related Problem Determine **Z** in Figure 15–38(a) if the frequency is doubled.

Conductance, Susceptance, and Admittance

Recall that **conductance, *G*,** is the reciprocal of resistance. The phasor expression for conductance is expressed as

$$\mathbf{G} = \frac{1}{R\angle 0°} = G\angle 0°$$

Two new terms are now introduced for use in parallel *RC* circuits. **Capacitive susceptance (B_C)** is the reciprocal of capacitive reactance. The phasor expression for capacitive susceptance is

$$\mathbf{B}_C = \frac{1}{X_C\angle -90°} = B_C\angle 90° = +jB_C$$

Admittance (Y) is the reciprocal of impedance. The phasor expression for admittance is

$$\mathbf{Y} = \frac{1}{Z\angle \pm\theta} = Y\angle \mp\theta$$

The unit of each of these terms is the siemens (S), which is the reciprocal of the ohm.

In working with parallel circuits, it is often easier to use conductance (G), capacitive susceptance (B_C), and admittance (Y) rather than resistance (R), capacitive reactance (X_C), and impedance (Z). In a parallel RC circuit, as shown in Figure 15–39, the total admittance is simply the phasor sum of the conductance and the capacitive susceptance.

Equation 15–19

$$\mathbf{Y} = G + jB_C$$

▶ **FIGURE 15–39**

Admittance in a parallel RC circuit.

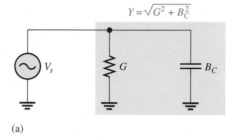

(a)

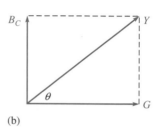

(b)

EXAMPLE 15–17

Determine the total admittance ($\mathbf{Y}$) and then convert it to total impedance ($\mathbf{Z}$) in Figure 15–40. Draw the admittance phasor diagram.

▶ **FIGURE 15–40**

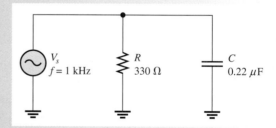

Solution From Figure 15–40, $R = 330\ \Omega$; thus $G = 1/R = 1/330\ \Omega = 3.03$ mS. The capacitive reactance is

$$X_C = \frac{1}{2\pi fC} = \frac{1}{2\pi(1000\ \text{Hz})(0.22\ \mu\text{F})} = 723\ \Omega$$

The capacitive susceptance magnitude is

$$B_C = \frac{1}{X_C} = \frac{1}{723\ \Omega} = 1.38\ \text{mS}$$

The total admittance is

$$\mathbf{Y}_{tot} = G + jB_C = \mathbf{3.03}\ \textbf{mS} + j\mathbf{1.38}\ \textbf{mS}$$

which can be expressed in polar form as

$$\mathbf{Y}_{tot} = \sqrt{G^2 + B_C^2}\,\angle\tan^{-1}\!\left(\frac{B_C}{G}\right)$$

$$= \sqrt{(3.03\ \text{mS})^2 + (1.38\ \text{mS})^2}\,\angle\tan^{-1}\!\left(\frac{1.38\ \text{mS}}{3.03\ \text{mS}}\right) = \mathbf{3.33}\angle\mathbf{24.5°}\ \textbf{mS}$$

The admittance phasor diagram is shown in Figure 15–41.

FIGURE 15–41

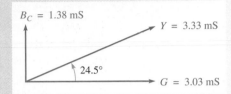

Convert total admittance to total impedance as follows:

$$\mathbf{Z}_{tot} = \frac{1}{\mathbf{Y}_{tot}} = \frac{1}{(3.33 \angle 24.5° \text{ mS})} = 300 \angle -24.5° \text{ } \Omega$$

Related Problem   Calculate the total admittance in Figure 15–40 if *f* is increased to 2.5 kHz.

**SECTION 15–5
REVIEW**

1. Define *conductance, capacitive susceptance,* and *admittance.*
2. If *Z* = 100 Ω, what is the value of Y?
3. In a certain parallel *RC* circuit, *R* = 47 kΩ and X_C = 75 kΩ. Determine **Y**.
4. In Question 3, what is **Z**?

15–6   ANALYSIS OF PARALLEL *RC* CIRCUITS

Ohm's law and Kirchhoff's current law are used in the analysis of *RC* circuits. Current and voltage relationships in a parallel *RC* circuit are examined.

After completing this section, you should be able to

◆ **Analyze a parallel *RC* circuit**

  ◆ Apply Ohm's law and Kirchhoff's current law to parallel *RC* circuits

  ◆ Express the voltages and currents as phasor quantities

  ◆ Show how impedance and phase angle vary with frequency

  ◆ Convert from a parallel circuit to an equivalent series circuit

For convenience in the analysis of parallel circuits, the Ohm's law formulas using impedance, previously stated, can be rewritten for admittance using the relation $Y = 1/Z$. Remember, the use of boldface nonitalic letters indicates phasor quantities.

$$\mathbf{V} = \frac{\mathbf{I}}{\mathbf{Y}}$$ **Equation 15–20**

$$\mathbf{I} = \mathbf{VY}$$ **Equation 15–21**

$$\mathbf{Y} = \frac{\mathbf{I}}{\mathbf{V}}$$ **Equation 15–22**

EXAMPLE 15–18

Determine the total current and phase angle in Figure 15–42. Draw a phasor diagram showing the relationship of V_s and I_{tot}.

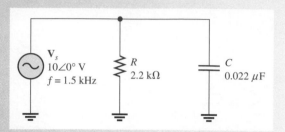

▲ FIGURE 15–42

Solution   The capacitive reactance is

$$X_C = \frac{1}{2\pi fC} = \frac{1}{2\pi(1.5\text{ kHz})(0.022\text{ }\mu F)} = 4.82\text{ k}\Omega$$

The capacitive susceptance magnitude is

$$B_C = \frac{1}{X_C} = \frac{1}{4.82\text{ k}\Omega} = 207\text{ }\mu S$$

The conductance magnitude is

$$G = \frac{1}{R} = \frac{1}{2.2\text{ k}\Omega} = 455\text{ }\mu S$$

The total admittance is

$$\mathbf{Y}_{tot} = G + jB_C = 455\text{ }\mu S + j207\text{ }\mu S$$

Converting to polar form yields

$$\mathbf{Y}_{tot} = \sqrt{G^2 + B_C^2}\angle\tan^{-1}\left(\frac{B_C}{G}\right)$$

$$= \sqrt{(455\text{ }\mu S)^2 + (207\text{ }\mu S)^2}\angle\tan^{-1}\left(\frac{207\text{ }\mu S}{455\text{ }\mu S}\right) = 500\angle 24.5°\text{ }\mu S$$

The phase angle is 24.5°.
   Use Ohm's law to determine the total current.

$$\mathbf{I}_{tot} = \mathbf{V}_s\mathbf{Y}_{tot} = (10\angle 0°\text{ V})(500\angle 24.5°\text{ }\mu S) = \mathbf{5.00\angle 24.5°\text{ mA}}$$

The magnitude of the total current is 5.00 mA, and it leads the applied voltage by **24.5°**, as the phasor diagram in Figure 15–43 indicates.

▶ FIGURE 15–43

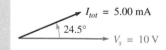

Related Problem   What is the total current (in polar form) if *f* is doubled?

Use Multisim file E15-18 to verify the calculated results in this example and to confirm your calculation for the related problem.

Phase Relationships of Currents and Voltages

Figure 15–44(a) shows all the currents in a basic parallel *RC* circuit. The total current, I_{tot}, divides at the junction into the two branch currents, I_R and I_C. The applied voltage, V_s, appears across both the resistive and the capacitive branches, so V_s, V_R, and V_C are all in phase and of the same magnitude.

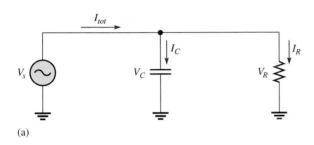
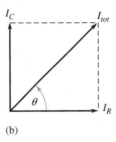

(a)   (b)

▲ FIGURE 15–44

Currents in a parallel *RC* circuit. The current directions shown in (a) are instantaneous and, of course, reverse when the source voltage reverses.

The current through the resistor is in phase with the voltage. The current through the capacitor leads the voltage, and thus the resistive current, by 90°. By Kirchhoff's current law, the total current is the phasor sum of the two branch currents, as shown by the phasor diagram in Figure 15–44(b). The total current is expressed as

$$\mathbf{I}_{tot} = I_R + jI_C$$

Equation 15–23

This equation can be expressed in polar form as

$$\mathbf{I}_{tot} = \sqrt{I_R^2 + I_C^2}\angle\tan^{-1}\left(\frac{I_C}{I_R}\right)$$

Equation 15–24

where the magnitude of the total current is

$$I_{tot} = \sqrt{I_R^2 + I_C^2}$$

and the phase angle between the resistor current and the total current is

$$\theta = \tan^{-1}\left(\frac{I_C}{I_R}\right)$$

Since the resistor current and the applied voltage are in phase, θ also represents the phase angle between the total current and the applied voltage. Figure 15–45 shows a complete current and voltage phasor diagram.

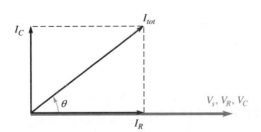

▲ FIGURE 15–45

Current and voltage phasor diagram for a parallel *RC* circuit (amplitudes are arbitrary).

EXAMPLE 15–19

Determine the value of each current in Figure 15–46, and describe the phase relationship of each with the applied voltage. Draw the current phasor diagram.

▶ **FIGURE 15–46**

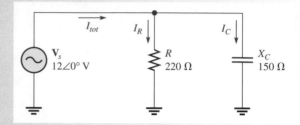

Solution The resistor current, the capacitor current, and the total current are expressed as follows:

$$\mathbf{I}_R = \frac{\mathbf{V}_s}{\mathbf{R}} = \frac{12\angle 0° \text{ V}}{220\angle 0° \ \Omega} = \mathbf{54.5\angle 0° \ mA}$$

$$\mathbf{I}_C = \frac{\mathbf{V}_s}{\mathbf{X}_C} = \frac{12\angle 0° \text{ V}}{150\angle -90° \ \Omega} = \mathbf{80\angle 90° \ mA}$$

$$\mathbf{I}_{tot} = I_R + jI_C = 54.5 \text{ mA} + j80 \text{ mA}$$

Converting $\mathbf{I}_{tot}$ to polar form yields

$$\mathbf{I}_{tot} = \sqrt{I_R^2 + I_C^2}\angle \tan^{-1}\left(\frac{I_C}{I_R}\right)$$

$$= \sqrt{(54.5 \text{ mA})^2 + (80 \text{ mA})^2}\angle \tan^{-1}\left(\frac{80 \text{ mA}}{54.5 \text{ mA}}\right) = \mathbf{96.8\angle 55.7° \ mA}$$

As the results show, the resistor current is 54.5 mA and is in phase with the voltage. The capacitor current is 80 mA and leads the voltage by 90°. The total current is 96.8 mA and leads the voltage by 55.7°. The phasor diagram in Figure 15–47 illustrates these relationships.

▶ **FIGURE 15–47**

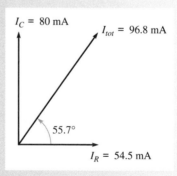

Related Problem In a parallel circuit, $\mathbf{I}_R = 100\angle 0°$ mA and $\mathbf{I}_C = 60\angle 90°$ mA. Determine the total current.

Conversion from Parallel to Series Form

For every parallel *RC* circuit, there is an equivalent series *RC* circuit for a given frequency. Two circuits are considered equivalent when they both present an equal impedance at their terminals; that is, the magnitude of impedance and the phase angle are identical.

To obtain the equivalent series circuit for a given parallel *RC* circuit, first find the impedance and phase angle of the parallel circuit. Then use the values of Z and θ to construct an impedance triangle, shown in Figure 15–48. The vertical and horizontal sides of the triangle represent the equivalent series resistance and capacitive reactance as indicated. These values can be found using the following trigonometric relationships:

$$R_{eq} = Z \cos \theta$$

Equation 15–25

$$X_{C(eq)} = Z \sin \theta$$

Equation 15–26

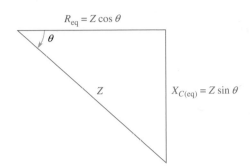

◀ **FIGURE 15–48**

Impedance triangle for the series equivalent of a parallel *RC* circuit. Z and θ are the known values for the parallel circuit. R_{eq} and $X_{C(eq)}$ are the series equivalent values.

EXAMPLE 15–20

Convert the parallel circuit in Figure 15–49 to a series form.

▶ **FIGURE 15–49**

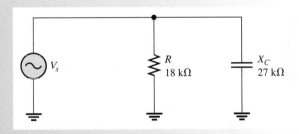

Solution First, find the admittance of the parallel circuit as follows:

$$G = \frac{1}{R} = \frac{1}{18 \text{ k}\Omega} = 55.6 \, \mu\text{S}$$

$$B_C = \frac{1}{X_C} = \frac{1}{27 \text{ k}\Omega} = 37.0 \, \mu\text{S}$$

$$\mathbf{Y} = G + jB_C = 55.6 \, \mu\text{S} + j37.0 \, \mu\text{S}$$

Converting to polar form yields

$$\mathbf{Y} = \sqrt{G^2 + B_C^2} \angle \tan^{-1}\left(\frac{B_C}{G}\right)$$

$$= \sqrt{(55.6 \, \mu\text{S})^2 + (37.0 \, \mu\text{S})^2} \angle \tan^{-1}\left(\frac{37.0 \, \mu\text{S}}{55.6 \, \mu\text{S}}\right) = 66.8 \angle 33.6° \, \mu\text{S}$$

Then, the total impedance is

$$\mathbf{Z}_{tot} = \frac{1}{\mathbf{Y}} = \frac{1}{66.8 \angle 33.6° \, \mu\text{S}} = 15.0 \angle -33.6° \text{ k}\Omega$$

Converting to rectangular form yields

$$\mathbf{Z}_{tot} = Z\cos\theta - jZ\sin\theta = R_{eq} - jX_{C(eq)}$$
$$= 15.0\,\text{k}\Omega\,\cos(-33.6°) - j15.0\,\text{k}\Omega\,\sin(-33.6°) = 12.5\,\text{k}\Omega - j8.31\,\text{k}\Omega$$

The equivalent series RC circuit is a 12.5 kΩ resistor in series with a capacitive reactance of 8.31 kΩ. This is shown in Figure 15–50.

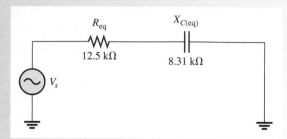

▲ FIGURE 15–50

Related Problem   The impedance of a parallel RC circuit is $\mathbf{Z} = 10\angle-26°\,\text{k}\Omega$. Convert to an equivalent series circuit.

**SECTION 15–6
REVIEW**

1. The admittance of an RC circuit is 3.50 mS, and the applied voltage is 6 V. What is the total current?
2. In a certain parallel RC circuit, the resistor current is 10 mA, and the capacitor current is 15 mA. Determine the magnitude and phase angle of the total current. This phase angle is measured with respect to what?
3. What is the phase angle between the capacitor current and the applied voltage in a parallel RC circuit?

OPTION 2 NOTE

Coverage of parallel reactive circuits continues in Chapter 16, Part 2, on page 691.

SERIES-PARALLEL CIRCUITS

15–7 ANALYSIS OF SERIES-PARALLEL *RC* CIRCUITS

The concepts studied with respect to series and parallel circuits are used to analyze circuits with combinations of both series and parallel *R* and *C* components.

After completing this section, you should be able to

◆ **Analyze series-parallel *RC* circuits**

  ◆ Determine total impedance

  ◆ Calculate currents and voltages

  ◆ Measure impedance and phase angle

The impedance of series components is most easily expressed in rectangular form, and the impedance of parallel components is best found by using polar form. The steps for analyzing a circuit with a series and a parallel component are illustrated in Example 15–21. First express the impedance of the series part of the circuit in rectangular form and the impedance of the parallel part in polar form. Next, convert the impedance of the parallel part to rectangular form and add it to the impedance of the series part. Once you determine the rectangular form of the total impedance, you can convert it to polar form in order to see the magnitude and phase angle and to calculate the current.

EXAMPLE 15–21

In the circuit of Figure 15–51, determine the following:

  (a) total impedance **(b)** total current **(c)** phase angle by which I_{tot} leads V_s

FIGURE 15–51

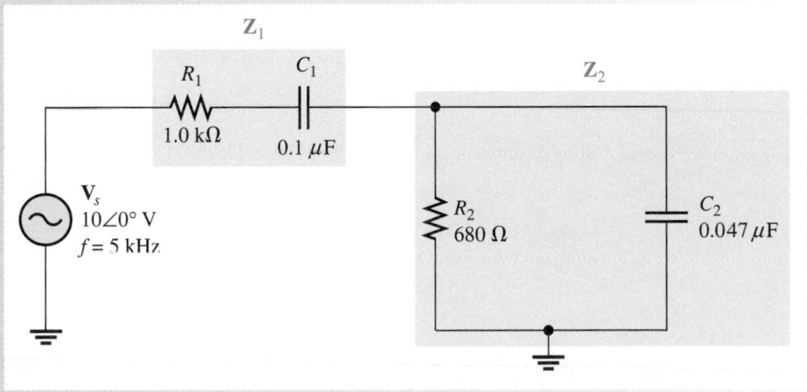

Solution **(a)** First, calculate the magnitudes of capacitive reactance.

$$X_{C1} = \frac{1}{2\pi fC} = \frac{1}{2\pi(5\text{ kHz})(0.1\ \mu\text{F})} = 318\ \Omega$$

$$X_{C2} = \frac{1}{2\pi fC} = \frac{1}{2\pi(5\text{ kHz})(0.047\ \mu\text{F})} = 677\ \Omega$$

One approach is to find the impedance of the series portion and the impedance of the parallel portion and combine them to get the total impedance. The impedance of the series combination of R_1 and C_1 is

$$\mathbf{Z}_1 = R_1 - jX_{C1} = 1.0\text{ k}\Omega - j318\ \Omega$$

To determine the impedance of the parallel portion, first determine the admittance of the parallel combination of R_2 and C_2.

$$G_2 = \frac{1}{R_2} = \frac{1}{680\ \Omega} = 1.47\text{ mS}$$

$$B_{C2} = \frac{1}{X_{C2}} = \frac{1}{677\ \Omega} = 1.48\text{ mS}$$

$$\mathbf{Y}_2 = G_2 + jB_{C2} = 1.47\text{ mS} + j1.48\text{ mS}$$

Converting to polar form yields

$$\mathbf{Y}_2 = \sqrt{G_2^2 + B_{C2}^2}\angle\tan^{-1}\left(\frac{B_{C2}}{G_2}\right)$$

$$= \sqrt{(1.47\text{ mS})^2 + (1.48\text{ mS})^2}\angle\tan^{-1}\left(\frac{1.48\text{ mS}}{1.47\text{ mS}}\right) = 2.09\angle45.2°\text{ mS}$$

Then, the impedance of the parallel portion is

$$\mathbf{Z}_2 = \frac{1}{\mathbf{Y}_2} = \frac{1}{2.09\angle45.2°\text{ mS}} = 478\angle-45.2°\ \Omega$$

Converting to rectangular form yields

$$\mathbf{Z}_2 = Z_2\cos\theta - jZ_2\sin\theta$$

$$= (478\ \Omega)\cos(-45.2°) - j(478\ \Omega)\sin(-45.2°) = 337\ \Omega - j339\ \Omega$$

The series portion and the parallel portion are in series with each other. Combine $\mathbf{Z}_1$ and $\mathbf{Z}_2$ to get the total impedance.

$$\mathbf{Z}_{tot} = \mathbf{Z}_1 + \mathbf{Z}_2$$

$$= (1.0\text{ k}\Omega - j318\ \Omega) + (337\ \Omega - j339\ \Omega) = 1337\ \Omega - j657\ \Omega$$

Expressing $\mathbf{Z}_{tot}$ in polar form yields

$$\mathbf{Z}_{tot} = \sqrt{Z_1^2 + Z_2^2}\angle-\tan^{-1}\left(\frac{Z_2}{Z_1}\right)$$

$$= \sqrt{(1338\ \Omega)^2 + (657\ \Omega)^2}\angle-\tan^{-1}\left(\frac{657\ \Omega}{1337\ \Omega}\right) = \mathbf{1.49\angle-26.2°\text{ k}\Omega}$$

(b) Use Ohm's law to determine the total current.

$$\mathbf{I}_{tot} = \frac{\mathbf{V}_s}{\mathbf{Z}_{tot}} = \frac{10\angle0°\text{ V}}{1.49\angle-26.2°\text{ k}\Omega} = \mathbf{6.71\angle26.2°\text{ mA}}$$

(c) The total current leads the applied voltage by **26.2°**.

Related Problem   Determine the voltages across $\mathbf{Z}_1$ and $\mathbf{Z}_2$ in Figure 15–51 and express in polar form.

Use Multisim file E15-21 to verify the calculated results in the (b) part of this example and to confirm your calculation for the related problem.

Example 15–22 shows two sets of series components in parallel. The approach is to first express each branch impedance in rectangular form and then convert each of these impedances to polar form. Next, calculate each branch current using polar notation. Once you know the branch currents, you can find the total current by adding the two branch currents in rectangular form. In this particular case, the total impedance is not required.

EXAMPLE 15–22   Determine all currents in Figure 15–52. Draw a current phasor diagram.

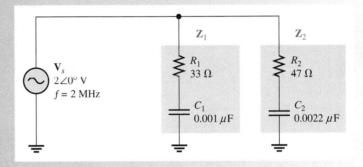

▲ **FIGURE 15–52**

Solution   First, calculate X_{C1} and X_{C2}.

$$X_{C1} = \frac{1}{2\pi fC} = \frac{1}{2\pi(2 \text{ MHz})(0.001 \ \mu\text{F})} = 79.6 \ \Omega$$

$$X_{C2} = \frac{1}{2\pi fC} - \frac{1}{2\pi(2 \text{ MHz})(0.0022 \ \mu\text{F})} = 36.2 \ \Omega$$

Next, determine the impedance of each of the two parallel branches.

$$\mathbf{Z}_1 = R_1 - jX_{C1} = 33 \ \Omega - j79.6 \ \Omega$$
$$\mathbf{Z}_2 = R_2 - jX_{C2} = 47 \ \Omega - j36.2 \ \Omega$$

Convert these impedances to polar form.

$$\mathbf{Z}_1 = \sqrt{R_1^2 + X_{C1}^2} \angle -\tan^{-1}\left(\frac{X_{C1}}{R_1}\right)$$

$$= \sqrt{(33 \ \Omega)^2 + (79.6 \ \Omega)^2} \angle -\tan^{-1}\left(\frac{79.6 \ \Omega}{33 \ \Omega}\right) = 86.2 \angle -67.5° \ \Omega$$

$$\mathbf{Z}_2 = \sqrt{R_2^2 + X_{C2}^2} \angle -\tan^{-1}\left(\frac{X_{C2}}{R_2}\right)$$

$$= \sqrt{(47 \ \Omega)^2 + (36.2 \ \Omega)^2} \angle -\tan^{-1}\left(\frac{36.2 \ \Omega}{47 \ \Omega}\right) = 59.3 \angle -37.6° \ \Omega$$

Calculate each branch current.

$$\mathbf{I}_1 = \frac{\mathbf{V}_s}{\mathbf{Z}_1} = \frac{2\angle 0° \text{ V}}{86.2\angle -67.5° \ \Omega} = \mathbf{23.2\angle 67.5° \text{ mA}}$$

$$\mathbf{I}_2 = \frac{\mathbf{V}_s}{\mathbf{Z}_2} = \frac{2\angle 0° \text{ V}}{59.3\angle -37.6° \ \Omega} = \mathbf{33.7\angle 37.6° \text{ mA}}$$

To get the total current, express each branch current in rectangular form so that they can be added.

$$\mathbf{I}_1 = 8.89 \text{ mA} + j21.4 \text{ mA}$$
$$\mathbf{I}_2 = 26.7 \text{ mA} + j20.6 \text{ mA}$$

The total current is

$$\mathbf{I}_{tot} = \mathbf{I}_1 + \mathbf{I}_2$$
$$= (8.89 \text{ mA} + j21.4 \text{ mA}) + (26.7 \text{ mA} + j20.6 \text{ mA}) = 35.6 \text{ mA} + j42.0 \text{ mA}$$

Converting $\mathbf{I}_{tot}$ to polar form yields

$$\mathbf{I}_{tot} = \sqrt{(35.6 \text{ mA})^2 + (42.0 \text{ mA})^2}\angle \tan^{-1}\left(\frac{42.0 \ \Omega}{35.6 \ \Omega}\right) = \mathbf{55.1\angle 49.7° \text{ mA}}$$

The current phasor diagram is shown in Figure 15–53.

▶ **FIGURE 15–53**

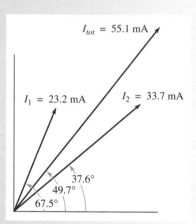

Related Problem Determine the voltages across each component in Figure 15–52 and draw a voltage phasor diagram.

 Use Multisim file E15-22 to verify the calculated results in this example and to confirm your calculations for the related problem.

Measurement of Z_{tot}

Now, let's see how the value of Z_{tot} for the circuit in Example 15–21 can be determined by measurement. First, the total impedance is measured as outlined in the following steps and as illustrated in Figure 15–54 (other ways are also possible):

Step 1. Using a sine wave generator, set the source voltage to a known value (10 V) and the frequency to 5 kHz. If your generator is not accurate, then it is advisable to check the voltage with an ac voltmeter and the frequency with a frequency counter rather than relying on the marked values on the generator controls.

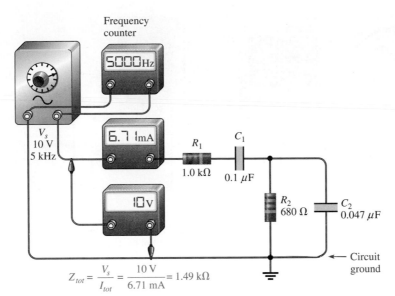

◀ FIGURE 15–54

Determining Z_{tot} by measurement of V_s and I_{tot}.

$$Z_{tot} = \frac{V_s}{I_{tot}} = \frac{10\text{ V}}{6.71\text{ mA}} = 1.49\text{ k}\Omega$$

Step 2. Connect an ac ammeter as shown in Figure 15–54, and measure the total current. Alternatively, you can measure the voltage across R_1 with a voltmeter and calculate the current.

Step 3. Calculate the total impedance by using Ohm's law.

Measurement of Phase Angle, θ

To measure the phase angle, the source voltage and the total current must be displayed on an oscilloscope screen in the proper time relationship. Two basic types of scope probes are available to measure the quantities with an oscilloscope: the voltage probe and the current probe. The current probe is a convenient device, but it is often not as readily available as a voltage probe. We will confine our phase measurement technique to the use of voltage probes in conjunction with the oscilloscope. Although there are special isolation methods, a typical oscilloscope voltage probe has two points that are connected to the circuit: the probe tip and the ground lead. Thus, all voltage measurements must be referenced to ground.

Since only voltage probes are to be used, the total current cannot be measured directly. However, for phase measurement, the voltage across R_1 is in phase with the total current and can be used to establish the phase angle of the current.

Before proceeding with the actual phase measurement, there is a problem with displaying V_{R1}. If the scope probe is connected across the resistor, as indicated in Figure 15–55(a), the ground lead of the scope will short point B to ground, thus bypassing the rest of the components and effectively removing them from the circuit electrically, as illustrated in Figure 15–55(b) (assuming that the scope is not isolated from power line ground).

To avoid this problem, you can switch the generator output terminals so that one end of R_1 is connected to the ground terminal, as shown in Figure 15–56(a). Now the scope can be connected across it to display V_{R1}, as indicated in Figure 15–56(b). The other probe is connected across the voltage source to display V_s as indicated. Now channel 1 of the scope has V_{R1} as an input, and channel 2 has V_s. The scope should be triggered from the source voltage (channel 2 in this case).

Before connecting the probes to the circuit, you should align the two horizontal lines (traces) so that they appear as a single line across the center of the scope screen. To do so, ground the probe tips and adjust the vertical position knobs to move the traces toward the center line of the screen until they are superimposed. This procedure ensures that both waveforms have the same zero crossing so that an accurate phase measurement can be made.

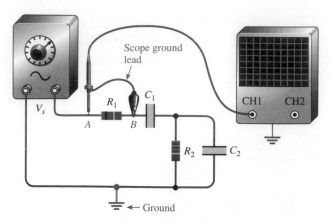
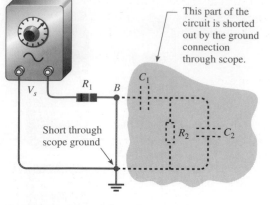

(a) Ground lead on scope probe grounds point B.

(b) The effect of grounding point B is to short out the rest of the circuit.

▲ FIGURE 15–55

Effects of measuring directly across a component when the instrument and the circuit are grounded.

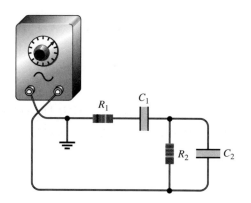
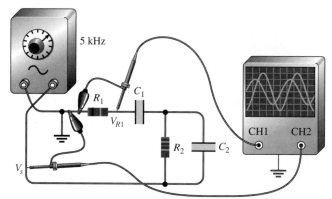

(a) Ground repositioned so that one end of R_1 is grounded.

(b) The scope displays V_{R1} and V_s. V_{R1} represents the phase of the total current.

▲ FIGURE 15–56

Repositioning ground so that a direct voltage measurement can be made with respect to ground without shorting out part of the circuit.

Once you have stabilized the waveforms on the scope screen, you can measure the period of the source voltage. Next, use the Volts/Div controls to adjust the amplitudes of the waveforms until they both appear to have the same amplitude. Now, spread the waveforms horizontally by using the Sec/Div control to expand the distance between them. This horizontal distance represents the time between the two waveforms. The number of divisions between the waveforms along any horizontal lines times the Sec/Div setting is equal to the time between them, Δt. Also, you can use the cursors to determine Δt if your oscilloscope has this feature.

Once you have determined the period, T, and the time between the waveforms, Δt, you can calculate the phase angle with the following equation:

Equation 15–27

$$\theta = \left(\frac{\Delta t}{T}\right)360°$$

An example screen display is shown in Figure 15–57. In this illustration, there are 1.5 horizontal divisions between the two waveforms, as indicated, and the Sec/Div control is set at 10 μs. The period of these waveforms is 200 μs and the Δt is

$$\Delta t = 1.5 \text{ divisions} \times 10\ \mu\text{s/division} = 15\ \mu\text{s}$$

The phase angle is

$$\theta = \left(\frac{\Delta t}{T}\right)360° = \left(\frac{15\ \mu\text{s}}{200\ \mu\text{s}}\right)360° = 27°$$

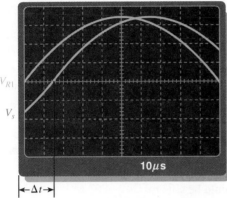

◀ FIGURE 15–57

Determining the phase angle on the oscilloscope.

10μs

←Δt→

$\Delta t = 1.5$ divisions $\times$ 10 μs/division = 15 μs

SECTION 15–7 **REVIEW**	1. What is the equivalent series *RC* circuit for the series-parallel circuit in Figure 15–51? 2. What is the total impedance in polar form of the circuit in Figure 15–52?

OPTION 2 NOTE

Coverage of series-parallel reactive circuits continues in Chapter 16, Part 3, on page 698.

15–8 POWER IN *RC* CIRCUITS

In a purely resistive ac circuit, all of the energy delivered by the source is dissipated in the form of heat by the resistance. In a purely capacitive ac circuit, all of the energy delivered by the source is stored by the capacitor during a portion of the voltage cycle and then returned to the source during another portion of the cycle so that there is no net energy conversion to heat. When there is both resistance and capacitance, some of the energy is alternately stored and returned by the capacitance and some is dissipated by the resistance. The amount of energy converted to heat is determined by the relative values of the resistance and the capacitive reactance.

After completing this section, you should be able to

- ◆ **Determine power in *RC* circuits**
    - ◆ Explain true and reactive power
    - ◆ Draw the power triangle
    - ◆ Define *power factor*
    - ◆ Explain apparent power
    - ◆ Calculate power in an *RC* circuit

When the resistance in a series *RC* circuit is greater than the capacitive reactance, more of the total energy delivered by the source is converted to heat by the resistance than is stored by the capacitor. Likewise, when the reactance is greater than the resistance, more of the total energy is stored and returned than is converted to heat.

The formulas for power in a resistor, sometimes called *true power* (P_{true}), and the power in a capacitor, called *reactive power* (P_r), are restated here. The unit of true power is the watt, and the unit of reactive power is the VAR (volt-ampere reactive).

Equation 15–28

$$P_{\text{true}} = I^2R$$

Equation 15–29

$$P_r = I^2X_C$$

The Power Triangle for *RC* Circuits

The generalized impedance phasor diagram for a series *RC* circuit is shown in Figure 15–58(a). A phasor relationship for the powers can also be represented by a similar diagram because the respective magnitudes of the powers, P_{true} and P_r, differ from R and X_C by a factor of I^2. This is shown in Figure 15–58(b).

The resultant power phasor, I^2Z, represents the **apparent power, P_a**. At any instant in time P_a is the total power that appears to be transferred between the source and the *RC*

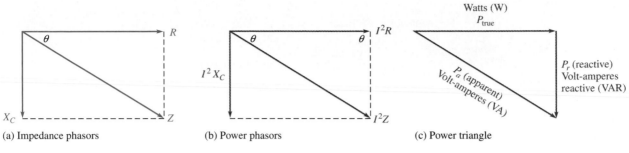

▲ FIGURE 15–58

Development of the power triangle for a series *RC* circuit.

circuit. The unit of apparent power is the volt-ampere, VA. The expression for apparent power is

$$P_a = I^2 Z$$

Equation 15–30

The power phasor diagram in Figure 15–58(b) can be rearranged in the form of a right triangle, as shown in Figure 15–58(c). This is called the *power triangle*. Using the rules of trigonometry, P_{true} can be expressed as

$$P_{\text{true}} = P_a \cos \theta$$

Since P_a equals $I^2 Z$ or VI, the equation for the true power dissipation in an *RC* circuit can be written as

$$P_{\text{true}} = VI \cos \theta$$

Equation 15–31

where V is the applied voltage and I is the total current.

For the case of a purely resistive current, $\theta = 0°$ and $\cos 0° = 1$, so P_{true} equals VI. For the case of a purely capacitive circuit, $\theta = 90°$ and $\cos 90° = 0$, so P_{true} is zero. As you already know, there is no power dissipation in an ideal capacitor.

The Power Factor

The term $\cos \theta$ is called the **power factor** and is stated as

$$PF = \cos \theta$$

Equation 15–32

As the phase angle between applied voltage and total current increases, the power factor decreases, indicating an increasingly reactive circuit. The smaller the power factor, the smaller the power dissipation.

The power factor can vary from 0 for a purely reactive circuit to 1 for a purely resistive circuit. In an *RC* circuit, the power factor is referred to as a leading power factor because the current leads the voltage.

EXAMPLE 15–23

Determine the power factor and the true power in the circuit of Figure 15–59.

▶ **FIGURE 15–59**

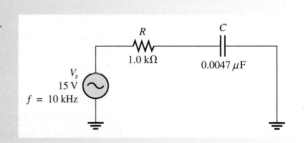

Solution   The capacitive reactance is

$$X_C = \frac{1}{2\pi fC} = \frac{1}{2\pi(10\text{ kHz})(0.0047\ \mu\text{F})} = 3.39\text{ k}\Omega$$

The total impedance of the circuit in rectangular form is

$$\mathbf{Z} = R - jX_C = 1.0\text{ k}\Omega - j3.39\text{ k}\Omega$$

Converting to polar form yields

$$\mathbf{Z} = \sqrt{R^2 + X_C^2}\angle -\tan^{-1}\left(\frac{X_C}{R}\right)$$

$$= \sqrt{(1.0\text{ k}\Omega)^2 + (3.39\text{ k}\Omega)^2}\angle -\tan^{-1}\left(\frac{3.39\text{ k}\Omega}{1.0\text{ k}\Omega}\right) = 3.53\angle -73.6°\text{ k}\Omega$$

The angle associated with the impedance is θ, the angle between the applied voltage and the total current; therefore, the power factor is

$$PF = \cos\theta = \cos(-73.6°) = \mathbf{0.282}$$

The current magnitude is

$$I = \frac{V_s}{Z} = \frac{15\text{ V}}{3.53\text{ k}\Omega} = 4.25\text{ mA}$$

The true power is

$$P_{\text{true}} = V_s I\cos\theta = (15\text{ V})(4.25\text{ mA})(0.282) = \mathbf{18.0\ mW}$$

Related Problem   What is the power factor if f is reduced by half in Figure 15–59?

Significance of Apparent Power

As mentioned, apparent power is the power that appears to be transferred between the source and the load, and it consists of two components—a true power component and a reactive power component.

In all electrical and electronic systems, it is the true power that does the work. The reactive power is simply shuttled back and forth between the source and load. Ideally, in terms of performing useful work, all of the power transferred to the load should be true power and none of it reactive power. However, in most practical situations the load has some reactance associated with it, and therefore you must deal with both power components.

In Chapter 14, the use of apparent power in relation to transformers was discussed. For any reactive load, there are two components of the total current: the resistive component and the reactive component. If you consider only the true power (watts) in a load, you are dealing with only a portion of the total current that the load demands from a source. In order to have a realistic picture of the actual current that a load will draw, you must consider apparent power (VA).

A source such as an ac generator can provide current to a load up to some maximum value. If the load draws more than this maximum value, the source can be damaged. Figure 15–60(a) shows a 120 V generator that can deliver a maximum current of 5 A to a load. Assume that the generator is rated at 600 W and is connected to a purely resistive load of 24 Ω (power factor of 1). The ammeter shows that the current is 5 A, and the wattmeter indicates that the power is 600 W. The generator has no problem under these conditions, although it is operating at maximum current and power.

Now, consider what happens if the load is changed to a reactive one with an impedance of 18 Ω and a power factor of 0.6, as indicated in Figure 15–60(b). The current is

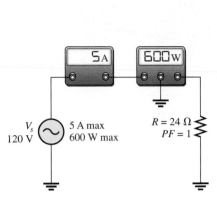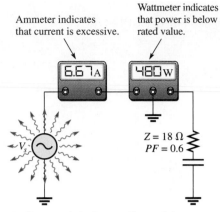

Ammeter indicates
that current is excessive.

Wattmeter indicates
that power is below
rated value.

(a) Generator operating at its limits with a
resistive load.

(b) Generator is in danger of internal damage
due to excess current, even though the
wattmeter indicates that the power is
below the maximum wattage rating.

▲ FIGURE 15–60

Wattage rating of a source is inappropriate when the load is reactive. The rating should be in VA
rather than in watts.

120 V/18 Ω = 6.67 A, which *exceeds* the maximum. Even though the wattmeter reads 480 W,
which is less than the power rating of the generator, the excessive current probably will
cause damage. This illustration shows that a true power rating can be deceiving and is inappropriate for ac sources. The ac generator should be rated at 600 VA, a rating that manufacturers generally use, rather than 600 W.

EXAMPLE 15–24

For the circuit in Figure 15–61, find the true power, the reactive power, and the apparent power.

▶ **FIGURE 15–61**

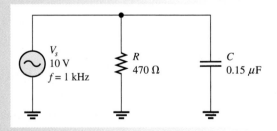

Solution The capacitive reactance and currents through *R* and *C* are

$$X_C = \frac{1}{2\pi f C} = \frac{1}{2\pi(1000 \text{ Hz})(0.15 \ \mu\text{F})} = 1061 \ \Omega$$

$$I_R = \frac{V_s}{R} = \frac{10 \text{ V}}{470 \ \Omega} = 21.3 \text{ mA}$$

$$I_C = \frac{V_s}{X_C} = \frac{10 \text{ V}}{1061 \ \Omega} = 9.43 \text{ mA}$$

The true power is

$$P_{true} = I_R^2 R = (21.3 \, \text{mA})^2 (470 \, \Omega) = \textbf{213 mW}$$

The reactive power is

$$P_r = I_C^2 X_C = (9.43 \, \text{mA})^2 (1061 \, \Omega) = \textbf{94.3 mVAR}$$

The apparent power is

$$P_a = \sqrt{P_{true}^2 + P_r^2} = \sqrt{(213 \, \text{mW})^2 + (94.3 \, \text{mVAR})^2} = \textbf{233 mVA}$$

Related Problem What is the true power in Figure 15–61 if the frequency is changed to 2 kHz?

SECTION 15–8 REVIEW

1. To which component in an *RC* circuit is the power dissipation due?
2. The phase angle, θ, is 45°. What is the power factor?
3. A certain series *RC* circuit has the following parameter values: $R = 330 \, \Omega$, $X_C = 460 \, \Omega$, and $I = 2 \, \text{A}$. Determine the true power, the reactive power, and the apparent power.

15–9 BASIC APPLICATIONS

RC circuits are found in a variety of applications, often as part of a more complex circuit. Three applications are phase shift oscillators, frequency-selective circuits, (filters) and ac coupling.

After completing this section, you should be able to

◆ **Discuss some basic *RC* applications**

  ◆ Discuss how the *RC* circuit is used as an oscillator

  ◆ Discuss how the *RC* circuit operates as a filter

  ◆ Discuss ac coupling

The Phase Shift Oscillator

As you know, a series *RC* circuit will shift the phase of the output voltage by an amount that depends on the values of *R* and *C* and the frequency of the signal. This ability to shift phase depending on frequency is vital in certain feedback oscillator circuits. An **oscillator** is a circuit that generates a periodic waveform and is an important circuit for many electronic systems. You will study oscillators in devices courses, so the focus here is on the application of *RC* circuits for shifting phase. The requirement is that a fraction of the output of the oscillator is returned to the input (called "feedback") in the proper phase to reinforce the input and sustain oscillations. Generally, the requirement is to feed back the signal with a total of 180° of phase shift.

A single *RC* circuit is limited to phase shifts that are smaller than 90°. The basic *RC* lag circuit discussed in Section 15–4 can be "stacked" to form a complex *RC* network as shown in Figure 15–62, which shows a specific circuit called a phase-shift oscillator. The phase shift oscillator typically uses three equal-component *RC* circuits that produce the required 180° phase shift at a certain frequency, which will be the frequency at which the oscillator

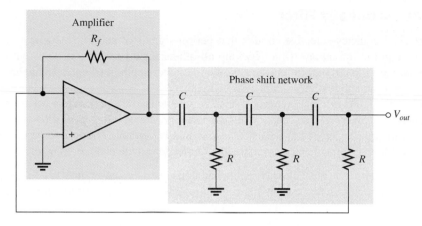

Phase shift oscillator.

works. The output of the amplifier is phase shifted by the RC network and returned to the input of the amplifier, which provides sufficient gain to maintain oscillations.

The process of putting several RC circuits together results in a loading effect, so the overall phase shift is not the same as simply adding the phase shifts of the individual RC circuits. The detailed calculation for this circuit is shown in Appendix B. With equal components, the frequency at which a 180° phase shift occurs is given by the equation

$$f_r = \frac{1}{2\pi\sqrt{6}\,RC}$$

Equation 15–33

It also turns out that the RC network attenuates (reduces) the signal from the amplifier by a factor of 29; the amplifier must make up for this attenuation by having a gain of -29 (the minus sign takes into account the phase shift).

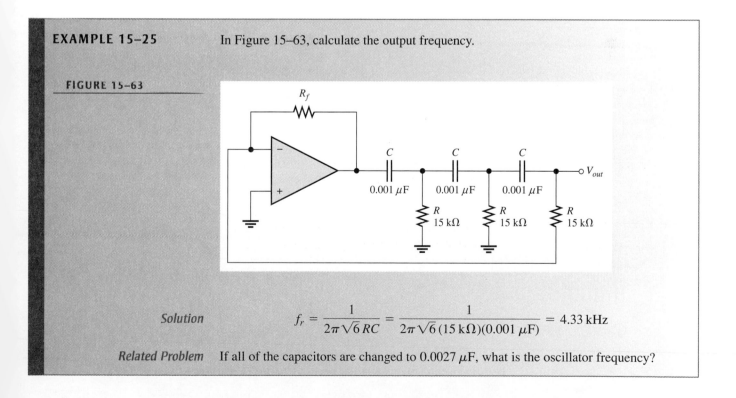

EXAMPLE 15–25   In Figure 15–63, calculate the output frequency.

FIGURE 15–63

Solution $f_r = \dfrac{1}{2\pi\sqrt{6}\,RC} = \dfrac{1}{2\pi\sqrt{6}\,(15\text{ k}\Omega)(0.001\ \mu\text{F})} = 4.33\text{ kHz}$

Related Problem   If all of the capacitors are changed to 0.0027 μF, what is the oscillator frequency?

The *RC* Circuit as a Filter

Filters are frequency-selective circuits that permit signals of certain frequencies to pass from the input to the output while blocking all others. That is, all frequencies but the selected ones are filtered out. Filters are covered in greater depth in Chapter 18 but are introduced here as an application example.

Series *RC* circuits exhibit a frequency-selective characteristic and therefore act as basic filters. There are two types. The first one that we examine, called a **low-pass filter**, is realized by taking the output across the capacitor, just as in a lag circuit. The second type, called a **high-pass filter**, is implemented by taking the output across the resistor, as in a lead circuit.

Low-Pass Filter You have already seen what happens to the output magnitude and phase angle in a lag circuit. In terms of its filtering action, we are interested primarily in the variation of the output magnitude with frequency.

Figure 15–64 shows the filtering action of a series *RC* circuit using specific values for illustration. In part (a) of the figure, the input is zero frequency (dc). Since the capacitor blocks constant direct current, the output voltage equals the full value of the input voltage because there is no voltage dropped across *R*. Therefore, the circuit passes all of the input voltage to the output (10 V in, 10 V out).

In Figure 15–64(b), the frequency of the input voltage has been increased to 1 kHz, causing the capacitive reactance to decrease to 159 Ω. For an input voltage of 10 V rms, the

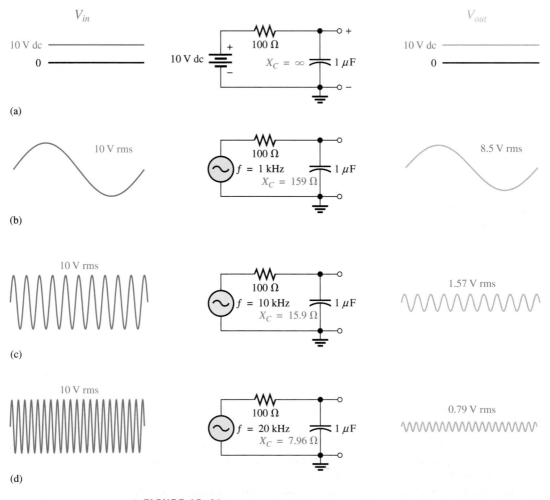

▲ **FIGURE 15–64**

Low-pass filtering action (phase shifts are not indicated).

output voltage is approximately 8.5 V rms, which can be calculated using the voltage-divider approach or Ohm's law.

In Figure 15–64(c), the input frequency has been increased to 10 kHz, causing the capacitive reactance to decrease further to 15.9 Ω. For a constant input voltage of 10 V rms, the output voltage is now 1.57 V rms.

As the input frequency is increased further, the output voltage continues to decrease and approaches zero as the frequency becomes very high, as shown in Figure 15–64(d).

A description of the circuit action is as follows: As the frequency of the input increases, the capacitive reactance decreases. Because the resistance is constant and the capacitive reactance decreases, the voltage across the capacitor (output voltage) also decreases according to the voltage-divider principle. The input frequency can be increased until it reaches a value at which the reactance is so small compared to the resistance that the output voltage can be neglected because it is very small compared to the input voltage. At this value of frequency, the circuit is essentially completely blocking the input signal.

As shown in Figure 15–64, the circuit passes dc (zero frequency) completely. As the frequency of the input increases, less of the input voltage is passed through to the output; that is, the output voltage decreases as the frequency increases. It is apparent that the lower frequencies pass through the circuit much better than the higher frequencies. This *RC* circuit is therefore a very basic form of low-pass filter.

The **frequency response** of the low-pass filter circuit in Figure 15–64 is shown in Figure 15–65 with a graph of output voltage magnitude versus frequency. This graph, called a *response curve,* indicates that the output decreases as the frequency increases.

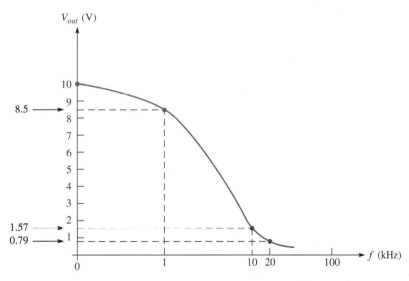

▲ **FIGURE 15–65**

Frequency response curve for the low-pass filter in Figure 15–64.

High-Pass Filter   Figure 15–66 illustrates high-pass filtering action, where the output is taken across the resistor, just as in a lead circuit. When the input voltage is dc (zero frequency) in part (a), the output is zero volts because the capacitor blocks direct current; therefore, no voltage is developed across *R*.

In Figure 15–66(b), the frequency of the input signal has been increased to 100 Hz with an rms value of 10 V. The output voltage is 0.63 V rms. Thus, only a small percentage of the input voltage appears on the output at this frequency.

In Figure 15–66(c), the input frequency is increased further to 1 kHz, causing more voltage to be developed across the resistor because of the further decrease in the capacitive reactance. The output voltage at this frequency is 5.32 V rms. As you can see, the output

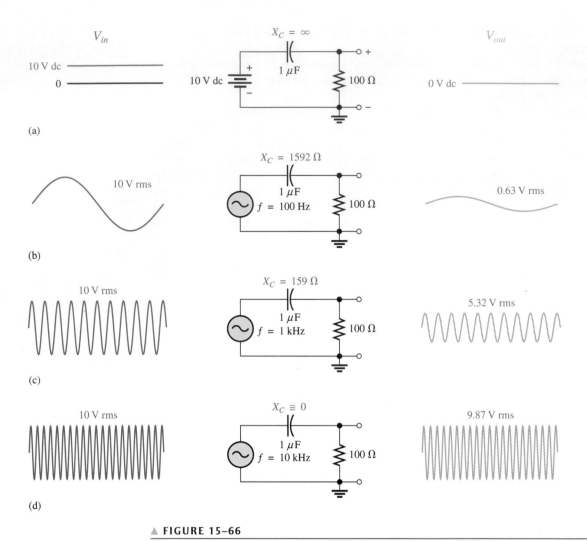

▲ **FIGURE 15–66**

High-pass filtering action (phase shifts are not indicated).

voltage increases as the frequency increases. A value of frequency is reached at which the reactance is negligible compared to the resistance, and most of the input voltage appears across the resistor, as shown in Figure 15–66(d).

As illustrated, this circuit tends to prevent lower frequencies from appearing on the output but allows higher frequencies to pass through from input to output. Therefore, this *RC* circuit is a basic form of high-pass filter.

The frequency response of the high-pass filter circuit in Figure 15–66 is shown in Figure 15–67 with a graph of output voltage magnitude versus frequency. This response curve shows that the output increases as the frequency increases and then levels off and approaches the value of the input voltage.

The Cutoff Frequency and the Bandwidth of a Filter   The frequency at which the capacitive reactance equals the resistance in a low-pass or high-pass *RC* filter is called the **cutoff frequency** and is designated f_c. This condition is expressed as $1/(2\pi f_c C) = R$. Solving for f_c results in the following formula:

Equation 15–34

$$f_c = \frac{1}{2\pi RC}$$

At f_c, the output voltage of the filter is 70.7% of its maximum value. It is standard practice to consider the cutoff frequency as the limit of a filter's performance in terms of

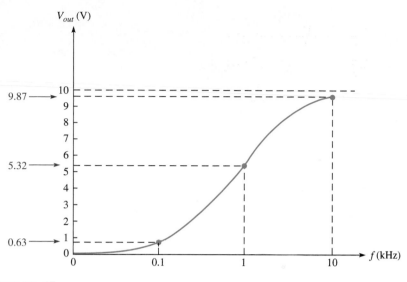

▲ FIGURE 15–67

Frequency response curve for the high-pass filter in Figure 15–66.

passing or rejecting frequencies. For example, in a high-pass filter, all frequencies above f_c are considered to be passed by the filter, and all those below f_c are considered to be rejected. The reverse is true for a low-pass filter.

The range of frequencies that is considered to be passed by a filter is called the **bandwidth**. Figure 15–68 illustrates the bandwidth and the cutoff frequency for a low-pass filter.

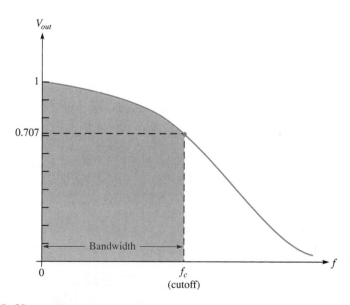

▲ FIGURE 15–68

Normalized general response curve of a low-pass filter showing the cutoff frequency and the bandwidth.

Coupling an AC Signal into a DC Bias Circuit

Figure 15–69 shows an RC circuit that is used to create a dc voltage level with an ac voltage superimposed on it. This type of circuit is commonly found in amplifiers in which the dc voltage is required to **bias** the amplifier to the proper operating point and the signal voltage to be

▶ FIGURE 15–69

Amplifier bias and signal-coupling circuit.

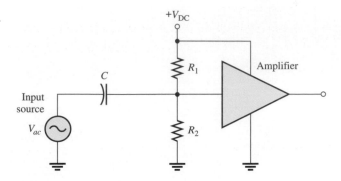

▶ FIGURE 15–69

Amplifier bias and signal-coupling circuit.

amplified is coupled through a capacitor and superimposed on the dc level. The capacitor prevents the low internal resistance of the signal source from affecting the dc bias voltage.

In this type of application, a relatively large value of capacitance is selected so that for the frequencies to be amplified, the reactance is very small compared to the resistance of the bias network. When the reactance is very small (ideally zero), there is practically no phase shift or signal voltage dropped across the capacitor. Therefore, all of the signal voltage passes from the source to the input of the amplifier.

Figure 15–70 illustrates the application of the superposition principle to the circuit in Figure 15–69. In part (a), the ac source has been effectively removed from the circuit and replaced with a short to represent its ideal internal resistance (actual generators typically have 50 Ω or 600 Ω of internal resistance). Since C is open to dc, the voltage at point A is determined by the voltage-divider action of R_1 and R_2 and the dc voltage source.

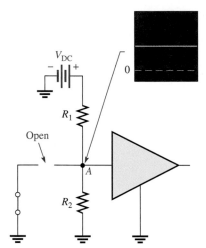

(a) dc equivalent: ac source replaced by short. C is open to dc. R_1 and R_2 act as dc voltage divider.

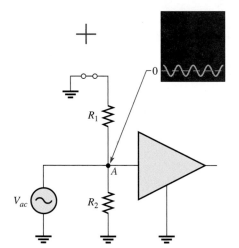

(b) ac equivalent: dc source is replaced by short. C is short to ac. All of V_{ac} is coupled to point A.

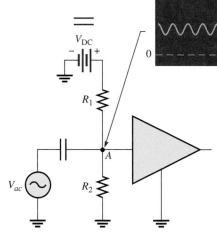

(c) dc + ac: Voltages are superimposed at point A.

▲ **FIGURE 15–70**

The superposition of dc and ac voltages in an *RC* bias and coupling circuit.

In Figure 15–70(b), the dc source has been effectively removed from the circuit and replaced with a short to represent its ideal internal resistance. Since C appears as a short at the frequency of the ac, the signal voltage is coupled directly to point A and appears across the parallel combination of R_1 and R_2.

Figure 15–70(c) illustrates that the combined effect of the superposition of the dc and the ac voltages results in the signal voltage "riding" on the dc level.

SECTION 15–9
REVIEW

SECTION 15–9
REVIEW

1. How much phase shift in produced by the *RC* circuit in a phase shift oscillator?
2. When an *RC* circuit is used as a low-pass filter, across which component is the output taken?

15–10 TROUBLESHOOTING

Typical component failures or degradation have an effect on the frequency response of basic *RC* circuits.

After completing this section, you should be able to

* **Troubleshoot *RC* circuits**
  * Find an open resistor or open capacitor
  * Find a shorted capacitor
  * Find a leaky capacitor

Effect of an Open Resistor It is easy to see how an open resistor affects the operation of a basic series *RC* circuit, as shown in Figure 15–71. Obviously, there is no path for current, so the capacitor voltage remains at zero; thus, the total voltage, V_s, appears across the open resistor.

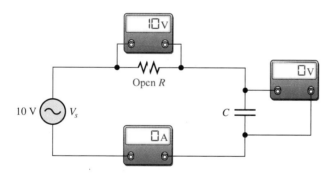

▲ FIGURE 15–71

Effect of an open resistor.

Effect of an Open Capacitor When the capacitor is open, there is no current; thus, the resistor voltage remains at zero. The total source voltage is across the open capacitor, as shown in Figure 15–72.

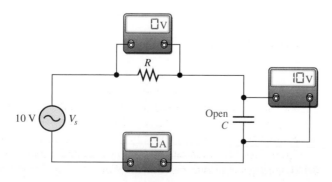

◀ FIGURE 15–72

Effect of an open capacitor.

Effect of a Shorted Capacitor Capacitors rarely short; but when a capacitor does short out, the voltage across it is zero, the current equals V_s/R, and the total voltage appears across the resistor, as shown in Figure 15–73.

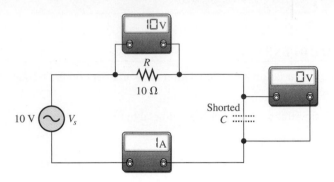

▲ **FIGURE 15–73**

Effect of a shorted capacitor.

Effect of a Leaky Capacitor When a large electrolytic capacitor exhibits a high leakage current, the leakage resistance effectively appears in parallel with the capacitor, as shown in Figure 15–74(a). When the leakage resistance is comparable in value to the circuit resistance, R, the circuit response is drastically affected. The circuit, looking from the capacitor toward the source, can be thevenized, as shown in Figure 15–74(b). The Thevenin equivalent resistance is R in parallel with R_{leak} (the source appears as a short), and the Thevenin equivalent voltage is determined by the voltage-divider action of R and R_{leak}.

$$R_{th} = R \| R_{leak} = \frac{R R_{leak}}{R + R_{leak}}$$

$$V_{th} = \left(\frac{R_{leak}}{R + R_{leak}} \right) V_s$$

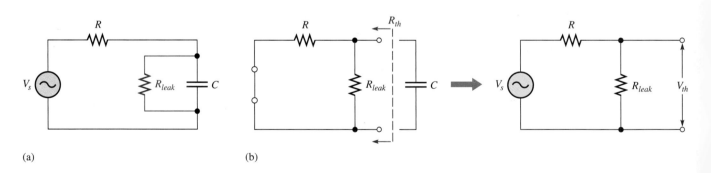

(a) (b)

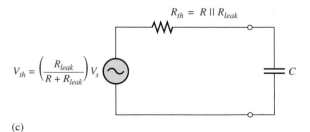

(c)

▲ **FIGURE 15–74**

Effect of a leaky capacitor.

As you can see, the voltage across the capacitor is reduced since $V_{th} < V_s$. Also, the circuit time constant is reduced, and the current is increased. The Thevenin equivalent circuit is shown in Figure 15–74(c).

EXAMPLE 15–26

Assume that the capacitor in Figure 15–75 is degraded to a point where its leakage resistance is $10 \text{ k}\Omega$. Determine the output voltage under the degraded condition.

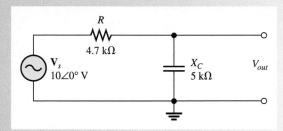

▲ **FIGURE 15–75**

Solution   The effective circuit resistance is

$$R_{th} = \frac{R R_{leak}}{R + R_{leak}} = \frac{(4.7 \text{ k}\Omega)(10 \text{ k}\Omega)}{14.7 \text{ k}\Omega} = 3.20 \text{ k}\Omega$$

To determine the output voltage, find the Thevenin equivalent voltage.

$$V_{th} = \left(\frac{R_{leak}}{R + R_{leak}} \right) V_s = \left(\frac{10 \text{ k}\Omega}{14.7 \text{ k}\Omega} \right) 10 \text{ V} = 6.80 \text{ V}$$

Then,

$$V_{out} = \left(\frac{X_C}{\sqrt{R_{th}^2 + X_C^2}} \right) V_{th} = \left(\frac{5 \text{ k}\Omega}{\sqrt{(3.2 \text{ k}\Omega)^2 + (5 \text{ k}\Omega)^2}} \right) 6.80 \text{ V} = \mathbf{5.73 \text{ V}}$$

Related Problem   What would the output voltage be if the capacitor were not leaky?

Other Troubleshooting Considerations

So far, you have learned about specific component failures and the associated voltage measurements. Many times, however, the failure of a circuit to work properly is not the result of a faulty component. A loose wire, a bad contact, or a poor solder joint can cause an open circuit. A short can be caused by a wire clipping or solder splash. Things as simple as not plugging in a power supply or a function generator happen more often than you might think. Wrong values in a circuit (such as an incorrect resistor value), the function generator set at the wrong frequency, or the wrong output connected to the circuit can cause improper operation.

When you have problems with a circuit, always check to make sure that the instruments are properly connected to the circuits and to a power outlet. Also, look for obvious things such as a broken or loose contact, a connector that is not completely plugged in, or a piece of wire or a solder bridge that could be shorting something out.

The point is that you should consider all possibilities, not just faulty components, when a circuit is not working properly. The following example illustrates this approach with a simple circuit using the APM (analysis, planning, and measurement) method.

EXAMPLE 15–27

The circuit represented by the schematic in Figure 15–76 has no output voltage, which is the voltage across the capacitor. You expect to see about 7.4 V at the output. The circuit is physically constructed on a protoboard. Use your troubleshooting skills to find the problem.

▶ **FIGURE 15–76**

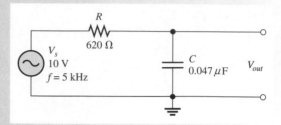

Solution   Apply the APM method to this troubleshooting problem.

Analysis: First think of the possible causes for the circuit to have no output voltage.

1. There is no source voltage or the frequency is so high that the capacitive reactance is almost zero.

2. There is a short between the output terminals. Either the capacitor could be internally shorted, or there could be some physical short in the circuit.

3. There is an open between the source and the output. This would prevent current and thus cause the output voltage to be zero. The resistor could be open, or the conductive path could be open due to a broken or loose connecting wire or a bad protoboard contact.

4. There is an incorrect component value. The resistor could be so large that the current and, therefore, the output voltage are negligible. The capacitor could be so large that its reactance at the input frequency is near zero.

Planning: You decide to make some visual checks for problems such as the function generator power cord not plugged in or the frequency set at an incorrect value. Also, broken leads, shorted leads, as well as an incorrect resistor color code or capacitor label often can be found visually. If nothing is discovered after a visual check, then you will make voltage measurements to track down the cause of the problem. You decide to use a digital oscilloscope and a DMM to make the measurements.

Measurement: Assume that you find that the function generator is plugged in and the frequency setting appears to be correct. Also, you find no visible opens or shorts during your visual check, and the component values are correct.

The first step in the measurement process is to check the voltage from the source with the scope. Assume a 10 V rms sine wave with a frequency of 5 kHz is observed at the circuit input as shown in Figure 15–77(a). The correct voltage is present, so *the first possible cause has been eliminated.*

Next, check for a shorted capacitor by disconnecting the source and placing a DMM (set on the ohmmeter function) across the capacitor. If the capacitor is good, an open will be indicated by an OL (overload) in the meter display after a short charging time. Assume the capacitor checks okay, as shown in Figure 15–77(b). *The second possible cause has been eliminated.*

Since the voltage has been "lost" somewhere between the input and the output, you must now look for the voltage. Reconnect the source and measure the voltage across the resistor with the DMM (set on the voltmeter function) from one resistor lead to the other. The voltage across the resistor is zero. This means there is no current, which indicates an open somewhere in the circuit.

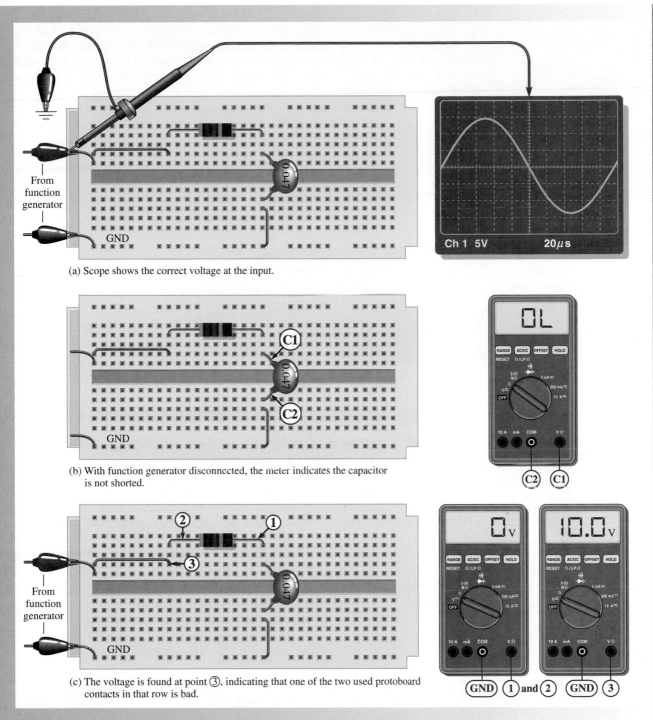

(a) Scope shows the correct voltage at the input.

(b) With function generator disconnected, the meter indicates the capacitor is not shorted.

(c) The voltage is found at point ③, indicating that one of the two used protoboard contacts in that row is bad.

FIGURE 15–77

Now, begin tracing the circuit back toward the source looking for the voltage (you could also start from the source and work forward). You can use either the scope or the DMM but decide to use the multimeter with one lead connected to ground and the other used to probe the circuit. As shown in Figure 15–77(c), the voltage on the right lead of the resistor, point ①, reads zero. Since you already have measured zero voltage across the resistor, the voltage on the left resistor lead at point ② must be zero as the meter indicates. Next, moving the meter probe to point ③, you read 10 V. You have found the voltage! Since there is zero volts on the left resistor lead, and there is 10 V at point ③, one of the two contacts in the protoboard hole into which the wire leads are inserted is bad. It could be that the small contacts were pushed in too far and were bent or broken so that the circuit lead does not make contact.

Move either or both the resistor lead and the wire to another hole in the same row. Assume that when the resistor lead is moved to the hole just above, you have voltage at the output of the circuit (across the capacitor).

Related Problem   Suppose you had measured 10 V across the resistor before the capacitor was checked. What would this have indicated?

SECTION 15–10 REVIEW

1. Describe the effect of a leaky capacitor on the response of an *RC* circuit.
2. In a series *RC* circuit, if all of the applied voltage appears across the capacitor, what is the problem?
3. What faults can cause 0 V across a capacitor in a series *RC* circuit if the source is functioning properly?

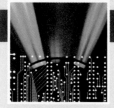

A Circuit Application

In Chapter 12, you studied the capacitively coupled input to an amplifier with voltage-divider bias. In this circuit application, you will check the output voltage and phase lag of a similar amplifier's input circuit to determine how they change with frequency. If too much voltage is dropped across the coupling capacitor, the overall performance of the amplifier is adversely affected.

As you learned in Chapter 12, the coupling capacitor (C_1) in Figure 15–78 passes the input signal voltage to the input of the amplifier (point A to point B) without affecting the dc level at

point B produced by the resistive voltage divider (R_1 and R_2). If the input frequency is high enough so that the reactance of the coupling capacitor is negligibly small, essentially no ac signal voltage is dropped across the capacitor. As the signal frequency is reduced, the capacitive reactance increases and more of the signal voltage is dropped across the capacitor. This lowers the overall voltage gain of the amplifier and thus degrades its performance.

The amount of signal voltage that is coupled from the input source (point A) to the amplifier input (point B) is determined by the values of the capacitor and the dc bias resistors (assuming the amplifier has no loading effect) in Figure 15–78. These

▶ **FIGURE 15–78**

A capacitively coupled amplifier.

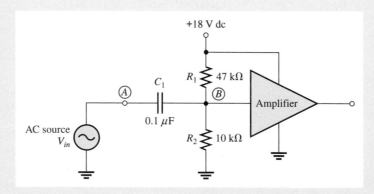

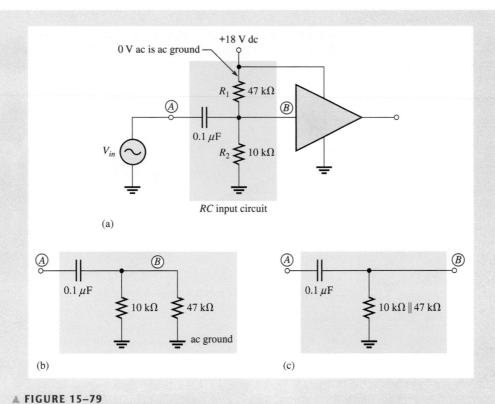

+18 V dc

0 V ac is ac ground

R_1 47 kΩ

Ⓐ

0.1 μF

Ⓑ

V_{in}

R_2 10 kΩ

RC input circuit

(a)

Ⓐ 0.1 μF Ⓑ

10 kΩ 47 kΩ

ac ground

(b)

Ⓐ 0.1 μF Ⓑ

10 kΩ ∥ 47 kΩ

(c)

▲ FIGURE 15–79

The RC input circuit acts effectively like a high-pass RC filter.

components actually form a high-pass RC filter, as shown in Figure 15–79. The voltage-divider bias resistors are effectively in parallel with each other as far as the ac source is concerned because the power supply has zero internal resistance. The lower end of R_2 goes to ground and the upper end of R_1 goes to the dc supply voltage as shown in Figure 15–79(a). Since there is no ac voltage at the +18 V dc terminal, the upper end of R_1 is at 0 V ac, which is referred to as *ac ground*. The development of the circuit into an effective high-pass RC filter is shown in parts (b) and (c).

The Amplifier Input Circuit

◆ Determine the value of the equivalent resistance of the input circuit. Assume the amplifier (shown inside the white dashed lines in Figure 15–80) has no loading effect on the input circuit.

The Response at Frequency f_1

Refer to Figure 15–80. The input signal voltage is applied to the amplifier circuit board and displayed on channel 1 of the oscilloscope, and channel 2 is connected to a point on the circuit board.

◆ Determine to what point on the circuit the channel 2 probe is connected, the frequency, and the voltage that should be displayed.

The Response at Frequency f_2

Refer to Figure 15–81 and the circuit board in Figure 15–80. The input signal voltage displayed on channel 1 of the oscilloscope is applied to the amplifier circuit board.

◆ Determine the frequency and the voltage that should be displayed on channel 2.

◆ State the difference between the channel 2 waveforms determined for f_1 and f_2. Explain the reason for the difference.

The Response at Frequency f_3

Refer to Figure 15–82 and the circuit board in Figure 15–80. The input signal voltage displayed on channel 1 of the oscilloscope is applied to the amplifier circuit board.

◆ Determine the frequency and the voltage that should be displayed on channel 2.

◆ State the difference between the channel 2 waveforms determined for f_2 and f_3. Explain the reason for the difference.

Response Curve for the Amplifier Input Circuit

◆ Determine the frequency at which the signal voltage at point B in Figure 15–78 is 70.7% of its maximum value.

◆ Plot the response curve using this voltage value and the values at frequencies f_1, f_2, and f_3.

◆ How does this curve show that the input circuit acts as a high-pass filter?

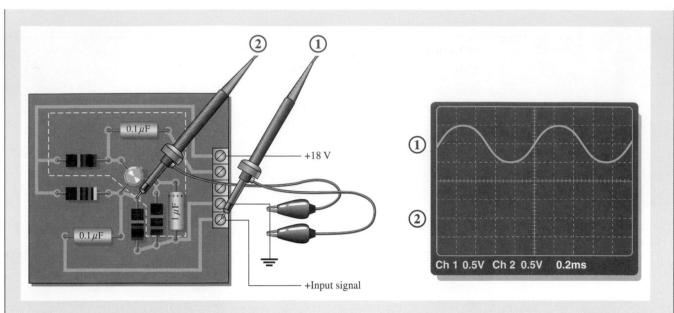

▲ FIGURE 15-80

Measuring the input circuit response at frequency f_1. Circled numbers relate scope inputs to the probes. The channel 1 waveform is shown.

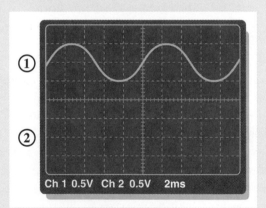

▲ FIGURE 15-81

Measuring the input circuit response at frequency f_2. The channel 1 waveform is shown.

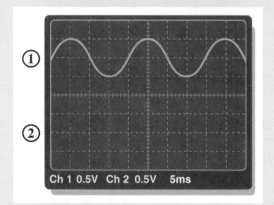

▲ FIGURE 15-82

Measuring the input circuit response at frequency f_3. The channel 1 waveform is shown.

◆ What can you do to the circuit to lower the frequency at which the voltage is 70.7% of maximum without affecting the dc bias voltage?

Review

1. Explain the effect on the response of the amplifier input circuit of reducing the value of the coupling capacitor.

2. What is the voltage at point *B* in Figure 15–78 if the coupling capacitor opens when the ac input signal is 10 mV rms?

3. What is the voltage at point *B* in Figure 15–78 if resistor R_1 is open when the ac input signal is 10 mV rms?

OPTION 2 NOTE

Coverage of special topics continues in Chapter 16, Part 4, on page 702.

SUMMARY

◆ A complex number represents a phasor quantity.

◆ The rectangular form of a complex number consists of a real part and a j part of the form $A + jB$.

◆ The polar form of a complex number consists of a magnitude and an angle of the form $C\angle\pm\theta$.

◆ Complex numbers can be added, subtracted, multiplied, and divided.

◆ When a sinusoidal voltage is applied to an RC circuit, the current and all the voltage drops are also sine waves.

◆ Total current in a series or parallel RC circuit always leads the source voltage.

◆ The resistor voltage is always in phase with the current.

◆ The capacitor voltage always lags the current by 90°.

◆ In a lag circuit, the output voltage lags the input voltage in phase.

◆ In a lead circuit, the output voltage leads the input voltage.

◆ In an RC circuit, the impedance is determined by both the resistance and the capacitive reactance combined.

◆ Impedance is expressed in units of ohms.

◆ The circuit phase angle is the angle between the total current and the applied (source) voltage.

◆ The impedance of a series RC circuit varies inversely with frequency.

◆ The phase angle (θ) of a series RC circuit varies inversely with frequency.

◆ For each parallel RC circuit, there is an equivalent series circuit for any given frequency.

◆ For each series RC circuit, there is an equivalent parallel circuit for any given frequency.

◆ The impedance of a circuit can be determined by measuring the applied voltage and the total current and then applying Ohm's law.

◆ In an RC circuit, part of the power is resistive and part reactive.

◆ The phasor combination of resistive power (true power) and reactive power is called *apparent power*.

◆ Apparent power is expressed in volt-amperes (VA).

◆ The power factor (PF) indicates how much of the apparent power is true power.

◆ A power factor of 1 indicates a purely resistive circuit, and a power factor of 0 indicates a purely reactive circuit.

◆ A filter passes certain frequencies and rejects others.

◆ A phase shift oscillator uses an RC network to produce a 180° phase shift.

KEY TERMS

Key terms and other bold terms in the chapter are defined in the end-of-book glossary.

Admittance (Y)   A measure of the ability of a reactive circuit to permit current; the reciprocal of impedance. The unit is the siemens (S).

Apparent power (P_a)   The phasor combination of resistive power (true power) and reactive power. The unit is the volt-ampere (VA).

Bandwidth   The range of frequencies that is considered to be passed by a filter.

Capacitive susceptance (B_C)   The ability of a capacitor to permit current; the reciprocal of capacitive reactance. The unit is the siemens (S).

Complex plane   An area consisting of four quadrants on which a quantity containing both magnitude and direction can be represented.

Cutoff frequency   The frequency at which the output voltage of a filter is 70.7% of the maximum output voltage.

Filter   A type of circuit that passes certain frequencies and rejects all others.

Frequency response   In electric circuits, the variation in the output voltage (or current) over a specified range of frequencies.

Imaginary number   A number that exists on the vertical axis of the complex plane.

Impedance   The total opposition to sinusoidal current expressed in ohms.

Polar form   One form of a complex number made up of a magnitude and an angle.

Power factor   The relationship between volt-amperes and true power or watts. Volt-amperes multiplied by the power factor equals true power.

Real number   A number that exists on the horizontal axis of the complex plane.

Rectangular form   One form of a complex number made up of a real part and an imaginary part.

FORMULAS

Complex Numbers

15–1 $C = \sqrt{A^2 + B^2}$

15–2 $\theta = \tan^{-1}\left(\dfrac{\pm B}{A}\right)$

15–3 $\pm A \pm jB = C\angle \pm \theta$

15–4 $A = C \cos \theta$

15–5 $B = C \sin \theta$

15–6 $C\angle \theta = A + jB$

Series *RC* Circuits

15–7 $\mathbf{Z} = R - jX_C$

15–8 $\mathbf{Z} = \sqrt{R^2 + X_C^2}\angle - \tan^{-1}\left(\dfrac{X_C}{R}\right)$

15–9 $\mathbf{V} = \mathbf{IZ}$

15–10 $\mathbf{I} = \dfrac{\mathbf{V}}{\mathbf{Z}}$

15–11 $\mathbf{Z} = \dfrac{\mathbf{V}}{\mathbf{I}}$

15–12 $\mathbf{V}_s = V_R - jV_C$

15–13 $\mathbf{V}_s = \sqrt{V_R^2 + V_C^2}\angle - \tan^{-1}\left(\dfrac{V_C}{V_R}\right)$

Lag Circuit

15–14 $\phi = -\tan^{-1}\left(\dfrac{R}{X_C}\right)$

15–15 $V_{out} = \left(\dfrac{X_C}{\sqrt{R^2 + X_C^2}}\right)V_{in}$

Lead Circuit

15–16 $\phi = \tan^{-1}\left(\dfrac{X_C}{R}\right)$

15–17 $V_{out} = \left(\dfrac{R}{\sqrt{R^2 + X_C^2}}\right)V_{in}$

Parallel *RC* Circuits

15–18 $\mathbf{Z} = \left(\dfrac{RX_C}{\sqrt{R^2 + X_C^2}}\right)\angle -\tan^{-1}\left(\dfrac{R}{X_C}\right)$

15–19 $\mathbf{Y} = G + jB_C$

15–20 $\mathbf{V} = \dfrac{\mathbf{I}}{\mathbf{Y}}$

15–21 $\quad \mathbf{I} = \mathbf{VY}$

15–22 $\quad \mathbf{Y} = \dfrac{\mathbf{I}}{\mathbf{V}}$

15–23 $\quad \mathbf{I}_{tot} = I_R + jI_C$

15–24 $\quad \mathbf{I}_{tot} = \sqrt{I_R^2 + I_C^2}\angle\tan^{-1}\left(\dfrac{I_C}{I_R}\right)$

15–25 $\quad R_{eq} = Z\cos\theta$

15–26 $\quad X_{C(eq)} = Z\sin\theta$

15–27 $\quad \theta = \left(\dfrac{\Delta t}{T}\right)360°$

Power in *RC* Circuits

15–28 $\quad P_{true} = I^2 R$

15–29 $\quad P_r = I^2 X_C$

15–30 $\quad P_a = I^2 Z$

15–31 $\quad P_{true} = VI\cos\theta$

15–32 $\quad PF = \cos\theta$

Applications

15–33 $\quad f_r = \dfrac{1}{2\pi\sqrt{6}RC}$

15–34 $\quad f_c = \dfrac{1}{2\pi RC}$

SELF-TEST

Answers are at the end of the chapter.

1. A positive angle of 20° is equivalent to a negative angle of
   (a) −160° (b) −340° (c) −70° (d) −20°
2. In the complex plane, the number $3 + j4$ is located in the
   (a) first quadrant (b) second quadrant (c) third quadrant (d) fourth quadrant
3. In the complex plane, $12 - j6$ is located in the
   (a) first quadrant (b) second quadrant (c) third quadrant (d) fourth quadrant
4. The complex number $5 + j5$ is equivalent to
   (a) $5\angle45°$ (b) $25\angle0°$ (c) $7.07\angle45°$ (d) $7.07\angle135°$
5. The complex number $35\angle60°$ is equivalent to
   (a) $35 + j35$ (b) $35 + j60$ (c) $17.5 + j30.3$ (d) $30.3 + j17.5$
6. $(4 + j7) + (-2 + j9)$ is equal to
   (a) $2 + j16$ (b) $11 + j11$ (c) $-2 + j16$ (d) $2 - j2$
7. $(16 - j8) - (12 + j5)$ is equal to
   (a) $28 - j13$ (b) $4 - j13$ (c) $4 - j3$ (d) $-4 + j13$
8. $(5\angle45°)(2\angle20°)$ is equal to
   (a) $7\angle65°$ (b) $10\angle25°$ (c) $10\angle65°$ (d) $7\angle25°$
9. $(50\angle10°)/(25\angle30°)$ is equal to
   (a) $25\angle40°$ (b) $2\angle40°$ (c) $25\angle-20°$ (d) $2\angle-20°$
10. In a series *RC* circuit, the voltage across the resistance is
   (a) in phase with the source voltage (b) lagging the source voltage by 90°
   (c) in phase with the current (d) lagging the current by 90°

11. In a series *RC* circuit, the voltage across the capacitor is
    (a) in phase with the source voltage   (b) lagging the resistor voltage by 90°
    (c) in phase with the current   (d) lagging the source voltage by 90°

12. When the frequency of the voltage applied to a series *RC* circuit is increased, the impedance
    (a) increases   (b) decreases   (c) remains the same   (d) doubles

13. When the frequency of the voltage applied to a series *RC* circuit is decreased, the phase angle
    (a) increases   (b) decreases   (c) remains the same   (d) becomes erratic

14. In a series *RC* circuit when the frequency and the resistance are doubled, the impedance
    (a) doubles   (b) is halved
    (c) is quadrupled   (d) cannot be determined without values

15. In a series *RC* circuit, 10 V rms is measured across the resistor and 10 V rms is also measured across the capacitor. The rms source voltage is
    (a) 20 V   (b) 14.14 V   (c) 28.28 V   (d) 10 V

16. The voltages in Question 15 are measured at a certain frequency. To make the resistor voltage greater than the capacitor voltage, the frequency
    (a) must be increased   (b) must be decreased
    (c) is held constant   (d) has no effect

17. When $R = X_C$, the phase angle is
    (a) 0°   (b) +90°   (c) −90°   (d) 45°

18. To decrease the phase angle below 45°, the following condition must exist:
    (a) $R = X_C$ (b) $R < X_C$ (c) $R > X_C$ (d) $R = 10X_C$

19. When the frequency of the source voltage is increased, the impedance of a parallel *RC* circuit
    (a) increases   (b) decreases   (c) does not change

20. In a parallel *RC* circuit, there is 1 A rms through the resistive branch and 1 A rms through the capacitive branch. The total rms current is
    (a) 1 A   (b) 2 A   (c) 2.28 A   (d) 1.414 A

21. A power factor of 1 indicates that the circuit phase angle is
    (a) 90°   (b) 45°   (c) 180°   (d) 0°

22. For a certain load, the true power is 100 W and the reactive power is 100 VAR. The apparent power is
    (a) 200 VA   (b) 100 VA   (c) 141.4 VA   (d) 141.4 W

23. Energy sources are normally rated in
    (a) watts   (b) volt-amperes   (c) volt-amperes reactive   (d) none of these

CIRCUIT DYNAMICS QUIZ

Answers are at the end of the chapter.

Refer to Figure 15–86.

1. If *C* opens, the voltage across it
   (a) increases   (b) decreases   (c) stays the same

2. If *R* opens, the voltage across *C*
   (a) increases   (b) decreases   (c) stays the same

3. If the frequency is increased, the voltage across *R*
   (a) increases   (b) decreases   (c) stays the same

Refer to Figure 15–87.

4. If R_1 opens, the voltage across R_2
   (a) increases   (b) decreases   (c) stays the same

5. If C_2 is increased to 0.47 μF, the voltage across it
   - **(a)** increases   **(b)** decreases   **(c)** stays the same

Refer to Figure 15–93.

6. If R becomes open, the voltage across the capacitor
   - **(a)** increases   **(b)** decreases   **(c)** stays the same

7. If the source voltage increases, X_C
   - **(a)** increases   **(b)** decreases   **(c)** stays the same

Refer to Figure 15–98.

8. If R_2 opens, the voltage from the top of R_2 to ground
   - **(a)** increases   **(b)** decreases   **(c)** stays the same

9. If C_2 shorts out, the voltage across C_1
   - **(a)** increases   **(b)** decreases   **(c)** stays the same

10. If the frequency of the source voltage is increased, the current through the resistors
    - **(a)** increases   **(b)** decreases   **(c)** stays the same

11. If the frequency of the source voltage is decreased, the current through the capacitors
    - **(a)** increases   **(b)** decreases   **(c)** stays the same

Refer to Figure 15–103.

12. If C_3 opens, the voltage from point B to ground
    - **(a)** increases   **(b)** decreases   **(c)** stays the same

13. If C_2 opens, the voltage from point B to ground
    - **(a)** increases   **(b)** decreases   **(c)** stays the same

14. If a short develops from point C to ground, the voltage from point A to ground
    - **(a)** increases   **(b)** decreases   **(c)** stays the same

15. If capacitor C_3 opens, the voltage from B to D
    - **(a)** increases   **(b)** decreases   **(c)** stays the same

16. If the source frequency increases, the voltage from point C to ground
    - **(a)** increases   **(b)** decreases   **(c)** stays the same

17. If the source frequency increases, the current from the source
    - **(a)** increases   **(b)** decreases   **(c)** stays the same

18. If R_2 shorts out, the voltage across C_1
    - **(a)** increases   **(b)** decreases   **(c)** stays the same

PROBLEMS

More difficult problems are indicated by an asterisk (*).
Answers to odd-numbered problems are at the end of the book.

PART 1: SERIES CIRCUITS

SECTION 15–1   The Complex Number System

1. What are the two characteristics of a quantity indicated by a complex number?

2. Locate the following numbers on the complex plane:
   - **(a)** +6   **(b)** −2   **(c)** +j3   **(d)** −j8

3. Locate the points represented by each of the following coordinates on the complex plane:
   - **(a)** 3, j5   **(b)** −7, j1   **(c)** −10, −j10

*4. Determine the coordinates of each point having the same magnitude but located 180° away from each point in Problem 3.

*5. Determine the coordinates of each point having the same magnitude but located 90° away from those in Problem 3.

6. Points on the complex plane are described below. Express each point as a complex number in rectangular form:

   (a) 3 units to the right of the origin on the real axis, and up 5 units on the j axis.

   (b) 2 units to the left of the origin on the real axis, and 1.5 units up on the j axis.

   (c) 10 units to the left of the origin on the real axis, and down 14 units on the $-j$ axis.

7. What is the value of the hypotenuse of a right triangle whose sides are 10 and 15?

8. Convert each of the following rectangular numbers to polar form:

   (a) $40 - j40$　　(b) $50 - j200$　　(c) $35 - j20$　　(d) $98 + j45$

9. Convert each of the following polar numbers to rectangular form:

   (a) $1000\angle-50°$　　(b) $15\angle160°$　　(c) $25\angle-135°$　　(d) $3\angle180°$

10. Express each of the following polar numbers using a negative angle to replace the positive angle:

    (a) $10\angle120°$　　(b) $32\angle85°$　　(c) $5\angle310°$

11. Identify the quadrant in which each point in Problem 8 is located.

12. Identify the quadrant in which each point in Problem 10 is located.

13. Write the polar expressions using positive angles for each phasor in Figure 15–83.

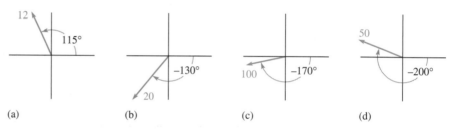

▲ FIGURE 15–83

14. Add the following sets of complex numbers:

    (a) $9 + j3$ and $5 + j8$　　　　(b) $3.5 - j4$ and $2.2 + j6$

    (c) $-18 + j23$ and $30 - j15$　　(d) $12\angle45°$ and $20\angle32°$

    (e) $3.8\angle75°$ and $1 + j1.8$　　(f) $50 - j39$ and $60\angle-30°$

15. Perform the following subtractions:

    (a) $(2.5 + j1.2) - (1.4 + j0.5)$　　(b) $(-45 - j23) - (36 + j12)$

    (c) $(8 - j4) - 3\angle25°$　　　　　(d) $48\angle135° - 33\angle-60°$

16. Multiply the following numbers:

    (a) $4.5\angle48°$ and $3.2\angle90°$　　(b) $120\angle-220°$ and $95\angle200°$

    (c) $-3\angle150°$ and $4 - j3$　　　　(d) $67 + j84$ and $102\angle40°$

    (e) $15 - j10$ and $-25 - j30$　　　(f) $0.8 + j0.5$ and $1.2 - j1.5$

17. Perform the following divisions:

    (a) $\dfrac{8\angle50°}{2.5\angle39°}$　　(b) $\dfrac{63\angle-91°}{9\angle10°}$　　(c) $\dfrac{28\angle30°}{14 - j12}$　　(d) $\dfrac{40 - j30}{16 + j8}$

18. Perform the following operations:

    (a) $\dfrac{2.5\angle65° - 1.8\angle-23°}{1.2\angle37°}$　　　　(b) $\dfrac{(100\angle15°)(85 - j150)}{25 + j45}$

    (c) $\dfrac{(250\angle90° + 175\angle75°)(50 - j100)}{(125 + j90)(35\angle50°)}$　　(d) $\dfrac{(1.5)^2(3.8)}{1.1} + j\left(\dfrac{8}{4} - j\dfrac{4}{2}\right)$

SECTION 15–2　Sinusoidal Response of Series RC Circuits

19. An 8 kHz sinusoidal voltage is applied to a series RC circuit. What is the frequency of the voltage across the resistor? Across the capacitor?

20. What is the wave shape of the current in the circuit of Problem 19?

SECTION 15–3   Impedance of Series *RC* Circuits

21. Express the total impedance of each circuit in Figure 15–84 in both polar and rectangular forms.

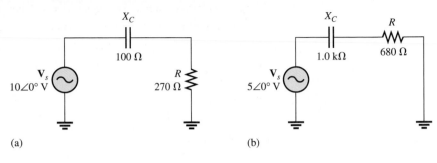

(a)   (b)

▲ **FIGURE 15–84**

22. Determine the impedance magnitude and phase angle in each circuit in Figure 15–85.

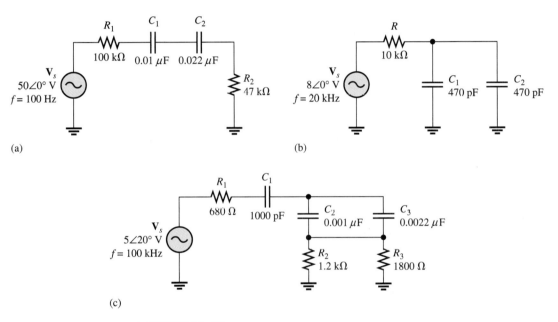

▲ **FIGURE 15–85**

23. For the circuit of Figure 15–86, determine the impedance expressed in rectangular form for each of the following frequencies:

   (a) 100 Hz   **(b)** 500 Hz   **(c)** 1 kHz   **(d)** 2.5 kHz

24. Repeat Problem 23 for $C = 0.0047\ \mu\text{F}$.

25. Determine the values of R and X_C in a series *RC* circuit for the following values of total impedance:

   (a) $\mathbf{Z} = 33\ \Omega - j50\ \Omega$ **(b)** $\mathbf{Z} = 300\angle{-25°}\ \Omega$

   (c) $\mathbf{Z} = 1.8\angle{-67.2°}\ \text{k}\Omega$ **(d)** $\mathbf{Z} = 789\angle{-45°}\ \Omega$

SECTION 15–4   Analysis of Series *RC* Circuits

26. Express the current in polar form for each circuit of Figure 15–84.

27. Calculate the total current in each circuit of Figure 15–85 and express in polar form.

28. Determine the phase angle between the applied voltage and the current for each circuit in Figure 15–85.

29. Repeat Problem 28 for the circuit in Figure 15–86, using $f = 5$ kHz.

30. For the circuit in Figure 15–87, draw the phasor diagram showing all voltages and the total current. Indicate the phase angles.

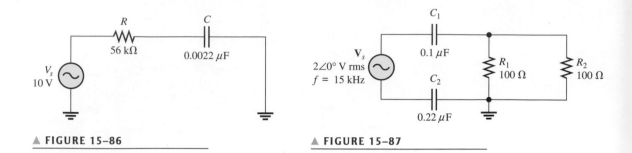

▲ FIGURE 15–86   ▲ FIGURE 15–87

31. For the circuit in Figure 15–88, determine the following in polar form:

   (a) **Z**   (b) **I**$_{tot}$ (c) **V**$_R$ (d) **V**$_C$

▶ FIGURE 15–88

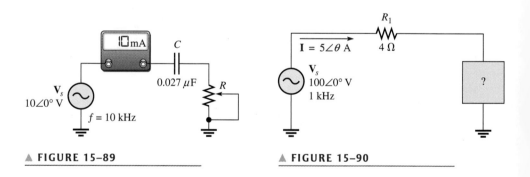

*32. To what value must the rheostat be set in Figure 15–89 to make the total current 10 mA? What is the resulting angle?

*33. Determine the series element or elements that must be installed in the block of Figure 15–90 to meet the following requirements: $P_{\text{true}} = 400$ W and there is a leading power factor (I_{tot} leads V_s).

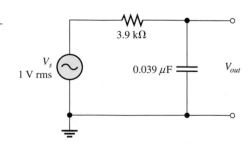

▲ FIGURE 15–89   ▲ FIGURE 15–90

34. For the lag circuit in Figure 15–91, determine the phase shift between the input voltage and the output voltage for each of the following frequencies:

   (a) 1 Hz   (b) 100 Hz   (c) 1 kHz   (d) 10 kHz

▶ FIGURE 15–91

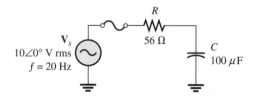

35. The lag circuit in Figure 15–91 also acts as a low-pass filter. Draw a response curve for this circuit by plotting the output voltage versus frequency for 0 Hz to 10 kHz in 1 kHz increments.

36. Repeat Problem 34 for the lead circuit in Figure 15–92.

▶ **FIGURE 15–92**

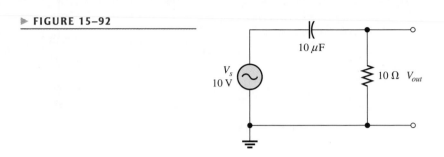

37. Plot the frequency response curve of the output amplitude for the lead circuit in Figure 15–92 for a frequency range of 0 Hz to 10 kHz in 1 kHz increments.

38. Draw the voltage phasor diagram for the circuit in Figure 15–91 for a frequency of 5 kHz with $V_s = 1$ V rms.

39. Repeat Problem 38 for the circuit in Figure 15–92. $V_s = 10$ V rms and $f = 1$ kHz.

PART 2: PARALLEL CIRCUITS

SECTION 15–5   Impedance and Admittance of Parallel *RC* Circuits

40. Determine the impedance and express it in polar form for the circuit in Figure 15–93.

41. Determine the impedance magnitude and phase angle in Figure 15–94.

42. Repeat Problem 41 for the following frequencies:

    (a) 1.5 kHz   **(b)** 3 kHz   **(c)** 5 kHz   **(d)** 10 kHz

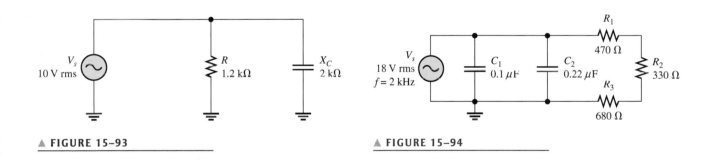

▲ **FIGURE 15–93**   ▲ **FIGURE 15–94**

SECTION 15–6   Analysis of Parallel *RC* Circuits

43. For the circuit in Figure 15–95, find all the currents and voltages in polar form.

▶ **FIGURE 15–95**

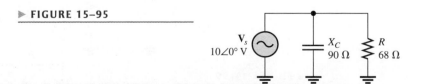

44. For the parallel circuit in Figure 15–96, find the magnitude of each branch current and the total current. What is the phase angle between the applied voltage and the total current?

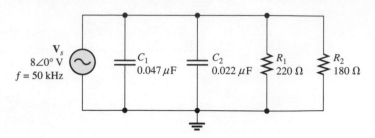

▲ FIGURE 15–96

45. For the circuit in Figure 15–97, determine the following:

    (a) Z   (b) I_R (c) $I_{C(tot)}$ (d) I_{tot} (e) θ

▶ FIGURE 15–97

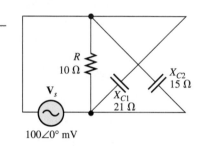

46. Repeat Problem 45 for $R = 5.6\,k\Omega$, $C_1 = 0.047\,\mu F$, $C_2 = 0.022\,\mu F$, and $f = 500$ Hz.

*47. Convert the circuit in Figure 15–98 to an equivalent series form.

▶ FIGURE 15–98

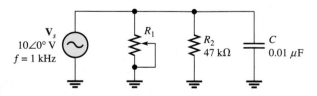

*48. Determine the value to which R_1 must be adjusted to get a phase angle of 30° between the source voltage and the total current in Figure 15–99.

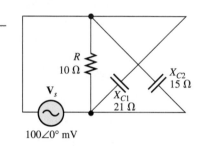

▲ FIGURE 15–99

PART 3: SERIES-PARALLEL CIRCUITS

SECTION 15–7 Analysis of Series-Parallel *RC* Circuits

49. Determine the voltages in polar form across each element in Figure 15–100. Draw the voltage phasor diagram.

50. Is the circuit in Figure 15–100 predominantly resistive or predominantly capacitive?

51. Find the current through each branch and the total current in Figure 15–100. Express the currents in polar form. Draw the current phasor diagram.

▶ **FIGURE 15–100**

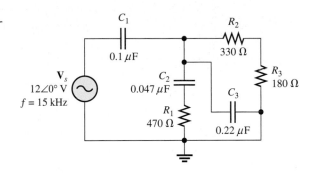

52. For the circuit in Figure 15–101, determine the following:

    (a) I_{tot} **(b)** θ **(c)** V_{R1} **(d)** V_{R2} **(e)** V_{R3} **(f)** V_C

*53. Determine the value of C_2 in Figure 15–102 when $V_A = V_B$.

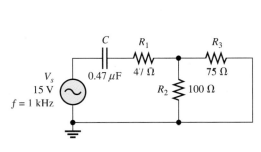

▲ **FIGURE 15–101**

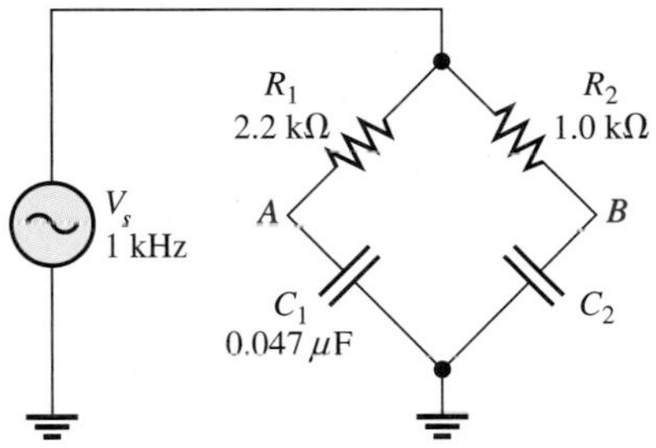

▲ **FIGURE 15–102**

*54. Determine the voltage and its phase angle at each point labeled in Figure 15–103.

*55. Find the current through each component in Figure 15–103.

*56. Draw the voltage and current phasor diagram for Figure 15–103.

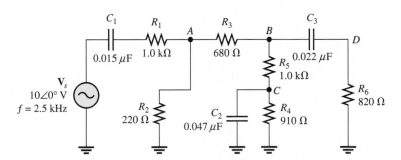

▲ **FIGURE 15–103**

PART 4: SPECIAL TOPICS

SECTION 15–8 Power in *RC* Circuits

57. In a certain series *RC* circuit, the true power is 2 W, and the reactive power is 3.5 VAR. Determine the apparent power.

58. In Figure 15–88, what is the true power and the reactive power?

59. What is the power factor for the circuit of Figure 15–98?

60. Determine P_{true}, P_r, P_a, and *PF* for the circuit in Figure 15–101. Draw the power triangle.

*61. A single 240 V, 60 Hz source drives two loads. Load *A* has an impedance of 50 Ω and a power factor of 0.85. Load *B* has an impedance of 72 Ω and a power factor of 0.95.

    (a) How much current does each load draw?

    (b) What is the reactive power in each load?

    (c) What is the true power in each load?

    (d) What is the apparent power in each load?

    (e) Which load has more voltage drop along the lines connecting it to the source?

SECTION 15–9 Basic Applications

62. Calculate the frequency of oscillation for the circuit in Figure 15–62 if all *C*s are 0.0022 *μ*F and all *R*s are 10 kΩ.

*63. What value of coupling capacitor is required in Figure 15–104 so that the signal voltage at the input of amplifier 2 is at least 70.7% of the signal voltage at the output of amplifier 1 when the frequency is 20 Hz?

▶ **FIGURE 15–104**

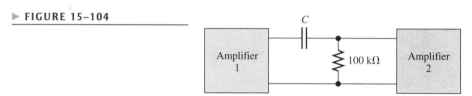

64. The rms value of the signal voltage out of amplifier *A* in Figure 15–105 is 50 mV. If the input resistance to amplifier *B* is 10 kΩ, how much of the signal is lost due to the coupling capacitor when the frequency is 3 kHz?

▶ **FIGURE 15–105**

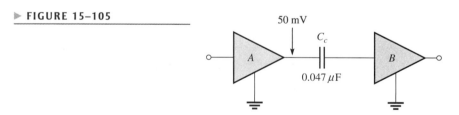

SECTION 15–10 Troubleshooting

65. Assume that the capacitor in Figure 15–106 is excessively leaky. Show how this degradation affects the output voltage and phase angle, assuming that the leakage resistance is 5 kΩ and the frequency is 10 Hz.

▶ **FIGURE 15–106**

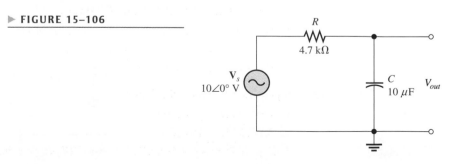

*66. Each of the capacitors in Figure 15–107 has developed a leakage resistance of 2 kΩ. Determine the output voltages under this condition for each circuit.

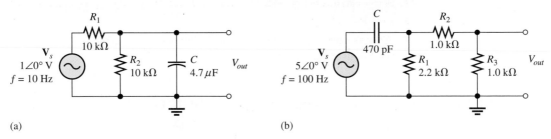

(a)   (b)

▲ **FIGURE 15–107**

67. Determine the output voltage for the circuit in Figure 15–107(a) for each of the following failure modes, and compare it to the correct output:

(a) R_1 open   (b) R_2 open   (c) C open   (d) C shorted

68. Determine the output voltage for the circuit in Figure 15–107(b) for each of the following failure modes, and compare it to the correct output:

(a) C open   (b) C shorted   (c) R_1 open   (d) R_2 open   (e) R_3 open

Multisim Troubleshooting and Analysis

These problems require your Multisim CD-ROM.

69. Open file P15-69 and determine if there is a fault. If so, find the fault.

70. Open file P15-70 and determine if there is a fault. If so, find the fault.

71. Open file P15-71 and determine if there is a fault. If so, find the fault.

72. Open file P15-72 and determine if there is a fault. If so, find the fault.

73. Open file P15-73 and determine if there is a fault. If so, find the fault.

74. Open file P15-74 and determine if there is a fault. If so, find the fault.

75. Open file P15-75 and determine the frequency response for the filter.

76. Open file P15-76 and determine the frequency response for the filter.

ANSWERS

SECTION REVIEWS

SECTION 15–1 The Complex Number System

1. $2.828 \angle 45°$; first
2. $3.54 - j3.54$, fourth
3. $4 + j1$
4. $3 + j7$
5. $16 \angle 110°$
6. $5 \angle 15°$

SECTION 15–2 Sinusoidal Response of Series RC Circuits

1. The voltage frequency is 60 Hz. The current frequency is 60 Hz.
2. The capacitive reactance causes the phase shift.
3. The phase angle is closer to 0°.

SECTION 15–3 Impedance of Series RC Circuits

1. $R = 150\ \Omega$; $X_C = 220\ \Omega$
2. $\mathbf{Z} = 33\ \mathrm{k}\Omega - j50\ \mathrm{k}\Omega$
3. $Z = \sqrt{R^2 + X_C^2} = 59.9\ \mathrm{k}\Omega$; $\theta = -\tan^{-1}(X_C/R) = -56.6°$

SECTION 15–4 **Analysis of Series *RC* Circuits**

1. $V_s = \sqrt{V_R^2 + V_C^2} = 7.21$ V
2. $\theta = -\tan^{-1}(X_C/R) = -56.3°$
3. $\theta = 90°$
4. When f increases, X_C decreases, Z decreases, and θ decreases.
5. $\phi = -90° + \tan^{-1}(X_C/R) = -62.8°$
6. $V_{out} = (R/\sqrt{R^2 + X_C^2})V_{in} = 8.90$ V rms

SECTION 15–5 **Impedance and Admittance of Parallel *RC* Circuits**

1. Conductance is the reciprocal of resistance, capacitive susceptance is the reciprocal of capacitive reactance, and admittance is the reciprocal of impedance.
2. $Y = 1/Z = 1/100\ \Omega = 10$ mS
3. $\mathbf{Y} = 1/\mathbf{Z} = 25.1\angle 32.1°\ \mu$S
4. $\mathbf{Z} = 39.8\angle -32.1°$ kΩ

SECTION 15–6 **Analysis of Parallel *RC* Circuits**

1. $I_{tot} = V_s Y = 21$ mA
2. $I_{tot} = \sqrt{I_R^2 + I_C^2} = 18$ mA; $\theta = \tan^{-1}(I_C/I_R) = 56.3°$; θ is with respect to applied voltage.
3. $\theta = 90°$

SECTION 15–7 **Analysis of Series-Parallel *RC* Circuits**

1. See Figure 15–108.
2. $\mathbf{Z}_{tot} = \mathbf{V}_s/\mathbf{I}_{tot} = 36.9\angle -51.6°\ \Omega$

▶ **FIGURE 15–108**

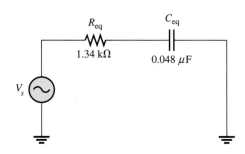

R_{eq}
1.34 kΩ

C_{eq}
0.048 μF

V_s

SECTION 15–8 **Power in *RC* Circuits**

1. Power dissipation is due to resistance.
2. $PF = \cos\theta = 0.707$
3. $P_{true} = I^2 R = 1.32$ kW; $P_r = I^2 X_C = 1.84$ kVAR; $P_a = I^2 Z = 2.26$ kVA

SECTION 15–9 **Basic Applications**

1. $180°$
2. The output is across the capacitor.

SECTION 15–10 **Troubleshooting**

1. The leakage resistance acts in parallel with C, which alters the circuit time constant.
2. The capacitor is open.
3. An open series resistor or the capacitor shorted will result in 0 V across the capacitor.

A Circuit Application

1. A lower value coupling capacitor will increase the frequency at which a significant drop in voltage occurs.

2. $V_B = 3.16\,\text{V dc}$

3. $V_B = 10\,\text{mV rms}$

RELATED PROBLEMS FOR EXAMPLES

15–1 **(a)** 1st **(b)** 4th **(c)** 3rd **(d)** 2nd

15–2 $29.2\angle 52°$

15–3 $70.1 - j34.2$

15–4 $-1 - j8$

15–5 $-13.5 - j4.5$

15–6 $1500\angle -50°$

15–7 $4\angle -42°$

15–8 $114.612390255\angle -60.75\ldots$
See Figure 15–109.

▶ **FIGURE 15–109**

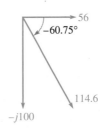

15–9 $\mathbf{V}_s = 2.56\angle -38.5°\,\text{V}$

15–10 $\mathbf{I} = 3.80\angle 33.4°\,\text{mA}$

15–11 $Z = 15.9\,\text{k}\Omega, \theta = -86.4°$

15–12 The phase lag increases.

15–13 The output voltage decreases.

15–14 The phase lead decreases.

15–15 The output voltage increases.

15–16 $\mathbf{Z} = 24.3\angle -76.0°\,\Omega$

15–17 $\mathbf{Y} = 4.60\angle 48.8°\,\text{mS}$

15–18 $\mathbf{I} = 6.16\angle 42.4°\,\text{mA}$

15–19 $\mathbf{I}_{tot} - 117\angle 31.0°\,\text{mA}$

15–20 $R_{eq} = 8.99\,\text{k}\Omega, X_{C(eq)} = 4.38\,\text{k}\Omega$

15–21 $\mathbf{V}_1 = 7.05\angle 8.9°\,\text{V}, \mathbf{V}_2 = 3.21\angle -18.9°\,\text{V}$

15–22 $\mathbf{V}_{R1} = 766\angle 67.5°\,\text{mV}; \mathbf{V}_{C1} = 1.85\angle -22.5°\,\text{V}; \mathbf{V}_{R2} = 1.58\angle 37.6°\,\text{V};$
$\mathbf{V}_{C2} = 1.22\angle -52.4°\,\text{V};$ See Figure 15–110.

▶ **FIGURE 15–110**

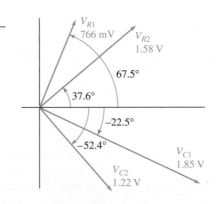

15–23 $PF = 0.146$

15–24 $P_{\text{true}} = 213\,\text{mW}$

15–25 1.60 kHz

15–26 $V_{out} = 7.29\,\text{V}$

15–27 Resistor open

SELF-TEST

1. (b)   **2.** (a)   **3.** (d)   **4.** (c)   **5.** (c)   **6.** (a)   **7.** (b)

8. (c)   **9.** (d)   **10.** (c)   **11.** (b)   **12.** (b)   **13.** (a)   **14.** (d)

15. (b)   **16.** (a)   **17.** (d)   **18.** (c)   **19.** (b)   **20.** (d)   **21.** (d)

22. (c)   **23.** (b)

CIRCUIT DYNAMICS QUIZ

1. (a)   **2.** (b)   **3.** (a)   **4.** (a)   **5.** (b)   **6.** (c)   **7.** (c)   **8.** (a)

9. (a)   **10.** (c)   **11.** (b)   **12.** (a)   **13.** (a)   **14.** (b)   **15.** (a)   **16.** (b)

17. (a)   **18.** (a)

RL CIRCUITS

16

CHAPTER OBJECTIVES

PART 1: SERIES CIRCUITS

- Describe the relationship between current and voltage in a series *RL* circuit
- Determine the impedance of a series *RL* circuit
- Analyze a series *RL* circuit

PART 2: PARALLEL CIRCUITS

- Determine impedance and admittance in a parallel *RL* circuit
- Analyze a parallel *RL* circuit

PART 3: SERIES-PARALLEL CIRCUITS

- Analyze series-parallel *RL* circuits

PART 4: SPECIAL TOPICS

- Determine power in *RL* circuits
- Describe two examples of *RL* circuit applications
- Troubleshoot *RL* circuits

KEY TERMS

- Inductive reactance
- Inductive susceptance (B_L)

A CIRCUIT APPLICATION PREVIEW

In the circuit application, you will use your knowledge of *RL* circuits to determine, based on parameter measurements, the type of filter circuits that are encapsulated in sealed modules and their component values.

VISIT THE COMPANION WEBSITE

Study aids for this chapter are available at http://www.prenhall.com/floyd

INTRODUCTION

In this chapter you will study series and parallel *RL* circuits. The analyses of *RL* and *RC* circuits are similar. The major difference is that the phase responses are opposite; inductive reactance increases with frequency, while capacitive reactance decreases with frequency.

An *RL* circuit contains both resistance and inductance. In this chapter, basic series and parallel *RL* circuits and their responses to sinusoidal ac voltages are presented. Series-parallel combinations are also analyzed. True, reactive, and apparent power in *RL* circuits are discussed and some basic *RL* circuit applications are introduced. Applications of *RL* circuits include filters and switching regulators. Troubleshooting is also covered in this chapter.

COVERAGE OPTIONS

If you chose Option 1 to cover all of Chapter 15 on *RC* circuits, then all of this chapter should be covered next.

If you chose Option 2 to cover reactive circuits beginning in Chapter 15 on the basis of the four major parts, then the appropriate part of this chapter should be covered next, followed by the corresponding part in Chapter 17.

SERIES CIRCUITS

16–1 SINUSOIDAL RESPONSE OF SERIES *RL* CIRCUITS

As with the *RC* circuit, all currents and voltages in a series *RL* circuit are sinusoidal when the input voltage is sinusoidal. The inductance causes a phase shift between the voltage and the current that depends on the relative values of the resistance and the inductive reactance.

After completing this section, you should be able to

- ◆ **Describe the relationship between current and voltage in a series *RL* circuit**
  - ◆ Discuss voltage and current waveforms
  - ◆ Discuss phase shift

In an *RL* circuit, the resistor voltage and the current lag the source voltage. The inductor voltage leads the source voltage. Ideally, the phase angle between the current and the inductor voltage is always 90°. These generalized phase relationships are indicated in Figure 16–1. Notice how they differ from those of the *RC* circuit that was discussed in Chapter 15.

The amplitudes and the phase relationships of the voltages and current depend on the values of the resistance and the **inductive reactance**. When a circuit is purely inductive, the phase angle between the applied voltage and the total current is 90°, with the current lagging the voltage. When there is a combination of both resistance and inductive reactance in a circuit, the phase angle is somewhere between 0° and 90°, depending on the relative values of the resistance and the inductive reactance.

Recall that practical inductors have winding resistance, capacitance between windings, and other factors that prevent an inductor from behaving as an ideal component. In practical

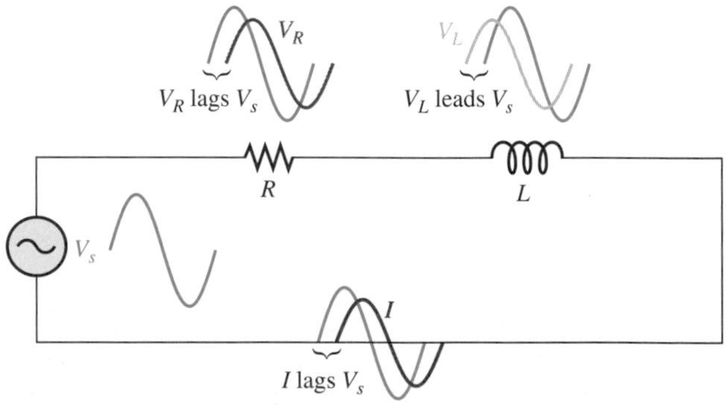

▲ FIGURE 16–1

Illustration of sinusoidal response with general phase relationships of V_R, V_L, and *I* relative to the source voltage. V_R and *I* are in phase, while V_R and V_L are 90° out of phase with each other.

circuits, these effects can be significant; however, for the purpose of isolating the inductive effects, we will treat inductors in this chapter as ideal (except in the Circuit Application).

1. A 1 kHz sinusoidal voltage is applied to an *RL* circuit. What is the frequency of the resulting current?

2. When the resistance in an *RL* circuit is greater than the inductive reactance, is the phase angle between the applied voltage and the total current closer to 0° or to 90°?

16–2 IMPEDANCE OF SERIES *RL* CIRCUITS

The impedance of a series *RL* circuit consists of resistance and inductive reactance and is the total opposition to sinusoidal current. Its unit is the ohm. The impedance also causes a phase difference between the total current and the source voltage. Therefore, the impedance consists of a magnitude component and a phase angle component.

After completing this section, you should be able to

- **Determine the impedance of a series *RL* circuit**

  - Express inductive reactance in complex form

  - Express total impedance in complex form

  - Calculate impedance magnitude and the phase angle

The impedance of a series *RL* circuit is determined by the resistance and the inductive reactance. Inductive reactance is expressed as a phasor quantity in rectangular form as

$$\mathbf{X}_L = jX_L$$

In the series *RL* circuit of Figure 16–2, the total impedance is the phasor sum of R and jX_L and is expressed as

$$\mathbf{Z} = R + jX_L$$

Equation 16–1

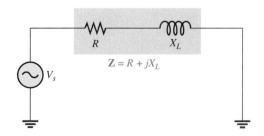

◀ FIGURE 16–2

Impedance in a series *RL* circuit.

In ac analysis, both R and X_L are as shown in the phasor diagram of Figure 16–3(a), with X_L appearing at a +90° angle with respect to R. This relationship comes from the fact that the inductor voltage leads the current, and thus the resistor voltage, by 90°. Since $\mathbf{Z}$ is the phasor sum of R and jX_L, its phasor representation is as shown in Figure 16–3(b). A repositioning of the phasors, as shown in part (c), forms a right triangle called the *impedance triangle*. The length of each phasor represents the magnitude of the quantity, and θ is the phase angle between the applied voltage and the current in the *RL* circuit.

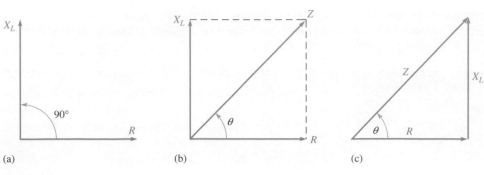

▲ FIGURE 16–3

Development of the impedance triangle for a series *RL* circuit.

The impedance magnitude of the series *RL* circuit can be expressed in terms of the resistance and reactance as

$$Z = \sqrt{R^2 + X_L^2}$$

The magnitude of the impedance is expressed in ohms.

The phase angle, θ, is expressed as

$$\theta = \tan^{-1}\left(\frac{X_L}{R}\right)$$

Combining the magnitude and the angle, the impedance can be expressed in polar form as

Equation 16–2

$$\mathbf{Z} = \sqrt{R^2 + X_L^2}\angle\tan^{-1}\left(\frac{X_L}{R}\right)$$

EXAMPLE 16–1

For each circuit in Figure 16–4, write the phasor expression for the impedance in both rectangular and polar forms.

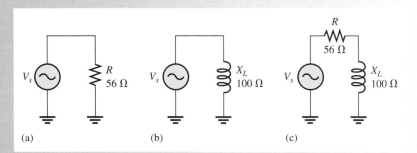

▲ FIGURE 16–4

Solution   For the circuit in Figure 16–4(a), the impedance is

$$\mathbf{Z} = R + j0 = R = \mathbf{56}\ \Omega \quad \text{in rectangular form } (X_L = 0)$$
$$\mathbf{Z} = R\angle 0° = \mathbf{56}\angle\mathbf{0}°\ \Omega \quad \text{in polar form}$$

The impedance is simply equal to the resistance, and the phase angle is zero because pure resistance does not introduce a phase shift.

For the circuit in Figure 16–4(b), the impedance is

$$\mathbf{Z} = 0 + jX_L = \mathbf{j100} \ \Omega \qquad \text{in rectangular form } (R = 0)$$

$$\mathbf{Z} = X_L \angle 90° = \mathbf{100 \angle 90°} \ \Omega \qquad \text{in polar form}$$

The impedance equals the inductive reactance in this case, and the phase angle is +90° because the inductance causes the current to lag the voltage by 90°.

For the circuit in Figure 16–4(c), the impedance in rectangular form is

$$\mathbf{Z} = R + jX_L = \mathbf{56} \ \Omega \ + \ \mathbf{j100} \ \Omega$$

The impedance in polar form is

$$\mathbf{Z} = \sqrt{R^2 + X_L^2} \angle \tan^{-1}\left(\frac{X_L}{R}\right)$$

$$= \sqrt{(56 \ \Omega)^2 + (100 \ \Omega)^2} \angle \tan^{-1}\left(\frac{100 \ \Omega}{56 \ \Omega}\right) = \mathbf{115 \angle 60.8°} \ \Omega$$

In this case, the impedance is the phasor sum of the resistance and the inductive reactance. The phase angle is fixed by the relative values of X_L and R.

*Related Problem** In a series *RL* circuit, $R = 1.8 \ \text{k}\Omega$ and $X_L = 950 \ \Omega$. Express the impedance in both rectangular and polar forms.

---

*Answers are at the end of the chapter.

**SECTION 16–2
REVIEW**

1. The impedance of a certain *RL* circuit is $150 \ \Omega + j220 \ \Omega$. What is the value of the resistance? The inductive reactance?
2. A series *RL* circuit has a total resistance of $33 \ \text{k}\Omega$ and an inductive reactance of $50 \ \text{k}\Omega$. Write the phasor expression for the impedance in rectangular form. Convert the impedance to polar form.

16–3 ANALYSIS OF SERIES *RL* CIRCUITS

In this section, Ohm's law and Kirchhoff's voltage law are used in the analysis of series *RL* circuits to determine voltage, current, and impedance. Also, *RL* lead and lag circuits are examined.

After completing this section, you should be able to

◆ **Analyze a series *RL* circuit**

   ◆ Apply Ohm's law and Kirchhoff's voltage law to series *RL* circuits

   ◆ Express the voltages and current as phasor quantities

   ◆ Show how impedance and phase angle vary with frequency

   ◆ Discuss and analyze the *RL* lead circuit

   ◆ Discuss and analyze the *RL* lag circuit

Ohm's Law

The application of Ohm's law to series *RL* circuits involves the use of the phasor quantities of **Z**, **V**, and **I**. The three equivalent forms of Ohm's law were stated in Chapter 15 for *RC* circuits. They apply also to *RL* circuits and are restated here:

$$\mathbf{V} = \mathbf{IZ} \qquad \mathbf{I} = \frac{\mathbf{V}}{\mathbf{Z}} \qquad \mathbf{Z} = \frac{\mathbf{V}}{\mathbf{I}}$$

Recall that since Ohm's law calculations involve multiplication and division operations, you should express the voltage, current, and impedance in polar form.

EXAMPLE 16–2

The current in Figure 16–5 is expressed in polar form as $\mathbf{I} = 0.2\angle0°$ mA. Determine the source voltage expressed in polar form, and draw a phasor diagram showing the relationship between the source voltage and the current.

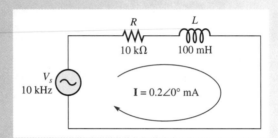

▲ FIGURE 16–5

Solution The magnitude of the inductive reactance is

$$X_L = 2\pi fL = 2\pi(10\,\text{kHz})(100\,\text{mH}) = 6.28\,\text{k}\Omega$$

The impedance in rectangular form is

$$\mathbf{Z} = R + jX_L = 10\,\text{k}\Omega + j6.28\,\text{k}\Omega$$

Converting to polar form yields

$$\mathbf{Z} = \sqrt{R^2 + X_L^2}\angle\tan^{-1}\left(\frac{X_L}{R}\right)$$

$$= \sqrt{(10\,\text{k}\Omega)^2 + (6.28\,\text{k}\Omega)^2}\angle\tan^{-1}\left(\frac{6.28\,\text{k}\Omega}{10\,\text{k}\Omega}\right) = \mathbf{11.8\angle32.1°\,k\Omega}$$

Use Ohm's law to determine the source voltage.

$$\mathbf{V}_s = \mathbf{IZ} = (0.2\angle0°\,\text{mA})(11.8\angle32.1°\,\text{k}\Omega) = \mathbf{2.36\angle32.1°\,V}$$

The magnitude of the source voltage is 2.36 V at an angle of 32.1° with respect to the current; that is, the voltage leads the current by 32.1°, as shown in the phasor diagram of Figure 16–6.

▶ FIGURE 16–6

Related Problem If the source voltage in Figure 16–5 were 5∠0° V, what would be the current expressed in polar form?

Use Multisim file E16-02 to verify the calculated results in this example and to confirm your calculation for the related problem.

Phase Relationships of Current and Voltages

In a series *RL* circuit, the current is the same through both the resistor and the inductor. Thus, the resistor voltage is in phase with the current, and the inductor voltage leads the current by 90°. Therefore, there is a phase difference of 90° between the resistor voltage, V_R, and the inductor voltage, V_L, as shown in the waveform diagram of Figure 16–7.

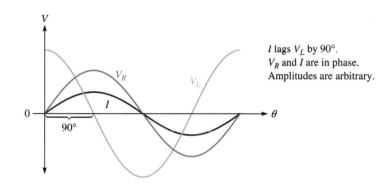

I lags V_L by 90°.
V_R and *I* are in phase.
Amplitudes are arbitrary.

◀ **FIGURE 16–7**

Phase relation of voltages and current in a series *RL* circuit.

From Kirchhoff's voltage law, the sum of the voltage drops must equal the applied voltage. However, since V_R and V_L are not in phase with each other, they must be added as phasor quantities with V_L leading V_R by 90°, as shown in Figure 16–8(a). As shown in part (b), $\mathbf{V}_s$ is the phasor sum of V_R and V_L.

$$\mathbf{V}_s = V_R + jV_L$$

Equation 16–3

This equation can be expressed in polar form as

$$\mathbf{V}_s = \sqrt{V_R^2 + V_L^2}\angle\tan^{-1}\left(\frac{V_L}{V_R}\right)$$

Equation 16–4

where the magnitude of the source voltage is

$$V_s = \sqrt{V_R^2 + V_L^2}$$

and the phase angle between the resistor voltage and the source voltage is

$$\theta = \tan^{-1}\left(\frac{V_L}{V_R}\right)$$

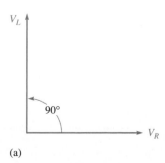

(a)

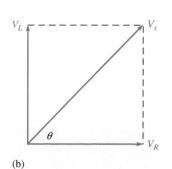

(b)

◀ **FIGURE 16–8**

Voltage phasor diagram for a series *RL* circuit.

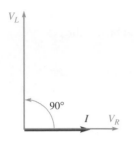

V_L

90°

I V_R

▲ FIGURE 16–9

Voltage and current phasor diagram for the waveforms in Figure 16–7.

Since the resistor voltage and the current are in phase, θ is also the phase angle between the source voltage and the current. Figure 16–9 shows a voltage and current phasor diagram that represents the waveform diagram of Figure 16–7.

Variation of Impedance and Phase Angle with Frequency

The impedance triangle is useful in visualizing how the frequency of the applied voltage affects the *RL* circuit response. As you know, inductive reactance varies directly with frequency. When X_L increases, the magnitude of the total impedance also increases; and when X_L decreases, the magnitude of the total impedance decreases. Thus, Z is directly dependent on frequency. The phase angle θ also varies directly with frequency because $\theta = \tan^{-1}(X_L/R)$. As X_L increases with frequency, so does θ, and vice versa.

The impedance triangle is used in Figure 16–10 to illustrate the variations in X_L, Z, and θ as the frequency changes. Of course, R remains constant. The main point is that because X_L varies directly with the frequency, so also do the magnitude of the total impedance and the phase angle. Example 16–3 illustrates this.

▶ FIGURE 16–10

As the frequency increases, X_L increases, Z increases, and θ increases. Each value of frequency can be visualized as forming a different impedance triangle.

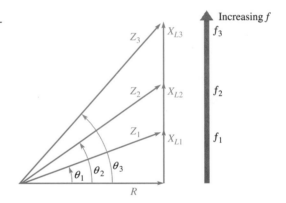

EXAMPLE 16–3

For the series *RL* circuit in Figure 16–11, determine the magnitude of the total impedance and the phase angle for each of the following frequencies:

(a) 10 kHz   (b) 20 kHz   (c) 30 kHz

▶ FIGURE 16–11

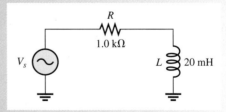

Solution   (a) For $f = 10$ kHz,

$$X_L = 2\pi fL = 2\pi(10 \text{ kHz})(20 \text{ mH}) = 1.26 \text{ k}\Omega$$

$$\mathbf{Z} = \sqrt{R^2 + X_L^2}\angle\tan^{-1}\left(\frac{X_L}{R}\right)$$

$$= \sqrt{(1.0 \text{ k}\Omega)^2 + (1.26 \text{ k}\Omega)^2}\angle\tan^{-1}\left(\frac{1.26 \text{ k}\Omega}{1.0 \text{ k}\Omega}\right) = 1.61\angle51.6° \text{ k}\Omega$$

Thus, $Z = \mathbf{1.61 \text{ k}\Omega}$ and $\theta = \mathbf{51.6°}$.

(b) For $f = 20$ kHz,

$$X_L = 2\pi(20 \text{ kHz})(20 \text{ mH}) = 2.51 \text{ k}\Omega$$

$$\mathbf{Z} = \sqrt{(1.0 \text{ k}\Omega)^2 + (2.51 \text{ k}\Omega)^2} \angle \tan^{-1}\left(\frac{2.51 \text{ k}\Omega}{1.0 \text{ k}\Omega}\right) = 2.70 \angle 68.3° \text{ k}\Omega$$

Thus, $Z = \mathbf{2.70 \text{ k}\Omega}$ and $\theta = \mathbf{68.3°}$.

(c) For $f = 30$ kHz,

$$X_L = 2\pi(30 \text{ kHz})(20 \text{ mH}) = 3.77 \text{ k}\Omega$$

$$\mathbf{Z} = \sqrt{(1.0 \text{ k}\Omega)^2 + (3.77 \text{ k}\Omega)^2} \angle \tan^{-1}\left(\frac{3.77 \text{ k}\Omega}{1.0 \text{ k}\Omega}\right) = \mathbf{3.90 \angle 75.1° \text{ k}\Omega}$$

Thus, $Z = \mathbf{3.90 \text{ k}\Omega}$ and $\theta = \mathbf{75.1°}$.

Notice that as the frequency increases, X_L, Z, and θ also increase.

Related Problem   Determine Z and θ in Figure 16–11 if f is 100 kHz.

The *RL* Lead Circuit

An *RL* lead circuit is a phase shift circuit in which the output voltage leads the input voltage by a specified amount. Figure 16–12(a) shows a series *RL* circuit with the output voltage taken across the inductor. Note that in the *RC* lead circuit, the output was taken across the resistor. The source voltage is the input, V_{in}. As you know, θ is the angle between the current and the input voltage; it is also the angle between the resistor voltage and the input voltage because V_R and I are in phase.

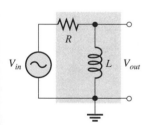

(a) A basic *RL* lead circuit

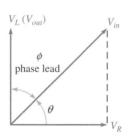

(b) Phasor voltage diagram showing V_{out} leading V_{in}

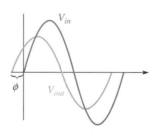

(c) Input and output voltage waveforms

◀ **FIGURE 16–12**

The *RL* lead circuit ($V_{out} = V_L$).

Since V_L leads V_R by 90°, the phase angle between the inductor voltage and the input voltage is the difference between 90° and θ, as shown in Figure 16–12(b). The inductor voltage is the output; it leads the input, thus creating a basic lead circuit.

The input and output voltage waveforms of the lead circuit are shown in Figure 16–12(c). The amount of phase difference, designated ϕ, between the input and the output is dependent on the relative values of the inductive reactance and the resistance, as is the magnitude of the output voltage.

Phase Difference Between Input and Output   The angle between V_{out} and V_{in} is designated ϕ (phi) and is developed as follows. The polar expressions for the input voltage and the current are $V_{in} \angle 0°$ and $I \angle -\theta$, respectively. The output voltage in polar form is

$$\mathbf{V}_{out} = (I \angle -\theta)(X_L \angle 90°) = IX_L \angle (90° - \theta)$$

This expression shows that the output voltage is at an angle of $90° - \theta$ with respect to the input voltage. Since $\theta = \tan^{-1}(X_L/R)$, the angle ϕ between the input and output is

$$\phi = 90° - \tan^{-1}\left(\frac{X_L}{R}\right)$$

Equivalently, this angle can be expressed as

Equation 16–5

$$\phi = \tan^{-1}\left(\frac{R}{X_L}\right)$$

The angle ϕ between the output and input is always positive, indicating that the output voltage leads the input voltage, as indicated in Figure 16–13.

▶ **FIGURE 16–13**

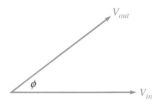

EXAMPLE 16–4

Determine the amount of phase lead from input to output in each lead circuit in Figure 16–14.

FIGURE 16–14

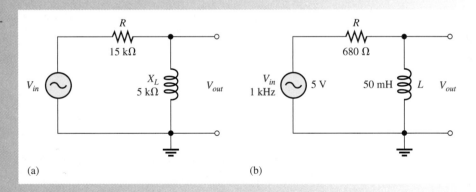

(a)   (b)

Solution For the lead circuit in Figure 16–14(a),

$$\phi = \tan^{-1}\left(\frac{R}{X_L}\right) = \tan^{-1}\left(\frac{15\,\text{k}\Omega}{5\,\text{k}\Omega}\right) = \mathbf{71.6°}$$

The output leads the input by 71.6°.

For the lead circuit in Figure 16–14(b), first determine the inductive reactance.

$$X_L = 2\pi f L = 2\pi(1\,\text{kHz})(50\,\text{mH}) = 314\,\Omega$$

$$\phi = \tan^{-1}\left(\frac{R}{X_L}\right) = \tan^{-1}\left(\frac{680\,\Omega}{314\,\Omega}\right) = \mathbf{65.2°}$$

The output leads the input by 65.2°.

Related Problem In a certain lead circuit, $R = 2.2\,\text{k}\Omega$ and $X_L = 1\,\text{k}\Omega$. What is the phase lead?

Use Multisim files E16-04A and E16-04B to verify the calculated results in this example and to confirm your calculation for the related problem.

Magnitude of the Output Voltage To evaluate the output voltage in terms of its magnitude, visualize the *RL* lead circuit as a voltage divider. A portion of the total input voltage is dropped across the resistor and a portion across the inductor. Because the output voltage is the voltage across the inductor, it can be calculated using either Ohm's law ($V_{out} = IX_L$) or the voltage-divider formula.

$$V_{out} = \left(\frac{X_L}{\sqrt{R^2 + X_L^2}} \right) V_{in}$$

Equation 16–6

The phasor expression for the output voltage of an *RL* lead circuit is

$$\mathbf{V}_{out} = V_{out} \angle \phi$$

EXAMPLE 16–5

For the lead circuit in Figure 16–14(b) (Example 16–4), determine the output voltage in phasor form when the input voltage has an rms value of 5 V. Draw the input and output voltage waveforms showing their peak values. The inductive reactance X_L (314 Ω) and ϕ (65.2°) were found in Example 16–4.

Solution The output voltage in phasor form is

$$\mathbf{V}_{out} = V_{out} \angle \phi = \left(\frac{X_L}{\sqrt{R^2 + X_L^2}} \right) V_{in} \angle \phi$$

$$= \left(\frac{314\ \Omega}{\sqrt{(680\ \Omega)^2 + (314\ \Omega)^2}} \right) 5 \angle 65.2°\ \text{V} = \mathbf{2.10} \angle \mathbf{65.2°}\ \textbf{V}$$

The peak values of voltage are

$$V_{in(p)} = 1.414 V_{in(rms)} = 1.414(5\ \text{V}) = 7.07\ \text{V}$$
$$V_{out(p)} = 1.414 V_{out(rms)} = 1.414(2.10\ \text{V}) = 2.97\ \text{V}$$

The waveforms with their peak values are shown in Figure 16–15. Notice that the output voltage leads the input voltage by 65.2°.

▶ **FIGURE 16–15**

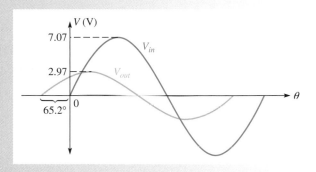

Related Problem In a lead circuit, does the output voltage increase or decrease when the frequency increases?

Use Multisim file E16-05 to verify the calculated results in this example and to confirm your calculation for the related problem.

The *RL* Lag Circuit

An *RL* lag circuit is a phase shift circuit in which the output voltage lags the input voltage by a specified amount. When the output of a series *RL* circuit is taken across the resistor rather than the inductor, as shown in Figure 16–16(a), it becomes a lag circuit.

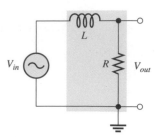

(a) A basic *RL* lag circuit

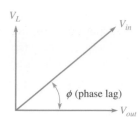

(b) Phasor voltage diagram
showing phase lag between
V_{in} and V_{out}

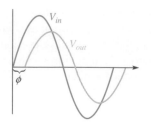

(c) Input and output waveforms

▲ FIGURE 16–16

The *RL* lag circuit ($V_{out} = V_R$).

Phase Difference Between Input and Output In a series *RL* circuit, the current lags the input voltage. Since the output voltage is taken across the resistor, the output lags the input, as indicated by the phasor diagram in Figure 16–16(b). The waveforms are shown in Figure 16–16(c).

As in the lead circuit, the amount of phase difference between the input and output and the magnitude of the output voltage in the lag circuit are dependent on the relative values of the resistance and the inductive reactance. When the input voltage is assigned a reference angle of 0°, the angle of the output voltage (ϕ) with respect to the input voltage equals θ because the resistor voltage (output) and the current are in phase with each other. The expression for the angle between the input voltage and the output voltage is

Equation 16–7

$$\phi = -\tan^{-1}\left(\frac{X_L}{R}\right)$$

This angle is negative because the output lags the input.

EXAMPLE 16–6

Calculate the output phase angle for each circuit in Figure 16–17.

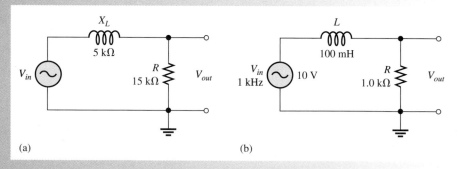

(a)

(b)

▲ FIGURE 16–17

Solution For the lag circuit in Figure 16–17(a).

$$\phi = -\tan^{-1}\left(\frac{X_L}{R}\right) = -\tan^{-1}\left(\frac{5\,k\Omega}{15\,k\Omega}\right) = -18.4°$$

The output lags the input by 18.4°.

For the lag circuit in Figure 16–17(b), first determine the inductive reactance.

$$X_L = 2\pi f L = 2\pi(1\text{ kHz})(100\text{ mH}) = 628\ \Omega$$

$$\phi = -\tan^{-1}\left(\frac{X_L}{R}\right) = -\tan^{-1}\left(\frac{628\ \Omega}{1.0\ \text{k}\Omega}\right) = -32.1°$$

The output lags the input by 32.1°.

Related Problem   In a certain lag circuit, $R = 5.6\ \text{k}\Omega$ and $X_L = 3.5\ \text{k}\Omega$. Determine the phase angle.

 Use Multisim files E16-06A and E16-06B to verify the calculated results in this example and to confirm your calculation for the related problem.

Magnitude of the Output Voltage   Since the output voltage of an *RL* lag circuit is taken across the resistor, the magnitude can be calculated using either Ohm's law ($V_{out} = IR$) or the voltage-divider formula.

$$V_{out} = \left(\frac{R}{\sqrt{R^2 + X_L^2}}\right)V_{in}$$

Equation 16–8

The expression for the output voltage in phasor form is

$$\mathbf{V}_{out} = V_{out}\angle\phi$$

EXAMPLE 16–7

The input voltage in Figure 16–17(b) (Example 16–6) has an rms value of 10 V. Determine the phasor expression for the output voltage. Draw the waveform relationships for the input and output voltages. The phase angle ($-32.1°$) and X_L (628 Ω) were found in Example 16–6.

Solution   The phasor expression for the output voltage is

$$\mathbf{V}_{out} = V_{out}\angle\phi = \left(\frac{R}{\sqrt{R^2 + X_L^2}}\right)V_{in}\angle\phi$$

$$= \left(\frac{1.0\ \text{k}\Omega}{1181\ \Omega}\right)10\angle-32.1°\ \text{V} = \mathbf{8.47\angle-32.1°\ V\ rms}$$

The waveforms are shown in Figure 16–18.

▶ **FIGURE 16–18**

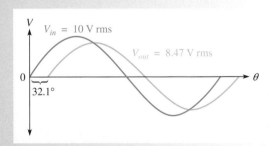

Related Problem   In a lag circuit, $R = 4.7\ \text{k}\Omega$ and $X_L = 6\ \text{k}\Omega$. If the rms input voltage is 20 V, what is the output voltage?

 Use Multisim file E16-07 to verify the calculated results in this example and to confirm your calculation for the related problem.

SECTION 16–3
REVIEW

1. In a certain series *RL* circuit, $V_R = 2\,V$ and $V_L = 3\,V$. What is the magnitude of the source voltage?

2. In Question 1, what is the phase angle between the source voltage and the current?

3. When the frequency of the applied voltage in a series *RL* circuit is increased, what happens to the inductive reactance? What happens to the magnitude of the total impedance? What happens to the phase angle?

4. A certain *RL* lead circuit consists of a 3.3 kΩ resistor and a 15 mH inductor. Determine the phase shift between input and output at a frequency of 5 kHz.

5. An *RL* lag circuit has the same component values as the lead circuit in Question 4. What is the magnitude of the output voltage at 5 kHz when the input is 10 V rms?

OPTION 2 NOTE

Coverage of series reactive circuits continues in Chapter 17, Part 1, on page 727.

PARALLEL CIRCUITS

16–4 IMPEDANCE AND ADMITTANCE OF PARALLEL *RL* CIRCUITS

In this section, you will learn how to determine the impedance and phase angle of a parallel *RL* circuit. The impedance consists of a magnitude component and a phase angle component. Also, inductive susceptance and admittance of a parallel *RL* circuit are introduced.

After completing this section, you should be able to

- **Determine impedance and admittance in a parallel *RL* circuit**
  - Express total impedance in complex form
  - Define and calculate *inductive susceptance* and *admittance*

Figure 16–19 shows a basic parallel *RL* circuit connected to an ac voltage source.

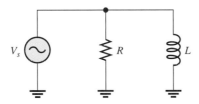

▲ **FIGURE 16–19**

Parallel *RL* circuit.

The expression for the total impedance of a two-component parallel *RL* circuit is developed as follows, using the product-over-sum rule.

$$\mathbf{Z} = \frac{(R\angle 0°)(X_L\angle 90°)}{R + jX_L} = \frac{RX_L\angle(0° + 90°)}{\sqrt{R^2 + X_L^2}\angle\tan^{-1}\left(\dfrac{X_L}{R}\right)}$$

$$= \left(\frac{RX_L}{\sqrt{R^2 + X_L^2}}\right)\angle\left(90° - \tan^{-1}\left(\frac{X_L}{R}\right)\right)$$

Equivalently, this equation can be expressed as

$$\mathbf{Z} = \left(\frac{RX_L}{\sqrt{R^2 + X_L^2}}\right)\angle\tan^{-1}\left(\frac{R}{X_L}\right)$$

Equation 16–9

EXAMPLE 16–8

For each circuit in Figure 16–20, determine the magnitude and phase angle of the total impedance.

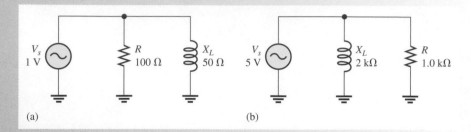

▲ FIGURE 16–20

Solution   For the circuit in Figure 16–20(a), the total impedance is

$$\mathbf{Z} = \left(\frac{RX_L}{\sqrt{R^2 + X_L^2}}\right)\angle\tan^{-1}\left(\frac{R}{X_L}\right)$$

$$= \left(\frac{(100\,\Omega)(50\,\Omega)}{\sqrt{(100\,\Omega)^2 + (50\,\Omega)^2}}\right)\angle\tan^{-1}\left(\frac{100\,\Omega}{50\,\Omega}\right) = \mathbf{44.7\angle63.4°\,\Omega}$$

Thus, $Z = 44.7\,\Omega$ and $\theta = 63.4°$.

For the circuit in Figure 16–20(b), the total impedance is

$$\mathbf{Z} = \left(\frac{(1.0\,k\Omega)(2\,k\Omega)}{\sqrt{(1.0\,k\Omega)^2 + (2\,k\Omega)^2}}\right)\angle\tan^{-1}\left(\frac{1.0\,k\Omega}{2\,k\Omega}\right) = \mathbf{894\angle26.6°\,\Omega}$$

Thus, $Z = 894\,\Omega$ and $\theta = 26.6°$.

Notice that the positive angle indicates that the voltage leads the current, as opposed to the RC case where the voltage lags the current.

Related Problem   In a parallel circuit, $R = 10\,k\Omega$ and $X_L = 14\,k\Omega$. Determine the total impedance in polar form.

Conductance, Susceptance, and Admittance

As you know from the previous chapter, conductance (G) is the reciprocal of resistance, susceptance (B) is the reciprocal of reactance, and admittance (Y) is the reciprocal of impedance.

For parallel RL circuits, the phasor expression for **inductive susceptance (B_L)** is

$$\mathbf{B}_L = \frac{1}{X_L\angle90°} = B_L\angle-90° = -jB_L$$

and the phasor expression for **admittance** is

$$\mathbf{Y} = \frac{1}{Z\angle\pm\theta} = Y\angle\mp\theta$$

In the basic parallel RL circuit shown in Figure 16–21, the total admittance is the phasor sum of the conductance and the inductive susceptance.

Equation 16–10
$$\mathbf{Y} = G - jB_L$$

As with the RC circuit, the unit for conductance (G), inductive susceptance (B_L), and admittance (Y) is the siemens (S).

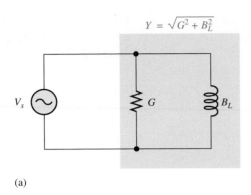

$$Y = \sqrt{G^2 + B_L^2}$$

(a)

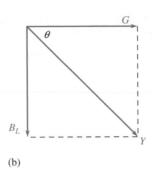

(b)

▲ **FIGURE 16–21**

Admittance in a parallel *RL* circuit.

EXAMPLE 16–9

Determine the total admittance and then convert it to total impedance in Figure 16–22. Draw the admittance phasor diagram.

▶ **FIGURE 16–22**

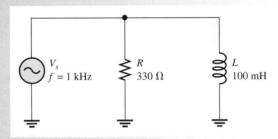

Solution   First, determine the conductance magnitude. $R = 330\ \Omega$; thus,

$$G = \frac{1}{R} = \frac{1}{330\ \Omega} = 3.03\ \text{mS}$$

Then, determine the inductive reactance.

$$X_L = 2\pi fL = 2\pi(1000\ \text{Hz})(100\ \text{mH}) = 628\ \Omega$$

The inductive susceptance magnitude is

$$B_L = \frac{1}{X_L} = \frac{1}{628\ \Omega} = 1.59\ \text{mS}$$

The total admittance is

$$\mathbf{Y}_{tot} = G - jB_L = 3.03\ \text{mS} - j1.59\ \text{mS}$$

which can be expressed in polar form as

$$\mathbf{Y}_{tot} = \sqrt{G^2 + B_L^2}\angle -\tan^{-1}\!\left(\frac{B_L}{G}\right)$$

$$= \sqrt{(3.03\ \text{mS})^2 + (1.59\ \text{mS})^2}\angle -\tan^{-1}\!\left(\frac{1.59\ \text{mS}}{3.03\ \text{mS}}\right) = \mathbf{3.42\angle -27.7°\ \text{mS}}$$

The admittance phasor diagram is shown in Figure 16–23.

▶ FIGURE 16–23

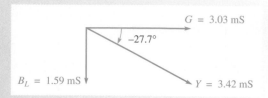

Convert total admittance to total impedance as follows:

$$\mathbf{Z}_{tot} = \frac{1}{\mathbf{Y}_{tot}} = \frac{1}{3.42\angle{-27.7°}\ \text{mS}} = \mathbf{292}\angle\mathbf{27.7°}\ \Omega$$

The positive phase angle indicates that the voltage leads the current.

Related Problem   What is the total admittance of the circuit in Figure 16–22 if *f* is increased to 2 kHz?

**SECTION 16–4
REVIEW**

1. If $Z = 500\ \Omega$, what is the value of the magnitude of the admittance Y?
2. In a certain parallel *RL* circuit, $R = 47\ \Omega$ and $X_L = 75\ \Omega$. Determine Y.
3. In the circuit of Question 2, does the total current lead or lag the applied voltage? By what phase angle?

16–5 ANALYSIS OF PARALLEL *RL* CIRCUITS

Ohm's law and Kirchhoff's current law are used in the analysis of *RL* circuits. Current and voltage relationships in a parallel *RL* circuit are examined.

After completing this section, you should be able to

◆ **Analyze a parallel *RL* circuit**

   ◆ Apply Ohm's law and Kirchhoff's current law to parallel *RL* circuits

   ◆ Express the voltages and currents as phasor quantities

The following example applies Ohm's law to the analysis of a parallel *RL* circuit.

EXAMPLE 16–10

Determine the total current and the phase angle in the circuit of Figure 16–24. Draw a phasor diagram showing the relationship of $\mathbf{V}_s$ and $\mathbf{I}_{tot}$.

▶ FIGURE 16–24

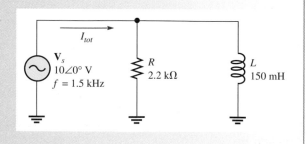

Solution   The inductive reactance is

$$X_L = 2\pi f L = 2\pi(1.5\,\text{kHz})(150\,\text{mH}) = 1.41\,\text{k}\Omega$$

The inductive susceptance magnitude is

$$B_L = \frac{1}{X_L} = \frac{1}{1.41\,\text{k}\Omega} = 709\,\mu\text{S}$$

The conductance magnitude is

$$G = \frac{1}{R} = \frac{1}{2.2\,\text{k}\Omega} = 455\,\mu\text{S}$$

The total admittance is

$$\mathbf{Y}_{tot} = G - jB_L = 455\,\mu\text{S} - j709\,\mu\text{S}$$

Converting to polar form yields

$$\mathbf{Y}_{tot} = \sqrt{G^2 + B_L^2}\angle-\tan^{-1}\!\left(\frac{B_L}{G}\right)$$

$$= \sqrt{(455\,\mu\text{S})^2 + (709\,\mu\text{S})^2}\angle-\tan^{-1}\!\left(\frac{709\,\mu\text{S}}{455\,\mu\text{S}}\right) = 842\angle-57.3°\,\mu\text{S}$$

The phase angle is −57.3°.
   Use Ohm's law to determine the total current.

$$\mathbf{I}_{tot} = \mathbf{V}_s\mathbf{Y}_{tot} = (10\angle0°\,\text{V})(842\angle-57.3°\,\mu\text{S}) = \mathbf{8.42\angle-57.3°\,mA}$$

The magnitude of the total current is 8.42 mA, and it lags the applied voltage by
57.3°, as indicated by the negative angle associated with it. The phasor diagram in
Figure 16–25 shows these relationships.

▶ **FIGURE 16–25**

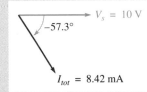

Related Problem   Determine the current in polar form if *f* is reduced to 800 Hz in Figure 16–24.

    Use Multisim file E16-10 to verify the calculated results in this example and to
confirm your calculation for the related problem.

Phase Relationships of Currents and Voltages

Figure 16–26(a) shows all the currents in a basic parallel *RL* circuit. The total current, I_{tot},
divides at the junction into the two branch currents, I_R and I_L. The applied voltage, V_s, appears across both the resistive and the inductive branches, so V_s, V_R, and V_L are all in phase
and of the same magnitude.

   The current through the resistor is in phase with the voltage. The current through the
inductor lags the voltage and the resistor current by 90°. By Kirchhoff's current law, the
total current is the phasor sum of the two branch currents, as shown by the phasor diagram
in Figure 16–26(b). The total current is expressed as

$$\mathbf{I}_{tot} = I_R - jI_L$$

Equation 16–11

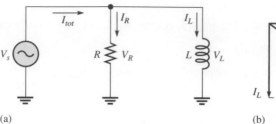
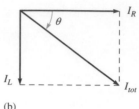

(a)   (b)

▲ FIGURE 16–26

Currents in a parallel *RL* circuit. The current directions shown in part (a) are instantaneous and, of course, reverse when the source voltage reverses during each cycle.

This equation can be expressed in polar form as

Equation 16–12

$$\mathbf{I}_{tot} = \sqrt{I_R^2 + I_L^2}\angle -\tan^{-1}\left(\frac{I_L}{I_R}\right)$$

where the magnitude of the total current is

$$I_{tot} = \sqrt{I_R^2 + I_L^2}$$

and the phase angle between the resistor current and the total current is

$$\theta = -\tan^{-1}\left(\frac{I_L}{I_R}\right)$$

Since the resistor current and the applied voltage are in phase, θ also represents the phase angle between the total current and the applied voltage. Figure 16–27 shows a complete current and voltage phasor diagram.

▶ FIGURE 16–27

Current and voltage phasor diagram for a parallel *RL* circuit (amplitudes are arbitrary).

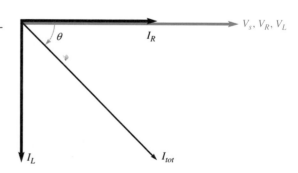

EXAMPLE 16–11

Determine the value of each current in Figure 16–28, and describe the phase relationship of each with the applied voltage. Draw the current phasor diagram.

▶ FIGURE 16–28

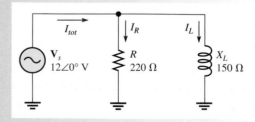

Solution   The resistor current, the inductor current, and the total current are expressed as follows:

$$\mathbf{I}_R = \frac{\mathbf{V}_s}{\mathbf{R}} = \frac{12\angle 0° \text{ V}}{220\angle 0° \ \Omega} = \mathbf{54.5\angle 0° \text{ mA}}$$

$$\mathbf{I}_L = \frac{\mathbf{V}_s}{\mathbf{X}_L} = \frac{12\angle 0° \text{ V}}{150\angle 90° \ \Omega} = \mathbf{80\angle -90° \text{ mA}}$$

$$\mathbf{I}_{tot} = I_R - jI_L = 54.5 \text{ mA} - j80 \text{ mA}$$

Converting $\mathbf{I}_{tot}$ to polar form yields

$$\mathbf{I}_{tot} = \sqrt{I_R^2 + I_L^2}\angle -\tan^{-1}\left(\frac{I_L}{I_R}\right)$$

$$= \sqrt{(54.5 \text{ mA})^2 + (80 \text{ mA})^2}\angle -\tan^{-1}\left(\frac{80 \text{ mA}}{54.5 \text{ mA}}\right) = \mathbf{96.8\angle -55.7° \text{ mA}}$$

As the results show, the resistor current is 54.5 mA and is in phase with the applied voltage. The inductor current is 80 mA and lags the applied voltage by 90°. The total current is 96.8 mA and lags the voltage by 55.7°. The phasor diagram in Figure 16–29 shows these relationships.

▶ **FIGURE 16–29**

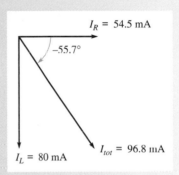

Related Problem   Find the magnitude of $\mathbf{I}_{tot}$ and the circuit phase angle if $X_L = 300 \ \Omega$ in Figure 16–28.

**SECTION 16–5
REVIEW**

1. The admittance of an *RL* circuit is 4 mS, and the applied voltage is 8 V. What is the total current?

2. In a certain parallel *RL* circuit, the resistor current is 12 mA, and the inductor current is 20 mA. Determine the magnitude and phase angle of the total current. This phase angle is measured with respect to what?

3. What is the phase angle between the inductor current and the applied voltage in a parallel *RL* circuit?

OPTION 2 NOTE

Coverage of series reactive circuits continues in Chapter 17, Part 2, on page 740.

16–6 ANALYSIS OF SERIES-PARALLEL *RL* CIRCUITS

The concepts studied with respect to series and parallel circuits are used to analyze circuits with combinations of both series and parallel *R* and *L* components.

After completing this section, you should be able to

♦ **Analyze series-parallel *RL* circuits**

    ♦ Determine total impedance

    ♦ Calculate currents and voltages

Recall from Section 15–1 that the impedance of series components is most easily expressed in rectangular form and that the impedance of parallel components is best found by using polar form. The steps for analyzing a circuit with a series and a parallel component are illustrated in Example 16–12. First express the impedance of the series part of the circuit in rectangular form and the impedance of the parallel part in polar form. Next, convert the impedance of the parallel part to rectangular form and add it to the impedance of the series part. Once you determine the rectangular form of the total impedance, you can convert it to polar form in order to see the magnitude and phase angle and to calculate the current.

EXAMPLE 16–12

In the circuit of Figure 16–30, determine the following values:

(a) $\mathbf{Z}_{tot}$ **(b)** $\mathbf{I}_{tot}$ **(c)** θ

FIGURE 16–30

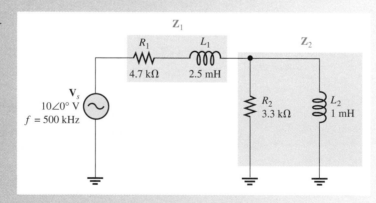

Solution **(a)** First, calculate the magnitudes of inductive reactance.

$$X_{L1} = 2\pi f L_1 = 2\pi(500\,\text{kHz})(2.5\,\text{mH}) = 7.85\,\text{k}\Omega$$

$$X_{L2} = 2\pi f L_2 = 2\pi(500\,\text{kHz})(1\text{mH}) = 3.14\,\text{k}\Omega$$

One approach is to find the impedance of the series portion and the impedance of the parallel portion and combine them to get the total impedance. The impedance of the series combination of R_1 and L_1 is

$$\mathbf{Z}_1 = R_1 + jX_{L1} = 4.7\,\text{k}\Omega + j7.85\,\text{k}\Omega$$

To determine the impedance of the parallel portion, first determine the admittance of the parallel combination of R_2 and L_2.

$$G_2 = \frac{1}{R_2} = \frac{1}{3.3\,\text{k}\Omega} = 303\,\mu\text{S}$$

$$B_{L2} = \frac{1}{X_{L2}} = \frac{1}{3.14\,\text{k}\Omega} = 318\,\mu\text{S}$$

$$\mathbf{Y}_2 = G_2 - jB_L = 303\,\mu\text{S} - j318\,\mu\text{S}$$

Converting to polar form yields

$$\mathbf{Y}_2 = \sqrt{G_2^2 + B_{L2}^2}\angle -\tan^{-1}\left(\frac{B_{L2}}{G_2}\right)$$

$$= \sqrt{(303\,\mu\text{S})^2 + (318\,\mu\text{S})^2}\angle -\tan^{-1}\left(\frac{318\,\mu\text{S}}{303\,\mu\text{S}}\right) = 439\angle -46.4°\,\mu\text{S}$$

Then, the impedance of the parallel portion is

$$\mathbf{Z}_2 = \frac{1}{\mathbf{Y}_2} = \frac{1}{439\angle -46.4°\,\mu\text{S}} = 2.28\angle 46.4°\,\text{k}\Omega$$

Converting to rectangular form yields

$$\mathbf{Z}_2 = Z_2\cos\theta + jZ_2\sin\theta$$
$$= (2.28\,\text{k}\Omega)\cos(46.4°) + j(2.28\,\text{k}\Omega)\sin(46.4°) = 1.57\,\text{k}\Omega + j1.65\,\text{k}\Omega$$

The series portion and the parallel portion are in series with each other. Combine $\mathbf{Z}_1$ and $\mathbf{Z}_2$ to get the total impedance.

$$\mathbf{Z}_{tot} = \mathbf{Z}_1 + \mathbf{Z}_2$$
$$= (4.7\,\text{k}\Omega + j7.85\,\text{k}\Omega) + (1.57\,\text{k}\Omega + j1.65\,\text{k}\Omega) = 6.27\,\text{k}\Omega + j9.50\,\text{k}\Omega$$

Expressing $\mathbf{Z}_{tot}$ in polar form yields

$$\mathbf{Z}_{tot} = \sqrt{Z_1^2 + Z_2^2}\angle \tan^{-1}\left(\frac{Z_2}{Z_1}\right)$$

$$= \sqrt{(6.27\,\text{k}\Omega)^2 + (9.50\,\text{k}\Omega)^2}\angle \tan^{-1}\left(\frac{9.50\,\text{k}\Omega}{6.27\,\text{k}\Omega}\right) = \mathbf{11.4\angle 56.6°\,k\Omega}$$

(b) Use Ohm's law to find the total current.

$$\mathbf{I}_{tot} = \frac{\mathbf{V}_s}{\mathbf{Z}_{tot}} = \frac{10\angle 0°\,\text{V}}{11.4\angle 56.6°\,\text{k}\Omega} = \mathbf{877\angle -56.6°\,\mu A}$$

(c) The total current lags the applied voltage by **56.6°**.

Related Problem **(a)** Determine the voltage across the series part of the circuit in Figure 16–30.

(b) Determine the voltage across the parallel part of the circuit in Figure 16–30.

Use Multisim file E16-12 to verify the calculated results in the (b) part of this example and to confirm your calculation for the related problem, part (b).

Example 16–13 shows two sets of series components in parallel. The approach is to first express each branch impedance in rectangular form and then convert each of these impedances to polar form. Next, calculate each branch current using polar notation. Once you know the branch currents, you can find the total current by adding the two branch currents in rectangular form. In this particular case, the total impedance is not required.

EXAMPLE 16–13

Determine the voltage across each component in Figure 16–31. Draw a voltage phasor diagram and a current phasor diagram.

▶ **FIGURE 16–31**

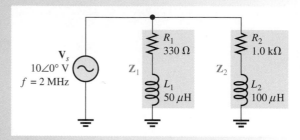

Solution First, calculate X_{L1} and X_{L2}.

$$X_{L1} = 2\pi f L_1 = 2\pi(2\,\text{MHz})(50\,\mu\text{H}) = 628\,\Omega$$

$$X_{L2} = 2\pi f L_2 = 2\pi(2\,\text{MHz})(100\,\mu\text{H}) = 1.26\,\text{k}\Omega$$

Next, determine the impedance of each branch.

$$\mathbf{Z}_1 = R_1 + jX_{L1} = 330\,\Omega + j628\,\Omega$$

$$\mathbf{Z}_2 = R_2 + jX_{L2} = 1.0\,\text{k}\Omega + j1.26\,\text{k}\Omega$$

Convert these impedances to polar form.

$$\mathbf{Z}_1 = \sqrt{R_1^2 + X_{L1}^2}\,\angle\tan^{-1}\!\left(\frac{X_{L1}}{R_1}\right)$$

$$= \sqrt{(330\,\Omega)^2 + (628\,\Omega)^2}\,\angle\tan^{-1}\!\left(\frac{628\,\Omega}{330\,\Omega}\right) = 709\angle62.3°\,\Omega$$

$$\mathbf{Z}_2 = \sqrt{R_2^2 + X_{L2}^2}\,\angle\tan^{-1}\!\left(\frac{X_{L2}}{R_2}\right)$$

$$= \sqrt{(1.0\,\text{k}\Omega)^2 + (1.26\,\text{k}\Omega)^2}\,\angle\tan^{-1}\!\left(\frac{1.26\,\text{k}\Omega}{1.0\,\text{k}\Omega}\right) = 1.61\angle51.6°\,\text{k}\Omega$$

Calculate each branch current.

$$\mathbf{I}_1 = \frac{\mathbf{V}_s}{\mathbf{Z}_1} = \frac{10\angle0°\,\text{V}}{709\angle62.3°\,\Omega} = 14.1\angle-62.3°\,\text{mA}$$

$$\mathbf{I}_2 = \frac{\mathbf{V}_s}{\mathbf{Z}_2} = \frac{10\angle0°\,\text{V}}{1.61\angle51.6°\,\text{k}\Omega} = 6.21\angle-51.6°\,\text{mA}$$

Now, use Ohm's law to get the voltage across each element.

$$\mathbf{V}_{R1} = \mathbf{I}_1\mathbf{R}_1 = (14.1\angle-62.3°\,\text{mA})(330\angle0°\,\Omega) = \mathbf{4.65\angle-62.3°\,V}$$

$$\mathbf{V}_{L1} = \mathbf{I}_1\mathbf{X}_{L1} = (14.1\angle-62.3°\,\text{mA})(628\angle90°\,\Omega) = \mathbf{8.85\angle27.7°\,V}$$

$$\mathbf{V}_{R2} = \mathbf{I}_1\mathbf{R}_2 = (6.21\angle-51.6°\,\text{mA})(1\angle0°\,\text{k}\Omega) = \mathbf{6.21\angle-51.6°\,V}$$

$$\mathbf{V}_{L2} = \mathbf{I}_2\mathbf{X}_{L2} = (6.21\angle-51.6°\,\text{mA})(1.26\angle90°\,\text{k}\Omega) = \mathbf{7.82\angle38.4°\,V}$$

The voltage phasor diagram is shown in Figure 16–32, and the current phasor diagram is shown in Figure 16–33.

▶ FIGURE 16–32

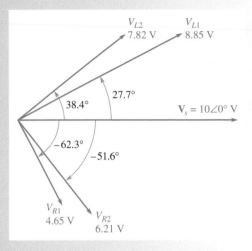

▶ FIGURE 16–33

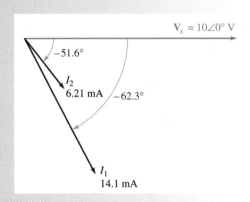

Related Problem What is the total current in polar form in Figure 16–31?

Use Multisim file E16-13 to verify the calculated results in this example and to confirm your calculation for the related problem.

SECTION 16–6 REVIEW

1. What is the total impedance in polar form of the circuit in Figure 16–31?
2. Determine the total current in rectangular form for the circuit in Figure 16–31.

OPTION 2 NOTE

Coverage of series-parallel reactive circuits continues in Chapter 17, Part 3, on page 749.

16–7 POWER IN *RL* CIRCUITS

In a purely resistive ac circuit, all of the energy delivered by the source is dissipated in the form of heat by the resistance. In a purely inductive ac circuit, all of the energy delivered by the source is stored by the inductor in its magnetic field during a portion of the voltage cycle and then returned to the source during another portion of the cycle so that there is no net energy conversion to heat. When there is both resistance and inductance, some of the energy is alternately stored and returned by the inductance and some is dissipated by the resistance. The amount of energy converted to heat is determined by the relative values of the resistance and the inductive reactance.

After completing this section, you should be able to

- ◆ **Determine power in *RL* circuits**
  - ◆ Explain true and reactive power
  - ◆ Draw the power triangle
  - ◆ Explain power factor correction

When the resistance in a series *RL* circuit is greater than the inductive reactance, more of the total energy delivered by the source is converted to heat by the resistance than is stored by the inductor. Likewise, when the reactance is greater than the resistance, more of the total energy is stored and returned than is converted to heat.

As you know, the power dissipation in a resistance is called the *true power*. The power in an inductor is reactive power and is expressed as

Equation 16–13
$$P_r = I^2 X_L$$

The Power Triangle for *RL* Circuits

The generalized power triangle for a series *RL* circuit is shown in Figure 16–34. The **apparent power, P_a,** is the resultant of the average power, P_{true}, and the reactive power, P_r.

▶ **FIGURE 16–34**

Power triangle for an *RL* circuit.

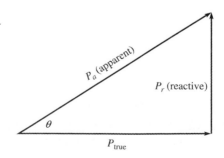

Recall that the power factor equals the cosine of θ ($PF = \cos\theta$). As the phase angle between the applied voltage and the total current increases, the power factor decreases, indicating an increasingly reactive circuit. A smaller power factor indicates less true power and more reactive power.

EXAMPLE 16–14

Determine the power factor, the true power, the reactive power, and the apparent power in the circuit in Figure 16–35.

▶ **FIGURE 16–35**

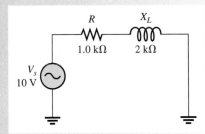

Solution The total impedance of the circuit in rectangular form is

$$\mathbf{Z} = R + jX_L = 1.0\,\text{k}\Omega + j2\,\text{k}\Omega$$

Converting to polar form yields

$$\mathbf{Z} = \sqrt{R_2^2 + X_L^2}\angle\tan^{-1}\!\left(\frac{X_L}{R}\right)$$

$$= \sqrt{(1.0\,\text{k}\Omega)^2 + (2\,\text{k}\Omega)^2}\angle\tan^{-1}\!\left(\frac{2\,\text{k}\Omega}{1.0\,\text{k}\Omega}\right) = 2.24\angle63.4°\,\text{k}\Omega$$

The current magnitude is

$$I = \frac{V_s}{Z} - \frac{10\,\text{V}}{2.24\,\text{k}\Omega} = 4.46\,\text{mA}$$

The phase angle, indicated in the expression for **Z**, is

$$\theta = 63.4°$$

The power factor is, therefore,

$$PF = \cos\theta = \cos(63.4°) = \mathbf{0.448}$$

The true power is

$$P_{\text{true}} = V_sI\cos\theta = (10\,\text{V})(4.46\,\text{mA})(0.448) = \mathbf{20\,mW}$$

The reactive power is

$$P_r = I^2X_L = (4.46\,\text{mA})^2(2\,\text{k}\Omega) = \mathbf{39.8\,mVAR}$$

The apparent power is

$$P_a = I^2Z = (4.46\,\text{mA})^2(2.24\,\text{k}\Omega) = \mathbf{44.6\,mVA}$$

Related Problem If the frequency in Figure 16–35 is increased, what happens to P_{true}, P_r, and P_a?

Significance of the Power Factor

As you learned in Chapter 15, the **power factor** (*PF*) is important in determining how much useful power (true power) is transferred to a load. The highest power factor is 1, which indicates that all of the current to a load is in phase with the voltage (resistive). When the power factor is 0, all of the current to a load is 90° out of phase with the voltage (reactive).

Generally, a power factor as close to 1 as possible is desirable because then most of the power transferred from the source to the load is the useful or true power. True power goes only one way—from source to load—and performs work on the load in terms of energy dissipation. Reactive power simply goes back and forth between the source and the load with no net work being done. Energy must be used in order for work to be done.

Many practical loads have inductance as a result of their particular function, and it is essential for their proper operation. Examples are transformers, electric motors, and speakers, to name a few. Therefore, inductive (and capacitive) loads are important considerations.

To see the effect of the power factor on system requirements, refer to Figure 16–36. This figure shows a representation of a typical inductive load consisting effectively of inductance and resistance in parallel. Part (a) shows a load with a relatively low power factor (0.75), and part (b) shows a load with a relatively high power factor (0.95). Both loads dissipate equal amounts of power as indicated by the wattmeters. Thus, an equal amount of work is done on both loads.

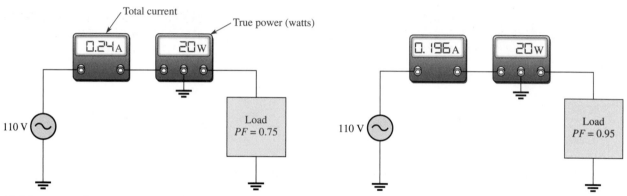

(a) A lower power factor means more total current for a given power dissipation (watts). A larger source VA rating is required to deliver the true power (watts).

(b) A higher power factor means less total current for a given power dissipation. A smaller source can deliver the same true power (watts).

▲ FIGURE 16–36

Illustration of the effect of the power factor on system requirements such as source rating (VA) and conductor size.

Although both loads are equivalent in terms of the amount of work done (true power), the low power factor load in Figure 16–36(a) draws more current from the source than does the high power factor load in Figure 16–36(b), as indicated by the ammeters. Therefore, the source in part (a) must have a higher VA rating than the one in part (b). Also, the lines connecting the source to the load in part (a) must be a larger wire gauge than those in part (b), a condition that becomes significant when very long transmission lines are required, such as in power distribution.

Figure 16–36 has demonstrated that a higher power factor is an advantage in delivering power more efficiently to a load.

Power Factor Correction

The power factor of an inductive load can be increased by the addition of a capacitor in parallel, as shown in Figure 16–37. The capacitor compensates for the phase lag of the total current by creating a capacitive component of current that is 180° out of phase with the inductive component. This has a canceling effect and reduces the phase angle (and power factor) as well as the total current, as illustrated in the figure.

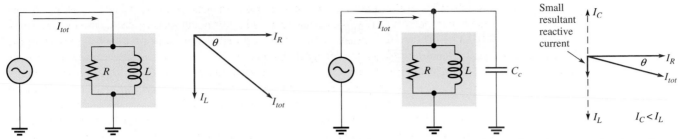

(a) Total current is the resultant of I_R and I_L.

(b) I_C subtracts from I_L, leaving only a small reactive current, thus decreasing I_{tot} and the phase angle.

▲ **FIGURE 16–37**

Example of how the power factor can be increased by the addition of a compensating capacitor.

SECTION 16–7
REVIEW

1. To which component in an *RL* circuit is the power dissipation due?
2. Calculate the power factor when $\theta = 50°$.
3. A certain *RL* circuit consists of a 470 Ω resistor and an inductive reactance of 620 Ω at the operating frequency. Determine P_{true}, P_r, and P_a when $I = 100$ mA.

16–8 BASIC APPLICATIONS

Two applications of *RL* circuits are covered in this section. The first application is a basic frequency selective (filter) circuit. The second application is the switching regulator, a widely used circuit in power supplies because of its high efficiency. The switching regulator uses other components, but the *RL* circuit is emphasized.

After completing this section, you should be able to

- ◆ **Describe two examples of *RL* circuit applications**
  - ◆ Discuss how the *RL* circuit operates as a filter
  - ◆ Discuss the advantage of an inductor in a switching regulator

The *RL* Circuit as a Filter

As with *RC* circuits, series *RL* circuits also exhibit a frequency-selective characteristic and therefore act as basic filters.

Low-Pass Filter   You have seen what happens to the output magnitude and phase angle in a lag circuit. In terms of the filtering action, the variation of the magnitude of the output voltage as a function of frequency is important.

Figure 16–38 shows the filtering action of a series *RL* circuit using specific values for purposes of illustration. In part (a) of the figure, the input is zero frequency (dc). Since the inductor ideally acts as a short to constant direct current, the output voltage equals the full value of the input voltage (neglecting the winding resistance). Therefore, the circuit passes all of the input voltage to the output (10 V in, 10 V out).

In Figure 16–38(b), the frequency of the input voltage has been increased to 1 kHz, causing the inductive reactance to increase to 62.83 Ω. For an input voltage of 10 V rms, the output voltage is approximately 8.47 V rms, which can be calculated using the voltage-divider approach or Ohm's law.

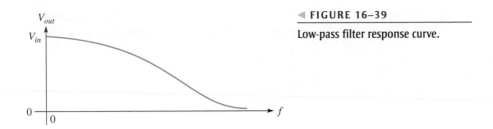

◀ FIGURE 16–39

Low-pass filter response curve.

In Figure 16–40(b), the frequency of the input signal has been increased to 100 Hz with an rms value of 10 V. The output voltage is 0.63 V rms. Thus, only a small percentage of the input voltage appears at the output at this frequency.

In Figure 16–40(c), the input frequency is increased further to 1 kHz, causing more voltage to be developed as a result of the increase in the inductive reactance. The output voltage at this frequency is 5.32 V rms. As you can see, the output voltage increases as the frequency increases. A value of frequency is reached at which the reactance is very large compared to the resistance and most of the input voltage appears across the inductor, as shown in Figure 16–40(d).

This circuit tends to prevent lower frequency signals from appearing on the output but permits higher frequency signals to pass through from input to output; thus, it is a basic form of high-pass filter.

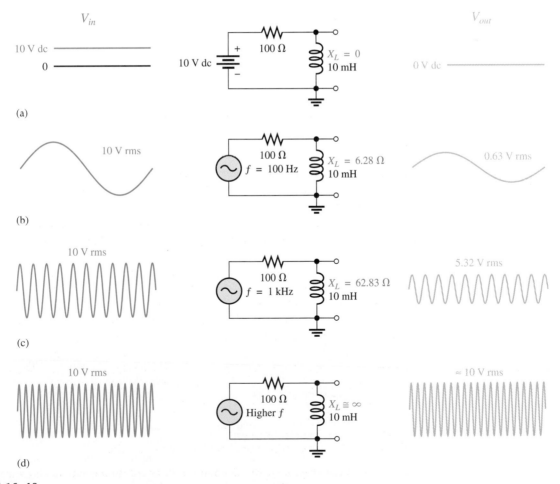

▲ FIGURE 16–40

High-pass filtering action of an *RL* circuit (phase shift from input to output is not indicated).

The response curve in Figure 16–41 shows that the output voltage increases and then levels off as it approaches the value of the input voltage as the frequency increases.

▶ **FIGURE 16–41**

High-pass filter response curve.

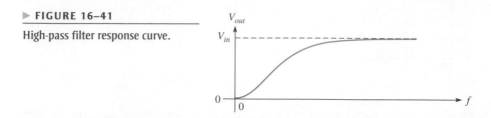

▶ **FIGURE 16–41**

High-pass filter response curve.

The Switching Regulator

In high-frequency switching power supplies, small inductors are used as an essential part of the filter section. A switching power supply is much more efficient at converting ac to dc than any other type of supply. For this reason it is widely used in computers and other electronic systems. A switching regulator precisely controls the dc voltage. One type of switching regulator is shown in Figure 16–42. It uses an electronic switch to change unregulated dc to high-frequency pulses. The output is the average value of the pulses. The pulse width is controlled by the pulse width modulator, which rapidly turns on and off a transistor switch and then is filtered by the filter section to produce regulated dc. (Ripple in the figure is exaggerated to show the cycle.) The pulse width modulator can increase pulse width if the output drops, or decrease it if the output rises, thus maintaining a constant average output for varying conditions.

▶ **FIGURE 16–42**

Switching regulator block diagram.

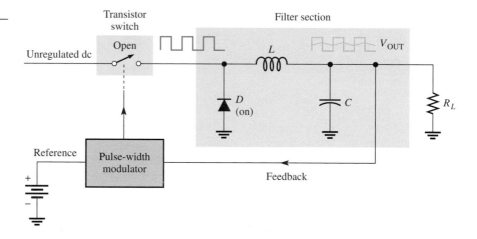

Figure 16–43 illustrates the basic filtering action. The filter consists of a diode, an inductor, and a capacitor. The diode is a one-way device for current that you will study in a devices course. In this application the diode acts as an on-off switch that allows current in only one direction.

An important component of the filter section is the inductor, which in this type of regulator will always have current in it. The average voltage and the load resistor determine the amount of current. Recall that Lenz's law states that an induced voltage is created across a coil that opposes a *change* in current. When the transistor switch is closed, the pulse is high and current is passed through the inductor and on to the load, as shown in Figure 16–43(a). The diode is off at this time. Notice that the inductor has an induced voltage across it that opposes a change in current. When the pulse goes low, as in Figure 16–43(b), the transistor is off and the inductor develops a voltage in the opposite direction than before. The diode acts as a closed switch, which provides a path for current. This action tends to keep the load current constant. The capacitor adds to this smoothing action by charging and discharging a small amount during the process.

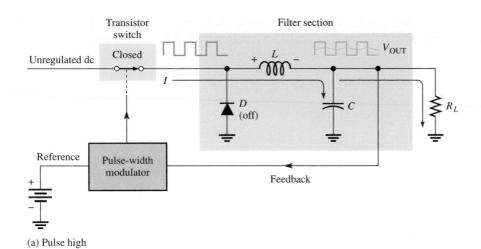

(a) Pulse high

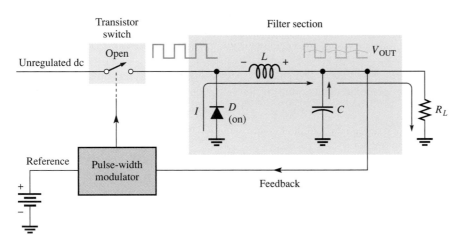

(b) Pulse low

▲ FIGURE 16–43

Switching regulator action.

<table>
<tr><td>SECTION 16–8
REVIEW</td><td>1. When an *RL* circuit is used as a low-pass filter, across which component is the output taken?
2. What is the major advantage of a switching regulator?
3. What happens to the pulse width of a switching regulator if the output voltage drops?</td></tr>
</table>

16–9 TROUBLESHOOTING

Typical component failures have an effect on the frequency response of basic *RL* circuits.

After completing this section, you should be able to

◆ **Troubleshoot *RL* circuits**

   ◆ Find an open inductor

   ◆ Find an open resistor

   ◆ Find an open in a parallel circuit

Effect of an Open Inductor The most common failure mode for inductors occurs when the winding opens as a result of excessive current or a mechanical contact failure. It is easy to see how an open coil affects the operation of a basic series *RL* circuit, as shown in Figure 16–44. Obviously, there is no current path; therefore, the resistor voltage is zero, and the total applied voltage appears across the inductor. If you suspect an open coil, remove one or both leads from the circuit and check continuity with an ohmmeter.

▶ **FIGURE 16–44**

Effect of an open coil.

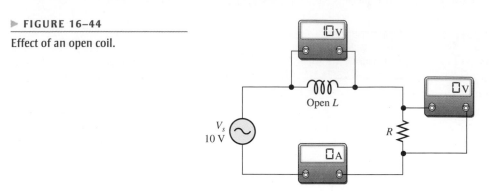

Effect of an Open Resistor When the resistor is open, there is no current and the inductor voltage is zero. The total input voltage is across the open resistor, as shown in Figure 16–45.

▶ **FIGURE 16–45**

Effect of an open resistor.

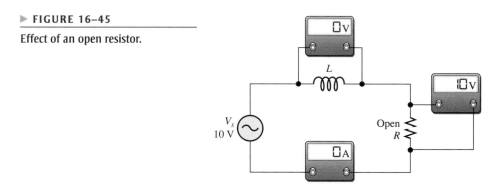

Open Components in Parallel Circuits In a parallel *RL* circuit, an open resistor or inductor will cause the total current to decrease because the total impedance will increase. Obviously, the branch with the open component will have zero current. Figure 16–46 illustrates these conditions.

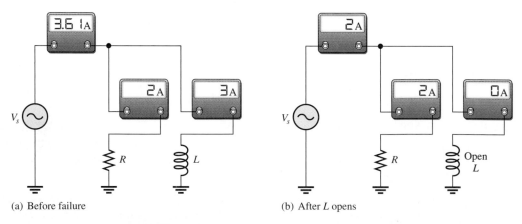

(a) Before failure   (b) After *L* opens

▲ **FIGURE 16–46**

Effect of an open component in a parallel circuit with V_s constant.

Effect of an Inductor with Shorted Windings   Although a very rare occurrence, it is possible for some of the windings of coils to short together as a result of damaged insulation. This failure mode is much less likely than the open coil and is difficult to detect. Shorted windings may result in a reduction in inductance because the inductance of a coil is proportional to the square of the number of turns. A short between windings effectively reduces the number of turns, which may or may not have an adverse effect on the circuit depending on the number of turns that are shorted.

Other Troubleshooting Considerations

As you have learned, the failure of a circuit to work properly is not always the result of a faulty component. A loose wire, a bad contact, or a poor solder joint can cause an open circuit. A short can be caused by a wire clipping or solder splash. Things as simple as not plugging in a power supply or a function generator happen more often than you might think. Wrong values in a circuit (such as an incorrect resistor value), the function generator set at the wrong frequency, or the wrong output connected to the circuit can cause improper operation.

Always check to make sure that the instruments are properly connected to the circuits and to a power outlet. Also, look for obvious things such as a broken or loose contact, a connector that is not completely plugged in, or a piece of wire or a solder bridge that could be shorting something out.

The following example illustrates a troubleshooting approach to a circuit containing inductors and resistors using the APM (analysis, planning, and measurement) method and half-splitting.

EXAMPLE 16–15   The circuit represented by the schematic in Figure 16–47 has no output voltage. The circuit is physically constructed on a protoboard. Use your trouble- shooting skills to find the problem.

FIGURE 16–47

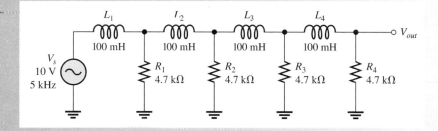

Solution   Apply the APM method to this troubleshooting problem.

Analysis: First think of the possible causes for the circuit to have no output voltage.

1. There is no source voltage or the frequency is so high that the inductors appear to be open because their reactances are extremely high compared to the resistance values.

2. There is a short between one of the resistors and ground. Either a resistor could be shorted, or there could be some physical short. A shorted resistor is not a common fault.

3. There is an open between the source and the output. This would prevent current and thus cause the output voltage to be zero. An inductor could be open, or the conductive path could be open due to a broken or loose connecting wire or a bad protoboard contact.

4. There is an incorrect component value. A resistor could be so small that the voltage across it is negligible. An inductor could be so large that its reactance at the input frequency is extremely high.

Planning: You decide to make some visual checks for problems such as the function generator power cord not plugged in or the frequency set at an incorrect value. Also, broken leads, shorted leads, as well as an incorrect resistor color code or inductor value often can be found visually. If nothing is discovered after a visual check, then

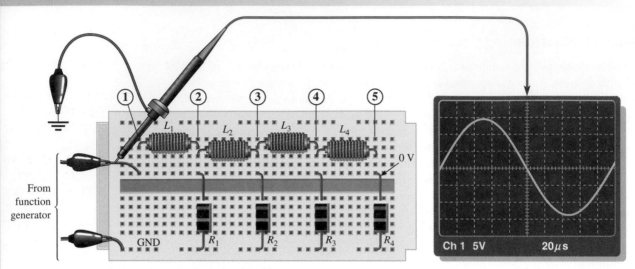

(a) Scope shows the correct voltage at the input. The scope probe ground lead is not shown.

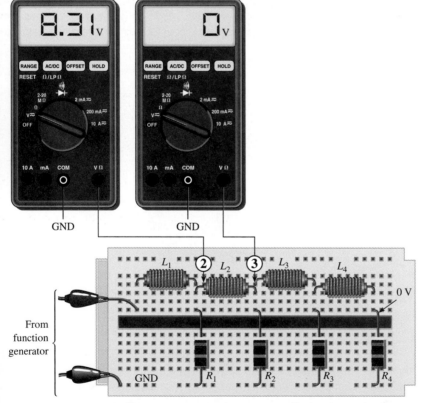

(b) A zero voltage at point ③ indicates the fault is between point ③ and the source. A reading of 10 V at point ② shows that L_2 is open.

▲ FIGURE 16–48

you will make voltage measurements to track down the cause of the problem. You decide to use a digital oscilloscope and a DMM to make the measurements using the half-splitting technique to more quickly isolate the fault.

Measurement: Assume that you find that the function generator is plugged in and the frequency setting appears to be correct. Also, you find no visible opens or shorts during your visual check, and the component values are correct.

The first step in the measurement process is to check the voltage from the source with the scope. Assume a 10 V rms sine wave with a frequency of 5 kHz is observed at the circuit input as shown in Figure 16–48(a). The correct ac voltage is present, so *the first possible cause has been eliminated.*

Next, check for a short by disconnecting the source and placing the DMM (set on the ohmmeter function) across each resistor. If any resistor is shorted (unlikely), the meter will read zero or a very small resistance. Assuming the meter readings are okay, *the second possible cause has been eliminated.*

Since the voltage has been "lost" somewhere between the input and the output, you must now look for the voltage. You reconnect the source and, using the half-splitting approach, measure the voltage at point ③ (the middle of the circuit) with respect to ground. The DMM test lead is placed on the right lead of inductor L_2, as indicated in Figure 16–48(b). Assume the voltage at this point is zero. This tells you that the part of the circuit to the right of point ③ is probably okay and the fault is in the circuit between point ③ and the source.

Now, you begin tracing the circuit back toward the source looking for the voltage (you could also start from the source and work forward). Placing the meter test lead on point ②, at the left lead of inductor L_2, results in a reading of 8.31 V as shown in Figure 16–48(b). This, of course, indicates that L_2 is open. Fortunately, in this case, a component, and not a contact on the board, is faulty. It is usually easier to replace a component than to repair a bad contact.

Related Problem Suppose you had measured 0 V at the left lead of L_2 and 10 V at the right lead of L_1. What would this have indicated?

**SECTION 16–9
REVIEW**

1. Describe the effect of an inductor with shorted windings on the response of a series *RL* circuit.
2. In the circuit of Figure 16–49, indicate whether I_{tot}, V_{R1}, and V_{R2} increase or decrease as a result of *L* opening.

▶ **FIGURE 16–49**

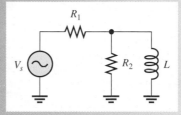

A Circuit Application

You are given two sealed modules that have been removed from a communications system that is being modified. Each module has three terminals and is labeled as an *RL* filter, but no specifications are given. You are asked to test the modules to determine the type of filters and the component values.

The sealed modules have three terminals labeled IN, GND, and OUT as shown in Figure 16–50. You will apply your knowledge of series *RL* circuits and some basic measurements to determine the internal circuit configuration and the component values.

Ohmmeter Measurement of Module 1

◆ Determine the arrangement of the two components and the values of the resistor and winding resistance for module 1 indicated by the meter readings in Figure 16–50.

AC Measurement of Module 1

◆ Determine the inductance value for module 1 indicated by the test setup in Figure 16–51.

▶ **FIGURE 16–50**

Ohmmeter measurements of module 1.

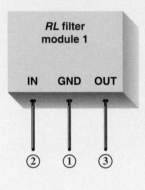

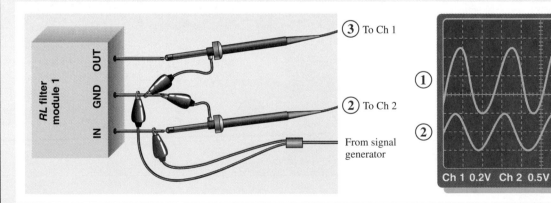

▲ **FIGURE 16–51**

AC measurements for module 1.

Ohmmeter Measurement of Module 2

◆ Determine the arrangement of the two components and the values of the resistor and the winding resistance for module 2 indicated by the meter readings in Figure 16–52.

AC Measurement of Module 2

◆ Determine the inductance value for module 2 indicated by the test setup in Figure 16–53.

Review

1. If the inductor in module 1 were open, what would you measure on the output with the test setup of Figure 16–51?

2. If the inductor in module 2 were open, what would you measure on the output with the test setup of Figure 16–53?

◀ **FIGURE 16–52**

Ohmmeter measurements of module 2.

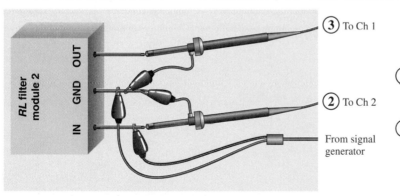
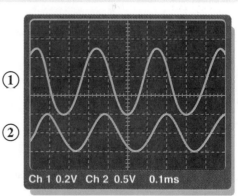

③ To Ch 1

② To Ch 2

From signal generator

Ch 1 0.2V   Ch 2 0.5V   0.1ms

▲ **FIGURE 16–53**

AC measurements for module 2.

OPTION 2 NOTE

Coverage of special topics continues in Chapter 17, Part 4, on page 757.

SUMMARY

♦ When a sinusoidal voltage is applied to an *RL* circuit, the current and all the voltage drops are also sine waves.

♦ Total current in a series or parallel *RL* circuit always lags the source voltage.

♦ The resistor voltage is always in phase with the current.

♦ In an ideal inductor, the voltage always leads the current by 90°.

♦ In a lag circuit, the output voltage lags the input voltage in phase.

♦ In a lead circuit, the output voltage leads the input voltage in phase.

♦ In an *RL* circuit, the impedance is determined by both the resistance and the inductive reactance combined.

♦ Impedance is expressed in units of ohms.

♦ The impedance of an *RL* circuit varies directly with frequency.

♦ The phase angle (θ) of a series *RL* circuit varies directly with frequency.

♦ You can determine the impedance of a circuit by measuring the applied voltage and the total current and then applying Ohm's law.

♦ In an *RL* circuit, part of the power is resistive and part reactive.

♦ The power factor indicates how much of the apparent power is true power.

♦ A power factor of 1 indicates a purely resistive circuit, and a power factor of 0 indicates a purely reactive circuit.

♦ A filter passes certain frequencies and rejects others.

KEY TERMS

Key terms and other bold terms in the chapter are defined in the end-of-book glossary.

Inductive reactance   The opposition of an inductor to sinusoidal current. The unit is the ohm.

Inductive susceptance (B_L)   The ability of an inductor to permit current; the reciprocal of inductive reactance. The unit is the siemens (S).

FORMULAS

Series *RL* Circuits

16–1 $\mathbf{Z} = R + jX_L$

16–2 $\mathbf{Z} = \sqrt{R^2 + X_L^2}\angle\tan^{-1}\left(\dfrac{X_L}{R}\right)$

16–3 $\mathbf{V}_s = V_R + jV_L$

16–4 $\mathbf{V}_s = \sqrt{V_R^2 + V_L^2}\angle\tan^{-1}\left(\dfrac{V_L}{V_R}\right)$

Lead Circuit

16–5 $\phi = \tan^{-1}\left(\dfrac{R}{X_L}\right)$

16–6 $\mathbf{V}_{out} = \left(\dfrac{X_L}{\sqrt{R^2 + X_L^2}}\right)V_{in}$

Lag Circuit

16–7 $\phi = -\tan^{-1}\left(\dfrac{X_L}{R}\right)$

16–8 $\mathbf{V}_{out} = \left(\dfrac{R}{\sqrt{R^2 + X_L^2}}\right)V_{in}$

Parallel *RL* Circuits

16–9 $\quad \mathbf{Z} = \left(\dfrac{RX_L}{\sqrt{R^2 + X_L^2}} \right) \angle \tan^{-1}\left(\dfrac{R}{X_L} \right)$

16–10 $\quad \mathbf{Y} = G - jB_L$

16–11 $\quad \mathbf{I}_{tot} = I_R - jI_L$

16–12 $\quad \mathbf{I}_{tot} = \sqrt{I_R^2 + I_L^2} \angle -\tan^{-1}\left(\dfrac{I_L}{I_R} \right)$

Power in *RL* Circuits

16–13 $\quad P_r = I^2 X_L$

SELF-TEST

Answers are at the end of the chapter.

1. In a series *RL* circuit, the resistor voltage
   - (a) leads the applied voltage
   - (b) lags the applied voltage
   - (c) is in phase with the applied voltage
   - (d) is in phase with the current
   - (e) answers (a) and (d)
   - (f) answers (b) and (d)

2. When the frequency of the voltage applied to a series *RL* circuit is increased, the impedance
   - (a) decreases
   - (b) increases
   - (c) does not change

3. When the frequency of the voltage applied to a series *RL* circuit is decreased, the phase angle
   - (a) decreases
   - (b) increases
   - (c) does not change

4. If the frequency is doubled and the resistance is halved, the impedance of a series *RL* circuit
   - (a) doubles
   - (b) halves
   - (c) remains constant
   - (d) cannot be determined without values

5. To reduce the current in a series *RL* circuit, the frequency should be
   - (a) increased
   - (b) decreased
   - (c) constant

6. In a series *RL* circuit, 10 V rms is measured across the resistor, and 10 V rms is measured across the inductor. The peak value of the source voltage is
   - (a) 14.14 V
   - (b) 28.28 V
   - (c) 10 V
   - (d) 20 V

7. The voltages in Problem 6 are measured at a certain frequency. To make the resistor voltage greater than the inductor voltage, the frequency is
   - (a) increased
   - (b) decreased
   - (c) doubled
   - (d) not a factor

8. When the resistor voltage in a series *RL* circuit becomes greater than the inductor voltage, the phase angle
   - (a) increases
   - (b) decreases
   - (c) is not affected

9. When the frequency of the source voltage is increased, the impedance of a parallel *RL* circuit
   - (a) increases
   - (b) decreases
   - (c) remains constant

10. In a parallel *RL* circuit, there are 2 A rms in the resistive branch and 2 A rms in the inductive branch. The total rms current is
    - (a) 4 A
    - (b) 5.656 A
    - (c) 2 A
    - (d) 2.828 A

11. You are observing two voltage waveforms on an oscilloscope. The time base (time/division) of the scope is adjusted so that one-half cycle of the waveforms covers the ten horizontal divisions. The positive-going zero crossing of one waveform is at the leftmost division, and the positive-going zero crossing of the other is three divisions to the right. The phase angle between these two waveforms is
    - (a) 18°
    - (b) 36°
    - (c) 54°
    - (d) 180°

12. Which of the following power factors results in less energy being converted to heat in an *RL* circuit?
    - (a) 1
    - (b) 0.9
    - (c) 0.5
    - (d) 0.1

13. If a load is purely inductive and the reactive power is 10 VAR, the apparent power is
    (a) 0 VA   (b) 10 VA   (c) 14.14 VA   (d) 3.16 VA

14. For a certain load, the true power is 10 W and the reactive power is 10 VAR. The apparent power is
    (a) 5 VA   (b) 20 VA   (c) 14.14 VA   (d) 100 VA

CIRCUIT DYNAMICS QUIZ

Answers are at the end of the chapter.

Refer to Figure 16–56.

1. If L opens, the voltage across it
   (a) increases   (b) decreases   (c) stays the same

2. If R opens, the voltage across L
   (a) increases   (b) decreases   (c) stays the same

3. If the frequency is increased, the voltage across R
   (a) increases   (b) decreases   (c) stays the same

Refer to Figure 16–63.

4. If L opens, the voltage across R
   (a) increases   (b) decreases   (c) stays the same

5. If f is increased, the current through R
   (a) increases   (b) decreases   (c) stays the same

Refer to Figure 16–69.

6. If R_1 becomes open, the current through L_1
   (a) increases   (b) decreases   (c) stays the same

7. If L_2 opens, the voltage across R_2
   (a) increases   (b) decreases   (c) stays the same

Refer to Figure 16–70.

8. If L_2 opens, the voltage from point B to ground
   (a) increases   (b) decreases   (c) stays the same

9. If L_1 opens, the voltage from point B to ground
   (a) increases   (b) decreases   (c) stays the same

10. If the frequency of the source voltage is increased, the current through R_1
    (a) increases   (b) decreases   (c) stays the same

11. If the frequency of the source voltage is decreased, the voltage from point A to ground
    (a) increases   (b) decreases   (c) stays the same

Refer to Figure 16–73.

12. If L_2 opens, the voltage across L_1
    (a) increases   (b) decreases   (c) stays the same

13. If R_1 opens, the output voltage
    (a) increases   (b) decreases   (c) stays the same

14. If R_3 becomes open, the output voltage
    (a) increases   (b) decreases   (c) stays the same

15. If a partial short develops in L_1, the source current
    (a) increases   (b) decreases   (c) stays the same

16. If the source frequency increases, the output voltage
    (a) increases   (b) decreases   (c) stays the same

PROBLEMS

More difficult problems are indicated by an asterisk (*).
Answers to odd-numbered problems are at the end of the book.

PART 1: SERIES CIRCUITS

SECTION 16–1 Sinusoidal Response of Series *RL* Circuits

1. A 15 kHz sinusoidal voltage is applied to a series *RL* circuit. What is the frequency of *I*, V_R, and V_L?

2. What are the wave shapes of *I*, V_R, and V_L in Problem 1?

SECTION 16–2 Impedance of Series *RL* Circuits

3. Express the total impedance of each circuit in Figure 16–54 in both polar and rectangular forms.

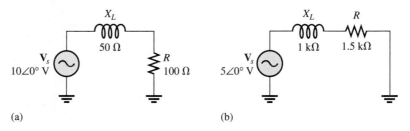

(a)   (b)

▲ FIGURE 16–54

4. Determine the impedance magnitude and phase angle in each circuit in Figure 16–55. Draw the impedance diagrams.

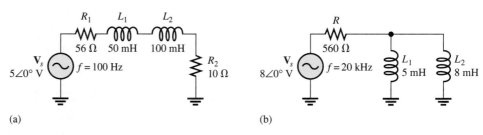

(a)   (b)

▲ FIGURE 16–55

5. In Figure 16–56, determine the impedance at each of the following frequencies:
   (a) 100 Hz   (b) 500 Hz   (c) 1 kHz   (d) 2 kHz

▶ FIGURE 16–56

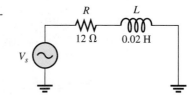

6. Determine the values of *R* and X_L in a series *RL* circuit for the following values of total impedance:
   (a) **Z** = 20 Ω + *j*45 Ω   (b) **Z** = 500∠35° Ω
   (c) **Z** = 2.5∠72.5° kΩ   (d) **Z** = 998∠45° Ω

7. Reduce the circuit in Figure 16–57 to a single resistance and inductance in series.

▶ **FIGURE 16–57**

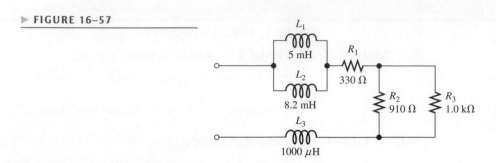

SECTION 16–3 Analysis of Series *RL* Circuits

8. A 5 V, 10 kHz sinusoidal voltage is applied to the circuit in Figure 16–57. Calculate the voltage across the total resistance found in Problem 7.

9. For the same applied voltage in Problem 8, determine the voltage across L_3 for the circuit in Figure 16–57.

10. Express the current in polar form for each circuit of Figure 16–54.

11. Calculate the total current in each circuit of Figure 16–55 and express in polar form.

12. Determine θ for the circuit in Figure 16–58.

▶ **FIGURE 16–58**

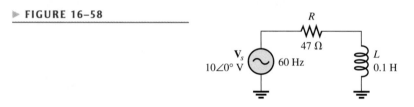

13. If the inductance in Figure 16–58 is doubled, does θ increase or decrease, and by how many degrees?

14. Draw the waveforms for $\mathbf{V}_s$, $\mathbf{V}_R$, and $\mathbf{V}_L$ in Figure 16–58. Show the proper phase relationships.

15. For the circuit in Figure 16–59, find $\mathbf{V}_R$ and $\mathbf{V}_L$ for each of the following frequencies:

   (a) 60 Hz   (b) 200 Hz   (c) 500 Hz   (d) 1 kHz

▶ **FIGURE 16–59**

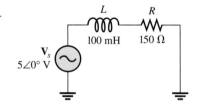

16. Determine the magnitude and phase angle of the source voltage in Figure 16–60.

▶ **FIGURE 16–60**

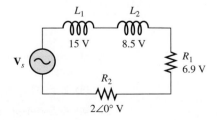

17. For the lag circuit in Figure 16–61, determine the phase lag of the output voltage with respect to the input for the following frequencies:

(a) 1 Hz (b) 100 Hz (c) 1 kHz (d) 10 kHz

18. Repeat Problem 17 for the lead circuit to find the phase lead in Figure 16–62.

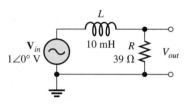

▲ FIGURE 16–61

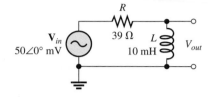

▲ FIGURE 16–62

PART 2: PARALLEL CIRCUITS

SECTION 16–4 Impedance and Admittance of Parallel *RL* Circuits

19. What is the impedance expressed in polar form for the circuit in Figure 16–63?

20. Repeat Problem 19 for the following frequencies:

(a) 1.5 kHz (b) 3 kHz (c) 5 kHz (d) 10 kHz

21. At what frequency does X_L equal R in Figure 16–63?

▶ FIGURE 16–63

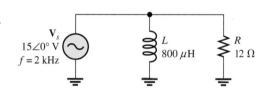

SECTION 16–5 Analysis of Parallel *RL* Circuits

22. Find the total current and each branch current in Figure 16–64.

▶ FIGURE 16–64

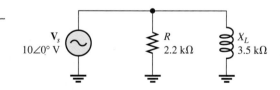

23. Determine the following quantities in Figure 16–65:

(a) **Z** (b) $\mathbf{I}_R$ (c) $\mathbf{I}_L$ (d) $\mathbf{I}_{tot}$ (e) θ

▶ FIGURE 16–65

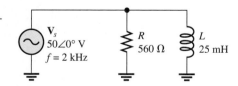

24. Repeat Problem 23 for $R = 56\ \Omega$ and $L = 330\ \mu\text{H}$.

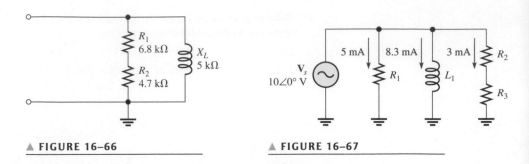

▲ FIGURE 16–66

▲ FIGURE 16–67

25. Convert the circuit in Figure 16–66 to an equivalent series form.

26. Find the magnitude and phase angle of the total current in Figure 16–67.

PART 3: SERIES-PARALLEL CIRCUITS

SECTION 16–6 Analysis of Series-Parallel *RL* Circuits

27. Determine the voltages in polar form across each element in Figure 16–68. Draw the voltage phasor diagram.

28. Is the circuit in Figure 16–68 predominantly resistive or predominantly inductive?

29. Find the current in each branch and the total current in Figure 16–68. Express the currents in polar form. Draw the current phasor diagram.

30. For the circuit in Figure 16–69, determine the following:

   (a) I_{tot} (b) θ (c) V_{R1} (d) V_{R2} (e) V_{R3} (f) V_{L1} (g) V_{L2}

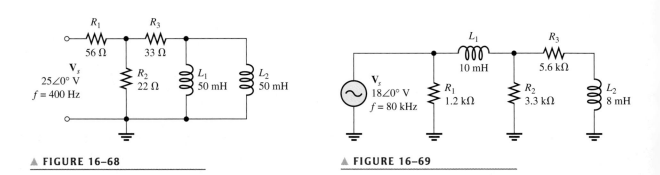

▲ FIGURE 16–68

▲ FIGURE 16–69

*31. For the circuit in Figure 16–70, determine the following:

   (a) I_{tot} (b) V_{L1} (c) V_{AB}

*32. Draw the phasor diagram of all voltages and currents in Figure 16–70.

▶ FIGURE 16–70

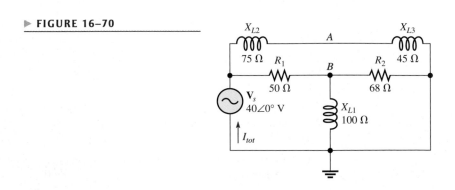

33. Determine the phase shift and attenuation (ratio of V_{out} to V_{in}) from the input to the output for the circuit in Figure 16–71.

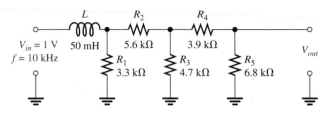

▲ FIGURE 16–71

*34. Determine the phase shift and attenuation from the input to the output for the ladder network in Figure 16–72.

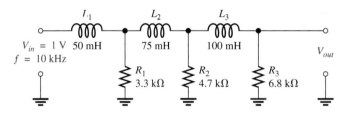

▲ FIGURE 16–72

35. Design an ideal inductive switching circuit that will provide a momentary voltage of 2.5 kV from a 12 V dc source when a switch is thrown instantaneously from one position to another. The drain on the source must not exceed 1 A.

PART 4: SPECIAL TOPICS

SECTION 16–7 Power in *RL* Circuits

36. In a certain *RL* circuit, the true power is 100 mW, and the reactive power is 340 mVAR. What is the apparent power?

37. Determine the true power and the reactive power in Figure 16–58.

38. What is the power factor in Figure 16–64?

39. Determine P_{true}, P_r, P_a, and *PF* for the circuit in Figure 16–69. Sketch the power triangle.

*40. Find the true power for the circuit in Figure 16–70.

SECTION 16–8 Basic Applications

41. Draw the response curve for the circuit in Figure 16–61. Show the output voltage versus frequency in 1 kHz increments from 0 Hz to 5 kHz.

42. Using the same procedure as in Problem 41, draw the response curve for Figure 16–62.

43. Draw the voltage phasor diagram for each circuit in Figures 16–61 and 16–62 for a frequency of 8 kHz.

SECTION 16–9   **Troubleshooting**

44. Determine the voltage across each component in Figure 16–73 if L_1 is open.

45. Determine the output voltage in Figure 16–73 for each of the following failure modes:

(a) L_1 open   (b) L_2 open   (c) R_1 open   (d) a short across R_2

▶ FIGURE 16–73

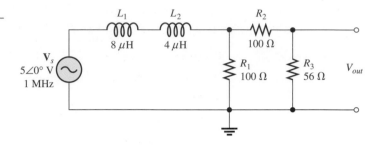

Multisim Troubleshooting and Analysis

These problems require your Multisim CD-ROM.

46. Open file P16-46 and determine if there is a fault. If so, find the fault.

47. Open file P16-47 and determine if there is a fault. If so, find the fault.

48. Open file P16-48 and determine if there is a fault. If so, find the fault.

49. Open file P16-49 and determine if there is a fault. If so, find the fault.

50. Open file P16-50 and determine if there is a fault. If so, find the fault.

51. Open file P16-51 and determine if there is a fault. If so, find the fault.

52. Open file P16-52 and determine the frequency response for the filter.

53. Open file P16-53 and determine the frequency response for the filter.

ANSWERS

SECTION REVIEWS

SECTION 16–1   **Sinusoidal Response of Series *RL* Circuits**

1. The current frequency is 1 kHz.

2. The phase angle is closer to 0°.

SECTION 16–2   **Impedance of Series *RL* Circuits**

1. $R = 150 \ \Omega$; $X_L = 220 \ \Omega$

2. $\mathbf{Z} = R + jX_L = 33 \ \text{k}\Omega + j50 \ \text{k}\Omega$; $\mathbf{Z} = \sqrt{R^2 + X_L^2} \angle \tan^{-1}(X_L/R) = 59.9 \angle 56.6° \ \text{k}\Omega$

SECTION 16–3   **Analysis of Series *RL* Circuits**

1. $V_s = \sqrt{V_R^2 + V_L^2} = 3.61 \ \text{V}$

2. $\theta = \tan^{-1}(V_L/V_R) = 56.3°$

3. When f increases, X_L increases, Z increases, and θ increases.

4. $\phi = 81.9°$

5. $V_{out} = 9.90 \ \text{V}$

SECTION 16–4   **Impedance and Admittance of Parallel *RL* Circuits**

1. $Y = \dfrac{1}{Z} = \dfrac{1}{\sqrt{R^2 + X_L^2}} = 2 \ \text{mS}$

2. $Y = \dfrac{1}{Z} = 25.1 \ \text{mS}$

3. $\mathbf{I}$ lags V_s; $\theta = 32.1°$

SECTION 16–5 **Analysis of Parallel *RL* Circuits**

 1. $I_{tot} = 32\,\text{mA}$
 2. $\mathbf{I}_{tot} = 23.3\angle{-59.0°}\,\text{mA}$; θ is with respect to the input voltage.
 3. $\theta = -90°$

SECTION 16–6 **Analysis of Series-Parallel *RL* Circuits**

 1. $\mathbf{Z} = 494\angle{59.0°}\,\Omega$
 2. $I_{tot} = 10.4\,\text{mA} - j17.3\,\text{mA}$

SECTION 16–7 **Power in *RL* Circuits**

 1. Power dissipation is due to resistance.
 2. $PF = 0.643$
 3. $P_{\text{true}} = 4.7\,\text{W}$; $P_r = 6.2\,\text{VAR}$; $P_a = 7.78\,\text{VA}$

SECTION 16–8 **Basic Applications**

 1. The output is across the resistor.
 2. It is more efficient than other types.
 3. It is adjusted by the pulse width modulator to be longer.

SECTION 16–9 **Troubleshooting**

 1. Shorted windings reduce L and thereby reduce X_L at any given frequency.
 2. I_{tot} decreases, V_{R1} decreases, V_{R2} increases.

A Circuit Application

 1. $V_{out} = 0\,\text{V}$
 2. $V_{out} = V_{in}$

RELATED PROBLEMS FOR EXAMPLES

16–1 $\mathbf{Z} = 1.8\,\text{k}\Omega + j950\,\Omega$; $\mathbf{Z} = 2.04\angle{27.8°}\,\text{k}\Omega$
16–2 $\mathbf{I} = 423\angle{-32.1°}\,\mu\text{A}$
16–3 $Z = 12.6\,\text{k}\Omega$; $\theta = 85.5°$
16–4 $\phi = 65.6°$
16–5 V_{out} increases.
16–6 $\phi = -32°$
16–7 $V_{out} = 12.3\,\text{V rms}$
16–8 $\mathbf{Z} = 8.14\angle{35.5°}\,\text{k}\Omega$
16–9 $\mathbf{Y} = 3.03\,\text{mS} - j0.796\,\text{mS}$
16–10 $\mathbf{I} = 14.0\angle{-71.1°}\,\text{mA}$
16–11 $I_{tot} = 67.6\,\text{mA}$; $\theta = 36.3°$
16–12 (a) $\mathbf{V}_1 = 8.04\angle{2.52°}\,\text{V}$ (b) $\mathbf{V}_2 = 2.00\angle{-10.2°}\,\text{V}$
16–13 $\mathbf{I}_{tot} = 20.2\angle{-59.0°}\,\text{mA}$
16–14 P_{true}, P_r, and P_a decrease.
16–15 Open connection between L_1 and L_2

SELF-TEST

1. (f)	2. (b)	3. (a)	4. (d)	5. (a)	6. (d)	7. (b)	8. (b)
9. (a)	10. (d)	11. (c)	12. (d)	13. (b)	14. (c)		

CIRCUIT DYNAMICS QUIZ

1. (a)	2. (b)	3. (b)	4. (c)	5. (c)	6. (c)	7. (a)	8. (c)
9. (a)	10. (b)	11. (c)	12. (b)	13. (a)	14. (a)	15. (a)	16. (b)

17

RLC CIRCUITS AND RESONANCE

CHAPTER OBJECTIVES

PART 1: SERIES CIRCUITS

◆ Determine the impedance of a series *RLC* circuit
◆ Analyze series *RLC* circuits
◆ Analyze a circuit for series resonance

PART 2: PARALLEL CIRCUITS

◆ Determine the impedance of a parallel *RLC* circuit
◆ Analyze parallel *RLC* circuits
◆ Analyze a circuit for parallel resonance

PART 3: SERIES-PARALLEL CIRCUITS

◆ Analyze series-parallel *RLC* circuits

PART 4: SPECIAL TOPICS

◆ Determine the bandwidth of resonant circuits
◆ Discuss some applications of resonant circuits

KEY TERMS

◆ Series resonance
◆ Resonant frequency (f_r)
◆ Parallel resonance
◆ Tank circuit
◆ Half-power frequency
◆ Selectivity

A CIRCUIT APPLICATION PREVIEW

In the circuit application, you will work with the resonant tuning circuit in the RF amplifier of an AM radio receiver. The tuning circuit is used to select any desired frequency within the AM band so that a desired station can be tuned in.

VISIT THE COMPANION WEBSITE

Study aids for this chapter are available at
http://www.prenhall.com/floyd

INTRODUCTION

In this chapter, the analysis methods learned in Chapters 15 and 16 are extended to the coverage of circuits with combinations of resistive, inductive, and capacitive components. Series and parallel *RLC* circuits, plus series-parallel combinations, are studied.

Circuits with both inductance and capacitance can exhibit the property of resonance, which is important in many types of applications. Resonance is the basis for frequency selectivity in communication systems. For example, the ability of a radio or television receiver to select a certain frequency that is transmitted by a particular station and, at the same time, to eliminate frequencies from other stations is based on the principle of resonance. The conditions in *RLC* circuits that produce resonance and the characteristics of resonant circuits are covered in this chapter.

COVERAGE OPTIONS

If you chose Option 1 to cover all of Chapter 15 and all of Chapter 16, then all of this chapter should be covered next.

If you chose Option 2 to cover reactive circuits in Chapters 15 and 16 on the basis of the four major parts, then the appropriate part of this chapter should be covered next, followed by the next part in Chapter 15, if applicable.

SERIES CIRCUITS

17–1 IMPEDANCE OF SERIES *RLC* CIRCUITS

A series *RLC* circuit contains resistance, inductance, and capacitance. Since inductive reactance and capacitive reactance have opposite effects on the circuit phase angle, the total reactance is less than either individual reactance.

After completing this section, you should be able to

- ◆ **Determine the impedance of a series *RLC* circuit**
  - ◆ Calculate total reactance
  - ◆ Determine whether a circuit is predominately inductive or capacitive

A series *RLC* circuit is shown in Figure 17–1. It contains resistance, inductance, and capacitance.

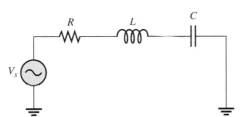

◀ **FIGURE 17–1**

Series *RLC* circuit.

As you know, inductive reactance ($\mathbf{X}_L$) causes the total current to lag the applied voltage. Capacitive reactance ($\mathbf{X}_C$) has the opposite effect: It causes the current to lead the voltage. Thus $\mathbf{X}_L$ and $\mathbf{X}_C$ tend to offset each other. When they are equal, they cancel, and the total reactance is zero. In any case, the magnitude of the total reactance in the series circuit is

$$X_{tot} = |X_L - X_C|$$

Equation 17–1

The term $|X_L - X_C|$ means the absolute value of the difference of the two reactances. That is, the sign of the result is considered positive no matter which reactance is greater. For example, $3 - 7 = -4$, but the absolute value is

$$|3 - 7| = 4$$

When $X_L > X_C$, the circuit is predominantly inductive, and when $X_C > X_L$, the circuit is predominantly capacitive.

The total impedance for the series *RLC* circuit is stated in rectangular form in Equation 17–2 and in polar form in Equation 17–3.

Equation 17–2

$$\mathbf{Z} = R + jX_L - jX_C$$

Equation 17–3

$$\mathbf{Z} = \sqrt{R^2 + (X_L - X_C)^2}\angle\pm\tan^{-1}\left(\frac{X_{tot}}{R}\right)$$

In Equation 17–3, $\sqrt{R^2 + (X_L - X_C)^2}$ is the magnitude and $\tan^{-1}(X_{tot}/R)$ is the phase angle between the total current and the applied voltage. If the circuit is predominately inductive, the phase angle is positive; and if predominately capacitive, the phase angle is negative.

EXAMPLE 17–1

For the series *RLC* circuit in Figure 17–2, determine the total impedance. Express it in both rectangular and polar forms.

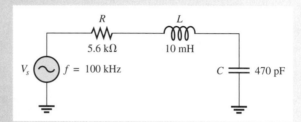

▲ **FIGURE 17–2**

Solution   First, find X_C and X_L.

$$X_C = \frac{1}{2\pi fC} = \frac{1}{2\pi(100\,\text{kHz})(470\,\text{pF})} = 3.39\,\text{k}\Omega$$

$$X_L = 2\pi fL = 2\pi(100\,\text{kHz})(10\,\text{mH}) = 6.28\,\text{k}\Omega$$

In this case, X_L is greater than X_C, and thus the circuit is more inductive than capacitive. The magnitude of the total reactance is

$$X_{tot} = |X_L - X_C| = |6.28\,\text{k}\Omega - 3.39\,\text{k}\Omega| = 2.89\,\text{k}\Omega \qquad \text{inductive}$$

The impedance in rectangular form is

$$\mathbf{Z} = R + (jX_L - jX_C) = 5.6\,\text{k}\Omega + (j6.28\,\text{k}\Omega - j3.39\,\text{k}\Omega) = \mathbf{5.6\,k\Omega + j2.89\,k\Omega}$$

The impedance in polar form is

$$\mathbf{Z} = \sqrt{R^2 + X_{tot}^2}\angle\tan^{-1}\left(\frac{X_{tot}}{R}\right)$$

$$= \sqrt{(5.6\,\text{k}\Omega)^2 + (2.89\,\text{k}\Omega)^2}\angle\tan^{-1}\left(\frac{2.89\,\text{k}\Omega}{5.6\,\text{k}\Omega}\right) = \mathbf{6.30\angle27.3°\,k\Omega}$$

The positive angle shows that the circuit is inductive.

Related Problem*   Determine **Z** in polar form if *f* is increased to 200 kHz.

---

*Answers are at the end of the chapter.

As you have seen, when the inductive reactance is greater than the capacitive reactance, the circuit appears inductive; so the current lags the applied voltage. When the capacitive reactance is greater, the circuit appears capacitive, and the current leads the applied voltage.

17–2 ANALYSIS OF SERIES *RLC* CIRCUITS

Recall that capacitive reactance varies inversely with frequency and that inductive reactance varies directly with frequency. In this section, the combined effects of the reactances as a function of frequency are examined.

After completing this section, you should be able to

◆ **Analyze series *RLC* circuits**

  ◆ Determine current in a series *RLC* circuit

  ◆ Determine the voltages in a series *RLC* circuit

  ◆ Determine the phase angle

Figure 17–3 shows that for a typical series *RLC* circuit the total reactance behaves as follows: Starting at a very low frequency, X_C is high, and X_L is low, and the circuit is predominantly capacitive. As the frequency is increased, X_C decreases and X_L increases until a value is reached where $X_C = X_L$ and the two reactances cancel, making the circuit purely resistive. This condition is **series resonance** and will be studied in Section 17–3. As the frequency is increased further, X_L becomes greater than X_C, and the circuit is predominantly inductive. Example 17–2 illustrates how the impedance and phase angle change as the source frequency is varied.

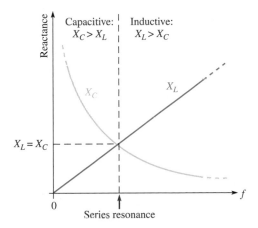

◀ **FIGURE 17–3**

How X_C and X_L vary with frequency.

The graph of X_L is a straight line, and the graph of X_C is curved, as shown in Figure 17–3. The general equation for a straight line is $y = mx + b$, where *m* is the slope of the line and *b* is the *y*-axis intercept point. The formula $X_L = 2\pi fL$ fits this general straight-line formula, where $y = X_L$ (a variable), $m = 2\pi L$ (a constant), $x = f$ (a variable), and $b = 0$ as follows: $X_L = 2\pi Lf + 0$.

The X_C curve is called a *hyperbola,* and the general equation of a hyperbola is $xy = k$. The equation for capacitive reactance, $X_C = 1/2\pi fC$, can be rearranged as $X_C f = 1/2\pi C$ where $x = X_C$ (a variable), $y = f$ (a variable), and $k = 1/2\pi C$ (a constant).

EXAMPLE 17–2

For each of the following input frequencies, find the impedance in polar form for the circuit in Figure 17–4. Note the change in magnitude and phase angle with frequency.

(a) $f = 1 \, \text{kHz}$ **(b)** $f = 2 \, \text{kHz}$ **(c)** $f = 3.5 \, \text{kHz}$ **(d)** $f = 5 \, \text{kHz}$

▶ **FIGURE 17–4**

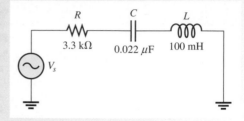

Solution   **(a)** At $f = 1 \, \text{kHz}$,

$$X_C = \frac{1}{2\pi fC} = \frac{1}{2\pi(1 \, \text{kHz})(0.022 \, \mu\text{F})} = 7.23 \, \text{k}\Omega$$

$$X_L = 2\pi fL = 2\pi(1 \, \text{kHz})(100 \, \text{mH}) = 628 \, \Omega$$

The circuit is clearly capacitive, and the impedance is

$$\mathbf{Z} = \sqrt{R^2 + (X_L - X_C)^2} \angle -\tan^{-1}\left(\frac{X_{tot}}{R}\right)$$

$$= \sqrt{(3.3 \, \text{k}\Omega)^2 + (628 \, \Omega - 7.23 \, \text{k}\Omega)^2} \angle -\tan^{-1}\left(\frac{6.60 \, \text{k}\Omega}{3.3 \, \text{k}\Omega}\right) = \mathbf{7.38} \angle -\mathbf{63.4°} \, \text{k}\Omega$$

The negative sign for the angle is used to indicate that the circuit is capacitive.

(b) At $f = 2 \, \text{kHz}$,

$$X_C = \frac{1}{2\pi(2 \, \text{kHz})(0.022 \, \mu\text{F})} = 3.62 \, \text{k}\Omega$$

$$X_L = 2\pi(2 \, \text{kHz})(100 \, \text{mH}) = 1.26 \, \text{k}\Omega$$

The circuit is still capacitive, and the impedance is

$$\mathbf{Z} = \sqrt{(3.3 \, \text{k}\Omega)^2 + (1.26 \, \text{k}\Omega - 3.62 \, \text{k}\Omega)^2} \angle -\tan^{-1}\left(\frac{2.36 \, \text{k}\Omega}{3.3 \, \text{k}\Omega}\right)$$

$$= \mathbf{4.06} \angle -\mathbf{35.6°} \, \text{k}\Omega$$

(c) At $f = 3.5 \, \text{kHz}$,

$$X_C = \frac{1}{2\pi(3.5 \, \text{kHz})(0.022 \, \mu\text{F})} = 2.07 \, \text{k}\Omega$$

$$X_L = 2\pi(3.5 \, \text{kHz})(100 \, \text{mH}) = 2.20 \, \text{k}\Omega$$

The circuit is very close to being purely resistive because X_C and X_L are nearly equal, but it is slightly inductive. The impedance is

$$\mathbf{Z} = \sqrt{(3.3 \, \text{k}\Omega)^2 + (2.20 \, \text{k}\Omega - 2.07 \, \text{k}\Omega)^2} \angle \tan^{-1}\left(\frac{0.13 \, \text{k}\Omega}{3.3 \, \text{k}\Omega}\right)$$

$$= \mathbf{3.3} \angle \mathbf{2.26°} \, \text{k}\Omega$$

(d) At $f = 5 \, \text{kHz}$,

$$X_C = \frac{1}{2\pi(5 \, \text{kHz})(0.022 \, \mu\text{F})} = 1.45 \, \text{k}\Omega$$

$$X_L = 2\pi(5 \, \text{kHz})(100 \, \text{mH}) = 3.14 \, \text{k}\Omega$$

The circuit is now predominantly inductive. The impedance is

$$\mathbf{Z} = \sqrt{(3.3\,\text{k}\Omega)^2 + (3.14\,\text{k}\Omega - 1.45\,\text{k}\Omega)^2}\angle\tan^{-1}\left(\frac{1.69\,\text{k}\Omega}{3.3\,\text{k}\Omega}\right)$$

$$= 3.71\angle 27.1°\,\text{k}\Omega$$

Notice how the circuit changed from capacitive to inductive as the frequency increased. The phase condition changed from the current leading to the current lagging as indicated by the sign of the angle. Note that the impedance magnitude decreased to a minimum equal to the resistance and then began increasing again. Also, notice that the negative phase angle decreased as the frequency increased and the angle became positive when the circuit became inductive and then increased with increasing frequency.

Related Problem   Determine **Z** in polar form for $f = 7$ kHz and draw a graph of impedance vs. frequency using the values in this example.

In a series *RLC* circuit, the capacitor voltage and the inductor voltage are always 180° out of phase with each other. For this reason, V_C and V_L subtract from each other, and thus the voltage across *L* and *C* combined is always less than the larger individual voltage across either element, as illustrated in Figure 17–5 and in the waveform diagram of Figure 17–6.

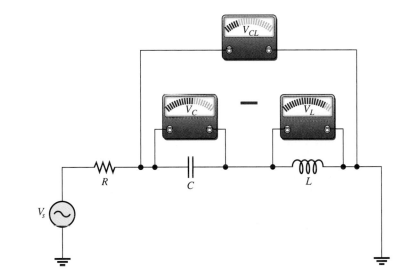

◀ **FIGURE 17–5**

The voltage across the series combination of *C* and *L* is always less than the larger individual voltage across either *C* or *L*.

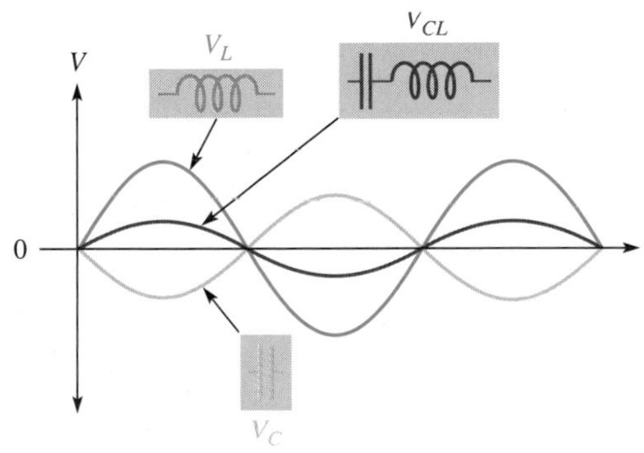

◀ **FIGURE 17–6**

V_{CL} is the algebraic sum of V_L and V_C. Because of the phase relationship, V_L and V_C effectively subtract.

In the next example, Ohm's law is used to find the current and voltages in a series *RLC* circuit.

EXAMPLE 17–3

Find the current and the voltages across each component in Figure 17–7. Express each quantity in polar form, and draw a complete voltage phasor diagram.

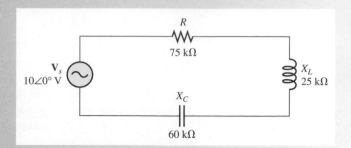

▲ FIGURE 17–7

Solution   First, find the total impedance.

$$\mathbf{Z} = R + jX_L - jX_C = 75 \, \text{k}\Omega + j25 \, \text{k}\Omega - j60 \, \text{k}\Omega = 75 \, \text{k}\Omega - j35 \, \text{k}\Omega$$

Convert to polar form for convenience in applying Ohm's law.

$$\mathbf{Z} = \sqrt{R^2 + X_{tot}^2}\angle -\tan^{-1}\left(\frac{X_{tot}}{R}\right)$$

$$= \sqrt{(75 \, \text{k}\Omega)^2 + (35 \, \text{k}\Omega)^2}\angle -\tan^{-1}\left(\frac{35 \, \text{k}\Omega}{75 \, \text{k}\Omega}\right) = 82.8\angle -25° \, \text{k}\Omega$$

where $X_{tot} = |X_L - X_C|$.

Apply Ohm's law to find the current.

$$\mathbf{I} = \frac{\mathbf{V}_s}{\mathbf{Z}} = \frac{10\angle 0° \, \text{V}}{82.8\angle -25° \, \text{k}\Omega} = \mathbf{121\angle 25.0° \, \mu A}$$

Now, apply Ohm's law to find the voltages across R, L, and C.

$$\mathbf{V}_R = \mathbf{IR} = (121\angle 25.0° \, \mu A)(75\angle 0° \, \text{k}\Omega) = \mathbf{9.08\angle 25.0° \, V}$$

$$\mathbf{V}_L = \mathbf{IX}_L = (121\angle 25.0° \, \mu A)(25\angle 90° \, \text{k}\Omega) = \mathbf{3.03\angle 115° \, V}$$

$$\mathbf{V}_C = \mathbf{IX}_C = (121\angle 25.0° \, \mu A)(60\angle -90° \, \text{k}\Omega) = \mathbf{7.26\angle -65.0° \, V}$$

The phasor diagram is shown in Figure 17–8. The magnitudes represent rms values. Notice that V_L is leading V_R by 90°, and V_C is lagging V_R by 90°. Also, there is a 180° phase difference between V_L and V_C. If the current phasor were shown, it would be at

▶ FIGURE 17–8

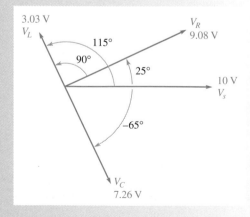

the same angle as V_R. The current is leading V_s, the source voltage, by 25°, indicating a capacitive circuit ($X_C > X_L$). The phasor diagram is rotated 25° from its usual position because the reference is the source voltage, V_s, which is shown oriented along the x-axis.

Related Problem   What will happen to the current as the frequency of the source voltage in Figure 17–7 is increased?

**SECTION 17–2
REVIEW**

1. The following voltages occur in a certain series *RLC* circuit. Determine the source voltage: $\mathbf{V}_R = 24\angle 30°$ V, $\mathbf{V}_L = 15\angle 120°$ V, and $\mathbf{V}_C = 45\angle -60°$ V.
2. When $R = 1.0\,\text{k}\Omega$, $X_C = 1.8\,\text{k}\Omega$, and $X_L = 1.2\,\text{k}\Omega$, does the current lead or lag the applied voltage?
3. Determine the total reactance in Question 2.

17–3 SERIES RESONANCE

In a series *RLC* circuit, series resonance occurs when $X_C = X_L$. The frequency at which resonance occurs is called the **resonant frequency** and is designated f_r.

After completing this section, you should be able to

◆ **Analyze a circuit for series resonance**

  ◆ Define *series resonance*

  ◆ Determine the impedance at resonance

  ◆ Explain why the reactances cancel at resonance

  ◆ Determine the series resonant frequency

  ◆ Calculate the current, voltages, and phase angle at resonance

Figure 17–9 illustrates the series resonant condition.

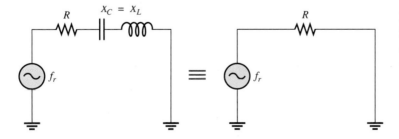

◀ **FIGURE 17–9**

Series resonance. X_C and X_L cancel each other resulting in a purely resistive circuit.

Resonance is a condition in a series *RLC* circuit in which the capacitive and inductive reactances are equal in magnitude; thus, they cancel each other and result in a purely resistive impedance. In a series *RLC* circuit, the total impedance was given in Equation 17–2 as

$$\mathbf{Z} = R + jX_L - jX_C$$

At resonance, $X_L = X_C$ and the j terms cancel; thus, the impedance is purely resistive. These resonant conditions are stated in the following equations:

$$X_L = X_C$$
$$Z_r = R$$

EXAMPLE 17–4

For the series *RLC* circuit in Figure 17–10, determine X_C and **Z** at resonance.

▶ **FIGURE 17–10**

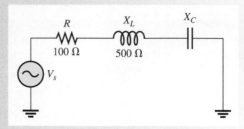

Solution $X_L = X_C$ at the resonant frequency. Thus, $X_C = X_L = \mathbf{500\ \Omega}$. The impedance at resonance is

$$\mathbf{Z}_r = R + jX_L - jX_C = 100\ \Omega + j500\ \Omega - j500\ \Omega = \mathbf{100\angle 0°\ \Omega}$$

This shows that the impedance at resonance is equal to the resistance because the reactances are equal in magnitude and therefore cancel.

Related Problem Just below the resonant frequency, is the circuit more inductive or more capacitive?

X_L and X_C Cancel at Resonance

At the series resonant frequency (f_r), the voltages across C and L are equal in magnitude because the reactances are equal. The same current is through both since they are in series ($IX_C = IX_L$). Also, V_L and V_C are always 180° out of phase with each other.

During any given cycle, the polarities of the voltages across C and L are opposite, as shown in parts (a) and (b) of Figure 17–11. The equal and opposite voltages across C and L cancel, leaving zero volts from point A to point B as shown. Since there is no voltage drop from A to B but there is still current, the total reactance must be zero, as indicated in part (c). Also, the voltage phasor diagram in part (d) shows that V_C and V_L are equal in magnitude and 180° out of phase with each other.

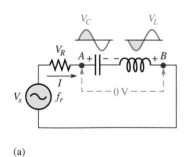

(a)

(b)

(c)

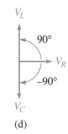

(d)

▲ **FIGURE 17–11**

At the resonant frequency, f_r, the voltages across C and L are equal in magnitude. Since they are 180° out of phase with each other, they cancel, leaving 0 V across the *LC* combination (point A to point B). The section of the circuit from A to B effectively looks like a short at resonance.

Series Resonant Frequency

For a given series *RLC* circuit, resonance occurs at only one specific frequency. A formula for this resonant frequency is developed as follows:

$$X_L = X_C$$

Substitute the reactance formulas.

$$2\pi f_r L = \frac{1}{2\pi f_r C}$$

Then, multiplying both sides by $f_r/2\pi L$,

$$f_r^2 = \frac{1}{4\pi^2 LC}$$

Take the square root of both sides. The formula for series resonant frequency is

$$f_r = \frac{1}{2\pi\sqrt{LC}}$$

Equation 17–4

EXAMPLE 17–5

Find the series resonant frequency for the circuit in Figure 17–12.

▶ **FIGURE 17–12**

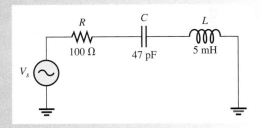

Solution   The resonant frequency is

$$f_r = \frac{1}{2\pi\sqrt{LC}} = \frac{1}{2\pi\sqrt{(5 \text{ mH})(47 \text{ pF})}} = \textbf{328 kHz}$$

Related Problem   If $C = 0.01\ \mu\text{F}$ in Figure 17–12, what is the resonant frequency?

Use Multisim file E17-05 to verify the calculated results in this example and to confirm your calculation for the related problem.

Current and Voltages in a Series *RLC* Circuit

At the series resonant frequency, the current is maximum ($I_{max} = V_s/R$). Above and below resonance, the current decreases because the impedance increases. A response curve showing the plot of current versus frequency is shown in Figure 17–13(a). The resistor voltage, V_R, follows the current and is maximum (equal to V_s) at resonance and zero at $f = 0$ and at $f = \infty$, as shown in Figure 17–13(b). The general shapes of the V_C and V_L curves are indicated in Figure 17–13(c) and (d). Notice that $V_C = V_s$ when $f = 0$, because the capacitor appears open. Also notice that V_L approaches V_s as f approaches infinity, because the inductor appears open. The voltage across the C and L combination decreases as the frequency increases below resonance, reaching a minimum of zero at the resonant frequency; then it increases above resonance, as shown in Figure 17–13(e).

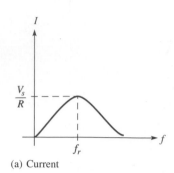

(a) Current

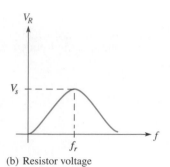

(b) Resistor voltage

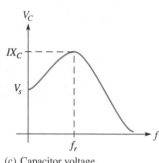

(c) Capacitor voltage

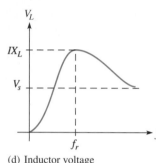

(d) Inductor voltage

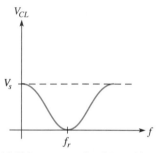

(e) Voltage across *C* and *L* combined

▲ **FIGURE 17–13**

Generalized current and voltage magnitudes as a function of frequency in a series *RLC* circuit. V_C and V_L can be much larger than the source voltage. The shapes of the graphs depend on specific circuit values.

The voltages are maximum at resonance but drop off above and below f_r. The voltages across *L* and *C* at resonance are exactly equal in magnitude but 180° out of phase; so they cancel. Thus, the total voltage across both *L* and *C* is zero, and $V_R = V_s$ at resonance, as indicated in Figure 17–14. Individually, V_L and V_C can be much greater than the source voltage, as you will see later. Keep in mind that V_L and V_C are always opposite in polarity regardless of the frequency, but only at resonance are their magnitudes equal.

▶ **FIGURE 17–14**

Series *RLC* circuit at resonance.

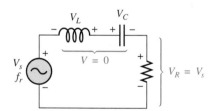

EXAMPLE 17–6

Find *I*, V_R, V_L, and V_C at resonance for the circuit in Figure 17–15. The resonant values of X_L and X_C are shown.

▶ **FIGURE 17–15**

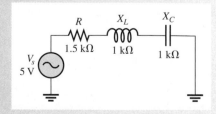

Solution   At resonance, I is maximum and equal to V_s/R.

$$I = \frac{V_s}{R} = \frac{5\,\text{V}}{1.5\,\text{k}\Omega} = \textbf{3.33 mA}$$

Apply Ohm's law to obtain the voltage magnitudes.

$$V_R = IR = (3.33\,\text{mA})(2.2\,\text{k}\Omega) = \textbf{7.33 V}$$
$$V_L = IX_L = (3.33\,\text{mA})(1\,\text{k}\Omega) = \textbf{3.33 V}$$
$$V_C = IX_C = (3.33\,\text{mA})(1\,\text{k}\Omega) = \textbf{3.33 V}$$

Notice that all of the source voltage is dropped across the resistor. Also, of course, V_L and V_C are equal in magnitude but opposite in phase. This causes these voltages to cancel, making the total reactive voltage zero.

Related Problem   What is the phase angle if the frequency is doubled?

Use Multisim file E17-06 to verify the calculated results in this example and to confirm your calculation for the related problem.

Series *RLC* Impedance

At frequencies below f_r, $X_C > X_L$; thus, the circuit is capacitive. At the resonant frequency, $X_C = X_L$, so the circuit is purely resistive. At frequencies above f_r, $X_L > X_C$; thus, the circuit is inductive.

The impedance magnitude is minimum at resonance ($Z = R$) and increases in value above and below the resonant point. The graph in Figure 17–16 illustrates how impedance changes with frequency. At zero frequency, both X_C and Z are infinitely large and X_L is zero because the capacitor looks like an open at 0 Hz and the inductor looks like a short. As the frequency increases, X_C decreases and X_L increases. Since X_C is larger than X_L at frequencies below f_r, Z decreases along with X_C. At f_r, $X_C = X_L$ and $Z = R$. At frequencies above f_r, X_L becomes increasingly larger than X_C, causing Z to increase.

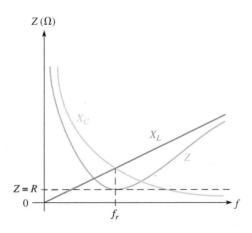

◀ **FIGURE 17–16**

Series *RLC* impedance as a function of frequency.

EXAMPLE 17–7

For the circuit in Figure 17–17, determine the impedance magnitude at the following frequencies:

(a) f_r **(b)** 1000 Hz below f_r **(c)** 1000 Hz above f_r

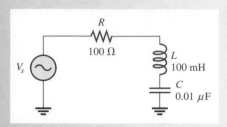

▲ FIGURE 17–17

Solution **(a)** At f_r, the impedance is equal to R.

$$Z = R = \textbf{100 } \mathbf{\Omega}$$

To determine the impedance above and below f_r, first calculate the resonant frequency.

$$f_r = \frac{1}{2\pi\sqrt{LC}} = \frac{1}{2\pi\sqrt{(100 \text{ mH})(0.01 \text{ } \mu\text{F})}} = 5.03 \text{ kHz}$$

(b) At 1000 Hz below f_r, the frequency and reactances are as follows:

$$f = f_r - 1 \text{ kHz} = 5.03 \text{ kHz} - 1 \text{ kHz} = 4.03 \text{ kHz}$$

$$X_C = \frac{1}{2\pi fC} = \frac{1}{2\pi(4.03 \text{ kHz})(0.01 \text{ } \mu\text{F})} = 3.95 \text{ k}\Omega$$

$$X_L = 2\pi fL = 2\pi(4.03 \text{ kHz})(100 \text{ mH}) = 2.53 \text{ k}\Omega$$

Therefore, the impedance at $f_r - 1 \text{ kHz}$ is

$$Z = \sqrt{R^2 + (X_L - X_C)^2} = \sqrt{(100 \text{ } \Omega)^2 + (2.53 \text{ k}\Omega - 3.95 \text{ k}\Omega)^2} = \textbf{1.42 k}\mathbf{\Omega}$$

(c) At 1000 Hz above f_r,

$$f = 5.03 \text{ kHz} + 1 \text{ kHz} = 6.03 \text{ kHz}$$

$$X_C = \frac{1}{2\pi(6.03 \text{ kHz})(0.01 \text{ } \mu\text{F})} = 2.64 \text{ k}\Omega$$

$$X_L = 2\pi(6.03 \text{ kHz})(100 \text{ mH}) = 3.79 \text{ k}\Omega$$

Therefore, the impedance at $f_r + 1 \text{ kHz}$ is

$$Z = \sqrt{(100 \text{ } \Omega)^2 + (3.79 \text{ k}\Omega - 2.64 \text{ k}\Omega)^2} = \textbf{1.15 k}\mathbf{\Omega}$$

In part (b) Z is capacitive, and in part (c) Z is inductive.

Related Problem What happens to the impedance magnitude if f is decreased below 4.03 kHz? Above 6.03 kHz?

The Phase Angle of a Series *RLC* Circuit

At frequencies below resonance, $X_C > X_L$, and the current leads the source voltage, as indicated in Figure 17–18(a). The phase angle decreases as the frequency approaches the resonant value and is 0° at resonance, as indicated in part (b). At frequencies above resonance, $X_L > X_C$, and the current lags the source voltage, as indicated in part (c). As the frequency goes higher, the phase angle approaches 90°. A plot of phase angle versus frequency is shown in part (d) of the figure.

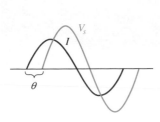

(a) Below f_r, I leads V_s.

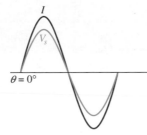

(b) At f_r, I is in phase with V_s.

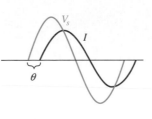

(c) Above f_r, I lags V_s.

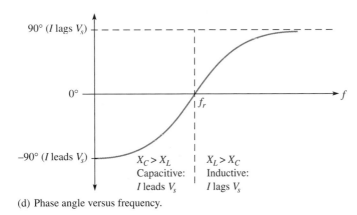

(d) Phase angle versus frequency.

▲ FIGURE 17–18

The phase angle as a function of frequency in a series *RLC* circuit.

SECTION 17–3 REVIEW	1. What is the condition for series resonance? 2. Why is the current maximum at the resonant frequency? 3. Calculate the resonant frequency for $C = 1000$ pF and $L = 1000$ μH. 4. In Question 3, is the circuit inductive or capacitive at 50 kHz?

OPTION 2 NOTE

This completes the coverage of series reactive circuits. Coverage of parallel reactive circuits begins in Chapter 15, Part 2, on page 626.

PARALLEL CIRCUITS

17–4 IMPEDANCE OF PARALLEL *RLC* CIRCUITS

In this section, you will study the impedance and phase angle of a parallel *RLC* circuit. Also, conductance, susceptance, and admittance of a parallel *RLC* circuit are covered.

After completing this section, you should be able to

♦ **Determine the impedance of a parallel *RLC* circuit**

  ♦ Calculate the conductance, susceptance, and admittance

  ♦ Determine whether a circuit is predominately inductive or capacitive

Figure 17–19 shows a parallel *RLC* circuit. The total impedance can be calculated using the reciprocal of the sum-of-reciprocals method, just as was done for circuits with resistors in parallel.

$$\frac{1}{\mathbf{Z}} = \frac{1}{R\angle 0°} + \frac{1}{X_L\angle 90°} + \frac{1}{X_C\angle -90°}$$

or

Equation 17–5

$$\mathbf{Z} = \cfrac{1}{\cfrac{1}{R\angle 0°} + \cfrac{1}{X_L\angle 90°} + \cfrac{1}{X_C\angle -90°}}$$

▶ **FIGURE 17–19**

Parallel *RLC* circuit.

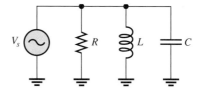

EXAMPLE 17–8

Find **Z** in polar form for the parallel *RLC* circuit in Figure 17–20.

▶ **FIGURE 17–20**

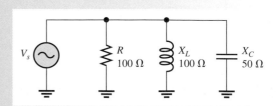

Solution   Use the sum-of-reciprocals formula.

$$\frac{1}{\mathbf{Z}} = \frac{1}{R\angle 0°} + \frac{1}{X_L \angle 90°} + \frac{1}{X_C \angle -90°} = \frac{1}{100\angle 0°\ \Omega} + \frac{1}{100\angle 90°\ \Omega} + \frac{1}{50\angle -90°\ \Omega}$$

Apply the rule for division of polar numbers.

$$\frac{1}{\mathbf{Z}} = 10\angle 0°\ \text{mS} + 10\angle -90°\ \text{mS} + 20\angle 90°\ \text{mS}$$

Recall that the sign of the denominator angle changes when dividing. Next, convert each term to its rectangular equivalent and combine.

$$\frac{1}{\mathbf{Z}} = 10\ \text{mS} - j10\ \text{mS} + j20\ \text{mS} = 10\ \text{mS} + j10\ \text{mS}$$

Take the reciprocal to obtain **Z** and then convert to polar form.

$$\mathbf{Z} = \frac{1}{10\ \text{mS} + j10\ \text{mS}} = \frac{1}{\sqrt{(10\ \text{mS})^2 + (10\ \text{mS})^2}\angle \tan^{-1}\left(\dfrac{10\ \text{mS}}{10\ \text{mS}}\right)}$$

$$= \frac{1}{14.14\angle 45°\ \text{mS}} = \mathbf{70.7\angle -45°\ \Omega}$$

The negative angle shows that the circuit is capacitive. This may surprise you because $X_L > X_C$. However, in a parallel circuit, the smaller quantity has the greater effect on the total current because its current is the greatest. Similar to the case of resistances in parallel, the smaller reactance draws more current and has the greater effect on the total Z. In this circuit, the total current leads the total voltage by a phase angle of 45°.

Related Problem   If the frequency in Figure 17–20 increases, does the impedance increase or decrease?

Conductance, Susceptance, and Admittance

The concepts of conductance (G), capacitive susceptance (B_C), inductive susceptance (B_L) and admittance (Y) were discussed in Chapters 15 and 16. The phasor formulas are restated here.

$$\mathbf{G} = \frac{1}{R\angle 0°} = G\angle 0° \qquad\qquad\qquad \textbf{Equation 17–6}$$

$$\mathbf{B}_C = \frac{1}{X_C \angle -90°} = B_C \angle 90° = jB_C \qquad\qquad \textbf{Equation 17–7}$$

$$\mathbf{B}_L = \frac{1}{X_L \angle 90°} = B_L \angle -90° = -jB_L \qquad\qquad \textbf{Equation 17–8}$$

$$\mathbf{Y} = \frac{1}{Z\angle \pm\theta} = Y\angle \mp\theta = G + jB_C - jB_L \qquad\qquad \textbf{Equation 17–9}$$

As you know, the unit of each of these quantities is the siemens (S).

EXAMPLE 17–9   For the *RLC* circuit in Figure 17–21 determine the conductance, capacitive susceptance, inductive susceptance, and total admittance. Also, determine the impedance.

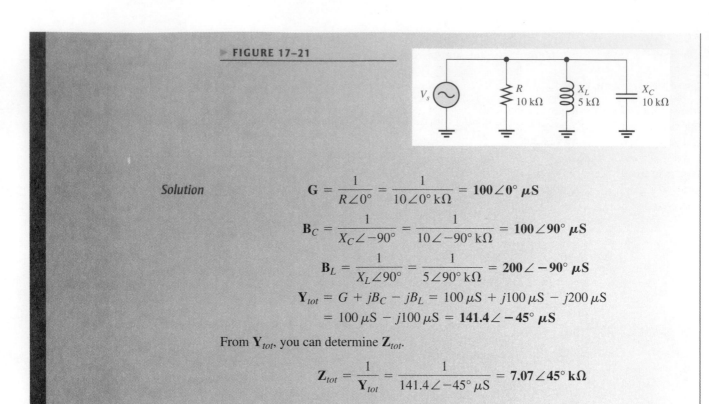

▶ FIGURE 17–21

Solution

$$G = \frac{1}{R\angle 0^\circ} = \frac{1}{10\angle 0^\circ \text{ k}\Omega} = 100\angle 0^\circ \ \mu\text{S}$$

$$B_C = \frac{1}{X_C\angle -90^\circ} = \frac{1}{10\angle -90^\circ \text{ k}\Omega} = 100\angle 90^\circ \ \mu\text{S}$$

$$B_L = \frac{1}{X_L\angle 90^\circ} = \frac{1}{5\angle 90^\circ \text{ k}\Omega} = 200\angle -90^\circ \ \mu\text{S}$$

$$\mathbf{Y}_{tot} = G + jB_C - jB_L = 100 \ \mu\text{S} + j100 \ \mu\text{S} - j200 \ \mu\text{S}$$
$$= 100 \ \mu\text{S} - j100 \ \mu\text{S} = 141.4\angle -45^\circ \ \mu\text{S}$$

From $\mathbf{Y}_{tot}$, you can determine $\mathbf{Z}_{tot}$.

$$\mathbf{Z}_{tot} = \frac{1}{\mathbf{Y}_{tot}} = \frac{1}{141.4\angle -45^\circ \ \mu\text{S}} = 7.07\angle 45^\circ \text{ k}\Omega$$

Related Problem Is the circuit in Figure 17–21 predominately inductive or predominately capacitive?

SECTION 17–4
REVIEW

1. In a certain parallel *RLC* circuit, the capacitive reactance is 60 Ω, and the inductive reactance is 100 Ω. Is the circuit predominantly capacitive or inductive?
2. Determine the admittance of a parallel circuit in which $R = 1.0 \text{ k}\Omega$, $X_C = 500 \ \Omega$, and $X_L = 1.2 \text{ k}\Omega$.
3. In Question 2, what is the impedance?

17–5 ANALYSIS OF PARALLEL *RLC* CIRCUITS

As you have learned, the smaller reactance in a parallel circuit dominates because it results in the larger branch current.

After completing this section, you should be able to

◆ **Analyze parallel *RLC* circuits**

◆ Explain how the currents are related in terms of phase

◆ Calculate impedance, currents, and voltages

Recall that capacitive reactance varies inversely with frequency and that inductive reactance varies directly with frequency. In a parallel *RLC* circuit at low frequencies, the inductive reactance is less than the capacitive reactance; therefore, the circuit is inductive. As the frequency is increased, X_L increases and X_C decreases until a value is reached where

$X_L = X_C$. This is the point of **parallel resonance**. As the frequency is increased further, X_C becomes smaller than X_L, and the circuit becomes capacitive.

Current Relationships

In a parallel *RLC* circuit, the current in the capacitive branch and the current in the inductive branch are *always* 180° out of phase with each other (neglecting any coil resistance). Because I_C and I_L add algebraically, the total current is actually the difference in their magnitudes. Thus, the total current into the parallel branches of *L* and *C* is always less than the largest individual branch current, as illustrated in Figure 17–22 and in the waveform diagram of Figure 17–23. Of course, the current in the resistive branch is always 90° out of phase with both reactive currents, as shown in the current phasor diagram of Figure 17–24.

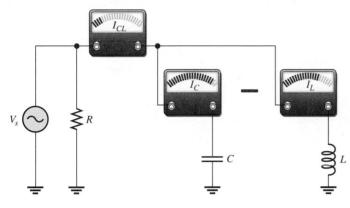

▲ **FIGURE 17–22**

The total current into the parallel combination of *C* and *L* is the difference of the two branch currents.

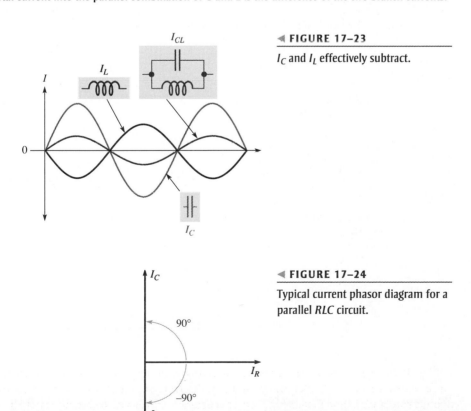

◀ **FIGURE 17–23**

I_C and I_L effectively subtract.

◀ **FIGURE 17–24**

Typical current phasor diagram for a parallel *RLC* circuit.

The total current can be expressed as

Equation 17–10

$$\mathbf{I}_{tot} = \sqrt{I_R^2 + (I_C - I_L)^2} \angle \tan^{-1}\left(\frac{I_{CL}}{I_R}\right)$$

where I_{CL} is $I_C - I_L$, the total current into the L and C branches.

EXAMPLE 17–10

For the circuit in Figure 17–25 find each branch current and the total current. Draw a diagram of their relationship.

▶ **FIGURE 17–25**

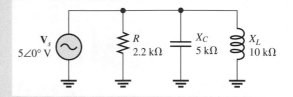

Solution   Use Ohm's law to find each branch current in phasor form.

$$\mathbf{I}_R = \frac{\mathbf{V}_s}{\mathbf{R}} = \frac{5\angle 0° \text{ V}}{2.2\angle 0° \text{ k}\Omega} = \mathbf{2.27\angle 0° \text{ mA}}$$

$$\mathbf{I}_C = \frac{\mathbf{V}_s}{\mathbf{X}_C} = \frac{5\angle 0° \text{ V}}{5\angle -90° \text{ k}\Omega} = \mathbf{1\angle 90° \text{ mA}}$$

$$\mathbf{I}_L = \frac{\mathbf{V}_s}{\mathbf{X}_L} = \frac{5\angle 0° \text{ V}}{10\angle 90° \text{ k}\Omega} = \mathbf{0.5\angle -90° \text{ mA}}$$

The total current is the phasor sum of the branch currents. By Kirchhoff's law,

$$\mathbf{I}_{tot} = \mathbf{I}_R + \mathbf{I}_C + \mathbf{I}_L$$
$$= 2.27\angle 0° \text{ mA} + 1\angle 90° \text{ mA} + 0.5\angle -90° \text{ mA}$$
$$= 2.27 \text{ mA} + j1 \text{ mA} - j0.5 \text{ mA} = 2.27 \text{ mA} + j0.5 \text{ mA}$$

Converting to polar form yields

$$\mathbf{I}_{tot} = \sqrt{I_R^2 + (I_C - I_L)^2} \angle \tan^{-1}\left(\frac{I_{CL}}{I_R}\right)$$

$$= \sqrt{(2.27 \text{ mA})^2 + (0.5 \text{ mA})^2} \angle \tan^{-1}\left(\frac{0.5 \text{ mA}}{2.27 \text{ mA}}\right) = \mathbf{2.32\angle 12.4° \text{ mA}}$$

The total current is 2.32 mA leading V_s by 12.4°. Figure 17–26 is the current phasor diagram for the circuit.

▶ **FIGURE 17–26**

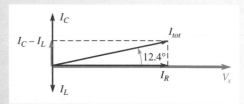

Related Problem   Will total current increase or decrease if the frequency in Figure 17–25 is increased?

17–6 PARALLEL RESONANCE

In this section, we will first look at the resonant condition in an ideal parallel LC circuit (no winding resistance). Then, we will examine the more realistic case where the resistance of the coil is taken into account.

After completing this section, you should be able to

- **Analyze a circuit for parallel resonance**
  - Describe parallel resonance in an ideal circuit
  - Describe parallel resonance in a nonideal circuit
  - Explain how impedance varies with frequency
  - Determine current and phase angle at resonance
  - Determine parallel resonant frequency

Condition for Ideal Parallel Resonance

Ideally, parallel resonance occurs when $X_C = X_L$. The frequency at which resonance occurs is called the *resonant frequency*, just as in the series case. When $X_C = X_L$, the two branch currents, I_C and I_L, are equal in magnitude, and, of course, they are always 180° out of phase with each other. Thus, the two currents cancel and the total current is zero, as shown in Figure 17–27.

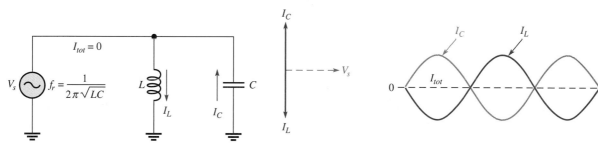

(a) Parallel circuit at resonance ($X_C = X_L$, $Z = \infty$)   (b) Current phasors   (c) Current waveforms

▲ **FIGURE 17–27**

An ideal parallel LC circuit at resonance.

Since the total current is zero, the impedance of the parallel LC circuit is infinitely large (∞). These ideal resonant conditions are stated as follows:

$$X_L = X_C$$
$$Z_r = \infty$$

Parallel Resonant Frequency

For an ideal (no resistance) parallel resonant circuit, the frequency at which resonance occurs is determined by the same formula as in series resonant circuits; that is,

$$f_r = \frac{1}{2\pi\sqrt{LC}}$$

Tank Circuit

The parallel resonant *LC* circuit is often called a **tank circuit**. The term *tank circuit* refers to the fact that the parallel resonant circuit stores energy in the magnetic field of the coil and in the electric field of the capacitor. The stored energy is transferred back and forth between the capacitor and the coil on alternate half-cycles as the current goes first one way and then the other when the inductor deenergizes and the capacitor charges, and vice versa. This concept is illustrated in Figure 17–28.

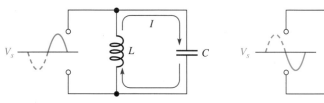

(a) The coil deenergizes as the capacitor charges. (b) The capacitor discharges as the coil energizes.

▲ **FIGURE 17–28**

Energy storage in an ideal parallel resonant tank circuit.

Variation of the Impedance with Frequency

Ideally, the impedance of a parallel resonant circuit is infinite. In practice, the impedance is maximum at the resonant frequency and decreases at lower and higher frequencies, as indicated by the curve in Figure 17–29.

▶ **FIGURE 17–29**

Generalized impedance curve for a parallel resonant circuit. The circuit is inductive below f_r, resistive at f_r, and capacitive above f_r.

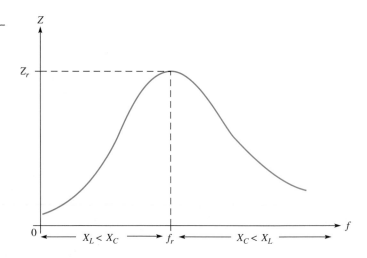

At very low frequencies, X_L is very small and X_C is very high, so the total impedance is essentially equal to that of the inductive branch. As the frequency goes up, the impedance also increases, and the inductive reactance dominates (because it is less than X_C) until the resonant frequency is reached. At this point, of course, $X_L \cong X_C$ (for $Q > 10$) and the impedance is at its maximum. As the frequency goes above resonance, the capacitive reactance dominates (because it is less than X_L) and the impedance decreases.

Current and Phase Angle at Resonance

In the ideal tank circuit, the total current from the source at resonance is zero because the impedance is infinite. In the nonideal case when the winding resistance is considered, there is some total current at the resonant frequency, and it is determined by the impedance at resonance.

$$I_{tot} = \frac{V_s}{Z_r}$$

Equation 17–11

The phase angle of the parallel resonant circuit is 0° because the impedance is purely resistive at the resonant frequency.

Effect of Winding Resistance on the Parallel Resonant Frequency

When the winding resistance is considered, the resonant condition can be expressed as

$$2\pi f_r L\left(\frac{Q^2 + 1}{Q^2}\right) = \frac{1}{2\pi f_r C}$$

where Q is the **quality factor** of the coil, X_L/R_W. Solving for f_r in terms of Q yields

$$f_r = \frac{1}{2\pi\sqrt{LC}}\sqrt{\frac{Q^2}{Q^2 + 1}}$$

Equation 17–12

When $Q \geq 10$, the term with the Q factors is approximately 1.

$$\sqrt{\frac{Q^2}{Q^2 + 1}} = \sqrt{\frac{100}{101}} = 0.995 \cong 1$$

Therefore, the parallel resonant frequency is approximately the same as the series resonant frequency as long as Q is equal to or greater than 10.

$$f_r \cong \frac{1}{2\pi\sqrt{LC}} \qquad \text{for } Q \geq 10$$

A precise expression for f_r in terms of the circuit component values is

$$f_r = \frac{\sqrt{1 - (R_W^2 C/L)}}{2\pi\sqrt{LC}}$$

Equation 17–13

This precise formula is seldom necessary and the simpler equation $f_r = 1/(2\pi\sqrt{LC})$ is sufficient for most practical situations. A derivation of Equation 17–13 is given in Appendix B.

EXAMPLE 17–11   Find the precise frequency and the value of Q at resonance for the circuit in Figure 17–30.

▶ **FIGURE 17–30**

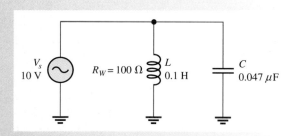

Solution Use Equation 17–13 to find the frequency.

$$f_r = \frac{\sqrt{1 - (R_W^2 C/L)}}{2\pi\sqrt{LC}} = \frac{\sqrt{1 - [(100\ \Omega)^2(0.047\ \mu F)/0.1\ H]}}{2\pi\sqrt{(0.047\ \mu F)(0.1\ H)}} = \mathbf{2.32\ kHz}$$

To calculate the quality factor, Q, first find X_L.

$$X_L = 2\pi f_r L = 2\pi(2.32\ kHz)(0.1\ H) = 1.46\ k\Omega$$

$$Q = \frac{X_L}{R_W} = \frac{1.46\ k\Omega}{100\ \Omega} = \mathbf{14.6}$$

Note that since $Q > 10$, the approximate formula, $f_r \cong 1/(2\pi\sqrt{LC})$, can be used.

Related Problem For a smaller R_W, will f_r be less than or greater than 2.32 kHz?

Use Multisim file E17-11 to verify the calculated results in this example and to confirm your answer for the related problem.

SECTION 17–6 REVIEW

1. Is the impedance minimum or maximum at parallel resonance?
2. Is the current minimum or maximum at parallel resonance?
3. For ideal parallel resonance, assume $X_L = 1500\ \Omega$. What is X_C?
4. A parallel tank circuit has the following values: $R_W = 4\ \Omega$, $L = 50\ mH$, and $C = 10\ pF$. Calculate f_r.
5. If $Q = 25$, $L = 50\ mH$, and $C = 1000\ pF$, what is f_r?
6. In Question 5, if $Q = 2.5$, what is f_r?

OPTION 2 NOTE

This completes the coverage of parallel reactive circuits. Coverage of series-parallel reactive circuits begins in Chapter 15, Part 3, on page 635.

SERIES-PARALLEL CIRCUITS

17–7 ANALYSIS OF SERIES-PARALLEL *RLC* CIRCUITS

In this section, series and parallel combinations of *R*, *L*, and *C* components are analyzed in specific examples. Also, conversion of a series-parallel circuit to an equivalent parallel circuit is covered and resonance in a nonideal parallel circuit is considered.

After completing this section, you should be able to

- ◆ **Analyze series-parallel *RLC* circuits**

  - ◆ Determine currents and voltages

  - ◆ Convert a series-parallel circuit to an equivalent parallel form

  - ◆ Analyze nonideal (with coil resistance) parallel circuits for parallel resonance

  - ◆ Examine the effect of a resistive load on a tank circuit

The following two examples illustrate an approach to the analysis of circuits with both series and parallel combinations of resistance, inductance, and capacitance.

EXAMPLE 17–12   In Figure 17–31, find the voltage across the capacitor in polar form. Is this circuit predominantly inductive or capacitive?

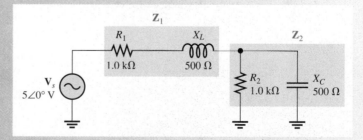

▲ FIGURE 17–31

Solution   Use the voltage-divider formula in this analysis. The impedance of the series combination of R_1 and X_L is called $\mathbf{Z}_1$. In rectangular form,

$$\mathbf{Z}_1 = R_1 + jX_L = 1000\ \Omega + j500\ \Omega$$

Converting to polar form yields

$$\mathbf{Z}_1 = \sqrt{R_1^2 + X_L^2}\angle\tan^{-1}\!\left(\frac{X_L}{R_1}\right)$$

$$= \sqrt{(1000\ \Omega)^2 + (500\ \Omega)^2}\angle\tan^{-1}\!\left(\frac{500\ \Omega}{1000\ \Omega}\right) = 1118\angle 26.6°\ \Omega$$

The impedance of the parallel combination of R_2 and X_C is called $\mathbf{Z}_2$. In polar form,

$$\mathbf{Z}_2 = \left(\frac{R_2 X_C}{\sqrt{R_2^2 + X_C^2}}\right)\angle -\tan^{-1}\!\left(\frac{R_2}{X_C}\right)$$

$$= \left[\frac{(1000\ \Omega)(500\ \Omega)}{\sqrt{(1000\ \Omega)^2 + (500\ \Omega)^2}}\right]\angle -\tan^{-1}\!\left(\frac{1000\ \Omega}{500\ \Omega}\right) = 447\angle -63.4°\ \Omega$$

Converting to rectangular form yields

$$\mathbf{Z}_2 = Z_2\cos\theta + jZ_2\sin\theta$$

$$= (447\ \Omega)\cos(-63.4°) + j447\sin(-63.4°) = 200\ \Omega - j400\ \Omega$$

The total impedance $\mathbf{Z}_{tot}$ in rectangular form is

$$\mathbf{Z}_{tot} = \mathbf{Z}_1 + \mathbf{Z}_2 = (1000\ \Omega + j500\ \Omega) + (200\ \Omega - j400\ \Omega) = 1200\ \Omega + j100\ \Omega$$

Converting to polar form yields

$$\mathbf{Z}_{tot} = \sqrt{(1200\ \Omega)^2 + (100\ \Omega)^2}\angle\tan^{-1}\!\left(\frac{100\ \Omega}{1200\ \Omega}\right) = 1204\angle 4.76°\ \Omega$$

Now apply the voltage-divider formula to get $\mathbf{V}_C$.

$$\mathbf{V}_C = \left(\frac{\mathbf{Z}_2}{\mathbf{Z}_{tot}}\right)\mathbf{V}_s = \left(\frac{447\angle -63.4°\ \Omega}{1204\angle 4.76°\ \Omega}\right)5\angle 0°\ \text{V} = \mathbf{1.86\angle -68.2°\ V}$$

Therefore, V_C is 1.86 V and lags V_s by 68.2°.

The $+j$ term in $\mathbf{Z}_{tot}$, or the positive angle in its polar form, indicates that the circuit is more inductive than capacitive. However, it is just slightly more inductive because the angle is small. This result may surprise you, because $X_C = X_L = 500\ \Omega$. However, the capacitor is in parallel with a resistor, so the capacitor actually has less effect on the total impedance than does the inductor. Figure 17–32 shows the phasor relationship of V_C and V_s. Although $X_C = X_L$, this circuit is not at resonance because the j term of the total impedance is not zero due to the parallel combination of R_2 and X_C. You can see this by noting that the phase angle associated with $\mathbf{Z}_{tot}$ is 4.76° and not zero.

▶ **FIGURE 17–32**

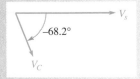

Related Problem   Determine the voltage across the capacitor in polar form if R_1 is increased to 2.2 kΩ.

EXAMPLE 17–13

For the reactive circuit in Figure 17–33, find the voltage at point *B* with respect to ground.

FIGURE 17–33

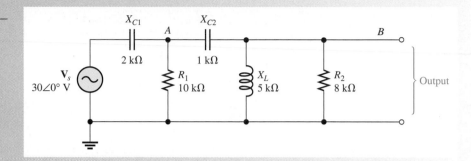

Solution   The voltage ($\mathbf{V}_B$) at point *B* is the voltage across the open output terminals. Use the voltage-divider approach. To do so, you must know the voltage ($\mathbf{V}_A$) at point *A* first; so you need to find the impedance from point *A* to ground as a starting point.

The parallel combination of X_L and R_2 is in series with X_{C2}. This combination is in parallel with R_1. Call this impedance from point *A* to ground, $\mathbf{Z}_A$. To find $\mathbf{Z}_A$, take the following steps. The impedance of the parallel combination of R_2 and X_L is called $\mathbf{Z}_1$.

$$\mathbf{Z}_1 = \left(\frac{R_2 X_L}{\sqrt{R_2^2 + X_L^2}}\right)\angle\tan^{-1}\left(\frac{R_2}{X_L}\right)$$

$$= \left(\frac{(8\,\text{k}\Omega)(5\,\text{k}\Omega)}{\sqrt{(8\,\text{k}\Omega)^2 + (5\,\text{k}\Omega)^2}}\right)\angle\tan^{-1}\left(\frac{8\,\text{k}\Omega}{5\,\text{k}\Omega}\right) = 4.24\angle 58.0°\,\text{k}\Omega$$

Next, combine $\mathbf{Z}_1$ in series with $\mathbf{X}_{C2}$ to get an impedance $\mathbf{Z}_2$.

$$\mathbf{Z}_2 = \mathbf{X}_{C2} + \mathbf{Z}_1$$
$$= 1\angle -90°\,\text{k}\Omega + 4.24\angle 58°\,\text{k}\Omega = -j1\,\text{k}\Omega + 2.25\,\text{k}\Omega + j3.6\,\text{k}\Omega$$
$$= 2.25\,\text{k}\Omega + j2.6\,\text{k}\Omega$$

Converting to polar form yields

$$\mathbf{Z}_2 = \sqrt{(2.25\,\text{k}\Omega)^2 + (2.6\,\text{k}\Omega)^2}\angle\tan^{-1}\left(\frac{2.6\,\text{k}\Omega}{2.25\,\text{k}\Omega}\right) = 3.44\angle 49.1°\,\text{k}\Omega$$

Finally, combine $\mathbf{Z}_2$ and $\mathbf{R}_1$ in parallel to get $\mathbf{Z}_A$.

$$\mathbf{Z}_A = \frac{\mathbf{R}_1\mathbf{Z}_2}{\mathbf{R}_1 + \mathbf{Z}_2} = \frac{(10\angle 0°\,\text{k}\Omega)(3.44\angle 49.1°\,\text{k}\Omega)}{10\,\text{k}\Omega + 2.25\,\text{k}\Omega + j2.6\,\text{k}\Omega}$$

$$= \frac{34.4\angle 49.1°\,\text{k}\Omega}{12.25\,\text{k}\Omega + j2.6\,\text{k}\Omega} = \frac{34.4\angle 49.1°\,\text{k}\Omega}{12.5\angle 12.0°\,\text{k}\Omega} = 2.75\angle 37.1°\,\text{k}\Omega$$

The simplified circuit is shown in Figure 17–34.

FIGURE 17–34

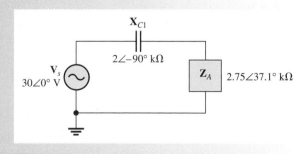

752 ◆ RLC CIRCUITS AND RESONANCE

Next, use the voltage-divider principle to find the voltage ($\mathbf{V}_A$) at point A in Figure 17–33. The total impedance is

$$\mathbf{Z}_{tot} = \mathbf{X}_{C1} + \mathbf{Z}_A$$
$$= 2\angle{-90°}\,\text{k}\Omega + 2.75\angle{37.1°}\,\text{k}\Omega = -j2\,\text{k}\Omega + 2.19\,\text{k}\Omega + j1.66\,\text{k}\Omega$$
$$= 2.19\,\text{k}\Omega - j0.340\,\text{k}\Omega$$

Converting to polar form yields

$$\mathbf{Z}_{tot} = \sqrt{(2.19\,\text{k}\Omega)^2 + (0.340\,\text{k}\Omega)^2}\angle{-\tan^{-1}\left(\frac{0.340\,\text{k}\Omega}{2.19\,\text{k}\Omega}\right)} = 2.22\angle{-8.82°}\,\text{k}\Omega$$

The voltage at point A is

$$\mathbf{V}_A = \left(\frac{\mathbf{Z}_A}{\mathbf{Z}_{tot}}\right)\mathbf{V}_s = \left(\frac{2.75\angle{37.1°}\,\text{k}\Omega}{2.22\angle{-8.82°}\,\text{k}\Omega}\right)30\angle{0°}\,\text{V} = 37.2\angle{45.9°}\,\text{V}$$

Next, find the voltage ($\mathbf{V}_B$) at point B by dividing $\mathbf{V}_A$ down, as indicated in Figure 17–35. $\mathbf{V}_B$ is the open terminal output voltage.

$$\mathbf{V}_B = \left(\frac{\mathbf{Z}_1}{\mathbf{Z}_2}\right)\mathbf{V}_A = \left(\frac{4.24\angle{58°}\,\text{k}\Omega}{3.44\angle{49.1°}\,\text{k}\Omega}\right)37.2\angle{45.9°}\,\text{V} = \mathbf{45.9\angle{54.8°}\,V}$$

Surprisingly, V_A is greater than V_s, and V_B is greater than V_A! This result is possible because of the out-of-phase relationship of the reactive voltages. Remember that X_C and X_L tend to cancel each other.

▶ FIGURE 17–35

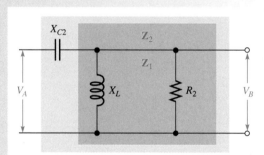

Related Problem   What is the voltage in polar form across C_1 in Figure 17–33?

Conversion of Series-Parallel to Parallel

The particular series-parallel configuration shown in Figure 17–36 is important because it represents a circuit having parallel L and C branches, with the winding resistance of the coil taken into account as a series resistance in the L branch.

▶ FIGURE 17–36

A series-parallel *RLC* circuit ($Q = X_L/R_W$).

It is helpful to view the series-parallel circuit in Figure 17–36 in an equivalent parallel form, as indicated in Figure 17–37.

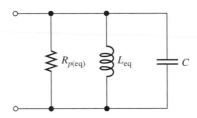

◀ **FIGURE 17–37**

Parallel equivalent form of the circuit in Figure 17–36.

The equivalent inductance, L_{eq}, and the equivalent parallel resistance, $R_{p(eq)}$, are given by the following formulas:

$$L_{eq} = L\left(\frac{Q^2 + 1}{Q^2}\right)$$

Equation 17–14

$$R_{p(eq)} = R_W(Q^2 + 1)$$

Equation 17–15

where Q is the quality factor of the coil, X_L/R_W. Derivations of these formulas are quite involved and thus are not given here. Notice in the equations that for a $Q \geq 10$, the value of L_{eq} is approximately the same as the original value of L. For example, if $L = 10$ mH and $Q = 10$, then

$$L_{eq} = 10\,\text{mH}\left(\frac{10^2 + 1}{10^2}\right) = 10\,\text{mH}(1.01) = 10.1\,\text{mH}$$

The equivalency of the two circuits means that at a given frequency, when the same value of voltage is applied to both circuits, the same total current is in both circuits and the phase angles are the same. Basically, an equivalent circuit simply makes circuit analysis more convenient.

EXAMPLE 17–14

Convert the series-parallel circuit in Figure 17–38 to an equivalent parallel form at the given frequency.

FIGURE 17–38

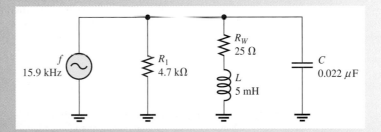

Solution Determine the inductive reactance.

$$X_L = 2\pi f L = 2\pi(15.9\,\text{kHz})(5\,\text{mH}) = 500\,\Omega$$

The Q of the coil is

$$Q = \frac{X_L}{R_W} = \frac{500\,\Omega}{25\,\Omega} = 20$$

Since $Q > 10$, then $L_{eq} \cong L = 5$ mH.

The equivalent parallel resistance is

$$R_{p(eq)} = R_W(Q^2 + 1) = (25\ \Omega)(20^2 + 1) = 10\ \text{k}\Omega$$

This equivalent resistance appears in parallel with R_1 as shown in Figure 17–39(a). When combined, they give a total parallel resistance ($R_{p(tot)}$) of 3.2 kΩ, as indicated in Figure 17–39(b).

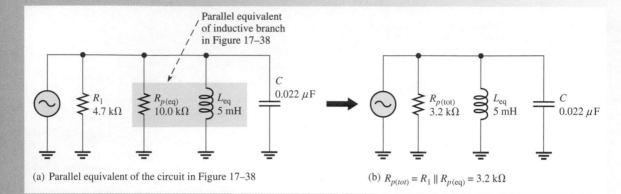

(a) Parallel equivalent of the circuit in Figure 17–38   (b) $R_{p(tot)} = R_1 \parallel R_{p(eq)} = 3.2\ \text{k}\Omega$

▲ **FIGURE 17–39**

Related Problem   Find the equivalent parallel circuit if $R_W = 10\ \Omega$ in Figure 17–38.

Parallel Resonant Conditions in a Nonideal Circuit

The resonance of an ideal parallel *LC* circuit was examined in Section 17–6. Now, let's consider resonance in a tank circuit with the resistance of the coil taken into account. Figure 17–40 shows a nonideal tank circuit and its parallel *RLC* equivalent.

Recall that the quality factor, *Q*, of the circuit at resonance is simply the *Q* of the coil.

$$Q = \frac{X_L}{R_W}$$

The expressions for the equivalent inductance and the equivalent parallel resistance were given in Equations 17–14 and 17–15 as

$$L_{eq} = L\left(\frac{Q^2 + 1}{Q^2}\right)$$

$$R_{p(eq)} = R_W(Q^2 + 1)$$

For $Q \geq 10, L_{eq} \cong L.$

▶ **FIGURE 17–40**

A practical treatment of parallel resonant circuits must include the coil resistance.

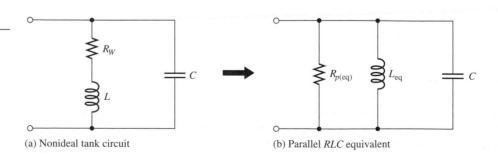

(a) Nonideal tank circuit   (b) Parallel *RLC* equivalent

At parallel resonance,

$$X_{L(eq)} = X_C$$

In the parallel equivalent circuit, $R_{p(eq)}$ is in parallel with an ideal coil and a capacitor, so the *L* and *C* branches act as an ideal tank circuit which has an infinite impedance at resonance as shown in Figure 17–41. Therefore, the total impedance of the nonideal tank circuit at resonance can be expressed as simply the equivalent parallel resistance.

$$Z_r = R_W(Q^2 + 1)$$

Equation 17–16

A derivation of Equation 17–16 is given in Appendix B.

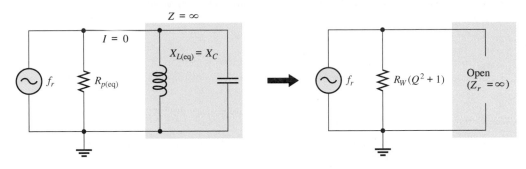

▲ **FIGURE 17–41**

At resonance, the parallel *LC* portion appears open and the source sees only $R_{p(eq)}$.

EXAMPLE 17–15

Determine the impedance of the circuit in Figure 17–42 at the resonant frequency ($f_r \cong 17{,}794$ Hz).

▶ **FIGURE 17–42**

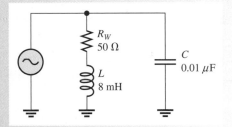

Solution   Before you can calculate the impedance using Equation 17–16, you must find the quality factor. To get *Q*, first find the inductive reactance.

$$X_L = 2\pi f_r L = 2\pi(17{,}794 \text{ Hz})(8 \text{ mH}) = 894 \ \Omega$$

$$Q = \frac{X_L}{R_W} = \frac{894 \ \Omega}{50 \ \Omega} = 17.9$$

$$Z_r = R_W(Q^2 + 1) = 50 \ \Omega(17.9^2 + 1) = \mathbf{16.1 \ k\Omega}$$

Related Problem   Determine Z_r for $R_W = 10 \ \Omega$.

An External Load Resistance Affects a Tank Circuit

In most practical situations, an external load resistance appears in parallel with a nonideal tank circuit, as shown in Figure 17–43(a). Obviously, the external resistor (R_L) will dissipate more of the energy delivered by the source and thus will lower the overall *Q* of the

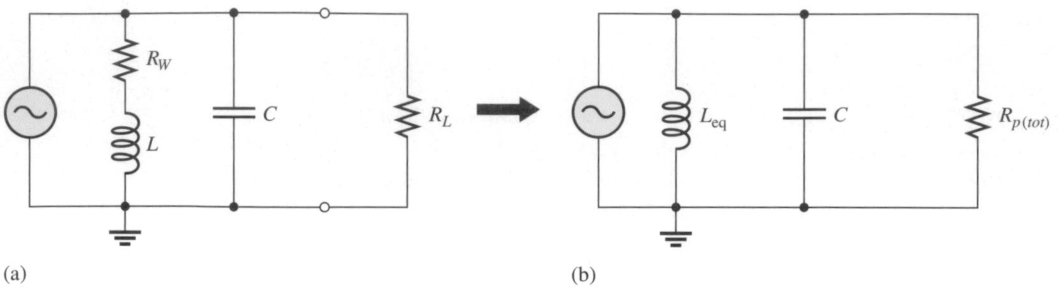

(a)   (b)

▲ FIGURE 17–43

Tank circuit with a parallel load resistor and its equivalent circuit.

circuit. The external resistor effectively appears in parallel with the equivalent parallel resistance of the coil, $R_{p(eq)}$, and both are combined to determine a total parallel resistance, $R_{p(tot)}$, as indicated in Figure 17–43(b).

$$R_{p(tot)} = R_L \parallel R_{p(eq)}$$

The overall Q, designated Q_O, for a parallel RLC circuit is expressed differently from the Q of a series circuit.

Equation 17–17

$$Q_O = \frac{R_{p(tot)}}{X_{L(eq)}}$$

As you can see, the effect of loading the tank circuit is to reduce its overall Q (which is equal to the coil Q when unloaded).

SECTION 17–7 REVIEW

1. A certain resonant circuit has a 100 μH inductor with a 2 Ω winding resistance in parallel with a 0.22 μF capacitor. If $Q = 8$, determine the parallel equivalent of this circuit.

2. Find the equivalent parallel inductance and resistance for a 20 mH coil with a winding resistance of 10 Ω at a frequency of 1 kHz.

OPTION 2 NOTE

This completes the coverage of series-parallel circuits. Coverage of special topics begins in Chapter 15, Part 4, on page 642.

17–8 BANDWIDTH OF RESONANT CIRCUITS

The current in a series *RLC* is maximum at the resonant frequency because the reactances cancel. The current in a parallel *RLC* is minimum at the resonant frequency because the inductive and capacitive currents cancel. This circuit behavior relates to a characteristic called bandwidth.

After completing this section, you should be able to

♦ **Determine the bandwidth of resonant circuits**

   ♦ Discuss the bandwidth of series and parallel resonant circuits

   ♦ State the formula for bandwidth

   ♦ Define *half-power frequency*

   ♦ Define *selectivity*

   ♦ Explain how the *Q* affects the bandwidth

Series Resonant Circuits

The current in a series *RLC* circuit is maximum at the resonant frequency (also known as *center frequency*) and drops off on either side of this frequency. Bandwidth, sometimes abbreviated *BW*, is an important characteristic of a resonant circuit. The bandwidth is the range of frequencies for which the current is equal to or greater than 70.7% of its resonant value.

Figure 17–44 illustrates bandwidth on the response curve of a series *RLC* circuit. Notice that the frequency f_1 below f_r is the point at which the current is $0.707I_{max}$ and is commonly called the *lower critical frequency*. The frequency f_2 above f_r, where the current is again $0.707I_{max}$, is the *upper critical frequency*. Other names for f_1 and f_2 are *−3 dB frequencies*, *cutoff frequencies*, and *half-power frequencies*. The significance of the latter term is discussed later in the chapter.

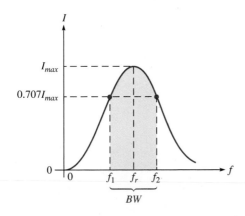

◀ **FIGURE 17–44**

Bandwidth on series resonant response curve for *I*.

EXAMPLE 17–16

A certain series resonant circuit has a maximum current of 100 mA at the resonant frequency. What is the value of the current at the critical frequencies?

Solution Current at the critical frequencies is 70.7% of maximum.

$$I_{f1} = I_{f2} = 0.707I_{max} = 0.707(100\,\text{mA}) = \textbf{70.7 mA}$$

Related Problem A certain series resonant circuit has a current of 25 mA at the critical frequencies. What is the current at resonance?

Parallel Resonant Circuits

For a parallel resonant circuit, the impedance is maximum at the resonant frequency; so the total current is minimum. The bandwidth can be defined in relation to the impedance curve in the same manner that the current curve was used in the series circuit. Of course, f_r is the frequency at which Z is maximum; f_1 is the lower critical frequency at which $Z = 0.707Z_{max}$; and f_2 is the upper critical frequency at which again $Z = 0.707Z_{max}$. The bandwidth is the range of frequencies between f_1 and f_2, as shown in Figure 17–45.

▶ **FIGURE 17–45**

Bandwidth of the parallel resonant response curve for Z_{tot}.

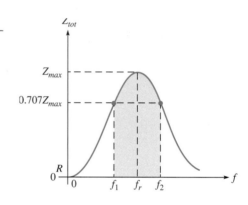

Formula for Bandwidth

The bandwidth for either series or parallel resonant circuits is the range of frequencies between the critical frequencies for which the response curve (I or Z) is 0.707 of the maximum value. Thus, the bandwidth is actually the difference between f_2 and f_1.

Equation 17–18

$$BW = f_2 - f_1$$

Ideally, f_r is the center frequency and can be calculated as follows:

Equation 17–19

$$f_r = \frac{f_1 + f_2}{2}$$

EXAMPLE 17–17

A resonant circuit has a lower critical frequency of 8 kHz and an upper critical frequency of 12 kHz. Determine the bandwidth and center (resonant) frequency.

Solution

$$BW = f_2 - f_1 = 12\,\text{kHz} - 8\,\text{kHz} = \textbf{4 kHz}$$

$$f_r = \frac{f_1 + f_2}{2} = \frac{12\,\text{kHz} + 8\,\text{kHz}}{2} = \textbf{10 kHz}$$

Related Problem If the bandwidth of a resonant circuit is 2.5 kHz and its center frequency is 8 kHz, what are the lower and upper critical frequencies?

Half-Power Frequencies

As previously mentioned, the upper and lower critical frequencies are sometimes called the **half-power frequencies**. This term is derived from the fact that the power from the source at these frequencies is one-half the power delivered at the resonant frequency. The following shows that this is true for a series circuit. The same end result also applies to a parallel circuit. At resonance,

$$P_{max} = I_{max}^2 R$$

The power at f_1 or f_2 is

$$P_{f1} = I_{f1}^2 R = (0.707 I_{max})^2 R = (0.707)^2 I_{max}^2 R = 0.5 I_{max}^2 R = 0.5 P_{max}$$

Selectivity

The response curves in Figures 17–44 and 17–45 are also called *selectivity curves*. **Selectivity** defines how well a resonant circuit responds to a certain frequency and discriminates against all others. *The narrower the bandwidth, the greater the selectivity.*

We normally assume that a resonant circuit accepts frequencies within its bandwidth and completely eliminates frequencies outside the bandwidth. Such is not actually the case, however, because signals with frequencies outside the bandwidth are not completely eliminated. Their magnitudes, however, are greatly reduced. The further the frequencies are from the critical frequencies, the greater is the reduction, as illustrated in Figure 17–46(a). An ideal selectivity curve is shown in Figure 17–46(b).

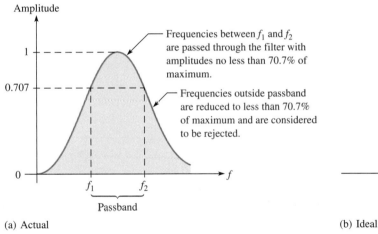

(a) Actual

(b) Ideal

▲ FIGURE 17–46

Generalized selectivity curve.

As you can see in Figure 17–46, another factor that influences selectivity is the sharpness of the slopes of the curve. The faster the curve drops off at the critical frequencies, the more selective the circuit is because it responds only to the frequencies within the bandwidth. Figure 17–47 shows a general comparison of three response curves with varying degrees of selectivity.

Q Affects Bandwidth

A higher value of circuit Q results in a narrower bandwidth. A lower value of Q causes a wider bandwidth. A formula for the bandwidth of a resonant circuit in terms of Q is stated in the following equation:

$$BW = \frac{f_r}{Q}$$

Equation 17–20

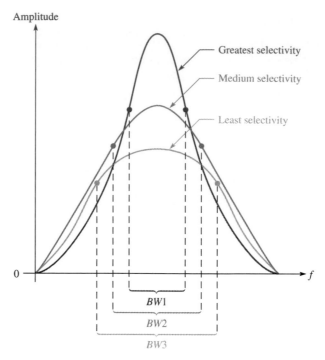

▲ FIGURE 17–47

Comparative selectivity curves.

EXAMPLE 17–18

What is the bandwidth of each circuit in Figure 17–48?

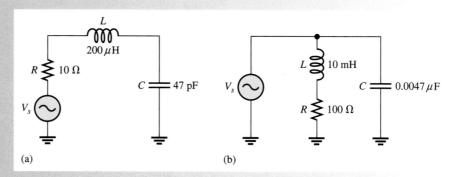

▲ FIGURE 17–48

Solution For the circuit in Figure 17–48(a), determine the bandwidth as follows:

$$f_r = \frac{1}{2\pi\sqrt{LC}} = \frac{1}{2\pi\sqrt{(200~\mu H)(47pF)}} = 1.64~\text{MHz}$$

$$X_L = 2\pi f_r L = 2\pi(1.64~\text{MHz})(200~\mu H) = 2.06~\text{k}\Omega$$

$$Q = \frac{X_L}{R} = \frac{2.06~\text{k}\Omega}{10~\Omega} = 206$$

$$BW = \frac{f_r}{Q} = \frac{1.64~\text{MHz}}{206} = \mathbf{7.96~kHz}$$

For the circuit in Figure 17–48(b),

$$f_r = \frac{\sqrt{1 - (R_W^2 \, C/L)}}{2\pi\sqrt{LC}} \cong \frac{1}{2\pi\sqrt{LC}} = \frac{1}{2\pi\sqrt{(10\text{ mH})(0.0047\ \mu F)}} = 23.2\text{ kHz}$$

$$X_L = 2\pi f_r L = 2\pi(23.2\text{ kHz})(10\text{ mH}) = 1.46\text{ k}\Omega$$

$$Q = \frac{X_L}{R} = \frac{1.46\text{ k}\Omega}{100\ \Omega} = 14.6$$

$$BW = \frac{f_r}{Q} = \frac{23.2\text{ kHz}}{14.6} = \textbf{1.59 kHz}$$

Related Problem   Change *C* in Figure 17–48(a) to 1000 pF and determine the bandwidth.

Use Multisim files E17-18A and E17-18B to verify the calculated results in this example and to confirm your calculation for the related problem.

**SECTION 17–8
REVIEW**

1. What is the bandwidth when $f_2 = 2.2$ MHz and $f_1 = 1.8$ MHz?
2. For a resonant circuit with the critical frequencies in Question 1, what is the center frequency?
3. The power dissipated at resonance is 1.8 W. What is the power at the upper critical frequency?
4. Does a larger *Q* mean a narrower or a wider bandwidth?

17–9 APPLICATIONS

Resonant circuits are used in a wide variety of applications, particularly in communication systems. In this section, we will look briefly at a few common communication systems applications to illustrate the importance of resonant circuits in electronic communication.

After completing this section, you should be able to

◆ **Discuss some applications of resonant circuits**

  ◆ Describe a tuned amplifier application

  ◆ Describe antenna coupling

  ◆ Describe tuned amplifiers

  ◆ Describe signal separation in a receiver

  ◆ Describe a radio receiver

Tuned Amplifiers

A *tuned amplifier* is a circuit that amplifies signals within a specified band. Typically, a parallel resonant circuit is used in conjunction with an amplifier to achieve the selectivity. In terms of the general operation, input signals with frequencies that range over a wide band are accepted on the amplifier's input and are amplified. The resonant circuit allows only a relatively narrow band of those frequencies to be passed on. The variable capacitor allows tuning over the range of input frequencies so that a desired frequency can be selected, as indicated in Figure 17–49.

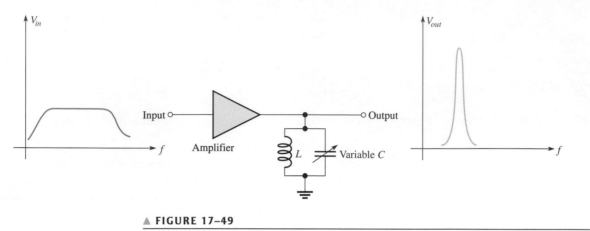

▲ **FIGURE 17–49**

A basic tuned band-pass amplifier.

Antenna Input to a Receiver

Radio signals are sent out from a transmitter via electromagnetic waves that propagate through the atmosphere. When the electromagnetic waves cut across the receiving antenna, small voltages are induced. Out of all the wide range of electromagnetic frequencies, only one frequency or a limited band of frequencies must be extracted. Figure 17–50 shows a typical arrangement of an antenna coupled to the receiver input by a transformer. A variable capacitor is connected across the transformer secondary to form a parallel resonant circuit.

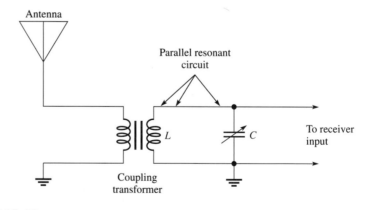

▲ **FIGURE 17–50**

Resonant coupling from an antenna.

Double-Tuned Transformer Coupling in a Receiver

In some types of communication receivers, tuned amplifiers are transformer-coupled together to increase the amplification. Capacitors can be placed in parallel with the primary and secondary windings of the transformer, effectively creating two parallel resonant band-pass filters that are coupled together. This technique, illustrated in Figure 17–51, can result in a wider bandwidth and steeper slopes on the response curve, thus increasing the selectivity for a desired band of frequencies.

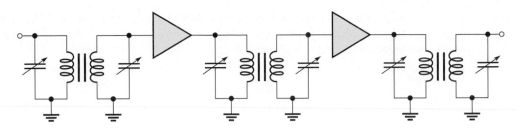

▲ FIGURE 17–51

Double-tuned amplifiers.

Signal Reception and Separation in a TV Receiver

A television receiver must handle both video (picture) signals and audio (sound) signals. Each TV transmitting station is allotted a 6 MHz bandwidth. Channel 2 is allotted a band from 54 MHz through 59 MHz, channel 3 is allotted a band from 60 MHz through 65 MHz, on up to channel 13 which has a band from 210 MHz through 215 MHz. You can tune the front end of the TV receiver to select any one of these channels by using tuned amplifiers. The signal output of the front end of the receiver has a bandwidth from 41 MHz through 46 MHz, regardless of the channel that is tuned in. This band, called the *intermediate frequency* (IF) band, contains both video and audio. Amplifiers tuned to the IF band boost the signal and feed it to the video amplifier.

Before the output of the video amplifier is applied to the picture tube, the audio signal is removed by a 4.5 MHz band-stop resonant circuit (called a *wave trap* filter), as shown in Figure 17–52. This trap keeps the sound signal from interfering with the picture. The video amplifier output is also applied to band-pass circuits that are tuned to the sound carrier frequency of 4.5 MHz. The sound signal is then processed and applied to the speaker as indicated in Figure 17–52.

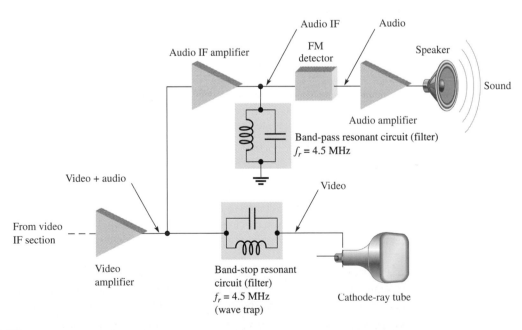

▲ FIGURE 17–52

A simplified portion of a TV receiver showing filter usage.

Superheterodyne Receiver

Another good example of resonant circuit (filter) applications is in the common AM (amplitude modulation) receiver. The AM broadcast band ranges from 535 kHz to 1605 kHz.

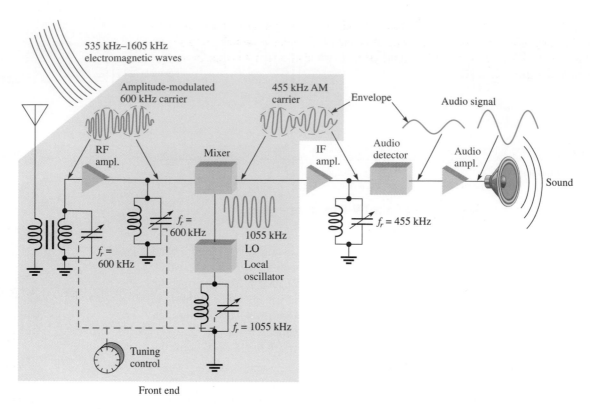

535 kHz–1605 kHz
electromagnetic waves

Amplitude-modulated
600 kHz carrier

455 kHz AM
carrier

Envelope

Audio signal

RF
ampl.

Mixer

IF
ampl.

Audio
detector

Audio
ampl.

Sound

$f_r =$
600 kHz

$f_r =$
600 kHz

1055 kHz
LO

Local
oscillator

$f_r = 455$ kHz

$f_r = 1055$ kHz

Tuning
control

Front end

▲ **FIGURE 17–53**

A simplified diagram of a superheterodyne AM radio broadcast receiver showing an example of the application of tuned resonant circuits.

Each AM station is assigned a 10 kHz bandwidth within that range. A simplified block diagram of a superheterodyne AM receiver is shown in Figure 17–53.

The tuned circuits are designed to pass only the signals from the desired radio station, rejecting all others. To reject stations outside the one that is tuned, the tuned circuits must be selective, passing on only the signals in the 10 kHz band and rejecting all others. Too much selectivity is not desirable either however. If the bandwidth is too narrow, some of the higher frequency modulated signals will be rejected, resulting in a loss of fidelity. Ideally, the resonant circuit must reject signals that are not in the desired passband.

In this system, there are basically three parallel resonant circuits in the front end of the receiver. Each of these resonant circuits is gang-tuned by capacitors; that is, the capacitors are mechanically or electronically linked together so that they change together as the tuning knob is turned. The front end is tuned to receive a desired station, for example, one that transmits at 600 kHz. The input resonant circuit from the antenna and the RF (radio frequency) amplifier resonant circuit select only a frequency of 600 kHz out of all the frequencies crossing the antenna.

The actual audio (sound) signal is carried by the 600 kHz carrier frequency by modulating the amplitude of the carrier so that it follows the audio signal as indicated. The variation in the amplitude of the carrier corresponding to the audio signal is called the *envelope*. The 600 kHz is then applied to a circuit called the *mixer*.

The *local oscillator* (LO) is tuned to a frequency that is 455 kHz above the selected frequency (1055 kHz, in this case). By a process called *heterodyning* or *beating,* the AM signal and the local oscillator signal are mixed together, and the 600 kHz AM signal is converted by the mixer to a 455 kHz AM signal (1055 kHz − 600 kHz = 455 kHz).

The 455 kHz is the intermediate frequency (IF) for standard AM receivers. No matter which station within the broadcast band is selected, its frequency is always converted to the

455 kHz IF. The amplitude-modulated IF is amplified by the IF amplifier which is tuned to 455 kHz. The output of the IF amplifier is applied to an *audio detector* which removes the IF, leaving only the envelope which is the audio signal. The audio signal is then amplified and applied to the speaker.

SECTION 17–9 REVIEW

1. Generally, why is a tuned filter necessary when a signal is coupled from an antenna to the input of a receiver?
2. What is a wave trap?
3. What is meant by *ganged tuning*?

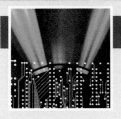

A Circuit Application

In the Chapter 11 circuit application, you worked with a receiver system to learn basic ac measurements. In this chapter, the receiver is again used to illustrate one application of resonant circuits. We will focus on a part of the "front end" of the receiver system that contains resonant circuits. Generally, the front end includes the RF amplifier, the local oscillator, and the mixer. In this circuit application, the RF amplifier is the focus. A knowledge of amplifier circuits is not necessary at this time.

A basic block diagram of an AM radio receiver is shown in Figure 17–54. In this particular system, the "front end" includes the circuitry used for tuning in a desired broadcasting station by

frequency selection and then converting that selected frequency to a standard intermediate frequency (IF). AM radio stations transmit in the frequency range from 535 kHz to 1605 kHz. The purpose of the RF amplifier is to take the signals picked up by the antenna, reject all but the signal from the desired station, and amplify it to a higher level.

A schematic of the RF amplifier is shown in Figure 17–55. The parallel resonant tuning circuit consists of L, C_1 and C_2. This particular RF amplifier does not have a resonant circuit on the output. C_1 is a varactor, which is a semiconductor device that you will learn more about in a later course. All that you need to know at this point is that the varactor is basically a variable capacitor whose capacitance is varied by changing the dc voltage

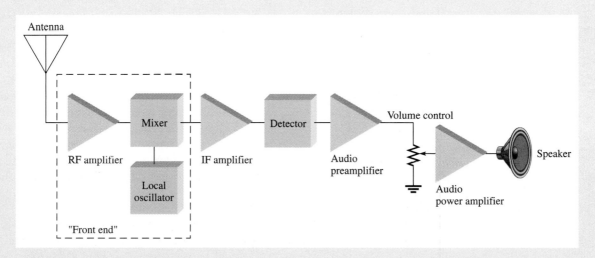

▲ **FIGURE 17–54**

Simplified block diagram of a basic radio receiver.

► FIGURE 17–55

Partial schematic of the RF amplifier showing the resonant tuning circuit.

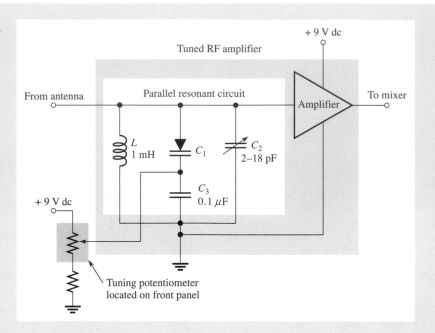

across it. In this circuit, the dc voltage comes from the wiper of the potentiometer used for tuning the receiver.

The voltage from the potentiometer can be varied from +1 V to +9 V. The particular varactor used in this circuit can be varied from 200 pF at 1 V to 5 pF at 9 V. The capacitor C_2 is a trimmer capacitor that is used for initially adjusting the resonant circuit. Once it is preset, it is left at that value. C_1 and C_2 are in parallel and their capacitances add to produce the total capacitance for the resonant circuit. C_3 has a minimal effect on the resonant circuit and can be ignored. The purpose of C_3 is to allow the dc voltage to be applied to the varactor while providing an ac ground.

In this circuit application, you will work with the RF amplifier circuit board in Figure 17–56. Although all of the amplifier components are on the board, the part that you are to focus on is the resonant circuit indicated by the highlighted area.

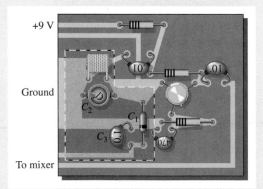

▲ FIGURE 17–56

RF amplifier circuit board.

Capacitance in the Resonant Circuit

◆ Calculate a capacitance setting for C_2 that will ensure a complete coverage of the AM frequency band as the varactor is varied over its capacitance range. C_3 can be ignored. The full range of resonant frequencies for the tuning circuit should more than cover the AM band, so that at the maximum varactor capacitance, the resonant frequency will be less than 535 kHz and at the minimum varactor capacitance, the resonant frequency will be greater than 1605 kHz.

◆ Using the value of C_2 that you have calculated, determine the values of the varactor capacitance that will produce a resonant frequency of 535 kHz and 1605 kHz, respectively.

Testing the Resonant Circuit

◆ Suggest a procedure for testing the resonant circuit using the instruments in the test bench setup of Figure 17–57. Develop a test setup by creating a point-to-point hook-up of the board and the instruments.

◆ Using the graph in Figure 17–58 that shows the variation in varactor capacitance versus varactor voltage, determine the resonant frequency for each indicated setting from the B outputs of the dc power supply (rightmost output terminals). The A output of the power supply is used to provide 9 V to the amplifier. The B output of the power supply is used to simulate the potentiometer voltage.

Review

1. What is the AM frequency range?
2. State the purpose of the RF amplifier.
3. How is a particular frequency in the AM band selected?

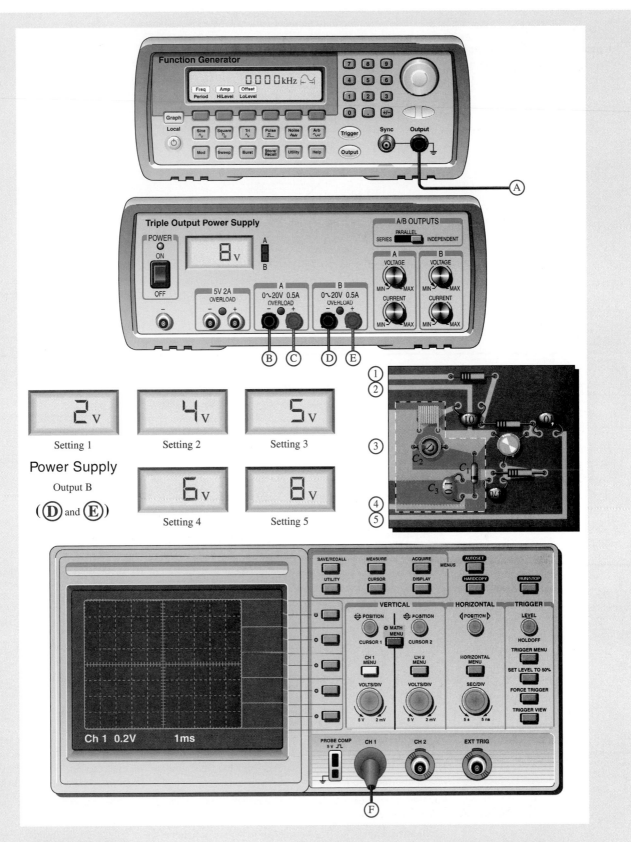

▲ FIGURE 17–57

Test bench setup.

▶ **FIGURE 17–58**

Varactor capacitance versus voltage.

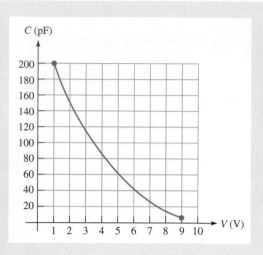

▶ **FIGURE 17–58**

Varactor capacitance versus voltage.

SUMMARY

◆ X_L and X_C have opposing effects in an *RLC* circuit.

◆ In a series *RLC* circuit, the larger reactance determines the net reactance of the circuit.

◆ At series resonance, the inductive and capacitive reactances are equal.

◆ The impedance of a series *RLC* circuit is purely resistive at resonance.

◆ In a series *RLC* circuit, the current is maximum at resonance.

◆ The reactive voltages V_L and V_C cancel at resonance in a series *RLC* circuit because they are equal in magnitude and 180° out of phase.

◆ In a parallel *RLC* circuit, the smaller reactance determines the net reactance of the circuit.

◆ In a parallel resonant circuit, the impedance is maximum at the resonant frequency.

◆ A parallel resonant circuit is commonly called a *tank circuit*.

◆ The impedance of a parallel *RLC* circuit is purely resistive at resonance.

◆ The bandwidth of a series resonant circuit is the range of frequencies for which the current is $0.707I_{max}$ or greater.

◆ The bandwidth of a parallel resonant circuit is the range of frequencies for which the impedance is $0.707Z_{max}$ or greater.

◆ The critical frequencies are the frequencies above and below resonance where the circuit response is 70.7% of the maximum response.

◆ A higher Q produces a narrower bandwidth.

KEY TERMS

Key terms and other bold terms in the chapter are defined in the end-of-book glossary.

Half-power frequency   The frequency at which the output power of a resonant circuit is 50% of the maximum (the output voltage is 70.7% of maximum); another name for *critical* or *cutoff frequency*.

Parallel resonance   A condition in a parallel *RLC* circuit in which the reactances ideally are equal and the impedance is maximum.

Resonant frequency   The frequency at which resonance occurs; also known as the *center frequency*.

Selectivity   A measure of how effectively a resonant circuit passes certain desired frequencies and rejects all others. Generally, the narrower the bandwidth, the greater the selectivity.

Series resonance A condition in a series *RLC* circuit in which the reactances ideally cancel and the impedance is minimum.

Tank circuit A parallel resonant circuit.

FORMULAS

Series *RLC* Circuits

17–1 $\quad X_{tot} = |X_L - X_C|$

17–2 $\quad \mathbf{Z} = R + jX_L - jX_C$

17–3 $\quad \mathbf{Z} = \sqrt{R^2 + (X_L - X_C)^2} \angle \pm \tan^{-1}\left(\dfrac{X_{tot}}{R}\right)$

Series Resonance

17–4 $\quad f_r = \dfrac{1}{2\pi\sqrt{LC}}$

Parallel *RLC* Circuits

17–5 $\quad \mathbf{Z} = \dfrac{1}{\dfrac{1}{R\angle 0°} + \dfrac{1}{X_L\angle 90°} + \dfrac{1}{X_C\angle -90°}}$

17–6 $\quad \mathbf{G} = \dfrac{1}{R\angle 0°} = G\angle 0°$

17–7 $\quad \mathbf{B}_C = \dfrac{1}{X_C\angle -90°} = B_C\angle 90° = jB_C$

17–8 $\quad \mathbf{B}_L = \dfrac{1}{X_L\angle 90°} = B_L\angle -90° = -jB_L$

17–9 $\quad \mathbf{Y} = \dfrac{1}{Z\angle \pm\theta} = Y\angle \mp\theta = G + jB_C - jB_L$

17–10 $\quad \mathbf{I}_{tot} = \sqrt{I_R^2 + (I_C - I_L)^2} \angle \tan^{-1}\left(\dfrac{I_{CL}}{I_R}\right)$

Parallel Resonance

17–11 $\quad I_{tot} = \dfrac{V_s}{Z_r}$

17–12 $\quad f_r = \dfrac{1}{2\pi\sqrt{LC}}\sqrt{\dfrac{Q^2}{Q^2+1}}$

17–13 $\quad f_r = \dfrac{\sqrt{1 - (R_W^2 C/L)}}{2\pi\sqrt{LC}}$

17–14 $\quad L_{eq} = L\left(\dfrac{Q^2+1}{Q^2}\right)$

17–15 $\quad R_{p(eq)} = R_W(Q^2+1)$

17–16 $\quad Z_r = R_W(Q^2+1)$

17–17 $\quad Q_O = \dfrac{R_{p(tot)}}{X_{L(eq)}}$

17–18 $\quad BW = f_2 - f_1$

17–19 $\quad f_r = \dfrac{f_1 + f_2}{2}$

17–20 $\quad BW = \dfrac{f_r}{Q}$

Answers are at the end of the chapter.

1. The total reactance of a series *RLC* circuit at resonance is
   (a) zero   (b) equal to the resistance   (c) infinity   (d) capacitive

2. The phase angle between the source voltage and current of a series *RLC* circuit at resonance is
   (a) $-90°$ (b) $+90°$ (c) $0°$ (d) dependent on the reactance

3. The impedance at the resonant frequency of a series *RLC* circuit with $L = 15$ mH, $C = 0.015$ μF, and $R_W = 80$ Ω is
   (a) 15 kΩ (b) 80 Ω (c) 30 Ω (d) 0 Ω

4. In a series *RLC* circuit that is operating below the resonant frequency, the current
   (a) is in phase with the applied voltage   (b) lags the applied voltage
   (c) leads the applied voltage

5. If the value of *C* in a series *RLC* circuit is increased, the resonant frequency
   (a) is not affected   (b) increases   (c) remains the same   (d) decreases

6. In a certain series resonant circuit, $V_C = 150$ V, $V_L = 150$ V, and $V_R = 50$ V. The value of the source voltage is
   (a) 150 V   (b) 300 V   (c) 50 V   (d) 350 V

7. A certain series resonant circuit has a bandwidth of 1 kHz. If the existing coil is replaced with one having a lower value of *Q*, the bandwidth will
   (a) increase   (b) decrease   (c) remain the same   (d) be more selective

8. At frequencies below resonance in a parallel *RLC* circuit, the current
   (a) leads the source voltage   (b) lags the source voltage
   (c) is in phase with the source voltage

9. The total current into the *L* and *C* branches of a parallel circuit at resonance is ideally
   (a) maximum   (b) low   (c) high   (d) zero

10. To tune a parallel resonant circuit to a lower frequency, the capacitance should be
    (a) increased   (b) decreased   (c) left alone   (d) replaced with inductance

11. The resonant frequency of a parallel circuit is approximately the same as a series circuit when
    (a) the *Q* is very low   (b) the *Q* is very high
    (c) there is no resistance   (d) either answer (b) or (c)

12. If the resistance in parallel with a parallel resonant circuit is reduced, the bandwidth
    (a) disappears   (b) decreases
    (c) becomes sharper   (d) increases

CIRCUIT DYNAMICS QUIZ

Answers are at the end of the chapter.

Refer to Figure 17–60.

1. If R_1 opens, the total current
   (a) increases   (b) decreases   (c) stays the same

2. If C_1 opens, the voltage across C_2
   (a) increases   (b) decreases   (c) stays the same

3. If L_2 opens, the voltage across it
   (a) increases   (b) decreases   (c) stays the same

Refer to Figure 17–63.

4. If *L* opens, the voltage across *R*
   (a) increases   (b) decreases   (c) stays the same

   5. If f is adjusted to its resonant value, the current through R

      (a) increases   **(b)** decreases   **(c)** stays the same

Refer to Figure 17–64.

   6. If L is increased to 100 mH, the resonant frequency

      (a) increases   **(b)** decreases   **(c)** stays the same

   7. If C is increased to 100 pF, the resonant frequency

      (a) increases   **(b)** decreases   **(c)** stays the same

   8. If L becomes open, the voltage across C

      (a) increases   **(b)** decreases   **(c)** stays the same

Refer to Figure 17–66.

   9. If R_2 becomes open, the voltage across L

      (a) increases   **(b)** decreases   **(c)** stays the same

  10. If C becomes shorted, the voltage across R_1

      (a) increases   **(b)** decreases   **(c)** stays the same

Refer to Figure 17–69.

  11. If L_1 opens, the voltage from point a to point b

      (a) increases   **(b)** decreases   **(c)** stays the same

  12. If the frequency of the source is increased, the voltage from a to b

      (a) increases   **(b)** decreases   **(c)** stays the same

  13. If the frequency of the source voltage is increased, the current through R_1

      (a) increases   **(b)** decreases   **(c)** stays the same

  14. If the frequency of the source voltage is decreased, the voltage across C

      (a) increases   **(b)** decreases   **(c)** stays the same

---

PROBLEMS

More difficult problems are indicated by an asterisk (*).
Answers to odd-numbered problems are at the end of the book.

PART 1: SERIES CIRCUITS

SECTION 17–1 **Impedance of Series *RLC* Circuits**

   1. A certain series *RLC* circuit has the following values: $R = 10\ \Omega$, $C = 0.047\ \mu\text{F}$, and $L = 5\ \text{mH}$. Determine the impedance in polar form. What is the net reactance? The source frequency is 5 kHz.

   2. Find the impedance in Figure 17–59, and express it in polar form.

   3. If the frequency of the source voltage in Figure 17–59 is doubled from the value that produces the indicated reactances, how does the magnitude of the impedance change?

   4. For the circuit of Figure 17–59, determine the net reactance that will make the impedance magnitude equal to 100 Ω.

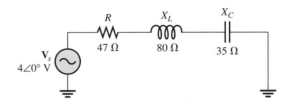

▲ **FIGURE 17–59**

SECTION 17–2 Analysis of Series *RLC* Circuits

5. For the circuit in Figure 17–59, find $\mathbf{I}_{tot}$, $\mathbf{V}_R$, $\mathbf{V}_L$, and $\mathbf{V}_C$ in polar form.

6. Draw the voltage phasor diagram for the circuit in Figure 17–59.

7. Analyze the circuit in Figure 17–60 for the following ($f = 25$ kHz):

   (a) $\mathbf{I}_{tot}$ (b) P_{true} (c) P_r (d) P_a

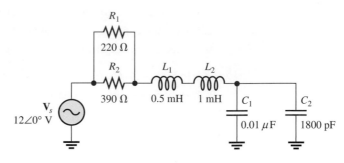

▲ FIGURE 17–60

SECTION 17–3 Series Resonance

8. For the circuit in Figure 17–59, is the resonant frequency higher or lower than the setting indicated by the reactance values?

9. For the circuit in Figure 17–61, what is the voltage across *R* at resonance?

10. Find X_L, X_C, Z, and I at the resonant frequency in Figure 17–61.

▶ FIGURE 17–61

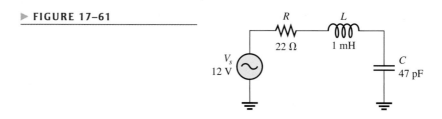

11. A certain series resonant circuit has a maximum current of 50 mA and a V_L of 100 V. The applied voltage is 10 V. What is Z? What are X_L and X_C?

12. For the *RLC* circuit in Figure 17–62, determine the resonant frequency.

13. What is the value of the current at the half-power points in Figure 17–62?

14. Determine the phase angle between the applied voltage and the current at the critical frequencies in Figure 17–62. What is the phase angle at resonance?

*15. Design a circuit in which the following series resonant frequencies are switch-selectable:

   (a) 500 kHz   (b) 1000 kHz   (c) 1500 kHz   (d) 2000 kHz

▶ FIGURE 17–62

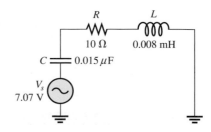

PART 2: PARALLEL CIRCUITS

SECTION 17–4   Impedance of Parallel *RLC* Circuits

16. Express the impedance of the circuit in Figure 17–63 in polar form.

17. Is the circuit in Figure 17–63 capacitive or inductive? Explain.

18. At what frequency does the circuit in Figure 17–63 change its reactive characteristic (from inductive to capacitive or vice versa)?

▶ FIGURE 17–63

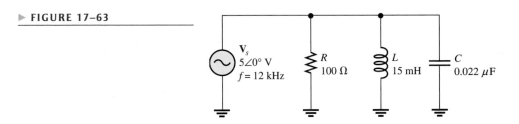

SECTION 17–5   Analysis of Parallel *RLC* Circuits

19. For the circuit in Figure 17–63, find all the currents and voltages in polar form.

20. Find the total impedance of the circuit in Figure 17–63 at 50 kHz.

21. Change the frequency to 100 kHz in Figure 17–63 and repeat Problem 19.

SECTION 17–6   Parallel Resonance

22. What is the impedance of an ideal parallel resonant circuit (no resistance in either branch)?

23. Find Z at resonance and f_r for the tank circuit in Figure 17–64.

24. How much current is drawn from the source in Figure 17–64 at resonance? What are the inductive current and the capacitive current at the resonant frequency?

25. Find P_{true}, P_r, and P_a in the circuit of Figure 17–64 at resonance.

▶ FIGURE 17–64

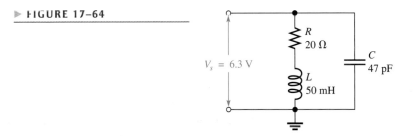

PART 3: SERIES-PARALLEL CIRCUITS

SECTION 17–7   Analysis of Series-Parallel *RLC* Circuits

26. Find the total impedance for each circuit in Figure 17–65.

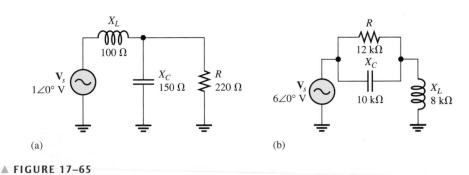

(a)   (b)

▲ FIGURE 17–65

27. For each circuit in Figure 17–65, determine the phase angle between the source voltage and the total current.

28. Determine the voltage across each element in Figure 17–66, and express each in polar form.

29. Convert the circuit in Figure 17–66 to an equivalent series form.

▶ FIGURE 17–66

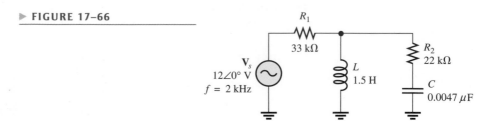

30. What is the current through R_2 in Figure 17–67?

31. In Figure 17–67, what is the phase angle between I_2 and the source voltage?

▶ FIGURE 17–67

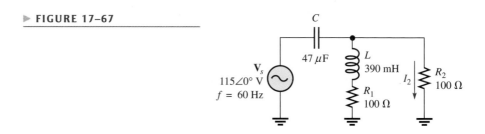

*32. Determine the total resistance and the total reactance in Figure 17–68.

*33. Find the current through each component in Figure 17–68. Find the voltage across each component.

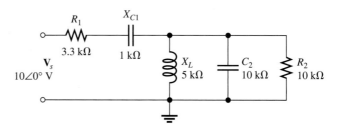

▲ FIGURE 17–68

34. Determine if there is a value of C that will make $V_{ab} = 0$ V in Figure 17–69. If not, explain.

*35. If the value of C is 0.22 μF, what is the current through a 100 Ω resistor connected from a to b in Figure 17–69?

▶ FIGURE 17–69

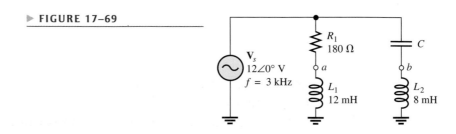

*36. How many resonant frequencies are there in the circuit of Figure 17–70? Why?

*37. Determine the resonant frequencies and the output voltage at each frequency in Figure 17–70.

▶ FIGURE 17–70

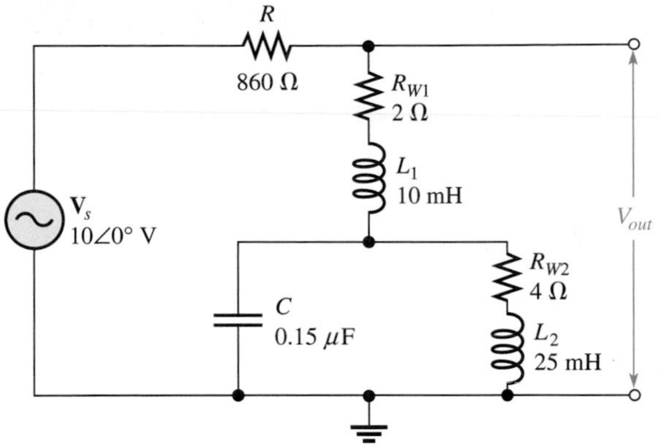

R
860 Ω

R_{W1}
2 Ω

L_1
10 mH

V_s
10∠0° V

V_{out}

R_{W2}
4 Ω

C
0.15 μF

L_2
25 mH

*38. Design a parallel-resonant network using a single coil and switch-selectable capacitors to produce the following resonant frequencies: 8 MHz, 9 MHz, 10 MHz, and 11 MHz. Assume a 10 μH coil with a winding resistance of 5 Ω.

PART 4: SPECIAL TOPICS

SECTION 17–8 Bandwidth of Resonant Circuits

39. At resonance, $X_L = 2\,\text{k}\Omega$ and $R_W = 25\,\Omega$ in a parallel *RLC* circuit. The resonant frequency is 5 kHz. Determine the bandwidth.

40. If the lower critical frequency is 2400 Hz and the upper critical frequency is 2800 Hz, what is the bandwidth? What is the resonant frequency?

41. In a certain *RLC* circuit, the power at resonance is 2.75 W. What is the power at the lower critical frequency?

*42. What values of *L* and *C* should be used in a tank circuit to obtain a resonant frequency of 8 kHz? The bandwidth must be 800 Hz. The winding resistance of the coil is 10 Ω.

43. A parallel resonant circuit has a *Q* of 50 and a *BW* of 400 Hz. If *Q* is doubled, what is the bandwidth for the same f_r?

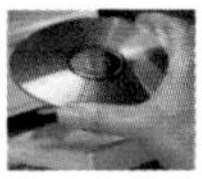

Multisim Troubleshooting and Analysis

These problems require your Multisim CD-ROM.

44. Open file P17-44 and determine if there is a fault. If so, find the fault.

45. Open file P17-45 and determine if there is a fault. If so, find the fault.

46. Open file P17-46 and determine if there is a fault. If so, find the fault.

47. Open file P17-47 and determine if there is a fault. If so, find the fault.

48. Open file P17-48 and determine if there is a fault. If so, find the fault.

49. Open file P17-49 and determine if there is a fault. If so, find the fault.

50. Open file P17-50 and determine the resonant frequency of the circuit.

51. Open file P17-51 and determine the resonant frequency of the circuit.

ANSWERS

SECTION REVIEWS

SECTION 17–1 Impedance of Series *RLC* Circuits

1. $X_{tot} = 70\,\Omega$; capacitive

2. $\mathbf{Z} = 84.3\angle{-56.1°}\,\Omega$; $Z = 84.3\,\Omega$; $\theta = -56.1°$; current is leading V_s.

SECTION 17–2 **Analysis of Series *RLC* Circuits**

    1. $\mathbf{V}_s = 38.4\angle-21.3° \text{ V}$

    2. Current leads the voltage.

    3. $X_{tot} = 600 \text{ } \Omega$

SECTION 17–3 **Series Resonance**

    1. For series resonance, $X_L = X_C$.

    2. The current is maximum because the impedance is minimum.

    3. $f_r = 159 \text{ kHz}$

    4. The circuit is capacitive.

SECTION 17–4 **Impedance of Parallel *RLC* Circuits**

    1. The circuit is capacitive.

    2. $\mathbf{Y} = 1.54\angle49.4° \text{ mS}$

    3. $\mathbf{Z} = 651\angle-49.4° \text{ } \Omega$

SECTION 17–5 **Analysis of Parallel *RLC* Circuits**

    1. $I_R = 80 \text{ mA}, I_C = 120 \text{ mA}, I_L = 240 \text{ mA}$

    2. The circuit is capacitive.

SECTION 17–6 **Parallel Resonance**

    1. Impedance is maximum at parallel resonance.

    2. The current is minimum.

    3. $X_C = 1500 \text{ } \Omega$

    4. $f_r = 225 \text{ kHz}$

    5. $f_r = 22.5 \text{ kHz}$

    6. $f_r = 20.9 \text{ kHz}$

SECTION 17–7 **Analysis of Series-Parallel *RLC* Circuits**

    1. $R_{p(eq)} = 130 \text{ } \Omega, L_{eq} = 101.6 \text{ } \mu\text{H}, C = 0.22 \text{ } \mu\text{F}$

    2. $L_{(eq)} = 20.1 \text{ mH}, R_{p(eq)} = 1.59 \text{ k}\Omega$

SECTION 17–8 **Bandwidth of Resonant Circuits**

    1. $BW = f_2 - f_1 = 400 \text{ kHz}$

    2. $f_r = 2 \text{ MHz}$

    3. $P_{f2} = 0.9 \text{ W}$

    4. Larger Q means narrower BW.

SECTION 17–9 **Applications**

    1. A tuned filter is used to select a narrow band of frequencies.

    2. A wave trap is a band-stop filter.

    3. Ganged tuning is done with several capacitors (or inductors) whose values can be varied simultaneously with a common control.

A Circuit Application

    1. The AM frequency range is 535 kHz to 1605 kHz.

    2. The RF amplifier rejects all signals but the one from the desired station. It then amplifies the selected signal.

    3. A particular AM frequency is selected by varying the varactor capacitance with a dc voltage.

RELATED PROBLEMS FOR EXAMPLES

17–1 $\mathbf{Z} = 12.7\angle 82.3°\,k\Omega$

17–2 $\mathbf{Z} = 4.72\angle 45.6°\,k\Omega$. See Figure 17–71.

▶ **FIGURE 17–71**

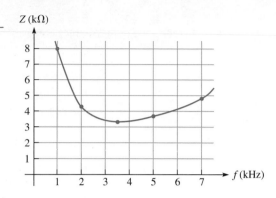

17–3 Current will increase with frequency to a certain point and then it will decrease.

17–4 The circuit is more capacitive.

17–5 $f_r = 22.5\,kHz$

17–6 45°

17–7 Z increases; Z increases.

17–8 Z decreases.

17–9 Inductive

17–10 I_{tot} increases.

17–11 Greater

17–12 $\mathbf{V}_C = 0.93\angle -65.8°\,V$

17–13 $\mathbf{V}_{C1} = 27.1\angle -81.1°\,V$

17–14 $R_{p(eq)} = 25\,k\Omega$, $L_{eq} = 5\,mH$; $C - 0.022\,\mu F$

17–15 $Z_r = 79.9\,k\Omega$

17–16 $I = 35.4\,mA$

17–17 $f_1 = 6.75\,kHz$; $f_2 = 9.25\,kHz$

17–18 $BW = 7.96\,kHz$

SELF-TEST

1. (a)   **2.** (c)   **3.** (b)   **4.** (c)   **5.** (d)   **6.** (c)   **7.** (a)   **8.** (b)

9. (d)   **10.** (a)   **11.** (b)   **12.** (d)

CIRCUIT DYNAMICS QUIZ

1. (b)   **2.** (a)   **3.** (a)   **4.** (c)   **5.** (c)   **6.** (b)   **7.** (b)   **8.** (c)

9. (a)   **10.** (a)   **11.** (a)   **12.** (a)   **13.** (b)   **14.** (a)

18

PASSIVE FILTERS

CHAPTER OUTLINE

CHAPTER OBJECTIVES

◆ Analyze the operation of *RC* and *RL* low-pass filters

◆ Analyze the operation of *RC* and *RL* high-pass filters

◆ Analyze the operation of band-pass filters

◆ Analyze the operation of band-stop filters

KEY TERMS

◆ Low-pass filter
◆ Passband
◆ Critical frequency (f_c)
◆ Roll-off
◆ Attenuation
◆ Decade

◆ Bode plot
◆ High-pass filter
◆ Band-pass filter
◆ Center frequency (f_0)
◆ Band-stop filter

A CIRCUIT APPLICATION PREVIEW

In the circuit application, you will plot the frequency responses of filters based on oscilloscope measurements and identify the types of filters.

VISIT THE COMPANION WEBSITE

Study aids for this chapter are available at
http://www.prenhall.com/floyd

INTRODUCTION

The concept of filters was introduced in Chapters 15, 16, and 17 to illustrate applications of *RC*, *RL*, and *RLC* circuits. This chapter is essentially an extension of the earlier material and provides additional coverage of the important topic of filters.

Passive filters are discussed in this chapter. Passive filters use various combinations of resistors, capacitors, and inductors. In a later course, you will study active filters that use passive components combined with amplifiers. You have already seen how basic *RC*, *RL*, and *RLC* circuits can be used as filters. Now, you will learn that passive filters can be placed in four general categories according to their response characteristics: low-pass, high-pass, band-pass, and band-stop. Within each category, there are several common types that will be examined.

18–1 LOW-PASS FILTERS

A **low-pass filter** allows signals with lower frequencies to pass from input to output while rejecting higher frequencies.

After completing this section, you should be able to

- ◆ **Analyze the operation of *RC* and *RL* low-pass filters**
  - ◆ Express the voltage and power ratios of a filter in decibels
  - ◆ Determine the critical frequency of a low-pass filter
  - ◆ Explain the difference between actual and ideal low-pass response curves
  - ◆ Define *roll-off*
  - ◆ Generate a Bode plot for a low-pass filter
  - ◆ Discuss phase shift in a low-pass filter

Figure 18–1 shows a block diagram and a general response curve for a low-pass filter. The range of frequencies passed by a filter within specified limits is called the **passband** of the filter. The point considered to be the upper end of the passband is at the critical frequency, f_c, as illustrated in Figure 18–1(b). The **critical frequency (f_c)** is the frequency at which the filter's output voltage is 70.7% of the maximum. The filter's critical frequency is also called the *cutoff frequency, break frequency,* or *−3 dB frequency* because the output voltage is down 3 dB from its maximum at this frequency. The term *dB (decibel)* is a commonly used unit in filter measurements.

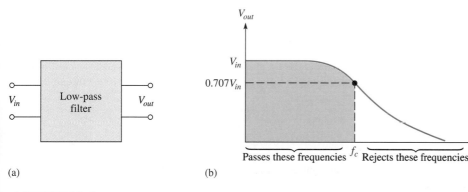

(a)   (b)

▲ **FIGURE 18–1**

Low-pass filter block diagram and general response curve.

Decibels

The basis for the decibel unit stems from the logarithmic response of the human ear to the intensity of sound. The **decibel** is a logarithmic measurement of the ratio of one power to another or one voltage to another, which can be used to express the input-to-output relationship of a filter. The following equation expresses a power ratio in decibels:

$$dB = 10 \log\left(\frac{P_{out}}{P_{in}}\right)$$

Equation 18–1

From the properties of logarithms, the following decibel formula for a voltage ratio is derived.

Equation 18–2

$$dB = 20 \log\left(\frac{V_{out}}{V_{in}}\right)$$

EXAMPLE 18–1

At a certain frequency, the output voltage of a filter is 5 V and the input is 10 V. Express the voltage ratio in decibels.

Solution

$$20 \log\left(\frac{V_{out}}{V_{in}}\right) = 20 \log\left(\frac{5\text{ V}}{10\text{ V}}\right) = 20 \log(0.5) = -6.02\text{ dB}$$

*Related Problem** Express the ratio $V_{out}/V_{in} = 0.85$ in decibels.

*Answers are at the end of the chapter.

RC Low-Pass Filter

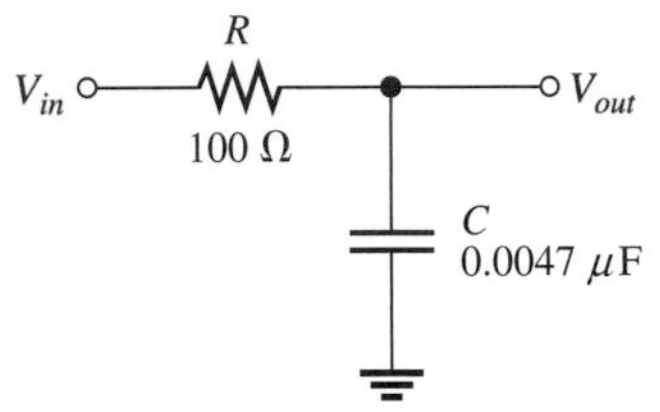

▲ FIGURE 18–2

A basic RC low-pass filter is shown in Figure 18–2. Notice that the output voltage is taken across the capacitor.

When the input is dc (0 Hz), the output voltage equals the input voltage because X_C is infinitely large. As the input frequency is increased, X_C decreases and, as a result, V_{out} gradually decreases until a frequency is reached where $X_C = R$. This is the critical frequency, f_c, of the filter.

$$X_C = \frac{1}{2\pi f_c C} = R$$

Solving for f_c,

Equation 18–3

$$f_c = \frac{1}{2\pi RC}$$

At any frequency, by application of the voltage-divider formula, the output voltage magnitude is

$$V_{out} = \left(\frac{X_C}{\sqrt{R^2 + X_C^2}}\right)V_{in}$$

Since $X_C = R$ at f_c, the output voltage at the critical frequency can be expressed as

$$V_{out} = \left(\frac{R}{\sqrt{R^2 + R^2}}\right)V_{in} = \left(\frac{R}{\sqrt{2R^2}}\right)V_{in} = \left(\frac{R}{R\sqrt{2}}\right)V_{in} = \left(\frac{1}{\sqrt{2}}\right)V_{in} = 0.707V_{in}$$

These calculations show that the output is 70.7% of the input when $X_C = R$. The frequency at which this occurs is, by definition, the critical frequency.

The ratio of output voltage to input voltage at the critical frequency can be expressed in decibels as follows:

$$V_{out} = 0.707V_{in}$$

$$\frac{V_{out}}{V_{in}} = 0.707$$

$$20 \log\left(\frac{V_{out}}{V_{in}}\right) = 20 \log(0.707) = -3\text{ dB}$$

EXAMPLE 18–2   Determine the critical frequency for the RC low-pass filter in Figure 18–2.

Solution
$$f_c = \frac{1}{2\pi RC} = \frac{1}{2\pi(100\ \Omega)(0.0047\ \mu F)} = \textbf{339 kHz}$$

The output voltage is 3 dB below V_{in} at this frequency (V_{out} has a maximum value of V_{in}).

Related Problem   A certain RC low-pass filter has $R = 1.0\ k\Omega$ and $C = 0.022\ \mu F$. Determine its critical frequency.

Roll-Off of the Response Curve

The blue line in Figure 18–3 shows an actual response curve for a low-pass filter. The maximum output is defined to be 0 dB as a reference. Zero decibels corresponds to $V_{out} = V_{in}$ because $20\log(V_{out}/V_{in}) = 20\log 1 = 0$ dB. The output drops from 0 dB to -3 dB at the critical frequency and then continues to decrease at a fixed rate. This pattern of decrease is called the **roll-off** of the frequency response. The red line shows an ideal output response that is considered to be "flat" out to the critical frequency. The output then decreases at the fixed rate.

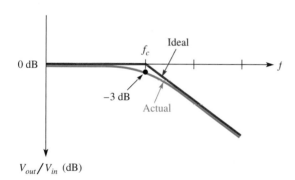

◀ **FIGURE 18–3**

Actual and ideal response curves for a low-pass filter.

As you have seen, the output voltage of a low-pass filter decreases by 3 dB when the frequency is increased to the critical value f_c. As the frequency continues to increase above f_c, the output voltage continues to decrease. In fact, for each tenfold increase in frequency above f_c, there is a 20 dB reduction in the output, as shown in the following steps.

Let's take a frequency that is ten times the critical frequency ($f = 10f_c$). Since $R = X_C$ at f_c, then $R = 10X_C$ at $10f_c$ because of the inverse relationship of X_C and f.

The **attenuation** is the reduction in voltage expressed as the ratio V_{out}/V_{in} and is developed as follows:

$$\frac{V_{out}}{V_{in}} = \frac{X_C}{\sqrt{R^2 + X_C^2}} = \frac{X_C}{\sqrt{(10X_C)^2 + X_C^2}}$$

$$= \frac{X_C}{\sqrt{100X_C^2 + X_C^2}} = \frac{X_C}{\sqrt{X_C^2(100 + 1)}} = \frac{X_C}{X_C\sqrt{101}} = \frac{1}{\sqrt{101}} \cong \frac{1}{10} = 0.1$$

The dB attenuation is

$$20\log\left(\frac{V_{out}}{V_{in}}\right) = 20\log(0.1) = -20\ dB$$

A tenfold change in frequency is called a **decade**. So, for an *RC* circuit, the output voltage is reduced by 20 dB for each decade increase in frequency. A similar result can be derived for a high-pass circuit. The roll-off is a constant -20 dB/decade for a basic *RC* or *RL* filter. Figure 18–4 shows the ideal frequency response plot on a semilog scale, where each interval on the horizontal axis represents a tenfold increase in frequency. This response curve is called a **Bode plot**.

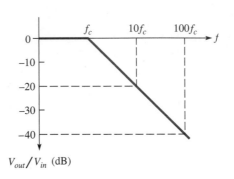

▲ FIGURE 18–4

Frequency roll-off for an *RC* low-pass filter (Bode plot).

EXAMPLE 18–3

Make a Bode plot for the filter in Figure 18–5 for three decades of frequency. Use semilog graph paper.

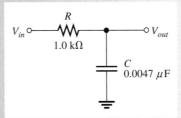

▲ FIGURE 18–5

Solution   The critical frequency for this low-pass filter is

$$f_c = \frac{1}{2\pi RC} = \frac{1}{2\pi(1.0 \text{ k}\Omega)(0.0047 \text{ }\mu\text{F})} = 33.9 \text{ kHz}$$

The idealized Bode plot is shown with the red line on the semilog graph in Figure 18–6. The approximate actual response curve is shown with the blue line. Notice first that the horizontal scale is logarithmic and the vertical scale is linear. The frequency is on the logarithmic scale, and the filter output in decibels is on the linear scale.

The output is flat below f_c (33.9 kHz). As the frequency is increased above f_c, the output drops at a -20 dB/decade rate. Thus, for the ideal curve, every time the frequency is increased by ten, the output is reduced by 20 dB. A slight variation from this occurs in actual practice. The output is actually at -3 dB rather than 0 dB at the critical frequency.

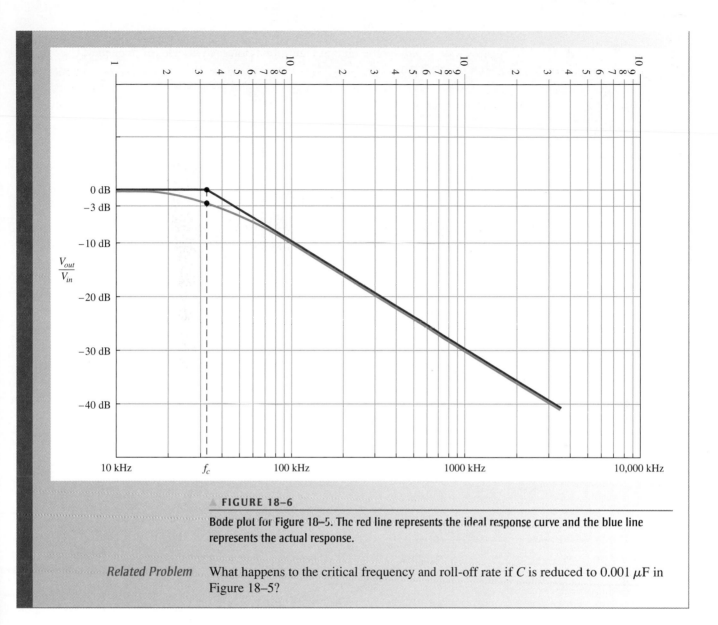

▲ FIGURE 18–6

Bode plot for Figure 18–5. The red line represents the ideal response curve and the blue line represents the actual response.

Related Problem   What happens to the critical frequency and roll-off rate if *C* is reduced to 0.001 μF in Figure 18–5?

RL Low-Pass Filter

A basic *RL* low-pass filter is shown in Figure 18–7. Notice that the output voltage is taken across the resistor.

When the input is dc (0 Hz), the output voltage ideally equals the input voltage because X_L is a short (if R_W is neglected). As the input frequency is increased, X_L increases and, as a result, V_{out} gradually decreases until the critical frequency is reached. At this point, $X_L = R$ and the frequency is

$$2\pi f_c L = R$$

$$f_c = \frac{R}{2\pi L}$$

$$f_c = \frac{1}{2\pi(L/R)}$$

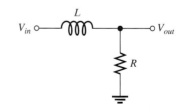

▲ FIGURE 18–7

RL low-pass filter.

Equation 18–4

Just as in the *RC* low-pass filter, $V_{out} = 0.707V_{in}$ and, thus, the output voltage is −3 dB below the input voltage at the critical frequency.

EXAMPLE 18–4

Make a Bode plot for the filter in Figure 18–8 for three decades of frequency. Use semilog graph paper.

▶ **FIGURE 18–8**

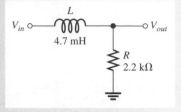

Solution

The critical frequency for this low-pass filter is

$$f_c = \frac{1}{2\pi(L/R)} = \frac{1}{2\pi(4.7 \text{ mH}/2.2 \text{ k}\Omega)} = 74.5 \text{ kHz}$$

The idealized Bode plot is shown with the red line on the semilog graph in Figure 18–9. The approximate actual response curve is shown with the blue line. Notice first that the horizontal scale is logarithmic and the vertical scale is linear. The frequency is on the logarithmic scale, and the filter output in decibels is on the linear scale.

▲ **FIGURE 18–9**

Bode plot for Figure 18–8. The red line is the ideal response curve and the blue line is the actual response.

The output is flat below f_c (74.5 kHz). As the frequency is increased above f_c, the output drops at a -20 dB/decade rate. Thus, for the ideal curve, every time the frequency is increased by ten, the output is reduced by 20 dB. A slight variation from this occurs in actual practice. The output is actually at -3 dB rather than 0 dB at the critical frequency.

Related Problem   What happens to the critical frequency and roll-off rate if L is reduced to 1 mH in Figure 18–8?

Use Multisim file E18-04 to verify the calculated results in this example and to confirm your calculations for the related problem.

Phase Shift in a Low-Pass Filter

The RC low-pass filter acts as a lag circuit. Recall from Chapter 15 that the phase shift from input to output is expressed as

$$\phi = -\tan^{-1}\left(\frac{R}{X_C}\right)$$

At the critical frequency, $X_C = R$ and, therefore, $\phi = -45°$. As the input frequency is reduced, ϕ decreases and approaches 0° when the frequency approaches zero. Figure 18–10, illustrates this phase characteristic.

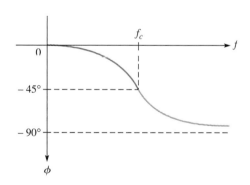

◀ FIGURE 18–10

Phase characteristic of a low-pass filter.

The RL low-pass filter also acts as a lag circuit. Recall from Chapter 16 that the phase shift is expressed as

$$\phi = -\tan^{-1}\left(\frac{X_L}{R}\right)$$

As in the RC filter, the phase shift from input to output is $-45°$ at the critical frequency and decreases for frequencies below f_c.

SECTION 18–1 REVIEW

Answers are at the end of the chapter.

1. In a certain low-pass filter, $f_c = 2.5$ kHz. What is its passband?
2. In a certain low-pass filter, $R = 100\ \Omega$ and $X_C = 2\ \Omega$ at a frequency, f_1. Determine V_{out} at f_1 when $V_{in} = 5\angle 0°$ V rms.
3. $V_{out} = 400$ mV, and $V_{in} = 1.2$ V. Express the ratio V_{out}/V_{in} in dB.

18–2 HIGH-PASS FILTERS

A **high-pass filter** allows signals with higher frequencies to pass from input to output while rejecting lower frequencies.

After completing this section, you should be able to

- ◆ **Analyze the operation of *RC* and *RL* high-pass filters**
  - ◆ Determine the critical frequency of a high-pass filter
  - ◆ Explain the difference between actual and ideal response curves
  - ◆ Generate a Bode plot for a high-pass filter
  - ◆ Discuss phase shift in a high-pass filter

Figure 18–11 shows a block diagram and a general response curve for a high-pass filter. The frequency considered to be the lower end of the passband is called the *critical frequency*. Just as in the low-pass filter, it is the frequency at which the output is 70.7% of the maximum, as indicated in the figure.

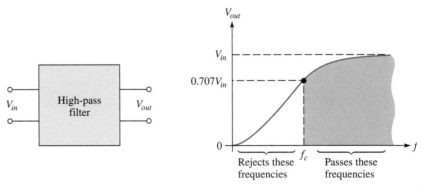

▲ FIGURE 18–11

High-pass filter block diagram and response curve.

RC High-Pass Filter

A basic *RC* high-pass filter is shown in Figure 18–12. Notice that the output voltage is taken across the resistor.

▶ FIGURE 18–12

RC high-pass filter.

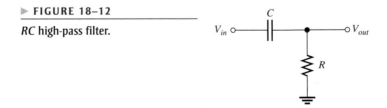

When the input frequency is at its critical value, $X_C = R$ and the output voltage is $0.707V_{in}$, just as in the case of the low-pass filter. As the input frequency increases above the critical frequency, X_C decreases and, as a result, the output voltage increases and approaches a value equal to V_{in}. The expression for the critical frequency of the high-pass filter is the same as for the low-pass filter.

$$f_c = \frac{1}{2\pi RC}$$

Below f_c, the output voltage decreases (rolls off) at a rate of -20 dB/decade. Figure 18–13 shows an actual and an ideal response curve for a high-pass filter.

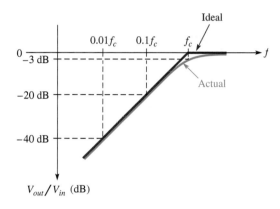

▲ FIGURE 18–13

Actual and ideal response curves for a high-pass filter.

EXAMPLE 18–5

Make a Bode plot for the filter in Figure 18–14 for three decades of frequency. Use semilog graph paper.

▶ FIGURE 18–14

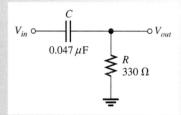

Solution   The critical frequency for this high-pass filter is

$$f_c = \frac{1}{2\pi RC} = \frac{1}{2\pi(330\ \Omega)(0.047\ \mu F)} = 10.3\ \text{kHz} \cong 10\ \text{kHz}$$

The idealized Bode plot is shown with the red line on the semilog graph in Figure 18–15. The approximate actual response curve is shown with the blue line. Notice first that the horizontal scale is logarithmic and the vertical scale is linear. The frequency is on the logarithmic scale, and the filter output in decibels is on the linear scale.

The output is flat above f_c (approximately 10 kHz). As the frequency is reduced below f_c, the output drops at a -20 dB/decade rate. Thus, for the ideal curve, every time the frequency is reduced by ten, the output is reduced by 20 dB. A slight variation from this occurs in actual practice. The output is actually at -3 dB rather than 0 dB at the critical frequency.

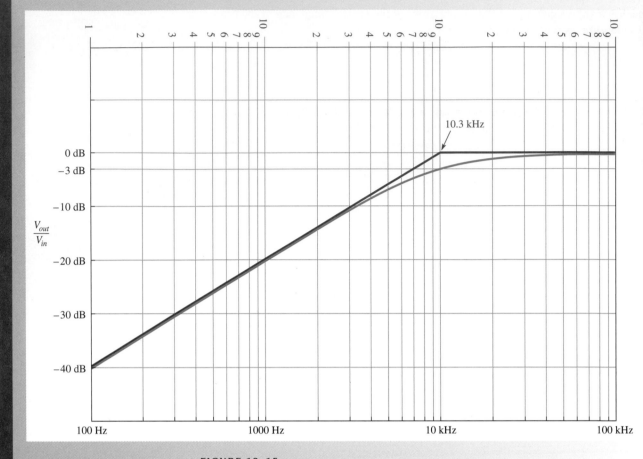

10.3 kHz

0 dB
−3 dB
−10 dB
$\dfrac{V_{out}}{V_{in}}$
−20 dB
−30 dB
−40 dB

100 Hz   1000 Hz   10 kHz   100 kHz

▲ FIGURE 18–15

Bode plot for Figure 18–14. The red line is the ideal response curve and the blue line is the actual response curve.

Related Problem   If the frequency for the high-pass filter is decreased to 10 Hz, what is the output to input ratio in decibels?

Use Multisim file E18-05 to verify the calculated results in this example and to confirm your calculation for the related problem.

RL High-Pass Filter

A basic *RL* high-pass filter is shown in Figure 18–16. Notice that the output is taken across the inductor.

When the input frequency is at its critical value, $X_L = R$, and the output voltage is $0.707V_{in}$. As the frequency increases above f_c, X_L increases and, as a result, the output voltage

▶ FIGURE 18–16

RL high-pass filter.

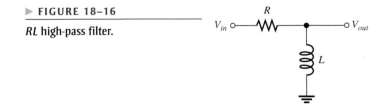

increases until it equals V_{in}. The expression for the critical frequency of the high-pass filter is the same as for the low-pass filter.

$$f_c = \frac{1}{2\pi(L/R)}$$

Phase Shift in a High-Pass Filter

Both the RC and the RL high-pass filters act as lead circuits. Recall from Chapters 15 and 16 that the phase shift from input to output for the RC lead circuit is

$$\phi = \tan^{-1}\left(\frac{X_C}{R}\right)$$

and the phase shift for the RL lead circuit is

$$\phi = \tan^{-1}\left(\frac{R}{X_L}\right)$$

At the critical frequency, $X_L = R$ and, therefore, $\phi = 45°$. As the frequency is increased, ϕ decreases toward $0°$, as shown in Figure 18–17.

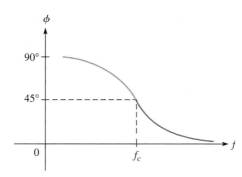

◀ FIGURE 18–17

Phase characteristic of a high-pass filter.

EXAMPLE 18–6

(a) In Figure 18–18, find the value of C so that X_C is ten times less than R at an input frequency of 10 kHz.

(b) If a 5 V sine wave with a dc level of 10 V is applied, what are the output voltage magnitude and the phase shift?

▸ FIGURE 18–18

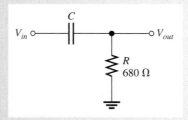

Solution   (a) Determine the value of C as follows:

$$X_C = 0.1R = 0.1(680 \ \Omega) = 68 \ \Omega$$

$$C = \frac{1}{2\pi f X_C} = \frac{1}{2\pi(10 \ \text{kHz})(68 \ \Omega)} = 0.234 \ \mu\text{F}$$

The nearest standard value of C is 0.22 μF.

(b) Determine the magnitude of the sinusoidal output as follows:

$$V_{out} = \left(\frac{R}{\sqrt{R^2 + X_C^2}}\right)V_{in} = \left(\frac{680\ \Omega}{\sqrt{(680\ \Omega)^2 + (68\ \Omega)^2}}\right)5\ V = \mathbf{4.98\ V}$$

The phase shift is

$$\phi = \tan^{-1}\left(\frac{X_C}{R}\right) = \tan^{-1}\left(\frac{68\ \Omega}{680\ \Omega}\right) = \mathbf{5.7°}$$

At $f = 10\ kHz$, which is a decade above the critical frequency, the sinusoidal output is almost equal to the input in magnitude, and the phase shift is very small. The 10 V dc level has been filtered out and does not appear at the output.

Related Problem   Repeat parts (a) and (b) of the example if R is changed to 220 Ω.

 Use Multisim file E18-06 to verify the calculated results in this example and to confirm your calculations for the related problem.

SECTION 18–2 REVIEW

1. The input voltage of a high-pass filter is 1 V. What is V_{out} at the critical frequency?
2. In a certain high-pass filter, $V_{in} = 10\angle 0°\ V$, $R = 1.0\ k\Omega$, and $X_L = 15\ k\Omega$. Determine V_{out}.

18–3 BAND-PASS FILTERS

A **band-pass filter** allows a certain band of frequencies to pass and attenuates or rejects all frequencies below and above the passband.

After completing this section, you should be able to

◆ **Analyze the operation of band-pass filters**

  ◆ Define *bandwidth*

  ◆ Show how a band-pass filter is implemented with low-pass and high-pass filters

  ◆ Explain the series-resonant band-pass filter

  ◆ Explain the parallel-resonant band-pass filter

  ◆ Calculate the bandwidth and output voltage of a band-pass filter

The bandwidth of a band-pass filter is the range of frequencies for which the current, and therefore the output voltage, is equal to or greater than 70.7% of its value at the resonant frequency.

As you know, bandwidth is often abbreviated *BW* and can be calculated as

$$BW = f_{c2} - f_{c1}$$

where f_{c1} is the lower cutoff frequency and f_{c2} is the upper cutoff frequency.
Figure 18–19 shows a typical band-pass response curve.

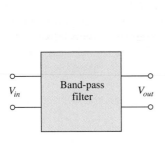 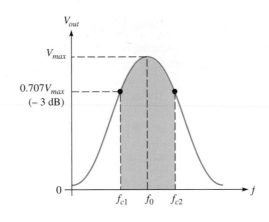

▲ FIGURE 18–19

Typical band-pass response curve.

Low-Pass/High-Pass Filter

A combination of a low-pass and a high-pass filter can be used to form a band-pass filter, as illustrated in Figure 18–20. The loading effect of the second filter on the first must be taken into account.

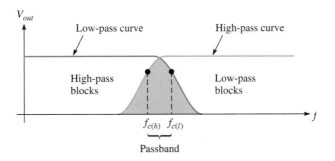

◀ FIGURE 18–20

A low-pass and a high-pass filter are used to form a band-pass filter.

If the critical frequency of the low-pass filter, $f_{c(l)}$, is higher than the critical frequency of the high-pass filter, $f_{c(h)}$, the responses overlap. Thus, all frequencies except those between $f_{c(h)}$ and $f_{c(l)}$ are eliminated, as shown in Figure 18–21.

◀ FIGURE 18–21

Overlapping response curves of a low-pass/high-pass filter.

EXAMPLE 18–7	A high-pass filter with $f_c = 2$ kHz and a low-pass filter with $f_c = 2.5$ kHz are used to construct a band-pass filter. Assuming no loading effect, what is the bandwidth of the passband?
Solution	$BW = f_{c(l)} - f_{c(h)} = 2.5 \text{ kHz} - 2 \text{ kHz} = \textbf{500 Hz}$
Related Problem	If $f_{c(l)} = 9$ kHz and the bandwidth is 1.5 kHz, what is $f_{c(h)}$?

Series Resonant Band-Pass Filter

A type of series resonant band-pass filter is shown in Figure 18–22. As you learned in Chapter 17, a series resonant circuit has minimum impedance and maximum current at the resonant frequency, f_r. Thus, most of the input voltage is dropped across the resistor at the resonant frequency. Therefore, the output across R has a band-pass characteristic with a maximum output at the frequency of resonance. The resonant frequency is called the **center frequency, f_0.** The bandwidth is determined by the quality factor, Q, of the circuit and the resonant frequency, as was discussed in Chapter 17. Recall that $Q = X_L/R$.

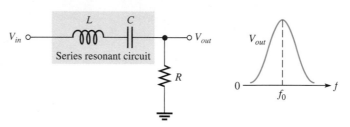

▲ FIGURE 18–22

Series resonant band-pass filter.

A higher value of Q results in a smaller bandwidth. A lower value of Q causes a larger bandwidth. A formula for the bandwidth of a resonant circuit in terms of Q is stated in the following equation:

Equation 18–5

$$BW = \frac{f_0}{Q}$$

EXAMPLE 18–8

Determine the output voltage magnitude at the center frequency (f_0) and the bandwidth for the filter in Figure 18–23.

▶ FIGURE 18–23

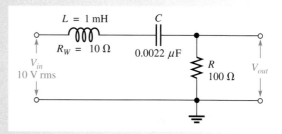

Solution At f_0, the impedance of the resonant circuit is equal to the winding resistance, R_W. By the voltage-divider formula,

$$V_{out} = \left(\frac{R}{R + R_W}\right)V_{in} = \left(\frac{100\ \Omega}{110\ \Omega}\right)10\ \text{V} = \textbf{9.09 V}$$

The center frequency is

$$f_0 = \frac{1}{2\pi\sqrt{LC}} = \frac{1}{2\pi\sqrt{(1\ \text{mH})(0.0022\ \mu\text{F})}} = 107\ \text{kHz}$$

At f_0, the inductive reactance is

$$X_L = 2\pi fL = 2\pi(107\,\text{kHz})(1\,\text{mH}) = 672\,\Omega$$

and the total resistance is

$$R_{tot} = R + R_W = 100\,\Omega + 10\,\Omega = 110\,\Omega$$

Therefore, the circuit Q is

$$Q = \frac{X_L}{R_{tot}} = \frac{672\,\Omega}{110\,\Omega} = 6.11$$

The bandwidth is

$$BW = \frac{f_0}{Q} = \frac{107\,\text{kHz}}{6.11} = \textbf{17.5 kHz}$$

Related Problem If a 1 mH coil with a winding resistance of 18 Ω replaces the existing coil in Figure 18–23, how is the bandwidth affected?

Use Multisim file E18-08 to verify the calculated results in this example and to confirm your answer for the related problem.

Parallel Resonant Band-Pass Filter

A type of band-pass filter using a parallel resonant circuit is shown in Figure 18–24. Recall that a parallel resonant circuit has maximum impedance at resonance. The circuit in Figure 18–24 acts as a voltage divider. At resonance, the impedance of the tank circuit is much greater than the resistance. Thus, most of the input voltage is across the tank circuit, producing a maximum output voltage at the resonant (center) frequency.

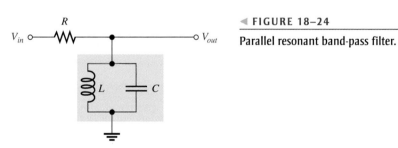

◀ **FIGURE 18–24**

Parallel resonant band-pass filter.

For frequencies above or below resonance, the tank circuit impedance drops off, and more of the input voltage is across R. As a result, the output voltage across the tank circuit drops off, creating a band-pass characteristic.

EXAMPLE 18–9

What is the center frequency of the filter in Figure 18–25? Assume $R_W = 0\,\Omega$.

▶ **FIGURE 18–25**

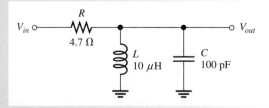

Solution The center frequency of the filter is its resonant frequency.

$$f_0 = \frac{1}{2\pi\sqrt{LC}} = \frac{1}{2\pi\sqrt{(10\,\mu H)(100\,pF)}} = \textbf{5.03 MHz}$$

Related Problem Determine f_0 in Figure 18–25 if C is changed to 1000 pF.

 Use Multisim file E18-09 to verify the calculated results in this example and to confirm your calculation for the related problem.

EXAMPLE 18–10

Determine the center frequency and bandwidth for the band-pass filter in Figure 18–26 if the inductor has a winding resistance of 15 Ω.

▶ **FIGURE 18–26**

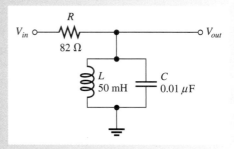

Solution Recall from Chapter 17 (Eq. 17–13) that the resonant (center) frequency of a nonideal tank circuit is

$$f_0 = \frac{\sqrt{1 - (R_W^2 C/L)}}{2\pi\sqrt{LC}} = \frac{\sqrt{1 - (15\,\Omega)^2(0.01\,\mu F)/50\,mH}}{2\pi\sqrt{(50\,mH)(0.01\mu F)}} = \textbf{7.12 kHz}$$

The Q of the coil at resonance is

$$Q = \frac{X_L}{R_W} = \frac{2\pi f_0 L}{R_W} = \frac{2\pi(7.12\,kHz)(50\,mH)}{15\,\Omega} = 149$$

The bandwidth of the filter is

$$BW = \frac{f_0}{Q} = \frac{7.12\,kHz}{149} = \textbf{47.8 Hz}$$

Note that since $Q > 10$, the simpler formula, $f_0 = 1/(2\pi\sqrt{LC})$, could have been used to calculate f_0.

Related Problem Knowing the value of Q, recalculate f_0 using the simpler formula.

SECTION 18–3 REVIEW

1. For a band-pass filter, $f_{c(h)} = 29.8$ kHz and $f_{c(l)} = 30.2$ kHz. What is the bandwidth?
2. A parallel resonant band-pass filter has the following values: $R_W = 15\,\Omega$, $L = 50\,\mu H$, and $C = 470$ pF. Determine the approximate center frequency.

18–4 BAND-STOP FILTERS

A band-stop filter is essentially the opposite of a band-pass filter in terms of the responses. A **band-stop filter** allows all frequencies to pass except those lying within a certain stopband.

After completing this section, you should be able to

- ◆ **Analyze the operation of band-stop filters**
  - ◆ Show how a band-stop filter is implemented with low-pass and high-pass filters
  - ◆ Explain the series-resonant band-stop filter
  - ◆ Explain the parallel-resonant band-stop filter
  - ◆ Calculate the bandwidth and output voltage of a band-stop filter

Figure 18–27 shows a general band-stop response curve.

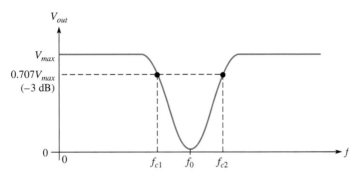

▲ **FIGURE 18–27**

General band-stop response curve.

Low-Pass/High-Pass Filter

A band-stop filter can be formed from a low-pass and a high-pass filter, as shown in Figure 18–28.

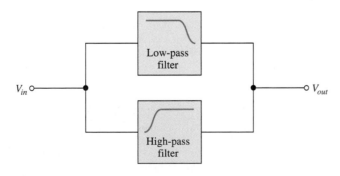

▲ **FIGURE 18–28**

A low-pass and a high-pass filter are used to form a band-stop filter.

If the low-pass critical frequency, $f_{c(l)}$, is set lower than the high-pass critical frequency, $f_{c(h)}$, a band-stop characteristic is formed as illustrated in Figure 18–29.

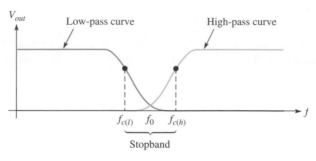

▲ FIGURE 18–29

Band-stop response curve.

Series Resonant Band-Stop Filter

A series resonant circuit used in a band-stop configuration is shown in Figure 18–30. Basically, it works as follows: At the resonant frequency, the impedance is minimum, and therefore the output voltage is minimum. Most of the input voltage is dropped across R. At frequencies above and below resonance, the impedance increases, causing more voltage across the output.

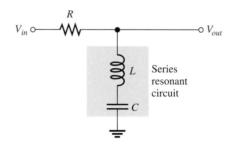

▲ FIGURE 18–30

Series resonant band-stop filter.

EXAMPLE 18–11

Find the output voltage magnitude at f_0 and the bandwidth in Figure 18–31.

▶ FIGURE 18–31

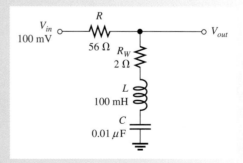

Solution Since $X_L = X_C$ at resonance, the output voltage is

$$V_{out} = \left(\frac{R_W}{R + R_W}\right)V_{in} = \left(\frac{2\,\Omega}{58\,\Omega}\right)100\,\text{mV} = \mathbf{3.45\,mV}$$

To determine the bandwidth, first calculate the center frequency and Q of the circuit.

$$f_0 = \frac{1}{2\pi\sqrt{LC}} = \frac{1}{2\pi\sqrt{(100 \text{ mH})(0.01 \ \mu\text{F})}} = 5.03 \text{ kHz}$$

$$Q = \frac{X_L}{R} = \frac{2\pi fL}{R} = \frac{2\pi(5.03 \text{ kHz})(100 \text{ mH})}{58 \ \Omega} = \frac{3.16 \text{ k}\Omega}{58 \ \Omega} = 54.5$$

$$BW = \frac{f_0}{Q} = \frac{5.03 \text{ kHz}}{54.5} = \mathbf{92.3 \text{ Hz}}$$

Related Problem   Assume $R_W = 10 \ \Omega$ in Figure 18–31. Determine V_{out} and the bandwidth.

Use Multisim file E18-11 to verify the calculated results in this example and to confirm your calculation for the related problem.

Parallel Resonant Band-Stop Filter

A parallel resonant circuit used in a band-stop configuration is shown in Figure 18–32. At the resonant frequency, the tank impedance is maximum, and so most of the input voltage appears across it. Very little voltage is across R at resonance. As the tank impedance decreases above and below resonance, the output voltage increases.

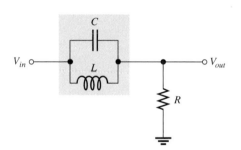

◀ **FIGURE 18–32**

Parallel resonant band-stop filter.

EXAMPLE 18–12

Find the center frequency of the filter in Figure 18–33. Draw the output response curve showing the minimum and maximum voltages.

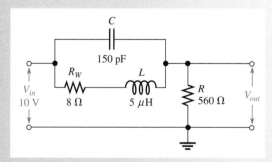

▲ **FIGURE 18–33**

Solution   The center frequency is

$$f_0 = \frac{\sqrt{1 - R_W^2 C/L}}{2\pi\sqrt{LC}} = \frac{\sqrt{1 - (8\ \Omega)^2(150\ \text{pF})/5\ \mu\text{H}}}{2\pi\sqrt{(5\ \mu\text{H})(150\ \text{pF})}} = \textbf{5.79 MHz}$$

At the center (resonant) frequency,

$$X_L = 2\pi f_0 L = 2\pi(5.79\ \text{MHz})(5\ \mu\text{H}) = 182\ \Omega$$

$$Q = \frac{X_L}{R_W} = \frac{182\ \Omega}{8\ \Omega} = 22.8$$

$$Z_r = R_W(Q^2 + 1) = 8\ \Omega(22.8^2 + 1) = 4.17\ \text{k}\Omega \quad \text{(purely resistive)}$$

Next, use the voltage-divider formula to find the minimum output voltage magnitude.

$$V_{out(min)} = \left(\frac{R}{R + Z_r}\right)V_{in} = \left(\frac{560\ \Omega}{4.73\ \text{k}\Omega}\right)10\ \text{V} = 1.18\ \text{V}$$

At zero frequency, the impedance of the tank circuit is R_W because $X_C = \infty$ and $X_L = 0\ \Omega$. Therefore, the maximum output voltage below resonance is

$$V_{out(max)} = \left(\frac{R}{R + R_W}\right)V_{in} = \left(\frac{560\ \Omega}{568\ \Omega}\right)10\ \text{V} = 9.86\ \text{V}$$

As the frequency increases much higher than f_0, X_C approaches $0\ \Omega$, and V_{out} approaches V_{in} (10 V). Figure 18–34 shows the response curve.

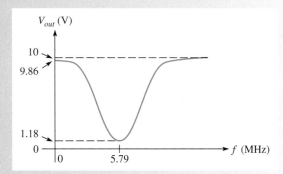

▲ FIGURE 18–34

Related Problem   What is the minimum output voltage if $R = 1.0\ \text{k}\Omega$ in Figure 18–33?

Use Multisim file E18-12 to verify the calculated results in this example and to confirm your calculation for the related problem.

SECTION 18–4
REVIEW

1. How does a band-stop filter differ from a band-pass filter?
2. Name three basic ways to construct a band-stop filter.

A Circuit Application

In this circuit application, you will plot the frequency responses of two types of filters based on a series of oscilloscope measurements and identify the type of filter in each case. The filters are contained in sealed modules as shown in Figure 18–35. You are concerned only with determining the filter response characteristics and not the types of internal components.

Filter Measurement and Analysis

◆ Refer to Figure 18–36. Based on the series of four oscilloscope measurements, create a Bode plot for the filter under test, specify applicable frequencies, and identify the type of filter.

◆ Refer to Figure 18–37. Based on the series of six oscilloscope measurements, create a Bode plot for the filter under test, specify applicable frequencies, and identify the type of filter.

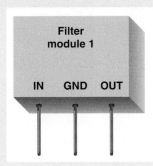

◀ **FIGURE 18–35**

Filter modules.

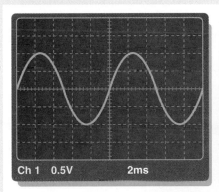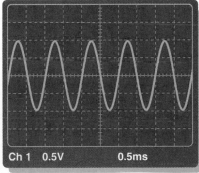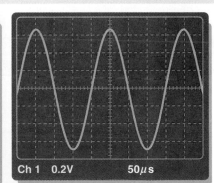

Ch 1   0.5V   2ms

Ch 1   0.5V   0.5ms

Ch 1   0.2V   50µs

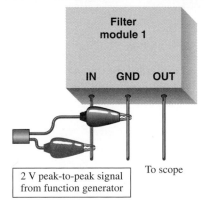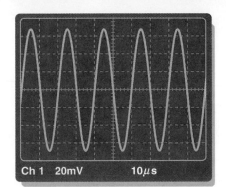

Filter module 1

IN   GND   OUT

To scope

2 V peak-to-peak signal from function generator

Ch 1   20mV   10µs

▲ **FIGURE 18–36**

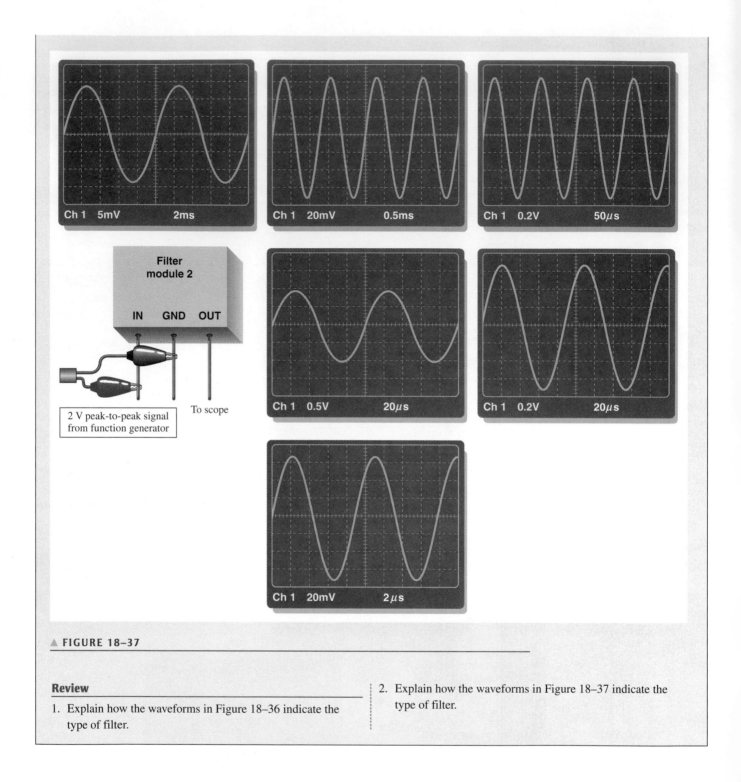

▲ FIGURE 18–37

Review

1. Explain how the waveforms in Figure 18–36 indicate the type of filter.

2. Explain how the waveforms in Figure 18–37 indicate the type of filter.

SUMMARY

◆ Four categories of passive filters according to their response characteristics are low-pass, high-pass, band-pass, and band-stop.

◆ In an *RC* low-pass filter, the output voltage is taken across the capacitor and the output lags the input.

◆ In an *RL* low-pass filter, the output voltage is taken across the resistor and the output lags the input.

◆ In an *RC* high-pass filter, the output is taken across the resistor and the output leads the input.

◆ In an *RL* high-pass filter, the output is taken across the inductor and the output leads the input.

◆ The roll-off rate of a basic *RC* or *RL* filter is −20 dB per decade.

◆ A band-pass filter passes frequencies between the lower and upper critical frequencies and rejects all others.

◆ A band-stop filter rejects frequencies between its lower and upper critical frequencies and passes all others.

◆ The bandwidth of a resonant filter is determined by the quality factor (*Q*) of the circuit and the resonant frequency.

◆ Critical frequencies are also called −3 dB frequencies.

◆ The output voltage is 70.7% of its maximum at the critical frequencies.

KEY TERMS

Key terms and other bold terms in the chapter are defined in the end-of-book glossary.

Attenuation A reduction of the output signal compared to the input signal, resulting in a ratio with a value of less than 1 for the output voltage to the input voltage of a circuit.

Band-pass filter A filter that passes a range of frequencies lying between two critical frequencies and rejects frequencies above and below that range.

Band-stop filter A filter that rejects a range of frequencies lying between two critical frequencies and passes frequencies above and below that range.

Bode plot The graph of a filter's frequency response showing the change in the output voltage to input voltage ratio expressed in dB as a function of frequency for a constant input voltage.

Center frequency (f_0) The resonant frequency of a band-pass or band-stop filter.

Critical frequency (f_c) The frequency at which a filter's output voltage is 70.7% of the maximum.

Decade A tenfold change in frequency or other parameter.

High-pass filter A type of filter that passes all frequencies above a critical frequency and rejects all frequencies below that critical frequency.

Low-pass filter A type of filter that passes all frequencies below a critical frequency and rejects all frequencies above that critical frequency.

Passband The range of frequencies passed by a filter.

Roll-off The rate of decrease of a filter's frequency response.

FORMULAS

18–1	$dB = 10 \log\left(\dfrac{P_{out}}{P_{in}}\right)$	Power ratio in decibels
18–2	$dB = 20 \log\left(\dfrac{V_{out}}{V_{in}}\right)$	Voltage ratio in decibels
18–3	$f_c = \dfrac{1}{2\pi RC}$	Critical frequency
18–4	$f_c = \dfrac{1}{2\pi(L/R)}$	Critical frequency
18–5	$BW = \dfrac{f_0}{Q}$	Bandwidth

SELF-TEST

Answers are at the end of the chapter.

1. The maximum output voltage of a certain low-pass filter is 10 V. The output voltage at the critical frequency is

   (a) 10 V   (b) 0 V   (c) 7.07 V   (d) 1.414 V

2. A sinusoidal voltage with a peak-to-peak value of 15 V is applied to an *RC* low-pass filter. If the reactance at the input frequency is zero, the output voltage is
(a) 15 V peak-to-peak   (b) zero
(c) 10.6 V peak-to-peak   (d) 7.5 V peak-to-peak

3. The same signal in Question 2 is applied to an *RC* high-pass filter. If the reactance is zero at the input frequency, the output voltage is
(a) 15 V peak-to-peak   (b) zero
(c) 10.6 V peak-to-peak   (d) 7.5 V peak-to-peak

4. At the critical frequency, the output of a filter is down from its maximum by
(a) 0 dB   (b) −3 dB   (c) −20 dB   (d) −6 dB

5. If the output of a low-pass *RC* filter is 12 dB below its maximum at $f = 1$ kHz, then at $f = 10$ kHz, the output is below its maximum by
(a) 3 dB   (b) 10 dB   (c) 20 dB   (d) 32 dB

6. In a passive filter, the ratio V_{out}/V_{in} is called
(a) roll-off   (b) gain
(c) attenuation   (d) critical reduction

7. For each decade increase in frequency above the critical frequency, the output of a low-pass filter decreases by
(a) 20 dB   (b) 3 dB   (c) 10 dB   (d) 0 dB

8. At the critical frequency, the phase shift through a high-pass filter is
(a) 90°   (b) 0°   (c) 45°   (d) dependent on the reactance

9. In a series resonant band-pass filter, a higher value of *Q* results in
(a) a higher resonant frequency   (b) a smaller bandwidth
(c) a higher impedance   (d) a larger bandwidth

10. At series resonance,
(a) $X_C = X_L$ (b) $X_C > X_L$ (c) $X_C < X_L$

11. In a certain parallel resonant band-pass filter, the resonant frequency is 10 kHz. If the bandwidth is 2 kHz, the lower critical frequency is
(a) 5 kHz   (b) 12 kHz   (c) 9 kHz   (d) not determinable

12. In a band-pass filter, the output voltage at the resonant frequency is
(a) minimum   (b) maximum
(c) 70.7% of maximum   (d) 70.7% of minimum

13. In a band-stop filter, the output voltage at the critical frequencies is
(a) minimum   (b) maximum
(c) 70.7% of maximum   (d) 70.7% of minimum

14. At a sufficiently high value of *Q*, the resonant frequency for a parallel resonant filter is ideally
(a) much greater than the resonant frequency of a series resonant filter
(b) much less than the resonant frequency of a series resonant filter
(c) equal to the resonant frequency of a series resonant filter

CIRCUIT DYNAMICS QUIZ

Answers are at the end of the chapter.

Refer to Figure 18–38(a).

1. If the frequency of the input voltage is increased, V_{out}
(a) increases   (b) decreases   (c) stays the same

2. If *C* is increased, the output voltage
(a) increases   (b) decreases   (c) stays the same

Refer to Figure 18–38(d).

3. If the frequency of the input voltage is increased, V_{out}

    (a) increases   **(b)** decreases   **(c)** stays the same

4. If L is increased, the output voltage

    (a) increases   **(b)** decreases   **(c)** stays the same

Refer to Figure 18–40.

5. If the switch is thrown from position 1 to position 2, the critical frequency

    (a) increases   **(b)** decreases   **(c)** stays the same

6. If the switch is thrown from position 2 to position 3, the critical frequency

    (a) increases   **(b)** decreases   **(c)** stays the same

Refer to Figure 18–41(a).

7. If the frequency of the input voltage is increased, V_{out}

    (a) increases   **(b)** decreases   **(c)** stays the same

8. If R is increased to 180 Ω, the output voltage

    (a) increases   **(b)** decreases   **(c)** stays the same

Refer to Figure 18–42.

9. If the switch is thrown from position 1 to position 2, the critical frequency

    (a) increases   **(b)** decreases   **(c)** stays the same

10. If the switch is in position 3 and R_5 opens, V_{out}

    (a) increases   **(b)** decreases   **(c)** stays the same

Refer to Figure 18–48.

11. If L_2 opens, the output voltage

    (a) increases   **(b)** decreases   **(c)** stays the same

12. If C becomes shorted, the output voltage

    (a) increases   **(b)** decreases   **(c)** stays the same

PROBLEMS

More difficult problems are indicated by an asterisk (*).
Answers to odd-numbered problems are at the end of the book.

SECTION 18–1 Low-Pass Filters

1. In a certain low-pass filter, $X_C = 500\ \Omega$ and $R = 2.2\ k\Omega$. What is the output voltage (V_{out}) when the input is 10 V rms?

2. A certain low-pass filter has a critical frequency of 3 kHz. Determine which of the following frequencies are passed and which are rejected:

    (a) 100 Hz   **(b)** 1 kHz   **(c)** 2 kHz   **(d)** 3 kHz   **(e)** 5 kHz

3. Determine the output voltage (V_{out}) of each filter in Figure 18–38 at the specified frequency when $V_{in} = 10$ V.

4. What is f_c for each filter in Figure 18–38? Determine the output voltage at f_c in each case when $V_{in} = 5$ V.

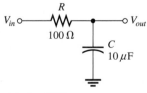

(a) $f = 60$ Hz

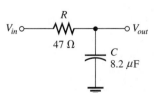

(b) $f = 400$ Hz

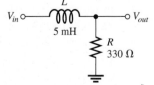

(c) $f = 1$ kHz

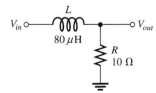

(d) $f = 2$ kHz

▲ **FIGURE 18–38**

5. For the filter in Figure 18–39, calculate the value of C required for each of the following critical frequencies:

   (a) 60 Hz   (b) 500 Hz   (c) 1 kHz   (d) 5 kHz

► FIGURE 18–39

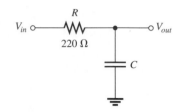

*6. Determine the critical frequency for each switch position on the switched filter network of Figure 18–40.

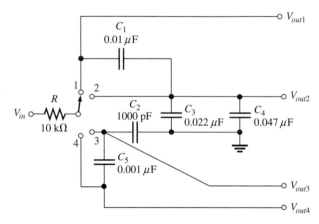

▲ FIGURE 18–40

7. Draw a Bode plot for each part of Problem 5.

8. For each following case, express the voltage ratio in dB:

   (a) $V_{in} = 1$ V, $V_{out} = 1$ V   (b) $V_{in} = 5$ V, $V_{out} = 3$ V

   (c) $V_{in} = 10$ V, $V_{out} = 7.07$ V   (d) $V_{in} = 25$ V, $V_{out} = 5$ V

9. The input voltage to a low-pass RC filter is 8 V rms. Find the output voltage at the following dB levels:

   (a) -1 dB   (b) -3 dB   (c) -6 dB   (d) -20 dB

10. For a basic RC low-pass filter, find the output voltage in dB relative to a 0 dB input for the following frequencies ($f_c = 1$ kHz):

   (a) 10 kHz   (b) 100 kHz   (c) 1 MHz

SECTION 18–2   High-Pass Filters

11. In a high-pass filter, $X_C = 500$ Ω and $R = 2.2$ kΩ. What is the output voltage ($\mathbf{V}_{out}$) when $V_{in} = 10$ V rms?

12. A high-pass filter has a critical frequency of 50 Hz. Determine which of the following frequencies are passed and which are rejected:

   (a) 1 Hz   (b) 20 Hz   (c) 50 Hz   (d) 60 Hz   (e) 30 kHz

13. Determine the output voltage of each filter in Figure 18–41 at the specified frequency when $V_{in} = 10$ V.

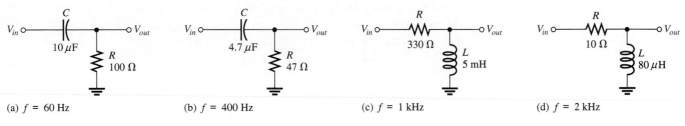

(a) $f = 60$ Hz

(b) $f = 400$ Hz

(c) $f = 1$ kHz

(d) $f = 2$ kHz

▲ FIGURE 18–41

14. What is f_c for each filter in Figure 18–41? Determine the output voltage at f_c in each case ($V_{in} = 10$ V).

15. Draw the Bode plot for each filter in Figure 18–41.

*16. Determine f_c for each switch position in Figure 18–42.

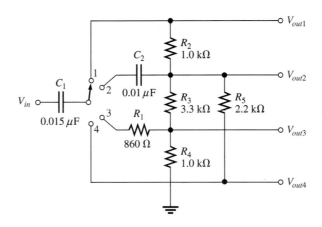

▲ FIGURE 18–42

SECTION 18–3 Band-Pass Filters

17. Determine the center frequency for each filter in Figure 18–43.

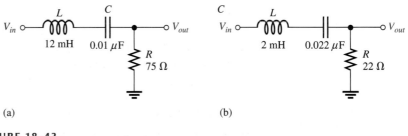

(a)

(b)

▲ FIGURE 18–43

18. Assuming that the coils in Figure 18–43 have a winding resistance of 10 Ω, find the bandwidth for each filter.

19. What are the upper and lower critical frequencies for each filter in Figure 18–43? Assume the response is symmetrical about f_0.

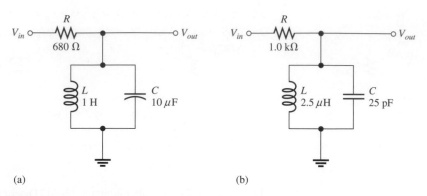

▲ FIGURE 18–44

20. For each filter in Figure 18–44, find the center frequency of the passband. Neglect R_W.

21. If the coils in Figure 18–44 have a winding resistance of 4 Ω, what is the output voltage at resonance when $V_{in} = 120$ V?

*22. Determine the separation of center frequencies for all switch positions in Figure 18–45. Do any of the responses overlap? Assume $R_W = 0$ Ω for each coil.

*23. Design a band-pass filter using a parallel resonant circuit to meet all of the following specifications: $BW = 500$ Hz; $Q = 40$; and $I_{C(max)} = 20$ mA, $V_{C(max)} = 2.5$ V.

▶ FIGURE 18–45

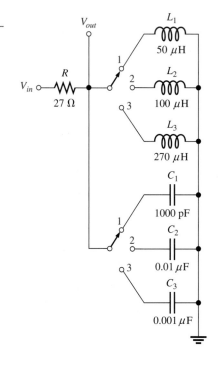

SECTION 18–4 Band-Stop Filters

24. Determine the center frequency for each filter in Figure 18–46.

▶ FIGURE 18–46

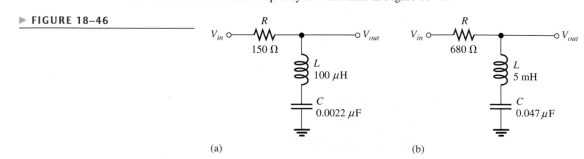

(a)   (b)

25. For each filter in Figure 18–47, find the center frequency of the stopband.

26. If the coils in Figure 18–47 have a winding resistance of 8 Ω, what is the output voltage at resonance when V_{in} = 50 V?

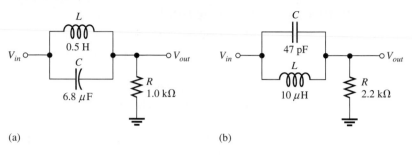

(a)   (b)

▲ **FIGURE 18–47**

*27. Determine the values of L_1 and L_2 in Figure 18–48 to pass a signal with a frequency of 1200 kHz and stop (reject) a signal with a frequency of 456 kHz.

▶ **FIGURE 18–48**

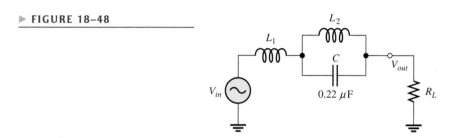

Multisim Troubleshooting and Analysis

These problems require your Multisim CD-ROM.

28. Open file P18-28 and determine if there is a fault. If so, find the fault.

29. Open file P18-29 and determine if there is a fault. If so, find the fault.

30. Open file P18-30 and determine if there is a fault. If so, find the fault.

31. Open file P18-31 and determine if there is a fault. If so, find the fault.

32. Open file P18-32 and determine if there is a fault. If so, find the fault.

33. Open file P18-33 and determine if there is a fault. If so, find the fault.

34. Open file P18-34 and determine the center frequency of the circuit.

35. Open file P18-35 and determine the bandwidth of the circuit.

ANSWERS

SECTION REVIEWS

SECTION 18–1 Low-Pass Filters

1. The passband is 0 Hz to 2.5 kHz.
2. $\mathbf{V}_{out} = 100 \angle -88.9°$ mV rms
3. $20 \log(V_{out}/V_{in}) = -9.54$ dB

SECTION 18–2 High-Pass Filters

1. $V_{out} = 0.707$ V
2. $\mathbf{V}_{out} = 9.98 \angle 3.81°$ V

SECTION 18–3 **Band-Pass Filters**

    1. $BW = 30.2\,\text{kHz} - 29.8\,\text{kHz} = 400\,\text{Hz}$

    2. $f_0 \cong 1.04\,\text{MHz}$

SECTION 18–4 **Band-Stop Filters**

    1. A band-stop filter rejects, rather than passes, a certain band of frequencies.

    2. High-pass/low-pass combination, series resonant circuit, and parallel resonant circuit

A Circuit Application

    1. The waveforms indicate that the output amplitude decreases with an increase in frequency as in a low-pass filter.

    2. The waveforms indicate that the output amplitude is maximum at 10 kHz and drops off above and below as in a band-pass filter.

RELATED PROBLEMS FOR EXAMPLES

18–1 $-1.41\,\text{dB}$

18–2   7.23 kHz

18–3 f_c increases to 159 kHz. Roll-off rate remains $-20\,\text{dB/decade}$.

18–4 f_c increases to 350 kHz. Roll-off rate remains $-20\,\text{dB/decade}$.

18–5 $-60\,\text{dB}$

18–6 $C = 0.723\,\mu\text{F}$; $V_{out} = 4.98\,\text{V}$; $\phi = 5.7°$

18–7   10.5 kHz

18–8 BW increases to 18.8 kHz.

18–9   1.59 MHz

18–10   7.12 kHz (no significant difference)

18–11 $V_{out} = 15.2\,\text{mV}$; $BW = 105\,\text{Hz}$

18–12   1.94 V

SELF-TEST

    1. (c)   **2.** (b)   **3.** (a)   **4.** (b)   **5.** (d)   **6.** (c)   **7.** (a)   **8.** (c)

    9. (b)   **10.** (a)   **11.** (c)   **12.** (b)   **13.** (c)   **14.** (c)

CIRCUIT DYNAMICS QUIZ

    1. (b)   **2.** (b)   **3.** (b)   **4.** (b)   **5.** (b)   **6.** (a)   **7.** (a)   **8.** (a)

    9. (a)   **10.** (c)   **11.** (b)   **12.** (a)

CIRCUIT THEOREMS IN AC ANALYSIS

19

CHAPTER OBJECTIVES

◆ Apply the superposition theorem to ac circuit analysis

◆ Apply Thevenin's theorem to simplify reactive ac circuits for analysis

◆ Apply Norton's theorem to simplify reactive ac circuits

◆ Apply the maximum power transfer theorem

KEY TERMS

◆ Superposition theorem

◆ Thevenin's theorem

◆ Equivalent circuit

◆ Norton's theorem

◆ Complex conjugate

A CIRCUIT APPLICATION PREVIEW

In the circuit application, you will evaluate a band-pass filter module to determine its internal component values. You will apply Thevenin's theorem to determine an optimum load impedance for maximum power transfer.

VISIT THE COMPANION WEBSITE

Study aids for this chapter are available at
http://www.prenhall.com/floyd

INTRODUCTION

Four important theorems were covered in Chapter 8 with emphasis on their applications in the analysis of dc circuits. This chapter is a continuation of that coverage with emphasis on applications in the analysis of ac circuits with reactive components.

The theorems in this chapter make analysis easier for certain types of circuits. These methods do not replace Ohm's law and Kirchhoff's laws, but they are normally used in conjunction with those laws in certain situations.

The superposition theorem helps you to deal with circuits that have multiple sources. Thevenin's and Norton's theorems provide methods for reducing a circuit to a simple equivalent form for easier analysis. The maximum power transfer theorem is used in applications where it is important for a given circuit to provide maximum power to a load.

19–1 THE SUPERPOSITION THEOREM

The superposition theorem was introduced in Chapter 8 for use in dc circuit analysis. In this section, the superposition theorem is applied to circuits with ac sources and reactive components.

After completing this section, you should be able to

◆ **Apply the superposition theorem to ac circuit analysis**

  ◆ State the superposition theorem

  ◆ List the steps in applying the theorem

The **superposition theorem** can be stated as follows:

The current in any given branch of a multiple-source circuit can be found by determining the currents in that particular branch produced by each source acting alone, with all other sources replaced by their internal impedances. The total current in the given branch is the phasor sum of the individual currents in that branch.

The procedure for the application of the superposition theorem is as follows:

Step 1. Leave one of the voltage (current) sources in the circuit, and replace all others with their internal impedance. For ideal voltage sources, the internal impedance is zero. For ideal current sources, the internal impedance is infinite. We will call this procedure *zeroing* the source.

Step 2. Find the current in the branch of interest produced by the one remaining source.

Step 3. Repeat Steps 1 and 2 for each source in turn. When complete, you will have a number of current values equal to the number of sources in the circuit.

Step 4. Add the individual current values as phasor quantities.

Example 19–1 illustrates this procedure for a circuit containing two ideal voltage sources, V_{s1} and V_{s2}.

EXAMPLE 19–1

Find the current in R of Figure 19–1 using the superposition theorem. Assume the internal source impedances are zero.

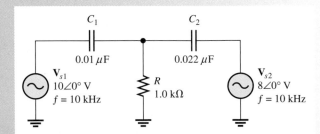

▲ **FIGURE 19–1**

Solution **Step 1.** Replace V_{s2} with its internal impedance (zero in this case), and find the current in R due to V_{s1}, as indicated in Figure 19–2.

$$X_{C1} = \frac{1}{2\pi f C_1} = \frac{1}{2\pi(10 \text{ kHz})(0.01 \text{ }\mu\text{F})} = 1.59 \text{ k}\Omega$$

$$X_{C2} = \frac{1}{2\pi f C_2} = \frac{1}{2\pi(10 \text{ kHz})(0.022 \text{ }\mu\text{F})} = 723 \text{ }\Omega$$

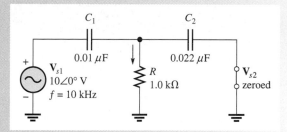

▲ **FIGURE 19–2**

Looking from V_{s1}, the impedance is

$$\mathbf{Z} = \mathbf{X}_{C1} + \frac{\mathbf{R}\mathbf{X}_{C2}}{\mathbf{R} + \mathbf{X}_{C2}} = 1.59\angle -90° \text{ k}\Omega + \frac{(1.0\angle 0° \text{ k}\Omega)(723\angle -90° \text{ }\Omega)}{1.0 \text{ k}\Omega - j723 \text{ }\Omega}$$

$$= 1.59\angle -90° \text{ k}\Omega + 588\angle -54.1° \text{ }\Omega$$

$$= -j1.59 \text{ k}\Omega + 345 \text{ }\Omega - j476 \text{ }\Omega = 345 \text{ }\Omega - j2.07 \text{ k}\Omega$$

Converting to polar form yields

$$\mathbf{Z} = 2.10\angle -80.5° \text{ k}\Omega$$

The total current from V_{s1} is

$$\mathbf{I}_{s1} = \frac{\mathbf{V}_{s1}}{\mathbf{Z}} = \frac{10\angle 0° \text{ V}}{2.10\angle -80.5° \text{ k}\Omega} = 4.76\angle 80.5° \text{ mA}$$

Use the current-divider formula. The current through R due to V_{s1} is

$$\mathbf{I}_{R1} = \left(\frac{X_{C2}\angle -90°}{R - jX_{C2}}\right)\mathbf{I}_{s1} = \left(\frac{723\angle -90° \text{ }\Omega}{1.0 \text{ k}\Omega - j723 \text{ }\Omega}\right)4.76\angle 80.5° \text{ mA}$$

$$= (0.588\angle -54.9° \text{ }\Omega)(4.76\angle 80.5° \text{ mA}) = 2.80\angle 25.6° \text{ mA}$$

Step 2. Find the current in R due to source V_{s2} by replacing V_{s1} with its internal impedance (zero), as shown in Figure 19–3.

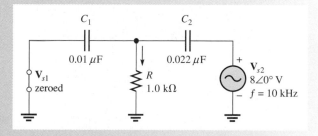

▲ **FIGURE 19–3**

Looking from V_{s2}, the impedance is

$$\mathbf{Z} = \mathbf{X}_{C2} + \frac{\mathbf{R}\mathbf{X}_{C1}}{\mathbf{R} + \mathbf{X}_{C1}} = 723\angle -90° \,\Omega + \frac{(1.0\angle 0° \,k\Omega)(1.59\angle -90° \,k\Omega)}{1.0\,k\Omega - j1.59\,k\Omega}$$

$$= 723\angle -90° \,\Omega + 847\angle -32.2° \,\Omega$$

$$= -j723\,\Omega + 717\,\Omega - j451\,\Omega = 717\,\Omega - j1174\,\Omega$$

Converting to polar form yields

$$\mathbf{Z} = 1376\angle -58.6° \,\Omega$$

The total current from V_{s2} is

$$\mathbf{I}_{s2} = \frac{\mathbf{V}_{s2}}{\mathbf{Z}} = \frac{8\angle 0° \,V}{1376\angle -58.6° \,\Omega} = 5.81\angle 58.6° \,mA$$

Use the current-divider formula. The current through R due to V_{s2} is

$$\mathbf{I}_{R2} = \left(\frac{X_{C1}\angle -90°}{R - jX_{C1}}\right)\mathbf{I}_{s2}$$

$$= \left(\frac{1.59\angle -90° \,k\Omega}{1.0\,k\Omega - j1.59\,k\Omega}\right)5.81\angle 58.6° \,mA = 4.91\angle 26.4° \,mA$$

Step 3. Convert the two individual resistor currents to rectangular form and add to get the total current through R.

$$\mathbf{I}_{R1} = 2.80\angle 25.6° \,mA = 2.53\,mA + j1.21\,mA$$

$$\mathbf{I}_{R2} = 4.91\angle 26.4° \,mA = 4.40\,mA + j2.18\,mA$$

$$\mathbf{I}_R = \mathbf{I}_{R1} + \mathbf{I}_{R2} = 6.93\,mA + j3.39\,mA = \mathbf{7.71\angle 26.1° \,mA}$$

Related Problem*   Determine $\mathbf{I}_R$ if $\mathbf{V}_{s2} = 8\angle 180° \,V$ in Figure 19–1.

 Use Multisim file E19-01 to verify the calculated results in this example and to confirm your calculation for the related problem.

*Answers are at the end of the chapter.

Example 19–2 illustrates the application of the superposition theorem for a circuit with two current sources, I_{s1} and I_{s2}.

EXAMPLE 19–2

Find the inductor current in Figure 19–4. Assume the current sources are ideal.

FIGURE 19–4

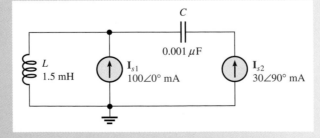

Solution **Step 1.** Find the current through the inductor due to current source I_{s1} by replacing source I_{s2} with an open, as shown in Figure 19–5. As you can see, the entire 100 mA from the current source I_{s1} is through the inductor.

▶ **FIGURE 19–5**

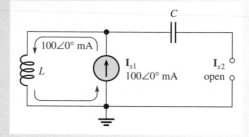

Step 2. Find the current through the inductor due to current source I_{s2} by replacing source I_{s1} with an open, as indicated in Figure 19–6. Notice that all of the 30 mA from source I_{s2} is through the inductor.

▶ **FIGURE 19–6**

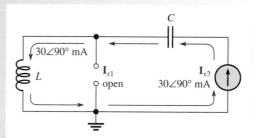

Step 3. To get the total inductor current, superimpose the two individual currents and add as phasor quantities.

$$\mathbf{I}_L = \mathbf{I}_{L1} + \mathbf{I}_{L2}$$
$$= 100\angle 0° \text{ mA} + 30\angle 90° \text{ mA} = 100 \text{ mA} + j30 \text{ mA}$$
$$= \mathbf{104\angle 16.7° \text{ mA}}$$

Related Problem Find the current through the capacitor in Figure 19–4.

Example 19–3 illustrates the analysis of a circuit with an ac voltage source and a dc voltage source. This situation is common in many amplifier applications.

EXAMPLE 19–3

Find the total current in the load resistor, R_L, in Figure 19–7. Assume the sources are ideal.

▶ **FIGURE 19–7**

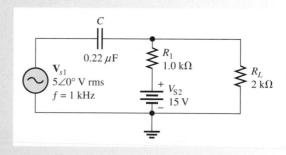

Solution   **Step 1.** Find the current through R_L due to the ac source V_{s1} by zeroing (replacing with its internal impedance) the dc source V_{S2}, as shown in Figure 19–8. Looking from V_{s1}, the impedance is

$$\mathbf{Z} = \mathbf{X}_C + \frac{\mathbf{R}_1\mathbf{R}_L}{\mathbf{R}_1 + \mathbf{R}_L}$$

$$X_C = \frac{1}{2\pi(1.0\,\text{kHz})(0.22\,\mu\text{F})} = 723\,\Omega$$

$$\mathbf{Z} = 723\angle{-90°}\,\Omega + \frac{(1.0\angle0°\,\text{k}\Omega)(2\angle0°\,\text{k}\Omega)}{3\angle0°\,\text{k}\Omega}$$

$$= -j723\,\Omega + 667\,\Omega = 984\angle{-47.3°}\,\Omega$$

The total current from the ac source is

$$\mathbf{I}_{s1} = \frac{\mathbf{V}_{s1}}{\mathbf{Z}} = \frac{5\angle0°\,\text{V}}{984\angle{-47.3°}\,\Omega} = 5.08\angle47.3°\,\text{mA}$$

Use the current-divider approach. The current in R_L due to V_{s1} is

$$\mathbf{I}_{RL(s1)} = \left(\frac{R_1}{R_1 + R_L}\right)\mathbf{I}_{s1} = \left(\frac{1.0\,\text{k}\Omega}{3\,\text{k}\Omega}\right)5.08\angle47.3°\,\text{mA} = 1.69\angle47.3°\,\text{mA}$$

▶ **FIGURE 19–8**

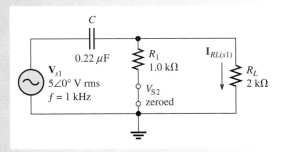

Step 2. Find the current in R_L due to the dc source V_{S2} by zeroing V_{s1} (replacing with its internal impedance), as shown in Figure 19–9. The impedance magnitude as seen by V_{S2} is

$$Z = R_1 + R_L = 3\,\text{k}\Omega$$

The current produced by V_{S2} is

$$I_{RL(S2)} = \frac{V_{S2}}{Z} = \frac{15\,\text{V}}{3\,\text{k}\Omega} = 5\,\text{mA dc}$$

▶ **FIGURE 19–9**

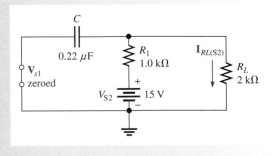

Step 3. By superposition, the total current in R_L is $1.69\angle47.3°$ mA riding on a dc level of 5 mA, as indicated in Figure 19–10.

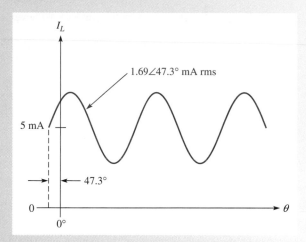

▲ FIGURE 19–10

Related Problem Determine the current through R_L if V_{S2} is changed to 9 V.

Use Multisim file E19-03 to verify the calculated results in this example and to confirm your calculation for the related problem.

SECTION 19–1
REVIEW
Answers are at the end of the chapter.

1. If two equal currents are in opposing directions at any instant of time in a given branch of a circuit, what is the net current at that Instant?

2. Why is the superposition theorem useful in the analysis of multiple-source circuits?

3. Using the superposition theorem, find the magnitude of the current through R in Figure 19–11.

▶ FIGURE 19–11

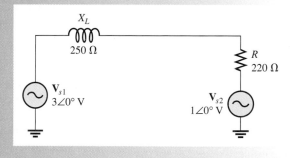

19–2 THEVENIN'S THEOREM

Thevenin's theorem, as applied to ac circuits, provides a method for reducing any circuit to an equivalent form that consists of an equivalent ac voltage source in series with an equivalent impedance.

After completing this section, you should be able to

◆ **Apply Thevenin's theorem to simplify reactive ac circuits for analysis**

    ◆ Describe the form of a Thevenin equivalent circuit

    ◆ Obtain the Thevenin equivalent ac voltage source

    ◆ Obtain the Thevenin equivalent impedance

    ◆ List the steps in applying Thevenin's theorem to an ac circuit

Equivalency

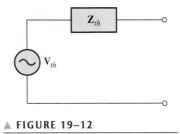

▲ FIGURE 19–12

Thevenin's equivalent circuit.

The form of Thevenin's **equivalent circuit** is shown in Figure 19–12. Regardless of how complex the original circuit is, it can always be reduced to this equivalent form. The equivalent voltage source is designated $\mathbf{V}_{th}$; the equivalent impedance is designated $\mathbf{Z}_{th}$ (lowercase italic subscript denotes ac quantity). Notice that the impedance is represented by a block in the circuit diagram. This is because the equivalent impedance can be of several forms: purely resistive, purely capacitive, purely inductive, or a combination of a resistance and a reactance.

Figure 19–13(a) shows a block diagram that represents an ac circuit of any given complexity. This circuit has two output terminals, A and B. A load impedance, $\mathbf{Z}_L$, is connected to the terminals. The circuit produces a certain voltage, $\mathbf{V}_L$, and a certain current, $\mathbf{I}_L$, as illustrated.

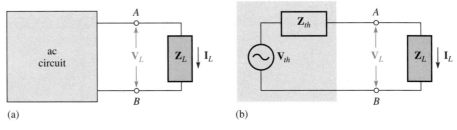

(a)   (b)

▲ FIGURE 19–13

An ac circuit of any complexity can be reduced to a Thevenin equivalent for analysis purposes.

By Thevenin's theorem, the circuit in the block can be reduced to an equivalent form, as indicated in the beige area of Figure 19–13(b). The term *equivalent* means that when the same value of load is connected to both the original circuit and Thevenin's equivalent circuit, the load voltages and currents are equal for both. Therefore, as far as the load is concerned, there is no difference between the original circuit and Thevenin's equivalent circuit. The load "sees" the same current and voltage regardless of whether it is connected to the original circuit or to the Thevenin equivalent. For ac circuits, the equivalent circuit is for one particular frequency. When the frequency is changed, the equivalent circuit must be recalculated.

Thevenin's Equivalent Voltage ($\mathbf{V}_{th}$)

As you have seen, the equivalent voltage, $\mathbf{V}_{th}$, is one part of the complete Thevenin equivalent circuit.

Thevenin's equivalent voltage is defined as the open circuit voltage between two specified terminals in a circuit.

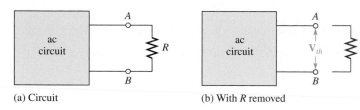

(a) Circuit   (b) With *R* removed

▲ **FIGURE 19–14**

How $\mathbf{V}_{th}$ is determined.

To illustrate, let's assume that an ac circuit of some type has a resistor connected between two specified terminals, *A* and *B*, as shown in Figure 19–14(a). We wish to find the Thevenin equivalent circuit for the circuit as "seen" by *R*. $\mathbf{V}_{th}$ is the voltage across terminals *A* and *B*, with *R* removed, as shown in part (b) of the figure. The circuit is viewed from the open terminals *A* and *B*, and *R* is considered external to the circuit for which the Thevenin equivalent is to be found.

The following three examples show how to find $\mathbf{V}_{th}$.

EXAMPLE 19–4

Refer to Figure 19–15. Determine $\mathbf{V}_{th}$ for the circuit within the beige box as viewed from terminals *A* and *B*.

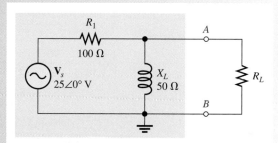

▲ **FIGURE 19–15**

Solution   Remove R_L and determine the voltage from *A* to *B* ($\mathbf{V}_{th}$). In this case, the voltage from *A* to *B* is the same as the voltage across X_L. This is determined using the voltage-divider method.

$$\mathbf{V}_L = \left(\frac{X_L \angle 90°}{R_1 + jX_L} \right) \mathbf{V}_s$$

$$= \left(\frac{50 \angle 90° \ \Omega}{100 \ \Omega + j50 \ \Omega} \right)$$

$$= \left(\frac{50 \angle 90° \ \Omega}{112 \angle 26.6° \ \Omega} \right) 25 \angle 0° \ \text{V} = 11.2 \angle 63.4° \ \text{V}$$

$$\mathbf{V}_{th} = \mathbf{V}_{AB} = \mathbf{V}_L = \mathbf{11.2 \angle 63.4° \ V}$$

Related Problem   Determine $\mathbf{V}_{th}$ if R_1 is changed to 47 Ω in Figure 19–15.

Use Multisim file E19-04 to verify the calculated results in this example and to confirm your calculation for the related problem.

EXAMPLE 19–5

Refer to Figure 19–16. Determine the Thevenin voltage for the circuit within the beige box as viewed from terminals A and B.

▶ FIGURE 19–16

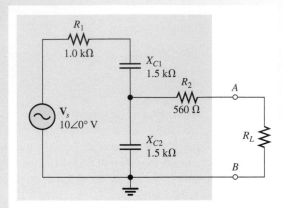

Solution

Thevenin's voltage for the circuit between terminals A and B is the voltage that appears across A and B with R_L removed from the circuit.

There is no voltage drop across R_2 because the open between terminals A and B prevents current through it. Thus, $\mathbf{V}_{AB}$ is the same as $\mathbf{V}_{C2}$ and can be found by the voltage-divider formula.

$$\mathbf{V}_{AB} = \mathbf{V}_{C2} = \left(\frac{X_{C2}\angle-90°}{R_1 - jX_{C1} - jX_{C2}}\right)\mathbf{V}_s = \left(\frac{1.5\angle-90°\,\text{k}\Omega}{1.0\,\text{k}\Omega - j3\,\text{k}\Omega}\right)10\angle0°\,\text{V}$$

$$= \left(\frac{1.5\angle-90°\,\text{k}\Omega}{3.16\angle-71.6°\,\text{k}\Omega}\right)10\angle0°\,\text{V} = 4.75\angle-18.4°\,\text{V}$$

$$\mathbf{V}_{th} = \mathbf{V}_{AB} = \mathbf{4.75\angle-18.4°\,V}$$

Related Problem

Determine $\mathbf{V}_{th}$ if R_1 is changed to 2.2 kΩ in Figure 19–16.

Use Multisim file E19-05 to verify the calculated results in this example and to confirm your calculation for the related problem.

EXAMPLE 19–6

Refer to Figure 19–17. Find $\mathbf{V}_{th}$ for the circuit within the beige box as viewed from terminals A and B.

▶ FIGURE 19–17

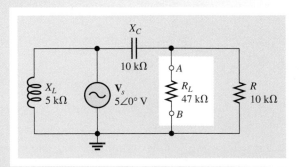

Solution First remove R_L and determine the voltage across the resulting open terminals, which is $\mathbf{V}_{th}$. Find $\mathbf{V}_{th}$ by applying the voltage-divider formula to X_C and R.

$$\mathbf{V}_{th} = \mathbf{V}_R = \left(\frac{R\angle 0°}{R - jX_C}\right)\mathbf{V}_s = \left(\frac{10\angle 0°\,\text{k}\Omega}{10\,\text{k}\Omega - j10\,\text{k}\Omega}\right)5\angle 0°\,\text{V}$$

$$= \left(\frac{10\angle 0°\,\text{k}\Omega}{14.1\angle -45°\,\text{k}\Omega}\right)5\angle 0°\,\text{V} = \mathbf{3.55\angle 45°\,V}$$

Notice X_L has no effect on the result, since the 5 V source appears across X_C and R in combination.

Related Problem Find $\mathbf{V}_{th}$ if R is 22 kΩ and R_L is 39 kΩ in Figure 19–17.

 Use Multisim file E19-06 to verify the calculated results in this example and to confirm your calculation for the related problem.

Thevenin's Equivalent Impedance (Z_{th})

The previous examples illustrated how to find $\mathbf{V}_{th}$. Now, let's determine the Thevenin equivalent impedance, $\mathbf{Z}_{th}$, the second part of a Thevenin equivalent circuit. As defined by Thevenin's theorem,

Thevenin's equivalent impedance is the total impedance appearing between two specified terminals in a given circuit with all sources replaced by their internal impedances.

To find $\mathbf{Z}_{th}$ between any two terminals in a circuit, replace all the voltage sources by a short (any internal impedance remains in series). Replace all the current sources by an open (any internal impedance remains in parallel). Then determine the total impedance between the two terminals. The following three examples illustrate how to find $\mathbf{Z}_{th}$.

EXAMPLE 19–7 Find $\mathbf{Z}_{th}$ for the part of the circuit in Figure 19–18 that is within the beige box as viewed from terminals A and B. This is the same circuit used in Example 19–4.

▶ **FIGURE 19–18**

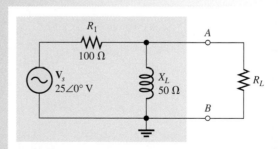

Solution First, replace $\mathbf{V}_s$ with its internal impedance (zero in this case), as shown in Figure 19–19. Looking in between terminals A and B, R_1 and X_L are in parallel. Thus,

$$\mathbf{Z}_{th} = \frac{(R_1\angle 0°)(X_L\angle 90°)}{R_1 + jX_L} = \frac{(100\angle 0°\,\Omega)(50\angle 90°\,\Omega)}{100\,\Omega + j50\,\Omega}$$

$$= \frac{(100\angle 0°\,\Omega)(50\angle 90°\,\Omega)}{112\angle 26.6°\,\Omega} = \mathbf{44.6\angle 63.4°\,\Omega}$$

▶ FIGURE 19–19

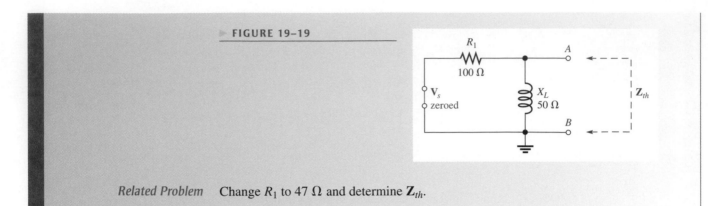

Related Problem Change R_1 to 47 Ω and determine $\mathbf{Z}_{th}$.

EXAMPLE 19–8

Refer to Figure 19–20. Determine $\mathbf{Z}_{th}$ for the circuit within the beige box as viewed from terminals A and B. This is the same circuit used in Example 19–5.

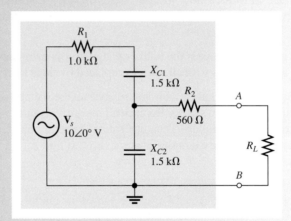

▲ FIGURE 19–20

Solution First, replace the voltage source with its internal impedance (zero in this case), as shown in Figure 19–21.

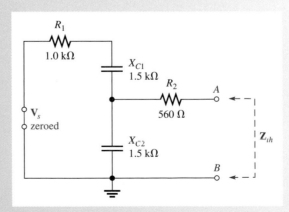

▲ FIGURE 19–21

Looking from terminals A and B, C_2 appears in parallel with the series combination of R_1 and C_1. This entire combination is in series with R_2. The calculation for $\mathbf{Z}_{th}$ is as follows:

$$\mathbf{Z}_{th} = R_2\angle 0° + \frac{(X_{C2}\angle -90°)(R_1 - jX_{C1})}{R_1 - jX_{C1} - jX_{C2}}$$

$$= 560\angle 0° \ \Omega + \frac{(1.5\angle -90° \ \text{k}\Omega)(1.0 \ \text{k}\Omega - j1.5 \ \text{k}\Omega)}{1.0 \ \text{k}\Omega - j3 \ \text{k}\Omega}$$

$$= 560\angle 0° \ \Omega + \frac{(1.5\angle -90° \ \text{k}\Omega)(1.8\angle -56.3° \ \text{k}\Omega)}{3.16\angle -71.6° \ \text{k}\Omega}$$

$$= 560\angle 0° \ \Omega + 854\angle -74.7° \ \Omega = 560 \ \Omega + 225 \ \Omega - j824 \ \Omega$$

$$= 785 \ \Omega - j824 \ \Omega = \mathbf{1138\angle -46.4° \ \Omega}$$

Related Problem Determine $\mathbf{Z}_{th}$ if R_1 is changed to 2.2 kΩ in Figure 19–20.

EXAMPLE 19–9

Refer to Figure 19–22. Determine $\mathbf{Z}_{th}$ for the portion of the circuit within the beige box as viewed from terminals A and B. This is the same circuit as in Example 19–6.

▶ **FIGURE 19–22**

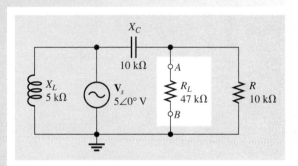

Solution With the voltage source replaced by its internal impedence (zero in this case), X_L is effectively out of the circuit. R and C appear in parallel when viewed from the open terminals, as indicated in Figure 19–23. $\mathbf{Z}_{th}$ is calculated as follows:

$$\mathbf{Z}_{th} = \frac{(R\angle 0°)(X_C\angle -90°)}{R - jX_C} = \frac{(10\angle 0° \ \text{k}\Omega)(10\angle -90° \ \text{k}\Omega)}{10 \ \Omega - j10 \ \text{k}\Omega}$$

$$= \frac{(10\angle 0° \ \text{k}\Omega)(10\angle -90° \ \text{k}\Omega)}{14.1\angle -45° \ \text{k}\Omega} = \mathbf{7.07\angle -45° \ k\Omega}$$

FIGURE 19–23

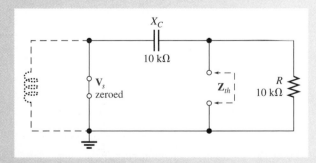

Related Problem Find $\mathbf{Z}_{th}$ if R is 22 kΩ and R_L is 39 kΩ in Figure 19–22.

Thevenin's Equivalent Circuit

The previous six examples have shown how to find the two equivalent components of a Thevenin circuit, $\mathbf{V}_{th}$ and $\mathbf{Z}_{th}$. Keep in mind that you can find $\mathbf{V}_{th}$ and $\mathbf{Z}_{th}$ for any circuit. Once you have determined these equivalent values, you must connect them in series to form the Thevenin equivalent circuit. The following three examples use the previous examples to illustrate this final step.

EXAMPLE 19–10

Refer to Figure 19–24. Draw the Thevenin equivalent for the circuit within the beige box as viewed from terminals *A* and *B*. This is the circuit used in Examples 19–4 and 19–7.

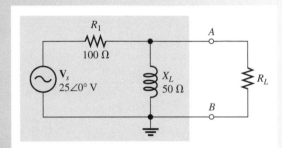

▲ FIGURE 19–24

Solution

From Examples 19–4 and 19–7, respectively, $\mathbf{V}_{th} = 11.2\angle63.4°$ V and $\mathbf{Z}_{th} = 44.6\angle63.4°\ \Omega$. In rectangular form, the impedance is

$$\mathbf{Z}_{th} = 20\ \Omega + j40\ \Omega$$

This form indicates that the impedance is a 20 Ω resistor in series with a 40 Ω inductive reactance. The Thevenin equivalent circuit is shown in Figure 19–25.

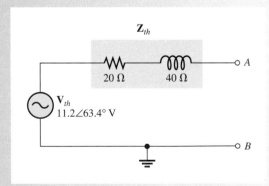

▲ FIGURE 19–25

Related Problem

Draw the Thevenin equivalent circuit for Figure 19–24 with $R_1 = 47\ \Omega$.

EXAMPLE 19–11

Refer to Figure 19–26. Draw the Thevenin equivalent for the circuit within the beige box as viewed from terminals A and B. This is the circuit used in Examples 19–5 and 19–8.

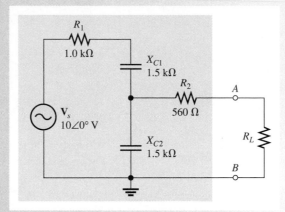

▲ FIGURE 19–26

Solution   From Examples 19–5 and 19–8, respectively, $\mathbf{V}_{th} = 4.75\angle-18.4°$ V and $\mathbf{Z}_{th} = 1138\angle-46.4°$ Ω. In rectangular form, the impedance is

$$\mathbf{Z}_{th} = 785\ \Omega - j824\ \Omega$$

The Thevenin equivalent circuit is shown in Figure 19–27.

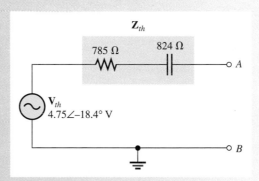

▲ FIGURE 19–27

Related Problem   Draw the Thevenin equivalent for the circuit in Figure 19–26 with $R_1 = 2.2\ k\Omega$.

EXAMPLE 19–12

Refer to Figure 19–28. Determine the Thevenin equivalent for the circuit within the beige box as viewed from terminals A and B. This is the circuit used in Examples 19–6 and 19–9.

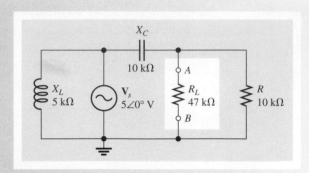

▲ FIGURE 19–28

Solution From Examples 19–6 and 19–9, respectively, $\mathbf{V}_{th} = 3.54\angle45°$ V, and $\mathbf{Z}_{th} = 7.07\angle-45°$ kΩ. The impedance in rectangular form is

$$\mathbf{Z}_{th} = 5\,\text{k}\Omega - j5\,\text{k}\Omega$$

Thus, the Thevenin equivalent circuit is as shown in Figure 19–29.

▶ FIGURE 19–29

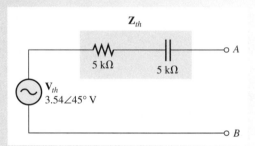

Related Problem Change R to 22 kΩ and R_L to 39 kΩ in Figure 19–28 and draw the Thevenin equivalent circuit.

Summary of Thevenin's Theorem

Remember that the Thevenin equivalent circuit is always a voltage source in series with an impedance regardless of the original circuit that it replaces. The significance of Thevenin's theorem is that the equivalent circuit can replace the original circuit as far as any external load is concerned. Any load connected between the terminals of a Thevenin equivalent circuit experiences the same current and voltage as if it were connected to the terminals of the original circuit.

A summary of steps for applying Thevenin's theorem follows.

Step 1. Open the two terminals between which you want to find the Thevenin circuit. This is done by removing the component from which the circuit is to be viewed.

Step 2. Determine the voltage across the two open terminals.

Step 3. Determine the impedance viewed from the two open terminals with ideal voltage sources replaced with shorts and ideal current sources replaced with opens (zeroed).

Step 4. Connect $\mathbf{V}_{th}$ and $\mathbf{Z}_{th}$ in series to produce the complete Thevenin equivalent circuit.

**SECTION 19–2
REVIEW**

1. What are the two basic components of a Thevenin equivalent ac circuit?
2. For a certain circuit, $\mathbf{Z}_{th} = 25\ \Omega - j50\ \Omega$, and $\mathbf{V}_{th} = 5\angle 0°$ V. Draw the Thevenin equivalent circuit.
3. For the circuit in Figure 19–30, find the Thevenin equivalent looking from terminals A and B.

▶ FIGURE 19–30

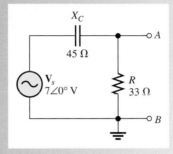

19–3 NORTON'S THEOREM

Like Thevenin's theorem, Norton's theorem provides a method of reducing a more complex circuit to a simpler, more manageable form for analysis. The basic difference is that **Norton's theorem** gives an equivalent current source (rather than a voltage source) in parallel (rather than in series) with an equivalent impedance.

After completing this section, you should be able to

◆ **Apply Norton's theorem to simplify reactive ac circuits**

◆ Describe the form of a Norton equivalent circuit

◆ Obtain the Norton equivalent ac current source

◆ Obtain the Norton equivalent impedance

The form of Norton's equivalent circuit is shown in Figure 19–31. Regardless of how complex the original circuit is, it can be reduced to this equivalent form. The equivalent current source is designated $\mathbf{I}_n$, and the equivalent impedance is $\mathbf{Z}_n$ (lowercase italic subscript denotes ac quantity).

Norton's theorem shows you how to find $\mathbf{I}_n$ and $\mathbf{Z}_n$. Once they are known, simply connect them in parallel to get the complete Norton equivalent circuit.

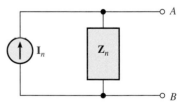

▲ FIGURE 19–31

Norton equivalent circuit.

Norton's Equivalent Current Source ($\mathbf{I}_n$)

$\mathbf{I}_n$ is one part of the Norton equivalent circuit; $\mathbf{Z}_n$ is the other part.

Norton's equivalent current is defined as the short-circuit current between two specified terminals in a given circuit.

Any load connected between these two terminals effectively "sees" a current source $\mathbf{I}_n$ in parallel with $\mathbf{Z}_n$.

To illustrate, let's suppose that the circuit shown in Figure 19–32 has a load resistor connected to terminals A and B, as indicated in part (a), and we wish to find the Norton equivalent

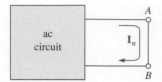

(a) Circuit with load resistor   (b) Load is replaced by a short, resulting in a short circuit current, $\mathbf{I}_n$.

▲ FIGURE 19–32

How $\mathbf{I}_n$ is determined.

for the circuit as viewed from terminals A and B. To find $\mathbf{I}_n$, calculate the current between terminals A and B with those terminals shorted, as shown in part (b). Example 19–13 shows how to find $\mathbf{I}_n$.

EXAMPLE 19–13

In Figure 19–33, determine $\mathbf{I}_n$ for the circuit as "seen" by the load resistor. The beige area identifies the portion of the circuit to be nortonized.

▶ FIGURE 19–33

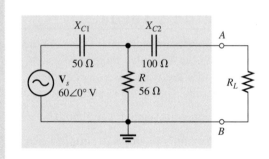

Solution   Short the terminals A and B, as shown in Figure 19–34.

▶ FIGURE 19–34

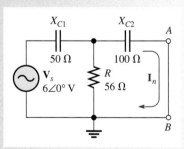

$\mathbf{I}_n$ is the current through the short and is calculated as follows. First, the total impedance viewed from the source is

$$\mathbf{Z} = \mathbf{X}_{C1} + \frac{\mathbf{R}\mathbf{X}_{C2}}{\mathbf{R} + \mathbf{X}_{C2}} = 50\angle-90°\ \Omega + \frac{(56\angle0°\ \Omega)(100\angle-90°\ \Omega)}{56\ \Omega - j100\ \Omega}$$

$$= 50\angle-90°\ \Omega + 48.9\angle-29.3°\ \Omega$$

$$= -j50\ \Omega + 42.6\ \Omega - j23.9\ \Omega = 42.6\ \Omega - j73.9\ \Omega$$

Converting to polar form yields

$$\mathbf{Z} = 85.3\angle-60.0°\ \Omega$$

Next, the total current from the source is

$$\mathbf{I}_s = \frac{\mathbf{V}_s}{\mathbf{Z}} = \frac{6\angle 0° \text{ V}}{85.3\angle -60.0° \ \Omega} = 70.3\angle 60.0° \text{ mA}$$

Finally, apply the current-divider formula to get $\mathbf{I}_n$ (the current through the short between terminals A and B).

$$\mathbf{I}_n = \left(\frac{\mathbf{R}}{\mathbf{R} + \mathbf{X}_{C2}}\right)\mathbf{I}_s = \left(\frac{56\angle 0° \ \Omega}{56 \ \Omega \ - \ j100 \ \Omega}\right)70.3\angle 60.0° \text{ mA} = \mathbf{34.4\angle 121° \text{ mA}}$$

This is the value for the equivalent Norton current source.

Related Problem   Determine $\mathbf{I}_n$ if $\mathbf{V}_s$ is changed to $2.5\angle 0°$ V and R is changed to 33 Ω in Figure 19–33.

Norton's Equivalent Impedance ($\mathbf{Z}_n$)

$\mathbf{Z}_n$ is defined the same as $\mathbf{Z}_{th}$: It is the total impedance appearing between two specified terminals of a given circuit viewed from the open terminals with all sources replaced by their internal impedances.

EXAMPLE 19–14   Find $\mathbf{Z}_n$ for the circuit in Figure 19–33 (Example 19–13) viewed from the open across terminals A and B.

Solution   First, replace $\mathbf{V}_s$ with its internal impedance (zero), as indicated in Figure 19–35.

▶ **FIGURE 19–35**

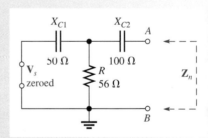

Looking in between terminals A and B, C_2 is in series with the parallel combination of R and C_1. Thus,

$$\mathbf{Z}_n = \mathbf{X}_{C2} + \frac{\mathbf{R}\mathbf{X}_{C1}}{\mathbf{R} + \mathbf{X}_{C1}} = 100\angle -90° \ \Omega + \frac{(56\angle 0° \ \Omega)(50\angle -90° \ \Omega)}{56 \ \Omega \ - \ j50 \ \Omega}$$

$$= 100\angle -90° \ \Omega + 37.3\angle -48.2° \ \Omega$$

$$= -j100 \ \Omega + 24.8 \ \Omega - j27.8 \ \Omega = \mathbf{24.8 \ \Omega - j128 \ \Omega}$$

The Norton equivalent impedance is a 24.8 Ω resistance in series with a 128 Ω capacitive reactance.

Related Problem   Find $\mathbf{Z}_n$ in Figure 19–33 if $R = 33 \ \Omega$.

Examples 19–13 and 19–14 showed how to find the two equivalent components of a Norton equivalent circuit. Keep in mind that you can find these values for any given ac circuit. Once you know $\mathbf{I}_n$ and $\mathbf{Z}_n$, connect them in parallel to form the Norton equivalent circuit, as Example 19–15 illustrates.

EXAMPLE 19–15

Show the complete Norton equivalent circuit for the circuit in Figure 19–33 (Example 19–13).

Solution   From Examples 19–13 and 19–14, respectively, $\mathbf{I}_n = 34.4\angle 121°$ mA and $\mathbf{Z}_n = 24.8\ \Omega - j128\ \Omega$. The Norton equivalent circuit is shown in Figure 19–36.

▶ FIGURE 19–36

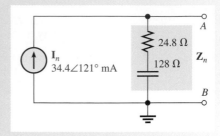

Related Problem   Show the Norton equivalent for the circuit in Figure 19–33 if $\mathbf{V}_s = 2.5\angle 0°$ V and $R = 33\ \Omega$.

Summary of Norton's Theorem

Any load connected between the terminals of a Norton equivalent circuit will have the same current through it and the same voltage across it as it would when connected to the terminals of the original circuit. A summary of steps for theoretically applying Norton's theorem is as follows:

Step 1. Replace the load connected to the two terminals between which the Norton circuit is to be determined with a short.

Step 2. Determine the current through the short. This is $\mathbf{I}_n$.

Step 3. Open the terminals and determine the impedance between the two open terminals with all sources replaced with their internal impedances. This is $\mathbf{Z}_n$.

Step 4. Connect $\mathbf{I}_n$ and $\mathbf{Z}_n$ in parallel.

SECTION 19–3 REVIEW

1. For a given circuit, $\mathbf{I}_n = 5\angle 0°$ mA, and $\mathbf{Z}_n = 150\ \Omega + j100\ \Omega$. Draw the Norton equivalent circuit.
2. Find the Norton circuit as seen by R_L in Figure 19–37.

▶ FIGURE 19–37

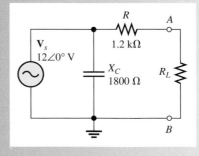

19–4 MAXIMUM POWER TRANSFER THEOREM

Maximum power is transferred to a load connected to a circuit when the load impedance is the complex conjugate of the circuit's output impedance.

After completing this section, you should be able to

♦ **Apply the maximum power transfer theorem**

 ♦ Explain the theorem

 ♦ Determine the value of load impedance for which maximum power is transferred from a given circuit

The **complex conjugate** of $R - jX_C$ is $R + jX_L$ and vice versa, where the resistances are equal in magnitude and the reactances are equal in magnitude but opposite in sign. The output impedance is effectively Thevenin's equivalent impedance viewed from the output terminals. When $\mathbf{Z}_L$ is the complex conjugate of $\mathbf{Z}_{out}$, maximum power is transferred from the circuit to the load with a power factor of 1. An equivalent circuit with its output impedance and load is shown in Figure 19–38.

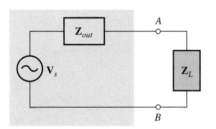

◀ **FIGURE 19–38**

Equivalent circuit with load.

Example 19–16 shows that maximum power occurs when the impedances are conjugately matched.

EXAMPLE 19–16

The circuit to the left of terminals A and B in Figure 19–39 provides power to the load $\mathbf{Z}_L$. This can be viewed as simulating a power amplifier delivering power to a complex load. It is the Thevenin equivalent of a more complex circuit. Calculate and plot a

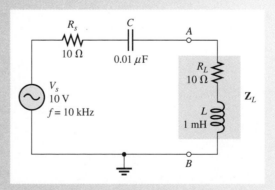

▲ **FIGURE 19–39**

graph of the power delivered to the load for each of the following frequencies: 10 kHz, 30 kHz, 50 kHz, 80 kHz, and 100 kHz.

Solution   For $f = 10$ kHz,

$$X_C = \frac{1}{2\pi fC} = \frac{1}{2\pi(10 \text{ kHz})(0.01 \text{ }\mu\text{F})} = 1.59 \text{ k}\Omega$$
$$X_L = 2\pi fL = 2\pi(10 \text{ kHz})(1 \text{ mH}) = 62.8 \text{ }\Omega$$

The magnitude of the total impedance is
$$Z_{tot} = \sqrt{(R_s + R_L)^2 + (X_L - X_C)^2} = \sqrt{(20 \text{ }\Omega)^2 + (1.53 \text{ k}\Omega)^2} = 1.53 \text{ k}\Omega$$

The current is
$$I = \frac{V_s}{Z_{tot}} = \frac{10 \text{ V}}{1.53 \text{ k}\Omega} = 6.54 \text{ mA}$$

The load power is
$$P_L = I^2R_L = (6.54 \text{ mA})^2(10 \text{ }\Omega) = \mathbf{428 \text{ }\mu W}$$

For $f = 30$ kHz,

$$X_C = \frac{1}{2\pi(30 \text{ kHz})(0.01 \text{ }\mu\text{F})} = 531 \text{ }\Omega$$
$$X_L = 2\pi(30 \text{ kHz})(1 \text{ mH}) = 189 \text{ }\Omega$$
$$Z_{tot} = \sqrt{(20 \text{ }\Omega)^2 + (342 \text{ }\Omega)^2} = 343 \text{ }\Omega$$
$$I = \frac{V_s}{Z_{tot}} = \frac{10 \text{ V}}{343 \text{ }\Omega} = 29.2 \text{ mA}$$
$$P_L = I^2R_L = (29.2 \text{ mA})^2(10 \text{ }\Omega) = \mathbf{8.53 \text{ mW}}$$

For $f = 50$ kHz,

$$X_C = \frac{1}{2\pi(50 \text{ kHz})(0.01 \text{ }\mu\text{F})} = 318 \text{ }\Omega$$
$$X_L = 2\pi(50 \text{ kHz})(1 \text{ mH}) = 314 \text{ }\Omega$$

Note that X_C and X_L are very close to being equal which makes the impedances approximately complex conjugates. The exact frequency at which $X_L = X_C$ is 50.3 kHz.

$$Z_{tot} = \sqrt{(20 \text{ }\Omega)^2 + (4 \text{ }\Omega)^2} = 20.4 \text{ }\Omega$$
$$I = \frac{V_s}{Z_{tot}} = \frac{10 \text{ V}}{20.4 \text{ }\Omega} = 490 \text{ mA}$$
$$P_L = I^2R_L = (490 \text{ mA})^2(10 \text{ }\Omega) = \mathbf{2.40 \text{ W}}$$

For $f = 80$ kHz,

$$X_C = \frac{1}{2\pi(80 \text{ kHz})(0.01 \text{ }\mu\text{F})} = 199 \text{ }\Omega$$
$$X_L = 2\pi(80 \text{ kHz})(1 \text{ mH}) = 503 \text{ }\Omega$$
$$Z_{tot} = \sqrt{(20 \text{ }\Omega)^2 + (304 \text{ }\Omega)^2} = 305 \text{ }\Omega$$
$$I = \frac{V_s}{Z_{tot}} = \frac{10 \text{ V}}{305 \text{ }\Omega} = 32.8 \text{ mA}$$
$$P_L = I^2R_L = (32.8 \text{ mA})^2(10 \text{ }\Omega) = \mathbf{10.8 \text{ mW}}$$

For $f = 100$ kHz,

$$X_C = \frac{1}{2\pi(100\text{ kHz})(0.01\ \mu\text{F})} = 159\ \Omega$$

$$X_L = 2\pi(100\text{ kHz})(1\text{ mH}) = 628\ \Omega$$

$$Z_{tot} = \sqrt{(20\ \Omega)^2 + (469\ \Omega)^2} = 469\ \Omega$$

$$I = \frac{V_s}{Z_{tot}} = \frac{10\text{ V}}{469\ \Omega} = 21.3\text{ mA}$$

$$P_L = I^2R_L = (21.3\text{ mA})^2(10\ \Omega) = \mathbf{4.54\ mW}$$

As you can see from the results, the power to the load peaks at the frequency (50 kHz) for which the load impedance is the complex conjugate of the output impedance (when the reactances are equal in magnitude). A graph of the load power versus frequency is shown in Figure 19–40. Since the maximum power is so much larger than the other values, an accurate plot is difficult to achieve without intermediate values.

▶ FIGURE 19–40

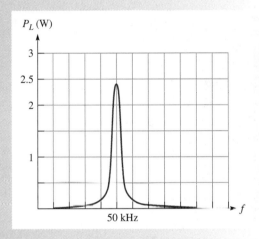

Related Problem If $R = 47\ \Omega$ and $C = 0.022\ \mu\text{F}$ in a series RC circuit, what is the complex conjugate of the impedance at 100 kHz?

 Use Multisim file E19-16 to verify the calculated results in this example.

Example 19–17 illustrates that the frequency at which maximum load power occurs is the value that makes the source and load impedances complex conjugates.

EXAMPLE 19–17

(a) Determine the frequency at which maximum power is transferred from the amplifier to the speaker in Figure 19–41(a). The amplifier and coupling capacitor are the source, and the speaker is the load, as shown in the equivalent circuit of Figure 19–41(b).

(b) How many watts of power are delivered to the speaker at this frequency if $V_s = 3.8$ V rms?

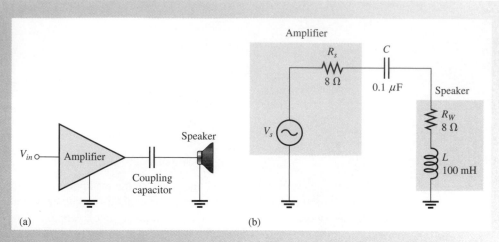

(a)

(b)

▲ FIGURE 19–41

Solution **(a)** When the power to the speaker is maximum, the source impedance $(R_s - jX_C)$ and the load impedance $(R_W + jX_L)$ are complex conjugates, so

$$X_C = X_L$$

$$\frac{1}{2\pi fC} = 2\pi fL$$

Solving for f,

$$f^2 = \frac{1}{4\pi^2 LC}$$

$$f = \frac{1}{2\pi\sqrt{LC}} = \frac{1}{2\pi\sqrt{(100\ \text{mH})(0.1\ \mu\text{F})}} \cong \mathbf{1.59\ kHz}$$

(b) Calculate the power to the speaker as follows:

$$Z_{tot} = R_s + R_W = 8\ \Omega + 8\ \Omega = 16\ \Omega$$

$$I = \frac{V_s}{Z_{tot}} = \frac{3.8\ \text{V}}{16\ \Omega} = 238\ \text{mA}$$

$$P_{max} = I^2 R_W = (238\ \text{mA})^2(8\ \Omega) = \mathbf{453\ mW}$$

Related Problem Determine the frequency at which maximum power is transferred from the amplifier to the speaker in Figure 19–41 if the coupling capacitor is 1 μF.

**SECTION 19–4
REVIEW**

1. If the output impedance of a certain driving circuit is 50 Ω − j10 Ω, what value of load impedance will result in the maximum power to the load?

2. For the circuit in Question 1, how much power is delivered to the load when the load impedance is the complex conjugate of the output impedance and when the load current is 2 A?

A Circuit Application

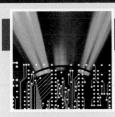

In this circuit application, you have a sealed band-pass filter module that has been removed from a system and two schematics. Both schematics indicate that the band-pass filter is implemented with a low-pass/high-pass combination. It is uncertain which schematic corresponds to the filter module, but one of them does. By certain measurements, you will determine which schematic represents the filter so that the filter circuit can be reproduced. Also, you will determine the proper load for maximum power transfer. The filter circuit contained in a sealed module and two schematics, one of which corresponds to the filter circuit, are shown in Figure 19–42.

Filter Measurement and Analysis

◆ Based on the oscilloscope measurement of the filter output shown in Figure 19–43, determine which schematic in Figure 19–42 represents the component values of the filter

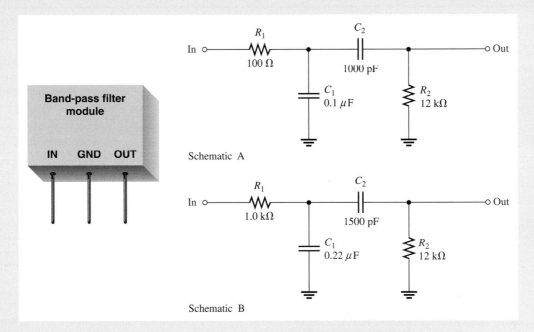

Schematic A

Schematic B

▲ **FIGURE 19–42**

Filter module and schematics.

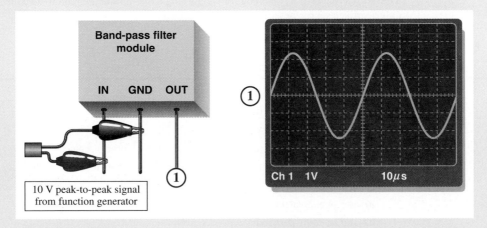

▲ **FIGURE 19–43**

circuit in the module. A 10 V peak-to-peak voltage is applied to the input.

◆ Based on the oscilloscope measurement in Figure 19–43, determine if the filter is operating at its approximate center frequency.

◆ Using Thevenin's theorem, determine the load impedance that will provide for maximum power transfer at the center frequency when connected to the output of the filter. Assume the source impedance is zero.

Review

1. Determine the peak-to-peak output voltage at the frequency shown in Figure 19–43 of the circuit in Figure 19–42 that was determined not to be in the module.

2. Find the center frequency of the circuit in Figure 19–42 that was determined not to be in the module.

SUMMARY

◆ The superposition theorem is useful for the analysis of both ac and dc multiple-source circuits.

◆ Thevenin's theorem provides a method for the reduction of any ac circuit to an equivalent form consisting of an equivalent voltage source in series with an equivalent impedance.

◆ The term *equivalency,* as used in Thevenin's and Norton's theorems, means that when a given load impedance is connected to the equivalent circuit, it will have the same voltage across it and the same current through it as when it is connected to the original circuit.

◆ Norton's theorem provides a method for the reduction of any ac circuit to an equivalent form consisting of an equivalent current source in parallel with an equivalent impedance.

◆ Maximum power is transferred to a load when the load impedance is the complex conjugate of the impedance of the driving circuit.

KEY TERMS

These key terms are also in the end-of-book glossary.

Complex conjugate   A complex number having the same real part and an oppositely signed imaginary part; an impedance containing the same resistance and a reactance opposite in phase but equal in magnitude to that of a given impedance.

Equivalent circuit   A circuit that produces the same voltage and current to a given load as the original circuit that it replaces.

Norton's theorem   A method for simplifying a two-terminal circuit to an equivalent circuit with only a current source in parallel with an impedance.

Superposition theorem   A method for the analysis of circuits with more than one source.

Thevenin's theorem   A method for simplifying a two-terminal circuit to an equivalent circuit with only a voltage source in series with an impedance.

SELF-TEST

Answers are at the end of the chapter.

1. In applying the superposition theorem,
    (a) all sources are considered simultaneously
    (b) all voltage sources are considered simultaneously
    (c) the sources are considered one at a time with all others replaced by a short
    (d) the sources are considered one at a time with all others replaced by their internal impedances
2. A Thevenin ac equivalent circuit always consists of an equivalent ac voltage source
    (a) and an equivalent capacitance
    (b) and an equivalent inductive reactance
    (c) and an equivalent impedance
    (d) in series with an equivalent capacitive reactance

3. One circuit is equivalent to another when
   (a) the same load has the same voltage and current when connected to either circuit
   (b) different loads have the same voltage and current when connected to either circuit
   (c) the circuits have equal voltage sources and equal series impedances
   (d) the circuits produce the same output voltage

4. The Thevenin equivalent voltage is
   (a) the open circuit voltage   (b) the short circuit voltage
   (c) the voltage across an equivalent load   (d) none of the above

5. The Thevenin equivalent impedance is the impedance looking from
   (a) the source with the output shorted
   (b) the source with the output open
   (c) any two specified open terminals with all sources replaced by their internal impedances
   (d) any two specified open terminals with all sources replaced by a short

6. A Norton ac equivalent circuit always consists of
   (a) an equivalent ac current source in series with an equivalent impedance
   (b) an equivalent ac current source in parallel with an equivalent reactance
   (c) an equivalent ac current source in parallel with an equivalent impedance
   (d) an equivalent ac voltage source in parallel with an equivalent impedance

7. The Norton equivalent current is
   (a) the total current from the source
   (b) the short circuit current
   (c) the current to an equivalent load
   (d) none of the above

8. The complex conjugate of $50\ \Omega\ +\ j100\ \Omega$ is
   (a) $50\ \Omega\ -\ j50\ \Omega$ (b) $100\ \Omega\ +\ j50\ \Omega$ (c) $100\ \Omega\ -\ j50\ \Omega$ (d) $50\ \Omega\ -\ j100\ \Omega$

9. In order to get maximum power transfer from a capacitive source, the load must
   (a) have a capacitance equal to the source capacitance
   (b) have an impedance equal in magnitude to the source impedance
   (c) be inductive
   (d) have an impedance that is the complex conjugate of the source impedance
   (e) answers (a) and (d)

CIRCUIT DYNAMICS QUIZ

Answers are at the end of the chapter.

Refer to Figure 19–47.

1. If the dc voltage source is shorted out, the voltage at point A with respect to ground
   (a) increases   (b) decreases   (c) stays the same

2. If C_2 opens, the voltage across R_5
   (a) increases   (b) decreases   (c) stays the same

3. If C_2 opens, the dc voltage across R_5
   (a) increases   (b) decreases   (c) stays the same

Refer to Figure 19–49(c).

4. If V_2 is reduced to 0 V, the voltage across R_L
   (a) increases   (b) decreases   (c) stays the same

5. If the frequency of the voltage sources is increased, the current through R_L
   (a) increases   (b) decreases   (c) stays the same

Refer to Figure 19–50.

6. If the frequency of the source voltage is increased, the voltage across R_1

    (a) increases   **(b)** decreases   **(c)** stays the same

7. If R_L opens, the voltage across it

    (a) increases   **(b)** decreases   **(c)** stays the same

Refer to Figure 19–51.

8. If the source frequency is increased, the voltage across R_3

    (a) increases   **(b)** decreases   **(c)** stays the same

9. If the capacitor value is reduced, the current from the source

    (a) increases   **(b)** decreases   **(c)** stays the same

Refer to Figure 19–54.

10. If R_2 opens, the current from the current source

    (a) increases   **(b)** decreases   **(c)** stays the same

11. If the frequency of the voltage source increase, X_{C2}

    (a) increases   **(b)** decreases   **(c)** stays the same

12. If the load is removed, the voltage across R_3

    (a) increases   **(b)** decreases   **(c)** stays the same

13. If the load is removed, the voltage across R_2

    (a) increases   **(b)** decreases   **(c)** stays the same

PROBLEMS

More difficult problems are indicated by an asterisk (*).
Answers to odd-numbered problems are at the end of the book.

SECTION 19–1   The Superposition Theorem

1. Using the superposition method, calculate the current through R_3 in Figure 19–44.

2. Use the superposition theorem to find the current in and the voltage across the R_2 branch of Figure 19–44.

▶ **FIGURE 19–44**

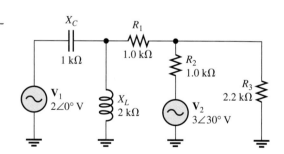

3. Using the superposition theorem, solve for the current through R_1 in Figure 19–45.

▶ **FIGURE 19–45**

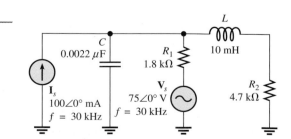

4. Using the superposition theorem, find the current through R_L in each circuit of Figure 19–46.

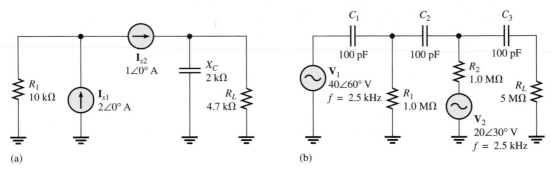

▲ FIGURE 19–46

*5. Determine the voltage at each point (A, B, C, D) in Figure 19–47. Assume $X_C = 0$ for all capacitors. Draw the voltage waveforms at each of the points.

*6. Use the superposition theorem to find the capacitor current in Figure 19–48.

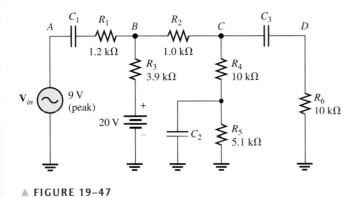

▲ FIGURE 19–47

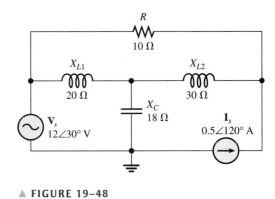

▲ FIGURE 19–48

SECTION 19–2 Thevenin's Theorem

7. For each circuit in Figure 19–49, determine the Thevenin equivalent circuit for the portion of the circuit viewed by R_L.

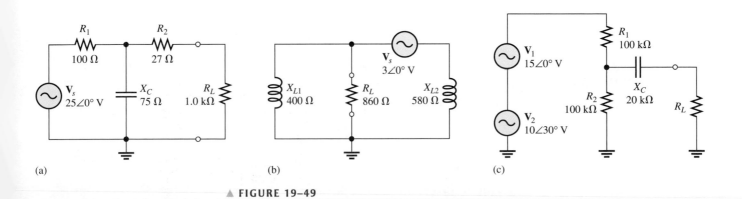

▲ FIGURE 19–49

8. Using Thevenin's theorem, determine the current through the load R_L in Figure 19–50.

*9. Using Thevenin's theorem, find the voltage across R_4 in Figure 19–51.

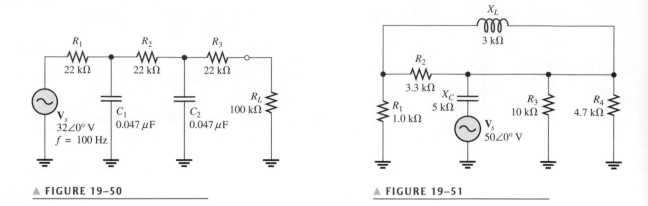

▲ FIGURE 19–50

▲ FIGURE 19–51

*10. Simplify the circuit external to R_3 in Figure 19–52 to its Thevenin equivalent.

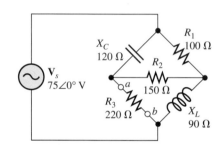

▲ FIGURE 19–52

SECTION 19–3 Norton's Theorem

11. For each circuit in Figure 19–49, determine the Norton equivalent as seen by R_L.

12. Using Norton's theorem, find the current through the load resistor R_L in Figure 19–50.

*13. Using Norton's theorem, find the voltage across R_4 in Figure 19–51.

SECTION 19–4 Maximum Power Transfer Theorem

14. For each circuit in Figure 19–53, maximum power is to be transferred to the load R_L. Determine the appropriate value for the load impedance in each case.

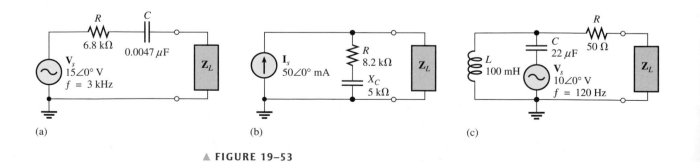

(a)   (b)   (c)

▲ FIGURE 19–53

*15. Determine $\mathbf{Z}_L$ for maximum power in Figure 19–54.

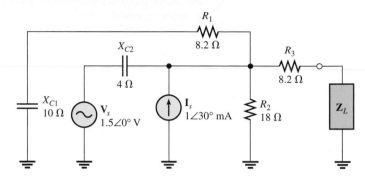

▲ FIGURE 19–54

*16. Find the load impedance required for maximum power transfer to Z_L in Figure 19–55. Determine the maximum true power.

*17. A load is to be connected in the place of R_2 in Figure 19–52 to achieve maximum power transfer. Determine the type of load, and express it in rectangular form.

▶ FIGURE 19–55

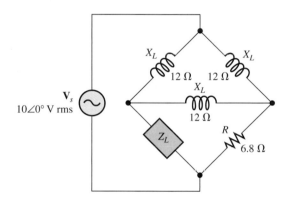

Multisim Troubleshooting and Analysis

These problems require your Multisim CD-ROM.

18. Open file P19-18 and determine if there is a fault. If so, find the fault.

19. Open file P19-19 and determine if there is a fault. If so, find the fault.

20. Open file P19-20 and determine if there is a fault. If so, find the fault.

21. Open file P19-21 and determine if there is a fault. If so, find the fault.

22. Open file P19-22 and determine the Thevenin equivalent circuit by measurement looking from Point A.

23. Open file P19-23 and determine the Norton equivalent circuit by measurement looking from Point A.

ANSWERS

SECTION REVIEWS

SECTION 19–1   The Superposition Theorem

1. The net current is zero.

2. The circuit can be analyzed one source at a time using superposition.

3. $I_R = 12\,\text{mA}$

SECTION 19–2   Thevenin's Theorem

1. The components of a Thevenin equivalent ac circuit are equivalent voltage source and equivalent series impedance.
2. See Figure 19–56.
3. $\mathbf{Z}_{th} = 21.5\ \Omega - j15.7\ \Omega$; $\mathbf{V}_{th} = 4.14\angle 53.8°$ V

▶ FIGURE 19–56

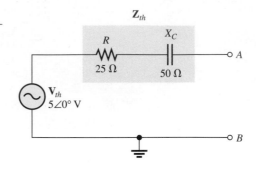

SECTION 19–3   Norton's Theorem

1. See Figure 19–57.
2. $\mathbf{Z}_n = R\angle 0° = 1.2\angle 0°$ kΩ; $\mathbf{I}_n = 10\angle 0°$ mA

▶ FIGURE 19–57

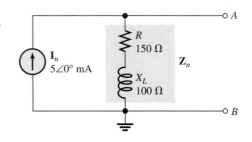

SECTION 19–4   Maximum Power Transfer Theorem

1. $\mathbf{Z}_L = 50\ \Omega + j10\ \Omega$
2. $P_L = 200$ W

A Circuit Application

1. $\mathbf{V}_{out} = 166\angle -66.1°$ mV pp
2. $f_0 = 4.76$ kHz

RELATED PROBLEMS FOR EXAMPLES

19–1 $2.11\angle -153°$ mA
19–2 $30\angle 90°$ mA
19–3 $1.69\angle 47.3°$ mA riding on a dc level of 3 mA
19–4 $18.2\angle 43.2°$ V
19–5 $4.03\angle -36.3°$ V
19–6 $4.55\angle 24.4°$ V
19–7 $34.3\angle 43.2°$ Ω
19–8 $1.37\angle -47.8°$ kΩ
19–9 $9.10\angle -65.6°$ kΩ
19–10   See Figure 19–58.
19–11   See Figure 19–59.
19–12   See Figure 19–60.

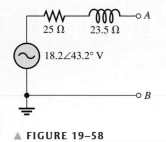

▲ FIGURE 19–58

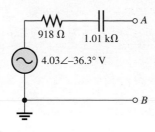

▲ FIGURE 19–59

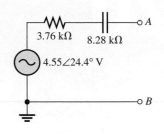

▲ FIGURE 19–60

19–13 $11.7\angle 135°$ mA

19–14 $117\angle -78.7°$ Ω

19–15 See Figure 19–61.

▶ FIGURE 19–61

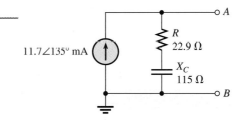

19–16 $47\ \Omega + j72.3\ \Omega$

19–17 503 Hz

SELF-TEST

1. (d)   **2.** (c)   **3.** (a)   **4.** (a)   **5.** (c)

6. (c)   **7.** (b)   **8.** (d)   **9.** (d)

CIRCUIT DYNAMICS QUIZ

1. (c)   **2.** (a)   **3.** (c)   **4.** (b)   **5.** (a)   **6.** (a)   **7.** (a)   **8.** (a)

9. (b)   **10.** (c)   **11.** (b)   **12.** (b)   **13.** (a)

20

TIME RESPONSE OF REACTIVE CIRCUITS

CHAPTER OBJECTIVES

◆ Explain the operation of an *RC* integrator
◆ Analyze an *RC* integrator with a single input pulse
◆ Analyze an *RC* integrator with repetitive input pulses
◆ Analyze an *RC* differentiator with a single input pulse
◆ Analyze an *RC* differentiator with repetitive input pulses
◆ Analyze the operation of an *RL* integrator
◆ Analyze the operation of an *RL* differentiator
◆ Explain the relationship of time response to frequency response
◆ Troubleshoot *RC* integrators and *RC* differentiators

KEY TERMS

◆ Integrator
◆ Time constant
◆ Transient time
◆ Steady-state
◆ Differentiator
◆ DC component

A CIRCUIT APPLICATION PREVIEW

In the circuit application, you will have to specify the wiring in a time-delay circuit. You will also determine component values to meet certain specifications and then determine instrument settings to properly test the circuit.

VISIT THE COMPANION WEBSITE

Study aids for this chapter are available at
http://www.prenhall.com/floyd

INTRODUCTION

In Chapters 15 and 16, the frequency response of *RC* and *RL* circuits was covered. In this chapter, the time response of *RC* and *RL* circuits with pulse inputs is examined. Before starting this chapter, you should review the material in Sections 12–5 and 13–4. Understanding exponential changes of voltages and currents in capacitors and inductors is crucial to the study of time response. Throughout this chapter, exponential formulas that were given in Chapters 12 and 13 are used.

With pulse inputs, the time responses of circuits are important. In the areas of pulse and digital circuits, technicians are often concerned with how a circuit responds over an interval of time to rapid changes in voltages or current. The relationship of the circuit time constant to the input pulse characteristics, such as pulse width and period, determines the wave shapes of voltages in the circuit.

Integrator and *differentiator,* terms used throughout this chapter, refer to mathematical functions that are approximated by these circuits under certain conditions. Mathematical integration is an averaging process, and mathematical differentiation is a process for establishing an instantaneous rate of change of a quantity.

20–1 THE *RC* INTEGRATOR

In terms of time response, a series *RC* circuit in which the output voltage is taken across the capacitor is known as an **integrator**. Recall that in terms of frequency response, it is a low-pass filter. The term *integrator* is derived from the mathematical process of integration, which this type of circuit approximates under certain conditions.

After completing this section, you should be able to

◆ **Explain the operation of an *RC* integrator**

   ◆ Describe how a capacitor charges and discharges

   ◆ Explain how a capacitor reacts to an instantaneous change in voltage or current

   ◆ Describe the basic output voltage waveform

Charging and Discharging of a Capacitor

When a pulse generator is connected to the input of an *RC* integrator, as shown in Figure 20–1, the capacitor will charge and discharge in response to the pulses. When the input goes from its low level to its high level, the capacitor charges toward the high level of the pulse through the resistor. This charging action is analogous to connecting a battery through a closed switch to the *RC* circuit, as illustrated in Figure 20–2(a). When the pulse goes from its high level back to its low level, the capacitor discharges back through the source. Compared to the resistance of the resistor, the resistance of the source is assumed to be negligible. This discharging action is analogous to replacing the source with a closed switch, as illustrated in Figure 20–2(b).

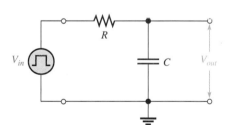

◀ **FIGURE 20–1**

An *RC* integrator with a pulse generator connected.

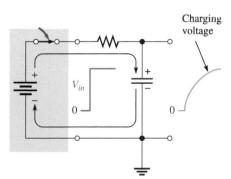

(a) When the input pulse goes HIGH, the source effectively acts as a battery in series with a closed switch, thereby charging the capacitor.

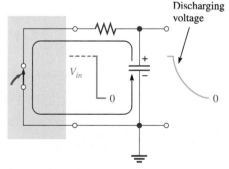

(b) When the input pulse goes back LOW, the source effectively acts as a closed switch, providing a discharge path for the capacitor.

▲ **FIGURE 20–2**

The equivalent action when a pulse source charges and discharges the capacitor.

As you learned in Chapter 12, a capacitor will charge and discharge following an exponential curve. Its rate of charging and discharging, of course, depends on the *RC* **time constant**, a fixed time interval determined by *R* and *C* ($\tau = RC$).

For an ideal pulse, both edges are considered to be instantaneous. Two basic rules of capacitor behavior help in understanding the response of *RC* circuits to pulse inputs.

1. The capacitor appears as a short to an instantaneous change in current and as an open to dc.

2. The voltage across the capacitor cannot change instantaneously—it can change only exponentially.

Capacitor Voltage

In an *RC* integrator, the output is the capacitor voltage. The capacitor charges during the time that the pulse is high. If the pulse is at its high level long enough, the capacitor will fully charge to the voltage amplitude of the pulse, as illustrated in Figure 20–3. The capacitor discharges during the time that the pulse is low. If the low time between pulses is long enough, the capacitor will fully discharge to zero, as shown in the figure. Then when the next pulse occurs, it will charge again.

▶ **FIGURE 20–3**

Illustration of a capacitor fully charging and discharging in response to a pulse input.

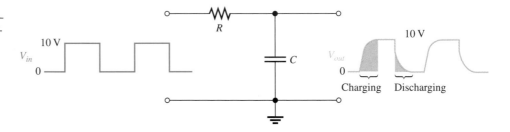

---

1. Define the term *integrator* in relation to an *RC* circuit.
2. What causes the capacitor in an *RC* circuit to charge and discharge?

20–2 RESPONSE OF AN *RC* INTEGRATOR TO A SINGLE PULSE

From the previous section, you have a general idea of how an *RC* integrator responds to an input pulse. In this section, the response to a single pulse is examined in detail.

After completing this section, you should be able to

◆ **Analyze an *RC* integrator with a single input pulse**

   ◆ Discuss the importance of the circuit time constant

   ◆ Define *transient time*

   ◆ Determine the response when the pulse width is equal to or greater than five time constants

   ◆ Determine the response when the pulse width is less than five time constants

Two conditions of pulse response must be considered:

1. When the input pulse width (t_W) is equal to or greater than five time constants ($t_W \geq 5\tau$)

2. When the input pulse width is less than five time constants ($t_W < 5\tau$)

Recall that five time constants is accepted as the time for a capacitor to fully charge or fully discharge; this time is often called the **transient time**. A capacitor will fully charge if the pulse width is equal to or greater than five time constants (5τ). This condition is expressed as $t_W \geq 5\tau$. At the end of the pulse, the capacitor fully discharges back through the source.

Figure 20–4 illustrates the output waveforms for various *RC* transient times and a fixed input pulse width. Notice that as the transient time becomes shorter, compared to the pulse width, the shape of the output pulse approaches that of the input. In each case, the output reaches the full amplitude of the input.

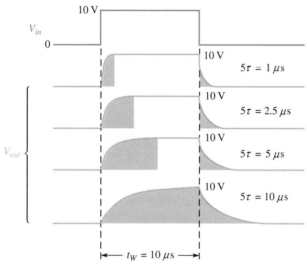

▲ FIGURE 20–4

Variation of an *RC* integrator's output pulse shape with time constant. The shaded areas indicate when the capacitor is charging or discharging.

Figure 20–5 shows how a fixed time constant and a variable input pulse width affect the integrator output. Notice that as the pulse width is increased, the shape of the output pulse approaches that of the input. Again, this means that the transient time is short compared to the pulse width.

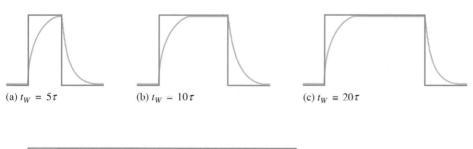

(a) $t_W = 5\tau$ (b) $t_W = 10\tau$ (c) $t_W = 20\tau$

◀ FIGURE 20–5

Variation of an *RC* integrator's output pulse shape with input pulse width (the time constant is fixed). Dark blue is input and light blue is output.

(d) $t_W = 40\tau$

Now let's examine the case in which the width of the input pulse is less than five time constants of the *RC* integrator. This condition is expressed as $t_W < 5\tau$. As you know, the capacitor charges for the duration of the pulse. However, because the pulse width is less than the time it takes the capacitor to fully charge (5τ), the output voltage will *not* reach the full input voltage before the end of the pulse. The capacitor only partially charges, as illustrated in Figure 20–6 for several values of *RC* time constants. Notice that for longer time constants, the output reaches a lower voltage because the capacitor cannot charge as much. Of course, in the examples with a single pulse input, the capacitor fully discharges after the pulse ends.

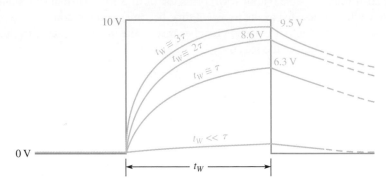

▲ FIGURE 20–6

Capacitor voltage for various time constants that are longer than the input pulse width. Dark blue is input and light blue is output.

When the time constant is much greater than the input pulse width, the capacitor charges very little, and, as a result, the output voltage becomes almost negligible, as indicated in Figure 20–6.

Figure 20–7 illustrates the effect of reducing the input pulse width for a fixed time constant value. As the input pulse width is reduced, the output voltage decreases because the capacitor has less time to charge. However, it takes the capacitor approximately the same length of time (5τ) to discharge back to zero for each condition after the pulse is removed.

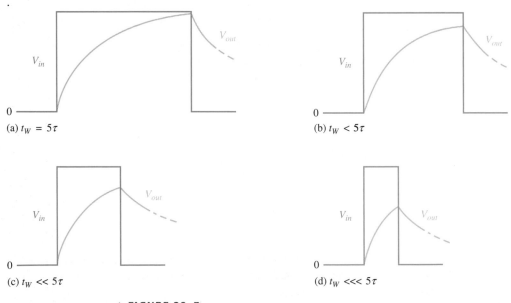

▲ FIGURE 20–7

The capacitor charges less and less as the input pulse width is reduced. The time constant is fixed.

EXAMPLE 20–1

A single 10 V pulse with a width of 100 μs is applied to the *RC* integrator in Figure 20–8.

(a) To what voltage will the capacitor charge?

(b) How long will it take the capacitor to discharge if the internal resistance of the pulse source is 50 Ω?

(c) Draw the output voltage waveform.

▶ **FIGURE 20–8**

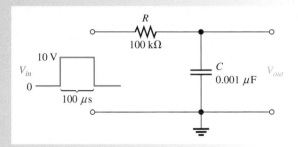

Solution   **(a)** The circuit time constant is

$$\tau = RC = (100\,\text{k}\Omega)(0.001\,\mu\text{F}) = 100\,\mu\text{s}$$

Notice that the pulse width is exactly equal to the time constant. Thus, the capacitor will charge approximately 63% of the full input amplitude in one time constant, so the output will reach a maximum voltage of

$$V_{out} = (0.63)10\,\text{V} = \textbf{6.3 V}$$

(b) The capacitor discharges back through the source when the pulse ends. You can neglect the 50 Ω source resistance in series with 100 kΩ. The total approximate discharge time, therefore, is

$$5\tau = 5(100\,\mu\text{s}) = \textbf{500}\,\boldsymbol{\mu}\textbf{s}$$

(c) The output charging and discharging curve is shown in Figure 20–9.

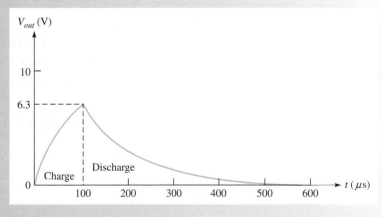

▲ **FIGURE 20–9**

Related Problem*   If the input pulse width in Figure 20–8 is increased to 200 μs, to what voltage will the capacitor charge?

*Answers are at the end of the chapter.

EXAMPLE 20–2

Determine how much the capacitor in Figure 20–10 will charge when the single pulse is applied to the input.

▶ **FIGURE 20–10**

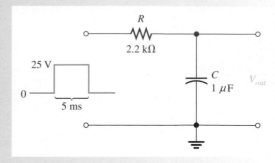

Solution Calculate the time constant.

$$\tau = RC = (2.2\ k\Omega)(1\ \mu F) = 2.2\ ms$$

Because the pulse width is 5 ms, the capacitor charges for 2.27 time constants (5 ms/2.2 ms = 2.27). Use the exponential formula from Chapter 12 (Eq. 12–19) to find the voltage to which the capacitor will charge. With V_F = 25 V and t = 5 ms, the calculation is as follows:

$$v = V_F(1 - e^{-t/RC})$$
$$= (25\ V)(1 - e^{-5ms/2.2ms}) = (25\ V)(1 - e^{-2.27})$$
$$= (25\ V)(1 - 0.103) = (25\ V)(0.897) = \mathbf{22.4\ V}$$

These calculations show that the capacitor charges to 22.4 V during the 5 ms duration of the input pulse. It will discharge back to zero when the pulse goes back to zero.

Related Problem Determine how much C will charge if the pulse width is increased to 10 ms.

SECTION 20–2 REVIEW

1. When an input pulse is applied to an *RC* integrator, what condition must exist in order for the output voltage to reach full amplitude?

2. For the circuit in Figure 20–11, which has a single input pulse, find the maximum output voltage and determine how long the capacitor will discharge.

3. For Figure 20–11, draw the approximate shape of the output voltage with respect to the input pulse.

4. If an integrator time constant equals the input pulse width, will the capacitor fully charge?

5. Describe the condition under which the output voltage has the approximate shape of a rectangular input pulse.

▶ **FIGURE 20–11**

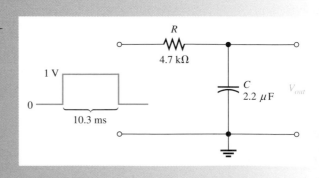

20–3 RESPONSE OF *RC* INTEGRATORS TO REPETITIVE PULSES

In electronic systems, you will encounter waveforms with repetitive pulses much more often than single pulses. However, an understanding of the integrator's response to single pulses is necessary in order to understand how these circuits respond to repeated pulses.

After completing this section, you should be able to

◆ **Analyze an *RC* integrator with repetitive input pulses**

  ◆ Determine the response when the capacitor does not fully charge or discharge

  ◆ Define *steady state*

  ◆ Describe the effect of an increase in time constant on circuit response

If a **periodic** pulse waveform is applied to an *RC* integrator, as shown in Figure 20–12, *the output waveshape depends on the relationship of the circuit time constant and the frequency (period) of the input pulses.* The capacitor, of course, charges and discharges in response to a pulse input. The amount of charge and discharge of the capacitor depends both on the circuit time constant and on the input frequency, as mentioned.

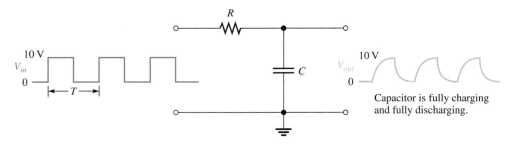

▲ **FIGURE 20–12**

RC integrator with a repetitive pulse waveform input ($T = 10\tau$).

If the pulse width and the time between pulses are each equal to or greater than five time constants, the capacitor will fully charge and fully discharge during each period (T) of the input waveform. This case is shown in Figure 20–12.

When the pulse width and the time between pulses are shorter than five time constants, as illustrated in Figure 20–13 for a square wave, the capacitor will *not* completely charge or discharge. We will now examine the effects of this situation on the output voltage of the *RC* integrator.

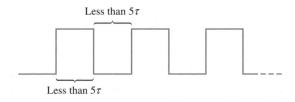

▲ **FIGURE 20–13**

Input waveform that does not allow full charge or discharge of the capacitor in an *RC* integrator.

For illustration, let's use an *RC* integrator with a charging and discharging time constant equal to the pulse width of a 10 V square wave input, as shown in Figure 20–14. This choice will simplify the analysis and will demonstrate the basic action of the integrator under these conditions. At this point, we do not care what the exact time constant value is because we know that an *RC* circuit charges approximately 63% during one time constant interval.

▶ **FIGURE 20–14**

RC integrator with a square wave input having a period equal to two time constants ($T = 2\tau$).

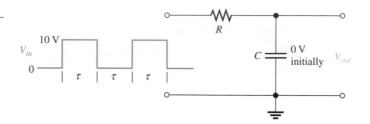

Let's assume that the capacitor in Figure 20–14 begins initially uncharged and examine the output voltage on a pulse-by-pulse basis. Figure 20–15 shows the charging and discharging shapes of five pulses.

▶ **FIGURE 20–15**

Input and output for the initially uncharged integrator in Figure 20–14.

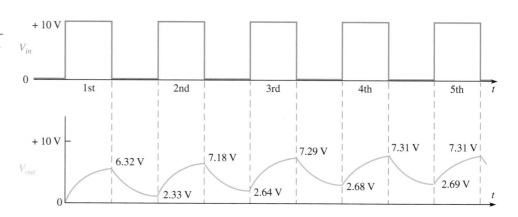

First pulse During the first pulse, the capacitor charges. The output voltage reaches 6.32 V (63.2% of 10 V), as shown in Figure 20–15.

Between first and second pulses The capacitor discharges, and the voltage decreases to 36.8% of the voltage at the beginning of this interval: 0.368(6.32 V) = 2.33 V.

Second pulse The capacitor voltage begins at 2.33 V and increases 63.2% of the way to 10 V. This calculation is as follows: The total charging range is 10 V − 2.33 V = 7.67 V. The capacitor voltage will increase an additional 63.2% of 7.67 V, which is 4.85 V. Thus, at the end of the second pulse, the output voltage is 2.33 V + 4.85 V = 7.18 V, as shown in Figure 20–15. Notice that the average is building up.

Between second and third pulses The capacitor discharges during this time, and therefore the voltage decreases to 36.8% of the initial voltage by the end of the second pulse: 0.368(7.18 V) = 2.64 V.

Third pulse At the start of the third pulse, the capacitor voltage begins at 2.64 V. The capacitor charges 63.2% of the way from 2.64 V to 10 V: 0.632(10 V − 2.64 V) = 4.65 V. Therefore, the voltage at the end of the third pulse is 2.64 V + 4.65 V = 7.29 V.

Between third and fourth pulses The voltage during this interval decreases due to capacitor discharge. It will decrease to 36.8% of its value by the end of the third pulse. The final voltage in this interval is 0.368(7.29 V) = 2.68 V.

Fourth pulse At the start of the fourth pulse, the capacitor voltage is 2.68 V. The voltage increases by 0.632(10 V − 2.68 V) = 4.63 V. Therefore, at the end of the fourth pulse, the capacitor voltage is 2.68 V + 4.63 V = 7.31 V. Notice that the values are leveling off as the pulses continue.

Between fourth and fifth pulses Between these pulses, the capacitor voltage drops to 0.368(7.31 V) = 2.69 V.

Fifth pulse During the fifth pulse, the capacitor charges 0.632(10 V − 2.69 V) = 4.62 V. Since it started at 2.69 V, the voltage at the end of the pulse is 2.69 V + 4.62 V = 7.31 V.

Steady-State Time Response

In the preceding discussion, the output voltage gradually built up and then began leveling off. It takes approximately 5τ for the output voltage to build up to a constant average value. This interval is the transient time of the circuit. Once the output voltage reaches the average value of the input voltage, a **steady-state** condition is reached that continues as long as the periodic input continues. This condition is illustrated in Figure 20–16 based on the values obtained in the preceding discussion.

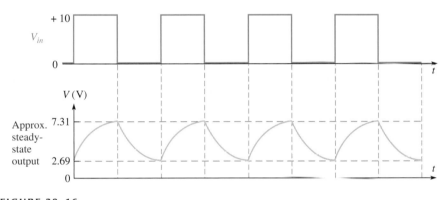

▲ FIGURE 20–16

Output reaches steady state after 5τ.

The transient time for our example circuit is the time from the beginning of the first pulse to the end of the third pulse. The reason for this interval is that the capacitor voltage at the end of the third pulse is 7.29 V, which is about 99% of the final voltage.

The Effect of an Increase in Time Constant

What happens to the output voltage if the *RC* time constant of the integrator is increased with a variable resistor, as indicated in Figure 20–17? As the time constant is increased, the capacitor charges less during a pulse and discharges less between pulses. The result is a

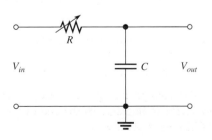

◀ FIGURE 20–17

Integrator with a variable *RC* time constant.

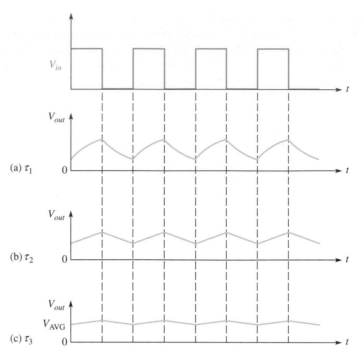

▲ FIGURE 20–18

Effect of longer time constants on the output of an RC integrator ($\tau_3 > \tau_2 > \tau_1$).

smaller fluctuation in the output voltage for increasing values of time constant, as shown in Figure 20–18.

As the time constant becomes extremely long compared to the pulse width, the output voltage approaches a constant dc voltage, as shown in Figure 20–18(c). This value is the average value of the input. For a square wave, it is one-half the amplitude.

EXAMPLE 20–3

Determine the output voltage waveform for the first two pulses applied to the *RC* integrator in Figure 20–19. Assume that the capacitor is initially uncharged and the rheostat is set to $5\,k\Omega$.

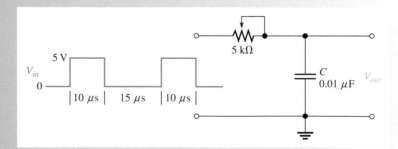

▲ FIGURE 20–19

Solution   First, calculate the circuit time constant.

$$\tau = RC = (5\,k\Omega)(0.01\,\mu F) = 50\,\mu s$$

Obviously, the time constant is much longer than the input pulse width or the interval between pulses (notice that the input is not a square wave). In this case, the exponential formulas must be applied, and the analysis is relatively difficult. Follow the solution carefully.

1. *Calculation for first pulse:* Use the equation for an increasing exponential because *C* is charging. Note that V_F is 5 V, and *t* equals the pulse width of 10 μs. Therefore,

$$v_C = V_F(1 - e^{-t/RC}) = (5\text{ V})(1 - e^{-10\mu s/50\mu s})$$
$$= (5\text{ V})(1 - 0.819) = \textbf{906 mV}$$

This result is plotted in Figure 20–20(a).

2. *Calculation for interval between first and second pulse:* Use the equation for a decreasing exponential because *C* is discharging. Note that V_i is 906 mV because *C* begins to discharge from this value at the end of the first pulse. The discharge time is 15 μs. Therefore,

$$v_C = V_i e^{-t/RC} = (906\text{ mV})e^{-15\mu s/50\mu s}$$
$$= (906\text{ mV})(0.741) = \textbf{671 mV}$$

This result is shown in Figure 20–20(b).

3. *Calculation for second pulse:* At the beginning of the second pulse, the output voltage is 671 mV. During the second pulse, the capacitor will again charge. In this case, it does not begin at zero volts. It already has 671 mV from the previous charge and discharge. To handle this situation, you must use the general exponential formula.

$$v = V_F + (V_i - V_F)e^{-t/\tau}$$

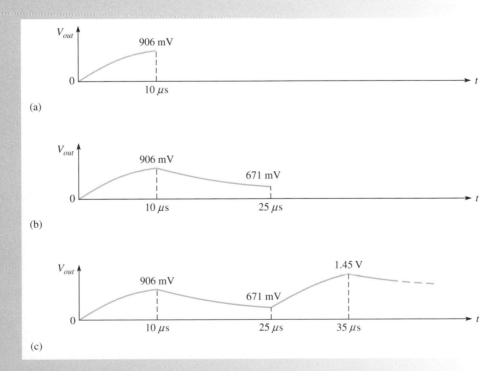

(a)

(b)

(c)

▲ FIGURE 20–20

Using this equation, you can calculate the voltage across the capacitor at the end of the second pulse as follows:

$$v_C = V_F + (V_i - V_F)e^{-t/RC}$$
$$= 5\,\text{V} + (671\,\text{mV} - 5\,\text{V})e^{-10\mu s/50\mu s}$$
$$= 5\,\text{V} + (-4.33\,\text{V})(0.819) = 5\,\text{V} - 3.55\,\text{V} = \mathbf{1.45\,V}$$

This result is shown in Figure 20–20(c).

Notice that the output waveform builds up on successive input pulses. After approximately 5τ, it will reach its steady state and will fluctuate between a constant maximum and a constant minimum, with an average equal to the average value of the input. You can see this pattern by carrying the analysis in this example further.

Related Problem   Determine V_{out} at the beginning of the third pulse.

    Use Multisim file E20-03 to verify the calculated results in this example and to confirm your calculation for the related problem.

SECTION 20–3 REVIEW

1. What conditions allow an *RC* integrator capacitor to fully charge and discharge when a periodic pulse waveform is applied to the input?

2. What will the output waveform look like if the circuit time constant is extremely short compared to the pulse width of a square wave input?

3. What is the time called that is required for the output voltage to build up to a constant average value when 5τ is greater than the pulse width of an input square wave?

4. Define *steady-state response*.

5. What does the average value of the output voltage of an integrator equal during steady state?

20–4   Response of an *RC* Differentiator to a Single Pulse

In terms of time response, a series *RC* circuit in which the output voltage is taken across the resistor is known as a **differentiator**. Recall that in terms of frequency response, it is a high-pass filter. The term *differentiator* is derived from the mathematical process of differentiation, which this type of circuit approximates under certain conditions.

After completing this section, you should be able to

◆ **Analyze an *RC* differentiator with a single input pulse**

 ◆ Describe the response at the rising edge of the input pulse

 ◆ Determine the response during and at the end of a pulse for various pulse width–time constant relationships

Figure 20–21 shows an *RC* differentiator with a pulse input. The same action occurs in a differentiator as in an integrator, except the output voltage is taken across the resistor rather than across the capacitor. The capacitor charges exponentially at a rate depending on the *RC* time constant. The shape of the differentiator's resistor voltage is determined by the charging and discharging action of the capacitor.

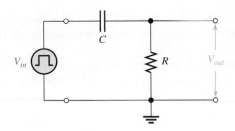

◀ FIGURE 20–21

An *RC* differentiator with a pulse generator connected.

Pulse Response

To understand how the output voltage is shaped by a differentiator, you must consider the following:

1. The response to the rising pulse edge

2. The response between the rising and falling edges

3. The response to the falling pulse edge

Response to the Rising Edge of the Input Pulse   Let's assume that the capacitor is initially uncharged prior to the rising pulse edge. Prior to the pulse, the input is zero volts. Thus, there are zero volts across the capacitor and also zero volts across the resistor, as indicated in Figure 20–22(a).

Let's also assume that a 10 V pulse is applied to the input. When the rising edge occurs, point *A* goes to +10 V. Recall that the voltage across a capacitor cannot change instantaneously, and thus the capacitor appears instantaneously as a short. Therefore, if point *A*

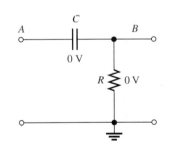

(a) Before pulse is applied

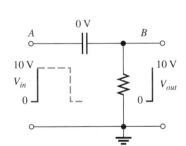

(b) At rising edge of input pulse

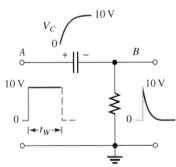

(c) During level part of pulse when $t_W \geq 5\tau$

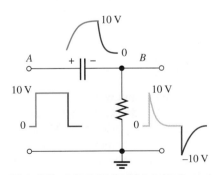

(d) At falling edge of pulse when $t_W \geq 5\tau$

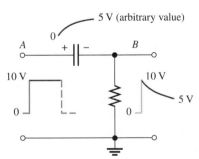

(e) During level part of pulse when $t_W < 5\tau$

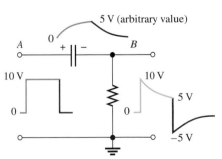

(f) At falling edge of pulse when $t_W < 5\tau$

▲ FIGURE 20–22

Examples of the response of an *RC* differentiator to a single input pulse under two conditions: $t_W \geq 5\tau$ and $t_W < 5\tau$.

goes instantly to +10 V, then point *B must* also go instantly to +10 V, keeping the capacitor voltage zero for the instant of the rising edge. The capacitor voltage is the voltage from point *A* to point *B*.

The voltage at point *B* with respect to ground is the voltage across the resistor (and the output voltage). Thus, the output voltage suddenly goes to +10 V in response to the rising pulse edge, as indicated in Figure 20–22(b).

Response During Pulse When $t_W \geq 5\tau$ While the pulse is at its high level between the rising edge and the falling edge, the capacitor is charging. When the pulse width is equal to or greater than five time constants ($t_W \geq 5\tau$), the capacitor has time to fully charge.

As the voltage across the capacitor builds up exponentially, the voltage across the resistor decreases exponentially until it reaches zero volts at the time the capacitor reaches full charge (+10 V in this case). This decrease in the resistor voltage occurs because the sum of the capacitor voltage and the resistor voltage at any instant must be equal to the applied voltage, in compliance with Kirchhoff's voltage law ($v_C + v_R = v_{in}$). This part of the response is illustrated in Figure 20–22(c).

Response to Falling Edge When $t_W \geq 5\tau$ Let's examine the case in which the capacitor is fully charged at the end of the pulse ($t_W \geq 5\tau$). Refer to Figure 20–22(d). On the falling edge, the input pulse suddenly goes from +10 V back to zero. An instant before the falling edge, the capacitor is charged to 10 V, so point *A* is +10 V and point *B* is 0 V. The voltage across a capacitor cannot change instantaneously, so when point *A* makes a transition from +10 V to zero on the falling edge, point *B must* also make a 10 V transition from zero to −10 V. This keeps the voltage across the capacitor at 10 V for the instant of the falling edge.

The capacitor now begins to discharge exponentially. As a result, the resistor voltage goes from −10 V to zero in an exponential curve, as indicated in red in Figure 20–22(d).

Response During Pulse When $t_W < 5\tau$ When the pulse width is less than five time constants ($t_W < 5\tau$), the capacitor does not have time to fully charge. Its partial charge depends on the relation of the time constant and the pulse width.

Because the capacitor does not reach the full +10 V, the resistor voltage will not reach zero volts by the end of the pulse. For example, if the capacitor charges to +5 V during the pulse interval, the resistor voltage will decrease to +5 V, as illustrated in Figure 20–22(e).

Response to Falling Edge When $t_W < 5\tau$ Now, let's examine the case in which the capacitor is only partially charged at the end of the pulse ($t_W < 5\tau$). For example, if the capacitor charges to +5 V, the resistor voltage at the instant before the falling edge is also +5 V because the capacitor voltage plus the resistor voltage must add up to +10 V, as illustrated in Figure 20–22(e).

When the falling edge occurs, point *A* goes from +10 V to zero. As a result, point *B* goes from +5 V to −5 V, as illustrated in Figure 20–22(f). This decrease occurs, of course, because the capacitor voltage cannot change at the instant of the falling edge. Immediately after the falling edge, the capacitor begins to discharge to zero. As a result, the resistor voltage goes from −5 V to zero, as shown.

Summary of *RC* Differentiator Response to a Single Pulse

A good way to summarize this section is to look at the general output waveforms of a differentiator as the time constant is varied from one extreme, when 5τ is much less than the pulse width, to the other extreme, when 5τ is much greater than the pulse width. These situations are illustrated in Figure 20–23. In part (a) of the figure, the output consists of narrow positive and negative "spikes." In part (e), the output approaches the shape of the input. Various conditions between these extremes are illustrated in parts (b), (c), and (d).

You may have observed a pulse that looks similar to Figure 20–23(e) when you ac couple a pulse to an oscilloscope. In this case the capacitor in the oscilloscope coupling circuit can act as an unwanted differentiating circuit, causing the pulse to droop. To avoid this, you can dc couple the scope and check the probe compensation.

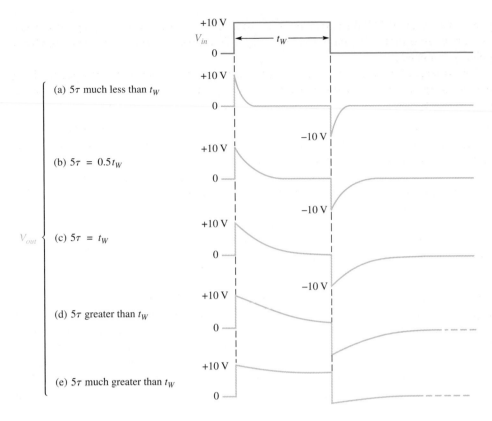

(a) 5τ much less than t_W

(b) $5\tau = 0.5t_W$

V_{out} (c) $5\tau = t_W$

(d) 5τ greater than t_W

(e) 5τ much greater than t_W

▲ **FIGURE 20–23**

Effects of a change in time constant on the shape of the output voltage of an *RC* differentiator.

EXAMPLE 20–4   Draw the output voltage for the *RC* differentiator in Figure 20–24.

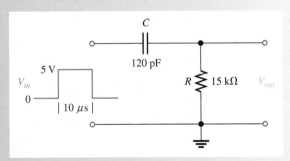

▲ **FIGURE 20–24**

Solution   First, calculate the time constant.

$$\tau = RC = (15\,\text{k}\Omega)(120\,\text{pF}) = 1.8\,\mu\text{s}$$

In this case, $t_W > 5\tau$, so the capacitor reaches full charge before the end of the pulse.
   On the rising edge, the resistor voltage jumps to $+5$ V and then decreases exponentially to zero by the end of the pulse. On the falling edge, the resistor voltage jumps to -5 V and then goes back to zero exponentially. The resistor voltage is, of course, the output, and its shape is shown in Figure 20–25.

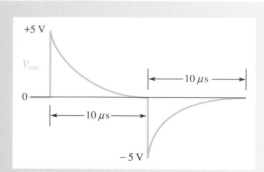

▲ FIGURE 20–25

Related Problem Draw the output voltage if C is changed to 12 pF in Figure 20–24.

 Use Multisim file E20-04 to verify the calculated results in this example and to confirm your answer for the related problem. To simulate a single pulse, specify a waveform with the given pulse width but a small duty cycle (long period).

EXAMPLE 20–5

Determine the output voltage waveform for the *RC* differentiator in Figure 20–26 with the rheostat set so that the total resistance of R_1 and R_2 is 2 kΩ.

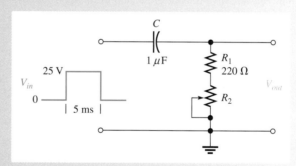

▲ FIGURE 20–26

Solution First, calculate the time constant.

$$\tau = R_{tot}C = (2\,k\Omega)(1\,\mu F) = 2\,ms$$

On the rising edge, the resistor voltage immediately jumps to +25 V. Because the pulse width is 5 ms, the capacitor charges for 2.5 time constants and therefore does not reach full charge. Thus, you must use the formula for a decreasing exponential in order to calculate to what voltage the output decreases by the end of the pulse.

$$v_{out} = V_i e^{-t/RC} = 25e^{-5ms/2ms} = 25(0.082) = 2.05\,V$$

where $V_i = 25$ V and $t = 5$ ms. This calculation gives the resistor voltage (v_{out}) at the end of the 5 ms pulse width interval.

On the falling edge, the resistor voltage immediately jumps from +2.05 V down to −22.95 V (a 25 V transition). The resulting waveform of the output voltage is shown in Figure 20–27.

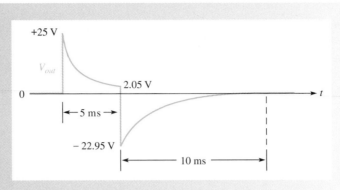

▲ FIGURE 20-27

Related Problem   Determine the voltage at the end of the pulse in Figure 20–26 if the rheostat is set so that the total resistance is 1.5 kΩ.

 Use Multisim file E20-05 to verify the calculated results in this example and to confirm your calculation for the related problem. To simulate a single pulse, specify a waveform with the given pulse width but a small duty cycle.

SECTION 20–4
REVIEW

1. Draw the output of a differentiator for a 10 V input pulse when $5\tau = 0.5t_W$.
2. Under what condition does the output pulse shape most closely resemble the input pulse for a differentiator?
3. What does the differentiator output look like when 5τ is much less than the pulse width of the input?
4. If the resistor voltage in a differentiating circuit is down to +5 V at the end of a 15 V input pulse, to what negative value will the resistor voltage go in response to the falling edge of the input?

20–5 RESPONSE OF *RC* DIFFERENTIATORS TO REPETITIVE PULSES

The *RC* differentiator response to a single pulse, covered in the last section, is extended in this section to repetitive pulses.

After completing this section, you should be able to

◆ **Analyze an *RC* differentiator with repetitive input pulses**

    ◆ Determine the response when the pulse width is less than five time constants

If a periodic pulse waveform is applied to an *RC* differentiating circuit, two conditions again are possible: $t_W \geq 5\tau$ or $t_W < 5\tau$. Figure 20–28 shows the output when $t_W = 5\tau$. As the time constant is reduced, both the positive and the negative portions of the output become narrower. Notice that the average value of the output is zero. An average value of zero means that the waveform has equal positive and negative portions. The average value of a waveform is its **dc component**. Because a capacitor blocks dc, the dc component of the input is prevented from passing through to the output, resulting in an average value of zero.

▶ FIGURE 20–28

Example of differentiator response
when $t_W = 5\tau$.

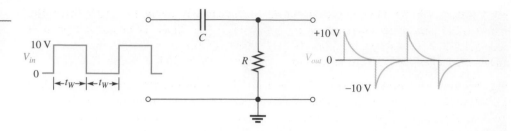

Figure 20–29 shows the steady-state output when $t_W < 5\tau$. As the time constant is increased, the positively and negatively sloping portions become flatter. For a very long time constant, the output approaches the shape of the input, but with an average value of zero.

▶ FIGURE 20–29

Example of differentiator response
when $t_W < 5\tau$.

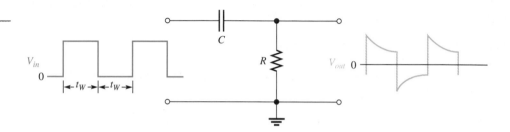

Analysis of a Repetitive Waveform

Like the integrator, the differentiator output takes time (5τ) to reach steady state. To illustrate the response, let's take an example in which the time constant equals the input pulse width. At this point, we do not care what the circuit time constant is because we know that the resistor voltage will decrease to approximately 37% of its maximum value during one pulse (1τ).

Let's assume that the capacitor in Figure 20–30 begins initially uncharged and then examine the output voltage on a pulse-by-pulse basis. The results of the analysis that follows are shown in Figure 20–31.

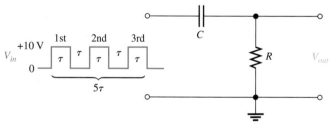

▲ FIGURE 20–30

RC differentiator with $\tau = t_W$.

▶ FIGURE 20–31

Differentiator output waveform
during transient time for the circuit
in Figure 20–30.

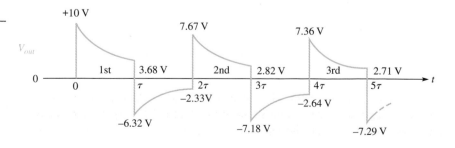

First pulse On the rising edge, the output instantaneously jumps to +10 V. Then the capacitor partially charges to 63.2% of 10 V, which is 6.32 V. Thus, the output voltage must decrease to 3.68 V, as shown in Figure 20–31. On the falling edge, the output instantaneously makes a negative-going 10 V transition to −6.32 V (−10 V + 3.68 V = −6.32 V).

Between first and second pulses The capacitor discharges to 36.8% of 6.32 V, which is 2.33 V. Thus, the resistor voltage, which starts at −6.32 V, must increase to −2.33 V. Why? Because at the instant prior to the next pulse, the input voltage is zero. Therefore, the sum of v_C and v_R must be zero (2.33 V − 2.33 V = 0). Remember that $v_C + v_R = v_{in}$ at all times, in accordance with Kirchhoff's voltage law.

Second pulse On the rising edge, the output makes an instantaneous, positive-going, 10 V transition from −2.33 V to 7.67 V. Then by the end of the pulse the capacitor charges $0.632 \times (10 \text{ V} - 2.33 \text{ V}) = 4.85$ V. Thus, the capacitor voltage increases from 2.33 V to 2.33 V + 4.85 V = 7.18 V. The output voltage drops to 0.368×7.67 V = 2.82 V.

On the falling edge, the output instantaneously makes a negative-going transition from 2.82 V to −7.18 V, as shown in Figure 20–31.

Between second and third pulses The capacitor discharges to 36.8% of 7.18 V, which is 2.64 V. Thus, the output voltage starts at −7.18 V and increases to −2.64 V because the capacitor voltage and the resistor voltage must add up to zero at the instant prior to the third pulse (the input is zero).

Third pulse On the rising edge, the output makes an instantaneous 10 V transition from −2.64 V to +7.36 V. Then the capacitor charges $0.632 \times (10 \text{ V} - 2.64 \text{ V}) = 4.65$ V to 2.64 V + 4.65 V = 7.29 V. As a result, the output voltage drops to 0.368×7.36 V = 2.71 V. On the falling edge, the output instantly goes from +2.71 V down to −7.29 V.

After the third pulse, five time constants have elapsed, and the output voltage is close to its steady state. Thus, it will continue to vary from a positive maximum of about +7.3 V to a negative maximum of about −7.3 V, with an average value of zero.

SECTION 20–5 REVIEW	1. What conditions allow an *RC* differentiator to fully charge and discharge when a periodic pulse waveform is applied to the input?
	2. What will the output waveform look like if the circuit time constant is extremely short compared to the pulse width of a square wave input?
	3. What does the average value of the differentiator output voltage equal during steady state?

20–6 RESPONSE OF *RL* INTEGRATORS TO PULSE INPUTS

A series *RL* circuit in which the output voltage is taken across the resistor is known as an integrator in terms of time response. Although only the response to a single pulse is discussed, it can be extended to repetitive pulses, as described for the *RC* integrator.

After completing this section, you should be able to

◆ **Analyze the operation of an *RL* integrator**

   ◆ Determine the response to a single input pulse

Figure 20–32 shows an *RL* integrator. The output waveform is taken across the resistor and, under equivalent conditions, is the same shape as that for the *RC* integrator. Recall that in the *RC* case, the output was across the capacitor.

► FIGURE 20–32

An *RL* integrator with a pulse generator connected.

As you know, each edge of an ideal pulse is considered to be instantaneous. Two basic rules for inductor behavior will aid in analyzing *RL* circuit responses to pulse inputs:

1. The inductor appears as an open to an instantaneous change in current and as a short (ideally) to dc.

2. The current in an inductor cannot change instantaneously—it can change only exponentially.

Response of the *RL* Integrator to a Single Pulse

When a pulse generator is connected to the input of the integrator and the voltage pulse goes from its low level to its high level, the inductor prevents a sudden change in current. As a result, the inductor acts as an open, and all of the input voltage is across it at the instant of the rising pulse edge. This situation is indicated in Figure 20–33(a).

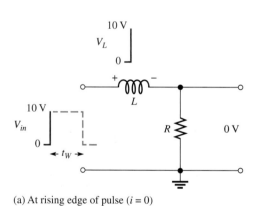

(a) At rising edge of pulse (*i* = 0)

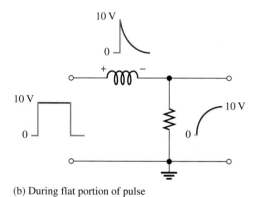

(b) During flat portion of pulse

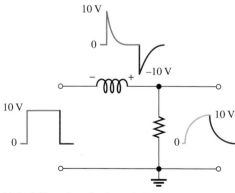

(c) At falling edge of pulse and after

▲ FIGURE 20–33

Illustration of the pulse response of an *RL* integrator ($t_W > 5\tau$).

After the rising edge, the current builds up, and the output voltage follows the current as it increases exponentially, as shown in Figure 20–33(b). The current can reach a maximum of V_p/R if the transient time is shorter than the pulse width ($V_p = 10$ V in this example).

When the pulse goes from its high level to its low level, an induced voltage with reversed polarity is created across the coil in an effort to keep the current equal to V_p/R. The output voltage begins to decrease exponentially, as shown in Figure 20–33(c).

The exact shape of the output depends on the L/R time constant as summarized in Figure 20–34 for various relationships between the time constant and the pulse width. You should note that the response of this *RL* circuit in terms of the shape of the output is identical to that of the *RC* integrator. The relationship of the L/R time constant to the input pulse width has the same effect as the *RC* time constant that was shown in Figure 20–4. For example, when $t_W < 5\tau$, the output voltage will not reach its maximum possible value.

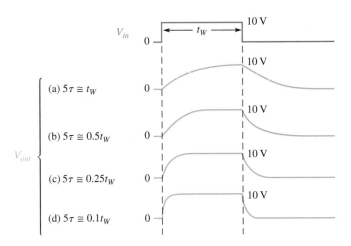

◀ FIGURE 20–34

Illustration of the variation in *RL* integrator output pulse shape with time constant.

EXAMPLE 20–6

Determine the maximum output voltage for the *RL* integrator in Figure 20–35 when a single pulse is applied as shown. The rheostat is set so that the total resistance is 50 Ω.

FIGURE 20–35

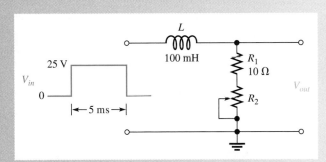

Solution   Calculate the time constant.

$$\tau = \frac{L}{R} = \frac{100\,\text{mH}}{50\,\Omega} = 2\,\text{ms}$$

Because the pulse width is 5 ms, the inductor charges for 2.5τ. Use the exponential formula to calculate the voltage.

$$v_{out(max)} = V_F(1 - e^{-t/\tau}) = 25(1 - e^{-5\text{ms}/2\text{ms}})$$
$$= 25(1 - e^{-2.5}) = 25(1 - 0.082) = 25(0.918) = \mathbf{22.9\,V}$$

To what resistance must the rheostat, R_2, be set for the output voltage to reach 25 V by the end of the pulse in Figure 20–35?

EXAMPLE 20–7

A pulse is applied to the *RL* integrator in Figure 20–36. Determine the complete wave-shapes and the values for *I*, V_R, and V_L.

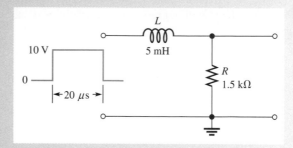

▲ FIGURE 20–36

Solution The circuit time constant is

$$\tau = \frac{L}{R} = \frac{5\,\text{mH}}{1.5\,\text{k}\Omega} = 3.33\,\mu\text{s}$$

Since $5\tau = 16.7\,\mu\text{s}$ is less than t_W, the current will reach its maximum value and remain there until the end of the pulse.

At the rising edge of the pulse,

$$i = 0\,\text{A}$$
$$v_R = 0\,\text{V}$$
$$v_L = 10\,\text{V}$$

The inductor initially appears as an open, so all of the input voltage appears across *L*.
During the pulse,

$$i\ \text{increases exponentially to}\ \frac{V_p}{R} = \frac{10\,\text{V}}{1.5\,\text{k}\Omega} = 6.67\,\text{mA in } 16.7\,\mu\text{s}$$

v_R increases exponentially to 10 V in 16.7 μs

v_L decreases exponentially to zero in 16.7 μs

At the falling edge of the pulse,

$$i = 6.67\,\text{mA}$$
$$v_R = 10\,\text{V}$$
$$v_L = -10\,\text{V}$$

After the pulse,

i decreases exponentially to zero in 16.7 μs

v_R decreases exponentially to zero in 16.7 μs

v_L increases exponentially to zero in 16.7 μs

The waveforms are shown in Figure 20–37.

FIGURE 20–37

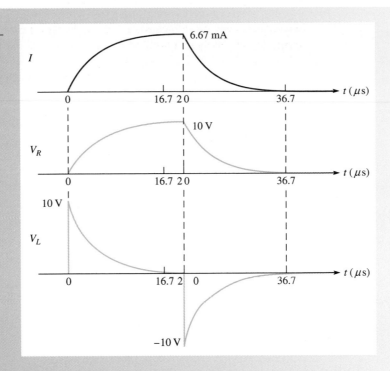

Related Problem   What will be the maximum output voltage if the amplitude of the input pulse is increased to 20 V in Figure 20–36?

   Use Multisim file E20-07 to verify the calculated results in this example and to confirm your calculation for the related problem. To simulate a single pulse, specify a waveform with the given pulse width but a small duty cycle.

EXAMPLE 20–8

A 10 V pulse with a width of 1 ms is applied to the *RL* integrator in Figure 20–38. Determine the voltage level that the output will reach during the pulse. If the source has an internal resistance of 30 Ω, how long will it take the output to decay to zero? Draw the output voltage waveform.

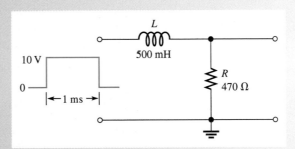

▲ **FIGURE 20–38**

Solution   The inductor charges through the 30 Ω source resistance plus the 470 Ω external resistor. The time constant is

$$\tau = \frac{L}{R_{tot}} = \frac{500\,\text{mH}}{470\,\Omega + 30\,\Omega} = \frac{500\,\text{mH}}{500\,\Omega} = 1\,\text{ms}$$

Notice that in this case the pulse width is exactly equal to τ. Thus, the output V_R will reach approximately 63% of the full input amplitude in 1τ. Therefore, the output voltage gets to **6.3 V** at the end of the pulse.

After the pulse is gone, the inductor discharges back through the 30 Ω source resistance and the 470 Ω resistor. The output voltage takes **5τ** to completely decay to zero.

$$5\tau = 5(1 \text{ ms}) = 5 \text{ ms}$$

The output voltage is shown in Figure 20–39.

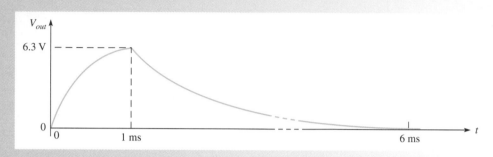

▲ **FIGURE 20–39**

Related Problem   To what value must R be changed to allow the output voltage to reach the input level during the pulse?

 Use Multisim file E20-08 to verify the calculated results in this example and to confirm your calculation for the related problem. To simulate a single pulse, specify a waveform with the given pulse width but a small duty cycle.

SECTION 20–6 REVIEW

1. In an *RL* integrator, across which component is the output voltage taken?
2. When a pulse is applied to an *RL* integrator, what condition must exist in order for the output voltage to reach the amplitude of the input?
3. Under what condition will the output voltage have the approximate shape of the input pulse?

20–7 RESPONSE OF *RL* DIFFERENTIATORS TO PULSE INPUTS

A series *RL* circuit in which the output voltage is taken across the inductor is known as a differentiator in terms of time response. Although only the response to a single pulse is discussed, it can be extended to repetitive pulses, as was described for the *RC* differentiator.

After completing this section, you should be able to

◆ **Analyze the operation of an *RL* differentiator**

   ◆ Determine the response to a single input pulse

Response of the *RL* Differentiator to a Single Pulse

Figure 20–40 shows an *RL* differentiator with a pulse generator connected to the input.

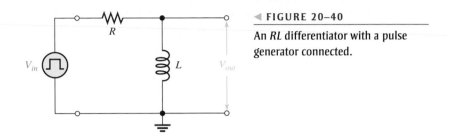

◀ **FIGURE 20–40**

An *RL* differentiator with a pulse generator connected.

Initially, before the pulse, there is no current in the circuit. When the input pulse goes from its low level to its high level, the inductor prevents a sudden change in current. It does so, as you know, with an induced voltage equal and opposite to the input. As a result, *L* looks like an open, and all of the input voltage appears across it at the instant of the rising edge, as shown in Figure 20–41(a) with a 10 V pulse.

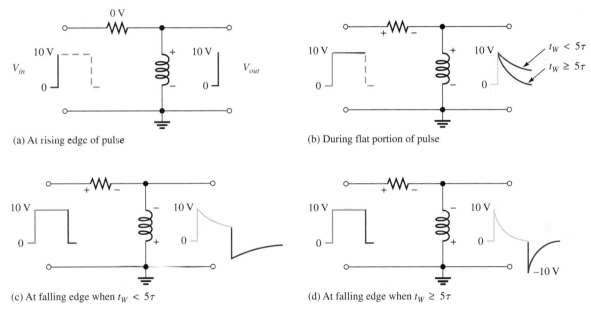

(a) At rising edge of pulse

(b) During flat portion of pulse

(c) At falling edge when $t_W < 5\tau$

(d) At falling edge when $t_W \geq 5\tau$

▲ **FIGURE 20–41**

Illustration of the response of an *RL* differentiator for both time constant conditions.

During the pulse, the current exponentially builds up. As a result, the inductor voltage decreases, as shown in Figure 20–41(b). The rate of decrease, as you know, depends on the L/R time constant. When the falling edge of the input occurs, the inductor reacts to keep the current as is, by creating an induced voltage in a direction as indicated in Figure 20–41(c). This reaction is seen as a sudden negative-going transition of the inductor voltage, as indicated in Figure 20–41(c) and (d).

Two conditions are possible, as indicated in Figure 20–41(c) and (d). In part (c), 5τ is greater than the input pulse width, and the output voltage does not have time to decay to zero. In part (d), 5τ is less than or equal to the pulse width, and so the output decays to zero before the end of the pulse. In this case a -10 V transition occurs at the trailing edge.

Keep in mind that as far as the input and output waveforms are concerned, the *RL* integrator and differentiator perform the same as their *RC* counterparts.

A summary of the *RL* differentiator response for relationships of various time constants and pulse widths is shown in Figure 20–42.

▶ FIGURE 20–42

Illustration of the variation in output pulse shape with the *RL* time constant.

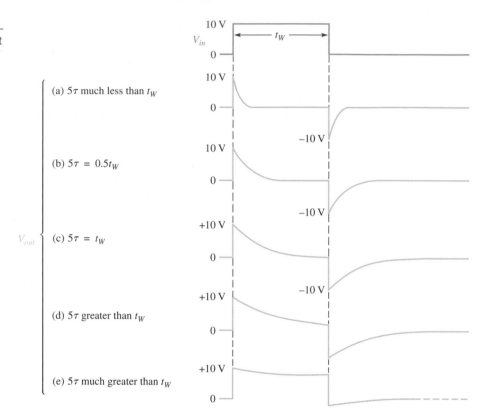

(a) 5τ much less than t_W

(b) 5τ = 0.5t_W

(c) 5τ = t_W

(d) 5τ greater than t_W

(e) 5τ much greater than t_W

V_{out}

EXAMPLE 20–9

Draw the output voltage for the *RL* differentiator in Figure 20–43.

▶ FIGURE 20–43

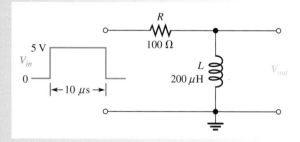

Solution First, calculate the time constant.

$$\tau = \frac{L}{R} = \frac{200\,\mu H}{100\,\Omega} = 2\,\mu s$$

In this case, $t_W = 5\tau$, so the output will decay to zero at the end of the pulse.

On the rising edge, the inductor voltage jumps to +5 V and then decays exponentially to zero. It reaches approximately zero at the instant of the falling edge. On the falling edge of the input, the inductor voltage jumps to −5 V and then goes back to zero. The output waveform is shown in Figure 20–44.

▶ **FIGURE 20–44**

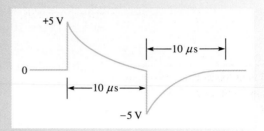

Related Problem   Draw the output voltage if the pulse width is reduced to 5 μs in Figure 20–43.

EXAMPLE 20–10

Determine the output voltage waveform for the *RL* differentiator in Figure 20–45.

▶ **FIGURE 20–45**

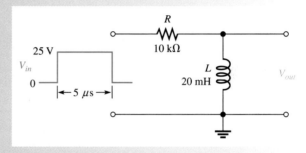

Solution   First, calculate the time constant.

$$\tau = \frac{L}{R} = \frac{20\,\text{mH}}{10\,\text{k}\Omega} = 2\,\mu s$$

On the rising edge, the inductor voltage immediately jumps to +25 V. Because the pulse width is 5 μs, the inductor charges for only 2.5τ, so you must use the formula for a decreasing exponential.

$$v_L = V_i e^{-t/\tau} = 25e^{-5\mu s/2\mu s} = 25e^{-2.5} = 25(0.082) = 2.05\,\text{V}$$

This result is the inductor voltage at the end of the 5 μs input pulse.

On the falling edge, the output immediately jumps from +2.05 V down to −22.95 V (a 25 V negative-going transition). The complete output waveform is shown in Figure 20–46.

▶ **FIGURE 20–46**

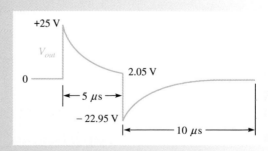

Related Problem   What must be the value of *R* for the output voltage to reach zero by the end of the pulse in Figure 20–45?

 Use Multisim file E20-10 to verify the calculated results in this example and to confirm your calculation for the related problem. To simulate a single pulse, specify a waveform with the given pulse width but a small duty cycle.

1. In an *RL* differentiator, across which component is the output taken?
2. Under what condition does the output pulse shape most closely resemble the input pulse?
3. If the inductor voltage in an *RL* differentiator is down to +2 V at the end of a +10 V input pulse, to what negative voltage will the output go in response to the falling edge of the input?

20–8 RELATIONSHIP OF TIME RESPONSE TO FREQUENCY RESPONSE

A definite relationship exists between time (pulse) response and frequency response. The fast rising and falling edges of a pulse waveform contain the higher frequency components. The flatter portions of the pulse waveform, which are the tops and the baseline of the pulses, represent slow changes or lower frequency components. The average value of the pulse waveform is its dc component.

After completing this section, you should be able to

◆ **Explain the relationship of time response to frequency response**

   ◆ Describe a pulse waveform in terms of its frequency components

   ◆ Explain how *RC* and *RL* integrators act as filters

   ◆ Explain how *RC* and *RL* differentiators act as filters

   ◆ State the formulas that relate rise and fall times to frequency

The relationships of pulse characteristics and frequency content of pulse waveforms are indicated in Figure 20–47.

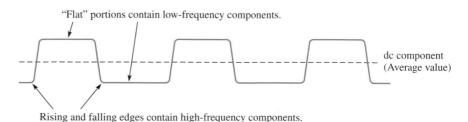

▲ **FIGURE 20–47**

Frequency content of a pulse waveform.

The Integrator

RC Integrator   In terms of frequency response, the RC integrator acts as a low-pass filter. As you learned, the RC integrator tends to exponentially "round off" the edges of the applied pulses. This rounding off occurs to varying degrees, depending on the relationship of the time constant to the pulse width and period. The rounding off of the edges indicates that the integrator tends to reduce the higher frequency components of the pulse waveform, as illustrated in Figure 20–48.

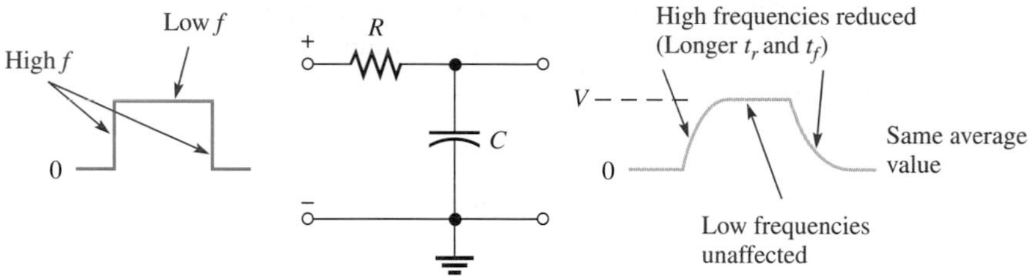

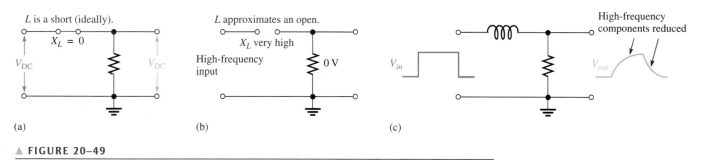

▲ **FIGURE 20–48**

Time and frequency response relationship in an RC integrator (one pulse in a repetitive waveform shown).

RL Integrator   Like the RC integrator, the RL integrator also acts as a basic low-pass filter because L is in series between the input and output. The inductive reactance, X_L, is small for low frequencies and offers little opposition. It increases with frequency, so at higher frequencies most of the total voltage is dropped across L and very little across R, the output. If the input is dc, L is like a short ($X_L = 0$). At high frequencies, L becomes like an open, as illustrated in Figure 20–49.

▲ **FIGURE 20–49**

Low-pass filtering action.

The Differentiator

RC Differentiator   In terms of frequency response, the RC differentiator acts as a high-pass filter. As you know, the differentiator tends to introduce tilt to the flat portion of a pulse. That is, it tends to reduce the lower frequency components of a pulse waveform. Also, it completely eliminates the dc component of the input and produces a zero average-value output. This action is illustrated in Figure 20–50.

RL Differentiator   Again like the RC differentiator, the RL differentiator also acts as a basic high-pass filter. Because L is connected across the output, less voltage is developed

▶ FIGURE 20–50

Time and frequency response relationship in an *RC* differentiator (one pulse in a repetitive waveform shown).

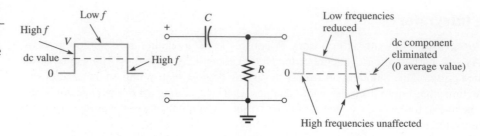

across it at lower frequencies than at higher ones. There are zero volts across the output for dc (ignoring winding resistance). For high frequencies, most of the input voltage is dropped across the output coil ($X_L = 0$ for dc; $X_L \cong$ open for high frequencies). Figure 20–51 shows high-pass filter action.

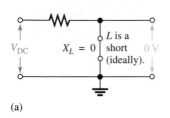

(a)

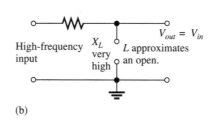

(b)

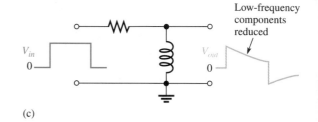

(c)

▲ FIGURE 20–51

High-pass filtering action.

Formulas Relating Time Response to Frequency Response

The fast transitions of a pulse (rise time, t_r, and fall time, t_f) are related to the highest frequency component, f_h, in that pulse by the following formula:

Equation 20–1

$$t_r = \frac{0.35}{f_h}$$

This formula also applies to fall time, and the fastest transition determines the highest frequency in the pulse waveform.

Equation 20–1 can be rearranged to give the highest frequency as follows:

Equation 20–2

$$f_h = \frac{0.35}{t_r}$$

also,

Equation 20–3

$$f_h = \frac{0.35}{t_f}$$

EXAMPLE 20–11

What is the highest frequency contained in a pulse that has rise and fall times equal to 10 nanoseconds (10 ns)?

Solution

$$f_h = \frac{0.35}{t_r} = \frac{0.35}{10 \times 10^{-9}\,\text{s}} = 0.035 \times 10^9\,\text{Hz}$$

$$= 35 \times 10^6\,\text{Hz} = \textbf{35 MHz}$$

Related Problem

What is the highest frequency in a pulse with $t_r = 20$ ns and $t_f = 15$ ns?

SECTION 20–8
REVIEW

1. What type of filter is an integrator?
2. What type of filter is a differentiator?
3. What is the highest frequency component in a pulse waveform having t_r and t_f equal to 1 μs?

20–9 TROUBLESHOOTING

In this section, *RC* circuits with pulse inputs are used to demonstrate the effects of common component failures in selected cases. The concepts can then be easily related to *RL* circuits.

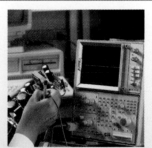

After completing this section, you should be able to

- ◆ **Troubleshoot *RC* integrators and *RC* differentiators**
  - ◆ Recognize the effect of an open capacitor
  - ◆ Recognize the effect of a leaky capacitor
  - ◆ Recognize the effect of a shorted capacitor
  - ◆ Recognize the effect of an open resistor

Open Capacitor

If the capacitor in an *RC* integrator opens, the output has the same waveshape as the input, as shown in Figure 20–52(a). If the capacitor in a differentiator opens, the output is zero because it is held at ground through the resistor, as illustrated in part (b).

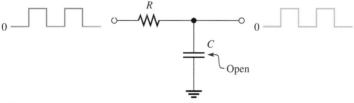

(a) Integrator

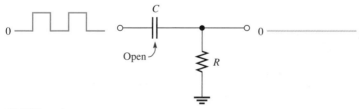

(b) Differentiator

▲ **FIGURE 20–52**

Examples of the effect of an open capacitor.

Leaky Capacitor

If the capacitor in an RC integrator becomes leaky, three things happen: (a) the time constant will be effectively reduced by the leakage resistance (when thevenized, looking from C it appears in parallel with R); (b) the waveshape of the output voltage (across C) is altered from normal by a shorter charging time; and (c) the amplitude of the output is reduced because R and R_{leak} effectively act as a voltage divider. These effects are illustrated in Figure 20–53(a).

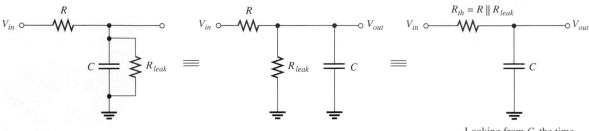

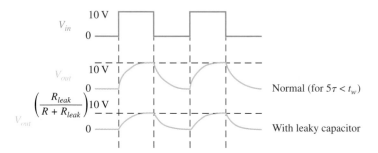

(a)

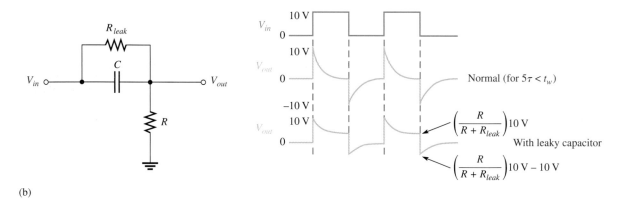

(b)

▲ **FIGURE 20–53**

Examples of the effect of a leaky capacitor.

If the capacitor in a differentiator becomes leaky, the time constant is reduced, just as in the integrator (they are both simply series RC circuits). When the capacitor reaches full charge, the output voltage (across R) is set by the effective voltage-divider action of R and R_{leak}, as shown in Figure 20–53(b).

Shorted Capacitor

If the capacitor in an *RC* integrator shorts, the output is at ground, as shown in Figure 20–54(a). If the capacitor in a *RC* differentiator shorts, the output is the same as the input, as shown in part (b).

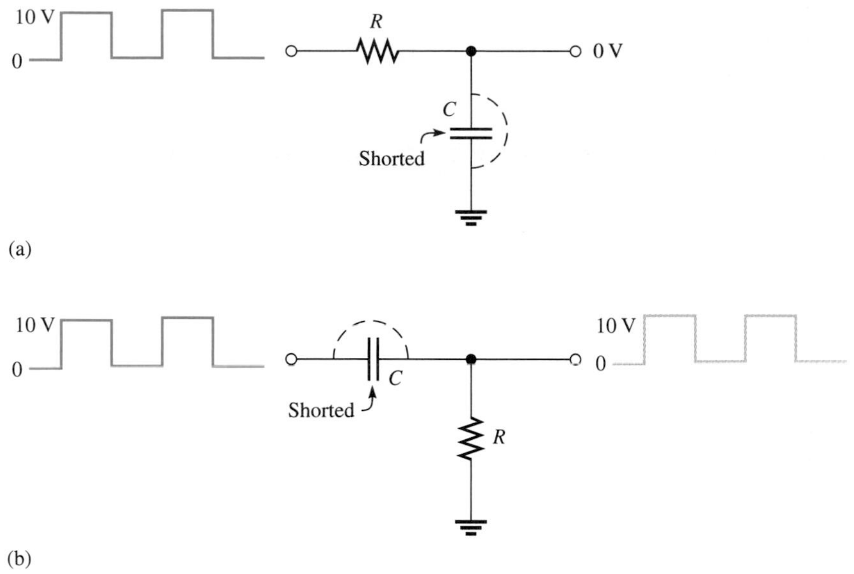

FIGURE 20–54

Examples of the effect of a shorted capacitor.

Open Resistor

If the resistor in an *RC* integrator opens, the capacitor has no discharge path, and, ideally, it will hold its charge. In an actual situation, the charge will gradually leak off or the capacitor will discharge slowly through a measuring instrument connected to the output. This is illustrated in Figure 20–55(a).

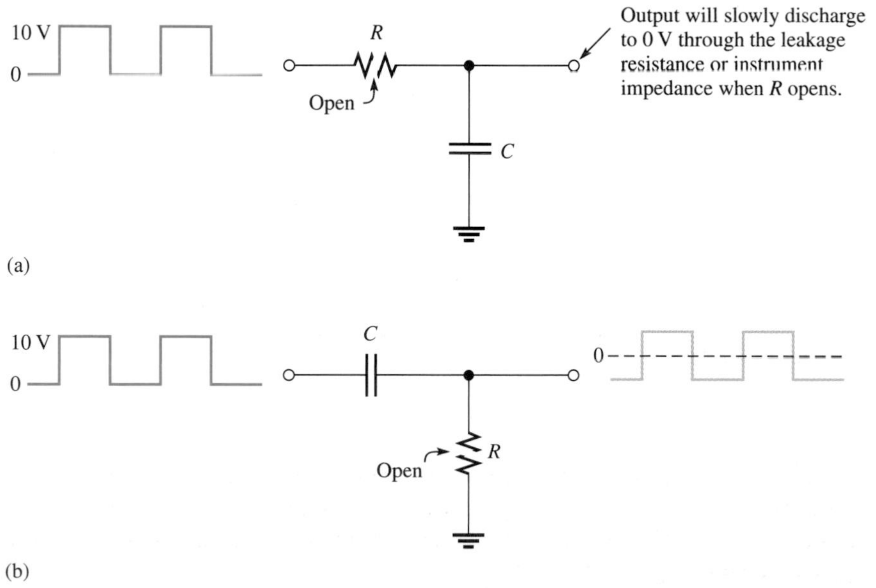

FIGURE 20–55

Examples of the effects of an open resistor.

If the resistor in a differentiator opens, the output looks like the input except for the dc level because the capacitor now must charge and discharge through the extremely high resistance of the oscilloscope, as shown in Figure 20–55(b).

SECTION 20–9 REVIEW

1. An *RC* integrator has a zero output with a square wave input. What are the possible causes of this problem?

2. If the capacitor in a differentiator is shorted, what is the output for a square wave input?

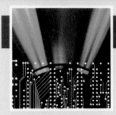

A Circuit Application

In this application, you are asked to build and test a time-delay circuit that will provide five switch-selectable delay times. An *RC* integrator is selected for this application. The input is a 5 V pulse of long duration, and the output goes to a threshold trigger circuit that is used to turn the power on to a portion of a system at any of the five selected time intervals after the occurrence of the original pulse.

A schematic of the selectable time-delay integrating circuit is shown in Figure 20–56. The *RC* integrator is driven by a pulse input; and the output is an exponentially increasing voltage that is used to trigger a threshold circuit at the 3.5 V level, which then turns power on to part of a system. The basic concept is shown in Figure 20–57. In this application, the delay time of the integrator is specified to be the time from the rising edge of the input pulse to the point where the output voltage reaches 3.5 V. The specified delay times are as listed in Table 20–1.

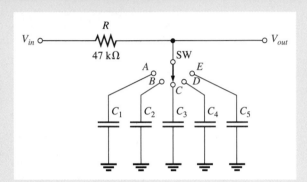

▲ **FIGURE 20–56**

Integrator delay circuit.

Capacitor Values

◆ Determine a value for each capacitor that will provide the specified delay times within 10%. Select from the following list of standard values (all are in μF): 0.1, 0.12, 0.15, 0.18,

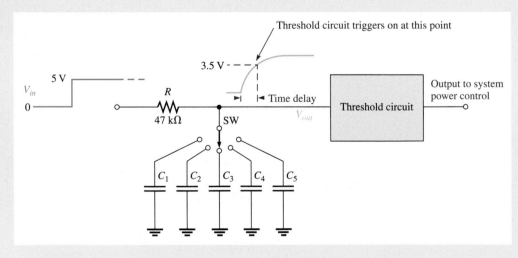

▲ **FIGURE 20–57**

Illustration of the time-delay application.

▼ TABLE 20–1

SWITCH POSITION	DELAY TIME
A	10 ms
B	25 ms
C	40 ms
D	65 ms
E	85 ms

0.22, 0.27, 0.33, 0.39, 0.47, 0.56, 0.68, 0.82, 1.0, 1.2, 1.5, 1.8, 2.2, 2.7, 3.3, 3.9, 4.7, 5.6, 6.8, 8.2.

Circuit Connections

Refer to Figure 20–58. The components for the *RC* integrator in Figure 20–56 are assembled, but not interconnected, on the circuit board.

♦ Using the circled numbers, develop a point-to-point wiring list to properly connect the circuit on the board.

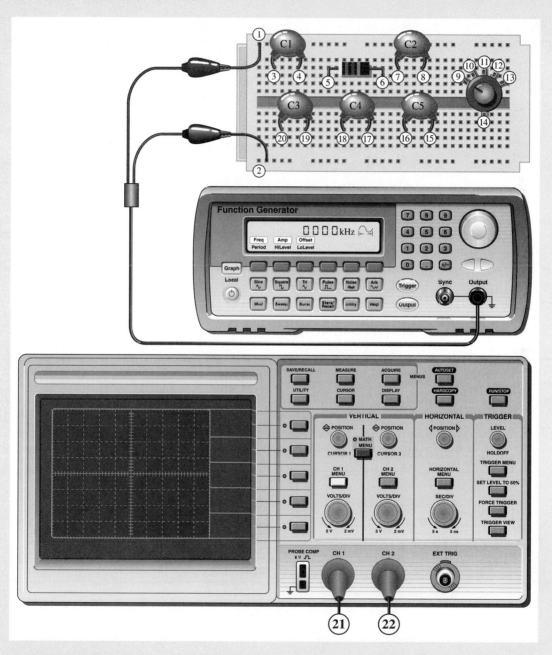

▲ FIGURE 20–58

◆ Indicate, using the appropriate circled numbers, how you would connect the instruments to test the circuit.

Test Procedure

◆ Specify the function, amplitude, and minimum frequency settings for the function generator in order to test all output delay times in Figure 20–58.

◆ Specify the minimum oscilloscope settings for measuring each of the specified delay times in Figure 20–58.

Review

1. To add an additional time delay to the circuit of Figure 20–57, what changes must be made?

2. An additional time delay of 100 ms is required for the time-delay circuit. Determine the capacitor value that should be added.

SUMMARY

◆ In an RC integrating circuit, the output voltage is taken across the capacitor.

◆ In an RC differentiating circuit, the output voltage is taken across the resistor.

◆ In an RL integrating circuit, the output voltage is taken across the resistor.

◆ In an RL differentiating circuit, the output voltage is taken across the inductor.

◆ In an integrator, when the pulse width (t_W) of the input is much less than the transient time, the output voltage approaches a constant level equal to the average value of the input.

◆ In an integrator, when the pulse width of the input is much greater than the transient time, the output voltage approaches the shape of the input.

◆ In a differentiator, when the pulse width of the input is much less than the transient time, the output voltage approaches the shape of the input but with an average value of zero.

◆ In a differentiator, when the pulse width of the input is much greater than the transient time, the output voltage consists of narrow, positive-going and negative-going spikes occurring on the leading and trailing edges of the input pulses.

◆ The rising and falling edges of a pulse waveform contain the higher frequency components.

◆ The flat portion of the pulse contains the lower frequency components.

KEY TERMS

Key terms and other bold terms in the chapter are defined in the end-of-book glossary.

DC component   The average value of a pulse waveform.

Differentiator   A circuit producing an output that approaches the mathematical derivative of the input.

Integrator   A circuit producing an output that approaches the mathematical integral of the input.

Steady state   The equilibrium condition of a circuit that occurs after an initial transient time.

Time constant   A fixed-time interval, set by R and C, or R and L values, that determines the time response of a circuit.

Transient time   An interval equal to approximately five time constants.

FORMULAS

20–1 $t_r = \dfrac{0.35}{f_h}$ Rise time

20–2 $f_h = \dfrac{0.35}{t_r}$ Highest frequency in relation to rise time

20–3 $f_h = \dfrac{0.35}{t_f}$ Highest frequency in relation to fall time

SELF-TEST

Answers are at the end of the chapter.

1. The output of an RC integrator is taken across the
   (a) resistor   (b) capacitor   (c) source   (d) coil

2. When a 10 V input pulse with a width equal to one time constant is applied to an RC integrator, the capacitor charges to
   (a) 10 V   (b) 5 V   (c) 6.3 V   (d) 3.7 V

3. When a 10 V pulse with a width equal to one time constant is applied to an RC differentiator, the capacitor charges to
   (a) 6.3 V   (b) 10 V   (c) 0 V   (d) 3.7 V

4. In an RC integrator, the output pulse closely resembles the input pulse when
   (a) τ is much larger than the pulse width   (b) τ is equal to the pulse width
   (c) τ is less than the pulse width   (d) τ is much less than the pulse width

5. In an RC differentiator, the output pulse closely resembles the input pulse when
   (a) τ is much larger than the pulse width   (b) τ is equal to the pulse width
   (c) τ is less than the pulse width   (d) τ is much less than the pulse width

6. The positive and negative portions of a differentiator's output voltage are equal when
   (a) $5\tau < t_W$ (b) $5\tau > t_W$
   (c) $5\tau = t_W$ (d) $5\tau > 0$
   (e) both (a) and (c)   (f) both (b) and (d)

7. The output of an RL integrator is taken across the
   (a) resistor   (b) coil   (c) source   (d) capacitor

8. The maximum possible current in an RL integrator is
   (a) $I = V_p/X_L$ (b) $I = V_p/Z$ (c) $I = V_p/R$

9. The current in an RL differentiator reaches its maximum possible value when
   (a) $5\tau = t_W$ (b) $5\tau < t_W$ (c) $5\tau > t_W$ (d) $\tau = 0.5t_W$

10. If you have an RC and an RL differentiator with equal time constants sitting side-by-side and you apply the same input pulse to both,
    (a) the RC has the widest output pulse
    (b) the RL has the most narrow spikes on the output
    (c) the output of one is an increasing exponential and the output of the other is a decreasing exponential
    (d) you can't tell the difference by observing the output waveforms

CIRCUIT DYNAMICS QUIZ

Answers are at the end of the chapter.

Refer to Figure 20–60.

1. If R_2 opens, the amplitude of the output voltage
   (a) increases   (b) decreases   (c) stays the same

2. If C doubled in value, the time constant
   (a) increases   (b) decreases   (c) stays the same

3. If R_1 is reduced in value, the output voltage amplitude
   (a) increases   (b) decreases   (c) stays the same

Refer to Figure 20–63.

4. If R_3 opens, the amplitude of the output voltage
   (a) increases   (b) decreases   (c) stays the same

5. If a constant dc voltage is applied to the input, the output voltage
   (a) increases   (b) decreases   (c) stays the same

6. If R_1 is 3.3 kΩ instead of 2.2 kΩ, the time constant
   (a) increases   (b) decreases   (c) stays the same

Refer to Figure 20–66.

7. If L is increased, the rise time of the output
   (a) increases   (b) decreases   (c) stays the same

8. If the width of the input pulse is increased to 5 ms, the amplitude of the output pulse
   (a) increases   (b) decreases   (c) stays the same

Refer to Figure 20–68.

9. If R_1 opens, the maximum amplitude of the output
   (a) increases   (b) decreases   (c) stays the same

10. If R_2 is shorted, the maximum amplitude of the output
    (a) increases   (b) decreases   (c) stays the same

PROBLEMS

Answers to odd-numbered problems are at the end of the book.

SECTION 20–1   The *RC* Integrator

1. An integrating circuit has $R = 2.2$ kΩ in series with $C = 0.047$ μF. What is the time constant?

2. Determine how long it takes the capacitor in an integrating circuit to reach full charge for each of the following series *RC* combinations:
   (a) $R = 56$ Ω, $C = 47$ μF
   (b) $R = 3300$ Ω, $C = 0.015$ μF
   (c) $R = 22$ kΩ, $C = 100$ pF
   (d) $R = 5.6$ MΩ, $C = 10$ pF

SECTION 20–2   Response of an *RC* Integrator to a Single Pulse

3. A 20 V pulse is applied to an *RC* integrator. The pulse width equals one time constant. To what voltage does the capacitor charge during the pulse? Assume that it is initially uncharged.

4. Repeat Problem 3 for the following values of t_W:
   (a) 2τ (b) 3τ (c) 4τ (d) 5τ

5. Draw the approximate shape of an integrator output voltage where 5τ is much less than the pulse width of a 10 V square-wave input. Repeat for the case in which 5τ is much larger than the pulse width.

6. Determine the output voltage for an *RC* integrator with a single input pulse, as shown in Figure 20–59. For repetitive pulses, how long will it take this circuit to reach steady state?

7. (a) What is τ in Figure 20–60?
   (b) Draw the output voltage.

8. Sketch the output voltage in Figure 20–60 if the pulse width is increased to 1.25 s.

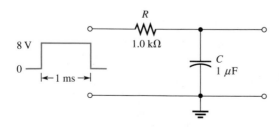

▲ FIGURE 20–59

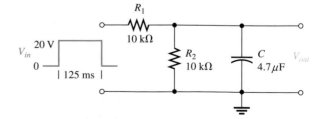

▲ FIGURE 20–60

SECTION 20–3   Response of *RC* Integrators to Repetitive Pulses

9. Draw the integrator output voltage in Figure 20–61, showing maximum voltages.

10. Sketch the output voltage if the pulse width of V_{in} in Figure 20–60 is changed to 47 ms and the frequency is the same.

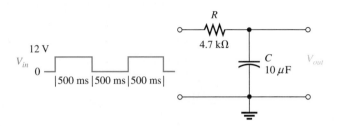

▲ FIGURE 20–61

11. A 1 V, 10 kHz pulse waveform with a duty cycle of 25% is applied to an integrator with $\tau = 25 \mu s$. Graph the output voltage for three initial pulses. *C* is initially uncharged.

12. What is the steady-state output voltage of the *RC* integrator with a square-wave input shown in Figure 20–62?

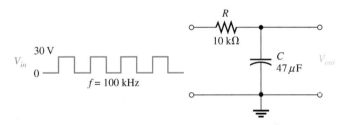

▲ FIGURE 20–62

SECTION 20–4   Response of an *RC* Differentiator to a Single Pulse

13. Repeat Problem 5 for an *RC* differentiator.

14. Redraw the circuit in Figure 20–59 to make it a differentiator, and repeat Problem 6.

15. (a) What is τ in Figure 20–63?

    (b) Draw the output voltage.

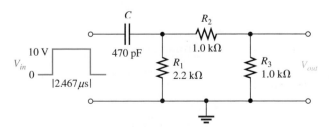

▲ FIGURE 20–63

SECTION 20–5 Response of *RC* Differentiators to Repetitive Pulses

16. Draw the differentiator output in Figure 20–64, showing maximum voltages.

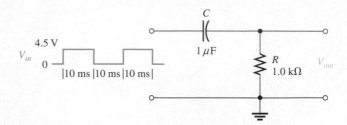

▲ **FIGURE 20–64**

17. What is the steady-state output voltage of the differentiator with the square-wave input shown in Figure 20–65?

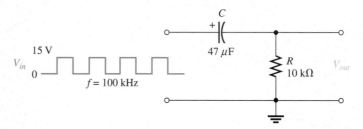

▲ **FIGURE 20–65**

SECTION 20–6 Response of *RL* Integrators to Pulse Inputs

18. Determine the output voltage for the circuit in Figure 20–66. A single input pulse is applied as shown.

▶ **FIGURE 20–66**

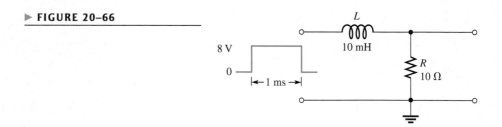

19. Draw the integrator output voltage in Figure 20–67, showing maximum voltages.

▶ **FIGURE 20–67**

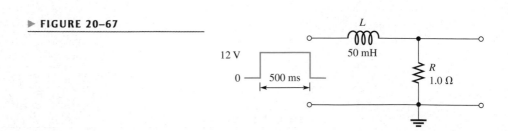

20. Determine the time constant in Figure 20–68. Is this circuit an integrator or a differentiator?

▶ FIGURE 20–68

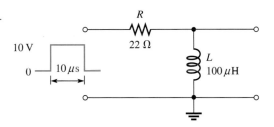

SECTION 20–7 Response of *RL* Differentiators to Pulse Inputs

21. **(a)** What is τ in Figure 20–69?   **(b)** Draw the output voltage.

▶ FIGURE 20–69

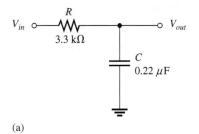

22. Draw the output waveform if a periodic pulse waveform with $t_W = 25\ \mu s$ and $T = 60\ \mu s$ is applied to the circuit in Figure 20–69.

SECTION 20–8 Relationship of Time Response to Frequency Response

23. What is the highest frequency component in the output of an integrator with $\tau = 10\ \mu s$? Assume that $5\tau < t_W$.

24. A certain pulse waveform has a rise time of 55 ns and a fall time of 42 ns. What is the highest frequency component in the waveform?

SECTION 20–9 Troubleshooting

25. Determine the most likely fault(s), in the circuit of Figure 20–70(a) for each set of waveforms in parts (b) through (d). V_{in} is a square wave with a period of 8 ms.

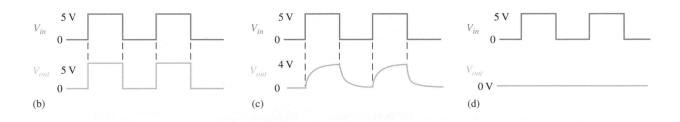

▲ FIGURE 20–70

26. Determine the most likely fault(s), if any, in the circuit of Figure 20–71(a) for each set of waveforms in parts (b) through (d). V_{in} is a square wave with a period of 8 ms.

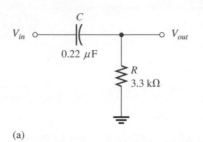

(a)

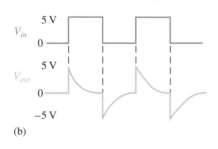

(b)

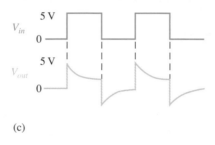

(c)

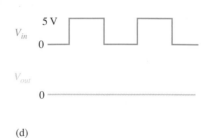

(d)

▲ **FIGURE 20–71**

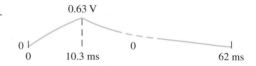

Multisim Troubleshooting and Analysis

These problems require your Multisim CD-ROM.

27. Open file P20-27 and determine if there is a fault. If so, find the fault.

28. Open file P20-28 and determine if there is a fault. If so, find the fault.

29. Open file P20-29 and determine if there is a fault. If so, find the fault.

30. Open file P20-30 and determine if there is a fault. If so, find the fault.

ANSWERS

SECTION REVIEWS

SECTION 20–1 **The *RC* Integrator**

1. An integrator is a series *RC* circuit in which the output is across the capacitor.

2. A voltage applied to the input causes the capacitor to charge. A short across the input causes the capacitor to discharge.

SECTION 20–2 **Response of an *RC* Integrator to a Single Pulse**

1. For the output of an integrator to reach amplitude, $5\tau \le t_W$.

2. $V_{out(max)} = 630$ mV; $t_{disch} = 51.7$ ms

3. See Figure 20–72.

▶ **FIGURE 20–72**

0.63 V

0 ⊦————————————————————⊦
0   10.3 ms   0   62 ms

4. No, *C* will not fully charge.

5. The output has approximately the shape of the input when $5\tau \ll t_W$ (5τ much less than t_W).

SECTION 20–3 **Response of *RC* Integrators to Repetitive Pulses**

    1. *C* will fully charge and discharge when $5\tau \leq t_W$ and $5\tau \leq$ time between pulses.

    2. When $\tau \ll t_W$, the output is approximately like the input.

    3. Transient time

    4. Steady-state response is the response after the transient time has passed.

    5. The average value of the output equals the average value of the input voltage.

SECTION 20–4 **Response of an *RC* Differentiator to a Single Pulse**

    1. See Figure 20–73.

    ▶ **FIGURE 20–73**

    2. The output resembles the input when $5\tau \gg t_W$.

    3. The output appears to be positive and negative spikes.

    4. V_R will go to -10 V.

SECTION 20–5 **Response of *RC* Differentiators to Repetitive Pulses**

    1. *C* will fully charge and discharge when $5\tau \leq t_W$ and $5\tau \leq$ time between pulses.

    2. The output appears to be positive and negative spikes.

    3. The average value is 0 V.

SECTION 20–6 **Response of *RL* Integrators to Pulse Inputs**

    1. The output is taken across the resistor.

    2. The output reaches the input amplitude when $5\tau \leq t_W$.

    3. The output has the approximate shape of the input when $5\tau \ll t_W$.

SECTION 20–7 **Response of *RL* Differentiators to Pulse Inputs**

    1. The output is taken across the inductor.

    2. The output has the approximate shape of the input when $5\tau \gg t_W$.

    3. V_L will go to -8 V.

SECTION 20–8 **Relationship of Time Response to Frequency Response**

    1. An integrator is a low-pass filter.

    2. A differentiator is a high-pass filter.

    3. $f_{max} = 350$ kHz

SECTION 20–9 **Troubleshooting**

    1. A 0 V output may be caused by an open resistor or shorted capacitor.

    2. If *C* is shorted, the output is the same as the input.

A Circuit Application

    1. A capacitor must be added and the switch changed to one with six positions.

    2. $C_6 = 100$ ms/$[(1.204)(47\,\text{k}\Omega)] = 1.77\,\mu\text{F}$ (use 1.8 μF)

RELATED PROBLEMS FOR EXAMPLES

20–1   8.65 V

20–2   24.7 V

20–3   1.08 V

20–4   See Figure 20–74.

▶ FIGURE 20–74

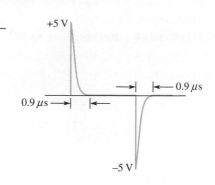

20–5   892 mV

20–6   Impossible with a 50 Ω rheostat

20–7   20 V

20–8   2.5 kΩ

20–9   See Figure 20–75.

▶ FIGURE 20–75

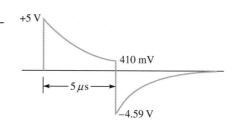

20–10 20 kΩ

20–11 23.3 MHz

SELF-TEST

  1. (b)   **2.** (c)   **3.** (a)   **4.** (d)   **5.** (a)   **6.** (e)   **7.** (a)   **8.** (c)

  9. (b)   **10.** (d)

CIRCUIT DYNAMICS QUIZ

  1. (b)   **2.** (a)   **3.** (c)   **4.** (a)   **5.** (b)   **6.** (a)   **7.** (a)   **8.** (a)

  9. (c)   **10.** (a)

THREE-PHASE SYSTEMS IN POWER APPLICATIONS

21

CHAPTER OBJECTIVES

◆ Describe basic three-phase machines
◆ Discuss the advantages of three-phase generators in power applications
◆ Analyze three-phase generator connections
◆ Analyze three-phase generators with three-phase loads
◆ Discuss power measurements in three-phase systems

KEY TERMS

◆ Rotor
◆ Field winding
◆ Stator
◆ Balanced load
◆ Phase voltage (V_θ)
◆ Phase current (I_θ)
◆ Line current (I_l)
◆ Line voltage (V_L)

VISIT THE COMPANION WEBSITE

Study aids for this chapter are available at
http://www.prenhall.com/floyd

INTRODUCTION

In the coverage of ac analysis in previous chapters, only single-phase sinusoidal sources were considered. In Chapter 11, you learned how a sinusoidal voltage can be generated by the rotation of a conductor at a constant velocity in a magnetic field, and the basic concepts of ac generators were introduced.

In this chapter, the basic generation of three-phase sinusoidal waveforms is examined. The advantages of three-phase systems in power applications are covered, and various types of three-phase connections and power measurement are introduced.

21–1 INTRODUCTION TO THREE-PHASE MACHINES

Three-phase generators simultaneously produce three sinusoidal voltages that are separated by certain constant phase angles. This multiple-phase generation is accomplished by multiple windings rotating through a magnetic field. Similarly, three-phase motors operate with three-phase sinusoidal inputs.

After completing this section, you should be able to

◆ **Describe basic three-phase machines**

　◆ Discuss a basic three-phase generator

　◆ Describe the construction of a three-phase generator

　◆ Describe a basic three-phase induction motor

The Generator

Figure 21–1(a) shows a generator with three separate windings placed at 120° intervals around the rotor. This configuration generates three sinusoidal voltages that are separated from each other by phase angles of 120°, as shown in part (b).

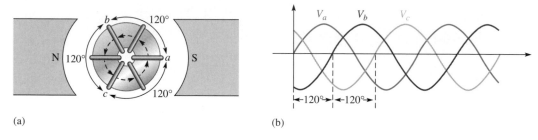

(a)　　　　　　　　　　　　　　　　　　　　　(b)

▲ **FIGURE 21–1**

Basic three-phase generator.

A two-pole, three-phase generator is shown in Figure 21–2. Most practical generators are of this form. Rather than using a permanent magnet in a fixed position, a rotating electromagnet is used. The electromagnet is created by passing direct current (I_F) through a winding around the **rotor**, as shown. This winding is called the **field winding**. The direct

▶ **FIGURE 21–2**

Two-pole, three-phase generator.

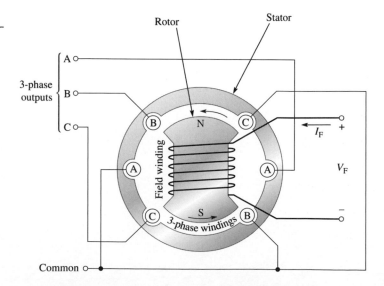

current is applied through a brush and slip ring assembly. The stationary outer portion of the generator is called the **stator**. Three separate windings are placed 120° apart around the stator; three-phase voltages are induced in these windings as the magnetic field rotates, as indicated in Figure 21–1(b).

The Motor

The most common type of ac motor is the three-phase induction motor. Basically, it consists of a stator with stator windings and a rotor assembly constructed as a cylindrical frame of metal bars arranged in a **squirrel-cage** type configuration. An end-view diagram is shown in Figure 21–3.

When the three-phase voltages are applied to the stator windings, a rotating magnetic field is established. As the magnetic field rotates, currents are induced in the conductors of the squirrel-cage rotor. The interaction of the induced currents and the magnetic field produces forces that cause the rotor to also rotate.

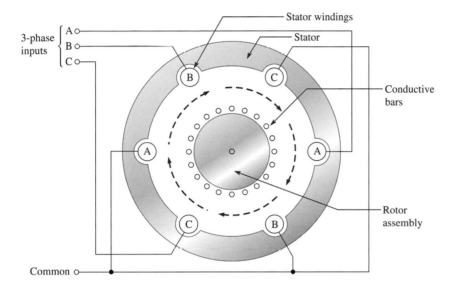

◀ FIGURE 21–3

Basic three-phase induction motor.

**SECTION 21–1
REVIEW**
Answers are at the end of the chapter.

1. Describe the basic principle used in ac generators.
2. How many separate armature windings are required in a three-phase generator?

21–2 GENERATORS IN POWER APPLICATIONS

There are advantages to using three-phase generators to deliver power to a load over using a single-phase machine.

After completing this section, you should be able to

◆ **Discuss the advantages of three-phase generators in power applications**

   ◆ Explain the copper advantage

   ◆ Compare single-phase and three-phase systems in terms of the copper advantage

   ◆ Explain the advantage of constant power

   ◆ Explain the advantage of a constant rotating magnetic field

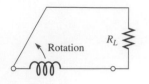

▲ **FIGURE 21–4**

Simplified representation of a single-phase generator connected to a resistive load.

The size of the copper wire required to carry current from a generator to a load can be reduced when a three-phase rather than a single-phase generator is used.

Figure 21–4 is a simplified representation of a single-phase generator connected to a resistive load. The coil symbol represents the generator winding.

For example, a single-phase sinusoidal voltage is induced in the winding and applied to a 60 Ω load, as indicated in Figure 21–5. The resulting load current is

$$\mathbf{I}_{RL} = \frac{120\angle 0°\ \text{V}}{60\angle 0°\ \Omega} = 2\angle 0°\ \text{A}$$

▶ **FIGURE 21–5**

Single-phase example.

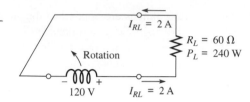

The total current that must be delivered by the generator to the load is $2\angle 0°$ A. This means that the two conductors carrying current to and from the load must each be capable of handling 2 A; thus, the total copper cross section must handle 4 A. (The copper cross section is a measure of the total amount of wire required based on its physical size as related to its diameter.) The total load power is

$$P_{L(tot)} = I_{RL}^2 R_L = 240\ \text{W}$$

Figure 21–6 shows a simplified representation of a three-phase generator connected to three 180 Ω resistive loads. An equivalent single-phase system would be required to feed three 180 Ω resistors in parallel, thus creating an effective load resistance of 60 Ω. The coils represent the generator windings separated by 120°.

▶ **FIGURE 21–6**

A simplified representation of a three-phase generator with each phase connected to a 180 Ω load.

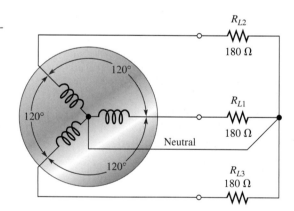

The voltage across R_{L1} is $120\angle 0°$ V, the voltage across R_{L2} is $120\angle 120°$ V, and the voltage across R_{L3} is $120\angle -120°$ V, as indicated in Figure 21–7(a). The current from each winding to its respective load is as follows:

$$\mathbf{I}_{RL1} = \frac{120\angle 0°\ \text{V}}{180\angle 0°\ \Omega} = 667\angle 0°\ \text{mA}$$

$$\mathbf{I}_{RL2} = \frac{120\angle 120°\ \text{V}}{180\angle 0°\ \Omega} = 667\angle 120°\ \text{mA}$$

$$\mathbf{I}_{RL3} = \frac{120\angle -120°\ \text{V}}{180\angle 0°\ \Omega} = 667\angle -120°\ \text{mA}$$

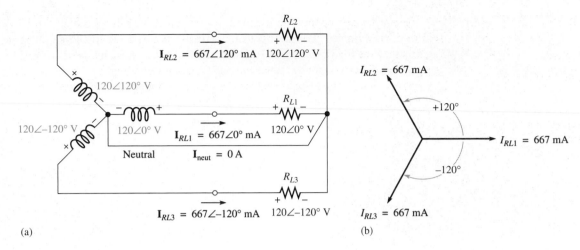

▲ FIGURE 21–7

Three-phase example.

The total load power is

$$P_{L(tot)} = I_{RL1}^2 R_{L1} + I_{RL2}^2 R_{L2} + I_{RL3}^2 R_{L3} = 240 \text{ W}$$

This is the same total load power as delivered by the single-phase system previously discussed.

Notice that four conductors, including the neutral, are required to carry the currents to and from the loads. The current in each of the three conductors is 667 mA, as indicated in Figure 21–7(a). The current in the neutral conductor is the phasor sum of the three load currents and is equal to zero, as shown in the following equation, with reference to the phasor diagram in Figure 21–7(b).

$$\begin{aligned}
\mathbf{I}_{RL1} + \mathbf{I}_{RL2} + \mathbf{I}_{RL3} &= 667\angle 0° \text{ mA} + 667\angle 120° \text{ mA} + 667\angle{-}120° \text{ mA} \\
&= 667 \text{ mA} - 333.5 \text{ mA} + j578 \text{ mA} - 333.5 \text{ mA} - j578 \text{ mA} \\
&= 667 \text{ mA} - 667 \text{ mA} = 0 \text{ A}
\end{aligned}$$

This condition, where all load currents are equal and the neutral current is zero, is called a **balanced load** condition.

The total copper cross section must handle 667 mA + 667 mA + 667 mA + 0 mA = 2 A. This result shows that considerably less copper is required to deliver the same load power with a three-phase system than is required for the single-phase system. The amount of copper is an important consideration in power distribution systems.

EXAMPLE 21–1

Compare the total copper cross sections in terms of current-carrying capacity for single-phase and three-phase 120 V systems with effective load resistances of 12 Ω.

Solution   *Single-phase system:*   The load current is

$$I_{RL} = \frac{120 \text{ V}}{12 \text{ Ω}} = 10 \text{ A}$$

The conductor to the load must carry 10 A, and the conductor from the load must also carry 10 A.

The total copper cross section, therefore, must be sufficient to handle 2 × 10 A = 20 A.

Three-phase system: For an effective load resistance of 12 Ω, the three-phase generator feeds three load resistors of 36 Ω each. The current in each load resistor is

$$I_{RL} = \frac{120 \text{ V}}{36 \text{ Ω}} = 3.33 \text{ A}$$

Each of the three conductors feeding the balanced load must carry 3.33 A, and the neutral current is zero.

Therefore, the total copper cross section must be sufficient to handle $3 \times 3.33 \text{ A} \cong 10 \text{ A}$. This is significantly less than for the single-phase system with an equivalent load.

Related Problem * Compare the total copper cross sections in terms of current-carrying capacity for single-phase and three-phase 240 V systems with effective load resistances of 100 Ω.

---

*Answers are at the end of the chapter.

A second advantage of three-phase systems over a single-phase system is that three-phase systems produce a constant amount of power in the load. As shown in Figure 21–8, the load power fluctuates as the square of the sinusoidal voltage divided by the resistance. It changes from a maximum of $V_{RL(\text{max})}^2/R_L$ to a minimum of zero at a frequency equal to twice that of the voltage.

▶ FIGURE 21–8

Single-phase load power (sin² curve).

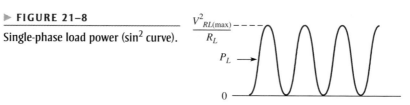

The power waveform across one of the load resistors in a three-phase system is 120° out of phase with the power waveforms across the other loads, as shown in Figure 21–9. Examination of the power waveforms shows that when three instantaneous values are added, the sum is always constant and equal to $V_{RL(\text{max})}^2/R_L$. A constant load power means a uniform conversion of mechanical to electrical energy, which is an important consideration in many power applications.

▶ FIGURE 21–9

Three-phase power ($P_L = V_{RL(\text{max})}^2/R_L$).

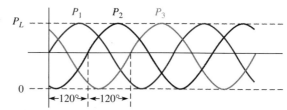

In many applications, ac generators are used to drive ac motors for conversion of electrical energy to mechanical energy in the form of shaft rotation in the motor. The original energy for operation of the generator can come from any of several sources, such as hydroelectric or steam. Figure 21–10 illustrates the basic concept.

When a three-phase generator is connected to the motor windings, a magnetic field is created within the motor that has a constant flux density and that rotates at the frequency of

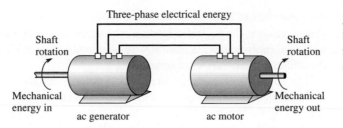

◀ FIGURE 21–10

Simple example of mechanical-to-electrical-to-mechanical energy conversion.

the three-phase sine wave. The motor's rotor is pulled around at a constant rotational velocity by the rotating magnetic field, producing a constant shaft rotation, which is an advantage of three-phase systems.

A single-phase system is unsuitable for many applications because it produces a magnetic field that fluctuates in flux density and reverses direction during each cycle without providing the advantage of constant rotation.

SECTION 21–2 REVIEW

1. List three advantages of three-phase systems over single-phase systems.
2. Which advantage is most important in mechanical-to-electrical energy conversions?
3. Which advantage is most important in electrical-to-mechanical energy conversions?

21–3 TYPES OF THREE-PHASE GENERATORS

In the previous sections, the Y-connection was used for illustration. In this section, the Y-connection is examined further and a second type, the Δ-connection, is introduced.

After completing this section, you should be able to

◆ **Analyze three-phase generator connections**

   ◆ Analyze the Y-connected generator

   ◆ Analyze the Δ-connected generator

The Y-Connected Generator

A Y-connected system can be either a three-wire or, when the neutral is used, a four-wire system, as shown in Figure 21–11, connected to a generalized load, which is indicated by

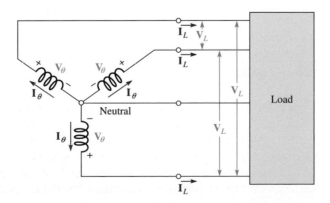

◀ FIGURE 21–11

Y-connected generator.

the green block. Recall that when the loads are perfectly balanced, the neutral current is zero; therefore, the neutral conductor is unnecessary. However, in cases where the loads are not equal (unbalanced), a neutral wire is essential to provide a return current path because the neutral current has a nonzero value.

The voltages across the generator windings are called **phase voltages (V_θ)**, and the currents through the windings are called **phase currents (I_θ)**. Also, the currents in the lines connecting the generator windings to the load are called **line currents (I_L)**, and the voltages across the lines are called the **line voltages (V_L)**. Note that the magnitude of each line current is equal to the corresponding phase current in the Y-connected circuit.

Equation 21–1

$$I_L = I_\theta$$

In Figure 21–12, the line terminations of the windings are designated a, b, and c, and the neutral point is designated n. These letters are added as subscripts to the phase and line currents to indicate the phase with which each is associated. The phase voltages are also designated in the same manner. Notice that the phase voltages are always positive at the terminal end of the winding and are negative at the neutral point. The line voltages are from one winding terminal to another, as indicated by the double-letter subscripts. For example, $\mathbf{V}_{L(ba)}$ is the line voltage from b to a.

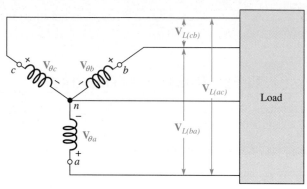

▲ FIGURE 21–12

Phase voltages and line voltages in a Y-connected system.

Figure 21–13(a) shows a phasor diagram for the phase voltages. By rotation of the phasors, as shown in part (b), $V_{\theta a}$ is given a reference angle of zero, and the polar expressions for the phasor voltages are as follows:

$$\mathbf{V}_{\theta a} = V_{\theta a}\angle 0°$$
$$\mathbf{V}_{\theta b} = V_{\theta b}\angle 120°$$
$$\mathbf{V}_{\theta c} = V_{\theta c}\angle -120°$$

▶ FIGURE 21–13

Phase voltage diagram.

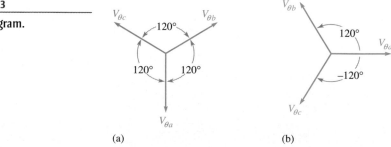

(a)　　　　　　　　　　　　　　　　(b)

There arc three line voltages: one between a and b, one between a and c, and another between b and c. It can be shown that the magnitude of each line voltage is equal to $\sqrt{3}$ times the magnitude of the phase voltage and that there is a phase angle of $30°$ between each line voltage and the nearest phase voltage.

$$V_L = \sqrt{3}V_\theta$$ **Equation 21–2**

Since all phase voltages are equal in magnitude,

$$\mathbf{V}_{L(ba)} = \sqrt{3}V_\theta \angle 150°$$
$$\mathbf{V}_{L(ac)} = \sqrt{3}V_\theta \angle 30°$$
$$\mathbf{V}_{L(cb)} = \sqrt{3}V_\theta \angle -90°$$

The line voltage phasor diagram is shown in Figure 21–14 superimposed on the phasor diagram for the phase voltages. Notice that there is a phase angle of $30°$ between each line voltage and the nearest phase voltage and that the line voltages are $120°$ apart.

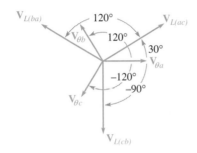

▲ **FIGURE 21–14**

Phase diagram for the phase voltages and line voltages in a Y-connected, three-phase system.

EXAMPLE 21–2

The instantaneous position of a certain Y-connected ac generator is shown in Figure 21–15. If each phase voltage has a magnitude of 120 V rms, determine the magnitude of each line voltage, and draw the phasor diagram.

▶ **FIGURE 21–15**

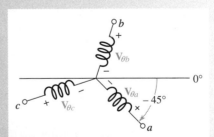

Solution   The magnitude of each line voltage is

$$V_L = \sqrt{3}V_\theta = \sqrt{3}(120 \text{ V}) = \mathbf{208 \text{ V}}$$

The phasor diagram for the given instantaneous generator position is shown in Figure 21–16.

▶ FIGURE 21–16

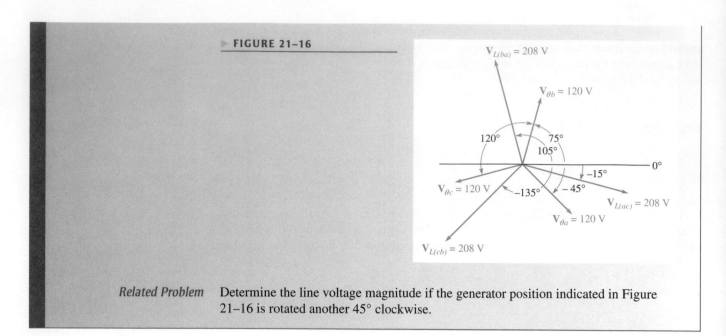

Related Problem Determine the line voltage magnitude if the generator position indicated in Figure 21–16 is rotated another 45° clockwise.

The Δ-Connected Generator

In the Y-connected generator, two voltage magnitudes are available at the terminals in the four-wire system: the phase voltage and the line voltage. Also, in the Y-connected generator, the line current is equal to the phase current. Keep these characteristics in mind as you examine the Δ-connected generator.

The windings of a three-phase generator can be rearranged to form a Δ-connected generator, as shown in Figure 21–17. By examination of this diagram, you can see that the magnitudes of the line voltages and phase voltages are equal, but the line currents do not equal the phase currents.

▶ FIGURE 21–17

Δ-connected generator.

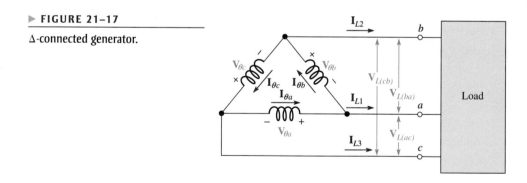

Since this is a three-wire system, only a single voltage magnitude is available, expressed as

Equation 21–3

$$V_L = V_\theta$$

All of the phase voltages are equal in magnitude; thus, the line voltages are expressed in polar form as follows:

$$\mathbf{V}_{L(ac)} = V_\theta \angle 0°$$
$$\mathbf{V}_{L(ba)} = V_\theta \angle 120°$$
$$\mathbf{V}_{L(cb)} = V_\theta \angle -120°$$

The phasor diagram for the phase currents is shown in Figure 21–18, and the polar expressions for each current are as follows:

$$\mathbf{I}_{\theta a} = I_{\theta a}\angle 0°$$
$$\mathbf{I}_{\theta b} = I_{\theta b}\angle 120°$$
$$\mathbf{I}_{\theta c} = I_{\theta c}\angle -120°$$

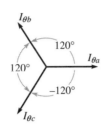

◀ **FIGURE 21–18**

Phase current diagram for the Δ-connected system.

It can be shown that the magnitude of each line current is equal to $\sqrt{3}$ times the magnitude of the phase current and that there is a phase angle of 30° between each line current and the nearest phase current.

$$I_L = \sqrt{3}I_\theta$$

Equation 21–4

Since all phase currents are equal in magnitude,

$$\mathbf{I}_{L1} = \sqrt{3}I_\theta\angle -30°$$
$$\mathbf{I}_{L2} = \sqrt{3}I_\theta\angle 90°$$
$$\mathbf{I}_{L3} = \sqrt{3}I_\theta\angle -150°$$

The current phasor diagram is shown in Figure 21–19.

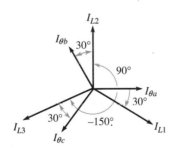

◀ **FIGURE 21–19**

Phasor diagram of phase currents and line currents.

EXAMPLE 21–3

The three-phase Δ-connected generator represented in Figure 21–20 is driving a balanced load such that each phase current is 10 A in magnitude. When $\mathbf{I}_{\theta a} = 10\angle 30°$ A, determine the following:

(a) The polar expressions for the other phase currents

(b) The polar expressions for each of the line currents

(c) The complete current phasor diagram

▶ FIGURE 21–20

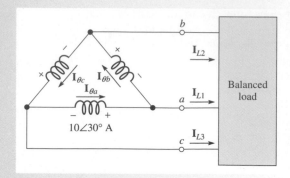

Solution   (a) The phase currents are separated by 120°; therefore,

$$\mathbf{I}_{\theta b} = 10\angle(30° + 120°) = \mathbf{10\angle 150°\ A}$$
$$\mathbf{I}_{\theta c} = 10\angle(30° - 120°) = \mathbf{10\angle -90°\ A}$$

(b) The line currents are separated from the nearest phase current by 30°; therefore,

$$\mathbf{I}_{L1} = \sqrt{3}I_{\theta a}\angle(30° - 30°) = \mathbf{17.3\angle 0°\ A}$$
$$\mathbf{I}_{L2} = \sqrt{3}I_{\theta b}\angle(150° - 30°) = \mathbf{17.3\angle 120°\ A}$$
$$\mathbf{I}_{L3} = \sqrt{3}I_{\theta c}\angle(-90° - 30°) = \mathbf{17.3\angle -120°\ A}$$

(c) The phasor diagram is shown in Figure 21–21.

▶ FIGURE 21–21

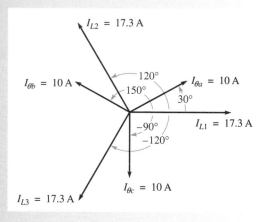

Related Problem   Repeat parts (a) and (b) of the example if $\mathbf{I}_{\theta a} = 8\angle 60°$ A.

SECTION 21–3 REVIEW

1. In a certain three-wire, Y-connected generator, the phase voltages are 1 kV. Determine the magnitude of the line voltages.
2. In the Y-connected generator mentioned in Question 1, all the phase currents are 5 A. What are the line current magnitudes?
3. In a Δ-connected generator, the phase voltages are 240 V. What are the line voltages?
4. In a Δ-connected generator, a phase current is 2 A. Determine the magnitude of the line current.

21–4 THREE-PHASE SOURCE/LOAD ANALYSIS

In this section, we look at four basic types of source/load configurations. As with the generator connections, a load can be either a Y or a Δ configuration.

After completing this section, you should be able to

◆ **Analyze three-phase generators with three-phase loads**

  ◆ Analyze the Y-Y source/load configuration

  ◆ Analyze the Y-Δ source/load configuration

  ◆ Analyze the Δ-Y source/load configuration

  ◆ Analyze the Δ-Δ source/load configuration

A Y-connected load is shown in Figure 21–22(a), and a Δ-connected load is shown in part (b). The blocks Z_a, Z_b, and Z_c represent the load impedances, which can be resistive, reactive, or both.

The four source/load configurations are:

1. Y-connected source driving a Y-connected load (Y-Y system)

2. Y-connected source driving a Δ-connected load (Y-Δ system)

3. Δ-connected source driving a Y-connected load (Δ-Y system)

4. Δ-connected source driving a Δ-connected load (Δ-Δ system)

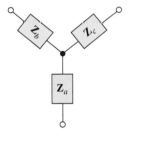

(a) Y-connected load

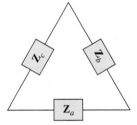

(b) Δ-connected load

▲ **FIGURE 21–22**

Three-phase loads.

The Y-Y System

Figure 21–23 shows a Y-connected source driving a Y-connected load. The load can be a balanced load, such as a three-phase motor where $Z_a = Z_b = Z_c$, or it can be three independent single-phase loads where, for example, Z_a is a lighting circuit, Z_b is a heater, and Z_c is an air-conditioning compressor.

An important feature of a Y-connected source is that two different values of three-phase voltage are available: the phase voltage and the line voltage. For example, in the standard power distribution system, a three-phase transformer can be considered a source of three-phase voltage supplying 120 V and 208 V. In order to utilize a phase voltage of 120 V, the loads are connected in the Y configuration. A Δ-connected load is used for the 208 V line voltages.

▶ FIGURE 21–23

A Y-connected source driving a
Y-connected load.

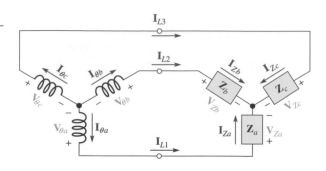

Notice in the Y-Y system in Figure 21–23 that the phase current, the line current, and the load current are all equal in each phase. Also, each load voltage equals the corresponding phase voltage. These relationships are expressed as follows and are true for either a balanced or an unbalanced load.

Equation 21–5

$$I_\theta = I_L = I_Z$$

Equation 21–6

$$V_\theta = V_Z$$

where V_Z and I_Z are the load voltage and current, respectively.

For a balanced load, all the phase currents are equal, and the neutral current is zero. For an unbalanced load, each phase current is different, and the neutral current is, therefore, nonzero.

EXAMPLE 21–4

In the Y-Y system of Figure 21–24, determine the following:

(a) Each load current   **(b)** Each line current   **(c)** Each phase current

(d) Neutral current   **(e)** Each load voltage

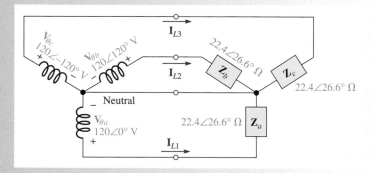

▲ FIGURE 21–24

Solution   This system has a balanced load. $\mathbf{Z}_a = \mathbf{Z}_b = \mathbf{Z}_c = 22.4\angle 26.6°\ \Omega$.

(a) The load currents are

$$\mathbf{I}_{Za} = \frac{\mathbf{V}_{\theta a}}{\mathbf{Z}_a} = \frac{120\angle 0°\ \text{V}}{22.4\angle 26.6°\ \Omega} = \mathbf{5.36}\angle\mathbf{-26.6°\ A}$$

$$\mathbf{I}_{Zb} = \frac{\mathbf{V}_{\theta b}}{\mathbf{Z}_b} = \frac{120\angle 120°\ \text{V}}{22.4\angle 26.6°\ \Omega} = \mathbf{5.36}\angle\mathbf{93.4°\ A}$$

$$\mathbf{I}_{Zc} = \frac{\mathbf{V}_{\theta c}}{\mathbf{Z}_c} = \frac{120\angle -120°\ \text{V}}{22.4\angle 26.6°\ \Omega} = \mathbf{5.36}\angle\mathbf{-147°\ A}$$

(b) The line currents are

$$\mathbf{I}_{L1} = \mathbf{5.36\angle{-26.6^\circ}\ A}$$
$$\mathbf{I}_{L2} = \mathbf{5.36\angle 93.4^\circ\ A}$$
$$\mathbf{I}_{L3} = \mathbf{5.36\angle{-147^\circ}\ A}$$

(c) The phase currents are

$$\mathbf{I}_{\theta a} = \mathbf{5.36\angle{-26.6^\circ}\ A}$$
$$\mathbf{I}_{\theta b} = \mathbf{5.36\angle 93.4^\circ\ A}$$
$$\mathbf{I}_{\theta c} = \mathbf{5.36\angle{-147^\circ}\ A}$$

(d) $\mathbf{I}_{\text{neut}} = \mathbf{I}_{Za} + \mathbf{I}_{Zb} + \mathbf{I}_{Zc}$
$= 5.36\angle{-26.6^\circ}\text{ A} + 5.36\angle 93.4^\circ\text{ A} + 5.36\angle{-147^\circ}\text{ A}$
$= (4.80\text{ A} - j2.40\text{ A}) + (-0.33\text{ A} + j5.35\text{ A}) + (-4.47\text{ A} - j2.95\text{ A}) = \mathbf{0\ A}$

If the load impedances were not equal (unbalanced load), the neutral current would have a nonzero value.

(e) The load voltages are equal to the corresponding source phase voltages.

$$\mathbf{V}_{Za} = \mathbf{120\angle 0^\circ\ V}$$
$$\mathbf{V}_{Zb} = \mathbf{120\angle 120^\circ\ V}$$
$$\mathbf{V}_{Zc} = \mathbf{120\angle{-120^\circ}\ V}$$

Related Problem Determine the neutral current if $\mathbf{Z}_a$ and $\mathbf{Z}_b$ are the same as in Figure 21–24, but $\mathbf{Z}_c = 50\angle 26.6^\circ\ \Omega$.

The Y-Δ System

Figure 21–25 shows a Y-connected source driving a Δ-connected load. An important feature of this configuration is that each phase of the load has the full line voltage across it.

$$V_Z = V_L \qquad \textbf{Equation 21–7}$$

The line currents equal the corresponding phase currents, and each line current divides into two load currents, as indicated. For a balanced load ($\mathbf{Z}_a = \mathbf{Z}_b = \mathbf{Z}_c$), the expression for the current in each load is

$$I_L = \sqrt{3}I_Z \qquad \textbf{Equation 21–8}$$

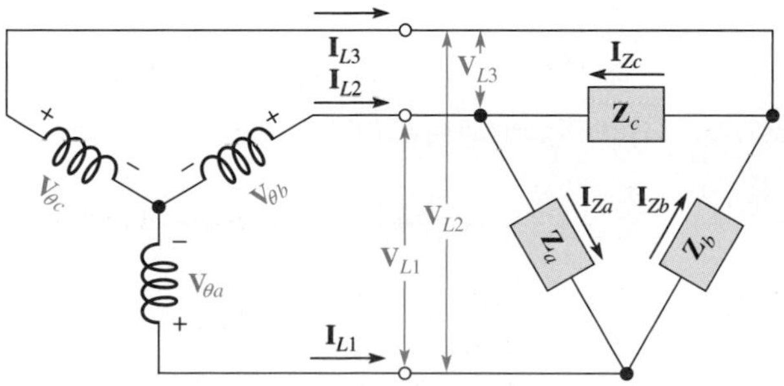

▲ **FIGURE 21–25**

A Y-connected source driving a Δ-connected load.

EXAMPLE 21–5

Determine the load voltages and load currents in Figure 21–26, and show their relationship in a phasor diagram.

FIGURE 21–26

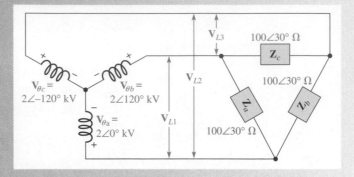

Solution

Using $V_L = \sqrt{3}V_\theta$ (Equation 21–2) and the fact that there is 30° between each line voltage and the nearest phase voltage, the load voltages are

$$\mathbf{V}_{Za} = \mathbf{V}_{L1} = 2\sqrt{3}\angle 150° \, \text{kV} = \mathbf{3.46}\angle\mathbf{150°}\,\textbf{kV}$$
$$\mathbf{V}_{Zb} = \mathbf{V}_{L2} = 2\sqrt{3}\angle 30° \, \text{kV} = \mathbf{3.46}\angle\mathbf{30°}\,\textbf{kV}$$
$$\mathbf{V}_{Zc} = \mathbf{V}_{L3} = 2\sqrt{3}\angle -90° \, \text{kV} = \mathbf{3.46}\angle\mathbf{-90°}\,\textbf{kV}$$

The load currents are

$$\mathbf{I}_{Za} = \frac{\mathbf{V}_{Za}}{\mathbf{Z}_a} = \frac{3.46\angle 150° \, \text{kV}}{100\angle 30° \, \Omega} = \mathbf{34.6}\angle\mathbf{120°}\,\textbf{A}$$

$$\mathbf{I}_{Zb} = \frac{\mathbf{V}_{Zb}}{\mathbf{Z}_b} = \frac{3.46\angle 30° \, \text{kV}}{100\angle 30° \, \Omega} = \mathbf{34.6}\angle\mathbf{0°}\,\textbf{A}$$

$$\mathbf{I}_{Zc} = \frac{\mathbf{V}_{Zc}}{\mathbf{Z}_c} = \frac{3.46\angle -90° \, \text{kV}}{100\angle 30° \, \Omega} = \mathbf{34.6}\angle\mathbf{-120°}\,\textbf{A}$$

The phasor diagram is shown in Figure 21–27.

▶ **FIGURE 21–27**

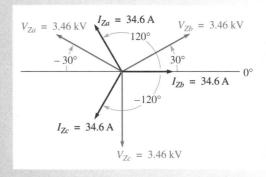

Related Problem

Determine the load currents in Figure 21–26 if the phase voltages have a magnitude of 240 V.

The Δ-Y System

Figure 21–28 shows a Δ-connected source driving a Y-connected balanced load. By examination of the figure, you can see that the line voltages are equal to the corresponding phase voltages of the source. Also, each phase voltage equals the difference of the corresponding load voltages, as you can see by the polarities.

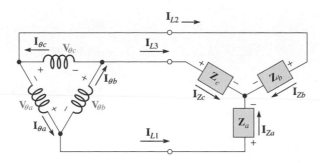

▲ FIGURE 21–28

A Δ-connected source driving a Y-connected load.

Each load current equals the corresponding line current. The sum of the load currents is zero because the load is balanced; thus, there is no need for a neutral return.

The relationship between the load voltages and the corresponding phase voltages (and line voltages) is

$$V_\theta = \sqrt{3}V_Z$$

Equation 21–9

The line currents and corresponding load currents are equal, and for a balanced load, the sum of the load currents is zero.

$$\mathbf{I}_L = \mathbf{I}_Z$$

Equation 21–10

As you can see in Figure 21–28, each line current is the difference of two phase currents.

$$\mathbf{I}_{L1} = \mathbf{I}_{\theta a} - \mathbf{I}_{\theta b}$$
$$\mathbf{I}_{L2} = \mathbf{I}_{\theta c} - \mathbf{I}_{\theta a}$$
$$\mathbf{I}_{L3} = \mathbf{I}_{\theta b} - \mathbf{I}_{\theta c}$$

EXAMPLE 21–6

Determine the currents and voltages in the balanced load and the magnitude of the line voltages in Figure 21–29.

FIGURE 21–29

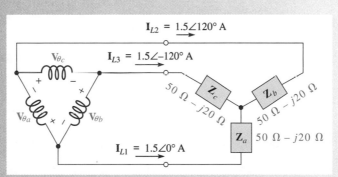

Solution The load currents equal the specified line currents.

$$\mathbf{I}_{Za} = \mathbf{I}_{L1} = \mathbf{1.5}\angle\mathbf{0°}\,\mathbf{A}$$
$$\mathbf{I}_{Zb} = \mathbf{I}_{L2} = \mathbf{1.5}\angle\mathbf{120°}\,\mathbf{A}$$
$$\mathbf{I}_{Zc} = \mathbf{I}_{L3} = \mathbf{1.5}\angle\mathbf{-120°}\,\mathbf{A}$$

The load voltages are

$$\mathbf{V}_{Za} = \mathbf{I}_{Za}\mathbf{Z}_a$$
$$= (1.5\angle0°\,\text{A})(50\,\Omega - j20\,\Omega)$$
$$= (1.5\angle0°\,\text{A})(53.9\angle-21.8°\,\Omega) = \mathbf{80.9}\angle\mathbf{-21.8°}\,\mathbf{V}$$

$$\mathbf{V}_{Zb} = \mathbf{I}_{Zb}\mathbf{Z}_b$$
$$= (1.5\angle120° \text{ A})(53.9\angle-21.8° \text{ }\Omega) = \mathbf{80.9}\angle\mathbf{98.2°} \text{ V}$$
$$\mathbf{V}_{Zc} = \mathbf{I}_{Zc}\mathbf{Z}_c$$
$$= (1.5\angle-120° \text{ A})(53.9\angle-21.8° \text{ }\Omega) = \mathbf{80.9}\angle\mathbf{-142°} \text{ V}$$

The magnitude of the line voltages is

$$V_L = V_\theta = \sqrt{3}V_Z = \sqrt{3}(80.9 \text{ V}) = \mathbf{140} \text{ V}$$

Related Problem   If the magnitudes of the line currents are 1 A, what are the load currents?

The Δ-Δ System

Figure 21–30 shows a Δ-connected source driving a Δ-connected load. Notice that the load voltage, line voltage, and source phase voltage are all equal for a given phase.

$$V_{\theta a} = V_{L1} = V_{Za}$$
$$V_{\theta b} = V_{L2} = V_{Zb}$$
$$V_{\theta c} = V_{L3} = V_{Zc}$$

▶ **FIGURE 21–30**

A Δ-connected source driving a Δ-connected load.

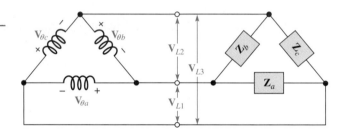

Of course, when the load is balanced, all the voltages are equal, and a general expression can be written

Equation 21–11

$$V_\theta = V_L = V_Z$$

For a balanced load and equal source phase voltages, it can be shown that

Equation 21–12

$$I_L = \sqrt{3}I_Z$$

EXAMPLE 21–7   Determine the magnitude of the load currents and the line currents in Figure 21–31.

FIGURE 21–31

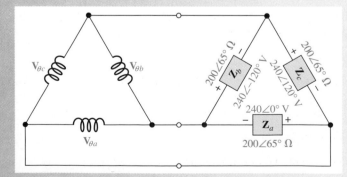

Solution

$$V_{Za} = V_{Zb} = V_{Zc} = 240 \text{ V}$$

The magnitude of the load currents is

$$I_{Za} = I_{Zb} = I_{Zc} = \frac{V_{Za}}{Z_a} = \frac{240 \text{ V}}{200 \text{ }\Omega} = \textbf{1.20 A}$$

The magnitude of the line currents is

$$I_L = \sqrt{3}I_Z = \sqrt{3}(1.20 \text{ A}) = \textbf{2.08 A}$$

Related Problem   Determine the magnitude of the load and line currents in Figure 21–31 if the magnitude of the load voltages is 120 V and the impedances are 600 Ω.

SECTION 21–4
REVIEW

1. List the four types of three-phase source/load configurations.
2. In a certain Y-Y system, the source phase currents each have a magnitude of 3.5 A. What is the magnitude of each load current for a balanced load condition?
3. In a given Y-Δ system, $V_L = 220$ V. Determine V_Z.
4. Determine the line voltages in a balanced Δ-Y system when the magnitude of the source phase voltages is 60 V.
5. Determine the magnitude of the load currents in a balanced Δ-Δ system having a line current magnitude of 3.2 A.

21–5 THREE-PHASE POWER

In this section, power in three-phase systems is studied and methods of power measurement are introduced.

After completing this section, you should be able to

◆ **Discuss power measurements in three-phase systems**

   ◆ Describe the three-wattmeter method
   ◆ Describe the two-wattmeter method

Each phase of a balanced three-phase load has an equal amount of power. Therefore, the total true load power is three times the power in each phase of the load.

$$P_{L(tot)} = 3V_Z I_Z \cos\theta$$

Equation 21–13

where V_Z and I_Z are the voltage and current associated with each phase of the load, and $\cos\theta$ is the power factor.

Recall that in a balanced Y-connected system, the line voltage and line current were

$$V_L = \sqrt{3}V_Z \qquad \text{and} \qquad I_L = I_Z$$

and in a balanced Δ-connected system, the line voltage and line current were

$$V_L = V_Z \qquad \text{and} \qquad I_L = \sqrt{3}I_Z$$

When either of these relationships is substituted into Equation 21–13, the total true power for both Y- and Δ-connected systems is

Equation 21–14

$$P_{L(tot)} = \sqrt{3}V_L I_L \cos\theta$$

EXAMPLE 21–8

In a certain Δ-connected balanced load, the line voltages are 250 V and the impedances are $50\angle30°\ \Omega$. Determine the total load power.

Solution   In a Δ-connected system, $V_Z = V_L$ and $I_L = \sqrt{3}I_Z$. The load current magnitudes are

$$I_Z = \frac{V_Z}{Z} = \frac{250\ \text{V}}{50\ \Omega} = 5\ \text{A}$$

and

$$I_L = \sqrt{3}I_Z = \sqrt{3}(5\ \text{A}) = 8.66\ \text{A}$$

The power factor is

$$\cos\theta = \cos30° = 0.866$$

The total load power is

$$P_{L(tot)} = \sqrt{3}V_L I_L \cos\theta = \sqrt{3}(250\ \text{V})(8.66\ \text{A})(0.866) = \textbf{3.25 kW}$$

Related Problem   Determine the total load power if $V_L = 120$ V and $\mathbf{Z} = 100\angle30°\ \Omega$.

Power Measurement

Power is measured in three-phase systems using wattmeters. The wattmeter uses a basic electrodynamometer-type movement consisting of two coils. One coil is used to measure the current, and the other is used to measure the voltage. Figure 21–32 shows a basic wattmeter schematic and the connections for measuring power in a load. The resistor in series with the voltage coil limits the current through the coil to a small amount proportional to the voltage across the coil.

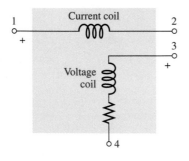

(a) Wattmeter schematic

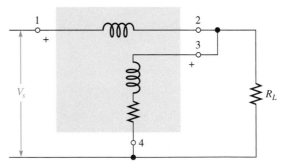

(b) Wattmeter connected to measure load power

▲ **FIGURE 21–32**

Three-Wattmeter Method   Power can be measured easily in a balanced or unbalanced three-phase load of either the Y or the Δ type by using three wattmeters connected as shown in Figure 21–33. This is sometimes known as the *three-wattmeter method*.

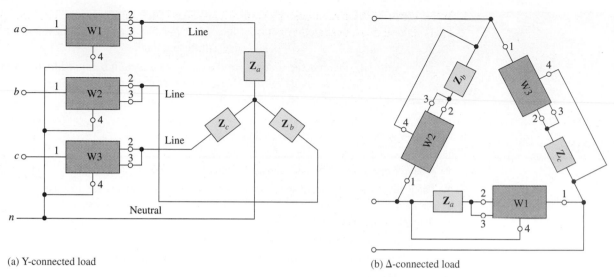

(a) Y-connected load

(b) Δ-connected load

▲ FIGURE 21–33

Three-wattmeter method of power measurement.

The total power is determined by summing the three wattmeter readings.

$$P_{tot} = P_1 + P_2 + P_3$$

Equation 21–15

If the load is balanced, the total power is simply three times the reading on any one wattmeter.

In many three-phase loads, particularly the Δ configuration, it is difficult to connect a wattmeter such that the voltage coil is across the load or such that the current coil is in series with the load because of inaccessibility of points within the load.

Two-Wattmeter Method   Another method of three-phase power measurement uses only two wattmeters. The connections for this two-wattmeter method are shown in Figure 21–34. Notice that the voltage coil of each wattmeter is connected across a line voltage and that the current coil has a line current through it. It can be shown that the algebraic sum of the two wattmeter readings equals the total power in the Y- or Δ-connected load.

$$P_{tot} = P_1 \pm P_2$$

Equation 21–16

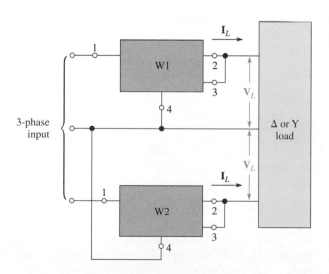

◀ FIGURE 21–34

Two-wattmeter method.

SECTION 21–5 REVIEW

1. $V_L = 30$ V, $I_L = 1.2$ A, and the power factor is 0.257. What is the total power in a balanced Y-connected load? In a balanced Δ-connected load?
2. Three wattmeters connected to measure the power in a certain balanced load indicate a total of 2678 W. How much power does each meter measure?

SUMMARY

- ◆ A simple three-phase generator consists of three conductive loops separated by 120°.
- ◆ Three advantages of three-phase systems over single-phase systems are a smaller copper cross section for the same power delivered to the load, constant power delivered to the load, and a constant, rotating magnetic field.
- ◆ In a Y-connected generator, $I_L = I_\theta$ and $V_L = \sqrt{3}V_\theta$.
- ◆ In a Y-connected generator, there is a 30° difference between each line voltage and the nearest phase voltage.
- ◆ In a Δ-connected generator, $V_L = V_\theta$ and $I_L = \sqrt{3}I_\theta$.
- ◆ In a Δ-connected generator, there is a 30° difference between each line current and the nearest phase current.
- ◆ A balanced load is one in which all the impedances are equal.
- ◆ Power is measured in a three-phase load using either the three-wattmeter method or the two-wattmeter method.

KEY TERMS

Key terms and other bold terms in the chapter are defined in the end-of-book glossary.

Balanced load A condition where all the load currents are equal and the neutral current is zero.

Field winding The winding on the rotor of an ac generator.

Line current (I_L) The current through a line feeding a load.

Line voltage (V_L) The voltage between lines feeding a load.

Phase current (I_θ) The current through a generator winding.

Phase voltage (V_θ) The voltage across a generator winding.

Rotor The rotating assembly in a generator or motor.

Stator The stationary outer part of a generator or motor.

FORMULAS

Y Generator

21–1 $I_L = I_\theta$

21–2 $V_L = \sqrt{3}V_\theta$

Δ Generator

21–3 $V_L = V_\theta$

21–4 $I_L = \sqrt{3}I_\theta$

Y-Y System

21–5 $I_\theta = I_L = I_Z$

21–6 $V_\theta = V_Z$

Y-Δ System

21–7 $V_Z = V_L$

21–8 $I_L = \sqrt{3}I_Z$

Δ-to-Y System

21–9 $V_\theta = \sqrt{3}V_Z$

21–10 $I_L = I_Z$

Δ-Δ System

21–11 $V_\theta = V_L = V_Z$

21–12 $I_L = \sqrt{3}I_Z$

Three-Phase Power

21–13 $P_{L(tot)} = 3V_Z I_Z \cos\theta$

21–14 $P_{L(tot)} = \sqrt{3}V_L I_L \cos\theta$

Three-Wattmeter Method

21–15 $P_{tot} = P_1 + P_2 + P_3$

Two-Wattmeter Method

21–16 $P_{tot} = P_1 \pm P_2$

SELF-TEST

Answers are at the end of the chapter.

1. In a three-phase system, the voltages are separated by
   (a) 90°   (b) 30°   (c) 180°   (d) 120°

2. The term *squirrel-cage* applies to a type of
   (a) three-phase ac generator
   (b) single-phase ac generator
   (c) a three-phase ac motor
   (d) a dc motor

3. Two major parts of an ac generator are
   (a) rotor and stator   (b) rotor and stabilizer
   (c) regulator and slip-ring   (d) magnets and brushes

4. Advantages of a three-phase system over a single-phase system are
   (a) smaller cross-sectional area for the copper conductors
   (b) slower rotor speed
   (c) constant power
   (d) smaller chance of overheating
   (e) both (a) and (c)
   (f) both (b) and (c)

5. The phase current produced by a certain 240 V, Y-connected generator is 12 A. The corresponding line current is
   (a) 36 A   (b) 4 A   (c) 12 A   (d) 6 A

6. A certain Δ-connected generator produces phase voltages of 30 V. The magnitude of the line voltages are
   (a) 10 V   (b) 30 V   (c) 90 V   (d) none of these

7. A certain Δ-Δ system produces phase currents of 5 A. The line currents are
   (a) 5 A   (b) 15 A   (c) 8.66 A   (d) 2.87 A

8. A certain Y-Y system produces phase currents of 15 A. Each line and load current is
   (a) 26 A   (b) 8.66 A   (c) 5 A   (d) 15 A

9. If the source phase voltages of a Δ-Y system are 220 V, the magnitude of the load voltages is
   (a) 220 V   (b) 381 V   (c) 127 V   (d) 73.3 V

<table>
<tr><td>PROBLEMS</td><td>More difficult problems are indicated by an asterisk (*).
Answers to odd-numbered problems are at the end of the book.</td></tr>
</table>

SECTION 21–1   Introduction to Three-Phase Machines

1. The output of an ac generator has a maximum value of 250 V. At what angle is the instantaneous value equal to 75 V?

2. A certain two-pole three-phase generator has a speed of rotation of 60 rpm. What is the frequency of each voltage produced by this generator? What is the phase angle between each voltage?

SECTION 21–2   Generators in Power Applications

3. A single-phase generator drives a load consisting of a 200 Ω resistor and a capacitor with a reactance of 175 Ω. The generator produces a voltage of 100 V. Determine the magnitude of the load current.

4. Determine the phase of the load current with respect to the generator voltage in Problem 3.

5. A certain three-phase unbalanced load in a four-wire system has currents of $2\angle 20°$ A, $3\angle 140°$ A, and $1.5\angle -100°$ A. Determine the current in the neutral line.

SECTION 21–3   Types of Three-Phase Generators

6. Determine the line voltages in Figure 21–35.

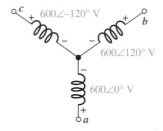

▲ FIGURE 21–35

7. Determine the line currents in Figure 21–36.

8. Develop a complete current phasor diagram for Figure 21–36.

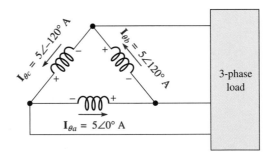

▲ FIGURE 21–36

SECTION 21–4   Three-Phase Source/Load Analysis

9. Determine the following quantities for the Y-Y system in Figure 21–37:
   (a) Line voltages   (b) Phase currents   (c) Line currents
   (d) Load currents   (e) Load voltages

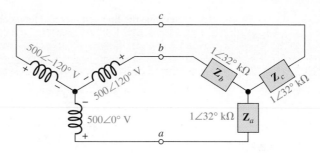

▲ FIGURE 21–37

10. Repeat Problem 9 for the system in Figure 21–38, and also find the neutral current.

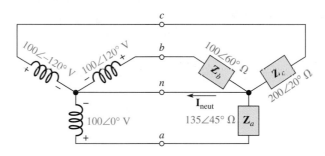

▲ FIGURE 21–38

11. Repeat Problem 9 for the system in Figure 21–39.

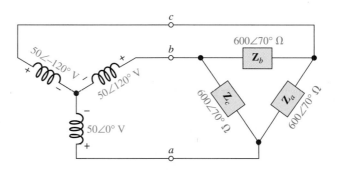

▲ FIGURE 21–39

12. Repeat Problem 9 for the system in Figure 21–40.

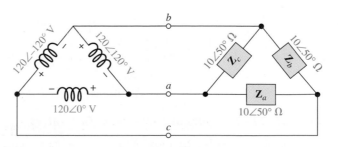

▲ FIGURE 21–40

13. Determine the line voltages and load currents for the system in Figure 21–41.

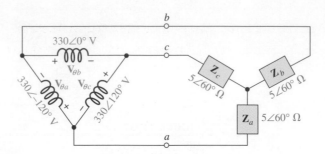

▲ **FIGURE 21–41**

SECTION 21–5 Three-Phase Power

14. The power in each phase of a balanced three-phase system is 1200 W. What is the total power?

15. Determine the load power in Figures 21–37 through 21–41.

16. Find the total load power in Figure 21–42.

▶ **FIGURE 21–42**

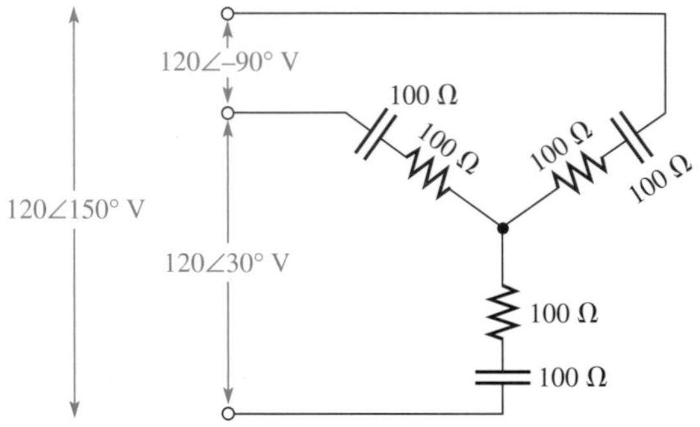

$120\angle-90°$ V

$100\,\Omega$

$100\,\Omega$

$100\,\Omega$

$100\,\Omega$

$120\angle150°$ V

$120\angle30°$ V

$100\,\Omega$

$100\,\Omega$

*__17.__ Using the three-wattmeter method for the system in Figure 21–42, how much power does each wattmeter indicate?

*__18.__ Repeat Problem 17 using the two-wattmeter method.

ANSWERS

SECTION REVIEWS

SECTION 21–1 Introduction to Three-Phase Machines

1. In ac generators, a sinusoidal voltage is induced when a conductive loop is rotated in a magnetic field at a constant speed.

2. Three armature windings

SECTION 21–2 Generators in Power Applications

1. The advantages of polyphase systems are less copper cross section to conduct current; constant power to load; and constant, rotating magnetic field.

2. Constant power

3. Constant magnetic field

SECTION 21–3 Types of Three-Phase Generators

1. $V_L = 1.73\,\text{kV}$

2. $I_L = 5\,\text{A}$

3. $V_L = 240$ V

4. $I_L = 3.46$ A

SECTION 21–4 **Three-Phase Source/Load Analysis**

1. The source/load configurations are Y-Y, Y-Δ, Δ-Y, and Δ-Δ.

2. $I_L = 3.5$ A

3. $V_Z = 220$ V

4. $V_L = 60$ V

5. $I_Z = 1.85$ A

SECTION 21–5 **Three-Phase Power**

1. $P_Y = 16.0$ W; $P_\Delta = 16.0$ W

2. $P = 893$ W

RELATED PROBLEMS FOR EXAMPLES

21–1 4.8 A total for single phase; 2.4 A total for three-phase

21–2 208 V

21–3 (a) $\mathbf{I}_{\theta b} = 8\angle 180°$ A, $\mathbf{I}_{\theta c} = 8\angle -60°$ A

  (b) $\mathbf{I}_{L1} = 13.9\angle 30°$ A, $\mathbf{I}_{L2} = 13.9\angle 150°$ A, $\mathbf{I}_{L3} = 13.9\angle -90°$ A

21–4 $2.96\angle 33.4°$ A

21–5 $\mathbf{I}_{Za} = 4.16\angle 120°$ A, $\mathbf{I}_{Zb} = 4.16\angle 0°$, $\mathbf{I}_{Zc} = 4.16\angle -120°$ A

21–6 $\mathbf{I}_{L1} = \mathbf{I}_{Za} = 1\angle 0°$ A, $\mathbf{I}_{L2} = \mathbf{I}_{Zb} = 1\angle 120°$ A, $\mathbf{I}_{L3} = \mathbf{I}_{Zc} = 1\angle -120°$ A

21–7 $I_Z = 200$ mA, $I_L = 346$ mA

21–8 374 W

SELF-TEST

1. (d)   2. (c)   3. (a)   4. (e)   5. (c)   6. (b)   7. (c)

8. (d)   9. (c)

Resistance Tolerance (±%)

0.1% 0.25% 0.5%	1%	2% 5%	10%	0.1% 0.25% 0.5%	1%	2% 5%	10%	0.1% 0.25% 0.5%	1%	2% 5%	10%	0.1% 0.25% 0.5%	1%	2% 5%	10%	0.1% 0.25% 0.5%	1%	2% 5%	10%	0.1% 0.25% 0.5%	1%	2% 5%	10%
10.0	10.0	10	10	14.7	14.7	—	—	21.5	21.5	—	—	31.6	31.6	—	—	46.4	46.4	—	—	68.1	68.1	68	68
10.1	—	—	—	14.9	—	—	—	21.8	—	—	—	32.0	—	—	—	47.0	—	47	47	69.0	—	—	—
10.2	10.2	—	—	15.0	15.0	15	15	22.1	22.1	22	22	32.4	32.4	—	—	47.5	47.5	—	—	69.8	69.8	—	—
10.4	—	—	—	15.2	—	—	—	22.3	—	—	—	32.8	—	—	—	48.1	—	—	—	70.6	—	—	—
10.5	10.5	—	—	15.4	15.4	—	—	22.6	22.6	—	—	33.2	33.2	33	33	48.7	48.7	—	—	71.5	71.5	—	—
10.6	—	—	—	15.6	—	—	—	22.9	—	—	—	33.6	—	—	—	49.3	—	—	—	72.3	—	—	—
10.7	10.7	—	—	15.8	15.8	—	—	23.2	23.2	—	—	34.0	34.0	—	—	49.9	49.9	—	—	73.2	73.2	—	—
10.9	—	—	—	16.0	—	16	—	23.4	—	—	—	34.4	—	—	—	50.5	—	—	—	74.1	—	—	—
11.0	11.0	11	—	16.2	16.2	—	—	23.7	23.7	—	—	34.8	34.8	—	—	51.1	51.1	51	—	75.0	75.0	75	—
11.1	—	—	—	16.4	—	—	—	24.0	—	24	—	35.2	—	—	—	51.7	—	—	—	75.9	—	—	—
11.3	11.3	—	—	16.5	16.5	—	—	24.3	24.3	—	—	35.7	35.7	—	—	52.3	52.3	—	—	76.8	76.8	—	—
11.4	—	—	—	16.7	—	—	—	24.6	—	—	—	36.1	—	36	—	53.0	—	—	—	77.7	—	—	—
11.5	11.5	—	—	16.9	16.9	—	—	24.9	24.9	—	—	36.5	36.5	—	—	53.6	53.6	—	—	78.7	78.7	—	—
11.7	—	—	—	17.2	—	—	—	25.2	—	—	—	37.0	—	—	—	54.2	—	—	—	79.6	—	—	—
11.8	11.8	—	—	17.4	17.4	—	—	25.5	25.5	—	—	37.4	37.4	—	—	54.9	54.9	—	—	80.6	80.6	—	—
12.0	—	12	12	17.6	—	—	—	25.8	—	—	—	37.9	—	—	—	56.2	—	—	—	81.6	—	—	—
12.1	12.1	—	—	17.8	17.8	—	—	26.1	26.1	—	—	38.3	38.3	—	—	56.6	56.6	56	56	82.5	82.5	82	82
12.3	—	—	—	18.0	—	18	18	26.4	—	—	—	38.8	—	—	—	56.9	—	—	—	83.5	—	—	—
12.4	12.4	—	—	18.2	18.2	—	—	26.7	26.7	—	—	39.2	39.2	39	39	57.6	57.6	—	—	84.5	84.5	—	—
12.6	—	—	—	18.4	—	—	—	27.1	—	27	27	39.7	—	—	—	58.3	—	—	—	85.6	—	—	—
12.7	12.7	—	—	18.7	18.7	—	—	27.4	27.4	—	—	40.2	40.2	—	—	59.0	59.0	—	—	86.6	86.6	—	—
12.9	—	—	—	18.9	—	—	—	27.7	—	—	—	40.7	—	—	—	59.7	—	—	—	87.6	—	—	—
13.0	13.0	13	—	19.1	19.1	—	—	28.0	28.0	—	—	41.2	41.2	—	—	60.4	60.4	—	—	88.7	88.7	—	—
13.2	—	—	—	19.3	—	—	—	28.4	—	—	—	41.7	—	—	—	61.2	—	—	—	89.8	—	—	—
13.3	13.3	—	—	19.6	19.6	—	—	28.7	28.7	—	—	42.2	42.2	—	—	61.9	61.9	62	—	90.9	90.9	91	—
13.5	—	—	—	19.8	—	—	—	29.1	—	—	—	42.7	—	—	—	62.6	—	—	—	92.0	—	—	—
13.7	13.7	—	—	20.0	20.0	20	—	29.4	29.4	—	—	43.2	43.2	43	—	63.4	63.4	—	—	93.1	93.1	—	—
13.8	—	—	—	20.3	—	—	—	29.8	—	—	—	43.7	—	—	—	64.2	—	—	—	94.2	—	—	—
14.0	14.0	—	—	20.5	20.5	—	—	30.1	30.1	30	—	44.2	44.2	—	—	64.9	64.9	—	—	95.3	95.3	—	—
14.2	—	—	—	20.8	—	—	—	30.5	—	—	—	44.8	—	—	—	65.7	—	—	—	96.5	—	—	—
14.3	14.3	—	—	21.0	21.0	—	—	30.9	30.9	—	—	45.3	45.3	—	—	66.5	66.5	—	—	97.6	97.6	—	—
14.5	—	—	—	21.3	—	—	—	31.2	—	—	—	45.9	—	—	—	67.3	—	—	—	98.8	—	—	—

NOTE: These values are generally available in multiples of 0.1, 1, 10, 100, 1 k, and 1 M.

DERIVATIONS

Equation 7–3 Output Voltage of Temperature Measuring Bridge

At balance $V_{OUT} = 0$ and all resistances have the value R. For a small imbalance:

$$V_B = \frac{V_S}{2} \text{ and } V_A = \left(\frac{R}{2R + \Delta R_{THERM}}\right)V_S$$

$$\Delta V_{OUT} = V_B - V_A = \frac{V_S}{2} - \left(\frac{R}{2R + \Delta R_{THERM}}\right)V_S$$

$$= \left(\frac{1}{2} - \frac{R}{2R + \Delta R_{THERM}}\right)V_S$$

$$= \left(\frac{2R + \Delta R_{THERM} - 2R}{2(2R + \Delta R_{THERM})}\right)V_S$$

$$= \left(\frac{\Delta R_{THERM}}{4R + 2\Delta R_{THERM}}\right)V_S$$

Assume $2\Delta R_{THERM} \ll 4R$, then

$$\Delta V_{OUT} \cong \left(\frac{\Delta R_{THERM}}{4R}\right)V_S - \Delta R_{THERM}\left(\frac{V_S}{4R}\right)$$

Equation 11–6 RMS (Effective) Value of a Sine Wave

The abbreviation "rms" stands for the root mean square process by which this value is derived. In the process, we first square the equation of a sine wave.

$$v^2 = V_p^2 \sin^2\theta$$

Next, we obtain the mean or average value of v^2 by dividing the area under a half-cycle of the curve by π (see Figure B–1). The area is found by integration and trigonometric identities.

$$V_{avg}^2 = \frac{\text{area}}{\pi} = \frac{1}{\pi}\int_0^\pi V_p^2 \sin^2\theta\, d\theta$$

$$= \frac{V_p^2}{2\pi}\int_0^\pi (1 - \cos 2\theta)d\theta = \frac{V_p^2}{2\pi}\int_0^\pi 1\, d\theta - \frac{V_p^2}{2\pi}\int_0^\pi (-\cos 2\theta)\, d\theta$$

$$= \frac{V_p^2}{2\pi}(\theta - \tfrac{1}{2}\sin 2\theta)_0^\pi = \frac{V_p^2}{2\pi}(\pi - 0) = \frac{V_p^2}{2}$$

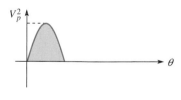

▲ FIGURE B–1

Finally, the square root of V_{avg}^2 is V_{rms}.

$$V_{rms} = \sqrt{V_{avg}^2} = \sqrt{V_p^2/2} = \frac{V_p}{\sqrt{2}} = 0.707 V_p$$

Equation 11–12   Average Value of a Half-Cycle Sine Wave

The average value of a sine wave is determined for a half-cycle because the average over a full cycle is zero.

The equation for a sine wave is

$$v = V_p \sin \theta$$

The average value of the half-cycle is the area under the curve divided by the distance of the curve along the horizontal axis (see Figure B–2).

$$V_{avg} = \frac{\text{area}}{\pi}$$

▶ FIGURE B–2

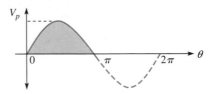

To find the area, we use integral calculus.

$$V_{avg} = \frac{1}{\pi} \int_0^\pi V_p \sin \theta \, d\theta = \frac{V_p}{\pi}(-\cos \theta)\Big|_0^\pi$$

$$= \frac{V_p}{\pi}[-\cos \pi - (-\cos 0)] = \frac{V_p}{\pi}[-(-1) - (-1)]$$

$$= \frac{V_p}{\pi}(2) = \frac{2}{\pi}V_p = 0.637 V_p$$

Equations 12–25 and 13–13   Reactance Derivations

Derivation of Capacitive Reactance

$$\theta = 2\pi ft = \omega t$$

$$i = C\frac{dv}{dt} = C\frac{d(V_p \sin \theta)}{dt} = C\frac{d(V_p \sin \omega t)}{dt} = \omega C(V_p \cos \omega t)$$

$$I_{rms} = \omega C V_{rms}$$

$$X_C = \frac{V_{rms}}{I_{rms}} = \frac{V_{rms}}{\omega C V_{rms}} = \frac{1}{\omega C} = \frac{1}{2\pi fC}$$

Derivation of Inductive Reactance

$$v = L\frac{di}{dt} = L\frac{d(I_p \sin \omega t)}{dt} = \omega L(I_p \cos \omega t)$$

$$V_{rms} = \omega L I_{rms}$$

$$X_L = \frac{V_{rms}}{I_{rms}} = \frac{\omega L I_{rms}}{I_{rms}} = \omega L = 2\pi fL$$

Equation 15–33

The feedback circuit in the phase-shift oscillator consists of three RC stages, as shown in Figure B–3. An expression for the attenuation is derived using the loop analysis method for the loop assignment shown. All Rs are equal in value, and all Cs are equal in value.

$$(R - j1/2\pi fC)I_1 - RI_2 + 0I_3 = V_{in}$$
$$-RI_1 + (2R - j1/2\pi fC)I_2 - RI_3 = 0$$
$$0I_1 - RI_2 + (2R - j1/2\pi fC)I_3 = 0$$

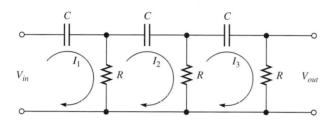

▲ FIGURE B–3

In order to get V_{out}, we must solve for I_3 using determinants:

$$I_3 = \frac{\begin{vmatrix} (R - j1/2\pi fC) & -R & V_{in} \\ -R & (2R - j1/2\pi fC) & 0 \\ 0 & -R & 0 \end{vmatrix}}{\begin{vmatrix} (R - j1/2\pi fC) & -R & 0 \\ -R & (2R - j1/2\pi fC) & -R \\ 0 & -R & (2R - j1/2\pi fC) \end{vmatrix}}$$

$$I_3 = \frac{R^2 V_{in}}{(R - j1/2\pi fC)(2R - j1/2\pi fC)^2 - R^2(2R - j1/2\pi fC) - R^2(R - 1/2\pi fC)}$$

$$\frac{V_{out}}{V_{in}} = \frac{RI_3}{V_{in}}$$

$$= \frac{R^3}{(R - j1/2\pi fC)(2R - j1/2\pi fC)^2 - R^3(2 - j1/2\pi fRC) - R^3(1 - 1/2\pi fRC)}$$

$$= \frac{R^3}{R^3(1 - j1/2\pi fRC)(2 - j1/2\pi fRC)^2 - R^3[(2 - j1/2\pi fRC) - (1 - j1/2\pi fRC)]}$$

$$= \frac{R^3}{R^3(1 - j1/2\pi fRC)(2 - j1/2\pi fRC)^2 - R^3(3 - j1/2\pi fRC)}$$

$$\frac{V_{out}}{V_{in}} = \frac{1}{(1 - j1/2\pi fRC)(2 - j1/2\pi fRC)^2 - (3 - j1/2\pi fRC)}$$

Expanding and combining the real terms and the j terms separately.

$$\frac{V_{out}}{V_{in}} = \frac{1}{\left(1 - \dfrac{5}{4\pi^2 f^2 R^2 C^2}\right) - j\left(\dfrac{6}{2\pi fRC} - \dfrac{1}{(2\pi f)^3 R^3 C^3}\right)}$$

For oscillation in the phase-shift amplifier, the phase shift through the RC circuit must equal 180°. For this condition to exist, the j term must be 0 at the frequency of oscillation f_r.

$$\frac{6}{2\pi f_r RC} - \frac{1}{(2\pi f_r)^3 R^3 C^3} = 0$$

$$\frac{6(2\pi)^2 f_r^2 R^2 C^2 - 1}{(2\pi)^3 f_r^3 R^3 C^3} = 0$$

$$6(2\pi)^2 f_r^2 R^2 C^2 - 1 = 0$$

$$f_r^2 = \frac{1}{6(2\pi)^2 R^2 C^2}$$

$$f_r = \frac{1}{2\pi\sqrt{6}RC}$$

Equation 17–13   Resonant Frequency for a Nonideal Parallel Resonant Circuit

$$\frac{1}{\mathbf{Z}} = \frac{1}{-jX_C} + \frac{1}{R_W + jX_L}$$

$$= j\left(\frac{1}{X_C}\right) + \frac{R_W - jX_L}{(R_W + jX_L)(R_W - jX_L)} = j\left(\frac{1}{X_C}\right) + \frac{R_W - jX_L}{R_W^2 + X_L^2}$$

The first term plus splitting the numerator of the second term yields

$$\frac{1}{\mathbf{Z}} = j\left(\frac{1}{X_C}\right) - j\left(\frac{X_L}{R_W^2 + X_L^2}\right) + \frac{R_W}{R_W^2 + X_L^2}$$

The j terms are equal.

$$\frac{1}{X_C} = \frac{X_L}{R_W^2 + X_L^2}$$

Thus,

$$R_W^2 = X_L^2 = X_L X_C$$

$$R_W^2 + (2\pi f_r L)^2 = \frac{2\pi f_r L}{2\pi f_r C}$$

$$R_W^2 + 4\pi^2 f_r^2 L^2 = \frac{L}{C}$$

$$4\pi^2 f_r^2 L^2 = \frac{L}{C} - R_W^2$$

Solving for f_r^2,

$$f_r^2 = \frac{\left(\dfrac{L}{C}\right) - R_W^2}{4\pi^2 L^2}$$

Multiply both numerator and denominator by C,

$$f_r^2 = \frac{L - R_W^2 C}{4\pi^2 L^2 C} = \frac{L - R_W^2 C}{L(4\pi^2 LC)}$$

Factoring an L out of the numerator and canceling gives

$$f_r^2 = \frac{1 - (R_W^2 C/L)}{4\pi^2 LC}$$

Taking the square root of both sides yields f_r,

$$f_r = \frac{\sqrt{1 - (R_W^2 C/L)}}{2\pi\sqrt{LC}}$$

Equation 17–16   Impedance of Nonideal Tank Circuit at Resonance

Begin with the following expression for $1/\mathbf{Z}$ that was developed in the derivation for Equation 17–13.

$$\frac{1}{\mathbf{Z}} = j\left(\frac{1}{X_C}\right) - j\left(\frac{X_L}{R_W^2 + X_L^2}\right) + \frac{R_W}{R_W^2 + X_L^2}$$

At resonance, $\mathbf{Z}$ is purely resistive; so it has no j part (the j terms in the last expression cancel). Thus, only the real part is left, as stated in the following equation for Z at resonance:

$$Z_r = \frac{R_W^2 + X_L^2}{R_W}$$

Splitting the denominator, we get

$$Z_p = \frac{R_W^2}{R_W} + \frac{X_L^2}{R_W} = R_W + \frac{X_L^2}{R_W}$$

Factoring out R_W gives

$$Z_r = R_W\left(1 + \frac{X_L^2}{R_W^2}\right)$$

Since $X_L^2/R_W^2 = Q^2$, then

$$Z_r = R_W(Q^2 + 1)$$

CAPACITOR COLOR CODING

Some capacitors have color-coded designations. The color code used for capacitors is basically the same as that used for resistors. Some variations occur in tolerance designation. The basic color codes are shown in Table C–1, and some typical color-coded capacitors are illustrated in Figure C–1.

▶ **TABLE C–1**

Typical composite color codes for capacitors (picofarads).

COLOR	DIGIT	MULTIPLIER	TOLERANCE
Black	0	1	20%
Brown	1	10	1%
Red	2	100	2%
Orange	3	1000	3%
Yellow	4	10000	
Green	5	100000	5% (EIA)
Blue	6	1000000	
Violet	7		
Gray	8		
White	9		
Gold		0.1	5% (JAN)
Silver		0.01	10%

NOTE: EIA stands for Electronic Industries Association, and JAN stands for Joint Army-Navy, a military standard.

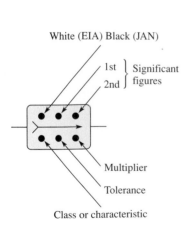

White (EIA) Black (JAN)

1st ⎱ Significant
2nd ⎰ figures

Multiplier

Tolerance

Class or characteristic

(a) Molded mica

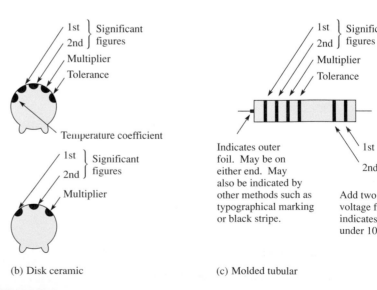

1st ⎱ Significant
2nd ⎰ figures

Multiplier

Tolerance

Temperature coefficient

1st ⎱ Significant
2nd ⎰ figures

Multiplier

(b) Disk ceramic

1st ⎱ Significant
2nd ⎰ figures

Multiplier

Tolerance

Indicates outer foil. May be on either end. May also be indicated by other methods such as typographical marking or black stripe.

1st ⎱ Significant
2nd ⎰ voltage figures

Add two zeros to significant voltage figures. One band indicates voltage ratings under 1000 volts.

(c) Molded tubular

▲ **FIGURE C–1**

Typical color-coded capacitors.

Marking Systems

A capacitor, as shown in Figure C–2, has certain identifying features.

- Body of one solid color (off-white, beige, gray, tan or brown).

- End electrodes completely enclose ends of part.

- Many different sizes:

    1. Type 1206: 0.125 inch long by 0.063 inch wide (3.2 mm × 1.6 mm) with variable thickness and color.

    2. Type 0805: 0.080 inch long by 0.050 inch wide (2.0 mm × 1.25 mm) with variable thickness and color.

    3. Variably sized with a single color (usually translucent tan or brown). Sizes range from 0.059 inch (1.5 mm) to 0.220 inch (5.6 mm) in length and in width from 0.032 inch (0.8 mm) to 0.197 inch (5.0 mm).

- Three different marking systems:

    1. Two place (letter and number only).

    2. Two place (letter and number or two numbers).

    3. One place (letter of varying color).

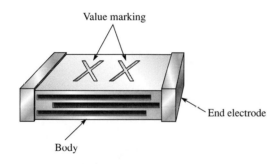

Value marking

End electrode

Body

▲ FIGURE C–2

Capacitor marking.

Standard Two-Place Code

Refer to Table C–2.

$$\mathbf{J3} \quad = 2.2 \times 10^3 = 2200 \text{ pF}$$

Multiplier (0–9)

Value (1st and 2nd significant digits)

Examples: S2 = 4.7 × 100 = 470 pF
b0 = 3.5 × 1.0 = 3.5 pF

► TABLE C–2

	VALUE*					MULTIPLIER
A	1.0	L	2.7	T	5.1	0 = ×1.0
B	1.1	M	3.0	U	5.6	1 = ×10
C	1.2	N	3.3	m	6.0	2 = ×100
D	1.3	b	3.5	V	6.2	3 = ×1000
E	1.5	P	3.6	W	6.8	4 = ×10000
F	1.6	Q	3.9	n	7.0	5 = ×100000
G	1.8	d	4.0	X	7.5	etc.
H	2.0	R	4.3	t	8.0	
J	2.2	e	4.5	Y	8.2	
K	2.4	S	4.7	y	9.0	
a	2.5	f	5.0	Z	9.1	

*Note uppercase and lowercase letters.

Alternate Two-Place Code

Refer to Table C–3.

   ◆ Values below 100 pF—Value read directly

   05 = 5 pF   **82** = 82 pF

   ◆ Values 100 pF and above—Letter/Number code

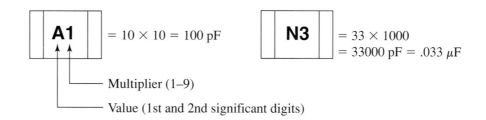

A1 = 10 × 10 = 100 pF

N3 = 33 × 1000
= 33000 pF = .033 μF

└── Multiplier (1–9)

└── Value (1st and 2nd significant digits)

► TABLE C–3

	VALUE*					MULTIPLIER
A	10	J	22	S	47	1 = ×10
B	11	K	24	T	51	2 = ×100
C	12	L	27	U	56	3 = ×1000
D	13	M	30	V	62	4 = ×10000
E	15	N	33	W	68	5 = ×100000
F	16	P	36	X	75	etc.
G	18	Q	39	Y	82	
H	20	R	43	Z	91	

*Note uppercase letters only.

Standard Single-Place Code

Refer to Table C–4.

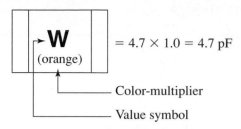

= 4.7 × 1.0 = 4.7 pF

—— Color-multiplier

—— Value symbol

Examples: R (Green) = 3.3 × 100 = 330 pF
7 (Blue) = 8.2 × 1000 = 8200 pF

◄ TABLE C–4

VALUE						MULTIPLIER (COLOR)
A	1.0	K	2.2	W	4.7	Orange = ×1.0
B	1.1	L	2.4	X	5.1	Black = ×10
C	1.2	N	2.7	Y	5.6	Green = ×100
D	1.3	O	3.0	Z	6.2	Blue = ×1000
E	1.5	R	3.3	3	6.8	Violet = ×10000
H	1.6	S	3.6	4	7.5	Red = ×100000
I	1.8	T	3.9	7	8.2	
J	2.0	V	4.3	9	9.1	

Answers to Odd-Numbered Problems

Chapter 1

1. (a) 3×10^3 (b) 7.5×10^4 (c) 2×10^6

3. (a) 8.4×10^3 (b) 9.9×10^4 (c) 2×10^5

5. (a) 3.2×10^4 (b) 6.8×10^{-3} (c) 8.7×10^{10}

7. (a) 0.0000025 (b) 500 (c) 0.39

9. (a) 4.32×10^7 (b) 5.00085×10^3
   (c) 6.06×10^{-8}

11. (a) 2.0×10^9 (b) 3.6×10^{14}
    (c) 1.54×10^{-14}

13. (a) 89×10^3 (b) 450×10^3
    (c) 12.04×10^{12}

15. (a) 345×10^{-6} (b) 25×10^{-3}
    (c) 1.29×10^{-9}

17. (a) 7.1×10^{-3} (b) 101×10^6
    (c) 1.50×10^6

19. (a) 22.7×10^{-3} (b) 200×10^6
    (c) 848×10^{-3}

21. (a) $345\,\mu A$ (b) 25 mA (c) 1.29 nA

23. (a) $3\,\mu F$ (b) 3.3 MΩ (c) 350 nA

25. (a) 7.5×10^{-12} A (b) 3.3×10^9 Hz
    (c) 2.8×10^{-7} W

27. (a) $5000\,\mu A$ (b) 3.2 mW
    (c) 5 MV (d) 10,000 kW

29. (a) 50.68 mA (b) 2.32 MΩ (c) $0.0233\,\mu F$

Chapter 2

1. 4.64×10^{-18} C

3. 80×10^{12} C

5. (a) 10 V (b) 2.5 V (c) 4 V

7. 20 V

9. 33.3 V

11. 0.2 A

13. 0.15 C

15. (a) 200 mS (b) 40 mS (c) 10 mS

17. dc power supply, solar cell, generator, battery

19. The power supply converts ac voltage to dc voltage.

21. (a) $27\,k\Omega \pm 5\%$ (b) $1.8\,k\Omega \pm 10\%$

23. 330 Ω: orange, orange, brown, gold
    2.2 kΩ: red, red, red, gold
    56 kΩ: green, blue, orange, gold
    100 kΩ: brown, black, yellow, gold
    39 kΩ: orange, white, orange, gold

25. (a) yellow, violet, silver, gold
    (b) red, violet, yellow, gold
    (c) green, brown, green, gold

27. (a) brown, yellow, violet, red, brown
    (b) orange, white, red, gold, brown
    (c) white, violet, blue, brown, brown

29. 4.7 kΩ

31. Through lamp 2

33. Circuit (b)

35. See Figure P–1.

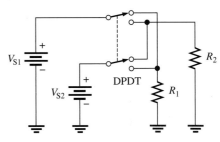

▲ FIGURE P–1

37. See Figure P–2.

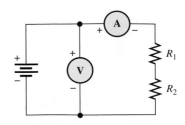

▲ FIGURE P–2

39. Position 1: V1 = 0 V, V2 = V_S

Position 2: V1 = V_S, V2 = 0 V

41. See Figure P–3.

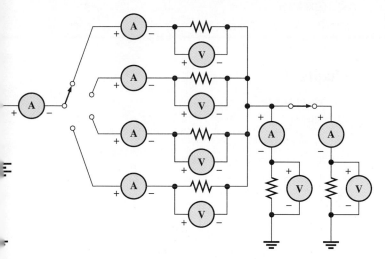

▲ **FIGURE P–3**

43. 250 V

45. (a) 20 Ω   (b) 1.50 MΩ   (c) 4500 Ω

47. See Figure P–4.

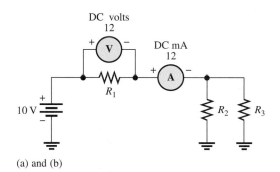

(a) and (b)

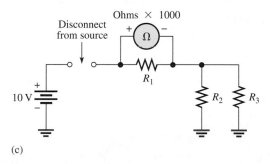

(c)

▲ **FIGURE P–4**

Chapter 3

1. (a) Current triples.   (b) Current is reduced 75%.

(c) Current is halved.   (d) Current increases 54%.

(e) Current quadruples.   (f) Current is unchanged.

3. $V = IR$

5. The graph is a straight line, indicating a linear relationship between V and I.

7. $R_1 = 0.5\ \Omega, R_2 = 1.0\ \Omega, R_3 = 2\ \Omega$

9. See Figure P–5.

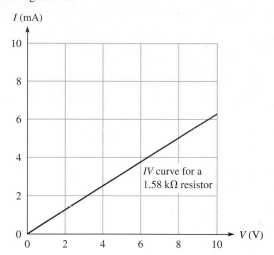

▲ **FIGURE P–5**

11. The voltage decreased by 4 V (from 10 V to 6 V).

13. See Figure P–6.

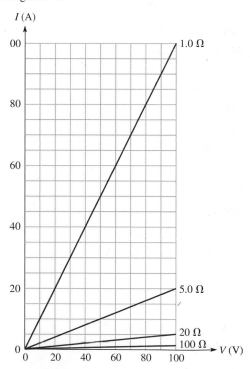

▲ **FIGURE P–6**

15. (a) 5 A   (b) 1.5 A   (c) 500 mA

(d) 2 mA   (e) 44.6 μA

17. 1.2 A

19. 532 μA

21. Yes. The current is now 0.642 A, which exceeds the rating of the fuse.

23. (a) 36 V (b) 280 V (c) 1700 V

 (d) 28.2 V (e) 56 V

25. 81 V

27. (a) 59.9 mA (b) 5.99 V (c) 4.61 mV

29. (a) 2 kΩ (b) 3.5 kΩ (c) 2 kΩ

 (d) 100 kΩ (e) 1.0 MΩ

31. 150 Ω

33. 133 Ω; 100 Ω; the source can be shorted if the rheostat is set to 0 Ω.

35. 95 Ω

37. Five

39. $R_A = 560\ k\Omega$; $R_B = 2.2\ M\Omega$;

 $R_C = 1.8\ k\Omega$; $R_D = 33\ \Omega$

41. $V = 18\ V$; $I = 5.455\ mA$;

 $R = 3.3\ k\Omega$

Chapter 4

1. volt = joule/coulomb

 amp = coulomb/second

 VI = joule/coulomb × coulomb/second = joule/sec = watt

3. 350 W

5. 20 kW

7. (a) 1 MW (b) 3 MW

 (c) 150 MW (d) 8.7 MW

9. (a) 2,000,000 μW

 (b) 500 μW

 (c) 250 μW

 (d) 6.67 μW

11. 8640 J

13. 2.02 kW/day

15. 0.00186 kWh

17. 37.5 Ω

19. 360 W

21. 100 μW

23. 40.2 mW

25. (a) 0.480 Wh (b) Equal

27. At least 12 W, to allow a safety margin of 20%

29. 7.07 V

31. 50,544 J

33. 8 A

35. 100 mW, 80%

37. 0.08 kWh

39. $V = 5\ V$; $I = 5\ mA$;

 $R = 1\ k\Omega$

Chapter 5

1. See Figure P–7.

3. 170 kΩ

5. R_1, R_7, R_8, and R_{10} are in series.

 R_2, R_4, R_6, and R_{11} are in series.

 R_3, R_5, R_9, and R_{12} are in series.

7. 5 mA

9. See Figure P–8.

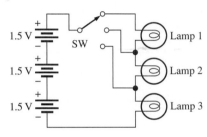

▲ FIGURE P–8

11. (a) 1560 Ω (b) 103 Ω

 (c) 13.7 kΩ (d) 3.671 MΩ

13. 67.2 kΩ

15. 3.9 kΩ

17. 17.8 MΩ

19. (a) 625 μA (b) 4.26 μA

21. (a) 34 mA (b) 16 V (c) 0.543 W

23. $R_1 = 330\ \Omega$, $R_2 = 220\ \Omega$, $R_3 = 100\ \Omega$, $R_4 = 470\ \Omega$

25. (a) 331 Ω

 (b) Position B: 9.15 mA

 Position C: 14.3 mA

 Position D: 36.3 mA

 (c) No

27. 14 V

29. (a) 23 V (b) 35 V (c) 0 V

31. 4 V

33. 22 Ω

▶ FIGURE P–7

(a)

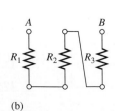

(b)

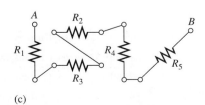

(c)

35. Position A: 4.0 V

Position B: 4.5 V

Position C: 5.4 V

Position D: 7.2 V

37. 4.82%

39. A output = 15 V

B output = 10.6 V

C output = 2.62 V

41. $V_R = 6$ V, $V_{2R} = 12$ V, $V_{3R} = 18$ V,

$V_{4R} = 24$ V, $V_{5R} = 30$ V

43. $V_2 = 1.79$ V, $V_3 = 1$ V, $V_4 = 17.9$ V

45. See Figure P–9.

▶ **FIGURE P–9**

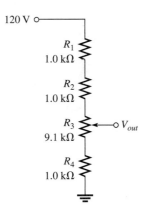

120 V

R_1
1.0 kΩ

R_2
1.0 kΩ

R_3 ——o V_{out}
9.1 kΩ

R_4
1.0 kΩ

47. 54.9 mW

49. 12.5 MΩ

51. $V_A = 100$ V, $V_B = 57.7$ V, $V_C = 15.2$ V, $V_D = 7.58$ V

53. $V_A = 14.82$ V, $V_B = 12.97$ V, $V_C = 12.64$ V, $V_D = 9.34$ V

55. (a) R_4 is open. **(b)** Short from A to B

57. Table 5–1 is correct.

59. Yes. There is a short between pin 4 and the upper side of R_{11}.

61. $R_T = 7.481$ kΩ

63. $R_3 = 22$ Ω

65. R_1 shorted

Chapter 6

1. See Figure P–10.

3. R_1, R_2, R_5, R_9, R_{10}, and R_{12} are in parallel.

R_4, R_6, R_7, and R_8 are in parallel.

R_3 and R_{11} are in parallel.

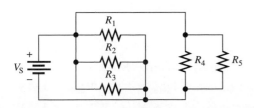

R_1

R_2

V_S $+$ $-$

R_3

R_4 R_5

▲ **FIGURE P–10**

5. 100 V

7. Position A:

$V_1 = 15$ V, $V_2 = 0$ V, $V_3 = 0$ V, $V_4 = 15$ V

Position B:

$V_1 = 15$ V, $V_2 = 0$ V, $V_3 = 15$ V, $V_4 = 0$ V

Position C:

$V_1 = 15$ V, $V_2 = 15$ V, $V_3 = 0$ V, $V_4 = 0$ V

9. 1.35 A

11. $R_2 = 22$ Ω, $R_3 = 100$ Ω, $R_4 = 33$ Ω

13. 11.4 mA

15. (a) 359 Ω   **(b)** 25.6 Ω   **(c)** 819 Ω   **(d)** 997 Ω

17. 567 Ω

19. 24.6 Ω

21. (a) 510 kΩ   **(b)** 245 kΩ   **(c)** 510 kΩ   **(d)** 193 kΩ

23. 10 A

25. 50 mA; When one bulb burns out, the others remain on.

27. 53.7 Ω

29. $I_2 = 167$ mA, $I_3 = 83.3$ mA, $I_T = 300$ mA,

$R_1 = 2$ kΩ, $R_2 = 600$ Ω

31. Position A: 2.25 mA

Position B: 4.75 mA

Position C: 7 mA

33. (a) $I_1 = 6.88$ μA, $I_2 = 3.12$ μA

(b) $I_1 = 5.25$ mA, $I_2 = 2.39$ mA, $I_3 = 1.59$ mA,

$I_4 = 772$ μA

35. $R_1 = 3.3$ kΩ, $R_2 = 1.8$ kΩ, $R_3 = 5.6$ kΩ, $R_4 = 3.9$ kΩ

37. (a) 1 mΩ   **(b)** 5 μA

39. (a) 68.8 μW   **(b)** 52.5 mW

41. $P_1 = 1.25$ W, $I_2 = 75$ mA, $I_1 = 125$ mA, $V_S = 10$ V,

$R_1 = 80$ Ω, $R_2 = 133$ Ω

43. 682 mA, 3.41 A

45. The 8.2 kΩ resistor is open.

47. Connect ohmmeter between the following pins:

Pins 1-2

Correct reading: $R = 1.0$ kΩ ‖ 3.3 kΩ = 767 Ω

R_1 open: $R = 3.3$ kΩ

R_2 open: $R = 1.0$ kΩ

Pins 3-4

Correct reading: $R = 270$ Ω ‖ 390 Ω = 159.5 Ω

R_3 open: $R = 390$ Ω

R_4 open: $R = 270$ Ω

Pins 5-6

Correct reading:

$R = 1.0$ MΩ ‖ 1.8 MΩ ‖ 680 kΩ ‖ 510 kΩ = 201 kΩ

R_5 open: $R = 1.8$ MΩ ‖ 680 kΩ ‖ 510 kΩ = 251 kΩ

R_6 open: $R = 1.0$ MΩ ‖ 680 kΩ ‖ 510 kΩ = 226 kΩ

R_7 open: $R = 1.0$ MΩ ‖ 1.8 MΩ ‖ 510 kΩ = 284 kΩ

R_8 open: $R = 1.0$ MΩ ‖ 1.8 MΩ ‖ 680 kΩ = 330 kΩ

49. Short between pins 3 and 4:

    (a) $R_{1-2} = (R_1 \parallel R_2 \parallel R_3 \parallel R_4 \parallel R_{11} \parallel R_{12})$

                $+ \ (R_5 \parallel R_6 \parallel R_7 \parallel R_8 \parallel R_9 \parallel R_{10}) = 940 \ \Omega$

    (b) $R_{2-3} = R_5 \parallel R_6 \parallel R_7 \parallel R_8 \parallel R_9 \parallel R_{10} = 518 \ \Omega$

    (c) $R_{2-4} = R_5 \parallel R_6 \parallel R_7 \parallel R_8 \parallel R_9 \parallel R_{10} = 518 \ \Omega$

    (d) $R_{1-4} = R_1 \parallel R_2 \parallel R_3 \parallel R_4 \parallel R_{11} \parallel R_{12} = 422 \ \Omega$

51. R_2 open

53. $V_S = 3.30$ V

Chapter 7

1. See Figure P–11.

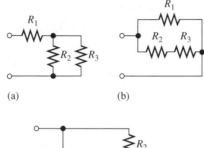

          (a)   (b)

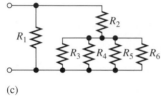

          (c)

▲ FIGURE P–11

3. **(a)** R_1 and R_4 are in series with the parallel combination of R_2 and R_3.

    (b) R_1 is in series with the parallel combination of R_2, R_3, and R_4.

    (c) The parallel combination of R_2 and R_3 is in series with the parallel combination of R_4 and R_5. This is all in parallel with R_1.

5. See Figure P–12.

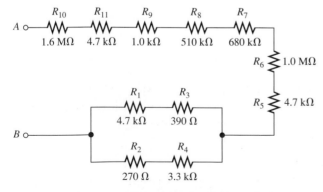

▲ FIGURE P–12

7. See Figure P–13.

9. **(a)** 133 Ω **(b)** 779 Ω **(c)** 852 Ω

▶ FIGURE P–13

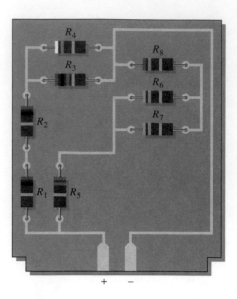

11. **(a)** $I_1 = I_4 = 11.3$ mA, $I_2 = I_3 = 5.64$ mA,

        $V_1 = 633$ mV, $V_2 = V_3 = 564$ mV,

        $V_4 = 305$ mV

    (b) $I_1 = 3.85$ mA, $I_2 = 563 \ \mu$A,

        $I_3 = 1.16$ mA, $I_4 = 2.13$ mA, $V_1 = 2.62$ V,

        $V_2 = V_3 = V_4 = 383$ mV

    (c) $I_1 = 5$ mA, $I_2 = 303 \ \mu$A,

        $I_3 = 568 \ \mu$A, $I_4 = 313 \ \mu$A,

        $I_5 = 558 \ \mu$A, $V_1 = 5$ V,

        $V_2 = V_3 = 1.88$ V, $V_4 = V_5 = 3.13$ V

13. SW1 closed, SW2 open: 220 Ω

    SW1 closed, SW2 closed: 200 Ω

    SW1 open, SW2 open: 320 Ω

    SW1 open, SW2 closed: 300 Ω

15. $V_A = 100$ V, $V_B = 61.5$ V, $V_C = 15.7$ V,

    $V_D = 7.87$ V

17. Measure the voltage at A with respect to ground and the voltage at B with respect to ground. The difference is V_{R2}.

19. 303 kΩ

21. **(a)** 110 kΩ **(b)** 110 mW

23. $R_{AB} = 1.32$ kΩ

    $R_{BC} = 1.32$ kΩ

    $R_{CD} = 0 \ \Omega$

25. 7.5 V unloaded, 7.29 V loaded

27. 47 kΩ

29. 8.77 V

31. $R_1 = 1000 \ \Omega$; $R_2 = R_3 = 500 \ \Omega$;

    lower tap loaded: $V_{lower} = 1.82$ V, $V_{upper} = 4.55$ V

    upper tap loaded: $V_{lower} = 1.67$ V, $V_{upper} = 3.33$ V

33. (a) $V_G = 1.75$ V, $V_S = 3.25$ V

(b) $I_1 = I_2 = 6.48\,\mu\text{A}$, $I_D = I_S = 2.17$ mA

(c) $V_{DS} = 2.55$ V, $V_{DG} = 4.05$ V

35. 1000 V

37. (a) 0.5 V range   (b) Approximately 1 mV

39. (a) 271 Ω   (b) 221 mA   (c) 58.7 mA   (d) 12 V

41. 621 Ω, $I_1 = I_9 = 16.1$ mA, $I_2 = 8.27$ mA,

$I_3 = I_8 = 7.84$ mA, $I_4 = 4.06$ mA, $I_5 = I_6 = I_7 = 3.78$ mA

43. 971 mA

45. (a) 9 V   (b) 3.75 V   (c) 11.25 V

47. 6 mV (right side positive with respect to left side)

49. No, it should be 4.39 V.

51. The 2.2 kΩ resistor (R_3) is open.

53. The 3.3 kΩ resistor (R_4) is open.

55. $R_T = 296.7$ Ω

57. $R_3 = 560$ kΩ

59. R_5 shorted

Chapter 8

1. $I_S = 6$ A, $R_S = 50$ Ω

3. 200 mΩ

5. $V_S = 720$ V, $R_S = 1.2$ kΩ

7. 845 μA

9. 1.6 mA

11. $V_{max} = 3.72$ V; $V_{min} = 1.32$ V

13. 90.7 V

15. $I_{S1} = 2.28$ mA, $I_{S2} = 1.35$ mA

17. 116 μA

19. $R_{TH} = 88.6$ Ω, $V_{TH} = 1.09$ V

21. 100 μA

23. (a) $I_N - 110$ mA, $R_N = 76.7$ Ω

(b) $I_N = 11.1$ mA, $R_N = 73$ Ω

(c) $I_N = 50\,\mu\text{A}$, $R_N = 35.9$ kΩ

(d) $I_N = 68.8$ mA, $R_N = 1.3$ kΩ

25. 17.9 V

27. $I_N = 953\,\mu\text{A}$, $R_N = 1175$ Ω

29. $I_N = -48.2$ mA, $R_N = 56.9$ Ω

31. 11.1 Ω

33. $R_{TH} = 48$ Ω, $R_4 = 160$ Ω

35. (a) $R_A = 39.8$ Ω, $R_B = 73$ Ω, $R_C = 48.7$ Ω

(b) $R_A = 21.2$ kΩ, $R_B = 10.3$ kΩ, $R_C = 14.9$ kΩ

37. R_1 leaky

39. $I_N = 0.383$ mA; $R_N = 9.674$ kΩ

41. $I_{AB} = 1.206$ mA; $V_{AB} = 3.432$ V;

$R_L = 2.846$ kΩ

Chapter 9

1. $I_1 = 371$ mA; $I_2 = 143$ mA

3. $I_1 = 0$ A, $I_2 = 2$ A

5. (a) −16,470   (b) −1.59

7. $I_1 = 1.24$ A, $I_2 = 2.05$ A, $I_3 = 1.89$ A

9. X1 = .371428571429 ($I_1 = 371$ mA)

X2 = −.142857142857 ($I_2 = -143$ mA)

11. $I_1 - I_2 - I_3 = 0$

13. $V_1 = 5.66$ V, $V_2 = 6.33$ V, $V_3 = 325$ mV

15. −1.84 V

17. $I_1 = -5.11$ mA, $I_2 = -3.52$ mA

19. $V_1 = 5.11$ V, $V_3 = 890$ mV, $V_2 = 2.89$ V

21. $I_1 = 15.6$ mA, $I_2 = -61.3$ mA, $I_3 = 61.5$ mA

23. −11.2 mV

25. Note: all Rs (coefficients) are in kΩ.

Loop A: $5.48I_A - 3.3I_B - 1.5I_C = 0$

Loop B: $-3.3I_A + 4.12I_B - 0.82I_C = 15$

Loop C: $-1.5I_A - 0.82I_B + 4.52I_C = 0$

27. $I_1 = 20.6$ mA, $I_3 = 193$ mA, $I_2 = -172$ mA

29. $V_A = 1.5$ V, $V_B = -5.65$ V

31. $I_1 = 193\,\mu\text{A}$, $I_2 = 370\,\mu\text{A}$, $I_3 = 179\,\mu\text{A}$,

$I_4 = 328\,\mu\text{A}$, $I_5 = 1.46$ mA, $I_6 = 522\,\mu\text{A}$,

$I_7 = 2.16$ mA, $I_8 = 1.64$ mA, $V_A = -3.70$ V,

$V_B = -5.85$ V, $V_C = 15.7$ V

33. No fault

35. R_4 open

37. Lower fuse open

39. R_4 open

Chapter 10

1. Decreases

3. 37.5 μWb

5. 1000 G

7. 597

9. 150 At

11. (a) Electromagnetic field   (b) Spring

13. Forces produced by the interaction of the electromagnetic field and the permanent magnetic field

15. Change the current.

17. Material A

19. The strength of the magnetic field, the length of the conductor exposed to the field, and the rotational rate of the conductor

21. Lenz's law defines the polarity of the induced voltage.

23. The commutator and brush arrangement electrically connect the loop to the external circuit.

25. Figure P–14.

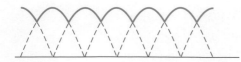

▲ FIGURE P–14

Chapter 11

1. (a) 1 Hz   (b) 5 Hz   (c) 20 Hz
 (d) 1 kHz   (e) 2 kHz   (f) 100 kHz

3. $2 \mu s$

5. 250 Hz

7. 200 rps

9. (a) 7.07 mA   (b) 0 A (full cycle), 4.5 mA (half-cycle)
 (c) 14.14 mA

11. (a) 0.524 or $\pi/6$ rad   (b) 0.785 or $\pi/4$ rad
 (c) 1.361 or $39\pi/90$ rad   (d) 2.356 or $3\pi/4$ rad
 (e) 3.491 or $10\pi/9$ rad   (f) 5.236 or $5\pi/3$ rad

13. 15°, A leading

15. See Figure P–15.

▶ FIGURE P–15

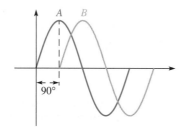

17. (a) 57.4 mA   (b) 99.6 mA   (c) −17.4 mA
 (d) −57.4 mA   (e) −99.6 mA   (f) 0 mA

19. 30°: 13.0 V
 45°: 14.5 V
 90°: 13.0 V
 180°: −7.5 V
 200°: −11.5 V
 300°: −7.5 V

21. 22.1 V

23. See Figure P–16.

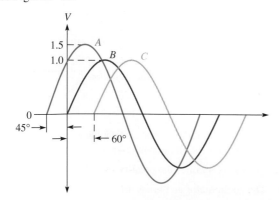

▲ FIGURE P–16

25. (a) 156 mV   (b) 1 V   (c) 0 V

27. $V_{1(avg)} = 40.5$ V, $V_{2(avg)} = 31.5$ V

29. $V_{max} = 39$ V, $V_{min} = 9$ V

31. −1 V

33. $t_r \cong 3.0$ ms, $t_f \cong 3.0$ ms, $t_W \cong 12.0$ ms, Ampl. $\cong 5$ V

35. 5.84 V

37. (a) −0.375 V   (b) 3.01 V

39. (a) 50 kHz   (b) 10 Hz

41. 75 kHz, 125 kHz, 175 kHz, 225 kHz, 275 kHz, 325 kHz

43. $V_p = 600$ mV, $T = 500$ ms

45. $V_{p(in)} = 4.44$ V, $f_{in} = 2$ Hz

47. $V_1 = 16.717$ V_{pp}; $V_1 = 5.911$ V_{rms};
 $V_2 = 36.766$ V_{pp}; $V_2 = 13.005$ V_{rms};
 $V_3 = 14.378$ V_{pp}; $V_3 = 5.084$ V_{rms}

49. No fault

51. $V_{min} = 2.000$ V_p; $V_{max} = 22.000$ V_p

Chapter 12

1. (a) $5 \mu F$ (b) $1 \mu C$ (c) 10 V

3. (a) $0.001 \mu F$ (b) $0.0035 \mu F$ (c) $0.00025 \mu F$

5. 125 J

7. (a) 8.85×10^{-12} F/m   (b) 35.4×10^{-12} F/m
 (c) 66.4×10^{-12} F/m   (d) 17.7×10^{-12} F/m

9. 983 pF

11. $0.0249 \mu F$

13. 12.5 pF increase

15. Ceramic

17. Aluminum, tantalum; they are polarized.

19. (a) $0.022 \mu F$ (b) $0.047 \mu F$
 (c) $0.001 \mu F$ (d) 220 pF

21. (a) $0.688 \mu F$ (b) 69.7 pF   (c) $2.64 \mu F$

23. $2 \mu F$

25. (a) 1057 pF   (b) $0.121 \mu F$

27. (a) $2.62 \mu F$ (b) 689 pF   (c) $1.6 \mu F$

29. (a) $0.411 \mu C$
 (b) $V_1 = 10.47$ V
 $V_2 = 1.54$ V
 $V_3 = 6.52$ V
 $V_4 = 5.48$ V

31. (a) 13.2 ms   (b) $247.5 \mu s$ (c) $11 \mu s$ (d) $280 \mu s$

33. (a) 9.20 V   (b) 1.24 V   (c) 0.458 V   (d) 0.168 V

35. (a) 17.9 V   (b) 12.8 V   (c) 6.59 V

37. $7.62 \mu s$

39. $3.00 \mu s$

41. See Figure P–17.

43. (a) 30.4Ω (b) $116 k\Omega$ (c) 49.7Ω

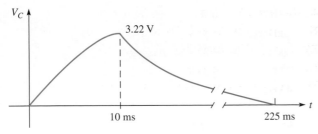

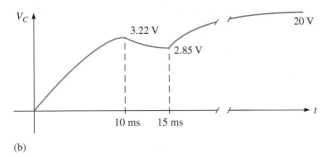

▲ FIGURE P-17

45. 200 Ω

47. 0 W, 3.39 mVAR

49. 0.00541 μF

51. The ripple is reduced.

53. 4.55 kΩ

55. V_1 = 3.103 V; V_2 = 6.828 V; V_3 = 2.069 V

57. I_C @ 1 kHz = 1.383 mA; I_C @ 500 Hz = 0.691 mA
I_C @ 2 kHz = 2.768 mA

59. C_4 shorted

Chapter 13

1. (a) 1000 mH   (b) 0.25 mH   (c) 0.01 mH   (d) 0.5 mH

3. 50 mV

5. 20 mV

7. 0.94 μJ

9. Inductor 2 has three-fourths the inductance of inductor 1.

11. 155 μH

13. 50.5 mH

15. 7.14 μH

17. (a) 4.33 H   (b) 50 mH   (c) 57.1 μH

19. (a) 1 μs   (b) 2.13 μs   (c) 2 μs

21. (a) 5.52 V   (b) 2.03 V   (c) 747 mV
(d) 275 mV   (e) 101 mV

23. (a) 12.3 V   (b) 9.10 V   (c) 3.35 V

25. 11.0 μs

27. 0.722 μs

29. 136 μA

31. (a) 144 Ω   (b) 10.1 Ω   (c) 13.4 Ω

33. (a) 55.5 Hz   (b) 796 Hz   (c) 597 Hz

35. 26.1 mA

37. V_1 = 12.953 V; V_2 = 11.047 V; V_3 = 5.948 V;
V_4 = 5.099 V; V_5 = 5.099 V

39. L_3 open

Chapter 14

1. 1.5 μH

3. 4; 0.25

5. (a) 100 V rms; in phase   (b) 100 V rms; out of phase
(c) 20 V rms; out of phase

7. 600 V

9. 0.25 (4:1)

11. 60 V

13. (a) 10 V   (b) 240 V

15. (a) 25 mA   (b) 50 mA   (c) 15 V   (d) 750 mW

17. 1.83

19. 9.76 W

21. 94.5 W

23. 0.98

25. 25 kVA

27. V_1 = 11.5 V, V_2 = 23.0 V, V_3 = 23.0 V, V_4 − 46.0 V

29. (a) 48 V   (b) 25 V

31. (a) V_{RL} = 35 V, I_{RL} = 2.92 A, V_C = 15 V, I_C = 1.5 A
(b) 34.5 Ω

33. Excessive primary current is drawn, potentially burning
out the source and/or the transformer unless the primary is
protected by a fuse.

35. Turns ratio 0.5

37. R_2 open

Chapter 15

1. Magnitude, angle

3. See Figure P-18.

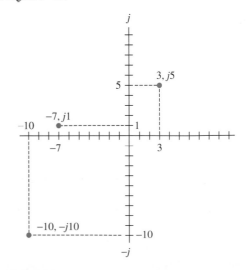

▲ FIGURE P-18

5. (a) $-5, +j3$ and $5, -j3$ (b) $-1, -j7$ and $1, +j7$

   (c) $-10, +j10$ and $10, -j10$

7. 18.0

9. (a) $643 - j766$ (b) $-14.1 + j5.13$

   (c) $-17.7 - j17.7$ (d) $-3 + j0$

11. (a) Fourth   (b) Fourth

    (c) Fourth   (d) First

13. (a) $12\angle115°$ (b) $20\angle230°$

    (c) $100\angle190°$ (d) $50\angle160°$

15. (a) $1.1 + j0.7$ (b) $-81 - j35$

    (c) $5.28 - j5.27$ (d) $-50.4 + j62.5$

17. (a) $3.2\angle11°$ (b) $7\angle-101°$

    (c) $1.52\angle70.6°$ (d) $2.79\angle-63.5°$

19. 8 kHz, 8 kHz

21. (a) $270\ \Omega - j100\ \Omega, 288\angle-20.3°\ \Omega$

    (b) $680\ \Omega - j1000\ \Omega, 1.21\angle-55.8°\ k\Omega$

23. (a) $56\ k\Omega - j723\ k\Omega$

    (b) $56\ k\Omega - j145\ k\Omega$

    (c) $56\ k\Omega - j72.3\ k\Omega$

    (d) $56\ k\Omega - j28.9\ k\Omega$

25. (a) $R = 33\ \Omega, X_C = 50\ \Omega$

    (b) $R = 272\ \Omega, X_C = 127\ \Omega$

    (c) $R = 698\ \Omega, X_C = 1.66\ k\Omega$

    (d) $R = 558\ \Omega, X_C = 558\ \Omega$

27. (a) $183\angle57.5°\ \mu A$

    (b) $611\angle40.3°\ \mu A$

    (c) $1.98\angle76.2°\ mA$

29. $-14.5°$

31. (a) $97.3\angle-54.9°\ \Omega$

    (b) $103\angle54.9°\ mA$

    (c) $5.76\angle54.9°\ V$

    (d) $8.18\angle-35.1°\ V$

33. $R_X = 12\ \Omega, C_X = 13.3\ \mu F$ in series.

35.
0 Hz	1 V
1 kHz	723 mV
2 kHz	464 mV
3 kHz	329 mV
4 kHz	253 mV
5 kHz	205 mV
6 kHz	172 mV
7 kHz	148 mV
8 kHz	130 mV
9 kHz	115 mV
10 kHz	104 mV

37.
0 Hz	0 V
1 kHz	5.32 V
2 kHz	7.82 V
3 kHz	8.83 V
4 kHz	9.29 V
5 kHz	9.53 V
6 kHz	9.66 V
7 kHz	9.76 V
8 kHz	9.80 V
9 kHz	9.84 V
10 kHz	9.87 V

39. See Figure P–19.

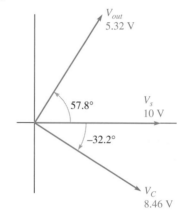

▲ FIGURE P–19

41. $245\ \Omega, -80.5°$

43. $\mathbf{V}_C = \mathbf{V}_R = 10\angle0°\ V$

    $\mathbf{I}_{tot} = 184\angle37.1°\ mA$

    $\mathbf{I}_R = 147\angle0°\ mA$

    $\mathbf{I}_C = 111\angle90°\ mA$

45. (a) $6.59\angle-48.8°\ \Omega$ (b) $10\angle0°\ mA$

    (c) $11.4\angle90°\ mA$ (d) $15.2\angle48.8°\ mA$

    (e) $-48.8°\ (I_{tot}$ leading $V_s)$

47. $18.4\ k\Omega$ resistor in series with 196 pF capacitor.

49. $\mathbf{V}_{C1} = 8.42\angle-2.9°\ V, \mathbf{V}_{C2} = 1.58\angle-57.5°\ V$

    $\mathbf{V}_{C3} = 3.65\angle6.8°\ V, \mathbf{V}_{R1} = 3.29\angle32.5°\ V$

    $\mathbf{V}_{R2} = 2.36\angle6.8°\ V, \mathbf{V}_{R3} = 1.29\angle6.8°\ V$

51. $\mathbf{I}_{tot} = 79.5\angle87.1°\ mA, \mathbf{I}_{C2R1} = 6.99\angle32.5°\ mA$

    $\mathbf{I}_{C3} = 75.7\angle96.8°\ mA, \mathbf{I}_{R2R3} = 7.16\angle6.8°\ mA$

53. $0.103\ \mu F$

55. $\mathbf{I}_{C1} = \mathbf{I}_{R1} = 2.27\angle74.5°\ mA$

    $\mathbf{I}_{R2} = 2.04\angle72.0°\ mA$

    $\mathbf{I}_{R3} = 246\angle84.3°\ \mu A$

    $\mathbf{I}_{R4} = 149\angle41.2°\ \mu A$

    $\mathbf{I}_{R5} = 180\angle75.1°\ \mu A$

    $\mathbf{I}_{R6} = \mathbf{I}_{C3} = 101\angle135°\ \mu A$

    $\mathbf{I}_{C2} = 101\angle131°\ \mu A$

57. 4.03 VA

59. 0.914

61. (a) $I_{LA} = 4.8$ A, $I_{LB} = 3.33$ A

(b) $P_{rA} = 606$ VAR, $P_{rB} = 250$ VAR

(c) $P_{trueA} = 979$ W, $P_{trueB} = 759$ W

(d) $P_{aA} = 1151$ VA, $P_{aB} = 799$ VA

(e) Load A

63. 0.0796 μF

65. Reduces V_{out} to 2.83 V and θ to $-56.7°$

67. (a) No output voltage   (b) $320\angle-71.3°$ mV

(c) $500\angle0°$ mV   (d) 0 V

69. No fault

71. R_1 open

73. No fault

75. 48.4 Hz

Chapter 16

1. 15 kHz

3. (a) $100\ \Omega + j50\ \Omega$; $112\angle26.6°\ \Omega$

(b) $1.5\ k\Omega + j1\ k\Omega$; $1.80\angle33.7°\ k\Omega$

5. (a) $17.4\angle46.4°\ \Omega$ (b) $64.0\angle79.2°\ \Omega$

(c) $127\angle84.6°\ \Omega$ (d) $251\angle87.3°\ \Omega$

7. $806\ \Omega$, 4.11 mH

9. 0.370 V

11. (a) $43.5\angle-55°$ mA   (b) $11.8\angle-34.6°$ mA

13. θ increases from 38.7° to 58.1°.

15. (a) $\mathbf{V}_R = 4.85\angle-14.1°$ V

$\mathbf{V}_L = 1.22\angle75.9°$ V

(b) $\mathbf{V}_R = 3.83\angle-40.0°$ V

$\mathbf{V}_L = 3.21\angle50.0°$ V

(c) $\mathbf{V}_R = 2.16\angle-64.5°$ V

$\mathbf{V}_L = 4.51\angle25.5°$ V

(d) $\mathbf{V}_R = 1.16\angle-76.6°$ V

$\mathbf{V}_L = 4.86\angle13.4°$ V

17. (a) $-0.0923°$ (b) $-9.15°$ (c) $-58.2°$ (d) $-86.4°$

19. $7.75\angle49.9°\ \Omega$

21. 2.39 kHz

23. (a) $274\angle60.7°\ \Omega$ (b) $89.3\angle0°$ mA

(c) $159\angle-90°$ mA   (d) $182\angle-60.7°$ mA

(e) 60.7° (I_{tot} lagging V_s)

25. $1.83\ k\Omega$ resistor in series with $4.21\ k\Omega$ inductive reactance

27. $\mathbf{V}_{R1} = 21.8\angle-3.89°$ V

$\mathbf{V}_{R2} = 7.27\angle9.61°$ V

$\mathbf{V}_{R3} = 3.38\angle-53.3°$ V

$\mathbf{V}_{L1} = \mathbf{V}_{L2} = 6.44\angle37.3°$ V

29. $\mathbf{I}_{R1} = I_T = 389\angle-3.89°$ mA

$\mathbf{I}_{R2} = 330\angle9.61°$ mA

$\mathbf{I}_{R3} = 102\angle-53.3°$ mA

$\mathbf{I}_{L1} = \mathbf{I}_{L2} = 51.3\angle-52.7°$ mA

31. (a) $588\angle-50.5°$ mA

(b) $22.0\angle16.1°$ V

(c) $8.63\angle-135°$ V

33. $\theta = 52.5°$ (V_{out} lags V_{in}), 0.143

35. See Figure P–20.

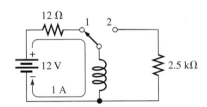

▲ FIGURE P–20

37. 1.29 W, 1.04 VAR

39. $P_{true} = 290$ mW; $P_r = 50.8$ mVAR;

$P_a = 296$ mVA; $PF = 0.985$

41. Use the formula, $V_{out} = \left(\dfrac{R}{Z_{tot}}\right)V_{in}$. See Figure P–21.

FREQUENCY (kHz)	X_L	Z_{tot}	V_{out}
0	$0\ \Omega$	$39.0\ \Omega$	1 V
1	$62.8\ \Omega$	$73.9\ \Omega$	528 mV
2	$126\ \Omega$	$132\ \Omega$	296 mV
3	$189\ \Omega$	$193\ \Omega$	203 mV
4	$251\ \Omega$	$254\ \Omega$	153 mV
5	$314\ \Omega$	$317\ \Omega$	123 mV

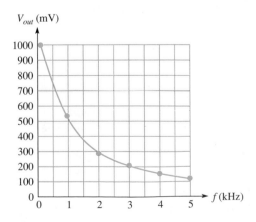

▲ FIGURE P–21

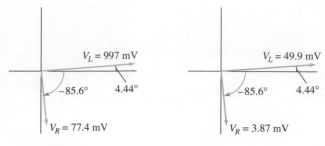

$V_L = 997$ mV $V_L = 49.9$ mV

$-85.6°$ $4.44°$ $-85.6°$ $4.44°$

$V_R = 77.4$ mV $V_R = 3.87$ mV

▲ FIGURE P–22

43. See Figure P–22.

45. (a) 0 V   (b) 0 V

    (c) $1.62\angle-25.8°$ V   (d) $2.15\angle-64.5°$ V

47. L_1 leaky

49. L_1 open

51. No fault

53. $f_c \approx 53.214$ kHz

Chapter 17

1. $520\angle-88.9°\ \Omega$; $520\ \Omega$ capacitive

3. Impedance increases to $150\ \Omega$

5. $\mathbf{I}_{tot} = 61.4\angle-43.8°$ mA

   $\mathbf{V}_R = 2.89\angle-43.8°$ V

   $\mathbf{V}_L = 4.91\angle46.2°$ V

   $\mathbf{V}_C = 2.15\angle-134°$ V

7. (a) $35.8\angle65.1°$ mA

   (b) 181 mW

   (c) 390 mVAR

   (d) 430 mVA

9. 12 V

11. $Z = 200\ \Omega$, $X_C = X_L = 2$ kΩ

13. 500 mA

15. See Figure P–23.

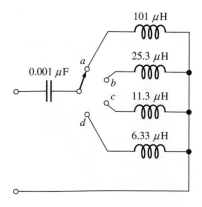

101 μH

0.001 μF

25.3 μH

a

b

c   11.3 μH

d

6.33 μH

▲ FIGURE P–23

17. The phase angle of $-4.43°$ indicates a slightly capacitive circuit.

19. $\mathbf{I}_R = 50\angle0°$ mA

    $\mathbf{I}_L = 4.42\angle-90°$ mA

    $\mathbf{I}_C = 8.29\angle90°$ mA

    $\mathbf{I}_{tot} = 50.2\angle4.43°$ mA

    $\mathbf{V}_R = \mathbf{V}_L = \mathbf{V}_C = 5\angle0°$ V

21. $\mathbf{I}_R = 50\angle0°$ mA, $\mathbf{I}_L = 531\angle-90°\ \mu$A,

    $\mathbf{I}_C = 69.1\angle90°\ \mu$A, $\mathbf{I}_{tot} = 84.9\angle53.9°$ mA

23. 53.5 MΩ, 104 kHz

25. $P_r = 0$ VAR, $P_a = 7.45\ \mu$VA, $P_{true} = 538$ mW

27. (a) $-1.97°$ (V_s lags I_{tot})

    (b) $23.0°$ (V_s leads I_{tot})

29. 49.1 kΩ resistor in series with 1.38 H inductor

31. $45.2°$ (I_2 leads V_s)

33. $\mathbf{I}_{R1} = \mathbf{I}_{C1} = 1.09\angle-25.7°$ mA

    $\mathbf{I}_{R2} = 767\angle19.3°\ \mu$A

    $\mathbf{I}_{C2} = 767\angle109.3°\ \mu$A

    $\mathbf{I}_L = 1.53\angle-70.7°$ mA

    $\mathbf{V}_{R2} = \mathbf{V}_{C2} = \mathbf{V}_L = 7.67\angle19.3°$ V

    $\mathbf{V}_{R1} = 3.60\angle-25.7°$ V

    $\mathbf{V}_{C1} = 1.09\angle-116°$ V

35. $52.2\angle126°$ mA

37. $f_{r(series)} = 4.11$ kHz

    $\mathbf{V}_{out} = 4.83\angle-61.0°$ V

    $f_{r(parallel)} = 2.6$ kHz

    $\mathbf{V}_{out} \cong 10\angle0°$ V

39. 62.5 Hz

41. 1.38 W

43. 200 Hz

45. C_1 leaky

47. C_1 leaky

49. No fault

51. $f_c \approx 338.698$ kHz

Chapter 18

1. $2.22\angle-77.2°$ V rms

3. (a) $9.36\angle-20.7°$ V

   (b) $7.18\angle-44.1°$ V

   (c) $9.96\angle-5.44°$ V

   (d) $9.95\angle-5.74°$ V

5. (a) $12.1\ \mu$F   (b) $1.45\ \mu$F

   (c) $0.723\ \mu$F   (d) $0.144\ \mu$F

7. See Figure P–24.

► FIGURE P–24

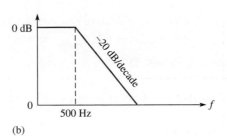

(a)　　　　　　　　　　　　(b)

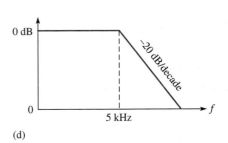

(c)　　　　　　　　　　　　(d)

9. (a) 7.13 V　(b) 5.67 V

(c) 4.01 V　(d) 0.800 V

11. 9.75∠12.8° V

13. (a) 3.53∠69.3° V

(b) 4.85∠61.0° V

(c) 947∠84.6° mV

(d) 995∠84.3° mV

15. See Figure P–25.

17. (a) 14.5 kHz

(b) 24.0 kHz

19. (a) 15.06 kHz, 13.94 kHz

(b) 25.3 kHz, 22.7 kHz

21. (a) 117 V　(b) 115 V

23. $C = 0.064\,\mu F$, $L = 989\,\mu H$, $f_r = 20$ kHz

25. (a) 86.3 Hz

(b) 7.34 MHz

27. $L_1 = 0.08\,\mu H$,

$L_2 = 0.554\,\mu H$

29. C_2 leaky

31. C_1 shorted

33. No fault

35. $BW \approx 88.93$ MHz

Chapter 19

1. 1.22∠28.6° mA

3. 81.0∠−11.9° mA

► FIGURE P–25

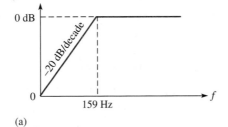

(a)　　　　　　　　　　　　(b)

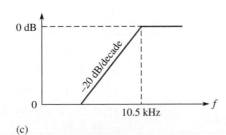

(c)　　　　　　　　　　　　(d)

5. $V_{A(dc)} = 0$ V, $V_{B(dc)} = 16.1$ V, $V_{C(dc)} = 15.1$ V,
$V_{D(dc)} = 0$ V, $V_{A(peak)} = 9$ V, $V_{B(peak)} = 5.96$ V,
$V_{C(peak)} = V_{D(peak)} = 4.96$ V

7. (a) $\mathbf{V}_{th} = 15\angle-53.1°$ V
  $\mathbf{Z}_{th} = 63\ \Omega - j48\ \Omega = 79.2\angle-37.3°\ \Omega$

 (b) $\mathbf{V}_{th} = 1.22\angle0°$ V
  $\mathbf{Z}_{th} = j237\ \Omega = 237\angle90°\ \Omega$

 (c) $\mathbf{V}_{th} = 12.1\angle11.9°$ V
  $\mathbf{Z}_{th} = 50\ k\Omega - j20\ k\Omega = 53.9\angle-21.8°\ k\Omega$

9. $16.9\angle88.2°$ V

11. (a) $\mathbf{I}_n = 189\angle-15.8°$ mA
  $\mathbf{Z}_n = 63\ \Omega - j48\ \Omega$

 (b) $\mathbf{I}_n = 5.15\angle-90°$ mA
  $\mathbf{Z}_n = j237\ \Omega$

 (c) $\mathbf{I}_n = 224\angle33.7°\ \mu A$
  $\mathbf{Z}_n = 50\ k\Omega - j20\ k\Omega$

13. $16.8\angle88.5°$ V

15. $9.18\ \Omega + j2.90\ \Omega$

17. $95.2\ \Omega + j42.7\ \Omega$

19. C_2 leaky

21. No fault

23. $\mathbf{I}_n = 30.142\angle-113.1°\ \mu A$
  $\mathbf{Z}_n = 30.3\angle64.28°\ k\Omega$

Chapter 20

1. $103\ \mu s$

3. 12.6 V

5. See Figure P–26.

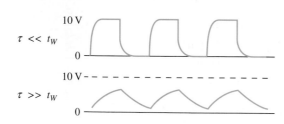

▲ FIGURE P–26

7. (a) 23.5 ms
 (b) See Figure P–27.

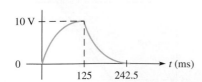

▲ FIGURE P–27

9. See Figure P–28.

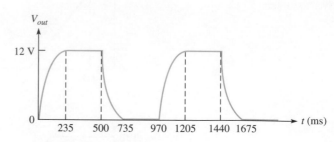

▲ FIGURE P–28

11. See Figure P–29.

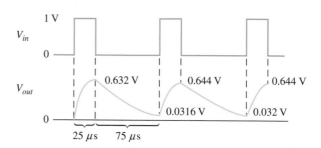

▲ FIGURE P–29

13. See Figure P–30.

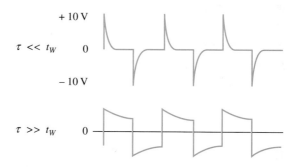

▲ FIGURE P–30

15. (a) 493.5 ns
 (b) See Figure P–31.

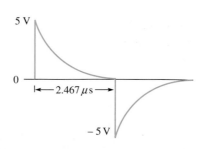

▲ FIGURE P–31

17. An approximate square wave with an average value of zero.

19. See Figure P–32.

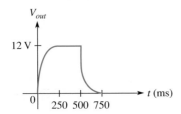

▲ **FIGURE P–32**

21. (a) 4.55 μs

    (b) See Figure P–33.

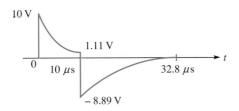

▲ **FIGURE P–33**

23. 15.9 kHz

25. (a) Capacitor open or R shorted.

    (b) C leaky or $R > 3.3 \text{ k}\Omega$ or $C > 0.22 \ \mu\text{F}$

    (c) Resistor open or capacitor shorted

27. C_1 open or R_1 shorted

29. R_1 or R_2 open

Chapter 21

1. 17.5°

3. 376 mA

5. 1.32∠121° A

7. $\mathbf{I}_{La} = 8.66\angle-30°$ A

    $\mathbf{I}_{Lb} = 8.66\angle90°$ A

    $\mathbf{I}_{Le} = 8.66\angle-150°$ A

9. (a) $\mathbf{V}_{L(ab)} = 866\angle-30°$ V

       $\mathbf{V}_{L(ca)} = 866\angle-150°$ V

       $\mathbf{V}_{L(bc)} = 866\angle90°$ V

    (b) $\mathbf{I}_{\theta a} = 500\angle-32°$ mA

       $\mathbf{I}_{\theta b} = 500\angle88°$ mA

       $\mathbf{I}_{\theta c} = 500\angle-152°$ mA

    (c) $\mathbf{I}_{La} = 500\angle-32°$ mA

       $\mathbf{I}_{Lb} = 500\angle88°$ mA

       $\mathbf{I}_{Lc} = 500\angle-152°$ mA

    (d) $\mathbf{I}_{Za} = 500\angle-32°$ mA

       $\mathbf{I}_{Zb} = 500\angle88°$ mA

       $\mathbf{I}_{Zc} = 500\angle-152°$ mA

    (e) $\mathbf{V}_{Za} = 500\angle0°$ V

       $\mathbf{V}_{Zb} = 500\angle120°$ V

       $\mathbf{V}_{Zc} = 500\angle-120°$ V

11. (a) $\mathbf{V}_{L(ab)} = 86.6\angle-30°$ V

       $\mathbf{V}_{L(ca)} = 86.6\angle-150°$ V

       $\mathbf{V}_{L(bc)} = 86.6\angle90°$ V

    (b) $\mathbf{I}_{\theta a} = 250\angle110°$ mA

       $\mathbf{I}_{\theta b} = 250\angle-130°$ mA

       $\mathbf{I}_{\theta c} = 250\angle-10°$ mA

    (c) $\mathbf{I}_{La} = 250\angle110°$ mA

       $\mathbf{I}_{Lb} = 250\angle-130°$ mA

       $\mathbf{I}_{Lc} = 250\angle-10°$ mA

    (d) $\mathbf{I}_{Za} = 144\angle140°$ mA

       $\mathbf{I}_{Zb} = 144\angle20°$ mA

       $\mathbf{I}_{Zc} = 144\angle-100°$ mA

    (e) $\mathbf{V}_{Za} = 86.6\angle-150°$ V

       $\mathbf{V}_{Zb} = 86.6\angle90°$ V

       $\mathbf{V}_{Zc} = 86.6\angle-30°$ V

13. $\mathbf{V}_{L(ab)} = 330\angle-120°$ V

    $\mathbf{V}_{L(ca)} = 330\angle120°$ V

    $\mathbf{V}_{L(bc)} = 330\angle0°$ V

    $\mathbf{I}_{Za} = 38.2\angle-150°$ A

    $\mathbf{I}_{Zb} = 38.2\angle-30°$ A

    $\mathbf{I}_{Zc} = 38.2\angle90°$ A

15. Figure 21–37:   636 W

    Figure 21–38:   149 W

    Figure 21–39:   12.8 W

    Figure 21–40:   2.78 kW

    Figure 21–41:   10.9 kW

17. 24.2 W

GLOSSARY

accuracy The degree to which a measured value represents the true or accepted value of a quantity.

admittance (Y) A measure of the ability of a reactive circuit to permit current; the reciprocal of impedance. The unit is the siemens (S).

ammeter An electrical instrument used to measure current.

ampere (A) The unit of electrical current.

ampere-hour (Ah) rating A number given in ampere-hours determined by multiplying the current (A) times the length of time (h) a battery can deliver that current to a load.

ampere-turn The current in a single loop (turn) of wire.

amplitude The maximum value of a voltage or current.

angular velocity The rotational rate of a phasor which is related to the frequency of the sine wave that the phasor represents.

apparent power The phasor combination of resistive power (true power) and reactive power. The unit is the volt-ampere (VA).

apparent power rating The method of rating transformers in which the power capability is expressed in volt-amperes (VA).

atom The smallest particle of an element possessing the unique characteristics of that element.

atomic number The number of protons in a nucleus.

attenuation A reduction of the output signal compared to the input signal, resulting in a ratio with a value of less than 1 for the output voltage to the input voltage of a circuit.

autotransformer A transformer in which the primary and secondary are in a single winding.

average value The average of a sine wave over one half-cycle. It is 0.637 times the peak value.

AWG American wire gauge; a standardization based on wire diameter.

balanced bridge A bridge circuit that is in the balanced state as indicated by 0 V across the output.

balanced load A condition where all the load currents are equal and the neutral current is zero.

band-pass filter A filter that passes a range of frequencies lying between two critical frequencies and rejects frequencies above and below that range.

band-stop filter A filter that rejects a range of frequencies lying between two critical frequencies and passes frequencies above and below that range.

bandwidth The range of frequencies for which the current (or output voltage) is equal to or greater than 70.7% of its value at the resonant frequency that is considered to be passed by a filter.

baseline The normal level of a pulse waveform; the voltage level in the absence of a pulse.

battery An energy source that uses a chemical reaction to convert chemical energy into electrical energy.

bias The application of a dc voltage to an electronic device to produce a desired mode of operation.

bleeder current The current left after the total load current is subtracted from the total current into the circuit.

Bode plot The graph of a filter's frequency response showing the change in the output voltage to input voltage ratio expressed in dB as a function of frequency for a constant input voltage.

branch One current path in a parallel circuit; a current path that connects two nodes.

branch current The actual current in a branch.

capacitance The ability of a capacitor to store electrical charge.

capacitive reactance The opposition of a capacitor to sinusoidal current. The unit is the ohm (Ω).

capacitive susceptance (B_C) The ability of a capacitor to permit current; the reciprocal of capacitive reactance. The unit is the siemens (S).

capacitor An electrical device consisting of two conductive plates separated by an insulating material and possessing the property of capacitance.

center frequency (f_0) The resonant frequency of a bandpass or band-stop filter.

center tap (CT) A connection at the midpoint of a winding in a transformer.

charge An electrical property of matter that exists because of an excess or a deficiency of electrons. Charge can be either positive or negative.

choke A type of inductor used to block or choke off high frequencies.

circuit An interconnection of electrical components designed to produce a desired result. A basic circuit consists of a source, a load, and an interconnecting current path.

circuit breaker A resettable protective device used for interrupting excessive current in an electric circuit.

circular mil (CM) A unit of the cross-sectional area of a wire.

closed circuit A circuit with a complete current path.

coefficient The constant number that appears in front of a variable.

coefficient of coupling (k) A constant associated with transformers that is the ratio of secondary magnetic flux to primary magnetic flux. The ideal value of 1 indicates that all the flux in the primary winding is coupled into the secondary winding.

common Reference ground.

complex conjugate A complex number having the same real part and an oppositely signed imaginary part; an impedance containing the same resistance and a reactance opposite in phase but equal in magnitude to that of a given impedance.

complex plane An area consisting of four quadrants on which a quantity containing both magnitude and direction can be represented.

conductance (G) The ability of a circuit to allow current; the reciprocal of resistance. The unit is the siemens (S).

conductor A material in which electric current is easily established. An example is copper.

core The physical structure around which the winding of an inductor is formed. The core material influences the electromagnetic characteristics of the inductor.

coulomb (C) The unit of electrical charge; the total charge possessed by 6.25×10^{18} electrons.

Coulomb's law A law that states a force exists between two charged bodies that is directly proportional to the product of the two charges and inversely proportional to the square of the distance between them.

critical frequency (f_c) The frequency at which a filter's output voltage is 70.7% of the maximum.

current The rate of flow of charge (electrons).

current divider A parallel circuit in which the currents divide inversely proportional to the parallel branch resistances.

current source A device that provides a constant current for a varying load.

cutoff frequency (f_c) The frequency at which the output voltage of a filter is 70.7% of the maximum output voltage; another term for critical frequency.

cycle One repetition of a periodic waveform.

DC component The average value of a pulse waveform.

decade A tenfold change in frequency or other parameter.

decibel A logarithmic measurement of the ratio of one power to another or one voltage to another, which can be used to express the input-to-output relationship of a filter.

degree The unit of angular measure corresponding to 1/360 of a complete revolution.

determinant The solution of a matrix consisting of an array of coefficients and constants for a set of simultaneous equations.

dielectric The insulating material between the plates of a capacitor.

dielectric constant A measure of the ability of a dielectric material to establish an electric field.

dielectric strength A measure of the ability of a dielectric material to withstand voltage without breaking down.

differentiator A circuit producing an output that approaches the mathematical derivative of the input.

digital multimeter An electronic instrument that combines meters for the measurement of voltage, current, and resistance.

DMM Digital multimeter; an electronic instrument that combines meters for measurement of voltage, current, and resistance.

duty cycle A characteristic of a pulse waveform that indicates the percentage of time that a pulse is present during a cycle; the ratio of pulse width to period, expressed as either a fraction or as a percentage.

effective value A measure of the heating effect of a sine wave; also known as the rms (root mean square) value.

efficiency The ratio of the output power delivered to a load to the input power to a circuit, usually expressed as a percentage.

electrical Related to the use of electrical voltage and current to achieve desired results.

electrical isolation The condition that exists when two coils are magnetically linked but have no electrical connection between them.

electrical shock The physical sensation resulting from electrical current through the body.

electromagnetic field A formation of a group of magnetic lines of force surrounding a conductor created by electrical current in the conductor.

electromagnetism The production of a magnetic field by current in a conductor.

electromagnetic induction The phenomenon or process by which a voltage is produced in a conductor when there is relative motion between the conductor and a magnetic or electromagnetic field.

electron A basic particle of electrical charge in matter. The electron possesses negative charge.

electronic Related to the movement and control of free electrons in semiconductors or vacuum devices.

electronic power supply A voltage source that converts the ac voltage from a wall outlet to a constant (dc) voltage at a level suitable for electronic components.

element One of the unique substances that make up the known universe. Each element is characterized by a unique atomic structure.

energy The ability to do work.

engineering notation A system for representing any number as a one-, two-, or three-digit number times a power of ten with an exponent that is a multiple of 3.

equivalent circuit A circuit that produces the same voltage and current to a given load as the original circuit that it replaces.

exponent The number to which a base number is raised.

falling edge The negative-going transition of a pulse.

fall time (t_f) The time interval required for a pulse to change from 90% to 10% of its amplitude.

farad (F) The unit of capacitance.

Faraday's law A law stating that the voltage induced across a coil of wire equals the number of turns in the coil times the rate of change of the magnetic flux.

field winding The winding on the rotor of an ac generator.

filter A type of circuit that passes certain frequencies and rejects all others.

free electron A valence electron that has broken away from its parent atom and is free to move from atom to atom within the atomic structure of a material.

frequency A measure of the rate of change of a periodic function; the number of cycles completed in 1 s. The unit of frequency is the hertz.

frequency response In electric circuits, the variation in the output voltage (or current) over a specified range of frequencies.

function generator An electronic instrument that produces electrical signals in the form of sine waves, triangular waves, and pulses.

fundamental frequency The repetition rate of a waveform.

fuse A protective device that burns open when there is excessive current in a circuit.

gauss (G) A CGS unit of flux density.

generator An energy source that produces electrical signals.

ground In electric circuits, the common or reference point.

half-power frequency The frequency at which the output power of a resonant circuit is 50% of the maximum (the output voltage is 70.7% of maximum); another name for critical or cutoff frequency.

half-splitting A troubleshooting procedure where one starts in the middle of a circuit or system and, depending on the first measurement, works toward the output or toward the input to find the fault.

harmonics The frequencies contained in a composite waveform, which are integer multiples of the repetition frequency (fundamental).

henry (H) The unit of inductance.

hertz (Hz) The unit of frequency. One hertz equals one cycle per second.

high-pass filter A type of filter that passes all frequencies above a critical frequency and rejects all frequencies below that critical frequency.

hysteresis A characteristic of a magnetic material whereby a change in magnetization lags the application of magnetic field intensity.

imaginary number A number that exists on the vertical axis of the complex plane.

impedance The total opposition to sinusoidal current expressed in ohms.

impedance matching A technique used to match a load resistance to a source resistance in order to achieve maximum transfer of power.

induced current (i_{ind}) A current induced in a conductor when the conductor moves through a magnetic field.

induced voltage (v_{ind}) Voltage produced as a result of a changing magnetic field.

inductance The property of an inductor whereby a change in current causes the inductor to produce a voltage that opposes the change in current.

inductive reactance The opposition of an inductor to sinusoidal current. The unit is the ohm (Ω).

inductive susceptance The ability of an inductor to permit current; the reciprocal of inductive reactance. The unit is the siemens (S).

inductor An electrical device formed by a wire wound around a core having the property of inductance; also known as *coil*.

instantaneous power The value of power in a circuit at any given instant of time.

instantaneous value The voltage or current value of a waveform at a given instant in time.

insulator A material that does not allow current under normal conditions.

integrator A circuit producing an output that approaches the mathematical integral of the input.

ion An atom that has a net positive or negative charge.

joule (J) The SI unit of energy.

junction A point at which two or more components are connected.

kilowatt-hour (kWh) A large unit of energy used mainly by utility companies.

Kirchhoff's current law A law stating that the total current into a node equals the total current out of the node. Equivalently, the algebraic sum of all the currents entering and leaving a node is zero.

Kirchhoff's voltage law A law stating that (1) the sum of the voltage drops around a single closed path equals the source voltage in that loop or (2) the algebraic sum of all the voltages (drops and sources) around a single closed path is zero.

lag Refers to a condition of the phase or time relationship of waveforms in which one waveform is behind the other in phase or time.

lead Refers to a condition of the phase or time relationship of waveforms in which one waveform is ahead of the other in phase or time; also, a wire or cable connection to a device or instrument.

leading edge The first step or transition of a pulse.

Lenz's law A law that states when the current through a coil changes, the polarity of the induced voltage created by the changing magnetic field is such that it always opposes the change in current that caused it. The current cannot change instantaneously.

linear Characterized by a straight-line relationship.

line current The current through a line feeding a load.

lines of force Magnetic flux lines in a magnetic field radiating from the north pole to the south pole.

line voltage The voltage between lines feeding a load.

load An element (resistor or other component) connected across the output terminals of a circuit that draws current from the source; an element in a circuit upon which work is done.

loop A closed current path in a circuit.

loop current A current assigned to a circuit purely for the purpose of mathematical analysis and not normally representing the actual physical current.

low-pass filter A type of filter that passes all frequencies below a critical frequency and rejects all frequencies above that critical frequency.

magnetic coupling The magnetic connection between two coils as a result of the changing magnetic flux lines of one coil cutting through the second coil.

magnetic field A force field radiating from the north pole to the south pole of a magnet.

magnetic field intensity The amount of mmf per unit length of magnetic material; also called *magnetizing force*.

magnetic flux The lines of force between the north and south poles of a permanent magnet or an electromagnet.

magnetic flux density The amount of flux per unit area perpendicular to the magnetic field.

magnetomotive force (mmf) The cause of a magnetic field, measured in ampere-turns.

magnitude The value of a quantity, such as the number of volts of voltage or the number of amperes of current.

matrix An array of numbers.

maximum power transfer A transfer of maximum power from a source to a load when the load resistance equals the internal source resistance.

metric prefix An affix that represents a power-of-ten number expressed in engineering notation.

multimeter An instrument that measures voltage, current, and resistance.

mutual inductance The inductance between two separate coils, such as in a transformer.

neutron An atomic particle having no electrical charge.

node A point in a circuit where two or more components are connected; also known as a *junction*.

Norton's theorem A method for simplifying a two-terminal linear circuit to an equivalent circuit with only a current source in parallel with a resistance or impedance.

nucleus The central part of an atom containing protons and neutrons.

Ohm (Ω) The unit of resistance.

Ohmmeter An instrument for measuring resistance.

Ohm's law A law stating that current is directly proportional to voltage and inversely proportional to resistance.

open A circuit condition in which there is not a complete current path.

open circuit A circuit in which there is not a complete current path.

oscillator An electronic circuit that produces a time-varying signal without an external input signal using positive feedback.

oscilloscope A measurement instrument that displays signal waveforms on a screen.

parallel The relationship in electric circuits in which two or more current paths are connected between two separate nodes.

parallel resonance A condition in a parallel *RLC* circuit in which the reactances ideally are equal and the impedance is maximum.

passband The range of frequencies passed by a filter.

peak-to-peak value The voltage or current value of a waveform measured from its minimum to its maximum points.

peak value The voltage or current value of a waveform at its maximum positive or negative points.

period (T) The time interval of one complete cycle of a periodic waveform.

periodic Characterized by a repetition at fixed-time intervals.

permeability The measure of ease with which a magnetic field can be established in a material.

phase The relative angular displacement of a time-varying quantity with respect to a given reference.

phase current (I_0) The current through a generator winding.

phase voltage (V_0) The voltage across a generator winding.

phasor A representation of a sine wave in terms of its magnitude (amplitude) and direction (phase angle).

photoconductive cell A type of variable resistor that is light-sensitive.

photovoltaic effect The process whereby light energy is converted directly into electrical energy.

piezoelectric effect The property of a crystal whereby a changing mechanical stress produces a voltage across the crystal.

polar form One form of a complex number made up of a magnitude and an angle.

potentiometer A three-terminal variable resistor.

power The rate of energy usage.

power factor The relationship between volt-amperes and true power or watts. Volt-amperes multiplied by the power factor equals true power.

power of ten A numerical representation consisting of a base of 10 and an exponent; the number 10 raised to a power.

power rating The maximum amount of power that a resistor can dissipate without being damaged by excessive heat buildup.

power supply A device that provides power to a load.

primary winding The input winding of a transformer; also called *primary*.

proton A positively charged atomic particle.

pulse A type of waveform that consists of two equal and opposite steps in voltage or current separated by a time interval.

pulse repetition frequency The fundamental frequency of a repetitive pulse waveform; the rate at which the pulses repeat expressed in either hertz or pulses per second.

pulse width (t_W) For a nonideal pulse, the time between the 50% points of the leading and trailing edges; the time interval between the opposite steps of an ideal pulse.

quality factor (Q) The ratio of true power to reactive power in a resonant circuit or the ratio of inductive reactance to winding resistance in an inductor.

radian A unit of angular measurement. There are 2π radians in one complete 360° revolution. One radian equals 57.3°.

ramp A type of waveform characterized by a linear increase or decrease in voltage or current.

RC time constant A fixed time interval set by the values of R and C that determines the time response of a series *RC* circuit. It equals the product of the resistance and the capacitance.

reactive power The rate at which energy is alternately stored and returned to the source by a capacitor or inductor. The unit is the VAR.

real number A number that exists on the horizontal axis of the complex plane.

rectangular form One form of a complex number made up of a real part and an imaginary part.

rectifier An electronic circuit that converts ac into pulsating dc; one part of a power supply.

reference ground A method of grounding whereby a large conducive area on a printed circuit board or the metal chassis that houses the assembly is used as the common or reference point.

reflected load The load as it appears to the source in the primary of a transformer.

reflected resistance The resistance in the secondary circuit reflected into the primary circuit.

relay An electromagnetically controlled mechanical device in which electrical contacts are opened or closed by a magnetizing current.

reluctance The opposition to the establishment of a magnetic field in a material.

resistance Opposition to current. The unit is the ohm (Ω).

resistor An electrical component specifically designed to have a certain amount of resistance.

resolution The smallest increment of a quantity that a DMM can measure.

resonance A condition in a series *RLC* circuit in which the capacitive and inductive reactances are equal in magnitude; thus, they cancel each other and result in a purely resistive impedance.

resonant frequency The frequency at which resonance occurs; also known as *center frequency.*

retentivity The ability of a material, once magnetized, to maintain a magnetized state without the presence of a magnetizing force.

rheostat A two-terminal variable resistor.

ripple voltage The variation in the dc voltage on the output of a filtered rectifier caused by the slight charging and discharging action of the filter capacitor.

rise time (t_r) The time interval required for a pulse to change from 10% to 90% of its amplitude.

rising edge The positive-going transition of a pulse.

RL time constant A fixed time interval set by the values of R and L that determines the time response of a circuit and is equal to L/R.

roll-off The rate of decrease of a filter's frequency response.

rms value The value of a sinusoidal voltage that indicates its heating effect, also known as the effective value. It is equal to 0.707 times the peak value. *rms* stands for root mean square.

rotor The rotating assembly in a generator or motor.

sawtooth waveform A type of electrical waveform composed of ramps; a special case of a triangular waveform in which one ramp is much shorter than the other.

schematic A symbolized diagram of an electrical or electronic circuit.

scientific notation A system for representing any number as a number between 1 and 10 times an appropriate power of ten.

secondary winding The output winding of a transformer; also called *secondary.*

Seebeck effect The generation of a voltage at the junction of two different materials that have a temperature difference between them.

selectivity A measure of how effectively a resonant circuit passes certain desired frequencies and rejects all others. Generally, the narrower the bandwidth, the greater the selectivity.

semiconductor A material that has a conductance value between that of a conductor and an insulator. Silicon and germanium are examples.

series In an electric circuit, a relationship of components in which the components are connected such that they provide a single current path between two points.

series resonance A condition in a series *RLC* circuit in which the reactances ideally cancel and the impedance is minimum.

shell The orbit in which an electron revolves.

short A circuit condition in which there is a zero or abnormally low resistance path between two points; usually an inadvertent condition.

SI Standardized international system of units used for all engineering and scientific work; abbreviation for French *Le Système International d'Unités.*

siemens (S) The unit of conductance.

simultaneous equations A set of n equations containing n unknowns, where n is a number with a value of 2 or more.

sine wave A type of waveform that follows a cyclic sinusoidal pattern defined by the formula $y = A \sin \theta$.

solenoid An electromagnetically controlled device in which the mechanical movement of a shaft or plunger is activated by a magnetizing current.

solenoid valve An electrically controlled valve for control of air, water, steam, oils, refrigerants, and other fluids.

source A device that produces electrical energy.

speaker An electromagnetic device that converts electrical signals to sound waves.

squirrel-cage A type of ac induction motor.

stator The stationary outer part of a generator or motor.

steady state The equilibrium condition of a circuit that occurs after an initial transient time.

step-down transformer A transformer in which the secondary voltage is less than the primary voltage.

step-up transformer A transformer in which the secondary voltage is greater than the primary voltage.

superposition theorem A method for the analysis of circuits with more than one source.

switch An electrical device for opening and closing a current path.

tank circuit A parallel resonant circuit.

tapered Nonlinear, such as a tapered potentiometer.

temperature coefficient A constant specifying the amount of change in the value of a quantity for a given change in temperature.

terminal equivalency The concept that when any given load resistance is connected to two sources, the same load voltage and load current are produced by both sources.

tesla (T) The SI unit of flux density.

thermistor A type of variable resistor that is temperature-sensitive.

thermocouple A thermoelectric type of voltage source commonly used to sense temperature.

Thevenin's theorem A method for simplifying a two-terminal linear circuit to an equivalent circuit with only a voltage source in series with a resistance or impedance.

time constant A fixed-time interval, set by R and C, or R and L values, that determines the time response of a circuit.

tolerance The limits of variation in the value of a component.

trailing edge The second step of transition of a pulse.

transformer An electrical device constructed of two or more coils (windings) that are electromagnetically coupled to each other to provide a transfer of power from one coil to another.

transient time An interval equal to approximately five time constants.

triangular waveform A type of electrical waveform that consists of two ramps.

trigger The activating signal for some electronic devices or instruments.

trimmer A small variable capacitor.

troubleshooting A systematic process of isolating, identifying, and correcting a fault in a circuit or system.

true power The power that is dissipated in a circuit, usually in the form of heat.

turns ratio (n) The ratio of turns in the secondary winding to turns in the primary winding.

unbalanced bridge A bridge circuit that is in the unbalanced state as indicated by a voltage across the bridge that is proportional to the amount of deviation from the balanced state.

valance Related to the outer shell or orbit of an atom.

valence electron An electron that is present in the outermost shell of an atom.

VAR (volt-ampere reactive) The unit of reactive power.

varactor A semiconductor device that exhibits a capacitance characteristic that is varied by changing the voltage across its terminals.

volt The unit of voltage or electromotive force.

voltage The amount of energy per charge available to move electrons from one point to another in an electric circuit.

voltage divider A circuit consisting of series resistors across which one or more output voltages are taken.

voltage drop The decrease in voltage across a resistor due to a loss of energy.

voltage source A device that provides a constant voltage for a varying load.

voltmeter An instrument used to measure voltage.

Watt (W) The unit of power. One watt is the power when 1 J of energy is used in 1 s.

Watt's law A law that states the relationships of power to current, voltage, and resistance.

waveform The pattern of variations of a voltage or current showing how the quantity changes with time.

weber The SI unit of magnetic flux, which represents 10^8 lines.

Wheatstone bridge A 4-legged type of bridge circuit with which an unknown resistance can be accurately measured using the balanced state of the bridge. Deviations in resistance can be measured using the unbalanced state.

winding The loops or turns of wire in an inductor.

wiper The sliding contact in a potentiometer.

INDEX